TELEPEN
16 0145756 5
AF742507
- 3 FEB 2000
CANCELLED
WITHDRAWN

GROUNDWATER IN ENGINEERING GEOLOGY

Proceedings of the 21st Annual Conference of the Engineering Group of the Geological Society held at The University of Sheffield 15th-19th September 1985

edited by

J.C. Cripps[1], F.G. Bell[2] and M.G. Culshaw[3]

[1] Department of Geology, University of Sheffield, Sheffield S1 3JD
[2] Department of Civil Engineering, Teesside Polytechnic, Middlesbrough TS1 3BA.
[3] Engineering Geology and Reservoir Rock Properties Research Group, British Geological Survey, Keyworth, Nottingham NG12 5GG

1986
Published by
The Geological Society
London

628.114
G

THE GEOLOGICAL SOCIETY

President Professor B.E. Leake
Secretaries Dr R.G. Park, Dr L.R.M. Cocks, Dr A.J. Martin
Foreign Secretary Professor J.B. Dawson
Treasurer Dr W.J. French

THE ENGINEERING GROUP OF THE GEOLOGICAL SOCIETY

Chairman Mr D.N. Holt
Vice-Chairman Dr R.K. Taylor
Secretary Mr J.W. Hooper
Treasurer Dr K.D. Privett
Committee Dr F.G. Bell, Dr C.R.I. Clayton, Dr J.C. Cripps, Mr M.G. Culshaw, Dr J.A. Dixon, Mr J.M. Head, Mrs S.M. Herbert, Mr B.A. Leach, Mr D. Norbury

FELLOWSHIP OF THE GEOLOGICAL SOCIETY MEMBERSHIP OF THE ENGINEERING AND HYDROGEOLOGICAL GROUPS

Membership is open to all Fellows of the Society. There are no rigid requirements for election to Fellowship, although it is usually expected that candidates will be graduates in geology or an allied subject, or have equivalent qualifications or experience. The specialist groups are open to all Fellows without additional charge. Enquiries concerning Fellowship of the Society and membership of the specialist groups should be sent to the Secretary, Geological Society, Burlington House, Piccadilly, London W1V 0JU. All Fellows of the Society are entitled to receive for their subscription one of the Society's three journals: the Journal of the Geological Society, Quarterly Journal of Engineering Geology or Marine and Petroleum Geology. On payment of an additional sum on the annual subscription, Fellows may obtain copies of additional journals.

CARDIFF
-3 AUG 1992

Published by

The Geological Society
Burlington House,
Piccadilly,
London W1V 0JU

First published 1986

© Copyright 1986 The Geological Society

All rights reserved. No part of this publication may be reproduced, stored in a retrieval system, or transmitted, in any form or by any means, electronic, mechanical, photocopying, recording or otherwise without the prior permission of the copyright owner.

Printed in The United Kingdom
by Acors Press, Billericay, Essex (England)

British Library Cataloguing in Publication Data

Geological Society of London. *Engineering Group. Conference (21st)*

Groundwater in engineering geology: proceedings of the 21st Annual Conference of the Engineering Group of the Geological Society. — (Engineering geology special publication, ISSN 0267-9914; no. 3)
1. Water, Under ground
I. Title II. Cripps, J.C. III. Bell, F.G.
IV. Culshaw, M.G. V. Series
627'.17 TC176

ISBN 0-903317-35-4
ISBN 0-903317-36-2 Pbk

Foreword

The inaugural meeting of the Engineering Group of the Geological Society was held on 24th June, 1964, and the first Regional meeting of the Group was held at Sheffield University in September, 1965.

This first Regional Meeting proceeded so well that the Group felt encouraged to organise further such meetings on a regular basis, with the result that regional meetings (now termed Annual Conferences) have been held every year since, each at a different locality within England, Ireland, Wales and Scotland.

An invitation from Sheffield to return there in 1985 for the 21st Conference was fortunate, well-timed and gratefully accepted by the Group. Thanks to Professor Dawson — Head of the Geology Department — and his staff, and particularly to Dr Cripps the local organiser, the Conference was excellently run and proceeded smoothly in every respect.

Over the 21 years which have passed since the first Sheffield Conference, the subject of which was Rock Mechanics, themes have been wide ranging. Titles of these themes have included Engineering Geology aspects of particular engineering operations such as tunnelling, or particular geological features such as soluble rocks or Quarternary deposits. Wider topics such as engineering geomorphology, land restoration, and redevelopment and the British Standard code of Practice on Site Investigation have also been chosen as Conference themes. Although groundwater is not specifically mentioned in any one of the listed past titles, the subject can hardly have escaped discussion at any single Conference. For 1985 the Group Committee therefore decided to give 'Groundwater in Engineering Geology' a platform in its own right in the 21st Conference year, and in retrospect this seems to have been fortunate since the response to the call for papers, the support from invited lead speakers, the attendance and the liveliness of the discussion were all very encouraging. Since then the Editors have put in much hard work on bringing all the contributions together in the form of these Proceedings.

The Group Committee is much aware of its indebtedness to all those organisers, delegates and participants (including trade exhibitors) who have supported this and past Conferences.

Having been so well launched into full adulthood in conference organising there can be no turning back, and so the Group aims to give its past supporters every incentive to return each year.

Newcomers also are essential to a healthy future, and the Office bearers, the Committee and the Group as a whole hope to see more and more of them as time goes by, not only as Conference delegates, but also as Fellows of the Society and as Engineering Group Members.

D.N. Holt
Chairman — Engineering Group
Freeman Fox & Partners
April, 1986

Preface

The twenty first Annual Conference of the Engineering Group of the Geological Society was held at Sheffield University between 16th and 19th September, 1985. The theme of the conference was 'Groundwater in Engineering Geology' being directed at the influence that groundwater has in construction operations.

The structure of this volume, and indeed that of the Conference, follows closely suggestions made by Dr M.H. de Freitas of Imperial College, London. By grouping topics by 'cause' in terms of groundwater rather than by their 'effect' on engineering structure, it has been possible to emphasise the importance of groundwater in ground engineering work without precluding reference to any particular type of structure or application. Although the application may be very relevant to the groundwater problems liable to arise, a further advantage of the approach adopted is that techniques hitherto regarded as appropriate in one type of engineering operation potentially become available in other fields of work.

Papers which originally formed a 615 page pre-print volume are presented here together with contributions submitted by lead speakers and discussion which took place at the meeting. In addition written discussion, enlargements to the verbal discussion and other contributions received after the meeting are included.

The Conference attracted a high attendance of delegates from industry, education and public service, both in the United Kingdom and overseas. It was gratifying to have among the delegates many people of International repute, among them Professor P.W. Rowe of Manchester University, Professor R.D.J. de Wiest of the State University of Florida, and Professor J.K. Mitchell of the University of California, Berkeley. The organisers are particularly indebted to Professor Mitchell for his contributions to the Conference.

F.G.Bell, J.C. Cripps, M.G. Culshaw
Sheffield, 1986

Organising Committee

Chairman	D.N. Holt
Local Secretary	J.C. Cripps
Committee	F.G. Bell M.G. Culshaw M.H. de Freitas S.M. Herbert J.W. Hooper R.J.W. McDermott K. Privett G. West

Acknowledgements

The preparation of these Proceedings has required the co-operation and assistance of many people. Special appreciation is expressed to Professor J.B. Dawson for the provision of secretarial, printing and draughting facilities in the Department of Geology at the University of Sheffield.

The editors are particularly indebted to Miss A. Hills, Miss R.J. Abbott (Geological Society) and Miss P. Mellor (University of Sheffield) for their patient and conscientious work in preparing this volume for publication. Miss P. Mellor and Miss A. Rutherford are also thanked for their assistance with the compilation of the pre-prints.

The authors are thanked for their co-operation in preparing the papers to the specified format at short notice. The editors are also grateful to the session chairmen and lead lecturers for their contributions, and would also like to thank the delegates for their active participation, especially during the discussion periods.

Contents

Groundwater in Engineering Geology, *London*, 1986, pp. 1-23 Printed in England

A review of the engineering behaviour of soils and rocks with respect to groundwater

F.G. Bell, J.C. Cripps & M.G. Culshaw

ABSTRACT: The effect of groundwater on the engineering behaviour of soils and rocks is of fundamental importance, indeed to paraphrase Tergazhi — without water there would be no soil mechanics. This paper reviews these effects in terms of the variation in the properties and behaviour of soils and rocks brought about by changes in moisture content, and associated changes, notably in effective stress, and by the dissolution of parts of the rock or soil mass.

Introduction

The principal source of groundwater is meteoric water, that is, precipitation (rain, sleet, snow and hail). Part of this water enters the ground by infiltration whilst the remainder is lost either as runoff on the surface or by evaporation and transpiration to the atmosphere. The main conditions which influence the relationship between infiltration, runoff, evaporation and transpiration are climate, topography, vegetation and geology.

Two other sources of water which are occasionally of some consequence are juvenile water which is derived from magmatic sources and connate water which represents the water trapped in the pore spaces of sedimentary rocks as they were deposited.

The retention of water in a soil depends upon the capillary force and the molecular attraction of the particles. As the pores in a soil become thoroughly wetted the capillary force declines so that gravity becomes more effective. In this way downward percolation can continue after infiltration has ceased but as the soil dries, so capillarity increases in importance. No further percolation occurs after the capillary and gravity forces are balanced. Thus, water percolates into the zone of saturation when the retention capacity is satisfied.

Many engineering properties of soils and rocks are very much influenced by their water content. But the water content in the ground can change above the water table and the water table itself can fluctuate in position. Such changes may be relatively rapid and are primarily influenced by the weather and climatic conditions as well as the retentivity of the ground. Accordingly, the effect of groundwater on the behaviour of the ground can change during the period of construction of a structure. Indeed changes can be brought about as a consequence of construction as, for example, in dewatering operations. It is therefore important prior to construction, to investigate what effects changes in the groundwater regime will have on the engineering performance of soil and rock masses. If adverse effects such as excessive ground movements due to swelling, shrinkage or settlement, are indicated as likely to occur, then precautionary measures should be taken.

Intrinsic properties

Porosity and permeability are the two most important factors governing the accumulation, migration and distribution of groundwater. However, both may change within a rock or soil mass in the course of its geological evolution. Furthermore, it is not uncommon to find changes in both porosity and permeability with depth due to variation in a number of features, including pore size distribution. The actual size of pores is significant since in narrow pores or capillaries, surface tension forces exert a control over the movement of fluids. In addition, chemical interaction may occur between, on the one hand water and dissolved gases, and on the other, certain rock or soil constituents, particularly clay — and soluble — minerals.

Porosity

Apart from particle size distribution, porosity is also affected by sorting, grain shape, fabric, degree of compaction and cementation, and mineralogical composition.

The highest porosity is commonly attained when all the grains are the same size. The addition of grains of different size to such an assemblage lowers its porosity and this is, within certain limits, directly proportional to the amount added. Irregularities in grain shape result in a larger possible range of porosity, as irregular forms may theoretically be packed either more tightly or loosely than spheres. Similarly angular grains may either cause an increase or a decrease in porosity.

From CRIPPS, J.C., BELL, F.G. & CULSHAW, M.G. (eds), 1986, *Groundwater in Engineering Geology*, Geological Society Engineering Geology Special Publication No. 3, pp. 1-23.

After a sediment has been buried and indurated, several additional factors help determine its porosity. The chief amongst these are closer spacing of grains, deformation and crushing of grains, recrystallisation, secondary growth of minerals, cementation and, in some cases, dissolution. Hence diagenetic change undergone by a rock may either increase or decrease its original porosity.

The porosity of a deposit does not necessarily provide an indication of the amount of water that can be obtained therefrom. Even though a rock or soil may be saturated, only a certain proportion of water can be removed by drainage under gravity or by pumping, the remainder being held in place by capillary or molecular forces.

The ratio of the volume of water retained to that of the total volume expressed as a percentage, is referred to as the specific retention. The amount of water retained varies directly in accordance with the surface area of the pores and indirectly with regard to the pore space. The relationship between porosity and specific yield on the one hand and particle size distribution on the other, is shown in Fig. 1. In soils the specific yield tends to decrease as the coefficient of uniformity increases. Examples of the specific yield of some common types of soil and rock are given in Table 1 (it must be appreciated that individual values of specific yield can vary considerably from those quoted).

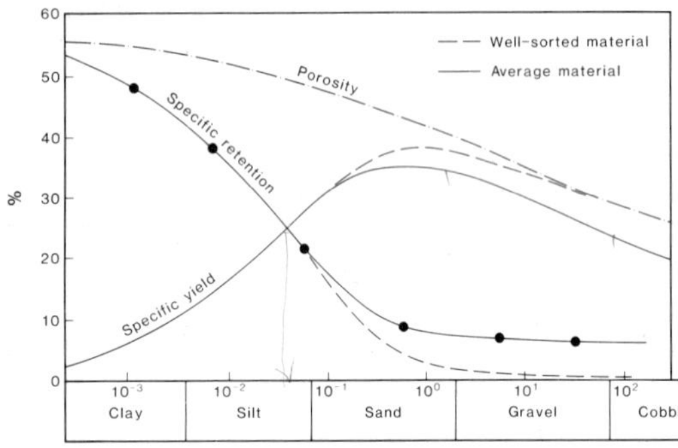

FIG. 1. Relationship between grain size, porosity, specific retention and specific yield (from Bear, J. 1979. *Hydraulics of Groundwater,* reproduced by permission of McGraw Hill, New York).

Permeability

In ordinary hydraulic usage a substance is termed permeable when it permits the passage of a measurable quantity of fluid in a finite period of time and impermeable when the rate at which it transmits that fluid is slow enough to be negligible under existing temperature-pressure conditions (Table 2).

TABLE 1. *Some examples of specific yield.*

Material	Specific yield (%)
Gravel	15 –30
Sand	10 –30
Dune sand	25 –35
Sand and gravel	15 –25
Loess	15 –20
Silt	5 –10
Clay	1 – 5
Till (silty)	4 – 7
Till (sandy)	12 –18
Sandstone	5 –25
Limestone	0.5 –10
Shale	0.5 – 5

The flow through a unit cross section of material is modified by temperature, hydraulic gradient and the permeability. The latter is affected by the uniformity and range of grain size, shape of the grains, stratification, the amount of consolidation and cementation undergone and the presence and nature of discontinuities. Temperature changes affect the flow rate of a fluid by changing its viscosity. The rate of flow is commonly assumed to be directly proportional to the hydraulic gradient but this is not always so in practice.

The permeability of a clastic material is also affected by the interconnections between the pore spaces. If these are highly tortuous then the permeability is accordingly reduced. Consequently tortuosity figures importantly in permeability, influencing the extent and rate of free water saturation. It can be defined as the ratio of the total path covered by a current flowing in the pore channels between two given points to the straight line distance between them.

Stratification in a formation varies within limits both vertically and horizontally. It is frequently difficult to predict what effect stratification has on the permeability of the beds. Nevertheless, in the great majority of cases where a directional difference in permeability exists, the greater permeability is parallel to the bedding. For example, the Permo-Triassic sandstones of the Mersey and Weaver Basins are notably anisotropic as far as permeability is concerned, the flow parallel to the bedding being higher than across it. Ratios of 5:1 are not uncommon and occasionally values of 100:1 have been recorded where fine marl partings occur.

The permeability of intact rock (primary permeability) is usually several orders less than the *in situ* permeability (secondary permeability). Although the secondary permeability is affected by the frequency, continuity and openness, and amount of infilling, of discontinuities, a rough estimate of the permeability can be obtained from their frequency (Table 3). Admittedly such estimates must be treated with caution and cannot be applied to rocks which are susceptible to solution.

TABLE 2. *Relative values of permeabilities.*

Rock types	Porosity: Primary (grain)	Porosity: Secondary (fracture)*	Permeability range (m/s): 10^0 Very high	10^-2 High	10^-4 Medium	10^-6 Low	10^-8 Very low	10^-10 Imperm-eable	Well yields: High	Medium	Low	Type of water-bearing unit
Sediments (unconsolidated)												
Gravel	30-40		—	—					—			Aquifer
Coarse sand	30-40			—	—				—	—		Aquifer
Medium to fine sand	25-35			—	—					—		Aquifer
Silt	40-50	Occasional				—	—				—	Aquiclude
Clay, till	45-55	Often fissured				—	—	—			—	Aquiclude
Sediments (consolidated)												
Limestones, dolostone	1-50	Solution joints, bedding planes	—	—	—	—	—	—	—	—	—	Aquifer or aquifuge
Coarse, medium sandstone	<20	Joints and bedding planes		—	—				—	—	—	Aquifer or aquiclude
Fine sandstone	<10	Joints and bedding planes			—	—					—	Aquifer or aquifuge
Shale, siltstone		Joints and bedding planes			—	—	—				—	Aquifuge or aquifer
Igneous rocks, Volcanic rocks *e.g. basalt*		Joints and bedding planes	—	—	—	—	—	—	—	—	—	Aquifer or aquifuge
Plutonic and metamorphic rocks		Weathering and joints decreasing as depth increases			—	—	—			—	—	Aquifuge or aquifer

* Rarely exceeds 10%

TABLE 3. *Estimation of secondary permeability from discontinuity frequency.*

Rock mass description	Interval (m)	Permeability: Term	Permeability: k (m/s)
Very closely to extremely closely spaced discontinuities	Less than 0.2	Highly permeable	10^{-2}-1
Closely to moderately widely spaced discontinuities	0.2-0.6	Moderately permeable	10^{-5}-10^{-2}
Widely to very widely spaced discontinuities	0.6-2.0	Slightly permeable	10^{-9}-10^{-5}
No discontinuities	Over 2.0	Effectively impermeable	Less than 10^{-9}

Water possesses three forms of energy, namely, potential energy attributable to its height, pressure energy owing to its pressure, and kinetic energy due to its velocity. The latter can usually be discounted in any assessment of flow through the soils. Energy in water is usually expressed in terms of head. The head possessed by water in soils or rocks is manifested by the height to which water will rise in a standpipe above a given datum. This height is usually referred to as the piezometric level and provides a measure of the total energy of the water. If at two different points within a continuous area of water there are different amounts of energy, then there will be a flow towards the point of lesser energy and the difference in head is expended in maintaining that flow. Other things being equal, the velocity of flow between two points is directly proportion to the difference in head between them.

The basic law concerned with flow is that enunciated by Darcy (1856) which states that the rate of flow per unit area is proportional to the gradient of the potential head. Darcy's law is valid as long as a laminar flow exists. Departures from Darcy's law, therefore, occur when the flow is turbulent, such conditions arise when the velocity of flow is high. Turbulent flow may occur in very permeable media, in which the Reynolds number attains values above four. Accordingly it is usually accepted that this law can be applied to those soils which have grain sizes finer than gravels. Furthermore, Darcy's law probably does not accurately represent the flow of water through a porous medium of extremely low permeability, because of the influence of surface and ionic phenomena and the presence of gases.

Apart from an increase in the mean velocity, the other factors which cause deviations from the linear laws of flow include, firstly, the non-uniformity of pore spaces, since differing porosity gives rise to differences in the seepage rates through pore channels. A second factor is an absence of a running-in section where the velocity profile can establish a steady state parabolic distribution. Lastly, such deviations may be developed by perturbations due to jet separation from wall irregularities.

Permeability also depends upon the density, and dynamic viscosity of the fluid, involved. In fact, permeability is directly proportional to the unit weight of the fluid concerned and is inversely proportional to its viscosity. The latter is very much influenced by temperature.

Ionic exchange on clay and colloid surfaces may bring about changes in mineral volume which, in turn, affect the shape and size of the pores. Moderate to high groundwater velocities are required to move colloids and clay particles. Solution and deposition may result from the pore fluids. Small changes in temperature and/or pressure may cause gas to come out of solution which may block pore spaces.

Capillarity and soil suction

Capillary movement in soil refers to the movement of moisture through the minute pores between the soil particles which act as capillaries. It takes place as a consequence of surface tension therefore moisture can rise from the water table. This movement, however, can occur in any direction, not just vertically upwards. It occurs whenever evaporation takes place from the surface of the soil, thus exerting a 'surface tension pull' on the moisture, the forces of surface tension increasing as evaporation proceeds. Accordingly capillary moisture is in hydraulic continuity with the water table and is raised against the force of gravity, the degree of saturation decreasing from the water table upwards. Equilibrium is attained when the forces of gravity and surface tension are balanced.

The boundary separating capillary moisture from the gravitational water in the zone of saturation is, as would be expected, ill-defined and cannot be determined accurately. That zone immediately above the water table which is saturated with capillary moisture is referred to as the closed capillary fringe, whilst above this, air and capillary moisture exist together in the pores of the open capillary fringe. The depth of the capillary fringe is largely dependent upon the particle size distribution and density of the soil mass, which in turn influence pore size. In other words the smaller the pore size, the greater is the depth. For example, capillary moisture can rise to great heights in clay soils (Table 4) but the movement is very slow. In soils which are poorly graded the height of the capillary fringe generally varies whereas in uniformly textured soils it attains roughly the same height. Where the water table is at shallow depth and the maximum capillary rise is large, moisture is continually attracted from the water table, due to evaporation from the ground surface, so that the uppermost soil is near saturation. For instance, under normal conditions peat deposits may be assumed to be within the zone of capillary saturation. This means that the height to which the water can rise in peat by capillary action is greater than the depth below ground to which the water table can be reduced by drainage. The coarse fibrous type of peat, containing appreciable sphagnum, may be an exception.

TABLE 4. *Capillary rises and pressures in soils.*

Soil	Capillary rise (mm)	Capillary pressure (kPa)
Fine gravel	up to 100	up to 1.0
Coarse gravel	100– 500	1.0– 1.5
Medium sand	150– 300	1.5– 3.0
Fine sand	1000– 10000	10.0– 100.0
Clay	over 10000	over 100.0

Drainage of capillary moisture cannot be effected by the installation of a drainage system within the capillary fringe as only that moisture in excess of that retained by surface tension can be removed, but it can be lowered by lowering the water table. The capillary ascent, however, can be interrupted by the installation of impermeable membranes or layers of coarse aggregate.

At each point where moisture menisci are in contact with soil particles the forces of surface tension are responsible for the development of capillary or suction pressure (Table 4). The air and water interfaces move into the smaller pores. In so doing the radii of curvature of the interfaces decrease and the soil suction increases. Hence the drier the soil, the higher is the soil suction.

Soil suction is a negative pressure and indicates the height to which a column of water could rise due to such suction. Since this height or pressure may be very large, a logarithmic scale has been adopted to express the relationship between soil suction and moisture content, the latter is referred to as the pF value (Table 5).

Soil suction tends to force soil particles together and these compressive stresses contribute towards the strength and stability of the soil. There is a particular suction pressure for a particular moisture content in a given soil, the magnitude of which is governed by whether it is becoming wetter or drier. In fact as clay soil dries out the soil suction may increase to the order of several thousands of kilopascals. However, the strength of a soil attributable to soil suction is only temporary and is destroyed upon saturation. At that point soil suction is zero.

TABLE 5. *Values of soil suction.*

pF Value	Equivalent suction	
	(mm water)	(kPa)
0	10	0.1
1	100	1.0
2	1000	10.0
3	10000	100.0
4	100000	1000.0
5	1000000	10000.0

Moisture content in clay soils

The engineering performance of clay deposits is very much affected by the total moisture content and by the energy with which this moisture is held. For instance, the moisture content influences their density, consistency and strength. The energy with which moisture is held influences their volume change characteristics since swelling, shrinkage and consolidation are affected by permeability and moisture migration. Furthermore moisture migration may give rise to differential movement in clay soils. The gradients which generate moisture migration in clays may arise from variations in temperature, extent of saturation, and chemical composition or concentration of pore solutions.

There tends to be a general reduction in porosity, and therefore also a reduction in moisture content, with increasing age and previous overburden. Mudrocks in Britain show a range of moisture content varying from about 3 to 30% (see Cripps & Taylor 1981) depending on these factors and also present overburden and weathering state.

The Atterberg or consistency limits of cohesive soils are founded on the concept that they can exist in any of four states depending on their water content. These limits are also influenced by the amount and character of the clay mineral content. In other words a cohesive soil is solid when dry but as water is added, it first turns to a semi-solid, then to a plastic, and finally to a liquid state. The water content at the boundaries between these states are referred to as the shrinkage limit, the plastic limit and the liquid limit respectively.

Schofield & Wroth (1968) indicated that the liquid and plastic limits for saturated remoulded soils represent critical states such that the plasticity index corresponds to an arbitrary increase in strength. Hence these limits can be regarded as being defined by water contents associated with particular strengths. Indeed the plasticity index has been redefined by Wroth & Wood (1978) as the change in water content giving rise to a one hundred-fold change in the strength of the soil.

Fully saturated clay soils often behave as incompressible materials when subjected to rapid loading. The amount of elasticity increases continuously as the water content is decreased. Elastic recovery may be immediate or may take place slowly. The linear relationship between stress and strain only applies to clays at low stresses.

Mineral hydration

Mineral hydration causes modifications to the engineering behaviour of soils and rocks by virtue of changes in volume or density, the interaction between mineral grains, and the physical properties of the materials involved.

Grim (1962) distinguished two hydration processes in clay soils, namely, intercrystalline and intracrystalline swelling. Intercrystalline swelling takes place when the uptake of moisture is restricted to the external crystal surfaces and void spaces between crystals. Such swelling may occur in all materials but it is most significant in fine grained ones, particularly clays. In relatively dry clays the particles are held together by relict water under tension from capillary forces. On wetting these forces are relaxed and the material expands. Such swelling occurs in any type of clay, irrespective of mineralogical composition, although the amount of swelling depends on a number of factors including mineral species and the type and concentration of cations present in the porewater. Intracrystalline swelling, on the other hand, is characteristic of the smectite family of clay minerals, in particular of montmorillonite. Vermiculite and some varieties of chlorite also display intracrystalline swelling behaviour. In swelling minerals the individual molecular layers of the mineral are weakly bonded so that on wetting water enters not only between the crystals, but also between the unit layers which comprise the crystals. Here the magnitude of swelling is a function of clay mineral type, the type of interlayer cations present in the mineral (see Taylor & Cripps 1984). Swelling in Na montmorillonite is the most notable and can amount to 800 to 1000 times the original volume, the clay then having formed a gel of dissociated platelets with dimensions similar to those of the unit cell (10Å).

Certain other minerals are susceptible to significant chemical hydration. Of particular importance in engineering geology are the hydration of anhydrite to gypsum and of oxides of calcium or magnesium present in some slags. These reactions and also dolomitisation are usually associated with volume increase. Volume decrease in consequence of the dehydration of gypsum may also occur in hot arid environments (see Hardie, 1967).

The hydration of anhydrite produces a theoretic swelling of up to about 63% which exerts pressures that have been variously estimated between 2 and 69 MPa. In fact Winkler & Wilhelm (1970) showed that the hydration pressure can be calculated from a consideration of the vapour pressures of the hydrated salt and atmospheric water.

When hydration occurs at shallow depths some of the gypsum formed may be removed in solution, leaving no net change in volume (see Holliday 1970). At greater depths anhydrite is effectively confined which results in a gradual build-up of pressure that may be liberated in a sudden or explosive ground movement. According to Brune (1965), in the United States such uplifts have occurred beneath reservoirs, these bodies of water providing a constant supply of hydration water. Hydration has been observed at depths as great as 152 m with an effective overburden pressure of about 2 MPa, although Brune (1965) considers that a pressure of 14 MPa would have been required to uplift and deform the overlying rockmass.

Swelling pressures attributable to the hydration of anhydrite encountered during the construction of tunnels in central Europe are similar to those given above. For example, Yuzer (1982) quoted values between 2 and 12 MPa. However, where swelling clays are associated with anhydrite it can be difficult to distinguish the cause of swelling. Serrano *et al.* (1981) attributed rapid swelling (days) to clay mineral effects and slower swelling (weeks) to the conversion of anhydrite in a gypsiferous marl formation in northern Spain.

Mineral solution and weathering

Surface and near surface environments can represent conditions in which certain rock and soil forming minerals are susceptible to chemical change within the context of weathering. The most important processes, namely, solution, oxidation and hydrolysis, may be controlled by the movement and composition of groundwater since it will act as a medium of transfer, both into and out of the reaction site, of active components and products respectively. Aspects of solutioning with respect to soluble rocks are dealt with on page 18 of this paper, but it is also pertinent that the removal of individual grains in heterogenous materials will lead to reductions in density and strength, together with increases in porosity and permeability. Commonly occurring minerals most at risk from solution effects include halite, carnalite and other evaporitic minerals including gypsum, anhydrite and to some extent calcite.

In dry air rocks decay very slowly. The presence of moisture hastens the rate tremendously, firstly, because water is itself an effective agent of weathering and, secondly, because it holds in solution substances which react with the component minerals of the rock. The most important of these substances are free oxygen, carbon dioxide, organic acids and acids of nitrogen.

Free oxygen is an important agent in the decay of all rocks which contain oxidisable substances, iron and sulphur being especially suspect. The rate of oxidation is quickened by the presence of water; indeed it may enter into the reaction itself, as for example, in the formation of hydrates. However, its role is chiefly that of a catalyst.

Carbonic acid is produced when carbon dioxide is dissolved in water and it may possess a pH value of about 5.7. The principal source of carbon dioxide is not the atmosphere, but the air contained in the pore spaces in soil where its content may be a hundred or so times greater than it is in the atmosphere. An abnormal concentration of carbon dioxide is released when organic material decays. Furthermore humic acids are formed by the decay of humus in soil waters; they ordinarily have pH values between 4.5 and 5.0, but occasionally they may be under 4.0. The nitrogen acids HNO = HNO_3 and HNO_2, are formed by organic decay or bacterial action in soils. They play only a minor part in weathering. In volcanic regions and in the oxidised zones of sulphide deposits, the sulphur acids H_2SO_3 and H_2SO_4, become important and in some localities their pH value may be lowered below 1.0.

Because of the hydrolysis of the dissolved metal ion, solutions forming from the oxidation of sulphides are acidic. For instance, when pyrite is initially oxidised, ferrous sulphate and sulphuric acid are formed. Further oxidation, assisted by autotropic bacteria, leads to the formation of ferric sulphate. Very insoluble ferric oxide or hydrated oxide is formed if highly acidic conditions are produced.

Experience with Canadian Ordovician black shales (see Gillot *et al.* 1974) and British examples (see Nixon, 1978) indicate that pyritiferous mudrocks can be subject to appreciable ground movements due to pyrite oxidation. Reaction of the sulphuric acid liberated during oxidation with soil and rock components or construction materials results in increased porosity and furthermore the products may be removed and precipitated elsewhere. These processes can also make a significant contribution to rock or soil degradation due to the dissolution of cement, creation of void space and crystallisation pressures that they entail.

Weathering of silicate minerals, including feldspar, augite, hornblende, mica and olivine, is primarily a process of hydrolysis. Much of the silica which is released by weathering forms silicic acid but where it is liberated in large quantities some of it may form colloidal or amorphous silica. Mafic silicates usually decay more rapidly than felsic silicates and in the process they release magnesium, iron and lesser amounts of calcium and alkalis. The process whereby feldspars are decomposed to form clay minerals is effected by the hydrolysing action of weakly carbonated water. The alkalis are removed in solution as carbonates and bicarbonates; some silica is hydrolysed to form silicic acid.

Clays are hydrated aluminium silicates and when they are subjected to severe chemical weathering in humic tropical regimes they break down to form laterite. The process involves the removal of siliceous material and this is again brought about by the action of carbonated waters. Intensive leaching of soluble mineral matter from surface rocks takes place during the wet season. During the subsequent dry season groundwater is drawn to the surface by capillary action and minerals are precipitated there as the water evaporates. The minerals generally consist of hydrated peroxides or iron, sometimes of aluminium, and very occasionally of manganese. The precipitation of these insoluble hydroxides gives rise to an impermeable lateritic soil which itself inhibits leaching to the extent ultimately that the formation of laterite ceases. As a consequence lateritic deposits are usually less than 7 m thick.

Effects of changes in moisture content

The amount of moisture contained within a soil or rock is a function of its porosity, the availability of water and the mineralogy of the material. The mechanical behaviour of some soils and rocks is very sensitive to pore space moisture such that significant changes of volume, porosity, strength and other parameters accompany wetting and drying. Moisture content changes occur for a variety of reasons; some of which are a consequence of deliberate drainage or recharge of groundwater, whilst others are either accidental or peripheral to other activities. Leakage from water supply or drainage systems and bodies of water is liable to result in an unintentional rise in moisture content. On the other hand, some engineering works have important side-effects with respect to groundwater regimes. The construction of dams and reservoirs usually causes a general rise in water levels, whereas tunnelling or mining operations may have the opposite effect. Significant changes in moisture content frequently relate to seasonal or longer term climatic variations. The nature of the actual ground surface also exerts a control by virtue of its influence over evaporation and transpiration of groundwater and the amount of run-off and infiltration which occurs during precipitation. Although the presence of vegetation generally increases infiltration at the expense of direct run-off; plants, particularly trees, with a high water demand cause a decrease in soil or rock moisture. A reduction may also result due to the effects of inadequately insulated hot process foundations or conduits.

Swelling and shrinkage of clay soils and rocks

One of the most notable characteristics of clays from an engineering point of view is their susceptibility to slow volume changes which can occur independently of loading. Differences in the period and magnitude of precipitation and evapo-transpiration, poor surface drainage, leakage and other causes of moisture content, change effect swelling and shrinkage behaviour. The density of a clay soil also influences the amount of swelling likely to occur. Expansive clay minerals adsorb water into their lattice structure, tending to expand into adjacent zones of looser soil before volume increase of the soil mass takes place. Once any void space has been filled and in densely packed soils, the soil mass has to swell to accommodate the volume change of the expansive clay particles.

Holtz & Gibbs (1952) showed that expansive clays can be recognised by their plasticity characteristics. Subsequently in a review of the various methods which have been used to determine the amount of swelling that an expansive clay is likely to undergo, O'Neill and Poormoayed (1980) supported the use of the United States Army Engineers Waterways Experimental Station (USAEWES) classification of potential swell. It can be seen from Table 6 that this also makes use of the plasticity characteristics of clay soils, as well as initial soil suction. Van der Merwe (1964) used the activity of a clay as a means of predicting its swell potential (see Fig. 2). As far as the maximum movement due to swelling beneath a building is concerned, Brackley (1980) suggested that this could be obtained from:

$$\text{Swell } (\%) = \frac{(PI - 10)}{10} \text{ Log}_{10}\, S/Q_{of}$$

where PI is the plasticity index, S is the soil suction at the time of construction (in kPa) and Q_{of} is the overburden pressure plus foundation pressure (kPa) acting on each layer of soil.

In the initial and residual stages of shrinkage the reduction in soil volume is less than the volume of water lost. Only during the normal shrinkage stage are the two equal. This can give rise to differential movement. Williams & Jennings (1977) found that soil structure has a major influence on the pattern of the shrinkage process, as well as on the total amount of shrinkage. They indicated that in clay soils with the same initial moisture content and density, the more random the particle arrangement, the less the total shrinkage.

Swelling of rocks is associated with weathering; shales, mudstones and marls being particularly prone although small amounts of swelling have been recorded in some sandstones. The clay mineral content plays a significant role in these rocks. For example, kaolinite is not expansive whilst montmorillonite is, Na montmorillonite being able to expand to many times its original volume. Since swelling is principally due to the ingress of water, the rock must be porous or fractured. If a rock has an intact unconfined compressive strength exceeding 40 MPa, it is not subject to swelling.

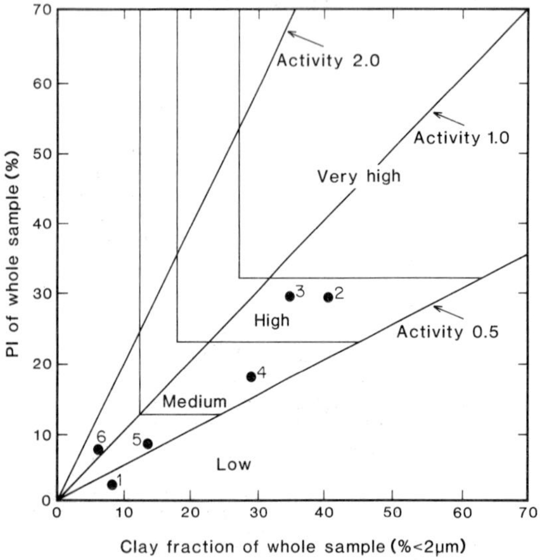

FIG. 2. Determination of potential expansiveness of soils (after Van der Merwe, 1965).

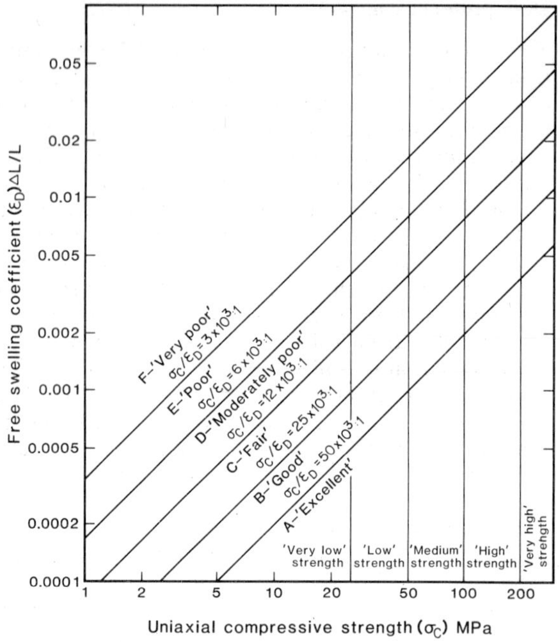

FIG. 3. Geodurability classification of the intact rock material (after Olivier, 1979).

Note:
1. ε_D determined from oven-dried (105°C) to 24-hour saturation condition.
2. ε_D plotted as the range and mean of the test results.
3. Strength ratings according to Deere and Miller (1966), as modified by Bieniawski (1973).

TABLE 6. *USAEWES classification of swell potential (from O'Neill and Poormoayed, 1980; reproduced by permission of American Society of Civil Engineers).*

Liquid limit (%)	Plastic limit (%)	Initial *in situ* suction (kPa)	Potential swell (%)	Classification
Less than 50	Less than 25	Less than 145	Less than 0.5	Low
50 to 60	25 to 35	145 to 385	0.5 to 1.5	Marginal
Over 60	Over 35	Over 385	Over 1.5	High

Failure of consolidated and poorly cemented rocks occurs during saturation when the swelling pressure (or internal saturation swelling stress, σ_s) developed by capillary suction pressures exceeds their tensile strength. An estimate of σ_s can be obtained from the modulus of deformation (E):

$$E = \sigma_s/\varepsilon_D$$

where ε_D is the free-swelling coefficient (i.e. the ratio of the change in length after swelling to initial length). Olivier (1979) proposed the geodurability classification (Fig. 3) which is based on the free-swelling coefficient and uniaxial compressive strength. This classification was developed primarily to assess the durability of mudrocks and poorly cemented sandstones during tunnelling operations, since the tendency of such rocks to deteriorate after exposure governs the stand-up time in tunnels.

Desiccation of clay soils

The volume change brought about by evaporation from a clay soil can be conservatively predicted by assuming the lower limit of the soil moisture content to be the shrinkage limit. Desiccation beyond this value cannot bring about further volume change. However, the effects of evaporation in semi-arid regions such as the highveld of South Africa may be reduced significantly by the presence of an inactive layer at the surface or by vegetation offering shade to the surface of the soil. On the other hand, transpiration from vegetative cover is a major cause of water loss from soils in semi-arid regions. Indeed the distribution of soil suction in the soil is primarily controlled by transpiration from vegetation. Very high values of soil suction have been recorded, for example, Williams & Pidgeon (1983) referred to a pF at a depth of 6 m being 4.8, whilst that at 15 m was 34.8.

The maximum soil suction which can be developed is governed by the ability of vegetation to extract moisture from the soil, that is, the water demand of the plants concerned. The field capacity of a soil refers to the point of desiccation, which defined in terms of soil suction is a pF value of approximately 2. The level at which moisture is no longer available to plants is termed the permanent wilting point and this corresponds to a pF value of about 4.2. In semi-arid regions the complete depth of active clay profiles usually does not become fully saturated during the wet season. For example, in the highveld even during the summer rainfall season the soil suction can still approach the wilting point where the soil is covered with vegetation. Nonetheless, the pF may fall to 3.2 during the wet season and changes in soil suction may be expected down to a depth of approximately 2m between the wet and dry seasons.

The moisture characteristic (moisture content v. soil suction; Fig. 4) of a soil provides valuable data concerning the moisture content corresponding to the field capacity, and the permanent wilting point as well as the rate at which changes in soil suction take place with variations in moisture content. This enables an assessment to be made of the range of soil suction and moisture content values which are likely to occur in the zone affected by seasonal changes of climate. The extent to which vegetation is able to increase the soil suction to the level associated with the shrinkage limit is obviously important. In fact the moisture content at the wilting point exceeds that of the shrinkage limit in soils with a high content of clay and is less in those possessing low clay contents. This explains why settlement resulting from the desiccating effects of trees is more notable in low to moderately expansive soils than in the more expansive ones.

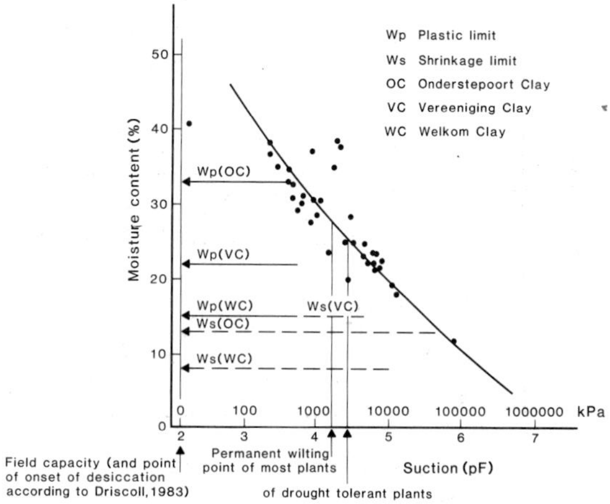

FIG. 4. Moisture content - soil suction relationships (the moisture characteristic) for some South African clays (modified after Williams & Pidgeon 1983).

TABLE 7. *Relative strength of various sandstones (Price, 1960).*

Rock type	Porosity (percent of volume)	Air dry pore water content as percentage of pore volume	Relative strengths for various pore water contents expressed as percentages of 'completely' dry rock.		
			Completely dry (σ_D)	Air dry (% of σ_D)	Saturated (% of σ_D)
Markham Sandstone (No. 21)	6.0	22 ± 5	100%	57	Not determined
Parkgate Rock (No. 23)	10.0	9 ± 1.5	100%	68	45
Pennant Sandstone (No. 26)	2.5	42 ± 3	100%	51	45
Darley Dale Sandstone (No. 27)	19.5	3 ± 1	100%	80	45

When vegetation is cleared from a site, its desiccating effect is also removed and the subsequent regain of moisture by clay soils can lead to swelling. Williams & Pidgeon (1983) noted that swelling movements on expansive clays in many areas of South Africa, associated with the removal of vegetation and the subsequent erection of buildings, amounted to about 150 mm. The largest upward movement recorded so far is 374 mm. Similar swelling movement was reported by O'Neill and Poormoayed (1980) from building sites in Texas which were located on areas of expansive clay cleared of vegetation. Uncleared areas did not swell to the same extent. In the latter areas surface movement was cyclical but irregular with an amplitude of up to 60 mm. Popescu (1979) noted that maximum seasonal changes in moisture content of the expansive clays of Rumania were around 20% at 0.5 m depth, 10% at 1.2 m and less than 5% at 1.8 m. The corresponding cyclic movements of the ground surface are about 100 to 120 mm. Desiccation of clays during the dry season leads to the soil cracking. The cracks can gape up to 150 mm and frequently extend to 2 m in depth. Popescu stated that the depth of the active zone in expansive clays in Rumania varies from about 2.0 to 2.5 m. In Australia, India, South Africa and the United States it may extend to 3 m depth, whereas in Israel depths of up to 6 m have been recorded.

The suction pressure associated with the onset of cracking is approximately pF 4.6. The presence of desiccation cracks enhances evaporation from the soil. Mahar & O'Neill (1983) assumed that such cracks led to a variable development of suction pressure, the highest suction occurring nearest the cracks. This, in turn, affects the preconsolidation pressure. Indeed it has been claimed that the effect of desiccation on clay soils is similar to that of heavy overconsolidation. For example Williams & Pidgeon (1983) suggested that if the pF of a soil rose to 4.5, then the state of stress therein approximates to a preconsolidation pressure of 3.1 MPa.

Sridharan & Allam (1982) found that repeated wetting and drying of clay soils in arid and semi-arid regions can bring about aggregation of soil particles due to cementation by Ca, Mg, Al and Fe compounds. This enhances the permeability of the clays and increases their shear strength. Such interparticle desiccation can reduce the swelling capacity of an expansive clay soil or may even render it non-expansive.

Many clay soils in Britain, especially in south-east England, possess a large potential for volume change. However, apart from periods of extreme drought, any significant deficits in soil moisture developed during the summer are confined to the upper 1.0 to 1.5 m of the soil and these are made good during the winter. Even so, deeper permanent deficits can be brought about by large trees. With this in mind, Driscoll (1983) suggested that desiccation could be regarded as commencing when the rate of change in moisture content (and therefore volume) with increasing soil suction, increases significantly. He proposed that this point approximates to a suction of pF about 2 (10 kPa). Similarly, notable suction could be assumed to have taken place if, on its disappearance, a low-rise building was uplifted due to the soil swelling. Driscoll maintained that this suction would have a pF value of about 3 (100 kPa). He went on to state that the moisture content (m%) of the soil at these two values of suction is approximately related to the liquid limit (LL) as follows:

(a) at pF = 2, m = 0.5 LL
(b) at pF = 3, m = 0.4 LL.

Hence these values of liquid limit provide crude estimates of the moisture contents at the beginning of desiccation and when it becomes significant.

Breakdown and slaking of argillaceous rocks

The natural moisture content of shales varies from less than 5% increasing to as high as 35% for some clayey shales. When the natural moisture content of shales exceeds 20% they frequently are suspect as they tend to develop potentially high pore water pressures. Usually the moisture content in the weathered zone is higher than in the unweathered shale beneath. According to Olivier (1980) the degradation processes in shales subjected to changes in the moisture content are related to partially irrevisible anisotropoic expansion and shrinkage of the rock as a result of capillary action and drying respectively.

Morgenstern & Eigenbrod (1974) carried out a series of compression softening tests on argillaceous materials. They found that the rate of softening of these materials when immersed in water largely depends upon their degree of induration. Poorly indurated materials soften very quickly and they may undergo a loss of up to 90% of their original strength within a few hours. Mudstones, at their natural water content, remain intact when immersed in water. However, they swell slowly, hence decreasing in bulk density and strength. This time-dependent loss in strength is a very significant engineering property of mudstones. A good correlation exists between their initial compressive strength and the amount of strength loss during softening.

Depending upon the relative humidity, many shales slake almost immediately when exposed to air (see Kennard *et al.* 1967). Desiccation of shale, following exposure, leads to the creation of negative pore water pressure and consequent tensile failure of the weak intercrystalline bonds. Hence shale particles of coarse sand and gravel size are produced. Alternate wetting and drying causes a rapid breakdown of compaction shales and low-grade varieties, in particular, undergo complete disintegration after a few cycles of drying and wetting.

On the other hand, well cemented shales are fairly resistant to slaking. The slaking behaviour of mudstones is often dominated by a tendency to breakage along irregular fracture patterns, which when well developed, can mean that these rocks disintegrate within one or two cycles of wetting and drying.

Russell (1984) examined the durability of two major shale formations in Ontario by means of the slake durability test. Although the shales appeared similar, one had a lower durability than the other. This was attributed to the difference in the nature of the microcracks and to the degree of cementation. In the less durable shale, the microcracks were more curved and it was also poorly cemented. Microcracks allow the penetration of water which leads to slaking and the more curved they are, the more they tend to meet on slaking thereby causing breakdown. Russell also noted that the durability of compaction shales was influenced by the amount of clay minerals present in the clastic fraction whereas this was not the case in cemented shales. Slaking in argillaceous sediments can bring about an increase in their plasticity index and augment their ability to swell. The air pressure in the pore spaces helps the development of the swell potential under cyclic wetting and drying conditions. On wetting the pore air pressure in a dry clay increases and it can become large enough to cause breakdown, which at times can be virtually explosive. The rate of wetting is important, slow wetting allows the air to diffuse through the pore water so that the pressure does not become large enough to cause disruption of the material.

Argillaceous materials are capable of undergoing appreciable suction before pore water is removed, drainage commencing when the necessary air-entry suction is achieved (about pF 2). Under increasing suction pressure the incoming air drives out water from a shale and some shrinkage takes place in the fabric before air can offer support. Generally as the natural moisture content and liquid limit increase so the effectiveness of soil suction declines.

Morgenstern & Eigenbrod (1974) used a water absorption test to assess the amount of slaking undergone by argillaceous material. This test measures the increase of water content in relation to the number of drying and wetting cycles undergone. They found that the maximum slaking water content increased linearly with increasing liquid limit and that during slaking all materials eventually reached a final water content equal to their liquid limit. Materials with medium to high liquid limits exhibited very substantial volume changes during each wetting stage, which caused large differential strains, resulting in complete destruction of the original structure. Thus materials characterised by high liquid limits are more severely weakened during slaking than materials with low liquid limits.

After a study of the disintegration of shales in water, Badger *et al.* (1956) concluded that this was brought about by two main processes, namely, air breakage and the dispersion of colloid material. It was noted that the former process only occurred in those shales which were mechanically weak whilst the latter appeared to be a general cause of disintegration. They also observed that the degree of disintegration of a shale when it was immersed in different liquids was governed by the manner in which those liquids affected air breakage and ionic dispersion forces. For example, in a liquid with a low dielectric constant little disintegration took place as a result of ionic forces because of the suppression of ionic dissociation from the shale colloids. It was found that the variation in disintegration of different shales in water was not usually connected with their total amount of clay colloid or the variation in the types of clay minerals present. It was rather controlled by the type of exchangeable cations attached to the clay particles and the accessibility of the latter to attack by water which, in turn, depended on the porosity of the shale. Air breakage could assist this process by presenting new surfaces of shale to water.

Nakano (1967) found that although some mudstones from Japan, when immersed in water swelled slowly and underwent a consequent decrease in bulk density and strength, they did not disintegrate even after immersion for a lengthy period of time. However, if they were dried and then wetted they disintegrated rapidly into small pieces. After conducting a series of slaking tests in vacuum, as well as in air, Nakano concluded that air breakage was not a significant mechanism in the breakdown of mudstones since he noticed no difference in the results of the slaking tests. He attributed the weakening of mudstone to chemical dissolution by hydrogen bonding of originally absorbed water molecules around clay particles with newly absorbed ones. He assumed that part of the free energy, which was evolved by water molecules when absorbed around clay particles, acted as the destructive force in slaking. It was observed that mudstones started to slake when drying in a relative humidity of 98%. In another instance, the relative humidity was 94%. In each case the drier the mudstone, compared with its natural state, the greater was the intensity of disintegration in water. This means that such mudstones (the clay fraction in these mudstones consisted of montmorillonite) may deteriorate readily in the zone of fluctuating water table or water vapour pressure. This does not mean that beneath these zones mudstones will not be affected, they can be softened, especially if heavily fractured.

After an investigation of the breakdown of Coal Measures rocks, Taylor & Spears (1970) found that some seatearths, notably the Brooch and Park seatearths, after desiccation, disintegrated very rapidly in water (the former was literally 'explosive' and the latter broke down in less then 30 min). Although the expandable clay content in these two rocks is high, and leads to intraparticle swelling, the authors maintained that this alone was not responsible for their rapid disintegration. In contrast to the work done by Nakano (1967), Taylor and Spears found that breakdown could be arrested by the removal of air from the samples under vacuum. Thus they concluded that air breakage was a principal disintegration mechanism in the weaker rocks. They suggested that during dry periods evaporation from the surfaces of rock fragments gives rise to high suction pressures which result in increased shearing resistance. With extreme desiccation most of the voids are filled with air, which, on immersion in water, becomes pressurised by the capillary pressures developed in the outer pore spaces. The mineral fabric may then fail along its weakest plane exposing an increased surface area to the same process. Taylor and Spears did not dismiss the physico-chemical ideas of Badger *et al.* (1956) or those of Nakano (1967), but suggested that such processes became progressively more important with time.

Strength of rocks

Much of the work on the effect of moisture on the compressive strength of rocks has concentrated on the variation of the uniaxial compressive strength between the dry and saturated state. Price (1960) tested a number of Coal Measure sandstones and found that the compressive strength for air-dried samples was between 51% and 80% of that for 'completely' dry samples and for saturated samples was 45% of the 'completely' dry strength. The loss of strength between the 'completely' dry and the air-dry state would appear to relate to the degree of saturation and the porosity (Table 7). In the semi-saturated state water tends to form a film on grain surfaces. As the ratio of grain surface area to pore volume is higher in low porosity rocks, the pore water content (i.e. the degree of saturation is higher for rocks of low porosity. It would seem, therefore, that for sandstones of low porosity in particular, much of the loss of strength on wetting takes place at very low degrees of saturation.

From his work on air-dry and saturated Fell Sandstone, Bell (1978) stated that the influence of moisture saturation on strength appears to become less effective as the strength of sandstones increases. This is only true when comparing the air-dry with the saturated strength, because, as the porosity decreases (and the strength increases), more of the strength loss on wetting takes place between the 'completely' dry and the air-dry states.

West (1979) demonstrated a loss of compressive strengths of 38% between the air-dry and saturated states for Bunter Sandstones from Warrington. From the results he derived a linear relationship between the uniaxial compressive strength and moisture content.

Colback and Wiid (1965) worked on quartzitic sandstones and shales from South Africa. A number of compression test were carried out at eight different moisture contents. These tests indicate that for both rock types the saturated strength is about half the 'completely' dry strength and that the relationship between strength and moisture content is only linear over the 'wetter' part of the curve (Fig. 5). Work on other rock types such as schist and gneiss by Feda (1966) and basalts, granite, gneiss and limestone by Ruiz (1966) shows similar loss of strength on saturation.

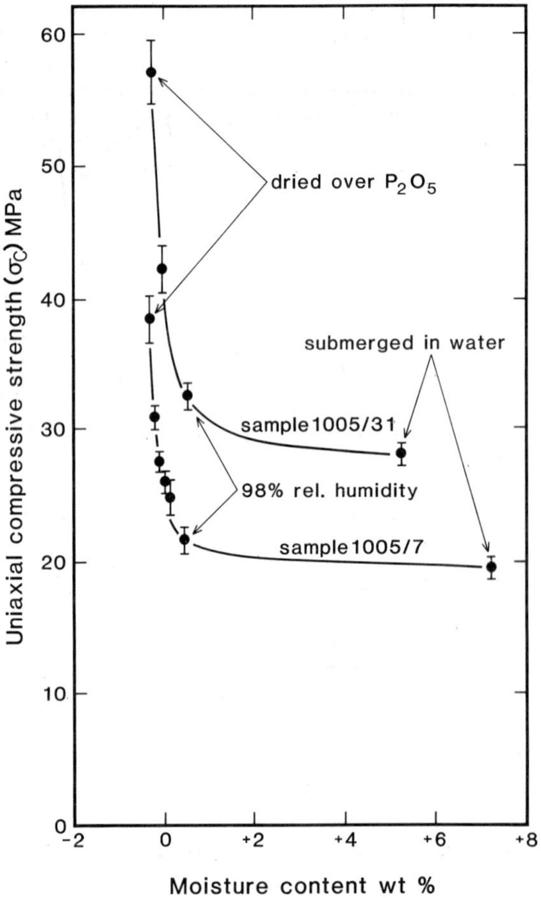

FIG. 5. Relationship between uniaxial compressive strength and moisture content for quartzitic sandstone specimens (after Colback & Wiid, 1965). Note: Moisture content zero datum defined as 50% relative humidity.

Colback & Wiid (1965) plotted the Mohr envelope for quartzite shale at two different moisture contents. From Fig. 6 it will be noted that the slopes of the Mohr envelopes are not sensibly different, indicating that the coefficient of internal friction is not significantly affected by changes in moisture content. Colback & Wiid therefore tentatively concluded that the reduction in strength witnessed with increasing moisture content, was primarily due to a lowering of the tensile strength, which is a function of the molecular cohesive strength of the material. Tests of specimens of quartzitic sandstone showed that their uniaxial compressive strength was inversely proportional to the surface tension of the different liquids into which they were placed. As the surface free energy of a solid submerged in a liquid is a function of the surface tension of the liquid, and since the uniaxial compressive strength is directly related to the uniaxial tensile strength, and this is to the molecular cohesive strength, it was postulated that the influence of the immersion liquid was to reduce the surface free energy of the rock and hence its strength. The authors therefore concluded that the reduction in strength from the dry to the saturated condition of predominantly quartzitic rocks was a constant which was governed by the reduction of the surface free energy of quartz due to the presence of any given liquid.

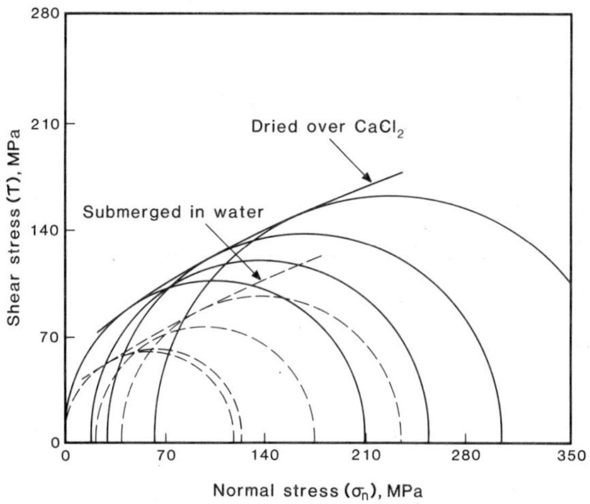

FIG. 6. Mohr envelope for quartzitic shale at two moisture contents (after Colback & Wiid 1965).

Volume change in loess, organic soils and fills

Instances of dramatic subsidence have been attributed to increases of moisture content of loess, a type of continental wind blown silt deposit, certain alluvial fan deposits formed under arid or semi-arid conditions and loosely compacted fills. In a dry state these materials may be sufficiently strong to resist void closure due to the overburden present. However, wetting weakens interparticle bonding to the extent that compaction takes place. Subsidence may also occur as a consequence of the drying out of organic soils.

Extensive damage to buildings, roads and services in the San Joaquin Valley of California, has been caused by hydrocompaction processes following an increase in moisture content of the underlying ground. Bull (1964) explained that materials susceptible to this type of behaviour comprise mudflow and other soils intermediate between waterlain and mudflow deposits. In the area in question, these contain about 20% clay (less than 0.004 mm) size particles and are in a desiccated condition with water contents between wilting and hydroscopic values. Under these circumstances the clay which coats the silt sized particles, develops high suction pressures that impart high strength to the material. Although many of these deposits also contain gypsum, it would appear that this mineral is not leached, so the principal cause of hydrocompaction is the loss of strength which accompanies a reduction in suction forces when the material is wetted. The problem is not prevalent in water lain deposits since in these the clay size fraction is concentrated into seams rather than it filling the interparticle void space or coating the sand and silt sized particles. Neither are deposits with small amounts of clay susceptible to hydrocompaction since they are insufficiently strong to resist void space closure as the sedimentary load increases. It is significant that in the alluvial fan deposits of the San Jaoquin Valley, the principal clay mineral is montmorillonite. In deposits with high clay size fractions little or no structural collapse occurs due to wetting due to the clay swelling, in fact volume increase may occur.

Besides compositional factors and the moisture condition of the material, the depth of overburden also directly controls the amount of subsidence consequent upon hydrocompaction. In the San Joaquin Valley most settlement occurs within the upper 60 m of deposits. Subsidence commonly attains 1.0 to 1.6 m with up to about 3.2 m in some small areas. The rate of compaction depends on the progress of the wetting front, but where continuous seepage of surface water has occurred, initial rates of about 75 mm per day have been recorded.

Ireland *et al.* (1984) describe pre-treatment carried out in connection with the California Aqueduct which crosses a number of areas susceptible to hydrocompaction. In these, ponds about 135 m wide were created along the alignment and in addition, infiltration wells were drilled on a 33 m grid over the central two-thirds of the ponds. During the 12 to 18 months period in which water depths of 0.5 to 0.6 m were maintained in the ponds, subsidence of the order of 0.8 to 1.6 m occurred.

Loess is a material which is liable to undergo appreciable volume reduction if wetted. Again collapse is attributed to the softening of clay bonding so that the soil settles under its own weight. Larionov (1965) recognised three micro-structures in loess namely, granular, where a filmy distribution of the fine dispersed fraction predominates; aggregate, consisting mainly of aggregates; and granular-aggregate, having an intermediate character. He suggested that generally loess soils with granular micro-structure are more likely to collapse on wetting than the aggregate types.

Several collapse criteria have been proposed which depend upon the void ratios at the liquid limit (e_l) and the plastic limit (e_p). According to Audric &

Bouquier (1976) collapse is probable when the natural void ratio is higher than a critical void ratio (e_c) which depends on e_l and e_p. They quoted the Denisov and Feda criteria as providing fairly good estimates of the likelihood of collapse:

$$e_c = e_l \text{ (Denisov)}$$
$$e_c = 0.85e_l + 0.15e_p \text{ (Feda).}$$

The main feature of the collapsible soil is the soil structure, which is formed by soil domains with a low number of grain contacts. Clevenger (1958) noted that on wetting, large settlements and low shearing resistance are encountered when the density of loess is below 1.28 Mg/m^3, whereas if the density exceeds 1.44 Mg/m^3, then settlement is small and shearing resistance fairly high. Moreover, the results of plate load tests have indicated that the bearing capacity of low-density loess may exceed 540 kPa when dry, and fall to as low as 27 kPa when wetted. The supporting capacity is reduced notably in low-density loess when the moisture content exceeds 15%. Indeed laboratory tests have shown that samples of low-density loess consolidate between 15 and 20% when prewetted (little consolidation occurs if there is no prewetting). It is not surprising therefore, that large settlements have occurred beneath footings in loess after it has been wetted.

Increased rates of settlement or a sudden resumption of settlement may occur due to the increased moisture content of domestic refuse, opencast mining and other fills. The problem has also been observed in compacted natural soils and rock fills, particularly if these materials are compacted dry of their optimum water contents.

Jennings & Burland (1962) referred to collapse phenomena due to inundation in laboratory tests on air dried and partly saturated silts. They attributed collapse which occurred even for small loads, to the loss of the bond effect produced by surface tension forces acting in water menisci at inter-particle points of contact. An alternative explanation is provided by Sowers *et al.* (1965) following laboratory tests. These demonstrated the significance of a reduction of rock strength at points of contact due to the entry of water into microfissures in the highly stressed contact zones.

The amount of settlement depends on many factors, including the depth, type and density of the fill. Often the exact nature of fill can be difficult to ascertain and the density depends on the method of emplacement. Leigh & Rainbow (1979) gave details of several backfill operations following opencast mining in which increased settlement has been attributed to a rise in groundwater levels. In one particular case in which the backfill consisted predominantly of shale excavated by scraper and drag line equipment, a large increase in the rate of settlement coincided with the cessation of groundwater pumping from an adjacent site. During the nine months following the completion of backfilling and prior to the cessation of pumping, the 21.8 m thickness of fill settled 24 mm. The accelerated settlement then gave rise to a total settlement of 168 mm in the succeeding two years.

Charles *et al.* (1977) described an opencast site at Horsley, Northumberland, in which 63.1 m of sandstone and mudstone fill was placed. During the initial 8-months during which a low water table was maintained, surface settlement at one particular measuring station amounted to 8 mm. Again a gradual rise in watertable was accompanied by an increased rate of settlement such that in the succeeding 32 months the total surface settlement amounted to 303 mm. The vertical strain measured depth ranges of 6 m within the fill was very variable. A maximum compression of 1.4% was recorded whereas extensions occurred in some places.

Although rising groundwater levels associated with the cessation of mining operations are commonly the cause of problems, ingress of water from the surface may also be responsible for the wetting. Charles *et al.* (1978) described a case in which surface water entered via trenches dug in the course of construction work. Many materials may be susceptible to this type of behaviour, for example, domestic refuse (see Sowers 1973), rock fill (see Sowers *et al.* 1965) and natural soils (see Clayton & Simons 1981). Settlement may occur even when a considerable time has elapsed between backfilling and wetting. Leigh & Rainbow (1978) describe a case in which structural damage to a building took place 24 years after backfilling. The cessation of pumping in the area 10 years before the damage occurred apparently caused partial inundation of the 10 to 18 m thickness of backfill which settled 1 to 2% as a result.

All organic soils subside when drained due to shrinkage upon drying out and consolidation due to the loss of the buoyant force of groundwater. For instance, the organic soils in the Everglades of Florida undergo an average subsidence of 25 mm per year; in the Sacramento-San Joaquin delta of California the average rate reported is 760 mm per year; and for the organic soils of Michigan the average rate was 91.6 mm per year. Stephens & Speir (1969) maintained that the subsidence rates for organic soils are positively correlated with the depth of groundwater -the higher the water table, the lower the rate of subsidence.

Peat is the most compressible of materials and is highly porous so that its water content may range up to 2000%. A classic example of subsidence of peat due to drainage is provided in the Fenlands. There drainage of the peat has taken place over the last 400 years. In some parts of the Fenlands the thickness of peat has almost been halved as a result. For example, in the period between 1848 and 1932 a total subsidence of 2.7 m was recorded by the Holme Post, the original thickness of peat being 6.7 m. Subsidence due to the reclamation of peat land has also taken place in the New Orleans area (see Traughber *et al.* 1979).

Effects of changes in effective stress

Below the water table the water contained in the pores of soil or rock is under normal hydrostatic load, the prewater pressure increasing with depth. Because these pressures exceed atmospheric pressure they are designated positive pressures. On the other hand the pressures existing in the capillary zone are less than atmospheric and so are termed negative pressure. Thus the water table is usually regarded as a datum of zero pressure between the positive pore water pressure below and the negative above.

Such water pressures have a significant influence on the engineering behaviour of most rock and soil masses, and their variations are responsible for changes in the stresses in these masses, which affect their deformation characteristics and failure.

The efficiency of a soil in supporting a structure is influenced by the effective or intergranular pressure, that is, the pressure between the particles of the soil which develops resistance to applied load. Because the moisture in the pores offers no resistance to shear, it is ineffective or neutral. Since the pore water or neutral pressure plus the effective pressure equals the total pressure, reduction in pore water pressure increases the effective pressure. Reduction of the porewater pressure by drainage consequently renders soil stronger and affords better conditions for carrying a proposed structure.

The presence of water in the voids of a granular soil does not usually produce significant changes in the value of the angle of shearing resistance. However, if stresses develop in the pore water they may bring about changes in the effective stresses between the particles whereupon the shear strength and the stress-strain relationships may be radically altered. Whether or not pore water pressures develop depends upon the drainage characteristics of the soil mass and its tendency to undergo volume changes when subjected to stress. If the pore water can readily drain from the soil mass during the application of stress, then the granular material behaves in a similar manner as it does when dry. On the other hand if loading takes place rapidly, particularly in fine grained sands which do not drain as easily, then pore water pressures are not dissipated. Since the water cannot escape readily from the voids of loosely packed, fine grained sands, no volume decrease can occur so the pressure increases in the pore water. If the sample is loose enough, almost the entire stress difference may be carried by the pore water so that very little increase occurs in the effective stress. In dense sands if the stress and drainage conditions are such that the water cannot flow into the sand as it is stressed, then the usual volume increase characteristic of dense dry sand does not occur and a negative pore water pressure develops.

There is some evidence which suggests that the law of effective stress as used in soil mechanics, in which the pore water pressure is subtracted from all direct stress components, holds true for some rocks. Those with low porosity may, at times, prove the exception. However, Serafim (1968) suggested that pore water pressures have no influence on brittle rocks. This is probably because the strength of such rocks is mainly attributable to the strength of the bonds between the component crystals or grains.

Pore water pressures also depend on the overall hydrological regime in which the soil or rock is situated. The movement of water through the ground is governed by a complex interaction between the permeability of the ground and the gravitational potential of sources and sinks within the system. Generally, unconsolidated or poorly cemented arenaceous or rudaceous materials or jointed rocks constitute the aquifers whilst aquicludes consist of fine-grained mudrocks or well-cemented or crystalline unjointed rocks. Water gaining access to an aquifer at a high level may become confined by an overlying aquiclude where the buried surface of the aquifer is lower than the water table in the recharge area. Under these circumstances there is an excess of pressure necessary to raise the water above the base of the confining layer. Where this pressure is sufficient to raise the water level above the ground level the conditions is described as artesian; otherwise it is known as sub-artesian and the water rises to a lower level. Artesian and sub-artesian conditions most commonly occur in the eroded cores of synclinal structures in systems of permeable and impermeable beds. They may also arise where impermeable igneous instrusions or impermeable fault gouge produce barriers that inhibit the movement of water through the ground.

Critical hydraulic gradient, quick conditions and hydraulic uplift

As water flows through the soil and loses head, its energy is transferred to the particles past which it is moving. This in turn creates a drag effect on the particles. If the drag effect is in the same direction as the force of gravity, then the effective pressure is increases and the soil is stable. Indeed the soil tends to become more dense. Conversely if water flows towards the ground surface, then the drag effect is counter to gravity thereby reducing the effective pressure between particles. In certain soils if the velocity of upward flow is sufficient it can buoy up the particles so that the effective pressure is reduced to zero. This represents a critical condition where the weight of the submerged soil is balanced by the upward acting seepage force. A critical condition sometimes occurs in silts and sands. If the upward velocity of flow increases beyond the critical hydraulic gradient a quick condition develops.

Quicksands, if subjected to deformation or disturbance, can undergo a spontaneous loss of strength. This loss of strength causes them to flow like viscous liquids. Terzaghi (1925) explained the quicksand phenomenon in the following terms.

Firstly, the sand or silt concerned must be saturated and loosely packed. Secondly, on disturbance the constituent grains become more closely packed which leads to an increase in pore water pressure, reducing the forces acting between the grains. This brings about a reduction in strength. If the pore water can escape very rapidly the loss in strength is momentary. Hence the third condition requires that pore water cannot escape readily. This is fulfilled if the sand has a low permeability and/or the seepage path is long. Casagrande (1936) demonstrated that a critical porosity existed above which a quick conditions could be developed. He maintained that many coarse-grained sands, even when loosely packed, have porosities approximately equal to the critical condition whilst medium- and fine-grained sands, especially if uniformly graded, exist well above the critical porosity when loosely packed. Accordingly fine sands tend to be potentially more unstable than coarse grained varieties. It must also be remembered that the finer sands generally have lower permeabilities.

As the velocity of the upward seepage force increases further from the critical gradient the soil begins to boil more and more violently. At such a point there is a complete loss of bearing capacity so that any supported structure sinks. Liquefaction of potential quicksands may be caused by sudden shocks such as the action of heavy machinery (notably pile driving), blasting and earthquakes. Such shocks increase the stress carried by the water, the neutral stress, and give rise to a decrease in the effective stress and shear strength of the soil. There is also a possibility of a quick condition developing in a layered soil sequence where the individual beds have different permeabilities. Hydraulic conditions are particularly favourable where water initially flows through a very permeable horizon with little loss of head, which means that flow takes place under a great hydraulic gradient.

When water percolates through heterogeneous soil masses it moves preferentially through the most permeable zones and it issues from the ground as springs. Piping refers to the erosive action of some such springs, where sediments are removed by seepage forces, so forming subsurface cavities and tunnels. In order that erosion tunnels may form, the soil must have some cohesion, the greater the cohesion, the wider the tunnel. In fact fine sands and silts are most susceptible to piping failures. Obviously the danger of piping occurs when the hydraulic gradient is high, that is, when there is a rapid loss of head over a short distance. As the pipe develops by backward erosion it nears the source of water supply so that eventually the water breaks into and rushes through the pipe. Ultimately, the hole, so produced, collapses from lack of support.

Hydraulic uplift phenomena occur under a wide range of geological conditions and at differing scales. For example, Moore & Longworth (1970) recorded a failure in a brick pit, 29 m in depth, excavated in Oxford Clay. The failure was brought about by a build-up of hydrostatic pressure in an underlying aquifer (either the Cornbrash or Blisworth Limestone located at depths of 6 and 11 m respectively below the surface of the pit). This initially gave rise to a heave of some 1500 mm, and then ruptured the surface clay, thereby allowing the rapid escape of approximately 7000 m^3 of water. The floor of the pit then settled up to 100 mm. Structures founded below ground water level are acted upon by uplift pressures. If the structure is weak this pressure can break it and, for example, cause a blow-out of a basement floor or collapse of a basement wall. If the structure is strong, but light, it may be lifted, that is, subjected to heave.

Consolidation and subsidence

As indicated in Table 8, subsidence attributable to the withdrawal of groundwater has occurred in numerous regions throughout the world. It has developed with most effect in those groundwater basins where there was, or is, intensive abstraction (see Poland, 1984). Such subsidence is attributed to the consolidation of sedimentary deposits as a result of increasing effective stress. The total overburden pressure of partially saturated or saturated deposits is borne by their granular structure and the pore water. When groundwater abstraction leads to a reduction in pore water pressure by draining water from the pores, there is a gradual transfer of stress from the pore water to the granular structure. For instance, if the water table is lowered by 1 m, then this gives rise to a corresponding increase in average effective overburden pressure of 10 kPa. As a result of having to carry this increased load the granular structure may deform in order to adjust to the new stress condition.

Consolidation may be elastic or non-elastic depending upon the character of the deposits involved and the range of stresses induced by a decline in the water level. In elastic deformation, stress and strain are proportional, and consolidation is independent of time and reversible. Non-elastic consolidation occurs when the granular structure of a deposit is rearranged to give a decrease in volume, that decrease being permanent. Generally recoverable consolidation represents compression in the pre-consolidation stress range, while irrecoverable consolidation represents compression due to stresses greater than the pre-consolidation pressure.

The amount of subsidence which occurs is governed by the increase in effective pressure, the thickness and compressibility of the deposits concerned, the length of time over which the increased loading is applied, and possibly the rate and type of stress applied. For example Deflache (1979) reported that the most noticeable subsidence in the Houston-Galveston region of Texas has occurred where the declines in head have been largest and where the thickness of clay in the aquifer system is greatest. Furthermore, the ratio between maximum subsidence and ground-

TABLE 8: *Areas of major land subsidence due to groundwater overdraft (from Poland 1972).*

Location	Depositional environment and age	Depth range of compacting beds (m)	Maximum subsidence (m)	Area of subsidence (km^2)	Time of principal occurrence	Remedial or protective measures taken
Japan, Osaka	Alluvial and shallow marine; Quarternary	9-400	3	192	1928-68	Reduced groundwater pumpage; built dikes, drainage pumping plants
Tokyo	As above	9-305	4.3	192	1920-70+	Reduced groundwater pumpage; built dikes, drainage pumps; pumping stations.
Mexico Mexico City	Alluvial and lacustrine; Late Cenozoic	9- 49	8.6	128	1938-70+	Reduced pumpage; imported water; built recharge wells.
Taiwan, Taipei Basin	Alluvial and lacustrine; Quaternary	9-244	1.3	128	1961-69+	Groundwater management code adopted; recharge planned.
United States, Arizona Central	Alluvial and lacustrine; late Cenozoic	91-550	2.3	640	1948-67	
California, Santa Clara Valley	Alluvial and shallow marine; late Cenozoic	55-305	4	640	1920-70	Built detention dams; increased local recharge; built dikes; imported water.
San Joaquin Valley (three subareas)	Alluvial and lacustrine; late Cenozoic	61-915	2.9-8.5	10750 (>0.3 m)	1935-70+	Built reservoirs and imported water to reduce groundwater pumpage.
Lancaster Area	Alluvial and lacustrine;	61-305	0.9	384	1955-67+	
Nevada, Las Vegas	Alluvial; late Cenozoic	61-305	0.9	512	1935-63	Moved well field away from fine-grained deposits; imported Colorado River water.
Texas, Houston Galveston area	Fluvial and shallow marine; late Cenozoic	61-610	0.9-1.5	6784 (>0.15 m)	1943-64+	Plans for surface-water imports under way
Louisiana, Baton Rouge	Fluvial and shallow marine; Miocene to Holocene	46-610	0.3	640	1934-65+	

water reservoir consolidation is related to the ratio between depth of burial and the lateral extent of the reservoir. In other words, small reservoirs which are deeply buried do not give rise to noticeable subsidence, even if subjected to considerable consolidation. By contrast, extremely large underground reservoirs may develop appreciable subsidence.

The rate at which consolidation occurs depends on the rate at which the pore water can drain from the system which in turn, is governed by its permeability. For instance, the low permeability and high specific storage of aquitards and aquicludes under virgin stress conditions means that the escape of water and resultant adjustment of pore water pressures is slow and time-dependent. Consequently, in fine-grained beds the increase in stress which accompanies the decline in head becomes effective only as rapidly as the pore water pressures are lowered towards equilibrium with the pressure in adjacent aquifers. The time required to reach this stage varies directly according to the specific storage and the square of the thickness of the zone from which drainage is occurring and inversely according to the vertical permeability of the aquitard. In fact, it may take months or years for fine-grained beds to adjust to increases in stress. Moreover, the rate of consolidation of slow-draining aquitards reduces with time and is usually small after a few years of loading.

FIG. 7. Historical faulting induced by groundwater level decline in Fremont Valley in southern California. Offset has occurred along a pre-existing geological fault (by courtesy of T.L. Holzer).

In addition to being the most prominent effect in subsiding groundwater basins, surface fissuring and faulting (Fig 7) may develop suddenly and therefore pose a greater potential threat to surface structures. In the USA such fissuring and faulting has occurred especially in the San Joaquin Valley, the Houston-Galveston region and in central Arizona. These fissures, and more particularly the faults, frequently occur along the periphery of the basin. The faults are high angled, normal faults, with the downthrow on the side towards the centre of the basin. Generally, displacements along the faults are not great, less than a metre, but movements may continue over a period of years. Holzer *et al.* (1979) argued that such faults did not extend beneath the zone where stresses due to lowering of the groundwater level occur. Hence their depth is shallow.

It is not only falling or low groundwater levels that cause problems, a rising or high water table can be equally troublesome. Since the mid 1960s, the rate of abstraction from the Chalk below London has decreased significantly so that groundwater levels are now increasing by as much as 1 m/year in some places (see Hurst & Wilkinson 1986). It must be borne in mind that natural groundwater levels in these areas had been lowered appreciably in the nineteenth century and first half of the twentieth by excessive abstraction.

A number of problems arise as a result of a rising water table. Obviously tunnels and deep basements are likely to be affected by leakage of groundwater into them, a problem now being faced in certain areas of Birmingham and Liverpool. One of the major changes is that in the pore water pressure regime. As the water table rises, saturation of the pore spaces in the soil or rock reduces the effective pressure, in the case of some soils by up to 50% of more. In turn, this lowers their strength and bearing capacity. Structures with deep foundations could be adversely affected in the future because of the decline in bearing capacity. The design for new structures will need to take these changing conditions into account.

In Louisville, Kentucky, increasing groundwater levels are causing concern over the possibility of structural settlement, damage to basement floors and the disruption of utility conduits. The control of groundwater levels has, therefore, an importance which extends beyond water supply considerations.

Effects of solution

The movement of groundwater within rock masses composed of a significant proportion of soluble minerals exerts a major control over their removal in solution. The actual site of dissolution is often controlled by the presence of rock mass discontinuities and other variations in mass permeability, including low permeability clay seams. However, besides removal from surfaces, reductions in bulk density or cavity formation may also be prevalent. The amount of mineral removed from the rockmass depends mainly on the rate of flow, quantity, temperature and chemistry of the water passing through it. Also of major significance is the solubility characteristics of the rock forming mineral. Natural processes liable to enhance the aggressiveness of water are considered in the context of weathering activity on page 6 of this paper. The effects of solution will be considered with reference to the commonly occurring rock formations most at risk from solution activity, namely calcareous rocks, gypsum and anhydrite and saliferous deposits.

Calcareous rocks

Limestones are chiefly composed of calcium carbonate and they are susceptible to acid attack because CO_3 readily combines with H to form the stable bicarbonate HCO_3.

The degree of aggressiveness of water to limestone can be assessed on the basis of the relationship between the dissolved carbonate content, the pH value and the temperature of the water. At any given pH value, the cooler the water the more aggressive it is. If solution continues its rate slackens and it eventually ceases when saturation is reached. Hence solution is greatest when the bicarbonate saturation is low. This occurs when water is circulating so that fresh supplies with low lime saturation are continually made available. As James & Kirkpatrick (1980) noted, a material dissolves at a rate and in a manner which is influenced by its solubility (C_s) and specific solution rate constant (K). The form a particular mineral adopts does not influence its solution rate. Not only is the rate of flow significant but the area of material exposed to flowing water is also important. Non-saline water can dissolve up to 400 ppm of calcium carbonate.

In water with a temperature of 25°C the solubility of calcium carbonate ranges from 0.01-0.05 g/l, depending upon the degree of saturation with carbon dioxide. Dolostone is somewhat less soluble than limestone. The solution of limestone is a very slow process. For instance, Kennard & Knill (1968) quoted mean rates of surface lowering of limestone areas in the British Isles which ranged from 0.041-0.099 mm annually. They also quoted experiments carried out in flowing non-saline water which produced an average solution rate of approximately 1 mm per year. Nevertheless solution may be accelerated by man-made changes in the groundwater conditions or by a change in the character of the surface water that drains into limestone. For instance, James & Kirkpatrick (1980) wrote that if dry fissured rocks are subjected to substantial hydraulic gradients then they undergo dissolution along the fissures, hence leading to rapidly accelerating seepage rates. From experimental work which they carried out on Portland Limestone they found that the values of solution rate constant (K) increased appreciably at flow velocities which corresponded with a transitional flow regime. They showed that such a flow regime occurred in fissures about 2.5 mm in width which experienced a hydraulic gradient of 0.2. According to these two authors solution takes place along a small fissure by retreat of the inlet face due to removal of soluble material. Dissolution of larger fissures gives rise to long tapered enlargements, that enable seepage rates to increase rapidly and runaway situations to develop.

Groundwater solution creates caves underground by secondary enlargement of primary fractures, and closed depressions with internal drainage are formed at the outcrop. These depressions are referred to as dolines, or, more commonly, sinkholes.

Caves and sinkholes are diagnostic features of the karst landscape found in many of the areas underlain by limestone. Sinkholes can range in size from 0.5 m to 500 m in both depth and radius, and can be classified genetically into solution, collapse, subsidence (or alluvial) and buried (or filled) sinkholes (Culshaw & Waltham 1986). Many have complex origins.

Solution sinkholes are the normal result of surface lowering in a karst landscape, their rate of formation being measured on a geological time scale. The existence of these features is dependent on drainage through caves and microcaves in the underlying bedrock. Like all sinkholes their location may be influenced by geological structure and they occur most commonly along the outcrop edge of impermeable rocks.

Collapse sinkholes are formed by surface collapse into caves. Because of the inherent strength of most cavernous limestones, and the relative scarcity of large caves they are rarer than is commonly thought. Progressive collapse of a heavily fissured zone of bedrock is more usual than catastrophic collapse of a cave roof. The collapse process is often contemporaneous with solution erosion, creating sinkholes of multiple origin.

Subsidence sinkholes form by the subsidence of a cover rock or soil into cavities in an underlying limestone. They are commonest in unconsolidated drift, which gives rise to their alternative name of alluvial sinkholes. They are also known as ravelling sinks (Beck 1984). There are two basic types of subsidence sinkhole in unconsolidated deposits. In sand the cover sediment is progressively sapped down into the limestone fissures. No large cavities can form and the result is slow surface subsidence. A clay cover is, however, capable of bridging a void in the sediment as the lower material is flushed into the limestone. The cavity enlarges by upward stoping until the surface fails suddenly in what is known as a drop-out. The locations of incipient dropouts are difficult to predict. Experience in Florida has shown that they form most commonly where the cover is less than 20 m thick (Steward, 1966) whilst in Missouri they frequently occur where soil thickness ranges from 12 to 30 m (Williams & Vineyard, 1976). A number of conditions accelerate the development of cavities in the soil and initiate collapse. Rapid changes in moisture content lead to aggravated slabbing or roofing in clays and flow in cohesionless sands. Lowering the water table increases the downward seepage gradient and accelerates downward erosion; reduces capillary attraction in sand and increases instability of flow through narrow openings. This also gives rise to shrinkage cracks in highly plastic clays which weakens the mass in dry weather and produces concentrated seepage during rains. Increased infiltration often initiates failure, particularly when it follows a period when the water table has been lowered.

Buried sinkholes occur in a limestone rockhead and are filled or buried by overlying sediment, commonly with no surface expression. They may form by subsurface solution, or as normal subaerial sinkholes subsequently filled by sediment as a result of environmental change. There may be associated cavities in either the soil cover or bedrock.

In chalk, solution sinkholes of significant size are rare and collapse sinkholes almost non-existent. Large subsidence sinkholes formed over geological lengths of time are known but active ones are on a small scale and are limited in distribution. In Britain they mostly occur in covers of unconsolidated sediment 1 to 10 m thick. Burial sinkholes are a more common feature in chalk and are chiefly found along the marginal outcrop close to the boundary with Tertiary deposits. They are mostly small features with diameters from 1 to 20 m and depths rarely more than 10 m.

An important effect of solution in some limestones is enlargement of the pores, which enhances water circulation thereby encouraging further solution. This brings about an increase in stress within the remaining rock framework which reduces the strength of the rockmass and leads to increasing stress corrosion. On loading the volume of the voids is reduced by fracture of the weakened cement between the particles and by the reorientation of the intact aggregations of rock that become separated by loss of bonding. Most of the resultant settlement takes place rapidly within a few days of the application of load. Sowers (1975) related a case of settlement of two- and three-storey reinforced concrete buildings founded on soft porous oolitic limestone in Florida, which had been subjected to dissolution. The limestone had been leached and was highly porous with little induration remaining 2.1 m below the rock surface. One of the buildings was surrounded by a crack in the more intact surface crust and settlement exceeding 100 mm was recorded.

Gypsum and anhydrite

Gypsum is more readily soluble than limestone, 2100 ppm can be dissolved in non-saline waters as compared with 400 ppm. Sinkholes and caverns can therefore develop in thick beds of gypsum (see Eck & Redfield, 1965) more rapidly than they can in limestone. Indeed in the United States such features have been known to form within a few years where beds of gypsum are located beneath dams. Extensive surface cracking and subsidence has occurred in certain areas of Oklahoma and New Mexico due to the collapse of cavernous gypsum (see Brune 1965). The problem is accentuated by the fact that gypsum is weaker than limestone and therefore collapses more readily. Kendal & Wroot (1924) quoted vivid, but highly exaggerated, accounts of subsidence due to the solution of gypsum which occurred in the Ripon district of Yorkshire in the eighteenth and nineteenth centuries. They wrote that wherever beds of gypsum approach the surface their presence can be traced by broad funnel-shaped craters formed by the collapse of overlying marl into areas from which gypsum has been removed by solution. Apparently these craters only took a matter of minutes to appear at the surface. However, where gypsum is effectively sealed from the ingress of water by overlying impermeable strata such as marl, dissolution does not occur (see Redfield 1968).

The solution rate of gypsum or anhydrite is principally controlled by the area of their surface in contact with water and the flow velocity of water associated with a unit area of the material. Hence the amount of fissuring in a rock mass, and whether it is enclosed by permeable or impermeable beds, is most important. Solution also depends on the sub-saturation concentration of calcium sulphate in solution. According to James & Lupton (1978) the concentration dependence for gypsum is linear whilst that for anhydrite is a square law. The salinity of the water also is influential. For example, the rates of solution or gypsum and anhydrite are increased by the presence of sodium chloride, carbonate and carbon dioxide in solution. It is therefore important to know the chemical composition of the groundwater.

Massive deposits of gypsum are usually less dangerous than those of anhydrite because gypsum tends to dissolve in a steady manner forming caverns or causing progressive settlements. For instance, if small fissures occur at less than 1 m intervals solution usually takes place by removal of gypsum as a front moving 'downstream' at less than 0.01 m/year. However, James & Lupton (1978) showed that if the following conditions were met:

i. the rock temperature was 10°C,
ii. the water involved contained no dissolved salts, and
iii. a hydraulic gradient of 0.2 was imposed,

then a fissure 0.2 mm in width and 100 m in length, in massive gypsum, would, in 100 years have widened by solution so that a block 1 m^3 in size could be accommodated in the entrance to the fissure. In other words a cavern would be formed. If the initial width of the fissure exceeds 0.6 mm, large caverns would form and a runaway situation would develop in a very short time. In long fissures the hydraulic gradient is low and the rate of flow is reduced so that solutions become saturated and little or no material is removed. Indeed James & Lupton implied that a flow rate of 10^{-3} m/s was rather critical in that if it was exceeded, extensive solution of gypsum could take place. Solution of massive gypsum is not likely to give rise to an accelerating deterioration in a foundation if precautions such as grouting are taken to keep seepage velocities low.

Massive anhydrite can be dissolved to produce uncontrollable runaway situations in which seepage flow rates increase in a rapidly accelerating manner. Even small fissures in massive anhydrite can prove dangerous. If anhydrite is taken in the above example not only is a cavern formed, but the fissure is enlarged as a long tapering section. Within about 13 years the flow rate increases to a runaway situation. However, if the fissure is 0.1 mm in width then the solution becomes supersaturated with calcium sulphate and gypsum is precipitated. This seals the outlet from the fissure and from that moment any anhydrite in contact with the water is hydrated to form gypsum. Accordingly 0.1 mm width seems to be a critical fissure size in anhydrite.

Anhydrite is less likely to undergo catastrophic solution in a fragmented or particulate form than gypsum. Conversion to gypsum is more characteristic of extensive deposits of permeable granular anhydrite.

Another point which should be borne in mind, and this particularly applies to conglomerate cemented with soluble material, is that when this is removed by solution the rock is reduced greatly in strength. A classic example of this is associated with the failure of the St. Francis Dam in California in 1928. One of its abutments was founded in conglomerate cemented with gypsum which was gradually dissolved, the rock losing strength, and ultimately the abutment failed.

Saliferous deposits

Salt is even more soluble than gypsum and the evidence of slumping, brecciation and collapse structures in rocks which overlie saliferous strata bear witness to the fact that salt has gone into solution in past geological times.

Considerable natural solution of salt in Cheshire means that it does not outcrop at the surface, it usually terminates in a solution surface at a depth ranging from 70-150 m. The regions beneath such surfaces, which are subjected to circulating groundwater and solution, are referred to as wet rock head. Where beds of rock salt are beyond the reach of circulating groundwater their original thickness is preserved and these regions are referred to as dry rock head. Natural subsidence in wet rock head areas are believed to have been responsible for the formation of the Cheshire Meres but generally speaking this type of subsidence, although it operates over large areas, takes place extremely slowly, it is perhaps imperceptible within the context of historic time. Some of the Meres are floored by peat deposits and it has been suggested that they may be pre-Pleistocene in age. Keuper Marl which generally overlies the saliferous beds is incompetent which means that large solution cavities have not been able to form since the collapse of the marl was probably contemporaneous with salt removal. These collapses produced linear subsidence depressions at the surface. Where a brine run occurs beneath glacial deposits cavities have formed in the latter. These then collapse to produce rather circular shaped, small hollows at the surface. They make their appearance almost instantaneously but fortunately, they only occur infrequently. It would seem that as the sand collapses it chokes the brine run, thereby preventing further subsidence.

It is generally believed, however, that in areas underlain by saliferous beds, measurable surface subsidence is unlikely to occur except where salt is being extracted (see Bell, 1975). Perhaps this is because equilibrium has been attained between the supply of unsaturated groundwater and the salt available for solution. Exceptionally cases have been recorded of rapid subsidence, such as the 'Meade salt sink' in Kansas. Johnson (1901) explained its formation as due to solution of beds of salt located at depth. This area of water, about 60 m in diameter, formed as a result of rapid subsidence in March 1879. At the same time, 64 km to the south-west, the railway station at Rosel and several buildings disappeared due to the sudden appearance of a sinkhole.

ACKNOWLEDGEMENTS: This paper is published by permission of the Director, British Geological Survey, Natural Environment Research Council.

References

AUDRIC, T. & BOUQUIER, L., 1976. Collapsing behaviour of some loess soils from Normandy. *Quarterly Journal of Engineering Geology*, **9**, 265-278.

BADGER, C.W., CUMMINGS, A.D. & WHITMORE, R.L., 1956. The disintegration of shale. *Journal of the Institute of Fuel*, **29**, 417-423.

BEAR, J., 1979. *Hydraulics of Groundwater*. McGraw Hill, New York.

BECK, B.F. (Editor) 1984. Sinkholes: their geology, engineering and environmental impact. *Proceedings of the First Multidisciplinary Conference on Sinkholes, Orlando, Florida,* October 1984. A.A. Balkema, Rotterdam, Boston.

BELL, F.G., 1975. Salt and subsidence in Cheshire, England. *Engineering Geology*, **9**, 237-247.

BELL, F.G., 1978. The physical and mechanical properties of the Fell Sandstone. *Engineering Geology*, **12**, 1-29..

BIENIAWSKI, Z., 1973. Engineering classification of jointed rock masses. *Transactions of the South African Institution of Civil Engineers*, **15**, 335-343.

BRACKLEY, I.J.A., 1980. Prediction of soil heave from suction measurements. *Proceedings of the 7th Regional African conference on Soil Mechanics, Accra*, **1**, 159-166.

BRUNE, G., 1965, Anhrydrite and gypsum problems in engineering geology. *Bulletin of the Association of Engineering Geologists*, **3**, 26-38.

BULL, W.B., 1964. Alluvial fans and near-surface subsidence in Western Fresno County, California. *US Geological Survey Professional paper 437-A*.

CASAGRANDE, A., 1936. Characteristics of cohesionless soils affecting the stability of slopes and earth fills. *Journal of the Boston Society of Civil Engineers*, **23**, 3-32.

CHARLES, J.A., NAISMITH, W.A. & BURFORD, D., 1977. Settlement of backfill at Horsley restored opencast coal mining site. *Large Ground Movements and Structures*. Geddes, J.D. (ed.), Pentech Press, London, 229-251.

CHARLES, J.A., EARLE, W. & BURFORD, D., 1978. Treatment and subsequent performance of cohesive fill left by opencast ironstone mining at Snatchill experimental housing site, Corby. *Proceedings of the Conference on Clay Fills, London*. Institution of Civil Engineers, 63-72.

CLAYTON, C.R.T. & SIMONS, N.E., 1980. The collapse of slabs on a compacted hoggin fill. *Ground movements and Structures*. Geddes, J.D. (ed.), Pentech Press, London, 655-670.

COLBACK, P.S.B. & WIID, B.L., 1965. Influence of moisture content on the compressive strength of rock. *Symposium of the Canadian Department of Mining Technology & Survey, Ottawa*, 65-83.

CRIPPS, J.C. & TAYLOR, R.K., 1981. The engineering properties of mudrocks. *Quarterly Journal of Engineering Geology*, **14**, 325-46.

CULSHAW, M.G. & WALTHAM, A.C., 1986. Natural and artificial cavities as ground engineering hazards, *Quarterly Journal of Engineering Geology*, **19** (in press).

DARCY, H., 1856. *Les Fontaines Publiques de la Ville de Dijon*. Dalmont, Paris.

DEERE, D.U. & MILLER, R.P., 1966. Engineering classification and Index Properties for Intact Rock. *Technical Report No. AFWL-TR-65-116, US Air Force Weapons Laboratory, Kirkland Air Base, New Mexico.*

DELFLACHE, A.P., 1979. Land subsidence versus head decline in Texas. *Evaluation and Prediction of Subsidence*. Saxena, S.K. (ed.), Proceedings of the Speciality Conference of ASCE, Gainsville, 320-331.

DRISCOLL, R., 1983. The influence of vegetation on the swelling and shrinkage of clay soils in Britain. *Geotechnique*, **33,** 93-105.

ECK, W. & REDFIELD, R.C, 1965. Engineering geology problems at Sanford Dam, Texas. *Bulletin of the Association of Engineering Geologists*, **3,** 15-25.

FEDA, J., 1966. The influence of water content on the behaviour of subsoil, formed by highly weathered rocks. *Proceedings of the 1st Congress of the International Society of Rock Mechanics., Lisbon*, 1966, **1,** 283-288.

GIDIGASU, M.D., 1976. *Laterite Soil Engineering*. Elsevier, Amsterdam.

GRIGGS, D.T., 1940. Experimental flow of rocks under conditions favouring recrystallisation. *Bulletin of the Geological Society America*, **51,** 1001-1022.

GRIM, R.E., 1962. *Applied Clay Mineralogy*. McGraw-Hill, New York

HARDY, L.A., 1967. The gypsum anhydrite equilibrium at one atmosphere pressure. *The American Mineralogist*, **62,** 171-200.

HARDY, R.M., 1966. Identification and performance of swelling soil types. *Canadian Geological Journal*, **2,** 141-53.

HOLTZ, W.G. & GIBBS, H.J., 1956. Engineering properties of expansive clays. *Transactions of the American Societies of Civil Engineers.*, **121,** 641-63.

HOLZER, T.L., DAVIS, S.N. & LOFGREN, B.N., 1979. Faulting caused by groundwater extraction in south-central Arizona. *Journal of Geophysical Research*, **84,** 603-612.

HURST, C.W. & WILKINSON, W.B., 1986. Rising groundwater levels in cities. *Groundwater in Engineering Geology, Proceedings of the 21st Annual Conference of Engineering Group of Geological Society*, The Geological Society, London, this volume, 75-80.

IRELAND, R.L., POLAND, J.F. & RILEY, F.S., 1984. Land subsidence in the San Joaquin Valley, California as of 1980. *US Geological Survey Professional Paper* **437-I.**

JAMES, A.N. & LUPTON, A.R.R., 1978. Gypsum and anhydrite in foundations of hydraulic structures. *Geotechnique*, **28,** 249-272.

JAMES, A.N. & KIRKPATRICK, I.M., 1980. Design of foundations of dams containing soluble rocks and soils. *Quarterly Journal of Engineering Geology*, **13,** 189-198.

JENNINGS, J.E.B. & BURLAND, J.B., 1962. Limitations to the use of effective stresses in partly saturated soils. *Geotechnique*, **12,** 125-144.

KENDALL, P.F. & WROOT, H.E., 1924. *The Geology of Yorkshire*. Printed Privately.

KENNARD, M.F. & KNILL, J.L., 1968. Reservoirs and limestones, with particular reference to the Cow Green Scheme. *Journal of the Institute of Water Engineers*. **23,** 87-113.

KENNARD, M.F., KNILL, J.L. & VAUGHAN, P.R., 1967. The geotechnical properties and behaviour of Carboniferous shale at Balderhead Dam. *Quarterly Journal of Engineering Geology*, **1,** 3-245.

LARIONOV, A.K., 1965. Structural characteristics of loess soils for evaluating their constructional properties. *Proceedings of the 6th International Conference on Soil Mechanics and Foundation Engineering. Montreal*, **1,** 64-8.

LEIGH, W.J.P. & RAINBOW, K.R., 1979. Observations on the settlement of restored backfill of opencast mine sites. *Proceedings of the Symposium on the Engineering Behaviour of Industrial and Urban Fill, Birmingham.* Midland Geotechnical Society, E99-E128.

LOFGREN, B.N., 1968. Analysis of stress causing land subsidence. *US Geological Survey Professional Paper*, 600-B, B219-225.

MAHAR, L.J. & O'NEILL, M.W., 1983. Geotechnical characterisation of desiccated clay. *Proceedings of ASCE Journal of the Geotechnical Engineering Division*, **109,** GT1, 56-71.

MEINZER, O.E., 1923. Outline of groundwater hydrology, with definitions. *US Geological Survey, Water Supply Paper 494*, Washington, D.C.

MOORE, J.F.A. & LONGWORTH, T., 1979. Hydraulic uplift at the base of a deep excavation in Oxford Clay. *Geotechnique*, **29,** 35-46.

MORGENSTERN, N.R. & EIGENBROD, K.D., 1974. Classification of argillaceous soils and rocks. *Proceedings of ASCE Journal of the Geotechnical Engineering Division*, **190,** GT10, 1137-1156.

NAKANO, R., 1967. On the weathering and change of Tertiary mudstone related to landslide. *Soil & Foundations*, **7,** 11-14.

OLIVIER, H.J., 1980. A new engineering-geological rock durability classification. *Engineering Geology*, **14,** 255-279.

OLIVIER, H.J., 1980. Some aspects of the influence of mineralogy and moisture redistribution on the weathering behaviour of mudrocks. *Proceedings of the 4th International Congress on Rock Mechanics, Montreaux*, **3,** Theme 1, 1-8.

O'NEILL, M.W. & POORMOAYED, A.M., 1980. Methodology for foundations on expansive clays. *Proceedings of ASCE Journal of the Geotechnical Engineering Division*, **106,** GT12, 1345-67.

POLAND, J.F., 1972. Subsidence and its control. *American Association of Petroleum Geologists, Memoir*, **18,** 50-71.

POLAND, J.F., (Ed.), 1984. Guidebook to studies of land subsidence due to groundwater withdrawal. *Studies and Reports in Hydrology*, No. 40, UNESCO, Paris.

POPESCU, M.E., 1979. Engineering problems associated with expansive clays from Romania. *Engineering Geology*, **14,** 43-53.

PRICE, N.J., 1960. The compressive strength of Coal Measures rocks. *Colliery Engineering*, **37,** 283-292.

REDFIELD, R.C., 1968. Brantley reservoir site - an investigation of evaporite and carbonate facies. *Bulletin of the Association of Engineering Geologists*, **6,** 14-30.

RUIZ, M.D., 1966. Some technological characteristics of twenty six Brazilian rock types. *Proceedings 1st Congress of the International Society of Rock Mechanics, Lisbon*, **1,** 115-119.

RUSSELL, D.J., 1981. Controls on slake durability: the response of two Ordovician shales in the slake durability test. *Canadian Geotechnical Journal*, **19,** 1-13.

SERRANO, A.A., OTEO, C., DAPENA, E. & VINAS, J.M., 1981. Analysis of swelling phenomena in a gypsum-marl formation. *Proceedings of the 10th International Conferences Soil Mechanics and Foundation Engineering, Stockholm*, **1**, 779-784.

SCHOFIELD, A.N. & WROTH, C.P., 1968. *Critical State Soil Mechanics*. McGraw-Hill, London.

SERAFIM, J.L., 1968. Influence of interstitial water of rock mass. *Rock Mechanics in Engineering Practice*. Stagg, K.G. and Zienkiewicz, O.C. (eds) Wiley, London, 55-97.

SOWERS, G.F., 1973. Settlement of waste disposal fills. *Proceedings of the 8th International Conference on Soil Mechanics and Foundation Engineering, Moscow*, **22**, 207-210.

SOWERS, G.F., 1975. Failure in limestones in the humid tropics. *Proceedings of the ASCE, Journal of the Geotechnical Engineering Division*, **101**, GT8, 777-784.

SOWERS, G.F., WILLIAMS, R.C. & WALLACE, T.S., 1965. Compressibility of broken rock and the settlement of rockfills. *Proceedings of the 6th International Conference of Soil Mechanics and Foundation Engineering, Montreal*, **2**, 561-565.

STEPHEN, J.L. & SPEIR, W.H., 1970. Subsidence of organic soils in the USA. *Proceedings of the 1st International Symposium on Land Subsidence*, UNESCO Publ. No. 88, International Association of Hydrological Scientists, Tokyo, **2**, 523-34.

STEWART, H.G., 1966. Groundwater Resources of Polk Country. *Florida Geological Survey, Report of Investigation*, Number 44, 170 pp.

SRIDHARAN, A. & ALLAM, M.M. 1982. Volume change behaviour of desiccated soils. *Proceedings of ASCE Journal of the Geotechnical Engineering Division*, **108**, GT8, 1057-71.

TAYLOR, R.K. & CRIPPS, J.C., 1984. Mineralogical controls on volume change. *Ground Movements and their Effects on Structures*. P.B. Attewell and R.K. Taylor (eds), Surrey University Press.

TAYLOR, R.K. & SPEARS, D.H., 1970. The breakdown of Coal Measures rocks. *International Journal of Rock Mechanics and Mining Science*, **7**, 481-501.

TERZAGHI, K., 1925. *Erdbaumechanik auf Bodenphysikalischer Grundlage*. Dueticke, Vienna.

TRAUGHBER, E.B., SNOWDEN, J.O. & SIMMONS, W.B., 1979. Differential subsidence on reclaimed marshland in metropolitan New Orleans, Louisiana. *Evaluation and Prediction of Subsidence*. Saxena, S.K. (ed.), Proceedings of the Speciality Conference, ASCE, Gainsville, 479-99.

VAN DER MERWE, D.H., 1964. The prediction of heave from the plasticity index and percentage clay fraction of soils. *The Civil Engineer in South Africa, Institute of Civil Engineers. S. Africa.*, **6**, 103-116.

WEST, G., 1979. Strength properties of Bunter Sandstone. *Tunnels and Tunnelling*, 7, No.7, 27-29

WILLIAMS, A.A.B. & JENNINGS, J.E., 1977. The *in situ* shear behaviour of fissured soils. *Proceedings of the 9th International Conference on Soil Mechanics and Foundation Engineering, Tokyo*, **2**, 169-176.

WILLIAMS, A.A.B. & PIDGEON, J.T., 1983. Evapotraspiration and heaving clays in South Africa. *Geotechnique*, **33**, 141-50.

WILLIAMS, J.H. & VINEYARD, J.D., 1976. Geologic indicators of catastrophic collapse in karst terrain in Missouri. *Transportation Research Record*, **612**, 31-37.

WINKLER, E.M. & WILHELM, E.J., 1970. Salt burst by hydration pressures in architectural stone in urban environments. *Bulletin of the Geological Society of America*, **81**, 567-572.

WROTH, C.P. & WOOD, D.M., 1978. The correlation of index properties with some basic engineering properties of soils. *Canadian Geotechnical Journal*, **15**, 137-145.

YUZER, E., 1982. Engineering properties and evaporitic formations of Turkey. *Bulletin of the International Association of Engineering Geology*, No. 25, 107-110.

F. G. BELL, Department of Civil Engineering, Teeside Polytechnic, Middlesbrough, T51 3BA, UK.
J. C. CRIPPS, Department of Geology, University of Sheffield, Sheffield S1 3JD, UK.
M.G. CULSHAW, Engineering Geology and Reservoir Rock Properties Research Group, British Geological Survey, Keyworth, Nottingham NG12 5GG, UK.

SESSION 1

ENGINEERING PROBLEMS POSED BY GROUNDWATER

The potentially latent dominance of groundwater in ground engineering

P.W. Rowe

ABSTRACT: Examples are given of unforeseen damage or difficulties arising during or after construction of buildings, wharves, foundations, tunnels, in-ground storage, pipe lines, and embankments. The underlying potentially latent mechanisms of groundwater are classified into four groups, namely water pressure changes, physical changes, erosion, and affects of earthworks, in order to assist early identification of some of the types of change in ground properties which may be induced by site operations and may lead to damage. It is concluded that otherwise unforeseen damage is more likely to be averted by a greater degree of investigation during construction into the consequences of the construction sequences, method and plant used.

Introduction

The combination of geology, groundwater levels and changes, ground works and methods of their construction are so extensive that nothwithstanding the known importance of comprehensive site investigations unexpected events do occur periodically. This implies that at some stage during the site investigation, design or construction the relevant potential mechanism was not foreseen by anyone connected with the works and in that sense the mechanism was hidden until the cause was diagnosed. Individual case histories provide a source whereby potential dangers can be formulated in similar situations.

Since the combinations of circumstances is rarely repeated it can sometimes prove difficult to extract the essence of one experience and apply it to another set of operations especially in the presence of continually developing site technique and plant. The present objective is to record brief but essential aspects of a number of widely differing cases so as to identify some of the types of mechanism whereby construction operations induced changes in groundwater pressures and properties in given geological conditions.

Groundwater pressure changes

Settlement due to pumping drawdown

The settlement of cities such as Mexico City, Venice, Bangkok and London, due to decrease in water pressure on water abstraction, is well known but the response rate of the underlying strata as a whole has not always been fully appreciated. At Venice (Fig. 1) industrial pumping from aquifers sunk to between 200 and 300 m depth on the mainland at Mestre and Marghera reduced water levels by a maximum of 9 m

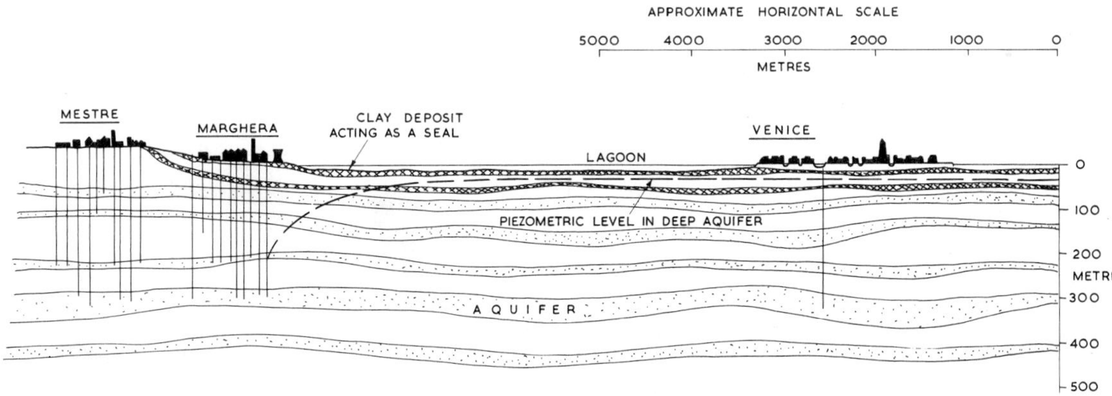

FIG. 1. Pumping drawdown spread below Venice.

From CRIPPS, J.C., BELL, F.G. & CULSHAW, M.G. (eds), 1986, *Groundwater in Engineering Geology,* Geological Society Engineering Geology Special Publication No. 3, pp. 27-42.

below Venice which is located some 6 km from the pumped wells. The geological structure of the clays and silts between the aquifers was such that little time lag occurred. The increase in settlement with time during 1900 to 1972 was directly due to the increasing demands of industry and by 1972 hardly at all due to past water lowering, (Rowe 1975). Once this was appreciated the settlement was largely arrested in 1974 by provision of an overland water supply to augment the ground source and avoidance of any increase in the rate of water abstraction. The response rate of geological deposits likewise affects whether or not temporary construction dewatering will do harm during the period of the works.

For Grangemouth New Lock (Fig. 2) deep pumping in glacial sands was essential to stabilise estuarine clays and silts having partings of fine sand. The presence of gravel at depth had not been identified during the site investigation and the dewatering wells had to penetrate the gravel to be effective. Once this was achieved the drawdown naturally took effect over a great distance laterally. The new lock, at its nearest point, was only 70 m from an existing old lock whose construction was such that its floor, founded on the silts, acted as a strut to the walls. The response rate of these silts would have allowed the old floor to consolidate during dewatering. The lateral thrust from the walls might

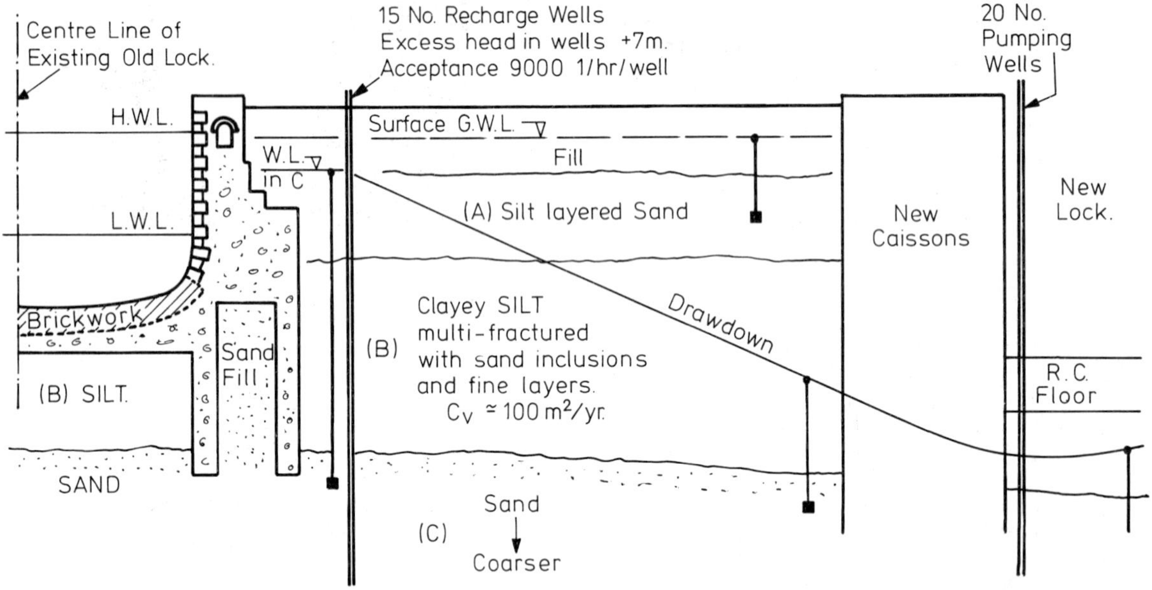

FIG. 2. Combined pumping and recharging at Grangemouth.

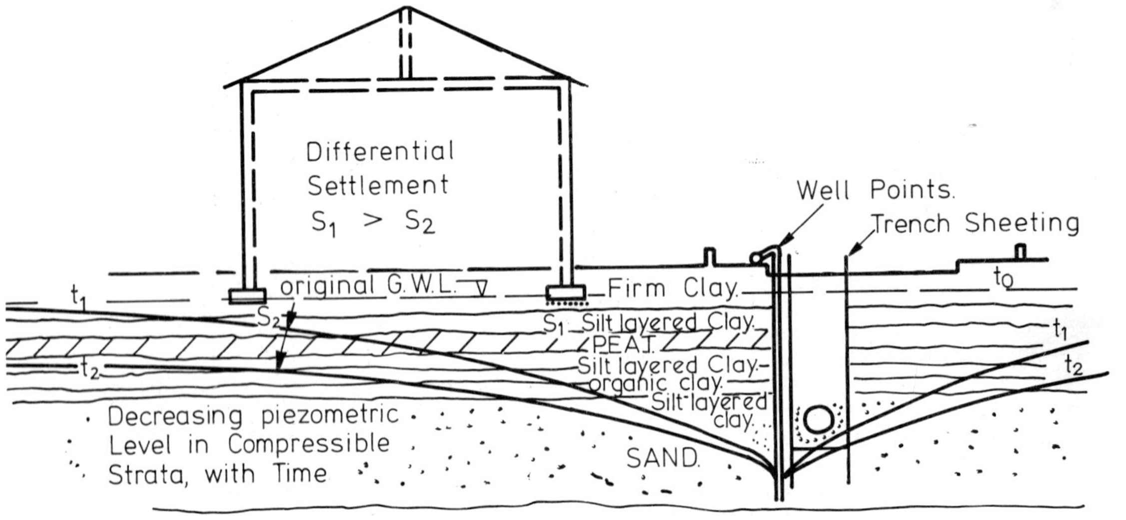

FIG. 3. A common case of short-term dewatering damage.

then have caused the floor to buckle and the walls to tilt inwards. The old lock had to be kept in operation until the new lock was completed. To prevent damage, recharge wells were installed alongside the old lock and purified water (to avoid clogging of the filters) was injected while pumping below the new lock took place. In the result, while the ground around the new lock settled some 300 mm that at the old lock hardly settled at all. The extensive measures that had to be taken could not have been identified from the original site investigation although in this case all that was needed was a better appreciation of the geology.

Sewer pipe installations in open cut are frequently located in building areas, and well point dewatering is often used to stabilise water bearing sands. Where adjacent properties are founded on compressible organic clays and silts with a high mass rate of response to drawdown via an underlying aquifer substantial settlements in excess of 100 mm have occurred causing extensive damage during the short period of pipe laying (Fig. 3).

Rise in water pressure causing instability

The 230 kV crossing of the Jamuna River, Bangladesh, (Chandler et al. 1984) located just north of its confluence with the Ganges, required the sinking of element caissons, 12.2 m diameter, to some 100 m depth within deep loose micaceous sand deposits in the flood plain of the Himalayas. The caissons (Fig. 4) were sunk from a level +10.7 m just below maximum flood level, the minimum river level being about +3 m. To prevent heave of the base during excavation, the water level inside the caissons was kept at about 2 m above river level. During design it was anticipated that a mud density of 1.2 Mg/m^3 would be required to support the loose sands to 100 m depth below flood level and that heavy minerals such as barytes would be used. In fact Caisson No. 11 was sunk successfully by March 1981 to a level −42.7 m with a bentonitic mud density of only 1.03 Mg/m^3, and when the excess head of the mud over the river level was only 4.5 m. This implied that the loose sand, acting radially over the circumference of 6.1 m radius was stabilised by arching and an equivalent lateral earth pressure coefficient K_a of only 0.11, compared to a figure of 0.32 applicable to an infinite plane. Since the caissons had to be sunk to level −104 m, and the river level was rising, an increase in mud density to a minimum of 1.15 Mg/m^3 was advised. This was not done. On 15 June 1981 when the caisson had reached −78.3 m and the river level had risen to +7.3 m, and with a sand plug 5.5 m thick inside the caisson plus an excess water head of 2.2 m inside the caisson, the ground around the caisson subsided over a radius of some 33 m and about 1000 m^3 of sand, bentonite and water, rose 23 m up inside the caisson (Fig. 5) flooding the steel fixers and the site with an almost equal volume of displaced water. The loose sand must have liquified

FIG. 4. Caisson.

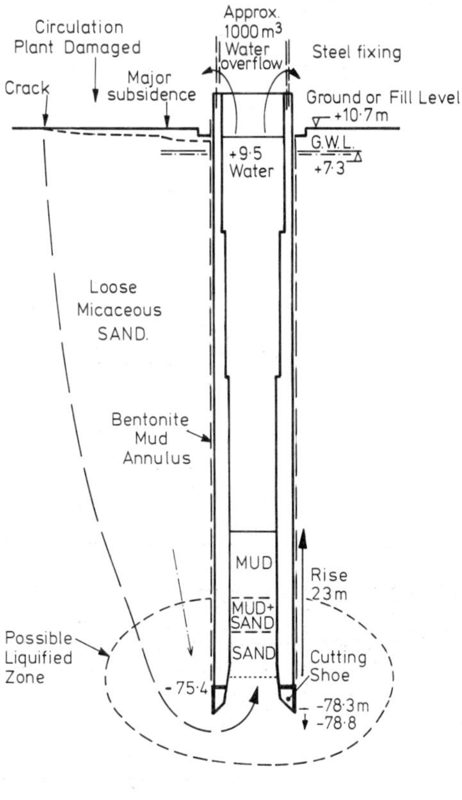

FIG. 5. Liquefaction of sand.

in shear around and below the caisson, but fortunately it sank only 0.5 m because once 530 m^3 of bentonite surround was lost the sand at its upper levels gripped the 10,000 t caisson in friction. Whereas in this case the problem had been foreseen, the depth of the caissons extended practical experience.

Fig. 6 shows a covered reservoir constructed over alternating bands of shale and limestone of the Lower Lias. Natural groundwater levels were low during the periods of site investigation and construction, and underdrainage was limited to that directly below the floor slabs. Subsequently exceptionally heavy rain caused a transient uplift in the confined limestone layers which lifted the reservoir floor at a time when the reservoir had been emptied for cleaning. The floor, restricted more at the sides than the centre rose in a dome shape over some 10,000 m^2, cracking all but the centre columns.

Natural hillside seepage pressures are raised if outlets are sealed off. Such was the case of the colliery tip flow disaster at Aberfan.

Rise of pore pressure due to construction fill is well appreciated but the long known effect of pore pressure spread can still be difficult to forecast. Fig. 7 shows a recent port construction in South America involving both cut and fill on either side of a new jetty. During the final stage of pile driving the piles were found to be moving slowly outwards by amounts which reached 400 mm. The sand fill settled at about 100 mm/month while the dredged sea bed rose 1.5 m. The positive excess pore pressures induced in the foundation layered clays below the fill, coupled with a reduction in pressure induced below the dredge level, caused lateral seepage and a change in the shape of the initially induced piezometric surface, with loss of stability followed by heave.

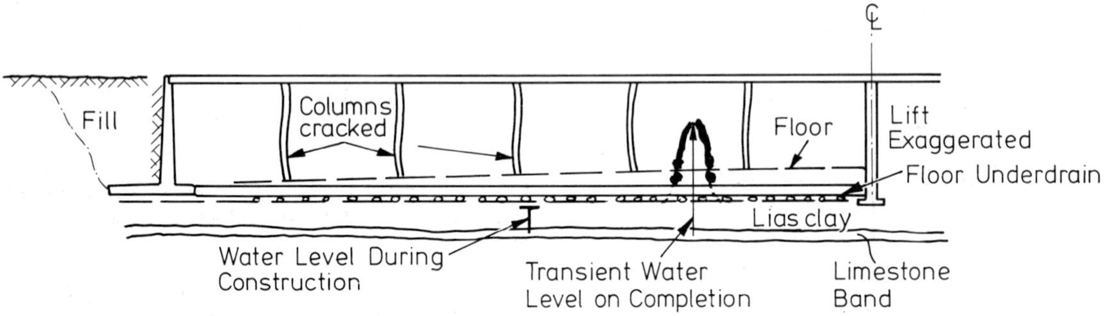

FIG. 6. Failure of concrete columns due to uplift.

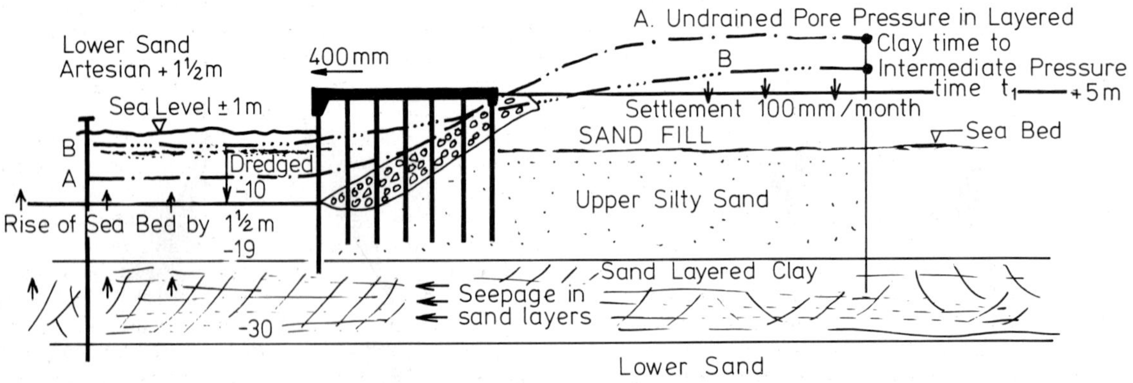

FIG. 7. Short-term pore pressure spread in layered clays.

Construction processes using grout or compressed air can induce rises in ground water pressure. Fig. 8 shows a case where cast *in situ* piles were formed alongside an existing building. The holes were drilled and cased and filled with preformed concrete units. The final stage involved the injection of cement grout under pressure as the casings were withdrawn. The fluid pressure caused a rise in ground water pressure within sand layers confined within the foundation clays of the adjacent building and the induced excess pore pressure spread below the building. The immediate settlement of the building was attributed to loss of bearing resistance. When the piling work was stopped the rate of settlement decreased with time in accordance with consolidation theory until the induced excess pore pressure had dissipated.

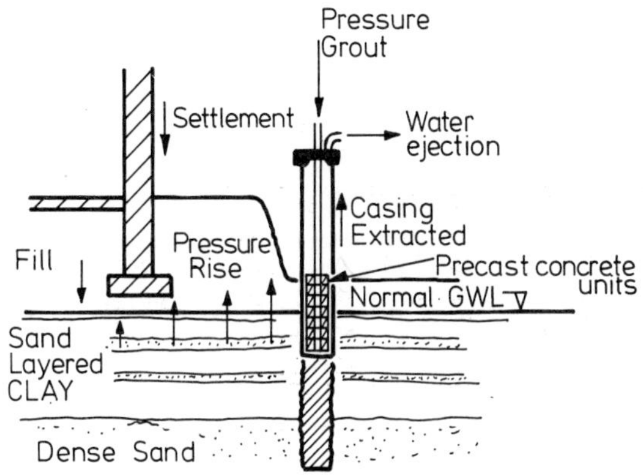

FIG. 8. Effect of grouting.

Fig. 9 shows a ground anchor installation to retain a temporarily sheeted excavation adjacent an existing building whose foundations were on sand underlain by a thin band of gravel overlying London Clay. The holes were cased and augered through the sand and gravel until a 'seal' was achieved in the clay. However a casing cannot be rotated over such a length concentrically and it would occupy an oval shaped hole. The holes were then advanced unlined and underreams were formed for the anchor length using compressed air to eject the cuttings. The anchor horizon chosen happened to coincide with a 2 m thick layer of silt unrevealed by the site investigation and the underreaming took up to an hour at each point. The anchors failed to achieve more than half their design load and the number of anchors was doubled. All this time compressed air was leaking past the 'seal' into the gravel and the water level was rising in the sand. At one stage sand and water burst through into the basement. The building foundations settled and eventually the building had to be demolished.

In both the above cases the geological detail from the site investigations was insufficient to enable the effect of construction on the ground to be seen at once even after the events and at a time when the construction processes were known.

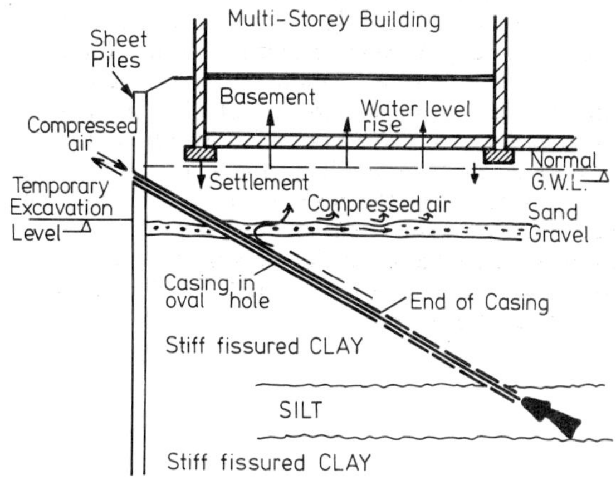

FIG. 9. Effect of drilling.

Pore pressure reduction due to excavation followed by rise due to construction-induced pressure gradients and seepage.

Fig. 10 shows a recent sheet pile wharf. During the final stages of dredging the wall moved seawards at a rate of 5 mm/day. A crack in the warehouse floor opened up gradually to 76 mm. The quay surface settled. Stone was hurriedly tipped back 3.5 m deep over the dredged length, yet during the next two months the wall moved outwards a further 56 mm at the level of the stone fill. Although the anchor length was clearly underdesigned the anchor held while the wall rotated anticlockwise about the anchor, indicating a process of progressive outward movement towards the pile toe. The piles had been driven into the Bracklesham Beds consisting of silty clays intercalated with thin silt and sand layers. The original pore pressure in these layers at depth was artesian relative to mean sea level, due to seepage from the land. On dredging, a suction component was induced in the layers below the dredge level. Sheet pile interlocks are permeable, and water under artesian pressure seeped through the interlocks along the confined permeable layers to negate the induced suction component and soften the clay. Just below the dredge level the clay may well have become virtually liquefied before the stone was placed. The wharf had been designed on the UK Code of Practice for Earth Retaining Structures (1952) using the well known $\phi = 0$ analysis together with 'softened' undrained strengths. This method is invalid where groundwater pressure changes are not directly related to change in total stress and the water content of the clays can change.

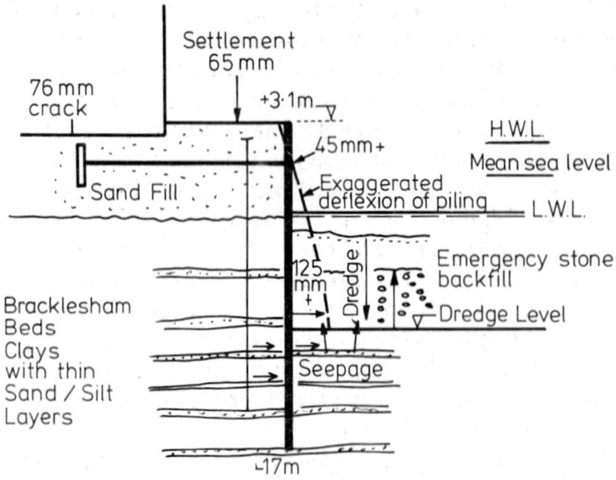

FIG. 10. Sheet pile wharf failure.

The concept that sheet piling acts as a cut-off for seepage is roughly true when the piles enter a uniform deposit and the main seepage flow is forced to rise from below. However the danger of percolation through the interlocks along confined layers is not always appreciated. Fig. 11 shows a cofferdam for a sewerage pumping station excavation where the sheet piles were driven to 6 m below formation. On excavating to some 3 m above formation a 'blow' occurred from the sand layer D just below the clay which lay beneath formation. For this fast uplift to occur there must have been a break in the interlocks at the corner near the 'blow'. The uplift problem was judged to have been solved once the pressure was relieved. The formation was reached and the base slab 1 m thick was cast while pumping from a sump at the location of the 'blow'. Subsequently, while attempting to form the walls, the slab rose differentially over 100 mm during a period of about 1 month, and cracked. Wellpoints jetted inside the cofferdam to layer C failed fully to arrest the rise until a well was sunk to intercept the sand layer B which may have had an indirect connection with the basal aquifer.

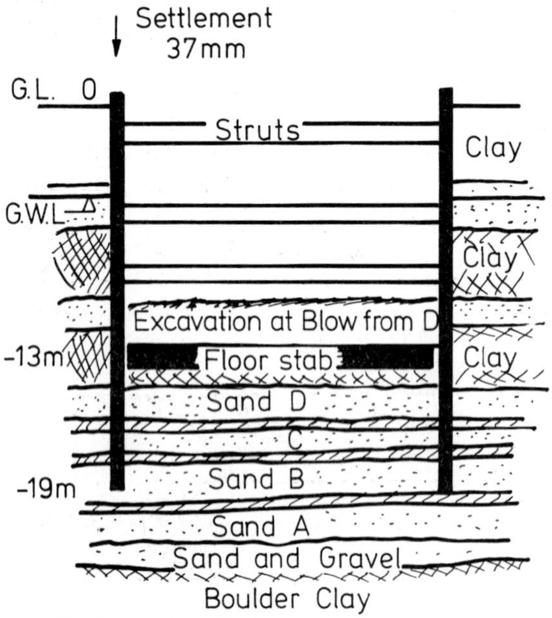

FIG. 11. Slab rise in cofferdam.

Fig. 12 shows the sequence of collapses experienced when forming a diaphragm wall through Keuper Marl by preboring staggered holes and using a grab on the dumpling in between. The Marl was described as weathered rock with SPT values increasing from 60 to above 100 within the upper 10.5 m. The walls had to penetrate to 15 m depth and the hardness below 10.5 m influence the choice of preboring at 1.5 m centres, initially by dry drilling. A normal grab tends to go off the vertical when meeting variations in rock hardness. In hole 1, collapses occurred below the water table due to very rapid softening of the Marl owing to its pedal silt fabric and the presence of thin sand layers. The hole then had to be lined to 13.5 m depth. Preboring was next attempted under bentonite, hole 2, but after successful completion of that hole, and while drilling hole 3 under bentonite, the dumpling in between these holes collapsed. Holes 4 and 5 were drilled under bentonite with the insertion also of a lining tube but this did not prevent collapse of the ground in between. The cause was attributed to an inward squeeze of the softening marl on to the rotary auger bucket. The wall was formed finally using a 4 m high hydraulically operated heavy grab under bentonite.

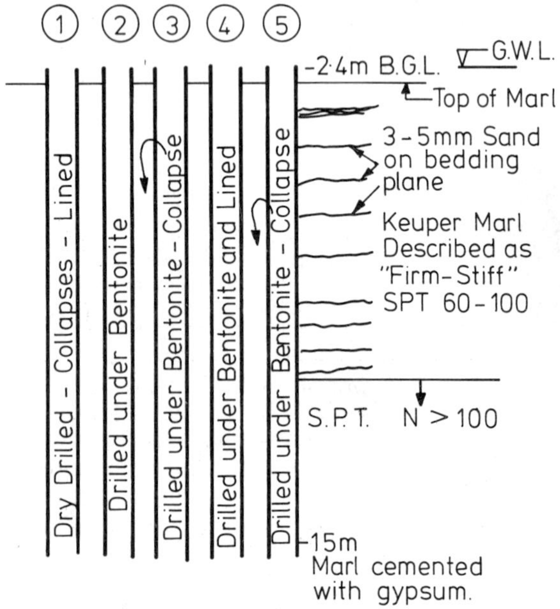

FIG. 12. Collapses on augering.

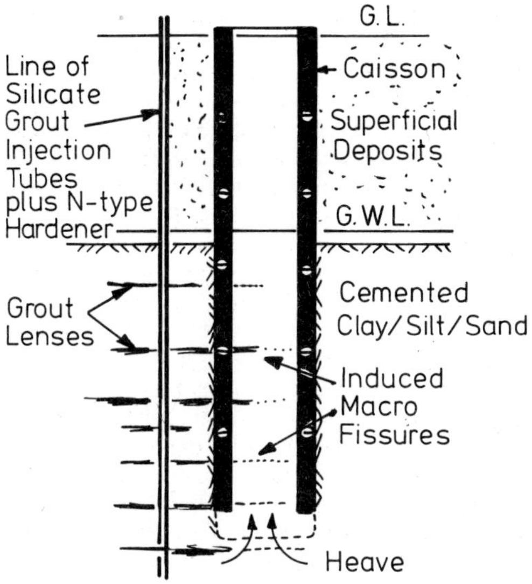

FIG. 13. Heave after grouting.

Grout injection is used widely to reduce the permeability of fissured rock, gravel or, with silicates, sands. However, if the deposit is granular but fine and weakly cemented, grout can enter only by hydrofracture to form discrete lenses. The ground is split and as when chopping wood with an axe, the split continues beyond the end of the grout lense (or axe) and to the sides of the intrusions. This can open up microfissures and actually increase permeability. Fig. 13 shows one of a series of caissons traditionally excavated to depths as much as 30 m below the water table successfully in the 'dry'. The permeability of the deposit in its natural state has to be sufficiently low to prevent undue softening during the period of excavation, when no prior groundwater lowering is required. Water pressures are lowered by the component of suction induced by removal of the soil weight. However, on one such site, silicate grout was injected on a line of holes 2.5 m to one side of a row of caissons 2.5 m diameter prior to commencement of excavation. Very wet and unusual conditions were met, increasing with depth, until the excavation had to cease in liquefied ground. Elsewhere on the site where prior line grouting had not been used, such wet conditions were not encountered. Limited intrusions of grout were found in the wet excavations and in later consecutive samples, with evidence of macrofissures beyond. Whereas blanket grouting over the whole area to be excavated can prove beneficial, line grouting can prove harmful.

When tunnelling in rocks of varying hardness and permeability below the water table, such as through sandstones, shales and mudstones, the tunnel face can be subjected to high seepage gradients through shale as sandstone is approached. To prevent the shale from softening excessively it is good practice to drive in the direction of the dip and to drill probe holes ahead to relieve the water pressure behind the face and below the advancing heading. Fig. 14 shows part of a drive through such materials for a tunnel of 6 m diameter using a Dosco Roadheader where, in parts of the drive only, excessive slurrying at the face when in shaley mudstone caused such a depth and volume of mud as to halt all mechanical operations of the machine and conveyor. Throughout the drive the shaley mudstones met were similar, exhibiting fine bedding planes on which silt or silt-sized floc of clay particles could be detected under a microscope. Some of the mudstones were fissured near fault zones or due to past mining operations at lower depth. Whereas all the strata over the drive were similar geologically, and all were below water, and probing ahead in the direction of dip was applied, the slurry

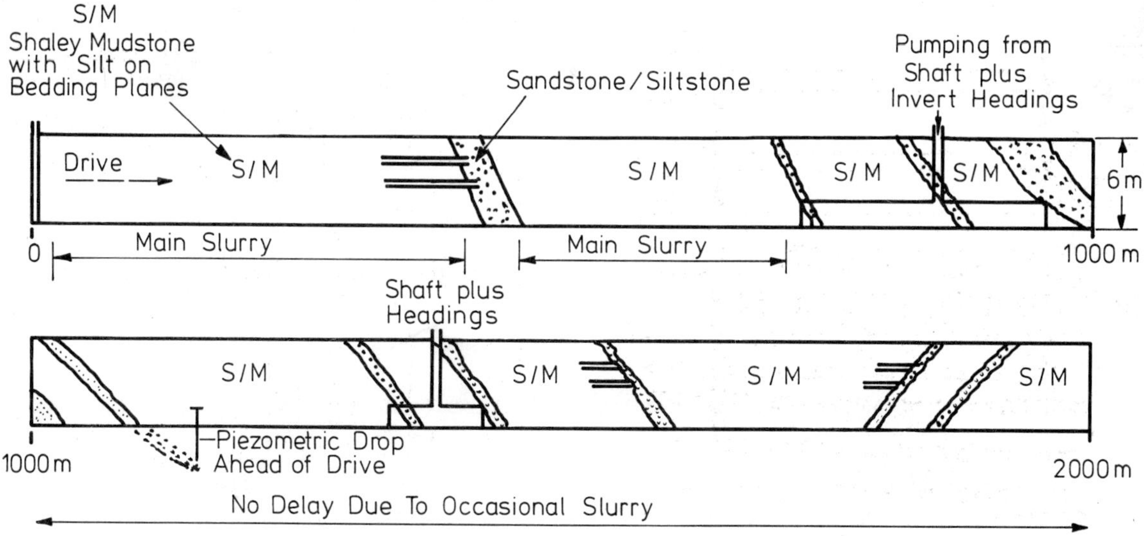

FIG. 14. Slurry problem with mechanical tunnelling in mudstone.

problem occurred more extensively over the initial lengths shown. This occurred partly because there was less frequency of sandstone/siltstone or fresh fissured rock and partly due to a greater frequency of faulting. The probe holes softened during drilling and yielded little or no water until a sandstone was met. The seepage gradients through the mudstone face were mainly high, causing the face to expand inwards and its water content to increase. The grinding action of the mechanical cutter turned the wet and fractured face of the finely-bedded shaley mudstone continually into slurry. In contrast, where the water pressure could be relieved ahead of the face the mudstone was encountered in a a relatively dry and hard condition suited to the machine operations. Whereas site investigations often fail to evaluate all features which may dictate the effects of construction operations this is particularly the case with long tunnel drives where boreholes tend to be relatively widely spaced.

Tunnel drives in Keuper Marl having thin sandstone layers do not suffer to the same extent as the advance drainage is more effective. However, the marl was formed by clay-sized particles settling in a hypersaline inland sea to form silt-sized flocs. Under the action of mechanical cutters and in the presence of moderate seepage without slurrying the flocculated structure collapses on to its interstitial water and the hard rock can come on the conveyor belt in the form of clay. If a chain conveyor is installed the clay can then clog the mechanical side rollers and bring the conveyor and all operations to a halt.

Groundwater control can be more difficult in variable glacial conditions where the effects of softening at the face of a tunnel may not be detected until too late. Fig. 15 shows part of a pneumatic spade drive, dug within a shield while working under compressed air. To prevent excessive loss of air into the ground overhead the pressure of the air was set to balance groundwater pressure at the midheight of the face. When in sands and gravels, seepage into the face emerged at the base of the face. When digging through glacial lake layered clays the water seepage remained confined within the layers at about midheight of the face, shown by a dashed line in Fig. 15. Hidden was the fact that as the drive advanced, from left to right on Fig. 15, the underside of the layered clays below the invert was rising over sand and gravel. The upward water gradients softened the thinning clays at the base of the heading and it passed unnoticed that this resulted in overexcavation at the base. After the primary linings were bolted in position, the back grouting did not travel far from the injection holes since the softened clay had squeezed on to the lining. While forming the secondary lining of cast-in-place concrete without the need for compressed air the weight of the concrete skips caused those rings over the region of the extruded clay to settle and crack. A large volume of silt and sand flowed into the tunnel causing subsidence of the tunnel rings, collapse of the ground above and damage to buildings. This illustrates again the importance of site investigation details in relation to the expected site operations.

Physical changes

Slag fill

Slag from steel making processes which utilise a basic flux are expansion prone due to the hydration of calcium and magnesium oxides. Such slags used as a backfill have caused floors to heave and walls to displace. Fresh slag is the more expansive because the calcium oxide hydrates quickly, but even after years of exposure to moisture the magnesium oxide can cause slag fill to expand at rates of 0.2%/yr at 20°C increasing to over 1%/yr at 30°C.

Heat source

In contrast the direct transmission of heat into ground has caused vaporisation of groundwater. In one case a pump room retaining wall continued to move inwards over a period of decades notwithstanding the insertion of counterfort walls in an attempt to arrest the movement. A borehole behind the wall revealed

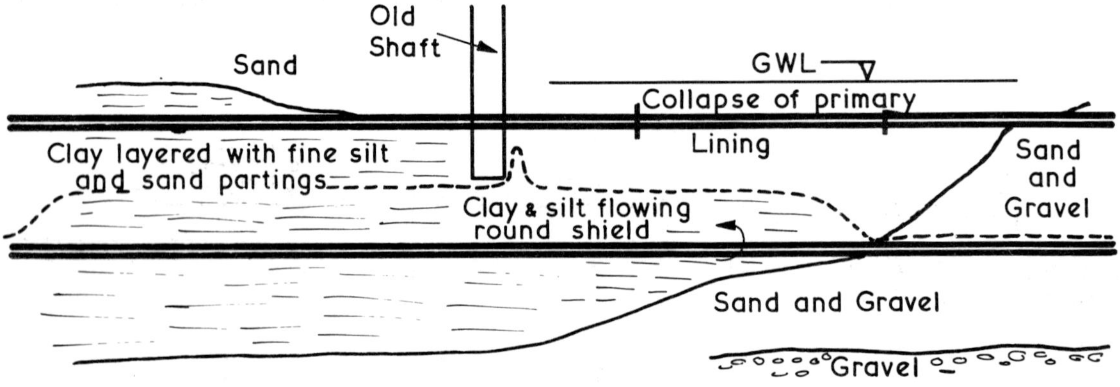

FIG. 15. Tunnel collapse after softening below invert.

the existence of water boiling in the ground with emission of steam due to the presence of a boiler located at ground level above the retaining wall. When the steam was vented the movements ceased.

Groundwater freezing

Ground heave as a consequence of freezing is well known but the dangers in respect of thinly layered soils may be less widely appreciated. An example of an inground storage reservoir for Liquid Natural Gas (LNG) is shown in Fig. 16. The reservoir walls and floor were formed by the normal brine circulation technique at above −10°C. The walls were 5.5 m thick and the minimum distance between the external surfaces of the ice walls of adjacent reservoirs was 23 m. The site consisted of alluvial clays having thin partings of silt and sand overlying gravel above a stiff fissured clay. After excavation of the inside of the frozen zone to form the 40 m diameter and 40 m deep reservoirs the walls were kept frozen while the roof was formed and the first LNG refrigerator ship arrived. The walls were then slowly cooled down by spraying with LNG which boils off at −160°C until the first reservoir could be gradually filled.

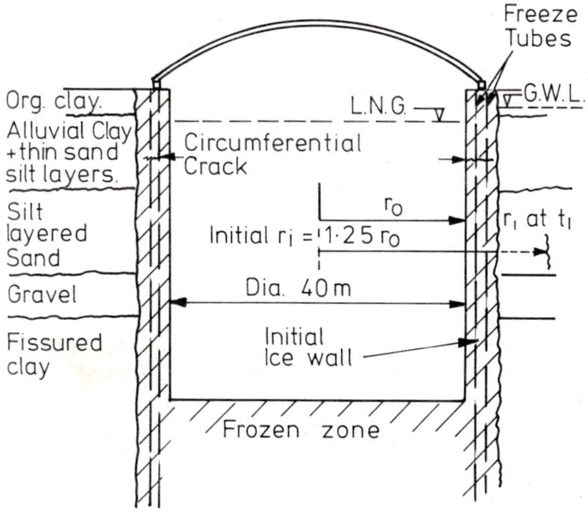

FIG. 16. LNG ice store.

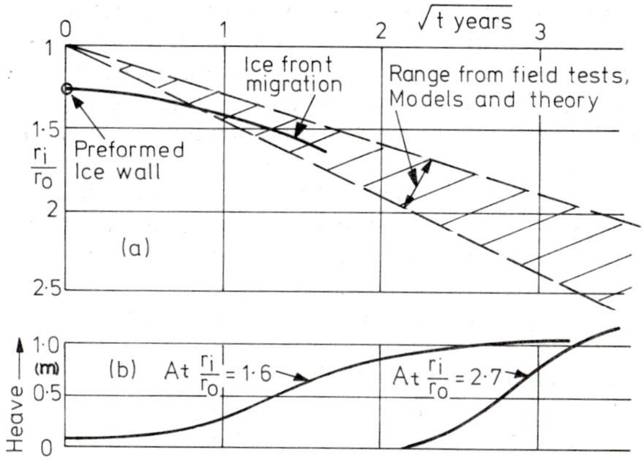

FIG. 17. Ice spread and heave.

It was known that the ice front would travel radially at a rate proportional to the square root of time (Fig. 17 a). The thermal conductivity of a wide range of soils from clays to sands lies in a relatively narrow range of about 4–33 m^2/yr unlike the analogous coefficient of consolidation for the flow of water which ranges over many orders of magnitude. The ice front progress was expected to cause ground heave due to water expansion *per se* but the unexpected event was that the thick reservoir ice walls cracked circumferentially at about 9 m depth, namely, towards the base of the alluvial layered clays, permitting the LNG to travel radially. The ground over a distance of more than twice the reservoir radius heaved excessively over a period of a few years, damaging the surface mechanical installations. Model experiments simulated the field observations which culminated in a heave of nearly 1 m over 4 years (Fig. 17 b) by which time the whole installation was abandoned. The physical mechanism (Fig. 18) was found from the models to be similar, in part, to that of the case of hydrofracture by grout injection shown in Fig. 13. As the ice front advances water expands on freezing and

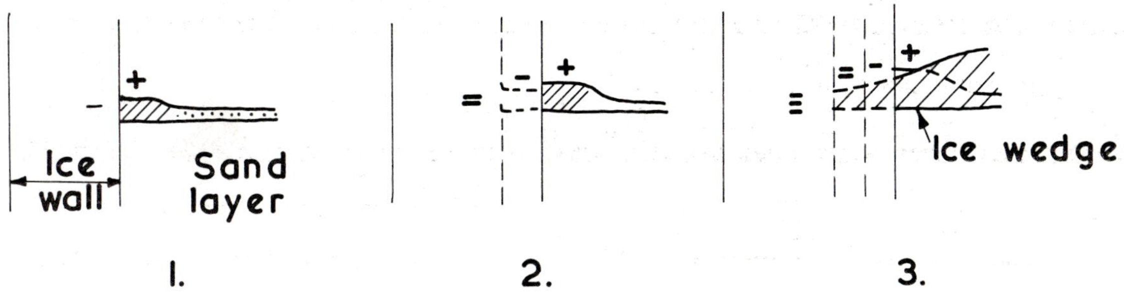

FIG. 18. Propogation mechanism of ice lenses.

the fine silty sand layers are forced apart. This sets up a component of pore suction just beyond the freezing water front which draws water radially inwards at a faster rate via the more permeable layers than in the clay. Consequently, as the ice lens penetrates, the thickness of the expanding water and that of the ice lenses increases. Not only did this cause heave by an amount greatly in excess of that of the expansion of the original pore water in absence of migration but it induced a tension in the ice wall which was the final cause of the crack and loss of LNG into the ground.

Chemical action

The basic interparticle angle of friction of minerals, which governs the effective strength parameters of soils, is a physical-chemical phenomenon which can be affected by changes in the acidity of the pore fluid. Recent attention has been drawn to the effect of chemical weathering of pyritic shale on shearing resistance (Steward & Cripps 1983).

Erosion

The process of scour round bridge piers and cofferdams, and the phenomenon of piping is well known. However it is possible for piping erosion to be induced by construction processes where none previously existed. An old dam had an enlarged draw-off pipe installed by thrust bore through the bank around the old pipe which was then removed. The ground around the new pipe was grouted by injection through the new pipe. The cement grout sealed natural coarse gravel passages which probably had acted as a natural filter to permit safe underseepage in the past. Within a year of the remedial work the reservoir emptied by piping through the silty sand adjacent to the grouted gravel.

In sewer pipe construction the pipe surround materials which are often recommended include stone graded 20–5 mm for 375 mm pipes and stone graded 63–10 mm for 1500 mm pipes, irrespective of the ground in which the pipes are laid. Fig. 19 shows one site where deep dewatering of a lower aquifer 10.5 m below ground level was necessary to stabilise the base of a pipe trench in silty sand at 7.5 m below ground level and some 5.5 m below the water table. The first pipes were laid on 20–5 mm gravel was reduced to 150 mm and the trench was deepened to allow dolomite stone 150–100 mm to be rammed into the base to a depth of 1000 mm below invert. After this appeared to be satisfactory the depth of the dolomite was reduced to 500 mm below invert and when a further seven pipe lengths had been laid and backfilled a temporary rise in groundwater level occurred due to a pump breakdown. The pipes then sank by amounts up to 150 mm. Re-excavation proved the sunken stone beds contained silty sand in their voids over a greater depth than to be expected from the ramming action. The reason was that as the water level rose it carried fines into the stone and gravel surround since the filter rules had not been applied. Once the fault was diagnosed the pipe line was completed satisfactorily using a bed of graded sandy gravel only 300 m thick.

A variation of this common occurrence is shown in Fig. 20 where a flexible 900 mm plastic outfall pipe was laid in open trench at times of low tide within a surround graded 12–0.6 mm on a formation of clayey silt and sand. It was backfilled and appeared to perform satisfactorily. The outfall led into a tidal estuary and six months after completion, just after a September high tide, the ground above the pipe subsided where the backfill was highest and remote from the manhole. Investigations showed that the pipe surround, subjected to tidal flow action, had gradually sunk into the soil fines below. The

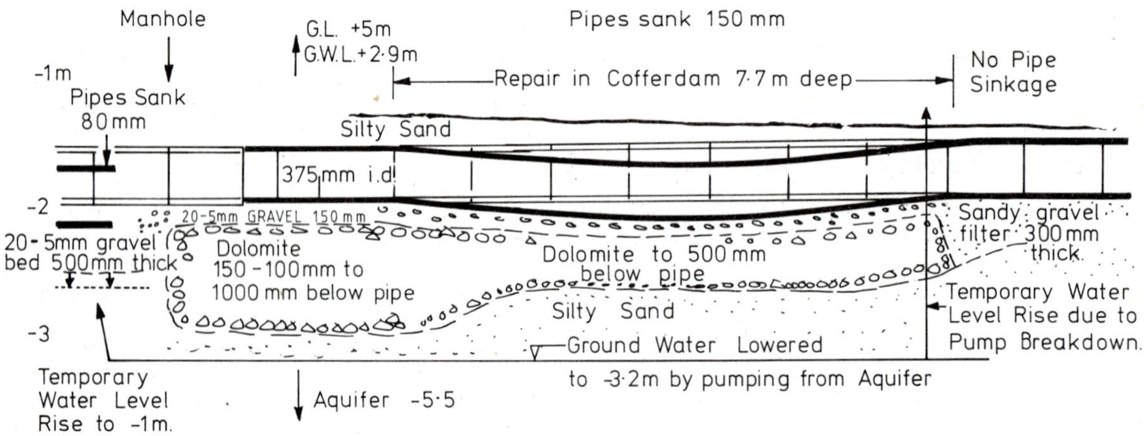

FIG. 19. Sinkage of sewer pipe after dewatering.

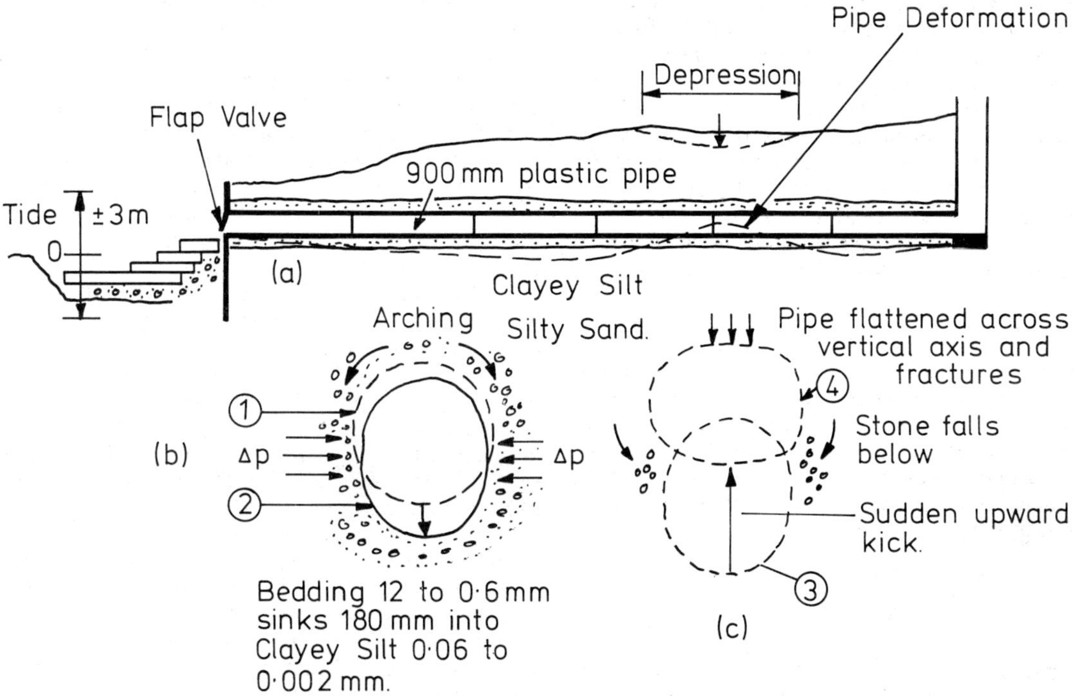

FIG. 20. Bedding settlement followed by pipe uplift.

downward movement would have left a void above the pipe and caused lateral arching. The pipe towards the outfall end (Fig. 20 a) was measured *in situ* to have settled and both decreased in width and increased in height (Fig. 20 b) by some 6%. There was a flap valve on the pipe so that either rising water did not enter or the outfall emptied ahead of the water under the pipe on a falling tide. With the pipe empty, and the water upthrust below, a stage was reached at which the vertical pressure above the pipe was no longer able to resist the uplift. The pipe was found to have been lifted upwards in the location shown in Fig. 20 a by over 40% of its diameter and deformed to the shape shown in Fig. 20 c, while the bedding fell below the pipe. The pipe cracked, soil was washed through and the ground above collapsed.

Effects of water during earthwork operations

Changes in fill properties between borrow pit and placement.

While it has been known that Keuper Marl, being rock for purposes of excavation, breaks down on stress relief continually into smaller particle size groups when transported, bladed and rolled in the presence of water and can become a plastic clay fill, the same applies to the 'unweathered Keuper Waterstone'. In two separate motorway contracts this material was specified as 'suitable' fill (in contrast to Keuper Marl specified as 'unsuitable') for placement below the water table at formation level, and for use initially in preference too imported stone fill. However, the Keuper Waterstone has sufficient of the silt-sized clay flocs of the marl in it to cause the hard excavated material to change to a plastic clay consistency during working, and in the presence of groundwater the subgrade rutted and was rejected. In this case the breakdown was indigenous to the material in the presence of groundwater and not a matter within the control of the contractor.

During another motorway construction the majority of the excavated material in cut was predicted to be 'suitable' for fill in embankment elsewhere. The cut passed through silt and sand layered clays to depths well below the original water table. As the cut was deepened the material below each successive excavation level softened due to lateral water pressure spreading into the zone of temporarily reduced pore pressure, and apart from plant traffic difficulty the 'suitable' fill became 'unsuitable', being too soft and wet for embanking. Subsequently there were over thirty slips of the cuttings before sufficient slope drainage was installed. This can also be a potential problem in borrow pit working for earth dams. In absence of advance drainage the early clay fill, placed in the more critical base of the dam, may be placed closer to the lower permissible limit than had been intended.

Difficulties of an opposite nature have arisen. The clay core of Foremark Reservoir embankment was being formed with weathered Moira Marl of suitable strength from an upsteam borrow pit in the floor of the valley when that source of clay ran out. Alternative supplies were obtained from a harder, less weathered, marl in the hillside. The ripped, transported and bladed, material still contained cobble sized lumps within a finer matrix. It was watered to bring the strength down to the specified range but the limit of water was reached when rutting tended to occur, with danger of plant shears. At this stage the samples tested showed strengths of 60 kN/m^2, well below the mean of 100 kN/m^2 specified but it was appreciated that it was not possible to obtain a uniform water content at that stage. After further raising it was found that the core pore water pressures were unexpectedly low. On drilling into the core several months after placement the samples gave strengths in excess of 200 kN/m^2, implying that the core was then too brittle. The reason was thought to be due to the lack of sufficient breakdown of the drier marl on excavation and working before watering. With time water could be sucked out of the wetter matrix into the drier lumps to give an average water content well below that consistent with the strength at placement.

Groundwater control during initial fill placement

Where excavation of natural surface material is undertaken prior to the formation of an embankment to levels below natural or modified groundwater level, pumping from deep wells (Foremark Reservoir) or sumps (Grimwith Reservoir) prevents water from rising upwards into the formation and provides a safeguard against possible softening of the formation and creation of plant shears in plastic clays. Once an effective drainage blanket is placed and the first layers of fill have been raised above equilibrium groundwater level at that stage of construction groundwater pumping is normally no longer needed.

FIG. 21. Embankment slip scarp.

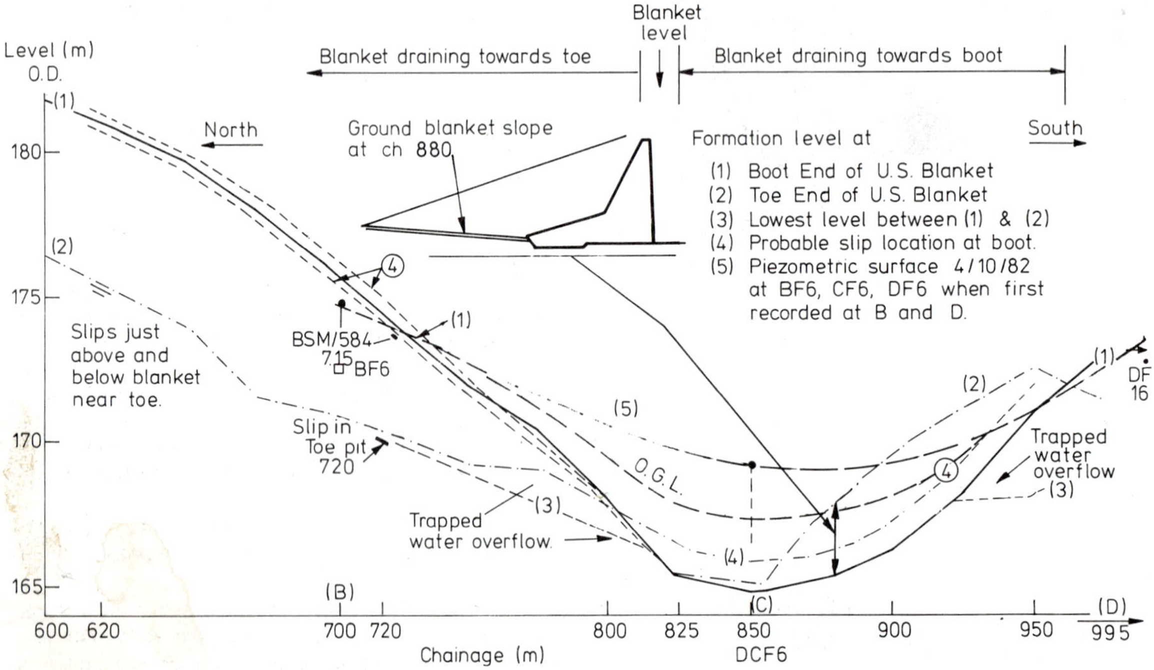

FIG. 22. Longitudinal section of formation and groundwater levels.

In this connexion it is relevant to examine the facts relating to the Carsington embankment construction slip which accelerated finally to failure from its north valley side in regard to the potential latent dominance of a seemingly small groundwater uplift pressure within the ground drainage blanket at the base of a 3.5 m deep excavation in the central width of the valley, over an initial minimum longitudinal distance of 80 m. The form of the scarp (Fig. 21) was similar over a length of some 500 m irrespective of differences in geological details at or just below formation, or of the extent of the excavation of the surface deposits to the south and whether or not the base of the slip passed through foundation or fill materials, and in particular it was very similar to that of a dam embankment slip over construction shears in Brazil (De Mello 1982).

The embankment was raised over basal Carboniferous shales whose permeability increased initially with depth. Prior to construction the groundwater pressure in the shales was at times artesian by 1–2 m in the bottom on the valley. After injection of a cement group curtain, and river diversion from a higher level upstream, and after completion of the clay cut-off, the groundwater pressure rose steadily with time to a level eventually some 5 m above the blanket level in the valley centre and probably to above blanket level over a chainage of 200 m. Line 5 of Fig. 22 shows the eventual piezometric surface on a longitudinal section for response points located 1–2 m below the ground blanket. Fig. 22 also shows that on the south valley side, from Ch 825 to 961 m, the upstream shoulder ground blanket sloped downwards towards the boot shaped core as shown by the difference in level between lines 1 and 2 and by the inserted cross section. In places on the north side the ground blanket dipped below the upstream outlet level.

The upstream toe trench which acted initially as a sump was filled with stone prior to completion of the adjacent ground blanket. Notwithstanding very dry weather the two filled rapidly with groundwater. This water was pumped only out of the top of the stone, as seen in Fig. 23. Site records noted upward seepage into the formation and thus likewise into the ground blanket.

FIG. 23. Water level at top of stone toe 29/7/82.

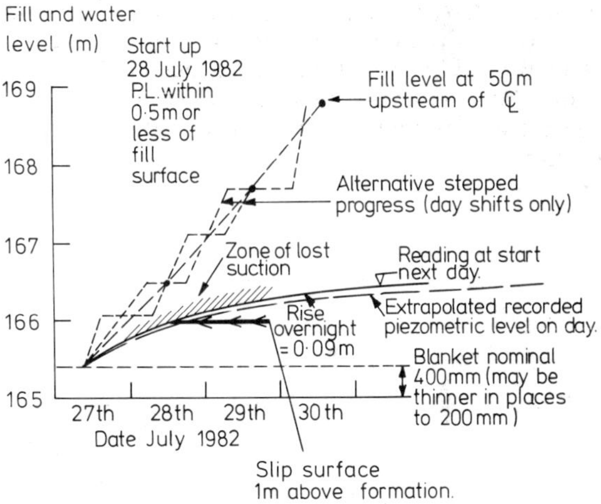

FIG. 24. Deduced water level in fill (28/7/82) and slip.

After the first layers of shale fill had been placed over the blanket on the first day in the centre of the valley the foundation piezometric level to within 1-2 m of the blanket rose both with response level near the boot and with time. The implied piezometric level in the blanket near the boot, consistent with the seepage gradients both vertically and horizontally to the toe, was within 0.5 m of the fill surface by the start of the second day's filling on 28 July 1982, as shown in Fig. 24. This level coincides with that of the main slip surface through the fill (Skempton & Coats 1985).

The present direct evidence is that after the slip and excavation into the foundation at the valley centre a number of adjacent levels of shear surfaces could be seen at about 300 mm distance apart vertically, namely, the fill layer thickness apart, in addition to the main shear plane. Furthermore, a sample taken from the main slip plane fell apart after later handing on a curved slickensided surface leading into the main shear plane (Fig. 25) indicative of overlapping plant shear. Plant shears of several metres length were found in the dark shale fill elsewhere at higher levels (Fig. 26) possibly due to later wet weather and lack of scarification, but the groundwater uplift in the blanket must have had a much more dominant effect and induced a greater frequency of plant shears just above the blanket than at higher levels. There the blanket sloped towards the core on the south side, such conditions can be envisaged as being repeated near the boot as the fill worked initially up the valley side. De Mello (1982) cites intensively laminated and sheared clay fill where compacted in the lower and wetter conditions of the clay blanket of the Brazilian dam which failed in a similar manner.

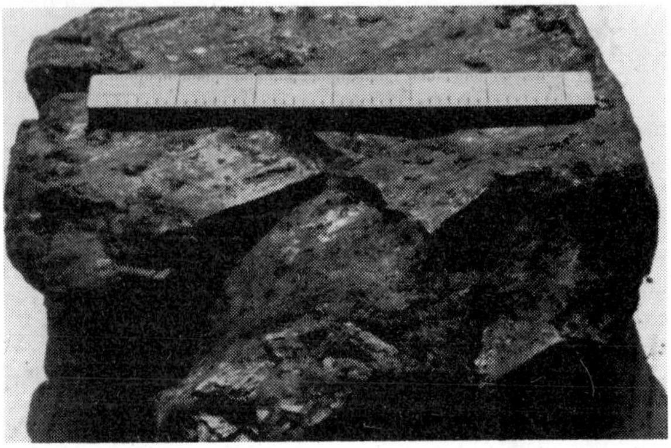

FIG. 25. Curved plant shear crossing main slip surface in fill.

FIG. 26. Typical plant shears in fill.

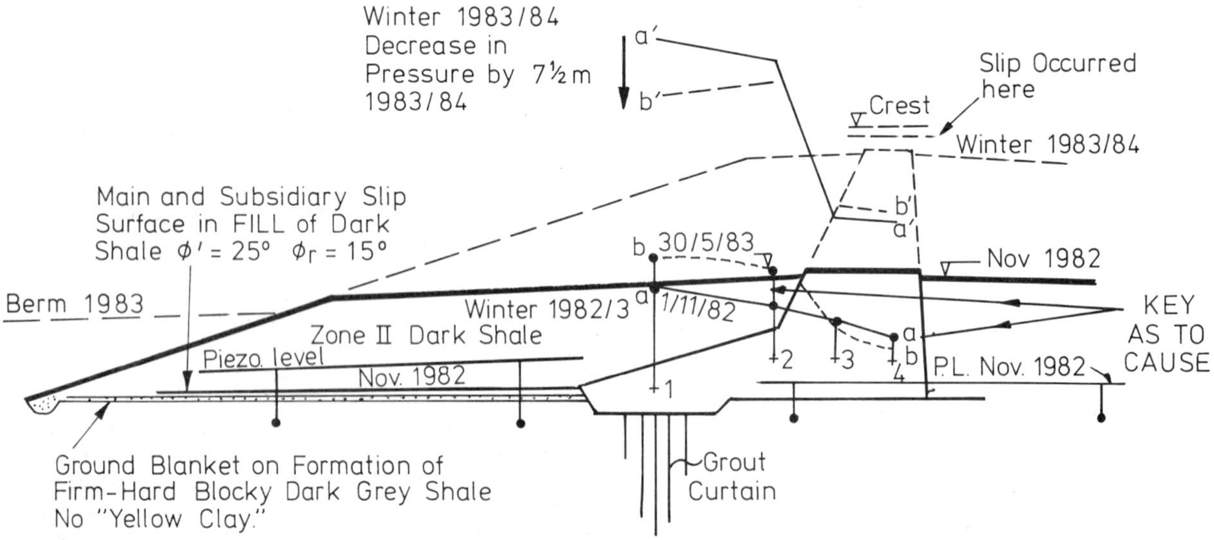

FIG. 27. Pore pressure changes in core and boot over rest periods.

The question as to whether the shear surfaces were built-in and caused the slip or whether the shear surface found was the result of the slip can be examined in another way. Under the thrust from the core the combination of sheared ($\phi_r = 15°$) and unsheared ($\phi' = 25°$) surfaces would be sensitive to progressive failure in the same way as perforations between postage stamps assist tearing. Given this condition the pore pressure records in the boot at Ch.850 can be interpreted as indicating that vertical total stress transfer occurred from the core to the boot, consistent with progressive shear in the base of the shoulder fill, even as early as the first winter rest period. This is seen in Fig. 27, comparing piezometric surface aa with bb, and in Fig. 28.

During the last winter rest period, 1983-84, the drop in piezometric level in the boot at Ch. 850 by some 7.5 m is consistent with stress transfer to the shoulder but as the latter was subject to the static water pressure in the blanket no response occurred from piezometers in the base of the fill or in the foundation. The embankment outward displacements,

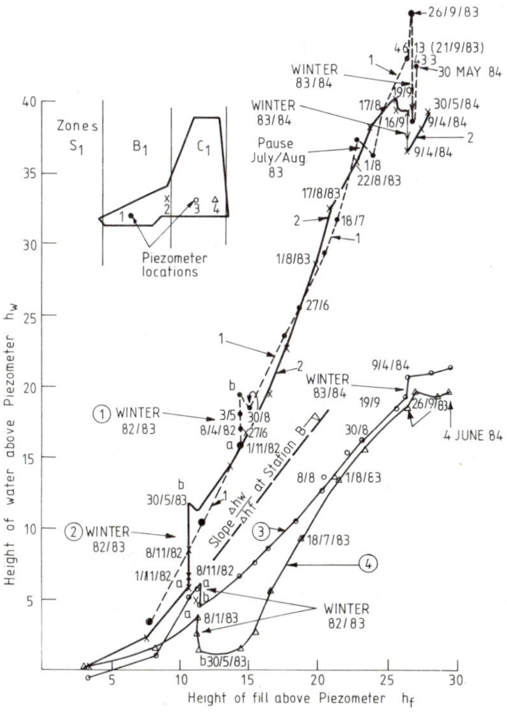

FIG. 28. Pore pressures Ch.850.

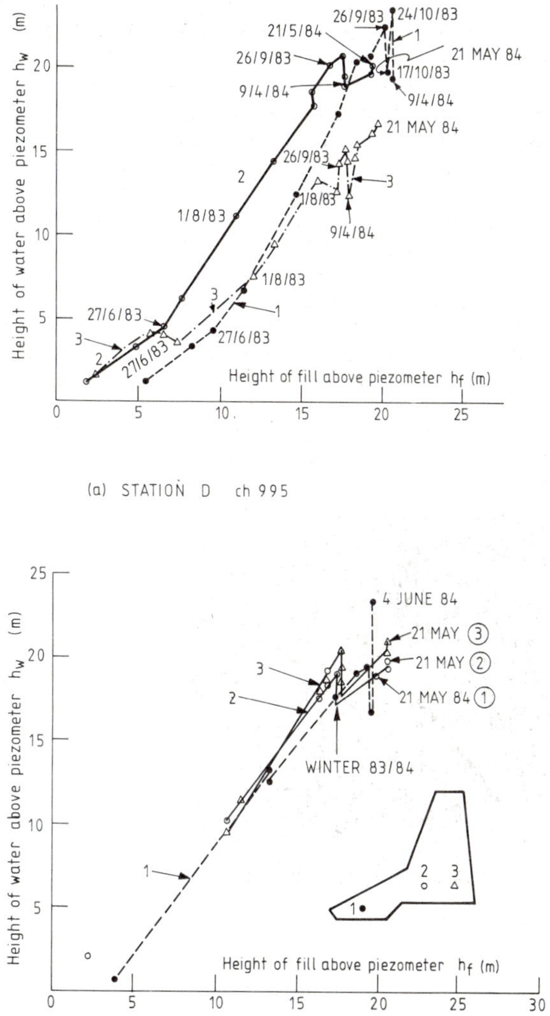

FIG. 29. Pore pressures Ch.995, Upper. Ch.705, Lower.

which led at the centre during construction over those at the valley sides, would almost certainly have been associated with high local shear strain over the plant shears at the base of the fill. Such tearing action could have spread progressively towards the valley sides at the boot for a similar mechanism of stress transfer to that at the centre can be picked up at the piezometers at Ch. 995 (Fig. 29 a) during the static period while the central berm was placed in 1983. The last instrumented location to show any drop in the core and boot pore pressures was that at Ch.705, where there is no evidence of uplift pressures in the ground blanket or reason to expect excessive plant shears in the fill. Here the bank was founded on a superficial head deposit of yellow clay whose peak strength was less than that of the fill and the slip passed through the foundation.

Embanking led to the crest from the North end and a factor of safety just below unity can be obtained at Ch.705 by inserting the residual, $\phi_r = 12°$ at the boot due to tearing action from the centre, increasing to peak $\phi' = 21°$ at the toe. At that stage the factor against slip through the base of the fill at the centre, (1.5 for intact fill), was just above unity in the presence of a then fully developed shear plane above the blanket. The average factor of safety at Ch.705 and Ch.850 was unity. This explains the extensive longitudinal propogation of the slip. The energy source and acceleration came from the region which had not previously progressed at its base to the residual. Further evidence may be obtained on reconstruction but correlation of the present reports with site events leads back to the groundwater levels at the start of embanking.

Conclusions

The cases illustrate that during and subsequent to a site investigation attention should be given to the possible temporary changes in ground properties arising from construction methods and sequences. A report of the original geology, groundwater and existing soil properties may not always enable an Engineer to formulate future events and when the report is included in the contract documents it may not always alert a Contractor sufficiently to the consequences of his proposed operations. Likewise no one concerned with the works may be enabled thereby readily to diagnose the true cause of unexpected events, even after an additional site investigation, unless that investigation is extended to reveal what it was that was built that failed. Plant and inground techniques are continually developing. Continuity of key personnel throughout all phases; continuity of site investigation and monitoring into the construction phase; and site discussions on what could possibly go wrong, are safeguards against damage.

References

CHANDLER, J.A., PERANINO, J. & ROWE, P.W. 1984. Jamuna river 230 kV crossing-Bangladesh.III. Construction of foundations. *Proceedings of the Institution of Civil Engineers, London, Part 1,* **76,** 965-984.

DE MELLO, V.F.B. 1982. A case history of a major construction period dam failure. *Amici et Alumini. Comite de'hommage au Professor E. de Beer. 1040 Bruxelles.* 63-78.

ROWE, P.W. 1975. Inherent difficulties in the application of goetechnical science. *Analysis of Soil Behaviour and its Application to Geotechnical Structures. University of New South Wales, Australia.* W.H. Sellen Pty Ltd. 3-29.

SKEMPTON, A.W. & COATS, D.J. 1985. Carsington dam failure. *Institution of Civil Engineers, London. Symposium on Failures in Earthworks.* Pre-print.

STEWARD, H.E. & CRIPPS, J.C. 1983. Some engineering implications of chemical weathering of pyritic shale. *Quarterly Journal of Engineering Geology,* **16,** 281-289.

P.W. ROWE, Styperson House, Adlington, Macclesfield, Cheshire, UK.

Discussion Session 1

Written Contribution:

L. Threadgold (Consulting Geotechnical Engineer)

In considering the groundwater problems of a site, it is important to appreciate that they can fall into three broad categories as follows:

1. Problems caused by the presence of the water itself, which can place major demands on pumping capacity or require under-water construction.

2. Problems caused by the flow of water, such as in erosion or in creating 'running sand' conditions in excavations.

3. Problems caused by water pressure as in slope stability for example.

I believe that this classification is the key to the formulation of the most satisfactory solution to the problems which groundwater can pose.

From Cripps, J.C., Bell, F.G & Culshaw, M.G (eds) 1986, *Groundwater in Engineering Geology*. Geological Society Engineering Geology Special Publication No. 3, p. 43.

SESSION 2

CHANGES IN GROUNDWATER CONDITIONS

The inter-relationship between changes in groundwater conditions and engineering construction

F.C. Brassington

ABSTRACT: The presence of groundwater and its effects are often major considerations in the design of civil engineering works. The engineering geologist is expected to provide reliable information on this subject by means of appropriate site investigation techniques. In many instances possible changes in groundwater levels beyond the normal seasonal variations are not considered. Large changes can occur, for example, a reduction in groundwater pumping will cause water tables to rise, flooding basements and tunnels and possibly changes in the load bearing properties of soils. Civil engineering construction can have a similarly significant effect on local water supplies especially in the case of road cuttings, tunnels and pipelines. The British Standard Code of Practice for Site Investigations (BS 5930: 1981) does not contain any recommendation that site investigations should include an examination of local groundwater abstraction. Nor does it suggest that surveys should be carried out to locate and identify local private water supplies where there is a danger that the proposed construction will affect them. It is recommended that the Code of Practice is amended to include both these omissions.

Introduction

To the engineering geologist, groundwater is a frequent source of trouble and groundwater conditions are a major element in all site investigations for new civil engineering works. Change in the water content of the soil is the greatest single cause of variation in shear strength, compressibility and permeability. Water chemistry is also important as dissolved constituents in groundwater determine its ability to corrode construction materials. The nature and extent of groundwater must be identified during the site investigation so that appropriate precautions are taken to contain it during construction and to prevent it affecting the long term stability of the new works.

The water engineer and hydrogeologist take a contrasting view and look on groundwater as a resource and asset. This difference in approach was recognised by Gray (1975) in his review of the scope of hydrogeology. In the circumstances, it is not surprising that both the effect on engineering works brought about by engineering construction can be overlooked. There is very little reference to this type of problem in the literature probably because when viewed against the engineering project as a whole, disruption of small water supplies is not very significant. It should not be forgotten, however, that these problems are of great importance to the people involved. A similar conflict of interests also exists between the extractive minerals industry and water supplies and has recently been examined in detail (Wood 1981; Brassington 1982).

Water supplies from groundwater

Groundwater plays an important role in providing water supplies in Britain. Public water supply authorities and major industrialists generally abstract water from deep boreholes whereas the smaller private supplies, mainly for agriculture and domestic use, are obtained from shallow boreholes, wells and springs. In England and Wales, some 32% of the water supplied by the ten regional water authorities is obtained from groundwater. In 1977 groundwater represented three-quarters of public water supplies provided by Southern Water and was about half the water supplied by the Anglian, Severn-Trent, Thames and Wessex Water Authorities. In the North West, South West and Yorkshire Water Authorities, the comparable figure was about 16% but is only 4% in Wales and less that 1% in Northumbria (Anon. 1978).

Private water supplies vary in size from 10 Ml/d (1 Ml/d = 1,000 m^3/d) to less than 1,000 l/d for supplies to individual households. In 1977 there were well over 27,000 licensed private groundwater abstractions in England and Wales (Anon. 1978). These represent only a small proportion of the total private water supplies however, because private domestic water supplies are exempt from licensing. The estimated total number of such sources in the United Kingdom is approaching 250,000 many of which are from groundwater (Brassington 1983; Parkinson & Reynolds 1985). When all agricultural supplies are taken into account the total number of abstractions from groundwater is likely to exceed half a million.

From CRIPPS, J.C., BELL, F.G. & CULSHAW, M.G. (eds), 1986 *Groundwater in Engineering Geology*, Geological Society Engineering Geology Special Publication No. 3, pp 47-50.

These statistics demonstrate the importance of groundwater supplies. This is recognised by water authorities who, besides making groundwater abstractions, have policies for the management and protection of groundwater resources (Hodkinson 1970; Sherlock 1973; Selby & Skinner 1978).

Monkhouse & Richards (1982) listed the geological formations which constitute aquifers in the United Kingdom. The majority of stratigraphical units are represented but mudrocks are defined as aquicludes rather than aquifers. Mudrocks in some circumstances, however, are capable of transmitting and storing water and are used for private water supplies (Tellam & Lloyd 1981). In Lower Palaeozoic shales groundwater is generally found in fissures in weathered rocks but in younger mudrock sequences groundwater tends to be found in arenaceous facies.

Engineering problems caused by changes in groundwater abstraction

Hodkinson (1970) described how large scale uncontrolled abstraction from the Permo-Triassic Sandstone aquifer in south Lancashire caused widespread lowering of the water table. Groundwater development on this scale with consequential lowering of the water table occurred in many parts of the country from the second half of the 19th century as industry grew in all major cities. Development of the Chalk aquifer in the London basin caused the water table to decline by some 45 m in central London between 1864 and 1932. Over the same period the resulting settlement caused ground levels to decline by as much as 0.2 m (Ineson 1957; Marsh & Davies 1983: Downing & Pontin 1984). This type of settlement problem has been experienced more recently in the well publicised case of Venice and the lesser known example of Bangkok.

Over recent years there has been a general reduction in the total quantity of water abstracted by industry. Between 1970 and 1977 in England and Wales the reduction was more than 15% (Anon. 1978). These changes have not occurred evenly over the country and in some places reductions have been substantially greater than this figure. In the Liverpool area, for example, abstractions were halved between 1960 and 1981 (Brassington 1984), largely caused by a decline in city centre industries, The water table in the area, particularly in the city centre, gradually rose causing flooding problems in deep basements and most seriously, in the British Rail Loop Tunnel. West (1983) shows the location and geology of this tunnel which was constructed in the early 1970s. The installed drainage system was not designed to cope with the quantities of groundwater now flowing into the tunnel and expensive remedial works are becoming necessary.

Flooded basements are a problem not only in Liverpool but have also been reported in Manchester and in Birmingham. In the latter example the water table has been rising in response to reductions in abstraction particularly in the northern part of the city since the early 1970s (Jackson & Lloyd 1983). In the London area no flooded basements have been reported at all but Marsh & Davies (1983) anticipate that rising groundwater in the Lower London Tertiary Series may significantly affect the construction techniques employed for deep-bored piled foundations and the design of deep underground structures.

Water supply problems caused by engineering construction

When a large volume of water bearing rock is removed from the ground the storage space available for groundwater is dramatically increased. The specific yield of aquifers varies widely, from less than 0.5% to about 30%. Typical values for gravels and coarse sands are < 20% whereas those for consolidated rocks are likely to be smaller and may be < 1% (Davis & DeWiest 1966). These values contrast with the 100% storage afforded by an excavation and a much larger volume of water is required to restore the hydraulic equilibrium in the aquifer than was originally contained in the rock before excavation. Flow is induced from the aquifer into the excavation thereby lowering the water table locally for periods which may extend over several years (Brassington 1982). If de-watering systems are installed to keep the excavation dry, the period with a lowered water table can be extended indefinitely. The consequential effect on small water supplies has been recognised by several workers, including Wood (1981) in relation to mining and quarrying.

Brassington (1983) warns of the potential impact on private water supplies caused by civil engineering construction, in particular of road cuttings, tunnels and pipelines. Carter & Mills (1976) report that a detailed survey was carried out along the route of the Kielder Tunnels to identify private water supplies which could have been affected by the tunnel construction. A detailed study was carried out involving the measurement of more than 40 springs by the Northumbrian Water Authority in addition to borehole and piezometer water levels and rainfall. These field measurements established that many springs are supported by perched water tables and isolated from the strata through which the tunnel was driven. They also permitted threatened supplies to be identified allowing an appropriate monitoring programme to be instituted.

The problems of groundwater inflows into deep motorway cuttings has been discussed by Buist *et al.* (1978) in relation to the Monk Wood cutting on the A61 in Derbyshire. This cutting was excavated to Lower Coal Measures strata which included a water bearing sandstone, the Silkstone Rock. Groundwater problems were anticipated but during construction these proved to be less severe than expected and only part of the designed system of pressure release holes was contracted. The usual french drains adjacent to the carriageway and a similar drainage system at the top of the cutting were installed but there is no report as to whether or not local springs were affected by this drainage.

Rogers (1972) describes how groundwater inflow problems were overcome during the excavation of a motorway cutting through a glacial sequence to the east of Coventry. Drainage during construction was carried out by longitudinal trenches excavated by dragline and the completed cutting was drained by standard french drains. This will have undoubtedly caused permanent lowering of the water table but no problems with private water supplies are reported. A deep cutting on the same motorway to the north-west of Coventry, however, did have a serious effect on two private water supplies. In this instance the cutting is excavated through sandstones of the South Warwickshire Coalfield and mudstones of the Corley Group (Shotton 1929). Groundwater contained in one of the sandstones is intercepted by the french drains adjacent to the carriageway. This substantially reduced the flow of a nearby spring and the yield of an adjacent borehole both of which were used for private water supplies.

Backfilled pipeline trenches are another potential threat to springs and other private water supplies, where the trench connects several permeable horizons. It is difficult to backfill trenches in a way which prevents water flowing through the fill and in any case, the modern practice of bedding pipes on granular fill makes groundwater movement along such trenches inevitable. Ham (1984) describes his experiences in laying a 400 mm diameter water main across 18 km of rich farmland in south Warwickshire. He recognises that this installation may disrupt field drains but does not mention problems with private water supplies.

Site Investigation

The British Standard Code of Practice for Site Investigations (BS 5930: 1981) covers in minute detail the planning and execution of site investigations for the purpose of assessing the suitability of such sites for civil engineering and building works. The problems which may be caused by groundwater to most construction works is fully recognised and appropriate methods are described for investigating its presence, quantity and quality. Dumbleton & West (1984, 1976a, 1976b) in their three publications covering site investigations also acknowledge that information on the presence or absence of groundwater can be the most critical factor in the design considerations of new works. None of these documents, however, fully recognises the relationship between groundwater abstraction and civil engineering construction.

Paragraph 3 of the Code of Practice (Anon. 1981) sets out the primary objectives of site investigations and sub-paragraph (d) states that:

Effect of changes

> To determine the changes that may arise in the ground and environmental conditions, either naturally or as a result of the works, and the effect of such changes of the works, on adjacent works, and on the environment in general.

It follows from this statement that it is appropriate for the Code to include reference to the interrelationship between civil engineering construction and groundwater supplies and in view of the importance of this relationship as illustrated earlier, it is suggested that the Code is amended.

Information regarding the use of the larger private water supplies may be obtained from water authorities in England and Wales who administer a system of abstraction licences set up under the provisions of the Water Resources Act, 1963. In other parts of the United Kingdom, where abstraction licences are not required, it may be possible to identify major industrial abstractors by contacting the local branch of the Confederation of British Industry or the local Chamber of Commerce. This will enable an assessment to be made of the extent to which the level of the water table is controlled by groundwater abstraction in the vicinity.

Most small abstractions are not required to be licensed, consequently the only way in which these can be identified is by a walk-over survey and liaison with local inhabitants. These small water supplies are likely to be the most vulnerable to any changes in groundwater conditions caused by the planned new works and it is important that they are found at an early date in any investigation.

Once all the local groundwater abstractors have been identified it should be possible to assess potential risks caused to an underground structure by changes in groundwater levels so that this can be taken into account in the design. Similarly an assessment can be made of the potential risks to the small private water supplies in the vicinity of the new structure and an appropriate monitoring programme established which uses field techniques described by Brassington (1983). It is recommended that all such small supplies are examined at an early date and their condition recorded. A common cause of failure of such water supplies is lack of maintenance yet frequently failures are attributed to other causes. Such a detailed record may avoid any confusion arising as to the reason for reduced yields.

Conclusions

Changes in groundwater abstraction may cause water tables to rise or fall causing problems to structures. The likelihood of these changes can be assessed at the site investigation stage and appropriate measures taken in the design.

Certain civil engineering construction particularly motorways, tunnels and pipelines can cause significant modification to groundwater conditions in their immediate vicinity and this may have a serious effect on small water supplies. These problems can also be anticipated by appropriate action at the site investigation stage. The British Standard Code of Practice for Site Investigations (BS 5930: 1981) should have appropriate minor amendments to cover an assessment of local groundwater development.

ACKNOWLEDGEMENTS: The Author is grateful to the General Manager and Divisional Engineer of the Rivers Division of North West Water for permission to publish this paper. The views expressed are those of the Author and not necessarily those of the Authority.

References

ANON., 1978. *Water Data 1977.* Water Data Unit, Reading

ANON., 1981 *Code of Practice for Site Investigations.* BS 5930. British Standards Institutions, London.

BRASSINGTON, F.C., 1982. Hydrogeological problems caused by mining and quarrying. *Transactions of the Institution of Mining and Metallurgy (Section: Applied Earth Sciences),* **91,** B21-B25.

BRASSINGTON, F.C., 1983. *Finding Water : A Guide to the Construction and Maintenance of Private Water Supplies.* Pelham Books, London.

BRASSINGTON, F.C., 1984. Problems caused by a rising water table in Liverpool. *British Geologist,* **10,** 8-9.

BUIST, D.S., BURNETT, A.D. & SANDERS, M.K., 1978. Engineering geological case study of the site investigation and design of a major trunk road cutting in Carboniferous rocks. *Quarterly Journal of Engineering Geology* **11,** 161-176.

DAVIS, S.N. & DEWIEST, R.J.M., 1966, *Hydrogeology,* John Wiley, New York, 463 pp.

DOWNING, R.A. & PONTIN, J.M.A., 1984. Discussion on March and Davies 1983. *Proceedings of the Institution of Civil Engineers,* **76,** 731-733.,

DUMBLETON, M.J. & WEST, G., 1974. Guidance on planning, directing and reporting site investigations. *TRRL Laboratory Report LR 625,* Transport and Road Research Laboratory, Crowthorne, Berks

DUMBLETON, M.J. & WEST, G., 1976a. Preliminary sources of information for site investigations in Britain. *TRRL Laboratory Report 403* (revised edition). Transport and Road Research Laboratory, Crowthorne, Berks.

DUMBLETON, M.J. & WEST, G., 1976b. A guide to site investigation procedures for tunnels. *TRRL Laboratory Report 740.* Transport and Road Research Laboratory, Crowthorne, Berks.

GRAY, D.A., 1975. The scope of hydrogeology. *Quarterly Journal of Engineering Geology,* **8,** 177-192.

HAM, B.V., 1984. Cross country pipelines and the landowner. *Journal of the Institution of Water Engineers & Scientists,* **38,** 467-474.

HODKINGSON, G.W., 1970. Management of an over-taxed aquifer, in WARREN, P.T. (ed) *Geological Aspects of Development and Planning in Northern England.* Yorkshire Geological Society.

INESON, J., 1957. *The movement of groundwater as influenced by geological factors, and its significance.* PhD thesis, University of London.

JACKSON, D & LLOYD, J.W., 1983. Groundwater chemistry of the Birmingham Triassic Sandstone aquifer and its relation to structure. *Quarterly Journal of Engineering Geology,* **16** 135-142.

MARSH, T.J. & DAVIES, P.A., 1983. The decline and partial recovery of groundwater levels below London. *Proceedings of the Institution of Civil Engineers,* **74,** 263-276.

MONKHOUSE, R.A. & RICHARDS, H.H., 1982. *Groundwater Resources of the United Kingdom.* Th. Schafer, Hanover, for the European Economic Community.

PARKINSON, N.H. & REYNOLDS, M.W., 1985. Private water supplies — a local authority viewpoint. In *COPA II and Drinking Water Directive, Preprint,* Institution of Water Engineers and Scientists, 43-60.

SELBY, K.H. & SKINNER, A.C., 1978. Aquifer protection in the Severn-Trent region: policy and practice. *Annual Conference, the Institute of Water Pollution Control 1978,* paper II.

SHERLOCK, D., 1973. The implementation of a policy for groundwater management in *2nd Symposium on River Basin Groundwater Development,* Institution of Water Engineers.

SHOTTON, F.W., 1929. The geology of the country around Kenilworth (Warwickshire), *Quarterly Journal of the Geological Society of London,* **85,** 167-222.

TELLAM, J.H. & LLOYD, J.W., 1981. A review of the hydrogeology of British onshore non-carbonate mudrocks. *Quarterly Journal of Engineering Geology,* **14,** 347-356.

WEST, G., 1983. Comparison between real and predicted geology in tunnels: examples from recent cases. *Quarterly Journal of Engineering Geology,* **16,** 113-126.

WOOD, P.A., 1981. Hydrological problems of surface mining. *Report number ICTIS/TR17,* IEA Coal Research, London.

F.C. BRASSINGTON, North West Water Authority, Rivers Division, PO Box 12, New Town House, Buttermarket Street, Warrington WA1 2QG, UK.

Groundwater conditions in the coastal landslides of the Isle of Sheppey

N. Dixon & E.N. Bromhead

ABSTRACT: The paper describes an investigation into the groundwater pressure distribution in two sections through the 46 m high London Clay cliffs at Warden Point, Isle of Sheppey. Over 50 piezometers were installed, and these show a groundwater pressure regime which indicates a significant degree of response to stress relief. The paper describes the influence of these pressures on the mechanics of failure of the cliffs, and the engineering implications of their presence.

Introduction

Pore water pressures in cut slopes may be depressed due to stress relief effects. This concept has been discussed by Vaughan & Walbancke (1973), Eigenbrod (1975) and Bromhead (1972), and is used to account for the delayed failure of cutting slopes in brown London Clay (Skempton 1977). Significant problems in the measurement of these pore water pressures arise because:

a. they are transient, i.e. change with time following the excavation,

b. they may be negative relative to atmospheric pressure at first, and

c. in a natural slope, complications may be caused by other degradation processes.

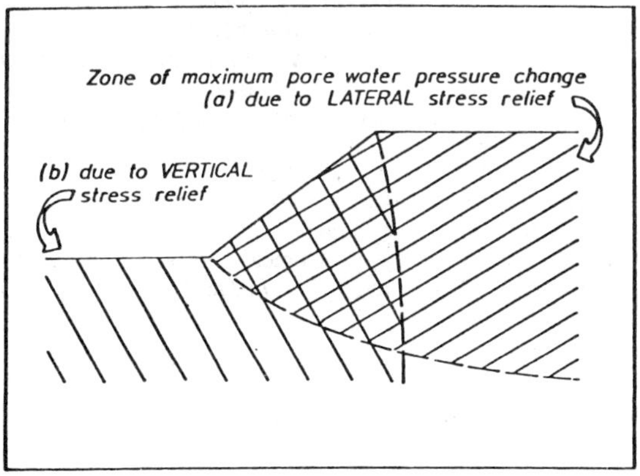

FIG. 1. Influence of stress relief on pore water pressures around an excavation.

Typically, in an unloaded artificial slope, there will be a depressed pore water pressure regime under the excavation and beneath the side slopes (Fig. 1). This will be modified in the case of a natural slope, by a mantle of shallow landslide (e.g. mudslide (Hutchinson 1967)) deposits, with a localised, perched and possibly artesian pore water pressure regime (Hutchinson & Bhandari 1971). If, in addition, it has been subject to deep seated slide activity, it may be grossly modified by pore water pressure distributions 'carried down' the slope, but otherwise unaltered.

To measure the magnitude and distribution of a depressed pore water pressure regime in a slope, the following criteria must be met:

a. an adequate number of measurements must be made,

b. the pore water pressure changes must be recent, and not partly equilibrated,

c. the magnitude of depression should be great so that inevitable inaccuracies in measurements are immaterial.

To meet objectives (b) and (c) requires a large slope which has been recently excavated, or is subject to rapid erosion (if natural). This is of particular importance due to uncertainty about field equilibration rates (Chandler 1984). Furthermore, a large slope makes it easier to install the requisite number of instruments (note (a)).

It is desirable to use simple, inexpensive and relatively foolproof instruments, especially where they are to be used on an exposed site or one subject to a variety of forms of mass movement and where a high rate of damage or loss is to be expected. This reinforces the need for an adequate number of instruments outlined in (a).

In the following, the investigation of the depressed pore water pressure regime in the coastal slopes at Warden Point on the Isle of Sheppey, Kent, is outlined. These slopes and their coastal landslides have been described elsewhere (Hutchinson 1965, 1973; Bromhead 1979). They meet the above criteria almost perfectly. Two sections were instrumented to allow the

From CRIPPS, J.C., BELL, F.G. & CULSHAW, M.G. (eds), 1986, *Groundwater in Engineering Geology,* Geological Society Engineering Geology Special Publication No. 3. pp. 51-58.

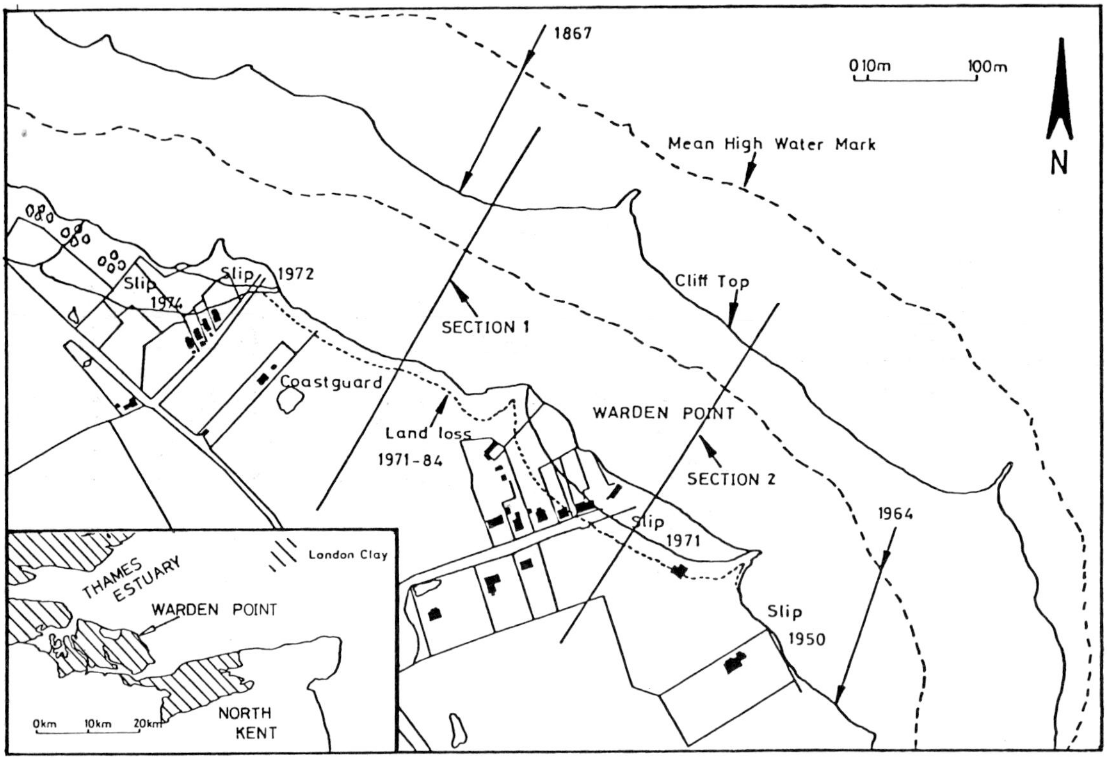

FIG. 2. Plan of the Warden Point area showing retreat of the cliff line since 1967 and location of the instrumented sections. (Based upon the 1867 and 1964 Ordnance Survey maps with permission of the Controller of Her Majesty's Stationery Office, Crown Copyright reserved.)

differentiation between pore water pressure changes simply resulting from stress relief, and those 'carried down' the cliff when one of the characteristic large and deep seated rotational slides takes place.

The main purpose of the investigation was to explore the pore water pressure regime. At the same time, however, the opportunity was taken to examine the impact of these pore water pressure conditions on the behaviour of the large coastal slide on Section 2. Ordinarily, these pore water pressures would be used, *via* a back analysis (Chandler 1977), to provide a fix on the field residual strength envelope. This envelope, for London Clay, is well defined (Skempton 1985), and agreement between the back-figured residual strength from this site and data from elsewhere may be taken as indirect corroboration of the broad correctness of the pore pressure measurements, as well as an extension to the *corpus* of knowledge.

Fieldwork

A total of 56 Casagrande standpipe-type piezometers have been installed at positions and depths shown on the site plan (Fig. 2) and the two cross sections, Figs 3 and 4. The cliff top piezometers were installed in the period November 1982 to April 1983, and the slope and foreshore instruments in September 1983. To facilitate access during the drilling and instrumentation programme on the slopes and foreshore, a track was built down each of the two cliff sections. The track on Section 2 also gave access to the beach, where instruments were installed between low and high tide marks. These tracks were built in approximately three days using a D6 bulldozer, which was retained on site and used to transport the drilling equipment onto and around the slopes.

A Traveller 30 rotary drilling machine was used to form 150 mm diameter boreholes by the continuous flight augering technique, a maximum depth capacity of 45 metres being available. In each borehole a simple standpipe piezometer was installed, comprising a low air entry porous tip connected to ground level by rigid PVC plastic tubing of approximately 19 mm diameter, the tip being surrounded by a nominally 600 mm long sand filter. Blockages in the tubing due to shearing or bending have been used to identify the position of zones of movement. Bentonite pellets were used to form a relatively impermeable seal above the filter, the remaining length of the borehole being backfilled with spoil in a matrix of bentonite-cement grout. The augering technique forms small pellets of clay which are ideal as a backfilling material when used in conjunction with a fairly liquid grout.

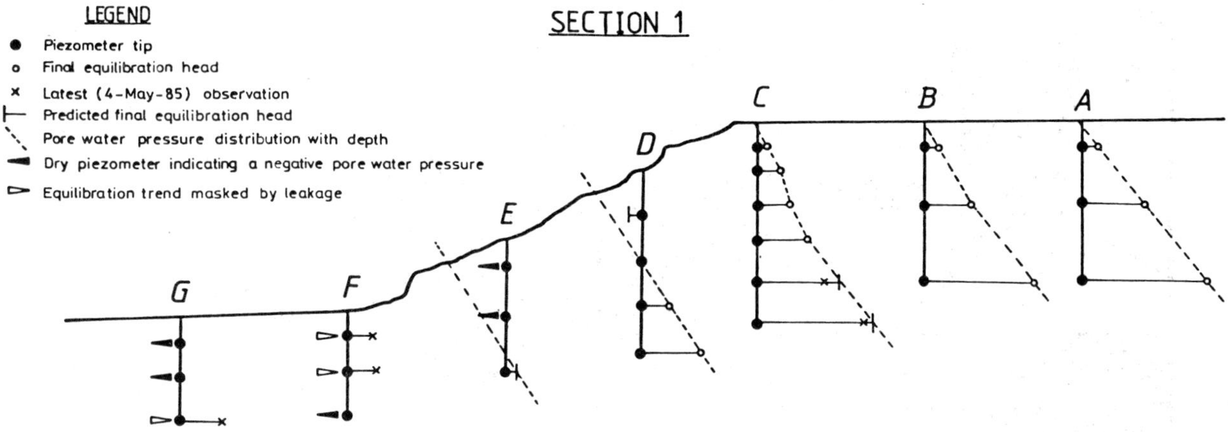

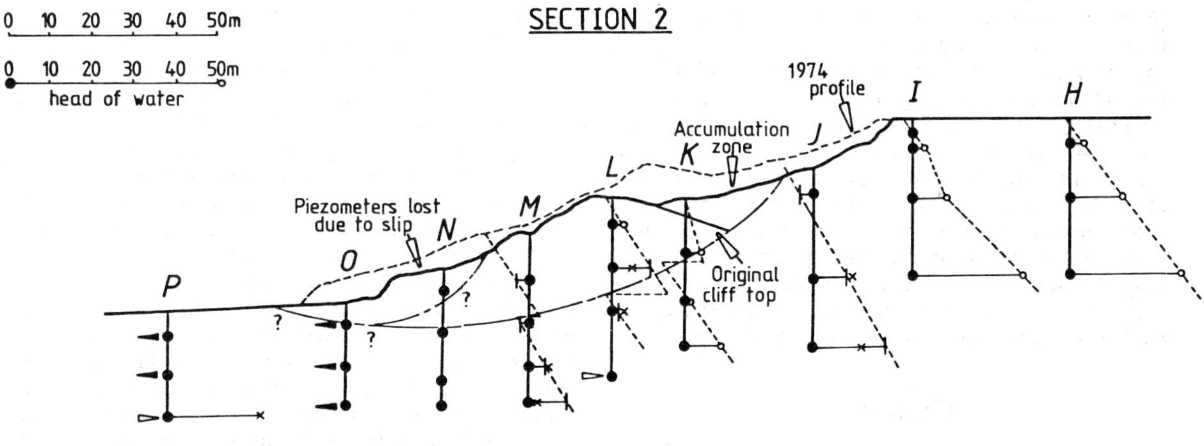

FIGS 3 (Upper) and 4 (Lower). Cross sections showing observed and predicted pore water pressures.

Protective covers were concreted in at the top of each instrument. For the beach piezometers, which are submerged at high tide, air- and water-tight covers were used, excess air pressure in the tubes being released each time the instruments are read.

Casagrande piezometers were chosen because a large number of instruments could be installed cheaply and quickly. They are also robust, easy to read and require little maintenance. However, because of their slow response time a special effort had to be made to gain information from the slope instruments before they became inaccessible or destroyed. In practice many peizometers have been lost due to shallow rotational slides and mudslides.

Instrumentation rates were on average 60 metres of borehole per day with three piezometers installed, including backfilling and construction of the protective caps. The total cost of the instrumentation programme, including materials and the services of the bulldozer was £7,200. Costed over the 1200 m of boring, this demonstrated the correct choice of technique, and resulted in an extremely economic cost of £6/metre inclusive.

Results

The cliff top piezometers have reached their final equilibrium level apart from the two deep instruments at 'C', which leak sporadically through their protective caps. These, by necessity, have been set flush with the ground, the site being regularly used for other purposes, and are inundated during rainstorms. This was not appreciated fully at the time of installation and capping, and as a result, the total ingress of surface water can take place at a faster rate than the natural equilibration of the piezometer. However, data from a number of segments of equilibration curves have been used to predict the *in situ* piezometric pressure which is now thought to be known with some confidence. Measurement of the final equilibration levels of the beach and slope piezometers has in many cases not been possible. These instruments fall into three categories, (i) those which were destroyed before equilibration was complete, (ii) piezometers suffering from regular leaks (this mainly applies to the beach instruments), (iii) instruments which become dry due to a negative

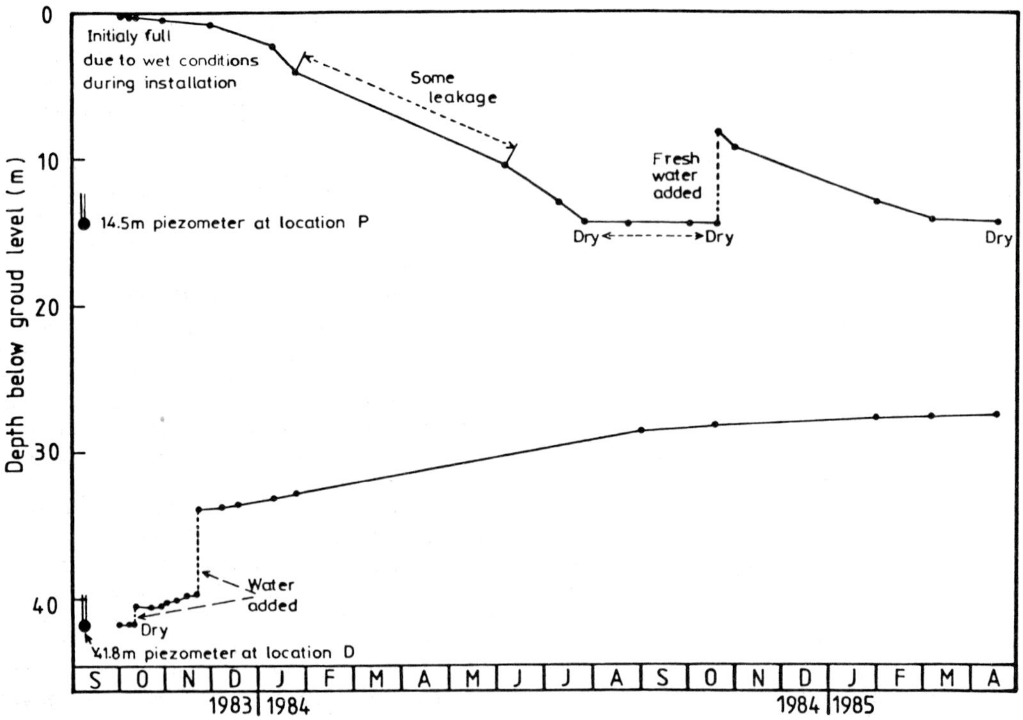

FIG. 5. Water level/time plots showing the equilibration response of typical beach and slope piezometers.

pore water pressure being present. However, for most of the instruments sufficient readings have been take to allow a prediction of the equilibrium pore water pressure to be made.

These predicted levels are derived from water level/time plots using Gibsons's (1963) variable head theory and a curve fitting technique. Where the piezometers have become dry, an estimate of the magnitude of the negative pore pressure is made using the above method.

Only eight of the piezometers have not given any useful information. A small secondary slide through the toe of the slope in Section 2 destroyed the four piezometers at location 'N' soon after installation. Four instruments have had their equilibration trend so masked by leakage that no predicted final level could be obtained. The measured/predicted pore water pressure variations with depth for each installation is shown on the sections (Figs 3 and 4). By way of example, Fig. 5 shows the response curve for the 41.8 m deep piezometer at location 'D', this is typical of many of the instruments. It also shows the response of the 14.5 m deep beach piezometer at 'P' (where suctions are present) to leakage initially, and then to filling with fresh water. The presence of suctions in this instrument are all the more remarkable in view of the fact that the backfilling of this hole had to be completed under water as the working site became inundated by the returning tide.

It has not been possible to explain the measured pore water pressure distribution in terms of a steady seepage regime, although a number of finite element analyses have been carried out using different permeability variations with depth, and using several different boundary conditions. The London Clay extends to a depth of 83 m below beach level at Warden Point (King *et al.* 1985), with a small amount of underdrainage taking place into the Oldhaven Beds. This base boundary is not critical, the slope being isolated from the underdrainage by decreasing permeability with depth (*vide* Bromhead & Vaughan 1980). The measured pore water pressures are consistently less than those predicted by seepage analyses. Taking into consideration the depressed pore pressures found in artificial cutting slopes as discussed previously, it seems likely that the low pore pressures are due to stress relief effects. The difference between the measured and predicted pore water pressures decreases as the distance from the cliff edge increases.

Pore water pressure versus depth relationships for the cliff top locations (Fig. 6) are comparable for the two sections with the exception of the two cliff edge installations. These are similar down to a depth of approximately 18 m, after which the pore water pressures are considerably greater for Section 2. The most probable explanation for this is the existence of the large landslide block which is situated 20 m down the slope, reducing the lateral stress relief below 20 m and leaving higher pore water pressures.

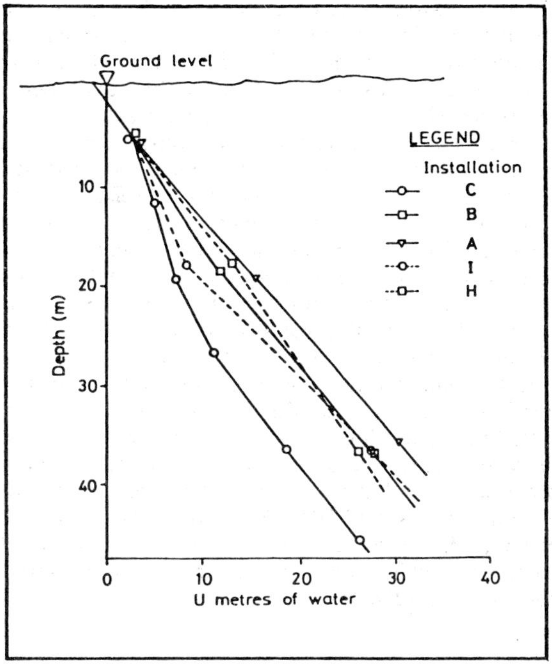

FIG. 6. Comparison of cliff top piezometric data between the two sections.

Installations 'D' and 'E' show a zone of depressed pore water pressures beneath the slope. Piezometers at 'D' do not record positive pore water pressures until a depth greater than 10 m below ground level is reached.

Although very few readings were taken at 'E', due to a shallow mudslide destroying the instruments in February 1984, a small head of water was measured in the 30 m piezometer and negative pore water pressures were indicated by the two shallower instruments. The zone of negative pore pressure would seem to extend to approximately 30 m below ground level, a depth similar to the amount of material which has been removed above this location. This pore water pressure distribution with depth is not that expected for steady seepage conditions.

On Section 2 the piezometers in the slipped mass have been placed at locations 'K', 'L', 'M', 'N' and 'O' with the lower-most piezometers penetrating into the intact clay below the landslide. At location 'J' the piezometers effectively cut the rear scarp and are placed in intact London Clay landward of the slide masses. Pore pressures in the slipped mass are usually positive but less than hydrostatic. It is thought that these are lateral stress influenced cliff-top pressures 'carried down' by the landslides. The piezometers which penetrate the intact clay show a zone of negative pore water pressure (indicated by the equilibration curves of piezometers therein), extending for several metres below the slip surface. Beneath this, pore water pressures are depressed, but positive in absolute terms. Readings taken so far at 'J' indicate pore pressures which are considerably depressed and of the same magnitude as those at position 'D' on Section 1; 'J' and 'D' have the same position relative to the cliff crest.

On the foreshore, piezometers at positions 'F', 'G', 'O' and 'P' are indicating a zone of negative pore water pressure which continues to a depth of over 24 m below beach level. As previously discussed, some of the beach piezometers are inoperable due to leakage, in others a small leakage of sea water into the piezometer can occur, causing an apparent reading in the 'dry' piezometer, but this vanishes when more care is taken over the securing of the protective cap, as seen in Fig. 5. The main problem associated with beach piezometers is if the water level is observed to rise, great care must be taken to ascertain whether the rise is due to equilibration or leakage, although a drop in level must certainly be equilibration.

Back Analysis — Section 2

A field residual strength envelope for London Clay based largely on data from the Herne Bay coastal landslides (Bromhead 1978), supplemented by a number of smaller slides, is given by Skempton (1985). He compares this to laboratory envelopes, based on slip surface tests and ring shear tests and comments that the ring shear test data tends to underestimate ϕ'_r by about 1.5°. A back analysis of the slide on Section 2 was undertaken to extend the field residual envelope. Surveys of the main cross section had been taken in 1974, some three years after the 'first time failure', and again in 1983. Significant changes had occurred. Naturally, the piezometric observations relate to the latter section, but their broad pattern has been extended back to the probable values for the earlier section, enabling two separate analyses to be made.

Having the two surveys meant that the shape of the slip surface, already defined by observations during the investigation and by sheared off instrumentation, could be further defined by reference to surface movement vectors. The movements that have occurred (8 m from 1974 to 1983) confirm that the shear strength is fully mobilised, or simply, F = 1.0, in both cases.

Analysis of the landslide has been carried out using Morgenstern and Price's method for non-circular slip surfaces with correction for side-shear.

Comparison of the back analysis results with a residual failure envelope based on ring shear tests from London Clay at Warden Point is shown in Fig. 7. The ring shear tests were carried out at Hatfield Polytechnic on the Bromhead ring shear apparatus, by arrangement with Mr E.A. James and his students whose help is gratefully acknowledged. Each point on the failure envelope is the average of 18 tests. Results of tests on natural slip surfaces from Warden Point

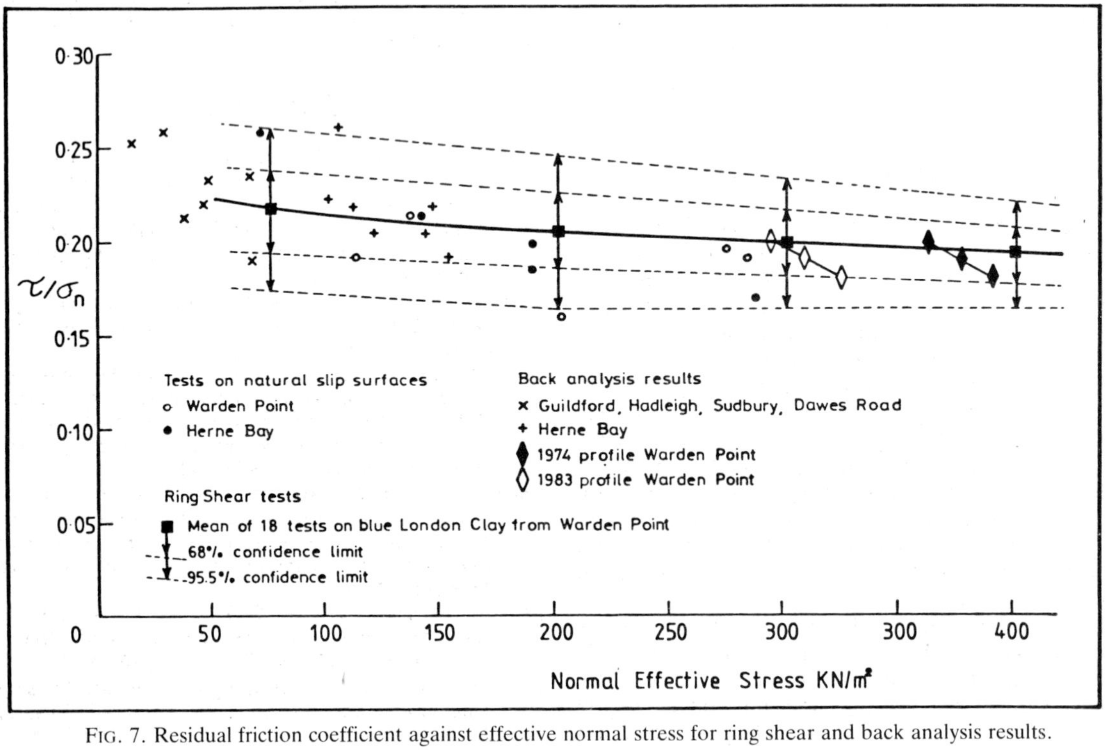

FIG. 7. Residual friction coefficient against effective normal stress for ring shear and back analysis results.

TABLE 1. *Stability analysis of Section 2.*

Profile	Average pore water pressure ratio r_u	ϕ'_r for equilibrium	Average shear stress kN/m^2	Average normal effective stress kN/m^3	Residual friction coefficient
1974	–0.35	10.3	71.5	392.6	0.182
	–0.29	10.7	71.6	378.1	0.189
	–0.22	11.1	71.6	363.6	0.197
1983	–0.66	10.1	58.4	325.6	0.179
	–0.54	10.7	58.5	310.4	0.189
	–0.42	11.3	58.6	295.1	0.199

(Bromhead & Curtis 1983) and Herne Bay (Bayley 1972) are included for comparison. Skempton's other data has been excluded, and only data from north Kent has been used, mainly for reasons of clarity in an otherwise congested diagram.

The back analysis has been carried out with three different pore pressure distributions in the toe region as some uncertainty still remains as to the size of the suctions in this area, higher pore pressures were used in the 1974 analysis to take into account the smaller amounts of stress relief. Results shown in Table 1 and plotted in Fig. 7 clearly show that stress relief induced pore water pressures are required for the back analysed residual shear strengths to agree with the laboratory defined failure envelope.

The close agreement between the field and laboratory test data confirms the correctness of the measured pore water pressures. Indeed, the differences noted earlier (Skempton 1985) between different methods of laboratory measurement and field strengths may well be more a matter of sampling bias on the one hand and shortage of data on the other.

Conclusions

This case record demonstrates conclusively that depressed pore water pressures in slopes subject to excavation or unloading are a significant effect. In contrast to the treatment of slopes where the pore water pressure regime is broadly in equilibrium with the hydraulic boundary conditions, the treatment of slopes containing significantly depressed pore pressures needs to be approached with caution. The long term tendency of the pore water pressures to rise will result in destabilisation. This is potentially capable of undermining the beneficial effect of stabilisation work such as regrading or certain types of drainage measures. The apparent rapid stabilisation of slopes with these measures may therefore prove to be an all too ephemeral effect.

The zone of depressed pore water pressure is carried inland with the cliff as the coastline retreats. It is local to the slope, but may extend a comparatively large distance inland. Equilibration rates in the London Clay are slow, as evidenced by the retention of suctions in the foreshore in areas first exposed in excess of 50 years ago. This slow equilibration rate prevents loss of depressed pore water pressure in the slope since soil is removed faster than moisture penetration takes place, although the general conclusion would not be applicable to smaller slopes retreating more slowly in more permeable strata than the London Clay.

Slide behaviour in these cliffs is dominated by the effects of the stress relief induced suctions.

ACKNOWLEDGEMENTS: A generous grant from the Science and Engineering Research Council has made the field work possible. Mr Dixon is supported financially by Kingston Polytechnic. The authors gratefully acknowledge the help in the field of Mr R. Sandman and from Dr M.P. Chandler, Dr K.L. Shepherd, Mr R.D. Curtis, Mr M. Dedic, Mr R. Thompson, Mr J. McEwen, Mr R. Ferrand, Mr A. Mancktelow, Mr N. Key, Mr M. Smith, Mr D. Williams and Mr S. Holsgrove, and at Hatfield, Messrs E.A. James, S. Williams and S. Sewell.

References

BAYLEY, M.J., 1972. Cliff stability at Herne Bay. *Civil Engineering and Public Works Review*, **67**, 788-793.

BROMHEAD, E.N., 1972. A study of some aspects of time dependent failure of cuttings in overconsolidated clays with special reference to transient pore water pressure effects. *MSc Thesis*, Imperial College.

BROMHEAD, E.N., 1978. Large landslides in London Clay at Herne Bay, Kent. *Quarterly Journal of Engineering Geology*, **11**, 291-304.

BROMHEAD, E.N.. 1979. Factors affecting the transition between various types of mass movement in coastal cliffs consisting largely of overconsolidated clay with special reference to Southern England. *Quarterly Journal of Engineering Geology* **12**, 291-300.

BROMHEAD, E.N. & VAUGHAN, P.R., 1980. Solutions for seepage in soils with an effective stress dependent permeability. *Proceedings of the first International Conference on Numerical Methods for Non-linear Problems*, 567-578.

BROMHEAD, E.N. & CURTIS, R.D., 1983. A comparison of alternative methods of measuring the residual strength of London Clay. *Ground Engineering*, **16**, 39-41.

CHANDLER, R.J., 1977. Back analysis techniques for slope stabilisation works: a case record. *Geotechnique*, **27**, 479-495.

CHANDLER, R.J., 1984. Recent European experience of landslides in over-consolidated clays and soft rocks. *Proceedings of the 4th International Symposium on Landslides, Toronto,* **1,** 61-81.

EIGENBROD, K.D., 1975. Analysis of the pore pressure changes following the excavation of a slope. *Canadian Geotechnical Journal,* **12,** 429-440.

GIBSON, R.E., 1963. An analysis of system flexibility and its effects on the time lag in pore water pressure measurement. *Geotechnique,* **13,** 1-11.

HUTCHINSON, J.N., 1965. Survey of the coastal landslides of Kent. Note No. EN35/36, Ministry of Technology, Building Research Station, Watford.

HUTCHINSON, J.N., 1967. The free degradation of London Clay cliffs. *Proceedings of the Geotechnical Conference, Oslo,* **1,** 113-118.

HUTCHINSON, J.N., 1973. The response of London Clay cliffs, to differing rates of toe erosion. *Geologia Applicata e Idrogeologia,* **8,** 221-239.

HUTCHINSON, J.N. & BHANDARI, R.K., 1971. Undrained loading, a fundamental mechanism of mudflows and other mass movements. *Geotechnique,* **21,** 353-358.

KING, C., SANDMAN, R.I., BROMHEAD, E.N. & DIXON, N. A faunal analysis and correlation of the London Clay formation for the Isle of Sheppey, Kent. in press.

SKEMPTON, A.W., 1977. Slope stability of cuttings in brown London Clay. *Proceedings of the 9th International Conference on Soil Mechanics, Tokyo,* **3,** 261-270.

SKEMPTON, A.W., 1985, Residual strength of clays in landslides, folded strata and the laboratory. *Geotechnique,* **1,** 3-18.

VAUGHAN, P.R. & WALBANCKE, H.J., 1973. Pore Pressure changes and the delayed failure of cutting slopes in overconsolidated clay. *Geotechnique,* **23,** 531-539.

N. DIXON & E.N. BROMHEAD, Department of Civil Engineering, Kingston Polytechnic, Canbury Park Centre, Canbury Park Road, Kingston-upon-Thames KT2 6LA, UK.

The effects and consequences of groundwater abstraction on foundations at Drax Power Station, North Yorkshire

M. Dobie & R.L. Newman

ABSTRACT: In order to provide an additional supply of water for Drax Power Station the CEGB proposed to abstract water from the Bunter Sandstone aquifer which underlies the site. A pumping test was carried out to assess the effects of abstraction, and this involved monitoring groundwater pressures and surface deformations.

The first part of the paper details the ground and groundwater conditions, and describes the design of the pumping test and installation of the instrumentation. Theoretical aspects of analysis and response of groundwater pressures to abstraction are also discussed.

The second part of the paper outlines and analyses the results of the pumping test. The results are presented as a series of graphs of drawdown, surface deformation, and duration and rate of pumping. The true aquifer characteristics were determined from the pumping test results using the analysis of Thiem incorporating Huisman's procedures for partially penetrating wells. Using the same analyses the effects of the pumping test are extrapolated to proposed 'licenced' rates. A discussion follows on the results and some interesting observed responses.

The final part of the paper assesses the consequences of surface deformations due to groundwater abstraction on existing and proposed structures. In particular, the effects of large differential settlements between buried services and piled structures, and also induced negative skin friction on vertical and raking piles are considered. Recommendations were subsequently given on the siting of the well which minimised its effects.

Introduction

In order to establish an alternative supply of process water to the river water that was being used, the CEGB wanted to exploit the groundwater in the Bunter Sandstone aquifer below the site. To extract this water a well was required so that the groundwater could be pumped out.

At the end of 1977, W.S. Atkins & Partners were asked for their comments on these proposals with respect to effects on the structures. The main conclusion, which was based upon initial analyses without the benefit of field data, was that pumping would cause a drop in water pressure in the clays that overlie the Bunter Sandstone and that settlement would occur. Not only would the ground settle, but also negative skin friction could be induced in the piles causing them to settle. The initial analyses indicated a likely drawdown curve that would be produced by pumping and from this information it was recommended that the well should be located at least 1000 m from any major structures.

In April 1978, under the instructions of the CEGB, a well was constructed at a location near the sedimentation tanks in the pretreatment area so that a field pumping test could be performed to verify the validity of the initial analysis. The site co-ordinates of the well are given in appendix A. In order to monitor the effects of pumping from this well, piezometers were installed to measure water pressures in the top of the sandstone and the base of the overlying clay, and ground deformations were recorded at various settlement stations. The position of the well, piezometers and settlement stations are shown in Fig. 1.

In this report, the results from the trial abstraction and associated monitoring are examined and analysed. The trial abstraction was carried out at a smaller rate and for a shorter period than the maxima envisaged. Thus it was necessary to extrapolate the measured response of the ground and structures caused by the trial to the likely response due to the intended production rate and duration. Therefore, the second section of this paper details the extrapolation based on the actual response and the theoretical response of the ground to pumping from the underlying aquifer.

Ground and groundwater conditions

The soil profile underlying the site consists of a series of clays overlying the Bunter Sandstone. Generally, the clays are laminated, with the frequency and thickness of partings varying. The silt and sand content of the clay increases with increasing depth, and at the base of the clay there is a layer of silt. The

From Cripps, J.C., Bell, F.G. & Culshaw, M.G. (eds), 1986, *Groundwater in Engineering Geology,* Geological Society Engineering Geology Special Publication No. 3, pp. 59–73.

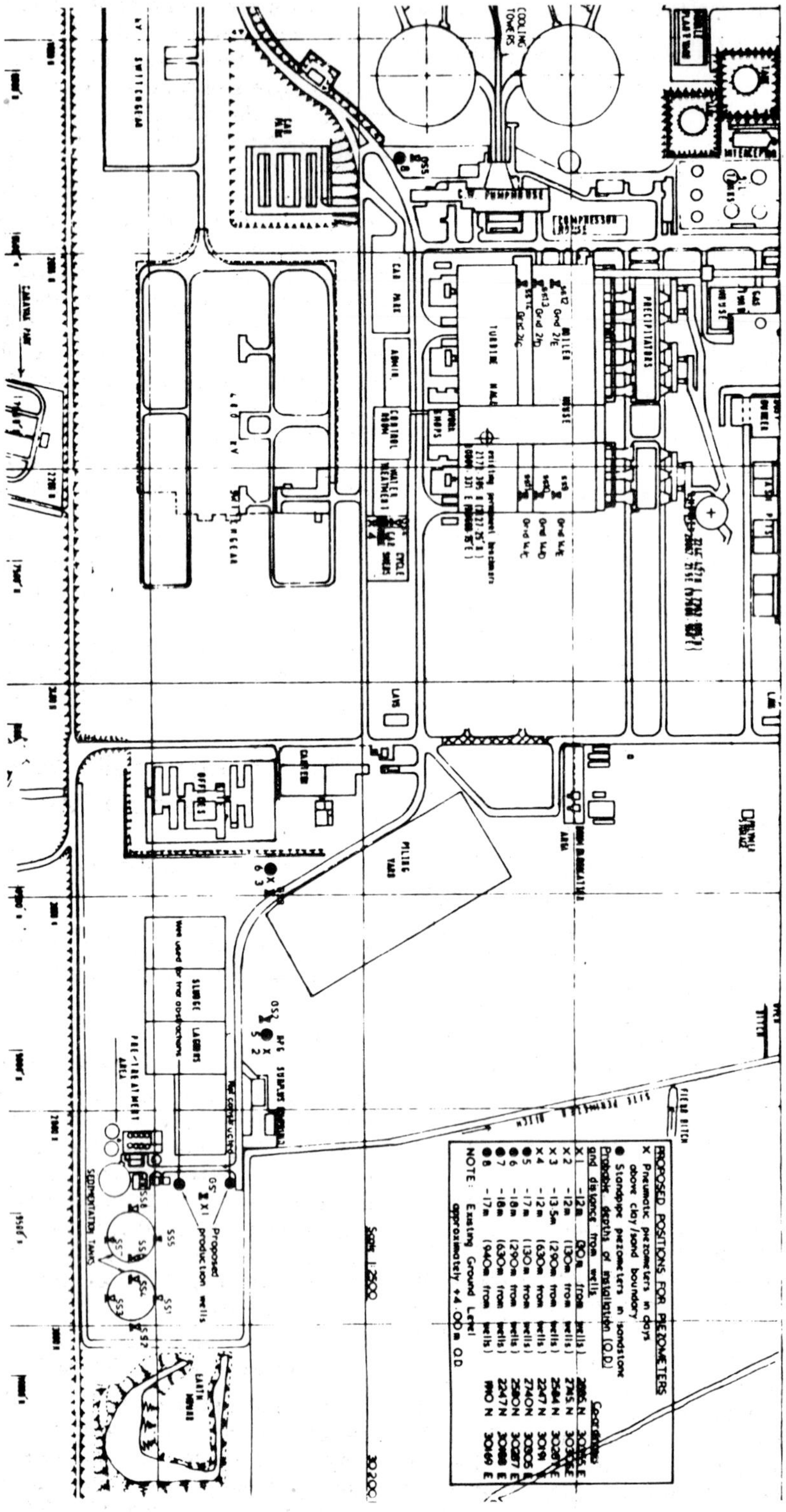

FIG. 1. Location of abstraction wells, settlement stations and piezometers.

Bunter Sandstone consists of sandstone with occasional bands of mudstone or siltstone, and is weathered at its junction with the overlying clays to an uncemented sand. A typical soil profile is as follows:

Soil profile	*Depth below ground surface (m)*
ALLUVIUM A firm grey brown mottled laminated silty clay.	0.0 to 5
GLACIAL LAKE CLAY A firm brown laminated silty clay with silt in partings and occasional bands of sand.	5.0 to 13.0
A grey and brown micaceous silt with fine laminations of clayey silt	13.0 to 16.0
BUNTER SANDSTONE A red-brown sand with occasional pebbles	16.0 to 20.0
A soft red-brown sandstone	20.0 to a considerable depth

The groundwater conditions are characterised by the standing water levels in the piezometers and the well, and are summarized in Appendix C.

Theoretical Response to Pumping

It is necessary at this stage to detail the theoretical response of the ground to pumping as the process and terminology involved will be used later in the paper.

The succession of soils on the site consists of clays of very low permeability overlying silts of higher permeability, which in turn overlie sands and sandstone of very much higher permeability. The groundwater is pumped from the sandstone.

Settlement of the ground occurs due to a drop in water pressure in the soil profile caused by pumping. The total stress at any point in a soil mass is made up to two portions:

(a) Stress carried by the soil skeleton which is termed the effective stress or average intergranular stress.

(b) Stress carried by the pore water which is termed the pore water pressure.

When considering the effects of the pumping test, the total stress remains essentially constant. Thus, if the water pressure is reduced, the effective stress increases and this results in consolidation settlements taking place. For consolidation to occur a small amount of water must drain out of the soil skeleton. Because there is no increase in total stress due to pumping, then there is no immediate elastic settlement as in the case of undrained loading by a foundation. Thus all settlements, whether they occur in the clays or in the sands, are consolidation settlements.

The rate of consolidation settlement is controlled by two time dependent processes. These are:

(a) The drawdown of pressure in the sandstone caused by pumping at the well.

(b) The drainage of the clays in response to the drop of pressure in the top of the sandstone.

The progress of these combined processes is shown in Fig. 2. Fig. 2a shows the initial state before pumping where there is an equilibrium condition. It is assumed that the ground surface is kept wet by rain, which was the case during most of the pumping test. Fig. 2a also shows the effect of a small amount of drawdown. The drop in pressure in the silt will be similar to that in the sandstone because it is able to drain relatively quickly. Very little pressure drop will have occurred in the clay due to its low permeability. Fig 2b shows the situation a short time after drawdown in the sandstone is complete. At this stage a small amount of drainage will have occurred in the clay. If pumping were continued for many years, consolidation would eventually be complete when a new equilibrium condition is reached, as shown in Fig. 2c. This situation will not be reached during the envisaged maximum pumping period. In each diagram the hatched area represents the decrease in water pressure that has occurred at that stage, and, therefore, reflects the progress of settlement.

Settlement could also be caused by extraction of fines from the soils by the action of pumping. This is only likely to occur close to the well.

Pumping test installation, arrangement and monitoring

In this section the installation and subsequent monitoring of a trial abstraction well situated in the north east corner of the site adjacent to the sedimentation tanks are described. Fig. 1 shows the locations of the settlement stations, piezometers and wells.

Abstraction well

The production wells consisted of two 450 mm diameter boreholes. One of these wells was utilised for the pumping test. The well for the pumping test was 92.0 m deep and extended approximately 73.0 m below the surface of the bedrock which, at the location of the well, was found to consist of a succession of marls and, predominantly, sandstones. The sandstone strata varied in thickness from 7.0 m to 44.0 m, whereas the thickness of the marl ranged from 2.0 m to 6.0 m.

Measurement of water table fluctuation

Four standpipe piezometers and four pneumatic piezometers were installed at intervals along an imaginary line between the trial abstraction well and the existing cooling tower area. The installation of the piezometers was completed in July 1978, and readings

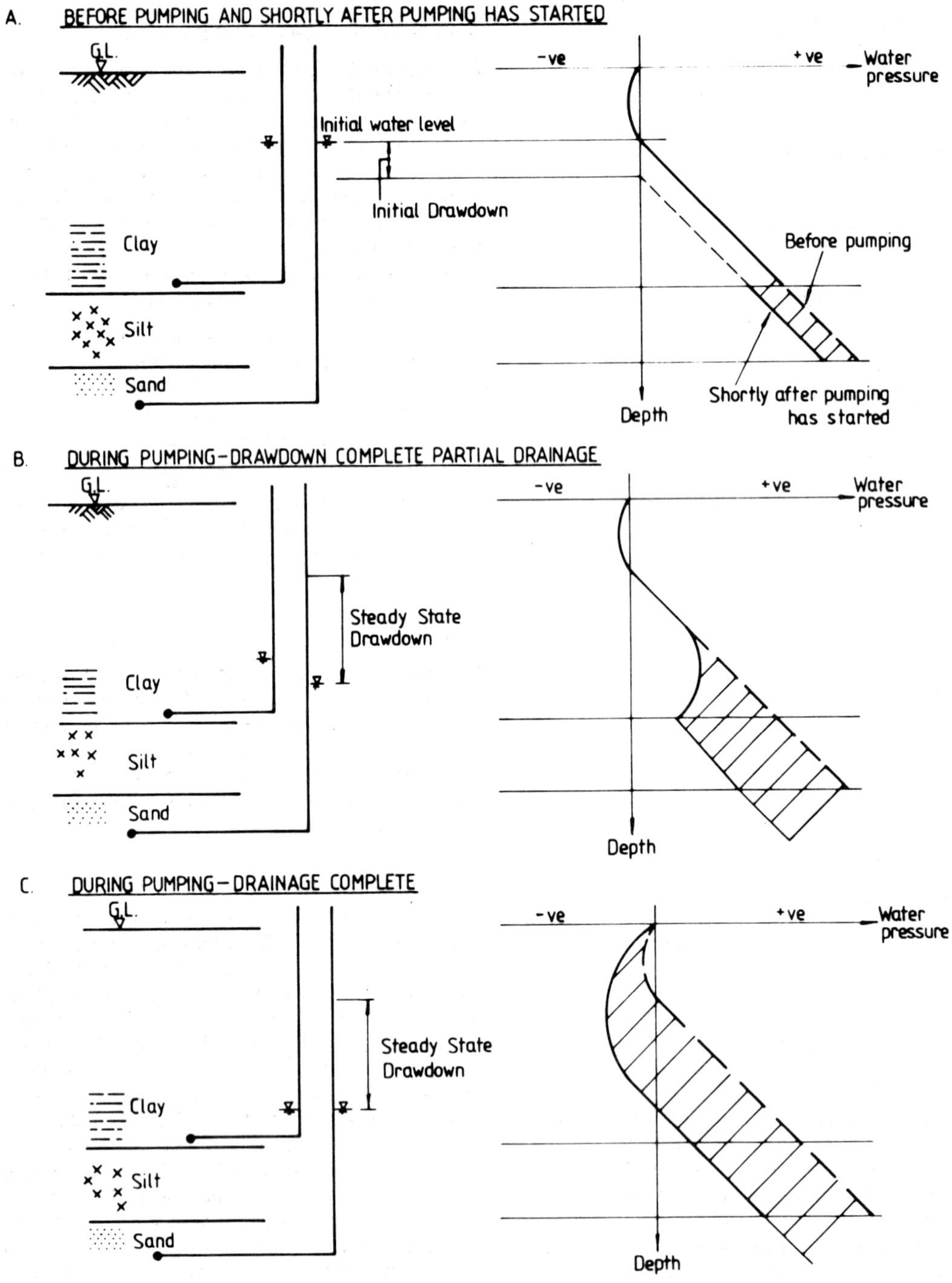

Notes
1. Dotted heavy line represents pressure before pumping
2. Solid heavy line represents pressure at time described
3. Hatched area represents drop in pressure at any time

FIG. 2. Process of Drainage in soils overlying sandstone.

were subsequently taken to establish 'at rest' or equilibrium conditions. Appendix 'A' details the depth and locations of the piezometers and Fig. 3 illustrates the positions of the piezometers with regard to the ground conditions and well location.

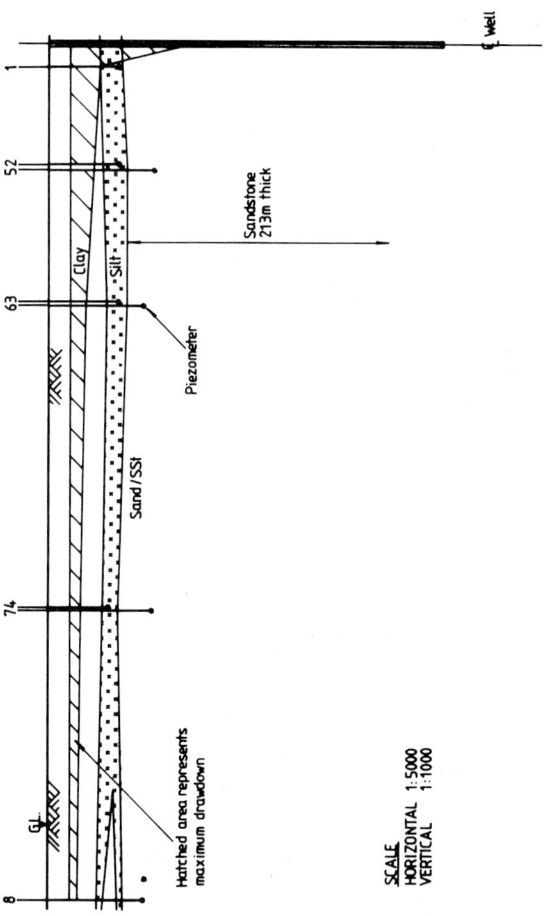

FIG. 3. Section through ground.

The pneumatic piezometers were installed within the silty clay/clayey silt deposits between 2.0 m to 3.0 m above the interface of this stratum and the underlying sand/sandstone deposits. Because of the relatively low permeability of the silt and clay deposits the response of pore pressure in these soils to changes in the head of water in the sand aquifer was correspondingly slow and it was for this reason that pumping had to be maintained as constant as possible and for as long as possible. Only on this basis could the long-term piezometric level during full scale process pumping be established. The standpipe piezometer tips were approximately 2.0 m below the surface of the sandstone bedrock and as they were installed in permeable material their response to porewater pressure changes was expected to be rapid.

Monitoring during and after pumping

The pneumatic piezometers were to be read as rapidly as possible during a two hour period immediately after the start of pumping or after any changes in pump discharge — including switching off of the pump at the end of the test. Hourly readings were required for the remaining time up to 48 hours after the start or change in discharge, with three daily readings this period.

The four standpipe piezometers were read simultaneously at the following frequency:

1 minute intervals to 15 minutes
5 minute intervals to 60 minutes
15 minutes intervals to 2 hours
30 minutes intervals to 6 hours
60 minutes intervals to 48 hours
then 3 to 4 times daily for the remainder of the test.

The frequency programme refers to (i) the start of pumping; (ii) each major change of pump discharge and (iii) switching off the pump at the end of the test.

General procedure

All readings commenced at time zero when pumping commenced. The exact time of each proposed change in pumping rate was made known to the entire team involved in the monitoring exercise and the time of such changes recorded by all concerned. A change in pump discharge required a change in reading frequency as outlined above.

Settlement stations

A total of five ground settlement stations were installed adjacent to each piezometer, or pair of piezometers, position. The ground settlement stations consist of 3 m long sections of steel reinforcing bar that are driven into the ground so that their tops are just proud of the existing ground surface. The bars were driven through a 1 m deep, 150 mm diameter hole and the top metre of the bar is sleeved in a grease filled length of scaffold tube. The tube was then surrounded by lean mix concrete and protected by a stop cock cover which prevented accidental damage to the top of the bar. The top portion of the bar was sleeved in this fashion in an attempt to isolate it from interference from near surface ground movements arising from temperature or moisture content changes. The top of the bars were domed to receive a levelling staff.

Fifteen structural settlement stations were established on the site and were positioned on the sedimentation tanks, the north and south ends of the boiler house/turbine hall and in the vicinity of the cooling water pumphouse. Structural settlement stations consisted of brass rivets (to receive levelling staffs) supported by brackets that were welded (in the case of steel) or anchor bolted (in the case of concrete) to the structural member concerned. One week was allowed after construction for local heave to stabilise and for 'interference' movements due to other factors apart from the proposed pumping; to be identified.

The reference bench mark for the settlement stations was situated at the main chimney where it was set into the top of the caisson wall which supports the chimney from bedrock level.

During pumping each settlement station was levelled once every day. The time and temperature at each station was also recorded.

Proposed pumping rates and well monitoring

The capacity of two 450 mm diameter boreholes was estimated to be 272,760 l/hour. For the trial abstraction from one borehole, water was abstracted at a rate of between 136,380 l and 159,110 l/hour. The pumping programme was to build up to this capacity over a period of two days with one-third capacity on the first day, two thirds capacity the second day, and full capacity at the start of the third. Full capacity was then proposed to be maintained for a minimum of 12 days. The measurement of the volume of water abstracted was by weir tanks. Pumping for any significant duration (i.e. to bring about a response in all the standpipe piezometers on the site) will furnish information on the flow characteristics of the aquifer irrespective of the amount of water removed. This would allow the derivation for use in any subsequent analysis or back analysis. The duration of pumping was intended to be variable in order to establish the pore pressure drop in the overlying clays and silts, and these are, by virtue of the soil particle size and structure, time dependent. Thus for any given constant flow after from the well it was important to continue pumping until steady state conditions were approached in the pneumatic piezometers.

At commencement and cessation of pumping, and at any change of pumping rate, the water level and pumping rate had been specified to be measured and recorded as follows:

every minute for the first 10 minutes,
every ten minutes for the next 50 minutes,
then hourly for the remainder of the test.

Actual pumping rates and well monitoring

The actual pumping rates and duration of pumping were not as envisaged as there were breakdowns which affected pumping. The nature and extent of the breakdowns need to be appreciated as clearly they affected the results.

The pumping test was commenced on 14th August, 1978, and was completed on 2nd September, 1978. The pumping rate was increased in three stages of 45,460 l/hour at 24 hour intervals until the maximum test rate of 30,000 l/hour was reached on 16th August, 1978. During the pumping several incidents occurred which disturbed the discharge rate. Firstly, a pipe broke which necessitated a shutdown of pumping of 10 minutes, between 1630-1640 hours on the 16th August, and 20 minutes, between 0920-0940 hours on 17th August and, secondly, a lower discharge rate due to a fault and a subsequent repair required cessation of pumping between 17th August at 2300 hours and 18th August at 1200 hours. Finally, on 2nd September 1978, the pump failed altogether and hence the test finished prematurely; it had been intended to complete the test on 4th September, 1978.

The eight piezometers installed to monitor draw-down were measured as intended. It was found from the first increment of the pumping rate that the high concentration of readings originally envisaged were not required so that the intervals between readings were increased for the second and third steps.

It was expected that the water levels in the clay would drop more slowly than in the sandstone. However, at the location of piezometers 2 and 5, initially the water level in the clay appeared to drop more quickly than in the sandstone. This was because the pneumatic piezometers installed in the clay reacted almost instantaneously to a change in pore water pressure. The standpipe piezometers in the sandstone, however, reacted slowly so that there was a time lag between the actual pressure drop and the same pressure drop being measured in the piezometer. As the test progressed the water level drop in the sandstone overtook that in the clay as the effect of time lag became less.

The maximum drawdown values measured in each piezometer are given in Appendix D. These are values which were recorded on 1st September. At the time the pump failed, the water levels were still dropping at a small, but appreciable rate. Due to the unexpected breakdown of the pump, it was not possible to get the high frequency of readings required following termination of the test and daily readings only were taken.

Levels before pumping started were established on 26th July 1978, by precise levelling techniques. The stations were monitored continuously by ordinary levelling techniques until after the start of pumping. As soon as movements were detected by ordinary levelling, a set of precise levels were taken. The results of the precise level surveys are shown in Appendix E.

Results

Test results

The drop in pressure is shown on the drawdown curves for each piezometer in Fig. 4. The maximum drawdowns for each piezometer are also summarised in Appendix D.

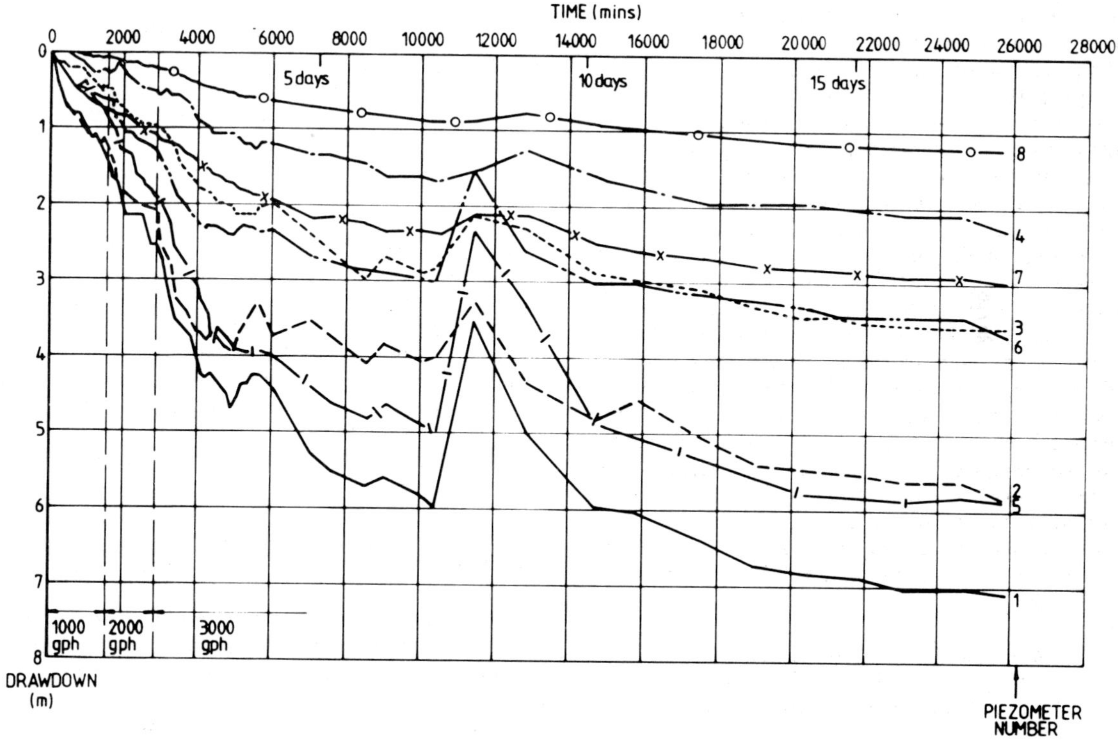

Notes
Nature of test: pumping
Starting: 14-8-1978 at 11.00 am

FIG. 4. Ground water levels.

The results from the precise levelling surveys are given in Appendix E. The levels of these stations were measured from time to time after the pumping test to check on fluctuations that may have occurred due to causes other than pumping. The accuracy of these values can be taken as ±0.5 mm. Settlement stations that were installed, but that are not shown in Appendix E, did not settle to a measurable extent during the pumping period.

Analysis of pumping test

The pumping test has been analysed assuming that the maximum drawdowns recorded represent steady state conditions. Although a genuine steady state was not reached, it is unlikely that this assumption will lead to significant inaccuracies. It was also assumed that the aquifer is confined, which is the case here, except for a zone close to the well where the water level dropped below the base of the clay. The drawdown level at the well was not used in the analysis. By making these assumptions it is possible to use the Thiem analysis to estimate permeability from the test results. Because the well partially penetrates the aquifer, corrections were made to the recorded drawdowns using Huisman's procedures. The calculated aquifer characteristics are given below:

Transmissivity	$= 4.7 \times 10^{-3}$ m²/s
Aquifer thickness	= 217 m (estimated)
Mean permeability	$= 2.17 \times 10^{-5}$ m/s

These results are reasonable for the Bunter Sandstone, and are similar to the values recorded at the Yorkshire Water Authority's Carlton Well which is near to Drax Power Station.

Extrapolation of results

The trial abstraction was carried out at a rate of 136,380 l/hour and over a period of 19 days. The CEGB had been granted a licence to pump 272,760 l/hour for a total of 6 weeks during any one year. It is possible that the 6 week pumping allowance could be used up in a continuous period if the need arose, for example, during a period of drought.

It was therefore important to assess the effects of the increased rate and duration of pumping, an extrapolition was made in two parts and these are discussed below.

Extrapolation due to increased pumping rate

The principal effect of increasing the pumping rate was to increase drawdown. The Thiem relationship, which was used to analyse the steady state pumping condition, is of the form:

$$Q = \frac{2\,k\,D\,(s_1 - s_2)}{(\ln r_1 - \ln r_2)} \qquad (1)$$

where: Q = pumping rate
k = permeability
D = depth of aquifer
s = drawdown
r = radius
ln = natural logarithm

If Q changes for a given well, the only variable on the right-hand side of the equation is the drawdown, s. Thus s is proportional to Q. In addition, corrections to drawdown due to partial penetration are also proportional to Q. Thus, for a particular well, equation (1) can be written in the form:

$$Q = Cs$$

where : C = constant

Therefore, because s is proportional to Q, if the pumping rate is doubled, then the drawdown is doubled. Fig. 5 shows the recorded surface settlements plotted against drawdown. The relationship established by this graph suggests that settlement is approximately proportional to drawdown. In conclusion, because drawdown is proportional to pumping rate, and settlement is approximately proportional to drawdown, it would be reasonable to double the settlements measured at 19 days to allow for twice the rate of pumping. The effects of this increase are shown in Appendix F. It should be stressed that these extrapolated settlements only apply to the rate of 272,760 l/hour after 19 days. The effect of increasing duration is discussed below.

Extrapolation due to increased duration of pumping

The principal effect of increasing the duration of pumping will be to increase the amount of time dependent consolidation settlement. The majority of settlement that occurred over the 19 day trial period was consolidation of the sands and silts. The piezometer measurements show that the pressure at the top of the silt remained similar to that in the sand beneath implying that consolidation of the silts was sensibly complete. This can be seen on the drawdown graphs for piezometers 2 and 5, or 3 and 6. Because the sand is likely to be of extremely low compressibility, then most consolidation will have taken place in the silt. Fig. 5 shows that some time dependent consolidation occurred during the test period, because, for a given drawdown, the measured deflection at 20 days is greater than that at 11 days.

To estimate the likely increase in settlement due to the increase in pumping duration, the laboratory measured coefficients of volume decrease were used to estimate a likely total settlement due to the measured drawdowns at each location. These total estimated settlements, which are given in Appendix G, would only occur on complete consolidation of the clays, as shown in Fig. 1c. By applying the laboratory measured coefficients of consolidation it is possible to estimate the proportion of the total settlement that will have occurred at various times. The result of doing this for 20 days and 42 days suggests that

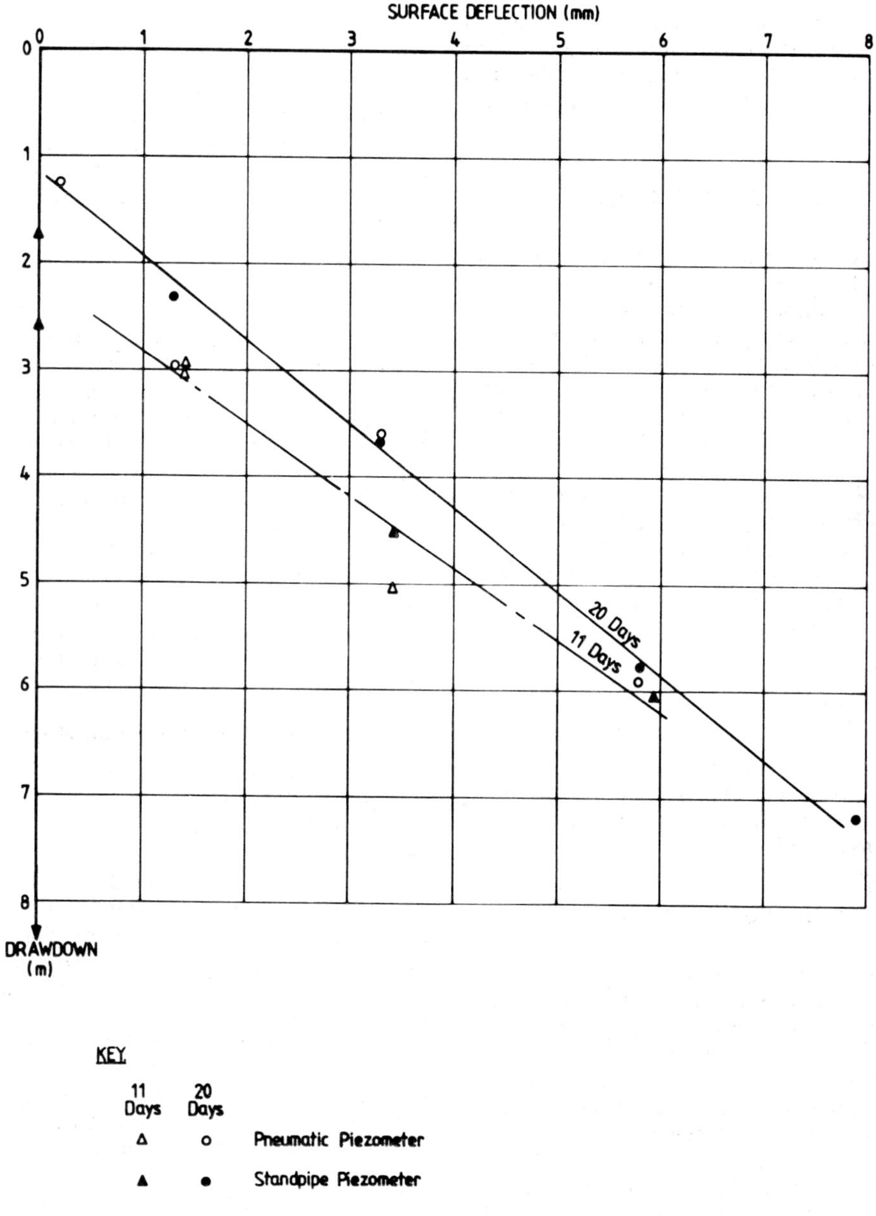

FIG. 5. Graph of drawdown against surface deflection.

consolidation settlement will occur of the order of 0.5 mm per 1.0 m drawdown at any point. The results of adding this consolidation settlement to the 19 day settlement value for 272,760 l/hour is given in brackets in the right hand column of Appendix F. The unbracketed figure has a factor of 2 applied to this addition. This factor has been applied due to the uncertainty of this part of the extrapolation. In addition, the extrapolation has been made for a single 6 week period of pumping. After pumping has ceased, settlements will reoccur, particularly in the silts and sands. Further pumping periods in subsequent years may cause settlements to progress beyond values in previous years, although only to a small extent. The factor also makes some allowance for this effect.

Discussion

The drawdown produced by pumping from the trial well is such that the radius of influence is in excess of 1000 m. The estimated settlements derived by extrapolating the measured results are considered below with relationship to existing and proposed structured and piles.

Power Station main building

The effects at the Power Station main building are not considered to be highly detrimental. The only problems which could occur would be differential settlements between turbine foundation blocks that are piled down to the sandstone and other foundations which are either unpiled or piled to shallower depths. This differential settlement is unlikely to be more than 5 mm.

Cooling towers

The cooling towers for Drax Completion were close to the current well location. They have piled foundations with a large proportion of piles which are raked. The implication of the effects of groundwater abstraction on the performance of piles is discussed in a subsequent section — 'Pile Design'. In addition, due to the large diameter of the towers, they could be subjected to appreciable differential settlement.

Pretreatment area

The closest structures to the well were in the pre-treatment area. The sedimentation tanks were monitored and recorded settlements are given in Appendix E. The settlements are relatively uniform with a small amount of tilt which is within the margins of possible survey accuracy. The relationship between

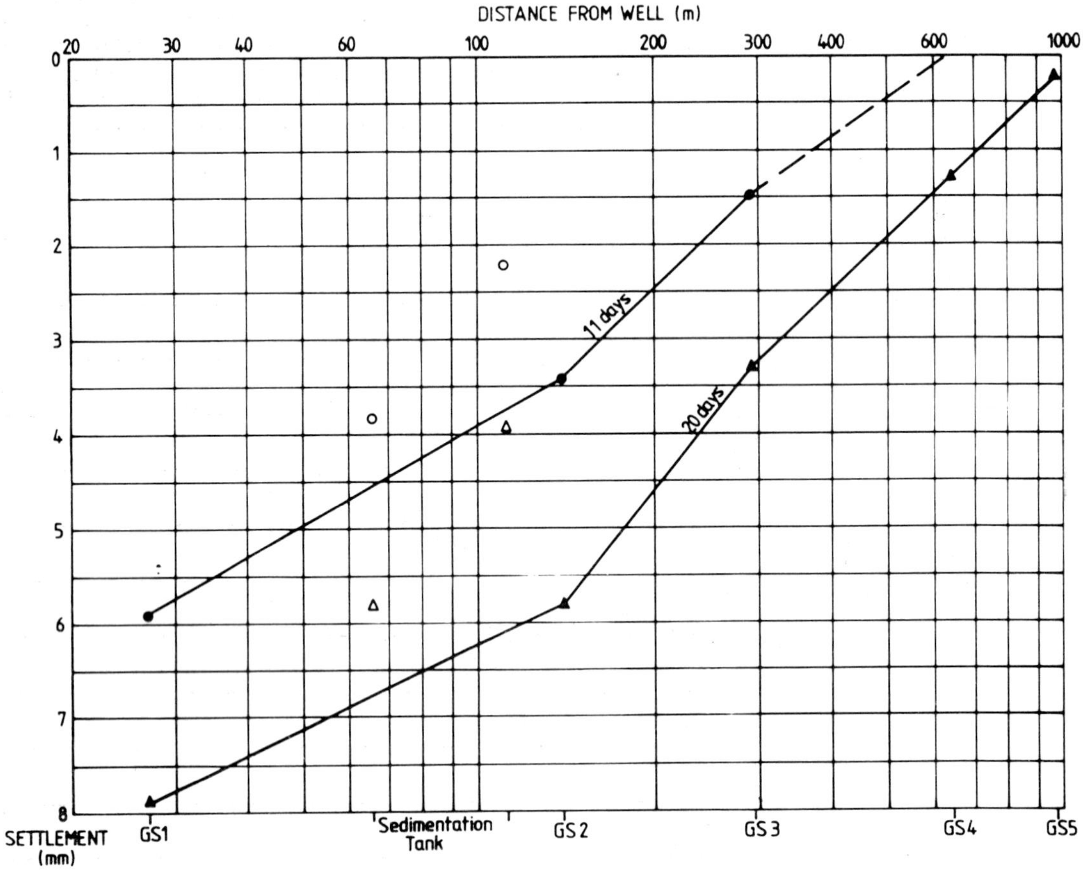

FIG. 6. Graph of settlement v. distance from well.

ground settlement and distance from the well is shown in Fig. 6, for 11 days and 20 days of pumping. Also shown, at their relevant locations, are the settlement of the sedimentation tanks. These are lower than the general trend of ground settlement, due to the restraining effect of the piled foundation. The extrapolated settlements for the sedimentation tanks are given in Appendix F. Also as the foundations are piled, the extrapolated values could be smaller than those estimated. One of the major problems that could be encountered by these structures is differential settlement between the structure and the ground which could lead to problems with pipe connections and possible opening of pipe joints. This differential settlement could be in the order of 10 mm under maximum pumping conditions. In addition, if tilt were to become excessive, the weir plates on the tanks may need adjusting.

Pile design

Initial recommendations, based on preliminary analyses and without the benefit of field pumping test data, concerning the location of the well, considered principally the effect of negative skin friction on piles. This effect still remained the main source of concern once the pumping test data had been analysed. Because most of the extrapolated settlements given in Appendix F will be caused by compression of the silts and the base of the clay, then for any pile founded in the sand or the sandstone the majority of the overlying strata will move downwards in relation to the pile. This will reduce any positive skin friction, and should the downward displacement of the soil relative to the pile exceed 10 mm, then the direction of friction could be reversed, and add to the load applied to the pile base. Based on the extrapolated settlement, up to a radius of 300 m from the well, skin friction could be reversed fully over most of the pile length for a pile founded in the sand or the sandstone. From 300 m to 600 m, the amount of negative skin friction would reduce, from almost full mobilisation at 300 m to zero at 600 m.

Values of negative skin friction for 100 t and 80 t piles have been assessed in Appendix H. These values are based on the following parameters:

Length of pile with fully mobilised negative skin friction	—	12 m
Average pile adhesion (mobilised)	—	35 kN/m^2

If the existing well location was to be used for the periodic abstraction considered, then the allowances for negative skin friction given in Appendix H should be made in pile design. Thus to determine ultimate bearing capacity, 170 t should be taken as the working load for a nominal 100 t pile, and 140 t for an 80 t pile. It was also found that for the design of Drax Completion piles, allowable concrete stress would be

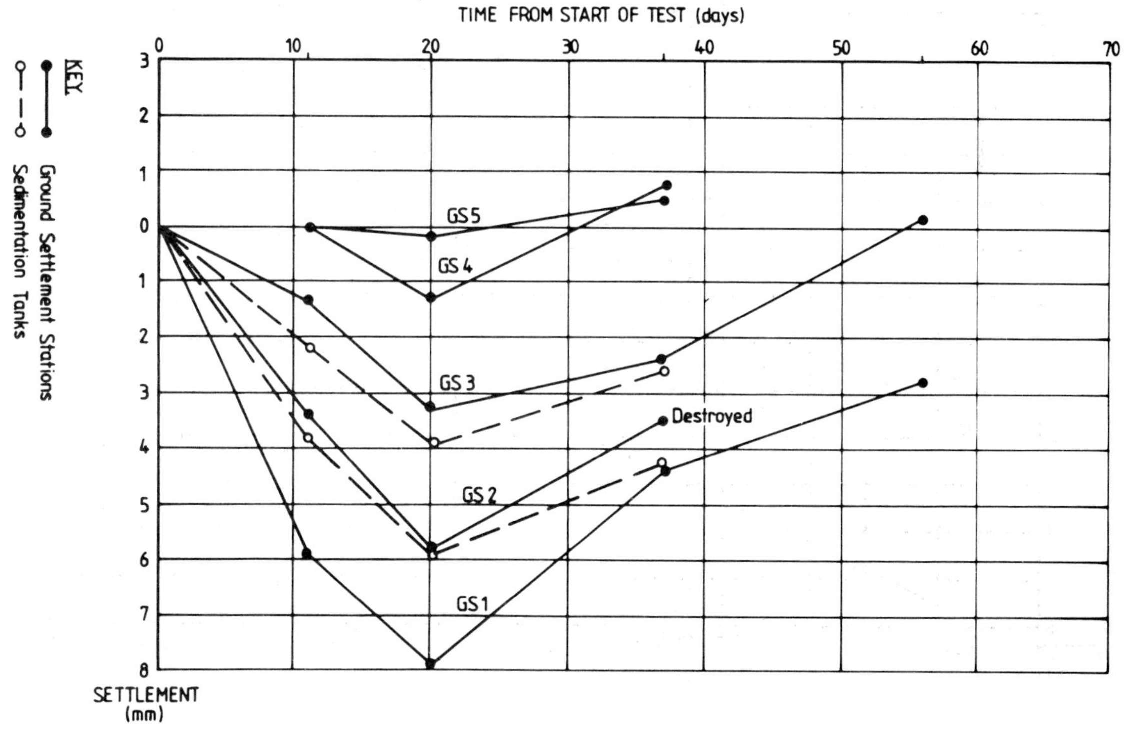

FIG. 7. Graph of settlement against time.

exceeded both for vertical piles and, to a greater extent, for raking piles. These results imply either that the piles would have to be redesigned, which could not have been considered at that stage, or that working loads of the Drax Completion piles would have needed to be down-rated which would have required considerable numbers of extra piles.

Extension of pumping time

There was a possibility that the current licence, at that time, to pump for 6 weeks in any one year could have been extended. If this occurred, then the effects on the ground, piles and structures would clearly have increased. Maximum credible ground settlements which are given in Appendix G, would have required full consolidation arising from groundwater abstraction for many years and are, therefore, unlikely to occur. However, these figures give some idea of the order of magnitude of ground settlements that could take place. As ground settlements increase, the differential settlement between the soil mass and piles would also increase. Thus negative skin friction could be mobilised at distances of 1000 m from a well. Large differential settlement between buried services and piled structures could result in considerable damage and fracturing of services.

Recommendations

Due to the possible danger of opening a major pipe joint at the sedimentation tanks and the implications to pile design, it was recommend that the production well should not be developed at the output allowed under the current licence at the location of the trial abstraction well. For these reasons, the well was recommended so as to be located at a minimum distance of 600 m from any piled structures. This minimum distance could have been reduced to 500 m if sufficient allowance for negative skin friction was made in pile design. If it is considered that, either due to reasons of economy or emergency, a substantial lengthening of the licensed pumping period might be sought and exploited, then the minimum distance of the well from piled structures should be increased to 1000 m.

To avoid any possible risk of over pumping in the future, and in view of the small difference in cost, it was recommended that the well should be located at least 1000 m from settlement structures sensitive to settlement.

Conclusion

This case history demonstrates how changes in the groundwater regime can give rise to detrimental ground deformations. The scale of the groundwater abstraction in this case produced wide reaching effects which were particularly severe, creating unacceptable differential settlements of settlement sensitive structures and unacceptable magnitudes of negative skin friction on piles under some major structures. It also demonstrates the benefits of field trials with associated instrumentation and the confidence it provides with predictions. As a general comment, it is stressed that whenever ground deformations are expected, reference stations for purposes of surface levelling should be installed, so that the extent and magnitude of the problem can be quickly assessed and appreciated.

ACKNOWLEDGEMENTS: The authors are grateful to the CEGB for their permission to publish this paper.

References

KRUSEMAN, G.P. & DE REDDER, N.A., 1970. Analysis and evaluation of pumping test data. *International Institute for Land Reclamation and Improvement Book II, Wagemenger, Netherlands.*

M. DOBIE, Dames & Moore, Booth House, 15-17 Church Street, Twickenham TW1 3NJ, UK. (Formerly at W.S. Atkins & Partners).
R.L. NEWMAN, W.S. Atkins & Partners, Woodcrote Grove, Ashley Road, Epsom, Surrey, UK.

Appendix A: Locations of well and piezometers

No.	Co-ordinates	Distance from well (m)	Depth (m)	Ground Level (mOD)
Well	30372.7 E 2865.2 N	-	92.0	5.3 (casing)
1	30355.0 E 2865.2 N	26.6	15.5	4.7
2	30355.0 E 2745.0 N	138.0	15.5	4.1
3	30287.0 E 2584.0 N	294.0	15.5	4.5
4	30186.0 E 2251.0 N	642.0	14.5	5.9
5	30305.0 E 2740.0 N	142.0	24.0	3.8
6	30287.0 E 2580.0 N	297.8	21.8	4.6
7	30183.0 E 2251.0 N	642.8	24.5	5.8
8	30163.0 E 1910.0 N	978.0	23.7	6.1

Appendix B: Locations of settlement stations

Type	No.	Location
Ground	GS1	Piezometer 1
	GS2	Piezometer 2 & 5
	GS3	Piezometer 3 & 6
	GS4	Piezometer 4 & 7
	GS5	Piezometer 8
Structures	SS1-554	Sedimentation tank nearest well
	SS5-558	Sedimentation tank furthest from well
	SS9	Grid 14 – E
	SS10	14 – D
	SS11	14 – C
	SS12	2 – E
	SS13	2 – D
	SS14	2 – C
	SS15	Cooling water pump house

Appendix C: Standing water levels

Piezometer	Ground Level (mOD)	Depth to Standing Water Leval (m)	Piezometer Level (mOD)
1	4.7	4.25	0.45
2	4.1	3.68	0.42
3	4.5	4.81	–0.31
4	5.9	5.50	0.40
5	3.8	3.29	0.51
6	4.6	4.75	–0.15
7	5.8	5.28	0.52
8	6.1	6.74	–0.64
Well	5.3	4.85	0.45

Appendix D: Values of drawdown for each piezometer

Piezometer	Maximum Redorded Drawdown (m)	Medium Level (mOD)
1	7.17	– 6.72
2	5.77	– 5.34
5	5.88	– 5.36
3	3.66	– 3.92
6	3.59	– 3.75
4	2.32	– 1.89
7	2.97	– 2.49
8	1.24	– 1.85
Well	25.88	–24.43

The drawdown curves for the piezometers are shown in the main text.

Appendix E: Results from precise levelling surveys

Station	Settlement (m)			
	25.8.78	3.9.78	20.9.78	9.10.78
GS1	5.9	7.9	4.4	2.8
GS2	3.4	5.8	3.5	Destroyed
GS3	1.4	3.3	2.4	–0.2
GS4	negligible	1.3	–0.7	–
GS5	negligible	0.2	–0.5	–
SS1	2.2	3.8	2.4	–
SS2	1.7	3.6	2.1	–
SS3	2.1	3.8	2.5	–
SS4	2.8	4.4	3.0	–
SS5	3.7	5.6	4.4	–
SS6	3.7	5.8	4.4	–
SS7	3.7	5.6	4.5	–
SS8	4.0	6.0	4.6	–

All settlements based on levels measured on 26th July, 1978.

Appendix F: Extrapolated settlements

Station	Measured settlement (mm)	Extrapolated settlement due to larger pumping rate (mm)	Extrapolated drawdown (m)	Extrapolated settlement due to larger pumping rate and duration (mm) *
GS1	7.9	15.8	14.4	30.2 (23.0)
GS2	5.8	11.6	11.8	23.4 (17.5)
GS3	3.3	6.68	7.3	13.9 (10.3)
GS4	1.3	2.68	5.8	8.4 (5.2)
GS5	0.2	0.48	2.5	2.9 (1.7)
SS1-SS4 (average)	3.9	7.8	13.2	24.8 (18.2)
SS5-SS8 (average	5.75	11.5	12.2	20.0 (13.9)

* The figures in brackets have no factor of safety on the extrapolation due to increased pumping duration. This is discussed in the text.

Appendix G: Estimated settlement after complete consolidation

Station	Total settlement (m)
GS1	160
GS2	143
GS3	75
GS4	56
GS5	21

The settlements assume total consolidation throughout the clay, as represented in Fig. 1c. This situation will not occur inside the envisaged maximum pumping duration.

Appendix H: Amount of negative skin friction

Radial distance from well (m)	Amount of negative skin friction (kN)	
	80 t pile	100 t pile
On to 300 m	600 kN	700 kN
300 m to 600 m	decreasing from 600 kN to zero	decreasing from 700 kN to zero
Beyond 600 m	zero	zero

Rising groundwater levels in cities

C.W. Hurst & W.B. Wilkinson

ABSTRACT: It is well documented that groundwater levels in a number of cities in this country and abroad have been rising over the past few years. Parts of central London are experiencing rises of one metre per year, while in Paris, previously dry basements are now affected by severe flooding.

The extraction of good quality water from city wells for a variety of industrial and domestic purposes led to a general lowering of groundwater levels from the early 1800s. During the period of depressed groundwater levels development within restricted city centre sites required higher buildings with deeper basements and foundations. Communication networks also moved underground to ease congestion at the surface. A large percentage of this work was carried out during a period when the groundwater was tens of metres below ground level; a situation accepted as being stable at the time. Consequently the effect of a higher water table was often not considered in the geotechnical design of structures. Within central London the water table is still very deep and may not cause problems for years to come. In Birmingham the situation is more critical and the rise in groundwater level is already causing problems.

Computer model predictions play a leading role in predicting tomorrow's groundwater levels. Evidence suggests that groundwater levels in the near future will attain those of natural conditions. The effect that these levels will have on current structures requires assessment. Simplified calculations for the London area show that the bearing capacity of deep basement foundations and piles could be reduced by 25-30% if the rise in groundwater level continues. High levels of sulphates have been recorded during recharge experiments and these will produce adverse effects on unprotected foundations. Problems with high artesian pressure during construction are also likely if the aquifer is confined. These is an urgent need to assess the geotechnical risk to engineering structures with respect to the rising groundwater levels. This should be carried out now so that problems can be properly anticipated and solutions developed in good time.

Introduction

Industrial development often relies heavily on the availability of good quality water. During the 19th and early 20th centuries the Chalk and Permo-Triassic sandstone aquifers in the UK were exploited extensively in this respect. When the demand on the aquifer surpasses the amount of natural recharge, water is then drawn from the aquifer storage. As a consequence of this the groundwater level will fall. This situation has occurred in many cities both in the UK and abroad. By the 1940s the groundwater level below several of the industrial cities in the UK had fallen be tens of metres.

During the period of depressed groundwater levels development within restricted city centre sites required higher buildings with deeper basements and foundations. In addition services were routed in tunnels constructed beneath the city centres. The basements, foundations, and tunnels would have been designed and constructed following site investigations which revealed a water table well below excavation level. Often records indicated that the water table had been stable in this level for decades. The assumption often made was that these levels would remain depressed and hence could be ignored in the geotechnical design of the particular structure.

However, over the last decade a rise in water levels has been observed below several major cities and there are indications that problems associated with this are beginning to occur. There are several factors to account for the rise in water levels. Primarily it can be attributed to the decline in public and private groundwater abstractions. The policy of industrial decentralisation has played a major role in this issue.

In some cases the rise in groundwater level has been dramatic. Central London has experienced rises of about 1 m per year. In Birmingham basement flooding has been reported and in Liverpool flooding of a British Rail tunnel has occurred (Slater 1983; Wilkinson 1984). The phenomena is not restricted to the UK. Several similar cases have been reported from France and West Germany (Bergeron *et al.* 1983; Anon. 1984). During the 1960s underground car parking, storage facilities and extensions to the Paris Metro were constructed in dewatered strata. Recovery of the levels has led to basement flooding. The New York Transit Authority is also reported to be experiencing some difficulties with the operation of its subway network due to rising water levels. At the Coptic Museum, Cairo, basements constructed during Roman times are now flooded for the first time in living memory (Pitts 1985).

From CRIPPS, J.C., BELL, F. G. & CULSHAW, M. G. (eds), 1986, *Groundwater in Engineering Geology,* Geological Society Engineering Geology Special Publication, No. 3, pp. 75-80.

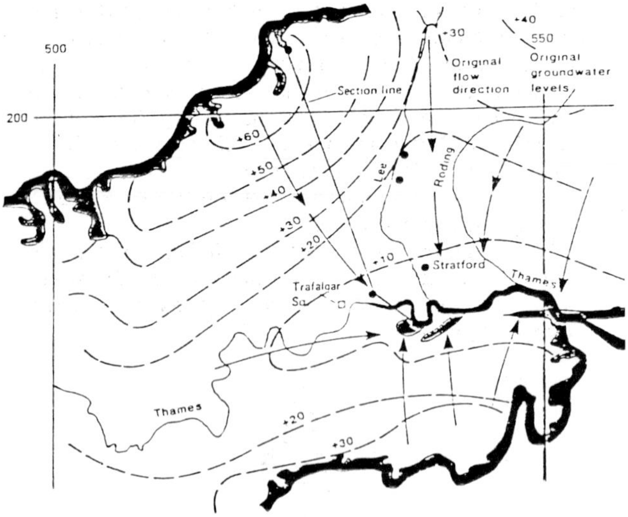

FIG. 1 Estimated groundwater levels and flow directions under natural conditions.

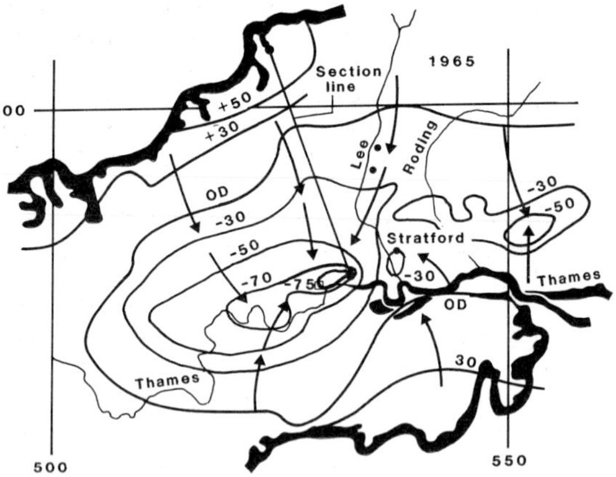

FIG. 3. Estimated groundwater levels and flow directions in 1965.

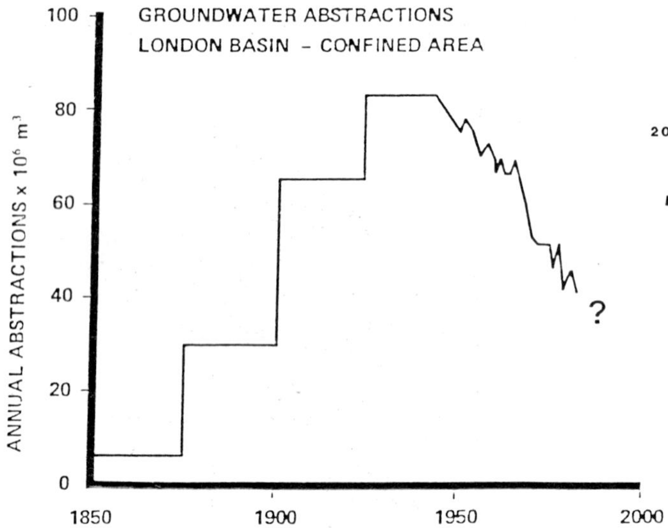

FIG. 2. Abstractions of groundwater from the confined area of the London Basin between 1850 and 1982.

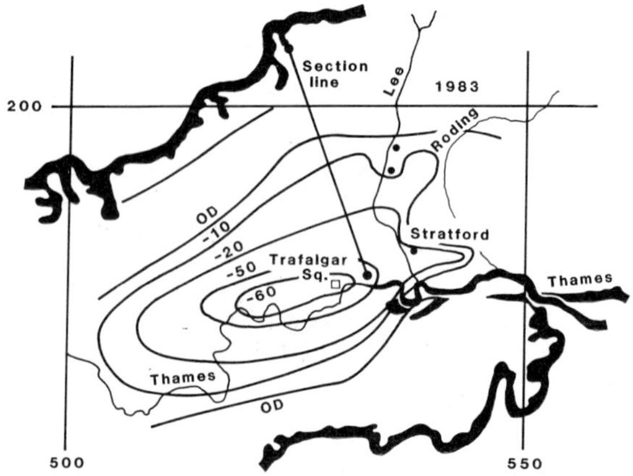

FIG. 4. Estimated groundwater level contours in 1983.

Changes in Groundwater Levels

London

On the basis of early records of water levels in wells within the London Basin the Water Resources Board (1972) prepared a map of original groundwater levels prior to any major abstraction (Fig. 1). The flow of water was essentially directed towards the River Thames around the Isle of Dogs where the Lower London Tertiaries and the Chalk are exposed.

The total groundwater abstraction in the confined area of the London Basin rose steadily from about 9 million m^3/year in the early 1800s to a peak of 83 million m^3/year in 1940. After this date the quantity abstracted has declined steadily to about 43 million m^3/year in 1983 (Fig. 2).

The cut back in abstractions during the 1950s and 1960s halted the fall in water levels which were probably at their lowest during the early 1960s. The flow in 1965 was directed towards a depression in Central London (Fig. 3) which is in marked contrast to the original flow path (Fig. 1). The Chalk outcrops around the Isle of Dogs became a zone of recharge. Reduction in groundwater abstraction has led to steady rise in groundwater levels since 1965. The groundwater level contours for 1983 are shown in Fig. 4.

Birmingham

From the examination of old well records a piezometric map has been constructed for the situation in the 1890s (Fig. 5). The directions of water flow were basically towards the River Tame to the north of Birmingham and easterly towards the major Birmingham Fault. In the latter case springs formed at the Fault boundary where the flow was blocked by Keuper Marl.

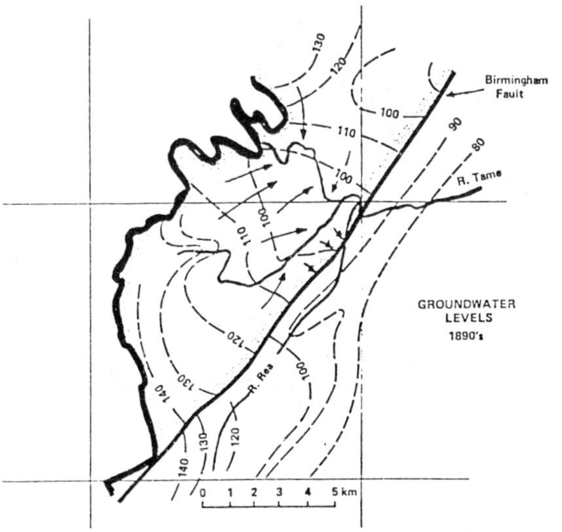

FIG. 5. Estimated groundwater level contours in the 1890s.

The steady rise in abstraction rates continued to a maximum around 1940 after which there has been a decline both to the east and west of the Birmingham Fault (Fig.6). Data plotted for 1968 (Fig. 7) reveals a cone of depression in central Birmingham, west of the fault, where the water level has dropped by about 40 m. An even more dramatic situation occurred to the east of the fault line. A substantial fall in water level was experienced as a result of pumping from the confined aquifer. In places the water level was drawn down to sea level and parts of the aquifer became unconfined. Water flow was now directed towards the cone of depression and away form the River Tame. Leakage across the fault line was also possible due to the high piezometric drop.

The situation since 1968 is similar to that in London. Reduced abstraction has allowed the water level to rise and by 1983 the recovery has matched the 1890s level in many places (Fig. 8).

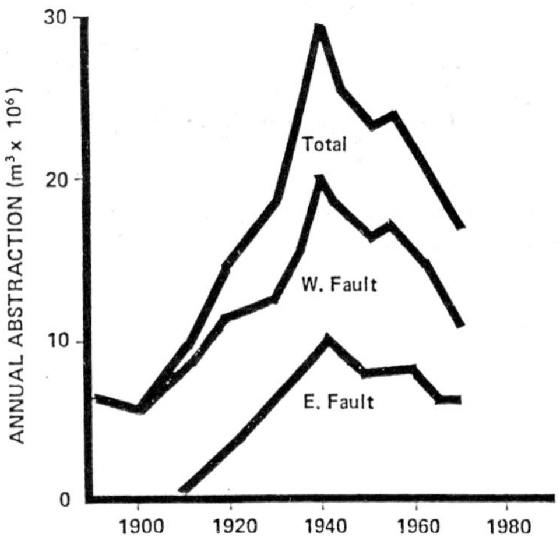

FIG. 6. Abstractions of groundwater from Birmingham area (1900–1970).

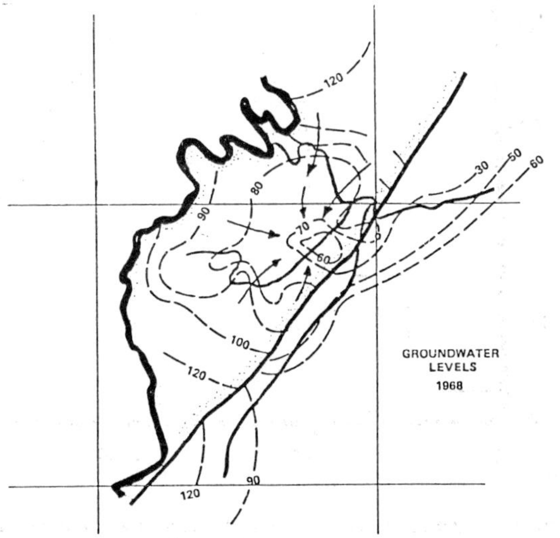

FIG. 7. Estimated groundwater level contours in 1968.

Water Level Predictions

From both a geotechnical and water resources standpoint it is important to predict the rate at which water levels are changing and also the expected final water levels that will be attained. At present the recovery of the aquifers in the London and Birmingham areas may be attributed principally to a decline in abstraction. In the case of Birmingham the number of wells pumping in excess of 180,000 m^3/year reduced by half in the period 1969–1983 (Slater 1983). Many abstractors in London are currently pumping a small proportion of their licensed entitlement. Whilst it looks likely that the cut back on abstraction will continue this cannot be guaranteed.

Models

Slater (1983) used a finite element model to predict groundwater levels in the Birmingham area. The model is three dimensional covering an area of 100 km$_2$. A total of 97 nodal points were selected. Results from the model indicate that by 1998 over 50% of the area would regain the groundwater levels of the 1890s (Fig. 9)

A simple finite difference model has been applied by Wilkinson (1985) to the London Basin. The model is only one dimensional and utilises 23 nodal points along a section extending from the northern Chalk outcrop to the Isle of Dogs. Although the model is very crude it does indicate the rates of rise likely to occur for two different abstraction rates (Fig. 10). The model results show that if the present (1983) abstraction rate is maintained the rate of rise will slow down and it may be 80 to 90 years before water levels approach the top of the Basal Sands in Central London. Reducing the abstraction rate causes the levels to rise much more rapidly so that within 40 years groundwater levels in parts of central London could be approaching the natural condition.

Geotechnical considerations

It is well established that a steady rise in groundwater levels is taking place in some areas. Once water levels rise to basement or foundation level their effects need to be considered in design. The principal effects on engineering structures would be from flooding, swelling of clays, reduction in bearing capacity, hydrostatic uplift pressures, chemical attack and difficulties during construction.

In central London groundwater levels are too low to pose any immediate leakage problems but in Birmingham, Liverpool and elsewhere in Europe basements and tunnels have been subject to severe flooding.

If water levels rise above the base of deep basements, then clearly, uplift pressures are developed on the basement slab and water pressure on the side retaining walls may be increased. Simplified calculations have shown that the bearing capacity of deep basement foundations and piles in central London may show a loss of bearing capacity of between 25 to 50% if water levels in the Basal Sands were to approach ground level (Chisholm 1984).

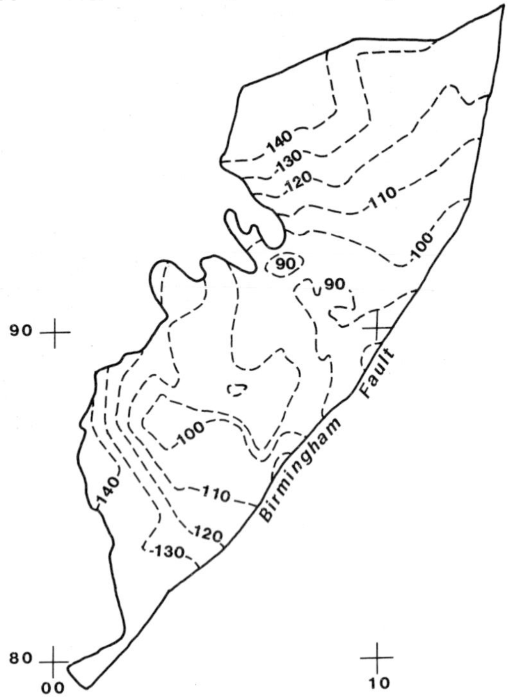

FIG. 8. Estimated groundwater level contours in 1983 (after Slater 1983).

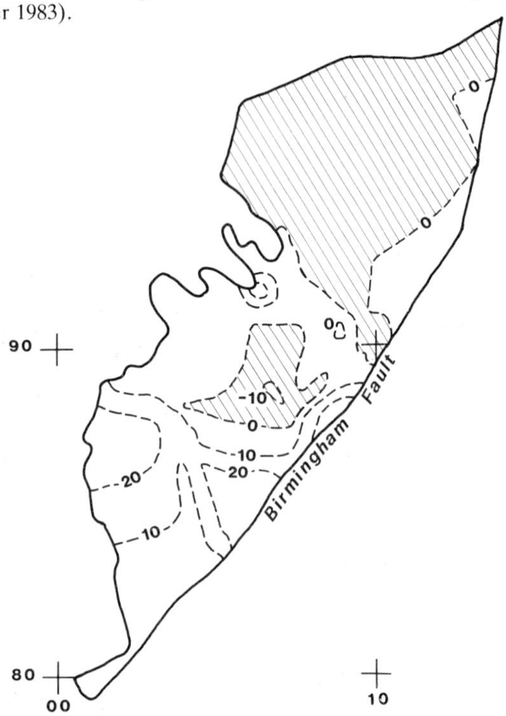

FIG. 9. Estimated difference in groundwater levels between 1898 situation and predicted 1998 situation (after Slater 1983).

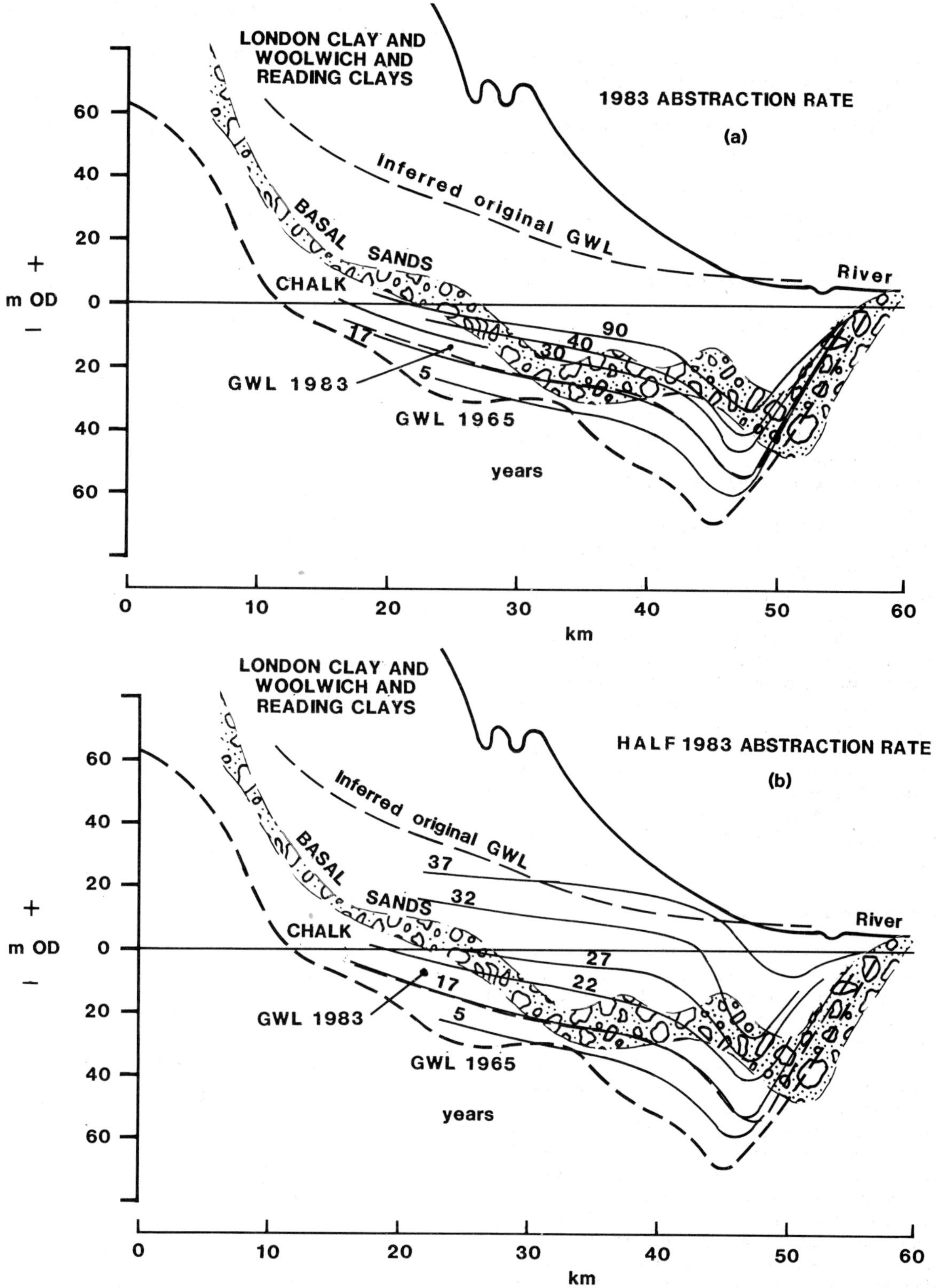

FIG. 10. One dimensional model showing the rise in water levels with the abstraction rate (a) held constant at the present value (b) halved.

Realisation that water levels are rising has led to a redesign of the foundations for the new British Library building in London. This has called for increases in the bearing capacity of the piles below basement level and a strengthening of the basement retaining walls. In addition, a drainage system below the basement has been incorporated to prevent a rise in water pressure above the basement slab level.

During the recharge experiments in the Lea Valley in the London Basin high sulphate concentrations of up to 2000 ppm were measured. Such high concentration coming into contact with foundations could lead to concrete attack and deterioration. This has already occurred in Birmingham where the concrete foundations of post-war houses in Northfield and Longbridge are being attacked by sulphates dissolved from the colliery shale hardcore as the water table has risen.

With respect to construction the situation developing in London calls for care. Once the aquifer achieves artesian pressure again, excavations in the clay close to the clay/Basal Sands interface could cause heave or blowout at the excavation base. Difficulties may also arise when constructing bored pile foundations in sandy strata or in clays with more permeable sand or silt layers.

Conclusions

In the London Basin it is apparent from preliminary modelling work that it will be several decades before groundwater achieves levels that have a detrimental effect on existing structures. For Birmingham that situation has already occurred. There is a need to assess the geotechnical risk to engineering structures in relation to the type of foundation, stratigraphy and changing groundwater levels. This should be carried out now so that problems can be properly anticipated and solutions developed in good time. It has been shown that changes in groundwater level is directly related to the rate of abstraction. At present no single organisation has the responsibility to evaluate an overall control policy through increased abstraction or other appropriate actions. It is suggested that this deficiency be rectified.

References

ANON, 1984. Renewed rise of groundwater following discontinuation of draw-down measures in the Federal Republic of Germany. *Federal Institute of Geo-Sciences and raw material. Archive No 95 678.* Hanover.

BERGERON, G., DEHAYS, M. and POINTER, T., 1983. Remontees des nappes d'eau sonterraine: cause et effets. Bureau de Recherches Geologique et minieres.

CHISHOLM, F.A., 1984. Rising groundwater levels under Greater London and their geotechnical significance. *MSc Thesis,* University of London.

MARSH, T. J. and DAVIES, P. A., 1984. The decline and partial recovery of groundwater levels below London. *Proceedings of the Institution of Civil Engineers.,* Part 1. **74,** 263-276.

PITTS, J., 1985. Personal communication.

SLATER, P. M., 1985. A finite element model and predictions of Birmingham groundwaters. *MSc Thesis,* University of London.

WATER RESOURCES BOARD, 1972. *Artificial recharge of the London Basin. I: Hydrogeology.* HMSO, London.

WILKINSON, W. B., 1984. Rising groundwater levels and geotechnical consequences. Informal discussion, Ground Engineering Group and Water Engineering Group. *Proceedings of the Institute of Civil Engineers., Part I,* **76,** 791-793.

WILKINSON, W. B., 1985. Rising groundwater levels in London and possible effects on engineering structures. *'Hydrogeology in the Service of Man' International Association of Hydrogeologists - 18th Congress Proceedings.* Cambridge 1985.

C. W. HURST & W. B. WILKINSON, School of Mechanical, Materials and Civil Engineering, Royal Military College of Science, Shrivenham, Swindon, Wilts, UK.

Tidal variations of groundwater level in an estuarine aquifer

M.S. Money

ABSTRACT: Groundwater levels in coastal aquifers respond to tides with amplitudes that decrease and time lags that increase inland. Analysis of tidal variations is complicated by the fact that long-term cycles are superimposed on the main diurnal cycle so that there are small but significant differences between successive tides and in the piezometric response.

It is shown that type curves can be constructed on the basis of simple theory to predict the piezometric response by using tidal factors as in navigational calculations.

Type curves have been drawn for piezometers on a site in the Tees estuary but these do not conform with the theoretical plots and it is suggested that the discrepancies are due to groundwater flow towards the coast. Recommendations are made for data collection.

Introduction

Groundwater levels in aquifers connected with tidal waters fluctuate in response to tides. As the sea or estuary level rises and falls a train of piezometric waves is generated in the aquifer, which travel inland and are progressively damped. The main tidal cycle is diurnal, but longer term cycles such as those which produce neap and spring tides also affect the range and period of the tide and these variations are reflected in the groundwater response.

The determination of groundwater behaviour on a large coastal site can present several problems. It can take a considerable time to read a number of piezometers unless they are connected to a central monitoring point, the passage of piezometric waves through the site means that levels may be rising in one part of a site and falling elsewhere, and variation of successive tides may render it difficult to correlate levels from one day to the next. Observations made during a site investigation may not be at the most critical time since it may be necessary to wait some weeks for spring tides or for some months for 'worst case' equinoctal tides to occur. A method is therefore proposed for standardizing groundwater levels so that predictions can, in theory, be made for any stage of any tide from a limited number of observations.

Tidal calculations

Daily variations in tides are computed for navigational purposes and nautical almanacs contain tables for major ports showing predicted times of high and low water and the heights of those tides above a local datum, chart datum, which is the level of the lowest astronomical tide. Admiralty charts and some almanacs show the relation between chart datum and Ordnance Datum. Tidal data for lesser ports are obtained by interpolation.

Depths (or elevations) of water at times between high and low tides are obtained by using Tidal Factors F_T, where

$$F_T = \frac{\text{rise of tide (above low water height)}}{\text{range of tide}}.$$

The range of each tide is the difference between high and low water heights. The tidal factor is thus zero at low water and one at high water for each tide, regardless of its range or period. Intermediate tidal factors may be obtained from curves or tables which are either standardized for general use or are specific to certain ports or areas. At some ports, for example, on estuaries, there are significant differences between flood and ebb tides and between spring and neap tides. In certain area, for example, the Solent, the tide cycle is more complicated and it must always be remembered that other factors such as floods in estuaries and strong winds may affect both the range and timing of water levels. The method does, however, enable predictions to be made which are sufficiently accurate for most practical purposes. Continuous records of tide levels are kept at some ports and can be used to check the predictions.

Groundwater response

The variation of piezometric levels in response to tides is illustrated in Fig. 1 which shows observations in piezometers on a site on the north bank of the River Tees, which is described below. These curves, which are based on observations at approximately hourly intervals show that whereas the mean piezometric level and the time lag are effectively constants for each piezometer, the amplitude of piezometric level varies with the range of the tide.

From Cripps, J.C., Bell, F.G. & Culshaw, M.G. (eds), 1986, *Groundwater in Engineering Geology,* Geological Society Engineering Geology Special Publication No. 3, pp. 81–85.

The ratio of piezometric range to tidal range, the tidal efficiency is, however, also constant. The following analysis shows how these factors may be used to predict the piezometric level corresponding to any tide level.

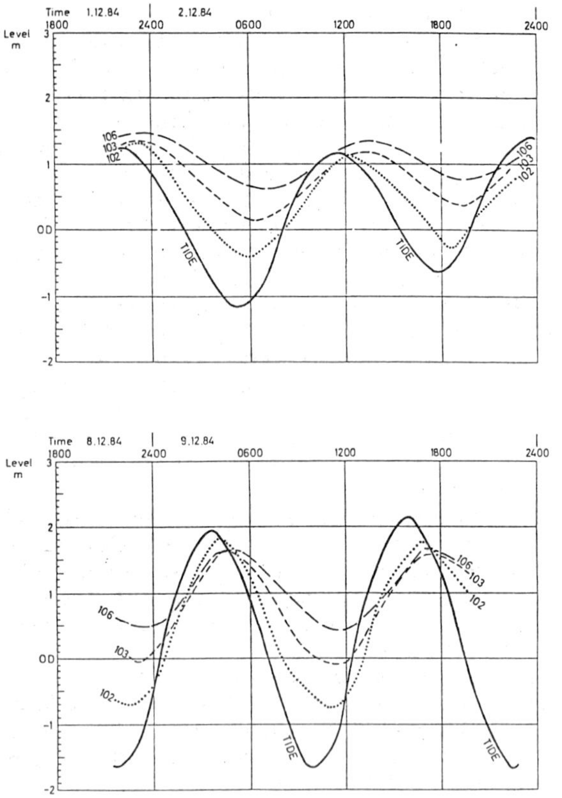

FIG. 1. The response of three piezometers to different tide ranges.

Theory

The relations between tide level and piezometric level have been analysed by several authors and are conveniently summarised by Todd (1980).

Consider a confined coastal aquifer in which flow occurs normal to the coast. Using the following:

h = rise or fall of piezometric level above or below mean level

h_o = amplitude of tide above or below mean level

h_x = amplitude of piezometric level relative to mean level at a distance x from the coastal outcrop

t = time from start of tide cycle (when $h_o = 0$)

t_o = period of tide cycle

t_L = tidal time lag of piezometric level

ω = angular velocity such that $\omega = \frac{2\pi}{t_o}$

S = storativity of aquifer.

T = transmissivity of aquifer.

The basic differential equation for flow in the aquifer is

$$\frac{\partial^2 h}{\partial x^2} = \frac{S}{T}\frac{\partial h}{\partial t} \quad (1)$$

for the following boundary conditions:

$$h = h_o \sin \omega t \text{ at } x = 0 \qquad h = 0 \text{ at } x = \infty$$

the solution of Equation (1) is:

$$h = h_o e^{-x\sqrt{\pi S/t_o T}} \sin\left(\frac{2\pi t}{t_o} - x\sqrt{\pi S/t_o T}\right). \quad (2)$$

It follows from this equation that the maximum and minimum piezometric levels occur when the sine term equals unity, hence

$$h_x = h_o\, e^{-x\sqrt{\pi S/t_o T}} \quad (3)$$

and

$$t_L = x\sqrt{t_o S/4\pi T} \quad (4)$$

Equations (2) and (3) can be combined to generate a piezometric factor, F_P, by analogy with the tidal factor, F_T, such that

$$F_p = \frac{\text{rise of piezometric level (above lowest level)}}{\text{range of piezometric level}}$$

$$\text{i.e. } F_p = \frac{h_x + h}{2hx} = {}^1\!/\!_2\left(1 + \frac{h}{hx}\right).$$

$$\text{Thus } F_p = {}^1\!/\!_2\left[1 + \sin\left(\frac{2\pi t}{t_o} - x\sqrt{\pi S/t_o T}\right)\right].$$

By substituting for x from Equation (4)

$$\text{then } F_p = {}^1\!/\!_2\left[1 + \sin 2\pi\left(\frac{t - t_L}{t_o}\right)\right].$$

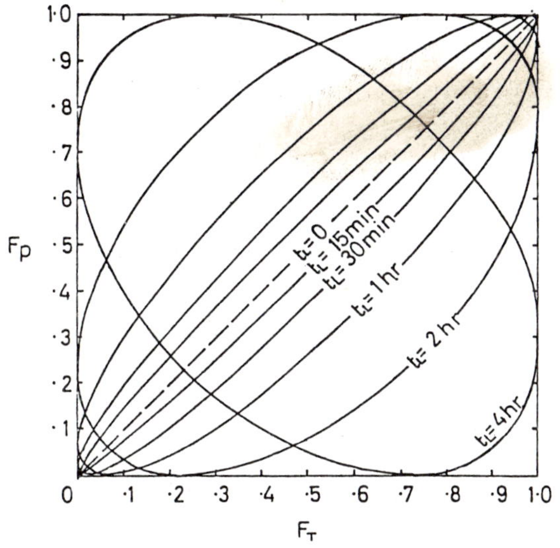

FIG. 2. Theoretical relationship between piezometric factor, F_p and tidal factor F_T. Tide lags are plotted for a tide cycle of 12.4 hours.

The factors F_T and F_p can now be plotted against each other for complete tide cycles for different values of t_L and will produce a series of Lissajous figures (Fig. 2).

If groundwater behaviour conforms with theory, then the response of each piezometer should be represented by a single curve on this plot corresponding to its tidal time lag. Variations in the length of the tide cycle can be accommodated by plotting in terms of t_L/t_o and thus making the curves dimensionless. It should be emphasized that this relationship is based on the assumptions that both tide and groundwater levels vary sinusoidally with time and that no account is taken of net groundwater flow towards the coast.

Type curves for individual piezometers can be constructed by obtaining timed readings during a flood and an ebb tide. It is, however, essential to obtain sufficient readings nears the maximum and minimum levels to define the range and the tidal time lag. The readings are then reduced to obtain the piezometric factor F_p and the simultaneous tidal factor F_T. The plotted curve may then be used to predict the piezometric level at any stage of any tide. Assessment of groundwater movement can then be carried out by 'freezing' the motion at specific times and comparing the instantaneous piezometric levels in relation to Ordnance or Chart Datum.

Observations

The validity of the theoretical curves has been tested using data from a site on the north bank of the River Tees where an extensive area of sand and mudflats has been reclaimed for industrial development. On the site in question several investigations including over 60 boreholes had been sunk for foundation design purposes but there were no observations of tidal variation except in a few holes sunk within the river. It became necessary to establish the groundwater behaviour in an undeveloped area of the site adjacent to the sea wall and for this purpose a special investigation was carried out.

The geology of the site is straightforward. The original mudflats, now represented by 1 to 2 m of organic clayey silty sand, overlie 6 to 8 m of uniform fine to medium sands. These form part of an extensive deposit of sands in the Tees estuary (Agar 1954) and constitute the aquifer investigated. The sands rest on 6–8 m of tills and laminated clays. Bedrock at the site consists of red mudstones of the Mercia Mudstone Formation with an approximate thickness near the river of about 100 m.

The area was reclaimed by constructing a sea wall, mainly of slag, and infilling with dredgings and some tipped material. Careful sampling of the fill on the site showed that it was predominantly clay, derived from glacial materials in the river channel. The fill, which is approximately 5 m thick, together with the organic material effectively confine the sand aquifer.

The site was investigated by sinking shell and auger boreholes at 13 locations. Six Casagrande standpipe piezometers and six observations wells were installed in the sands. A further 3 piezometers were installed at rockhead and 6 piezometers were installed in the fill.

Falling head permeability tests were carried out in the sands and indicate that below the organic layer the permeability decreases downwards in the range 3×10^{-5} to 6×10^{-6} m/s. The effective permeability is estimated to be 1.4×10^{-5} m/s.

All the piezometers and observation wells in the sand respond to tidal variations. Two of the three piezometers at rockhead failed during or soon after the investigation but all initially showed a slight tidal response with piezometric levels that were completely different from those in the sand. Piezometers in the fill show little or no tidal variation.

Results

Observations from three piezometers have been shown in Fig. 1. Fig. 3 shows the same data plotted as type curves. These curves have been derived graphically from smoothed plots and although there is some scatter of data there is reasonable agreement between the response from various tide cycles.

It will be seen that although standardized curves can be constructed, they do not conform to the theoretical type curves, which, for small values of t_L/t_o, are symmetrical about the $t_L/t_o = 0$ line.

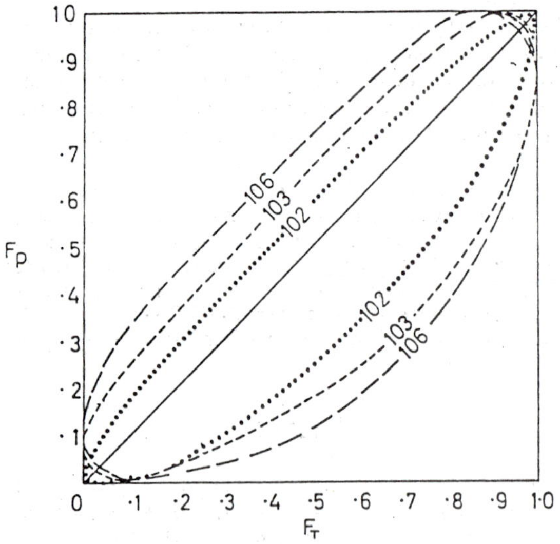

FIG. 3. Mean relationship between piezometric factor, F_P, and tidal factor, F_T, for the piezometers whose response is shown in Fig. 1.

In each case there is a significant lag in the rise of piezometric level in response to the flood tide and a comparatively rapid response as the tide falls, a relationship which is not immediately apparent from Fig. 1.

It is suggested that this distortion of the type curve is due to the coastward flow of groundwater across the site. The mean groundwater gradient towards the estuary is of the order of 1 in 200 and the net flow is calculated to be of the order of 40 litres/day per metre of river frontage.

The cyclic flow into and out of the sands due to tidal variation can be computed from the relationship given by Todd (1980) where V is the volume;

$$V = h_o \sqrt{2t_o ST/\pi}$$

For a medium range tide where $2\, h_o = 3.5$ m.

V is of the order of 20 litres per metre of coast per half cycle. This calculation is of course based on the simple theoretical relationship of tide and piezometric levels, nevertheless it suggests that on a rising tide the potential input to the aquifer from the estuary is at least of the same order of magnitude as the potential output due to the mean groundwater gradient. In simple terms therefore the effect of the rising tide is to impede and, near high water, to reverse the coastward seepage so that the main discharge of groundwater occurs with the ebbing tide.

The data from the individual piezometers has been used to assess the movement of groundwater across the site. The piezometric wave travels inland at approximately 100 m/hour and there are therefore significant tidal lags between piezometers. Specimen plots of piezometric contours are shown in Fig. 4.

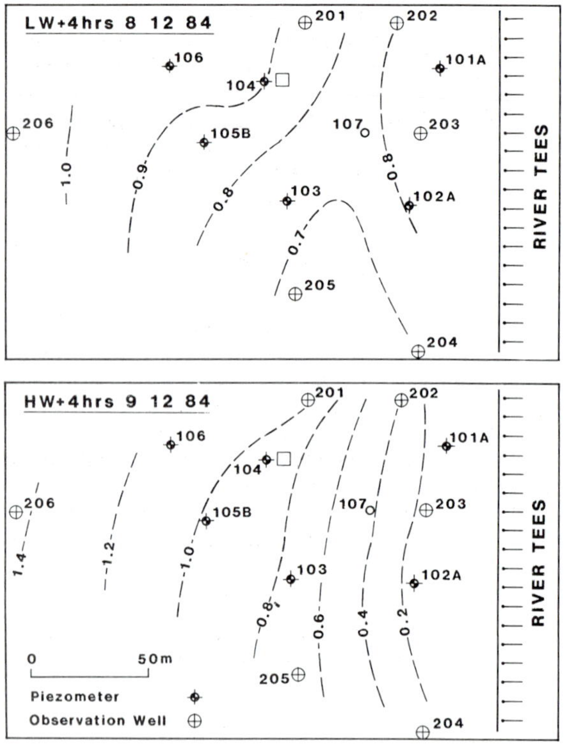

FIG. 4. Specimen plots of piezometric level in the Teeside site.

Practical considerations

A number of practical recommendations can be made as a result of carrying out this investigation.

(i) Conventional groundwater level observations in site investigation holes, for example, 'as struck' and 'start and finish of shift' are insufficient to detect or record tidal variations. Observations should be taken in completed or interrupted holes and preferably in properly constructed piezometers or observation wells.

(ii) Occasional 'spot' readings may not be sufficient to define variations and it is advisable to take readings systematically. If the type curves are constructed it should not be necessary to take readings throughout complete tide cycles or to wait for particular tide ranges.

(iii) The time of each piezometer reading should be recorded with appropriate accuracy. Reading the piezometers single-handed at the Teesside site took about 30 minutes. If only the nominal reading time is noted there can be substantial errors.

(iv) Tide times need to be checked carefully. Most navigational sources of actual and predicted tides relate to Greenwich Mean Time.

(v) It may be necessary to take account of the time lag of piezometers when assessing groundwater response. This point is discussed by Carr and Van der Kamp (1969) following Hvorslev (1951). At the Teesside site the Basic Time Lags were estimated to be 12 s for the piezometers and 90 s for the observation wells in the sands and were therefore ignored.

Conclusions

It has been shown that existing theory based on simple sinusoidal variations in tide and piezometric levels can be used to construct type curves to relate tidal and piezometric factors. Data from the site studied have been used to construct type curves and whilst these describe adequately the response of individual piezometers, and may therefore be used for prediction, they do not conform with the theoretical curves. It is suggested that the discrepancies are due to the net coastward flow of the groundwater which is impeded during the flood tide and accelerated during the ebb tide.

Some suggestions are made for good practice during investigations, in particular that considerable care should be taken over the timing of readings, of tides and of time lag in piezometers. Errors in timing may be more than the tidal lag of an instrument

ACKNOWLEDGEMENTS: This study was undertaken in conjunction with Simpson Coulson and Partners, Consulting Engineers, Stockton on Tees, and the author is grateful to Mr G.R. Simpson for his collaboration. The site investigation and installation of most of the piezometers was carried out by Soil Mechanics Ltd., Doncaster. Mr P. Matson took many of the piezometer readings, especially at night.

References

AGAR, R. 1954. Glacial and post-glacial geology of Middlesbrough and the Tees Estuary. *Proceedings of the Yorkshire Geological Society,* **29,** 237-253.

CARR, P.A. and G.S. VAN DER KAMP. 1969. Determining aquifer characteristics by the tidal method. *Water Resources Research,* **5** (5), 1023-1031.

HVORSLER, M.J. 1951. Time lag and soil permeability in groundwater observations. Bull. No. 36, Waterways Experimental Station, U.S. Army Corps. of Engineers, Vicksburg.

TODD, D.K. 1980. *Groundwater Hydrology* (2nd edn). John Wiley and Sons, New York.

M.S. MONEY, Department of Geotechnical Engineering, Drummond Building. The University, Newcastle Upon Tyne NE1 7RU, UK.

Landslides caused by rapid groundwater changes

J. Premchitt, E.W. Brand & H.B. Phillipson

ABSTRACT: A severe rainfall event occurs once every two years in Hong Kong, causing widespread landslides which result in human casualties and much damage. It has recently been established that these landslides are caused largely by short-period, high-intensity rainfall, with nearly all failures occurring within a few hours of peak intensity. The failures are therefore attributable mainly to rapid transient variations in groundwater conditions. These findings are in contradiction to previously accepted theories on failure mechanisms, and it is therefore necessary for these to be reconsidered.

A new theoretical model is now being examined in Hong Kong to explain the rapid landslide response following high-intensity rainfall. This is based on well-defined hillslope hydrological theories. Of particular significance in this context are thought to be subsurface erosion pipes and tunnels which have been found to exist in some of Hong Kong's hillslopes, especially those carpeted with colluvium.

Introduction

Landslides have been a major problem in Hong Kong, resulting in much damage and loss of life over the years (Lumb 1975, 1979; Brand 1984, 1985). Since the establishment of the Geotechnical Control Office in 1977, considerable efforts have been made to improve standards of safety for existing slopes and for the design and construction of new slopes. One aspect of this work has been to investigate more closely the mechanisms of slope failures, with a view to modifying the methods now used for the stability assessment and design of slopes.

The paper covers the background to Hong Kong's landslide problems and describes a recent study carried out to correlate time of landslide occurrence with the time of rainstorm event. A clear correlation is obtained which is not explained by the theory of failure which is traditionally accepted. Consideration is given to the possible major influence of rapid downhill subsurface flow in causing landslides.

Geography and geology

The rugged terrain of Hong Kong consists of many hills rising steeply from the sea to heights exceeding 400 m. Much of the terrain steeper than 30° is liable to become unstable under extreme rainfall conditions. Many types of slope feature are prone to landslides, including soil and rock cut slopes, earth fill slopes, retaining walls, natural slopes and boulders.

The geology of Hong Kong has been described by Ruxton (1960) and Allen & Stephens (1971). The predominant rock types are granite and volcanic rocks (Fig. 1). Granite predominates in those areas of the Territory which are heavily developed. Complete weathering of the parent rocks has resulted in the formation of mantles of residual soils, which are up to 50 m thick over the granite, and up to 20 m thick over the volcanics. These commonly contain corestones of less weathered material. Typical weathered profiles of Hong Kong have been described by Ruxton & Berry (1957) and Lumb (1975, 1983). A recent review of superficial deposits and weathering has been made by Bennett (1984).

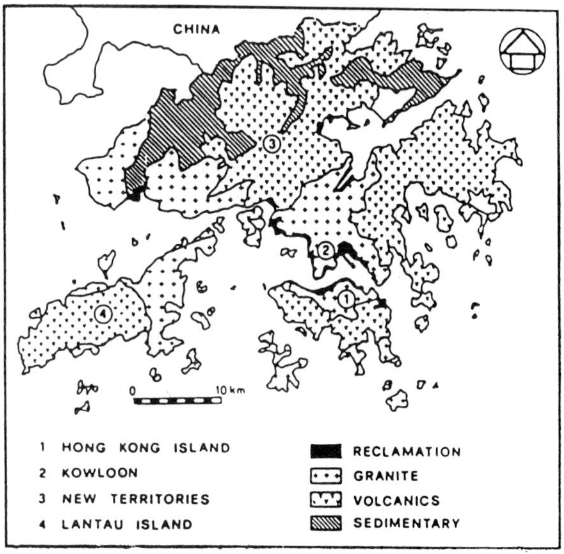

FIG. 1. Territory of Hong Kong showing the geology.

Of particular significance for the slope stability problems of Hong Kong is the presence of extensive bodies of colluvium which cover many of the footslopes. Colluvium is a slope debris material which

From CRIPPS, J.C., BELL, F.G. & CULSHAW, M.G. (eds), 1986, *Groundwater in Engineering Geology*, Geological Society Engineering Geology Special Publication No. 3, pp. 87–94

TABLE 1. *The seven major rainstorms since 1963.*

Date of rainstorm	Maximum rainfall, mm		Landslide consequences	
	24-hour	1-hour	No. failures* (newspaper)	No. people killed & injured
12 June 1966	525	157	100	35
16-17 June 1972	560	71	>15	21
17-18 June 1972	300	98	"dozens"	229
24-25 August 1976	500	82	314	57
29 May 1982	430	111	498	48
16 August 1982	370	95	62	9
17 June 1983	460	101	114	2

* The number of failures reported in newspapers represents the lower limit of the number of landslides that actually occurred.

is extremely variable in composition, but which most commonly consists of boulders, cobbles and gravel in a matrix of sand, silt and clay. It is often in a loose state, and it frequently gives rise to perched water-table conditions. The colluvium is up to 30 m thick, sometimes with a boulder content as high as 50%. From a geotechnical engineering point of view, it is a very problematical material, largely because of the difficulties of quantifying its engineering properties.

Rainfall

Average annual rainfall in Hong Kong is 2,225 mm, with 80% of this falling between May and September. Extreme rainfall events are associated either with tropical cyclones or troughs of low pressure. Rainfall intensities can be high; 24 hour rainfalls of more than 250 mm occur fairly frequently, and one-hour intensities of more than 50 mm are not uncommon. Table 1 lists chronologically the seven rainstorms which have caused the most failures in Hong Kong since 1963.

Rainfall distribution can vary significantly in rainstorm events. Fig. 2. shows contours for the major rainfall areas associated with the rainstorm of May 1983. The two main areas were separated by a three-hour period. This variation in rainfall intensity over small distances is typical for Hong Kong.

Hourly rainfall measurements have been made since 1884 at the Royal Observatory in Kowloon, with the exception of the war years from 1940 to 1946, and several autographic gauges have been in existence for many years to give a wide geographical coverage of detailed rainfall measurements. A much improved system of raingauges, however, has come into use since 1978, and this has provided information for the Geotechnical Control Office's landslide correlation studies. Forty-six automatic gauges are now in operation. These transmit rainfall measurements continuously through telephone lines to a central micro-computer which prints out rainfalls at five-minute intervals. The locations of all the existing continuously recording raingauges in Hong Kong are shown in Fig. 3. By international standards, the system is exceptional because of the concentration of gauges, the high recording frequency and the accuracy of rainfall timing.

The existence of the sophisticated system of automatic raingauges has enabled the Geotechnical Control Office to gather valuable data on rainfall distribution and intensity over the past few years. In particular, detailed information has been obtained on the two major rainstorms that occurred in May and August 1982, which together resulted in over 1,500 landslides, and on the rainstorm of June 1983, which caused more than 150 landslides.

Landslide data

Since the establishment of the Geotechnical Control Office in 1977, a considerable amount of valuable information has been collected, particularly in respect of landslide characteristics and mechanisms (Hencher *et al.* 1984) and in evaluating the susceptibility of terrain to landslides (Brand *et al.* 1982a, 1982b). Information on precise times of failure, however, has been difficult to obtain.

In collecting data for its landslide correlation studies, the Geotechnical Control Office has found the Hong Kong Fire Services Department to be an extremely reliable source of information on the occurrence of landslides over the past twenty years.

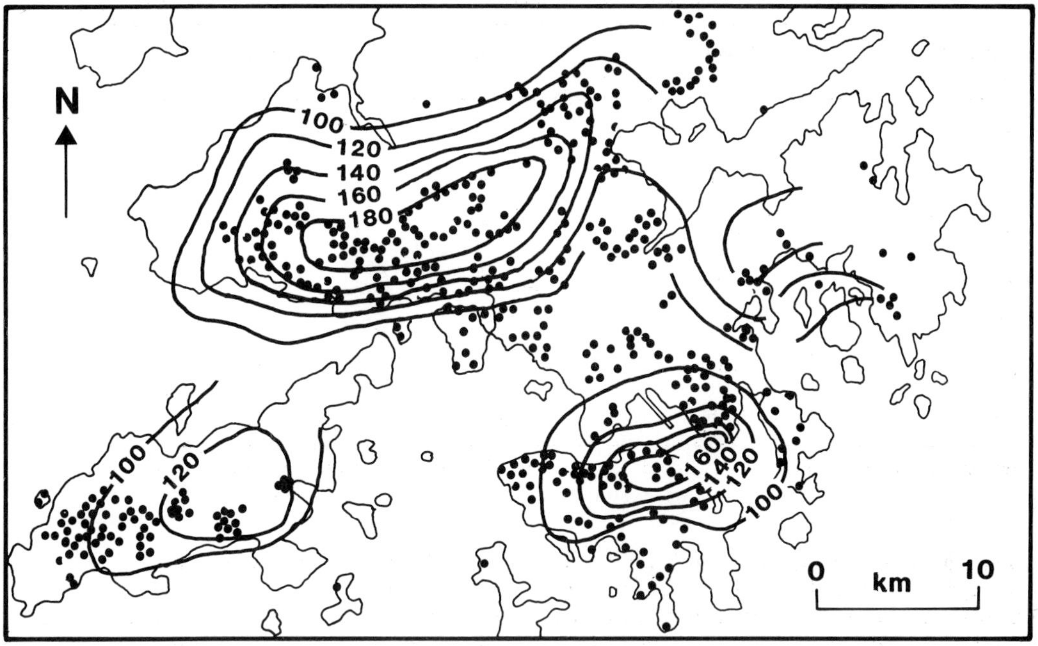

FIG. 2. Landslide sties in relation to isohyets of three-hour rainfalls (mm) on 29th May 1982.

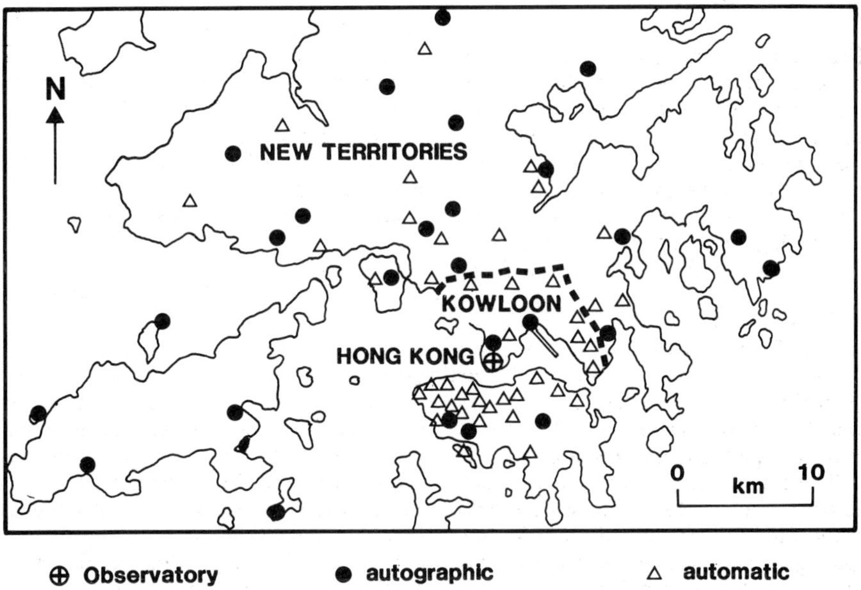

FIG. 3. Locations of continuously recording raingauges.

Although it is involved with only a small proportion of the total number of landslides that occur, this includes all the more serious incidents, and the Department's excellent records can be considered to be representative of the landslide events in Hong Kong as a whole. The data which relates to times of call-out is particularly good, and the fact that the Department's staff are usually called to a serious incident immediately after its occurrence enables the time of failures to be fixed fairly accurately.

Rainfall — landslide correlations

Correlations between rainfall and landslides for the period 1963 to 1983 have been undertaken by the Geotechnical Control Office. These have been based largely on the Fire Services Department records of the times of occurrence of landslides, and on the rainfall data from the principal Observatory raingauge (Fig. 3.) and the Geotechnical Control Office's automatic gauges. A full account of this correlation study has been given by Brand *et al.* (1984).

Fig. 4 illustrates the main findings of the correlation study for the May 1982 rainstorm. This figure shows the hourly and cumulative rainfall at the Observatory and at the location of measured maximum rainfall, compared with the occurrence of landslides as recorded by the Fire Services Department. The pattern of cumulative rainfall measured at the Observatory is quite similar to that measured at the location of maximum rainfall, although it is inevitably slightly lower.

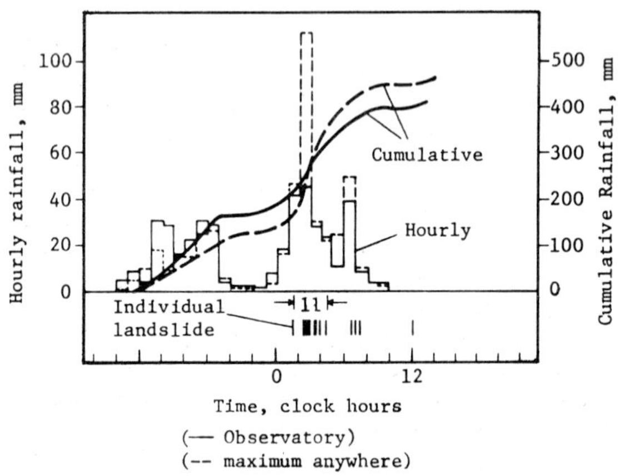

FIG. 4. Times of landslide reports on 29th May 1982.

In contrast, the hourly maximum rainfall is significantly higher than the corresponding one measured at the Observatory. Landslide timings for the rainstorm show very good correlations with hourly maximum rainfall, which suggests that high intensity rainfall of short duration is a major triggering factor for landslides.

The results presented in Fig. 5 for the period 1963 to 1983 show that the majority of the reported landslides occurred within four hours of peak rainfall intensity, with only 10% taking more than 16 hours to occur. This is in good agreement with the timing of landslides for the single rainstorm event of May 1982 illustrated in Fig. 4.

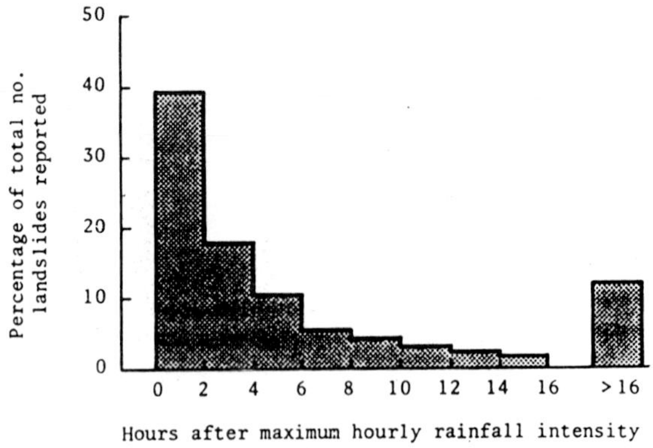

FIG. 5. Times between maximum rainfall intensity and landslide reports for period 1963 to 1983.

The conclusions to the landslide correlation study which are relevant to this paper are as follows :

(a) The large majority of landslides are induced by localised short-duration rainfalls of high intensity, and these landslides take place at about the same time as the peak hourly rainfall.
(b) Antecedent rainfall is not a major factor in landslide occurrence, except in cases of minor landslide events which take place under relatively low-intensity rainfalls of short duration.
(c) A rainfall intensity of about 70 mm/hour appears to be the threshold value above which landslides occur. The number of landslides and the severity of the consequences increase dramatically as the hourly intensity increases above this level.

Failure mechanisms

Some useful attempts at explaining rain-induced slope failure mechanisms have been made previously on the basis of rainfall infiltration, such as those by Lumb (1975, 1979), Eyles *et al.* (1978) and Cozier & Eyles (1980). These emphasized long-term build-up of groundwater and related pore pressures, and they considered antecedent rainfall over the past 15 days to be the major critical factor. Estimates of the time required to develop sufficient thickness of 'wetting band' were made by Lumb (1975) on the assumption of vertical rainfall infiltration into horizontal ground with an impermeable surface as a base. Subsurface

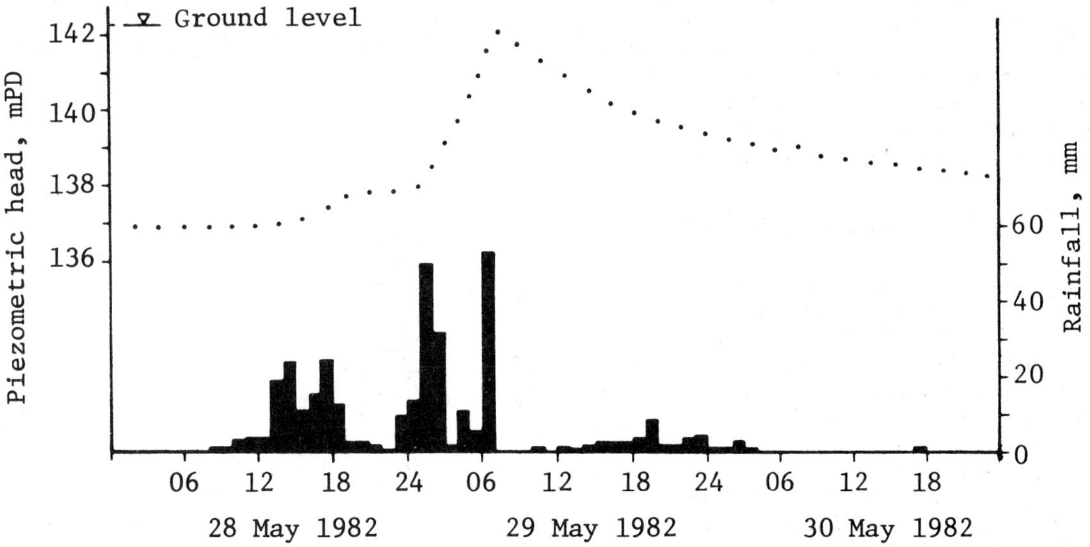

FIG. 6. Example of rapid piezometric response to storm rainfall in Hong Kong (Nash & Dale, 1984).

flow from the uphill area was not considered at all. With typical permeability values for decomposed granite and volcanic rocks of 8×10^{-6} to 1.5×10^{-6} m/s, it was estimated that the infiltration times required would be in the order of 14 hours to three days. It therefore seems unlikely that such a failure mechanism can apply to the majority of landslides considered in the Hong Kong correlation study. In most of these cases, it is probable that a rapid rise in groundwater table occurred during, or very soon after, the intensive rainfall.

Automatic piezometers have in fact recorded very rapid groundwater responses in Hong Kong, an example of which is shown in Fig. 6. Pope (1982) has described how, during a detailed hillside study in Hong Kong, rises in the perched groundwater level in colluvium occurred within a few hours of a rainfall, whereas these took several days to develop in the underlying decomposed granite. Cowland & Richards (1985) described a 4 m rise in the groundwater level within a few hours in sheeting joints of a Hong Kong granite slope.

Recent work by Leach & Herbert (1982) attempted to analyse subsurface flows in slopes by numerical models, and some useful discussions were presented for a particular hillside in Hong Kong. Their work indicated that, in order for their model to correctly reflect the measured groundwater changes, it would have been necessary for the sub-horizontal permeability in the upper 10 to 15 m zone to have been as high as 5×10^{-6} to 100×10^{-6} m/s, as compared with a vertical permeability of 1×10^{-6} m/sec. In this zone, subsurface horizontal flow would have been correspondingly greater than vertical infiltration, and under certain conditions could have led to the rapid build-up of perched groundwater levels, normally in the downslope area.

Whipkey (1965) carried out detailed measurements of subsurface horizontal flow in a forest hillslope. A narrow trench was dug in the slope to intercept the subsurface horizontal flow, with collecting gutters at various depths. Fig. 7. shows the results of this experiment, which was carried out on a natural slope of sandy loam, of 16° slope angle, using a simulated storm of 51 mm/hour lasting for two hours. The slope had previously drained for more than four days. The striking characteristics of the results are the rapid response of the subsurface flow and the restriction of flow to a limited depth. Small natural pipes near the ground surface were considered to have had a major influence on these characteristics.

Natural pipes, especially if large, can be a major agency for transmitting subsurface water downslope. Gilman & Newson (1980) investigated the discharge characteristics of such a pipe on a hillside in Wales. Special instruments were installed, and careful monitoring was carried out during rainstorms. Typical pipe-flow rates against rainfall are shown in Fig. 8. A very rapid symmetrical response was observed, indicating unimpeded flow through the pipe.

The influence of natural pipes on landslides has been discussed by several authors. Jones (1981) commented on this in his extensive review of the available research on the subject. Pierson (1983) described how soil pipes can improve the stability of slopes by rapidly draining subsurface water to stream courses. On the other hand, pipes are often subject to blockage, and this can cause a back-up of water, with a resulting build-up of pore water pressures which is much greater than that generated by vertical infiltration through the soil. Pierson concluded that such pore pressure increases can trigger landslides at sites that would otherwise be stable. Atkinson (1978) has also considered the contribution of natural pipes to groundwater flow regimes. In addition to the two possible extreme effects of pipes identified by Pierson (1983), it is possible that pipes could adequately drain the subsurface of a slope under normal rainfall conditions, but might not be able to do so during very intense rainstorms. Rapid pore pressure build-up might then result when the rainfall intensity exceeds a critical value.

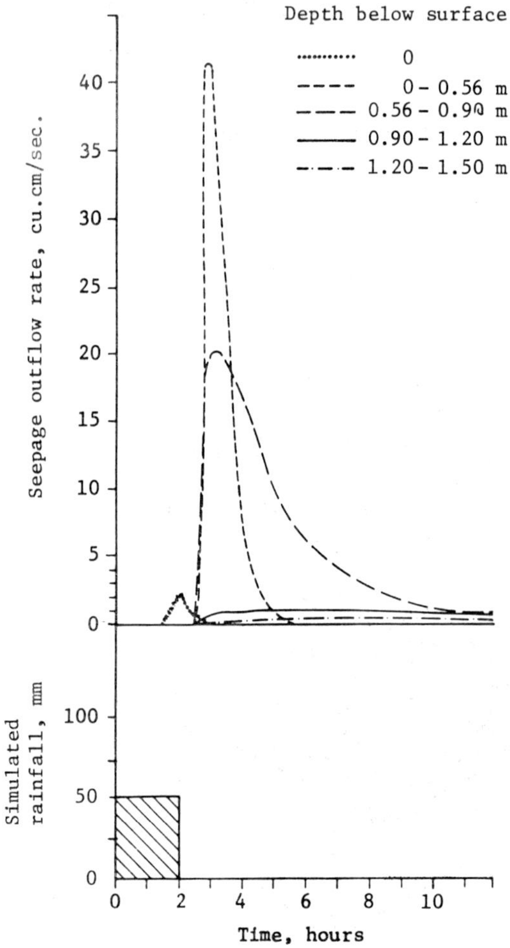

FIG. 7. Slope subsurface horizontal flow resulting from simulated rainfall (Whipkey 1965).

Natural pipes in Hong Kong

The movement of surface water during heavy rainstorms is known to be a landslide triggering agency, as is leakage from stormwater drains. It is considered, however, that a large proportion of slope failures in Hong Kong are better explained by the rapid transmission of groundwater through natural pipes and other preferential drainage paths causing sudden pore pressure increases.

High mass permeabilities of slope materials in Hong Kong can also be the result of agencies other than natural pipes. Zones of varying permeability roughly parallel to the slope are a common phenomenon in colluvium (Geotechnical Control Office 1982). Relict joints in the decomposed granitic and volcanic rocks also provide preferential drainage paths. Even for these situations, however, concentrations of flow with time lead eventually to the creation of natural pipes (Nash & Dale 1983).

The existence of natural pipes in Hong Kong is mentioned by several authors (Leach & Herbert 1982; Geotechnical Control Office 1982; Nash & Dale 1983; Brand *et al.* 1985). The paper by Nash & Dale (1983) describes in detail how such natural pipes can occur in Hong Kong, and examines the evidence which indicates their presence. For example, the piezometric response to a storm event shown in Fig. 6 is thought by them to have been caused by a natural pipe. The asymmetrical shape of the piezometric response curve was explained by them as the rapid build-up of hydrostatic head as the pipe filled, followed by the slow dissipation of this head as water seeped away from the blocked pipe.

Conclusions

It is the Authors' opinion that, although the influence of such agencies as direct surface water infiltration cannot be discounted, the rapid horizontal transmission of groundwater in natural pipes or along other preferential drainage paths is likely to be a major triggering mechanism for landslides in Hong Kong. It would satisfactorily explain why many landslides occur during heavy rainstorms or within a few hours of them.

A theoretical model, similar to the one described by Atkinson (1978) and illustrated in Fig. 9, should be devised for Hong Kong on the basis of the rainfall landslide correlation information and of the probable major significance of natural pipes. This should emphasize rapid infiltration and a large subsurface horizontal flow down a hillside during and following intense rainstorms leading to the build-up of pore water pressures. This model has not been considered in the more traditional explanations of rain-induced landslides, and no account is taken of it in current stability analysis procedures. It should be possible to formulate a simplified equation based on this model

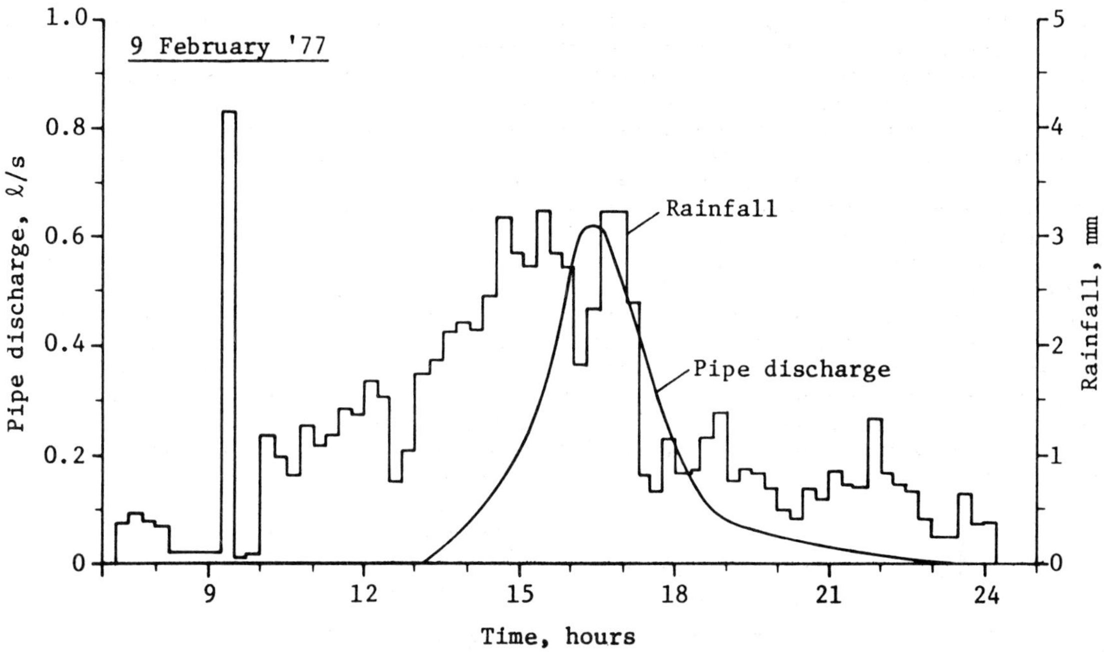

FIG. 8. Natural pipe discharge in response to rainfall (Gilman & Newson 1980).

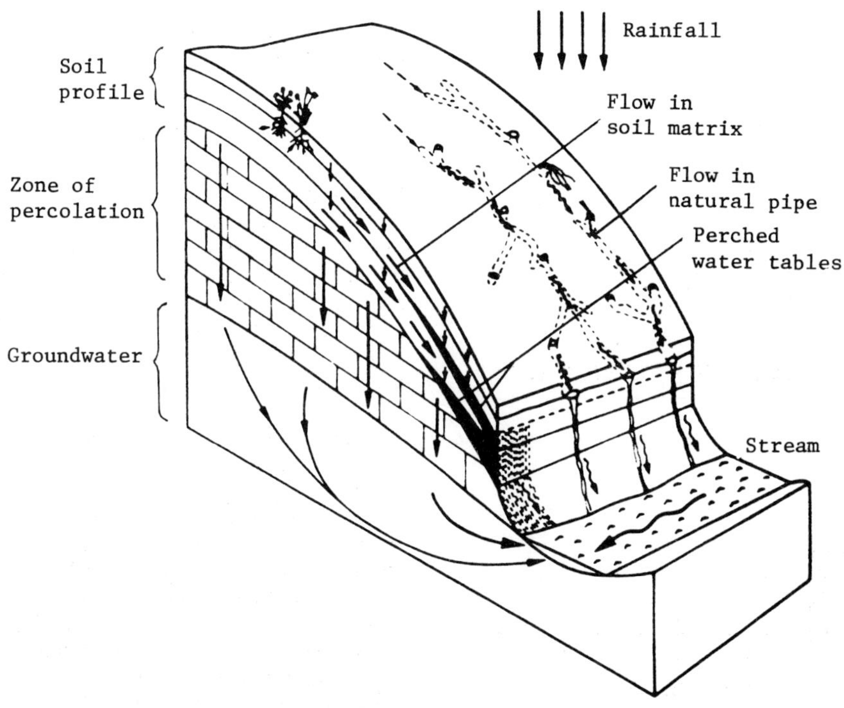

FIG. 9. Hillside stormwater flow model (Atkinson 1978).

for the estimation of the rise in groundwater pressures for subsequent use in slope stability calculation. This should include, among other factors, the critical rainfall intensity and its duration, the slope angle, the thicknesses and permeabilities of the various strata in the subsurface, and the catchment characteristics of the upper hillside area which generates the subsurface flow down the slope. Field studies would be necessary to confirm the validity of any such approach.

ACKNOWLEDGEMENTS: This paper is published with the permission of the Director of Engineering Development of the Hong Kong Government.

References

ALLEN, P.M. & STEPHENS, E.A., 1971. *Report on the Geological Survey of Hong Kong.* Hong Kong Government Press.

ATKINSON, T.C., 1978. Techniques for measuring subsurface flow on hillsides. *Hillslope Hydrology,* edited by M.J. Kirkby, 73-120. John Wiley, New York.

BENNETT, J.D., 1984. *Review of Superficial Deposits and Weathering in Hong Kong.* Geotechnical Control Office, Hong Kong, 51 pp.

BRAND, E.W., 1984. Landslides in Southeast Asia: a state-of-the-art report. *Proceedings of the 4th International Symposium on Landslides,* Toronto, **1,** 17-59. (Addendum, vol. 3, in press).

BRAND, E.W., 1985. Landslides in Hong Kong. *Proceedings of the 8th Southeast Asian Geotechnical Conference,* Kuala Lumpur, in press.

BRAND, E.W., DALE, M.J. & NASH, J.M., 1986. Discussion on 'Soil pipes and slope stability' by T.C. Pierson. *Quarterly Journal of Engineering Geology,* **19,** 301-304.

BRAND, E.W., PREMCHITT, J. & PHILLIPSON, H.B., 1984. Relationship between rainfall and landslides in Hong Kong. *Proceedings of the 4th International Symposium of Landslides, Toronto,* **1,** 377-384. (Discussion, Vol. 3, in press).

BRAND, E.W., BURNETT, A.D. & STYLES, K.A., 1982a. The Geotechnical Area Studies Programme in Hong Kong. *Proceedings of the 7th Southeast Asian Geotechnical Conference,* Hong Kong, **1,** 107-123.

BRAND, E.W., STYLES, K.A. & BURNETT, A.D., 1982b. Geotechnical land use maps for planning in Hong Kong. *Proceedings of the 4th Congress of the International Association of Engineering Geology,* New Delhi, **1,** 145-153.

COWLAND, J.W. & RICHARDS, L.R., 1985. Transient groundwater rises in sheeting joints in a Hong Kong granite slope. *Hong Kong Engineer,* **13** (2), 27-32.

CROZIER, M.J. & EYLES, R.J., 1980. Assessing the probability of rapid mass movement. *Proceedings of the 3rd Australian - New Zealand Conference on Geomechanics,* Wellington, **2,** 47-51.

EYLES, R.J., CROZIER, M.J. & WHEELER, R.H., 1978. Landslips in Wellington City. *New Zealand Geographer,* **34,** 58-74.

GEOTECHNICAL CONTROL OFFICE, 1982. *Mid-levels Study: Report on Geology, Hydrology and Soil Properties.* Geotechnical Manual for Slopes (2nd edition).

GEOTECHNICAL CONTROL OFFICE, 1984. *Geotechnical Manual for Slopes (2nd Edition).* Geotechnical Control Office, Hong Kong.

GILMAN, K. & NEWSON, M., 1980. *Soil Pipes and Pipeflow -A Hydrological Study in Upland Wales.* British Geomorphological Research Group Research Monograph, No. 1. Geo Books, Norwich, UK,

HENCHER, S.R., MASSEY, J.B. & BRAND, E.W., 1984, Application of back analysis to some Hong Kong landslides. *Proceedings of the 4th International Symposium on Landslides,* Toronto, **1,** 631-638.

JONES, J.A.A., 1981. *The Nature of Soil Piping - A Review of Research.* British Geomorphological Research Group Research Monograph No. 3, Geo Books, Norwich, UK,

LEACH, B. & HERBERT, R., 1982. The genesis of a numerical model for the study of the hydrogeology of a steep hillside in Hong Kong. *Quarterly Journal of Engineering Geology,* **15,** 243-259.

LUMB, P., 1975. Slope failures in Hong Kong. *Quarterly Journal of Engineering Geology.* **8,** 31-65.

LUMB, P., 1979. Statistics of natural disaster in Hong Kong, 1884-1976. *Proceedings of the 3rd International Conference on Applications of Statistics and Probability to Soil and Structural Engineering,* Sydney, **1,** 9-22.

LUMB, P., 1983. Engineering properties of fresh and decomposed igneous rocks from Hong Kong. *Engineering Geology,* **19,** 81-94.

NASH, J.M. & DALE, M.J., 1983. Geology and hydrogeology of natural tunnel erosion in superficial deposits in Hong Kong. *Proceedings of the Meeting on the Geology of Surficial Deposits in Hong Kong.* Hong Kong. 61-72.

PIERSON, T.C., 1983 Soil pipes and slope stability. *Quarterly Journal of Engineering Geology,* **16,** 1-11.

POPE, R.G., 1982. Measurement and prediction of groundwater levels. (Report by J.R.A. Gammon on a Geotechnical Group Meeting). *Hong Kong Engineer,* **10** (12), 37-38.

RUXTON, B.P., 1960. The geology of Hong Kong. *Quarterly Journal of the Geological Society of London,* **115,** 233-260.

RUXTON, B.P., & BERRY, L., 1957. Weathering of granite and associated erosional features in Hong Kong. *Bulletin of the Geological Society of America,* **68,** 1263-1291.

WHIPKEY, R.Z., 1965. Subsurface stormflow on forested slopes. *Bulletin of the International Association for Scientific Hydrology,* **10** (2), 74-85.

J. PREMCHITT, E.W. BRAND & H.B. PHILLIPSON, Geotechnical Control Office, 6th Floor, Empire Centre, 68 Mody Road, Tsim Sha Shui East, Kowloon, Hong Kong.

CRRDIS UWCC LIBRARY

Discussion Session 2

P.W. Rowe (Emeritus Professor, Manchester University) asked Professor De Wiest to comment on the possibility of the uplift being caused by the presence of thin bentonite seams. Also, does a rise in porewater pressure occur due to downward flowing groundwater meeting upward flowing groundwater?

R.D.J. De Wiest (State University of Florida) replied that it is generally assumed that thin bentonite seams at the base of the landslide have a thixo-tropic behaviour and that the slip occurs precisely in these seams. Downward flowing groundwater results in a decrease in porewater pressure along the path of flow, and hence it produces an increase in intergranular pressure or effective stress. This process favours increased stability. The opposite result, that is decreased stability, is achieved by upward flowing groundwater under artesian pressure.

J. Pontin (Hydraulic Research Ltd.) asked Mr. Hurst or Professor Wilkinson for clarification regarding the estimation of time for rising groundwater levels in view of the fact that the storage coefficient during extraction generally exceeds that for recharge.

Written reply:

C.W. Hurst & W.B. Wilkinson (Royal Military College of Science)

The simple finite difference model described in the paper used unconfined and confined storage coefficients of 0.02 and 0.001 for both the Chalk and Basal Sand aquifers. No allowance was made for differences in storage coefficients that may occur between an abstraction and a recharge phase. It must be stressed that the model was only developed to illustrate some of the principles involved and that this degree of refinement was quite unnecessary.

A detailed study of the geotechnical problems associated with rising water levels in the London Basin is now being undertaken by CIRIA. As part of this work Thames Water Authority in association with the Water Research Centre are developing a detailed groundwater model which can fully reflect the variations in hydrogeological properties of the aquifer.

D.W. Cox (Polytechnic of Central London) asked Mr. Hurst to clarify the position regarding reductions in bearing capacity caused by groundwater levels. Using standard bearing capacity theory and imposing a hydraulic gradient a reduction is predicted. Tests confirm that this occurs (See written contribution to discussion, this volume).

Written contribution further to verbal discourse:

C.W. Hurst & W.B. Wilkinson

We accept the point made by Cox, Armishaw and Mojabi (written contribution) that an imposed hydraulic gradient will contribute to the reduction in bearing capacity of a foundation. This is simply equivalent to reducing the effective stress. In the London Basin under pre-abstraction conditions the artesian pressures in most areas were relatively small in relation to the thickness of the London Clay. If, in the future, the water table attains its pre-abstraction level then a similar situation is expected and therefore the hydraulic gradient would be small. The authors consider it better under such conditions to work in terms of effective stress rather than hydraulic gradient.

A. Campbell (Ove Arup and Partners) asked Mr. Hurst about the uniformity of the 1 m per year rise in groundwater levels quoted for London.

C.W. Hurst replied that the 1 m per year figure is a maximum for the London area. The rise is highest at the centre of the basin and reduces radically outwards.

H.B. Phillipson (Geotechnical Control Office, Hong Kong) in answer to a questions regarding the depth of the landslips referred to in his paper stated that they are essentially shallow ones, normally not deeper than 2 to 3 metres.

P.W. Rowe asked Mr. Phillipson for clarification regarding the telephone/micro-computer system and hence the possibility of establishing an early warning system for slope failures in Hong Kong. Mr. Phillipson explained that this was not feasible because of the very large numbers of slopes involved and the complexity of the variables leading to failure. Also, most slips occur without warning, within half an hour of the local peak rainstorm event.

Written contribution further to verbal discourse:

D.W. Cox, J. armishaw & S. Mojabi (Polytechnic of Central London)

Effect of water level on bearing capacity

Further to Mr. Hurst and Professor Wilkinson's comment that water level only has a minor effect on ultimate bearing pressure, full scale experiments to show the relationship between water level and

From Cripps, J.C., Bell, F.G. & Culshaw, M.G. (eds), 1986, *Groundwater in Engineering Geology,* Geological Society Engineering Geology Special Publication No. 3, pp. 95-97

bearing pressure have shown the following characteristics.

Rising or falling water level under shallow formations

This creates a hydraulic gradient in the soil and the theoretical bearing pressure for shallow foundations is given by:

$$q = {}^1/_2 B N\gamma\ (\gamma_b - \gamma_w - i\gamma_w) \quad \text{where:}$$

q = ultimate bearing pressure

B = footing width

$N\gamma$ = bearing capacity factor (depending on the angle of friction of soil)

γ_b = bulk density of soil

γ_w = density of water

i = hydraulic gradient

Thirty tests on 0.5 m and 1 m diameter foundations on well graded sand fill using a series of constant flow conditions gave the results shown in Fig. 1. These confirm the theoretical assumptions that bearing pressure is inversely proportional to hydraulic gradient. At sites where large lateral soil pressures occur *in situ*, this might modify the relationship slightly. During a further series of three tests (see Mojabi 1984), in which the load was kept constant and the hydraulic gradient was varied, the sand fill failed when the gradient reached a value reasonably consistant with that in Fig. 1.

Change in water level below shallow foundations

The results of experiments shown in Fig. 1 also give the bearing pressure at zero hydraulic gradient with the water level constant at ground level. When the water was allowed to drain freely from the sand, leaving a partially saturated soil with a negative pore pressure, an increase of between times two and times five in ultimate bearing pressure over the value with water level at ground level (eight tests) was produced. Other experiments in which the water table was pressively lowered and held constant within a depth equal to the plate diameter indicated increases in bearing pressure of times two over the value with the water table at ground level (four tests).

Changes in water level around piles

Full scale load tests on twelve end bearing driven piles of different depths in sand and gravel were conducted with groundwater pressures controlled by pumping into or out of a ring of wells around the pile test area. The results of these tests are shown in Fig. 2. Further details are given in Armishaw & Cox (1979). In conclusion the loss in bearing pressures of shallow footings from the lowest to the highest water table is substantial and generally more than 50%. However, it should be noted that actual factors of safety of foundations on granular soils are usually very generous.

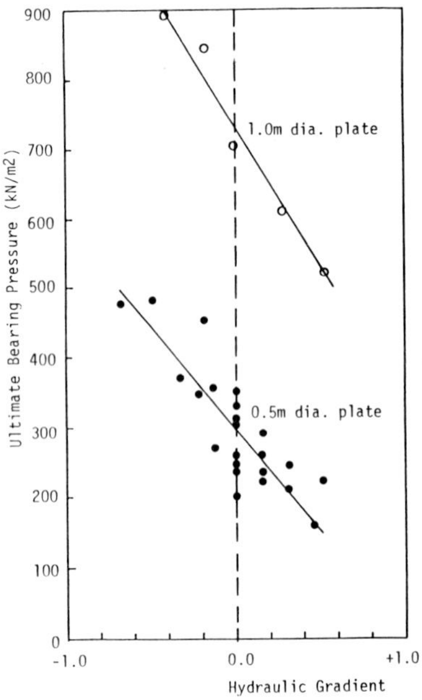

FIG. 1. Variation of ultimate bearing pressure with hydraulic gradient.

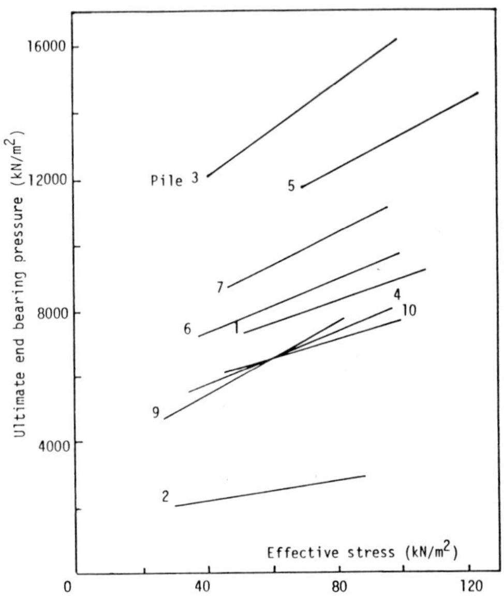

FIG. 2. Graph of ultimate end bearing pressure versus effective overburden pressure at the level of the base of the pile during testing.

ACKNOWLEDGEMENTS: Some of these finds follow undergraduate project work by Messrs Tan, Lan and Ng.

References

ARMISHAW, J. and COX, D.W. (1979). *Proceedings of a conference on recent developments in design and construction of piles.* Institution of Civil Engineers.

MOJABI, S. (1984). A study of the effect of pore water pressure on the ultimate bearing capacity of sand. *MPhil Thesis* Polytechnic of Central London.

SESSIONS 3 & 4

EFFECTS OF GROUNDWATER ON SOILS, ROCKS AND CONSTRUCTION MATERIALS

Effects of groundwater on soils, rocks and construction materials: an introduction

M.H. de Freitas

ABSTRACT: Although the effects of groundwater on soils, rocks and construction materials may appear to be slow to occur, there is increasing evidence to demonstrate that such effects progress at rates that are significant to engineering. A growing body of evidence that demonstrates the influence exerted by groundwater chemistry and flow upon the geotechnical character of the ground is also accumulating. With these thoughts in mind the papers submitted to this session of the conference are introduced and the work they describe reviewed.

Four papers have been submitted to this section of the conference and it is the intention of the introduction given here to place them within a wider context, so that the implications of the work they describe may be more readily appreciated.

When working with the phenomenon of groundwater flow it is always important to remember that such flow is a geological process and that geological processes, slow though many of them may seem, do operate at rates that are significant to engineering. This can be illustrated by observing the seepage through concrete dams, seen best in wintery weather when evaporation from the downstream concrete surface is at a minimum and makes visible the dampness (Fig. 1). Good concrete has a permeability in the order of 3 cm/year and no dam engineer would claim this figure is insignificant as its consequences are visible on concrete dams, yet 3 cm/year is a commonly occurring rate for continental drift. So, rates of geological, and hence hydrogeological,

FIG. 1. A modern gravity dam exhibiting dampness on its downstream face which betrays the presence of seepage through the concrete.

From CRIPPS, J.C., BELL, F.G. & CULSHAW, M.G. (eds), 1986, *Groundwater in Engineering Geology,* Geological Society Engineering Geology Special Publication No. 3, pp. 101–106 processes are not to be ignored because their values may easily be of a significant magnitude to engineering works.

The paper by Barton & Thompson is interesting in that it presents yet another case to demonstrate the importance of rates of groundwater response to geotechnical engineering. Their particular study concerns the rate at which equilibrium pore pressures are established on either side of a lithological boundary separating materials of different permeability.

At this point it is worth recording that a substantial number of delegates attending this conference come from colleges and research establishments, and it is therefore appropriate for this introduction to propose research areas that seem worthy of development. The first such area is the relationship between the change of groundwater conditions in a vertical sequence (and particularly the rate at which the change can occur) and other geotechnical characters and properties.

The paper by Barton & Thompson demonstrates the insight that is provided by such studies. It would be fascinating to see these parameters recorded with the same regularity against depth as moisture content, consistency limits and undrained shear strength in areas where groundwater is either suspected or known to make a significant contribution to the processes governing strength and deformation. Permeability is the great controller of rate of change and perhaps the time has come when a measurement of *in situ* permeability and its variation with depth should be more widely considered as an essential piece of information to be obtained from site investigation, and given greater status in ground investigation than it presently enjoys.

The next subject to be introduced in this session concerns the effect of groundwater flow upon the strength of ground. Groundwater has a chemistry of its own and a content of total dissolved solids that reflects its residence in the ground. From this it follows that the ground must impart its chemistry to the water flowing through it and lessen in some degree that may make it weaker. In short, the processes can be summarised by the general statement illustrated in Fig. 2: viz that the mass input to an element of ground plus the changes in mass that occur within the element will equal the mass output from the element. Honeycombed karstic limestone is as clear an example of this process as any.

The reverse action is also true when groundwater precipitates some part of its dissolved solids as mineral cement to bind and clog a porous media, and in so doing, to strengthen it. Fig. 3 shows two examples of coarse granular material that have been cemented by percolating groundwater. Similar occurrences have been revealed to exist in the granular filters of highways, converting them from the loose unbound aggregate they originally were to a bound aggregate similar in character to breccia, and having a much lower permeability than the filter design specified. Some of our largest engineering structures are made from granular fill, embankment dams being typical examples, and the changes which can occur in this fill consequent to groundwater flow deserve more attention from geologists than they appear to receive.

The contribution from Shirlaw is therefore both welcome and timely. His work forms part of a wide range of studies that investigate the geomechanical changes which occur in the presence of a continuing flux. A recent example of this, forming studies that

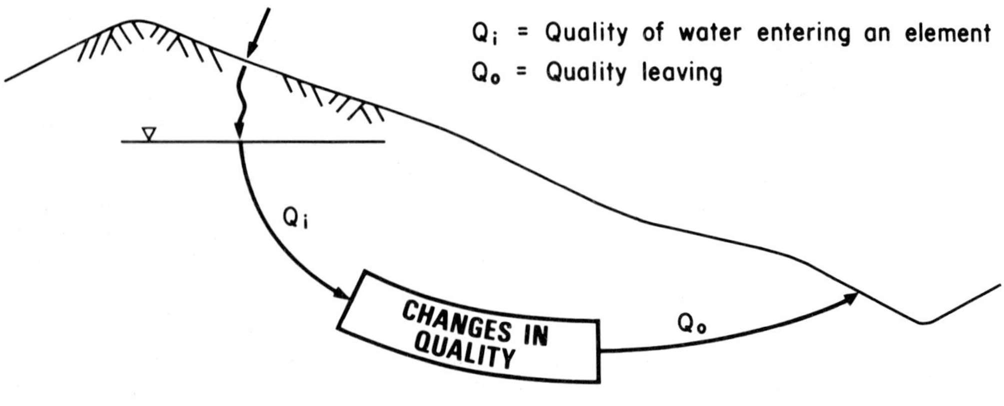

FIG. 2. A view of mass transfer, seen in terms of water quality and analogous to that commonly used for defining water quantity (ie. Vol. entering + change in storage = vol. leaving).

that are currently progressing, seeks to explain some remarkable structures found along the flanks of the Dead Sea. An example of these is shown in Fig. 4 and is one of a series of such structures found in the Lisan deposits; a sequence of fluvially derived clastics accumulated during the Upper Pleistocene by seasonal floods that discharged into Lake Lisan, which is considered to be a precursor of the Dead Sea, and mixed with the products of evaporation. The sediments dip at a very small angle and the structures within them tend to be concentrated in the region of lateral valleys draining to the Dead Sea (Fig. 5a).

FIG. 3. Fragments of (a) Jurassic limestone, and (b) Carboniferous sandstone, cemented by percolating groundwater: the limestone with $CaCO_3$ and the sandstone with Fe-oxides. Larger fragments in both cases, are approximately 5 cm in length.

The geometry of these structures implies that the sediments were of low strength and in a plastic condition when deformed, and shortened in a direction parallel to the dip of the sediments towards the axis of the lake. The stratigraphic relationship of the structures to undeformed horizons that lie immediately above and below them confirm that such deformation as formed them originated and was largely completed during the general period of deposition, that is, they were formed quickly.

FIG. 4. Folds in the Lisan deposits: vertical thickness of the band of folding is approximately 1m.

Numerous hypotheses have been proposed to explain these structures, including periodic seismicity associated within movement of the Rift Valley, but all have been discounted on the basis of field evidence. A new explanation currently under consideration is that the sediments, which were eventually accumulated in an extremely saline environment, were weakened in the vicinity of the valleys by the flushout of soluble evaporites within them during periods of sustained covergent discharge of comparatively fresh groundwater flowing towards the base of the valleys (Fig. 5b). Similar processes of flushing by less saline water are known to have greatly weakened the marine clays deposited marginal to the land masses of Scandinavia and N. America during the fall of sea levels and isostatic rise in land levels at the end of the Pleistocene glaciation. The quick-clays of these areas are well known and it seems possible from the occurrence of the fold structures in the Lisan deposits that such purging mechanisms as are recognised to have

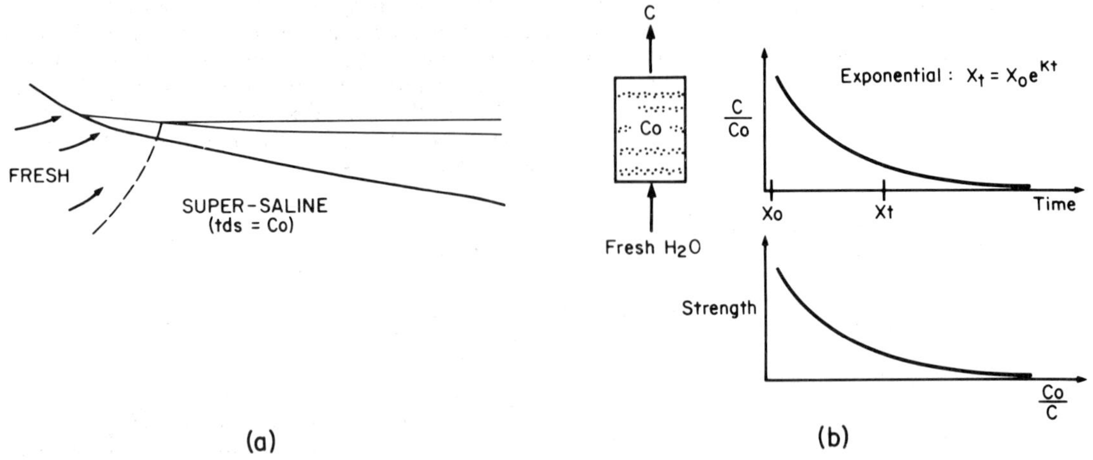

FIG. 5. (a) A sketch vertical cross-section along the axis of a lateral valley to the Dead Sea and into the sea itself illustrating the presence of a tongue of fresh groundwater following periods of inland recharge from rain. (b) The change in concentration of dissolved solids, with time and consequent change such a mechanism may have on the strength of the sediment.

occurred in the quick-clays are much more widespread in their occurrence than had previously been realised. If such a mechanism operated it would be expected to obey an exponential law with respect to time, as the soluble products are leached from storage: hence strength would be expected to decay gradually, and in the case of the Lisan deposits, to reduce sufficiently for gravitational slumping of the sediment to occur at intervals during deposition (Fig. 5b). It is therefore of great interest to note that the decay in strength recorded by Shirlaw follows exactly this behaviour and Fig. 6 illustrates the data presented by Shirlaw in his Fig. 2 as an arithmetic plot. This work underlines the need that presently exists to study more widely the relationship between the strength of a material and the flux through it, especially in recently accumulated sediments and man-made structures composed of accumulations of aggregate, and this is clearly another area worthy of research.

The third submission to this session of the conference, the paper by Hart, is not unrelated to the work involving the relationship between strength and flux, because it considers the extent to which groundwater influences ground strain in areas relaxed by excavation.

A mineralogical index (*XC*) is defined (Fig. 7) and its relationship to ground response examined, including the swell associated with an unrestricted access to water. The material studied has been obtained from two sources: Red Vein, Betws Colliery

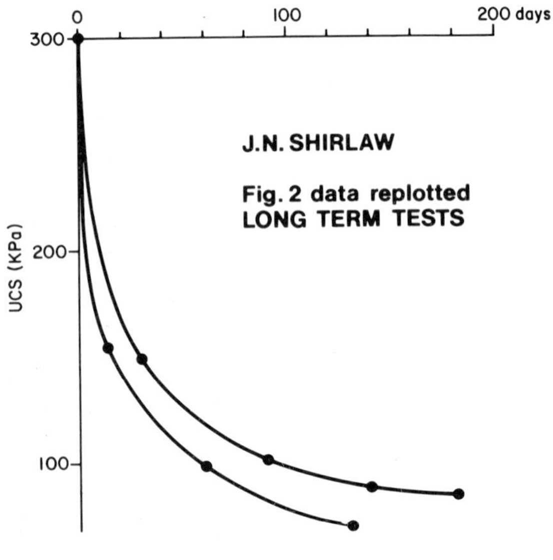

FIG. 6. The change in confined compressive strength as a function of time (or flux).

(Middle Coal Measures) and the Taff Merthyr '7 Foot' seam (Lower Coal Measures), both in S. Wales. Hart presents the swelling characters of the material from these areas (Hart, fig. 4) but unfortunately does not differentiate the origin of the specimens. Replotting this data as a semi-log graph clarifies a trend visible in Hart's fig. 4; viz. that there appear to be two different rate relationships (Fig. 7) and one is tempted to suggest that they reflect the different sources of material, there being one rate relationship for the Red Vein (MCM) and another for the '7 foot' (LCM). This would be in keeping with the response expected from ground when exposed to water, for such a response must reflect the geological history of the ground.

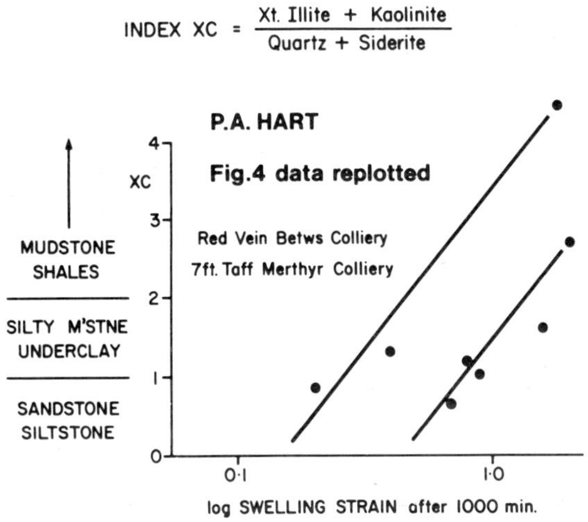

FIG. 7. The possible relationship between an index (XC), swelling and geological history.

This raises an interesting prospect: we are accustomed to measuring and interpreting the response of the ground to electrical currents, seismic accelerations, gamma and neutron bombardment and the like, as a normal part of geophysical studies, so why not also accept that a measure of ground response to water is another geophysical response that can be valuable and relevant to ground engineering, and should be measured more frequently than it is? Anyone who has had to engineer with soft rock, especially argillaceous materials, will appreciate how significant is the response of such material to water. It is just over 10 years since Morgentern & Eigenbrod (1974) based an important appreciation of argillaceous materials upon such a response to water. In this country we have earthworks curtailed by bad weather and overseas there are many areas where weak rocks, especially of volcanic association, display a dramatic change in properties when wetted. It is therefore comforting to know that Hart's contribution to the conference brings this aspect of hydogeology to our attention.

It is only a small step from Hart's work to that described in the fourth submission; the paper by Turk & Dearman. Twenty years ago Colback & Wiid (1965) brought to the attention of engineers and geologists alike, the amazing effect water has upon rock strength. The significance of moisture content must not be under-estimated, as recent work by Priest & Selvakumar (Imperial College report for Transport and Road Research Lab. 1982) demonstrates (Fig. 8): note how great can be the loss in strength within the first few percentages of moisture content. Turk & Dearman report the same trends (Turk & Dearman, fig. 3) with the strength of saturated specimens being consistently and significantly less than those for dry specimens.

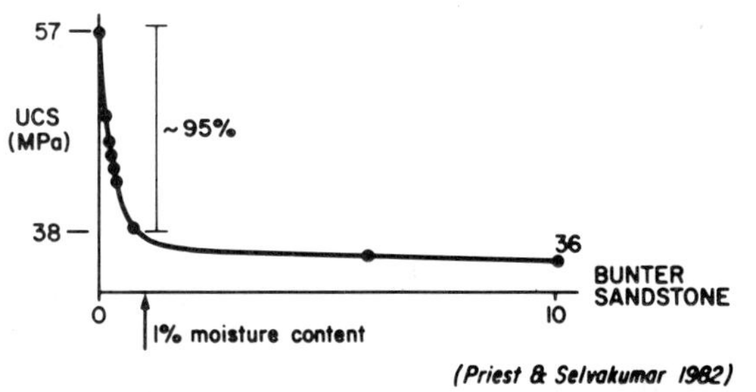

FIG. 8. The characteristic response of rock strength to the presence of water.

Many theories have been postulated to account for this dramatic behaviour. Some require a decrease in capillary tension between grains with increasing moisture content, others a rise in pore water pressure, whilst a number link the change to reductions in the coefficient of friction of solid surfaces. The latest theories rely heavily upon aspects of chemical deterioration of solid lattices in the presence of water and a consequent reduction in the energy required to fracture the lattice. The work of structural geologists in the field of fracture mechanics has confirmed that an explanation for this phenomenon may be possible in terms of fracture energy. Such a theory, written in these terms may unify the diverse theories based upon behaviour described by capillary tension, pore pressure, friction and the like, these being merely less direct ways of quantifying the charge on surfaces created during fracture in the presence of water.

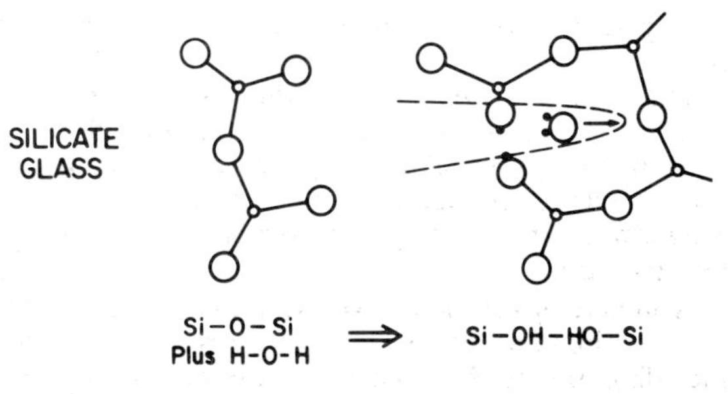

FIG. 9. A schematic representation of H-O-H invasion into the lattice of a silica glass.

The processes involved can be introduced with reference to work undertaken 10 years ago on the behaviour of silicate glass (Lawn & Wilshaw 1975). Fig. 9 illustrates the Si-O-Si lattice of such glass in a diagrammatic way. The lattice can easily be broken in the presence of H-O-H with its disjointed end-members satisfied by the available oxygen and hydrogen, leaving a window in the lattice through which further supplies of these elements can enter so that the process may continue.

Some remarkable relationships are being revealed between moisture content and strength. Last year Dobereiner (Dobereiner & de Freitas 1984) presented a relationship between saturated moisture content and saturated unconfined compressive strength, and it is clear from such relationships that moisture content somehow summarises all those parameters that significantly contribute to rock strength. Rocks when wet are weaker than they are when dry, and the fact is that we do not yet have an adequate explanation for this most basic of rock responses. Here, surely, is an area of research that needs to be pursued. There is much that can be learned from the work of the structural geologists and those physicists, and engineers, involved in fracture mechanics. The time seems right for such research to enjoy a renaissance.

The papers submitted to this session therefore represent enquiries into four major areas of concern, viz. rates of response, influence of flux, geological history, and strength, and so contribute to the value of this conference. But there remains an area of interest that has not attracted attention and deserves a mention: it concerns the movement of a wetting front through the ground and the effect this can have upon soils, rocks and construction materials. Perhaps this oversight arises from the fact that the UK rarely endures sustained periods when the rate of precipitation exceeds the infiltration capacity of the ground surface. This is not the case overseas and recent research by Zorich (1985) illustrates the phenomenon that can occur when attempts are made to reproduce in the laboratory the infiltration that results from intense rainfall.

Many areas of the world are exposed to weather patterns that bring with them prolonged periods of rain which include intense bursts when more than 20 mm of rain may fall in less than 1 hour, and that after a period when 300 mm may have fallen over the previous 30 days. The common consequence of this is widespread landsliding on shallow sliding surfaces sub-parallel to ground level. Fig. 10 illustrates the basic characters of a laboratory apparatus that was used to reproduce this environment. To the great surprise of observers, the wetting front succeeded in reaching the water table whilst at the same time enclosing a bubble of trapped air which on occasions was capable of splitting the sample horizontally; a form of pneumatic fracture. It is hoped to describe this work more fully elsewhere, but these preliminary tests demonstrated that in this process there exists a very simple mechanism for failing the ground in tension and thereby creating a surface that has little, if any cohesion. Shallow, planar failure surfaces oriented sub-parallel to ground surface can be seen on slopes all over the world, including our own motorway cuttings in the UK, and their existence seems evidence enough at present for the continued study of this phenomenon.

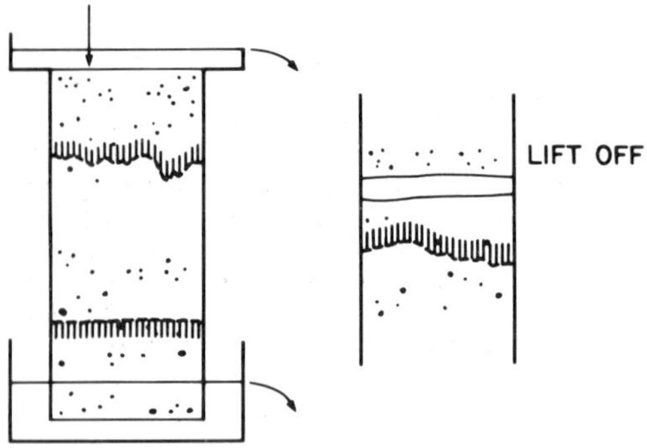

FIG. 10. A column of partially saturated ground sandwiched between a lower capillary fringe and an upper (and downward moving) wetting front.

In concluding this introduction it is pertinent to consider what the future may hold for the subjects considered during the session. In them there is research work for many. Those who favour field work will find much to study in the relationship between groundwater and ground response, seen as a function of geological history. Those preferring a mixture of work in the laboratory and the field will find in the subject of strength and associated rates of change of groundwater conditions, many fruitful areas of investigation. For the theoretician there remains the hidden secrets of fracture energy and the influence upon it of water. In short, there is something for everyone in this session alone and it is to be hoped that this conference with its proceedings, will be a catalyst that renews in the UK the pursuit of hydrogeological research.

References

COLBACK, P.S.B. & WIID, B.L. 1965. The influence of moisture content on the compressive strength of rocks. *Proceedings of the 3rd Canadian Rock Mechanics Symposium, Toronto, 65-83.*

DOBEREINER, C. & de FREITAS, M.H. 1984. *Proceedings of the 20th Regional Conference of the Engineering Group of the Geological Society, London* 143-156.

LAWN & WILSHAW 1975. *Fracture of Brittle Solids,* Cambridge University Press.

MORGERNTERN, N.R. & EIGENBORD, K 1974, Classification of argillaceous soils and rocks. *Proceedings of American Society of Civil Engineers, Journal of the Geotechnical Engineering Division,* **100,** GT10, 1137-1156.

ZORICH, E.A. 1985. Rain-induced failures in residual and colluvial soils on natural Slopes, *MSc thesis,* Imperial College (University of London).

M.H. DE FREITAS, Geology Department, Imperial College, London SW7 2AZ

Seepage characteristics and landsliding of the A3 zone of the Barton Beds

M.E. Barton & R.I. Thomson

ABSTRACT: The most characteristic feature of the Barton Clay cliffs of Christchurch Bay is the presence of a number of preferred bedding plane surfaces of shearing. Most of these are contained within clay horizons and the reason for their preferential use by the coastal landslides is unknown. One surface, however, is at the base of the 2.7 m thick A3 Zone which consists of inter-bedded sand and clays. This surface is well exposed and forms a distinct feature in the cliffs throughout nearly the whole of the 1.5 mile coastal outcrop of the A3 Zone.

The paper discusses the reason why a shear surface is preferentially developed at this stratigraphic location. Three possible causes are discussed under the headings of (i) pore pressure fluctuation, (ii) seepage erosion and piping and (iii) equilibration response time. The first two are well established theories but the third is a new hypothesis derived from the original Bishop and Bjerrum theory of delayed equilibration. It is considered that the discussion is useful in the general context of landsliding at sand/clay junctions in overconsolidated soils.

Preferred shear surfaces in the Barton Clay

The lithological character of the 46.4 m thick Barton Clay is indicated in Fig. 1. Certain of the bedding planes are preferentially used as basal surfaces of shearing in the compound-shaped (bench-form) landslides of the coastal cliffs. The four bedding planes which are used throughout nearly the whole of their outcrop, and which consequently exert a controlling influence on the undercliff form, are indicated on Fig. 1. They are notated according to the zone in which they occur or, to which they are adjacent if at the junction of two zones.

Of these four preferred bedding planes, only the 'A3' surface occurs at, or very near, the junction of a sand with a clay. The other three are contained within clay zones and there is no obvious evidence as to why these particular planes are preferred. The A3 Zone is the only zone in the Barton Clay which contains actual beds of sand. The suggestion made by Bayley (1972) that sand beds are responsible for the characteristic benched profile in the Christchurch Bay cliffs is not applicable to the Barton Clay cliffs other than in the vicinity of Zone A3. The problem of the other three preferred shear surfaces are very briefly discussed elsewhere (Barton 1973, 1977) and no further reference will be made to this here.

The purpose of this paper is to discuss the A3 Zone and to suggest why the associated 'A3' bedding plane shear surface develops at the A3/A2 junction. The exact stratigraphic position of the shear surface cannot be precisely located owing to the removal of the superincumbent strata during sliding. While in broad terms, it occurs at the lower boundary of the A3 Zone, it seems more probable that actual sliding develops along a clay/clay contact just below the lowest sand (perhaps at a distance of a few microns) rather than at a sand/clay contact. It is convenient, however, to refer the shear surface to the junction between the zones since the close proximity of the sand bed creates different conditions to those experienced by the surfaces located well within beds of clay. The cause, or causes, of landsliding along the 'A3' surface may be similar to other examples of landsliding at, or very close to, the junction of sands and clays in sequences of over-consolidated soil strata.

The A3 Zone

The A3 Zone contains interbedded sand and clay and is 2.7 m thick at Highcliffe Recreation Ground. The sand beds are generally thin (from 5 to 15 mm) but can be much thicker (up to 0.6 m in the central area but over 1 m at the western end of the outcrop). The sand beds consist of a pale silty fine sand (Fig. 2). The clay beds are a dark grey silty clay.

The upper boundary of the A3 Zone is a well marked junction to the bioturbated Zone B which contains very irregular lenses of sand in a sandy clay matrix. The lower boundary is drawn at the base of the lowest sand bed. Fig. 3 shows an exposure of the lowest sand bed which occurs at the extreme eastern limit of its outcrop, just above beach level. Cliff-face exposures of this lowest sand bed do not occur elsewhere since throughout the rest of the outcrop, the lower bedding plane is used as a basal shear surface. In consequence all that is seen is colluvium resting on A2 Zone clay (Fig. 4).

From CRIPPS, J.C., BELL, F.G. & CULSHAW, M.G. (eds), 1986, *Groundwater in Engineering Geology*, Geological Society Engineering Geology Special Publication No. 3, pp. 107–114

Hydrogeology

Detailed work with some sub-surface investigations have been carried out at two locations in the coastal cliff outcrop (Fig. 5). These are at Highcliffe Recreation Ground (Barton 1973) and Naish Farm (Barton & Coles 1984; Barton & Thomson 1984). Geological cross-sections for these two sites are given in Fig. 6.

The hydrogeological conditions at both sites are complex but the relatively high permeability of the A3 Zone makes it possible to consider this as a separate geohydrological domain. Unfortunately there are insufficient standpipe/piezometers to establish unequivocally the drawdown curve for this zone but an attempt has been made as shown in Fig. 6.

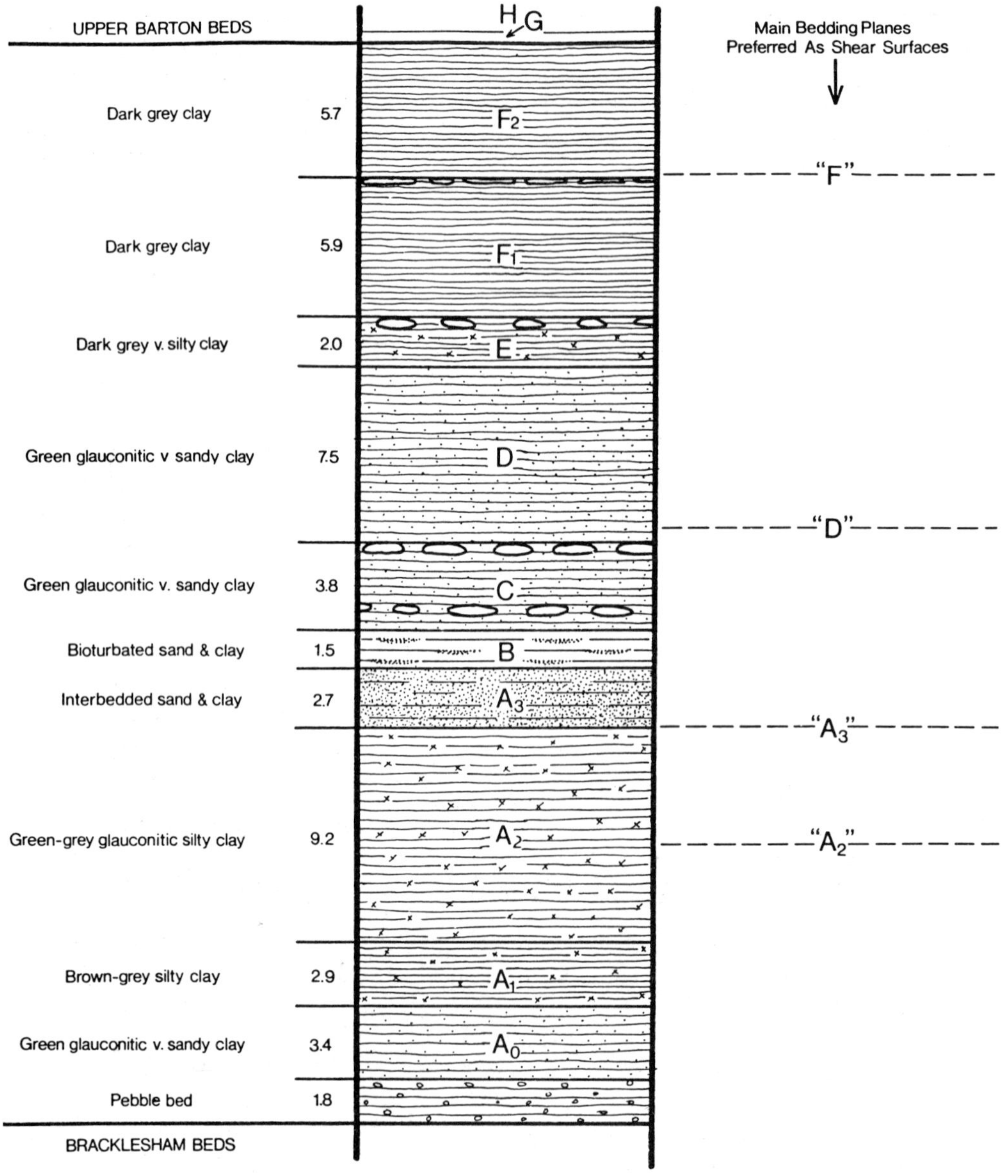

FIG. 1. Lithological descriptions of the zones in the Barton Clay and the stratigraphic location of the main preferred bedding plane shear surfaces. The 'A3' surface is unique in occurring at a sand/clay junction. The zonal sequence is based on the palaeontogical sequence described by Burton (1933) but slightly modified to take account of lithological characteristics (Barton, 1973). The Barton Clay includes the 'Lower and Middle Barton Beds' of current stratigraphic nomenclature.

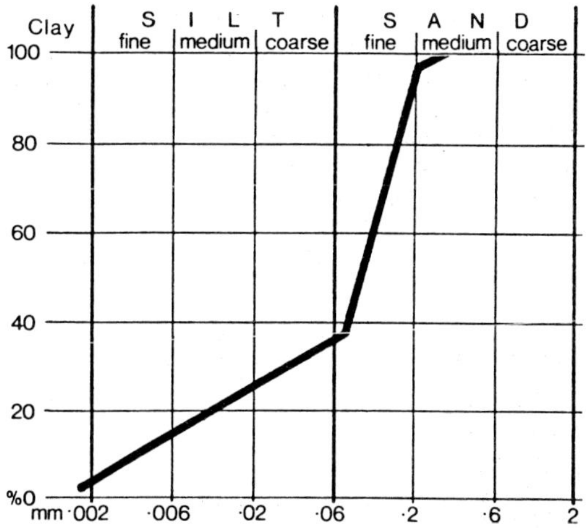

FIG. 2. Particle size distribution of the sand beds in the A3 Zone.

The colluvium on the bench formed by the sheared 'A3' surface contains much derived A3 and B Zone material and *en masse* is relatively permeable. While it probably differs in some degree from the *in situ* A3 sand, the permeability contrast has been ignored and a smooth curve drawn to link up the piezometric level recorded in standpipe/piezometers installed in the A3 (and B at Highcliffe Recreation Ground) from, or near, the cliff top and the level recorded in the 'A3' bench colluvium. The piezometric levels in the colluvium are markedly seasonal but only the lowest (summer) levels have been used to draw the curve as they seem the most likely to be controlled by the seepage through the A3 Zone. Using these levels, a total of three points for the drawdown curve on each section is obtained and a smooth curve fitted to the data (see Appendix).

FIG. 3. Exposure near beach level of the A2/A3 boundary. The visible base of the ranging rod marks the actual boundary. Above this is a 13 cm thick bed of sand at the base of the A3 zone. The sand bed has well developed vertical jointing. The hollows are most probably formed by wave erosion. The ranging rod is marked in feet - the dark band being 1ft (0.305 m).

FIG. 4. Exposure of the A3 shear surface (separating *in situ* clay from colluvium) in the unprotected cliffs east of Chewton Bunny. The lighter tone below is the *in situ* A2 clay forming a 4 m (foreground) to 3 m (background) high scarp at the back of the beach. The darker tone above (approximately 3 m high) is colluvium. A wide bench level occupies the middle of the cliffs (at the elevation of the lower end of the pipe seen near the top left of the photograph). The back scarp is formed in the D, E and lower F zones. The prominent A3 shear surface can be seen reducing in elevation to the east (away from the camera) since it confirms to the general dip of the Barton Clay (approximately $^1/_2$° along the cliff-line).

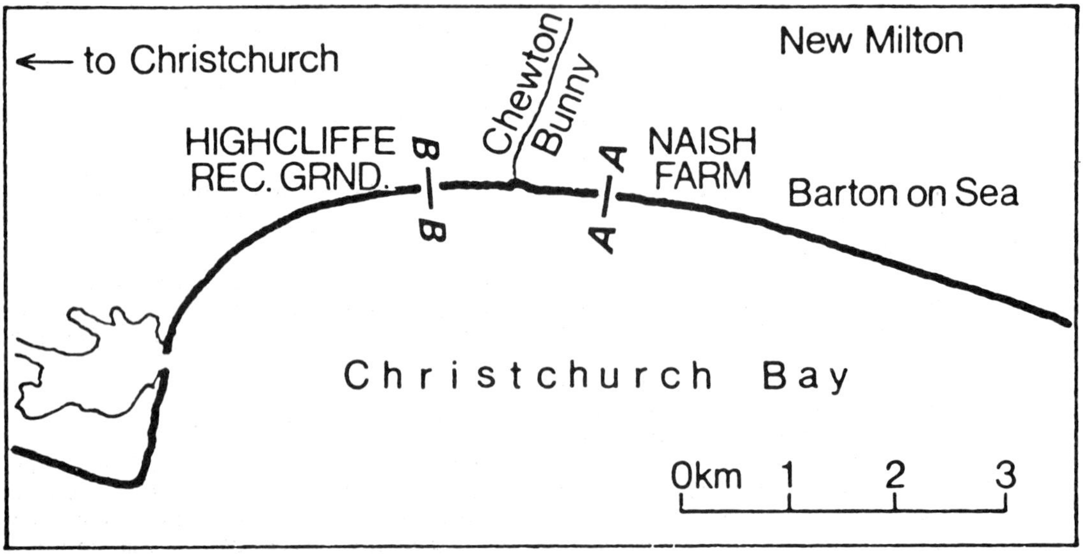

FIG. 5. Sketch-map of the coastal outcrop of the Barton Clay in Christchurch Bay. Geological sections along lines AA (Naish Farm) and BB (Highcliffe Recreation Ground) are shown in Fig. 6.

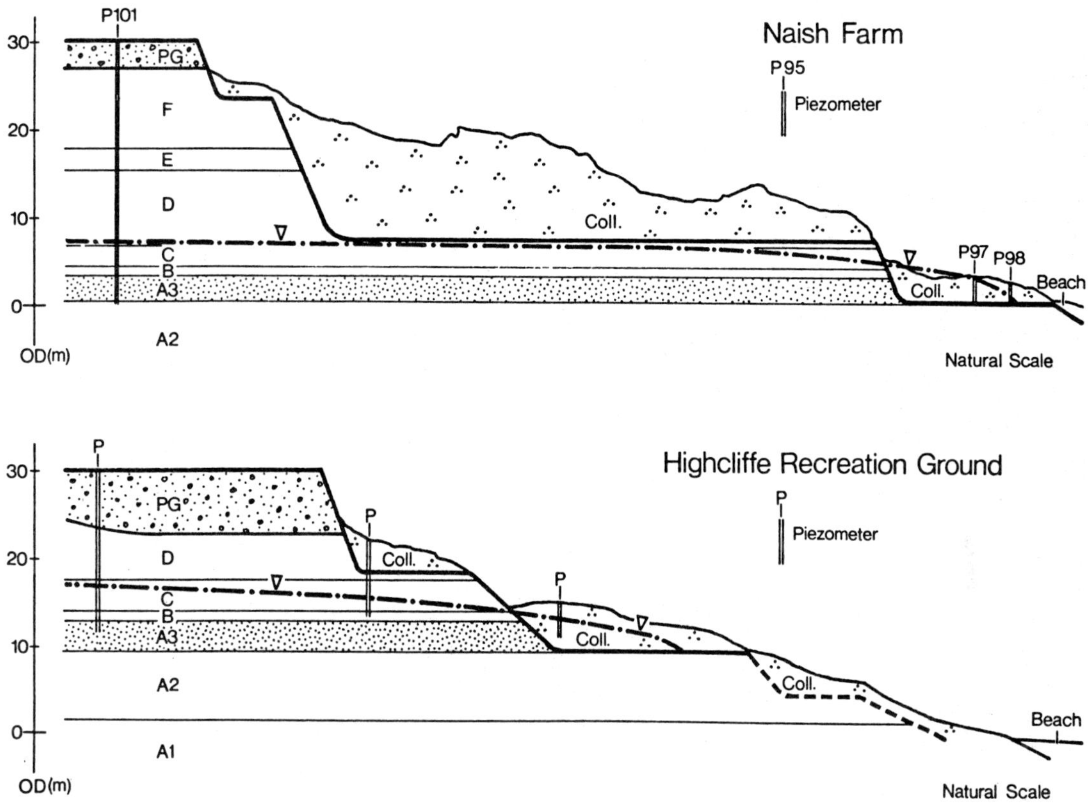

FIG. 6. Geological sections at Naish Farm (above) and Highcliffe Recreation Ground (below). The colluvial bench formed by sliding on the A3 surface is at a lower elevation at Naish Farm than at the Recreation Ground owing to the eastwards dip of the bedding. A piezometric surface for groundwater contained in the A3 zone and colluvium on the A3 bench is shown on both sections.

The curves indicate average hydraulic gradients, i, in the vicinity of the buried, *in situ* scarp face as follows:-

Location	i
Naish Farm	0.11
Highcliffe Recreation Ground	0.12

These values may be compared with the value of $i = 0.07$ given by Ward (in Hutchinson 1982) for a sand in the Woolwich and Reading Beds in the cliffs at Newhaven (Ward 1948).

During the winter half of the year, the increase in effective rainfall causes a marked increase in piezometric levels in the undercliff colluvium. Under these conditions the drawdown in the A3 Zone will be altered, reducing the hydraulic gradient in the vicinity of the buried scarp and increasing it at the seaward end of the colluvial bench.

The outflow from the A3 Zone introduces a significant quantity of water into the colluvium of the 'A3' bench. In the unprotected cliffs at Naish Farm, the resulting pore pressures cause continuous sliding of the colluvium throughout the year with very rapid rates in the winter (Barton & Coles 1984). In the stabilised cliffs at Highcliffe a drain has been constructed to collect the outflow from the A3 Zone and discharge it to the beach via sealed drains (Mockridge 1983).

Possible causes of landsliding on the 'A3' surface

Three possible causes will be discussed here under the headings of (i) pore pressure fluctuation, (ii) seepage erosion and piping and (iii) equilibration response time.

(i) Pore Pressure Fluctuation

The greater permeability of the sands ensures that the pore pressures developing at the sand/clay boundary are determined, almost exclusively, by the water flow in the sand bed (Henkel, 1967). Hence an increase in total head in the sand layer will be more rapidly observed as an increase in pore pressure at the top of the clay than at some depth. Seasonal fluctuations within the A3 Zone aquifer, as shown by successive readings in the standpipe/piezometers sunk from the cliff top, are small. More significant fluctuations are caused by the winter recharge to the colluvium, reducing the hydraulic gradient in the *in situ* A3 Zone. These fluctuations could be a significant factor in the development of sliding at the A3/A2 junction during renewed slumping of the scarp slope behind the colluvial bench.

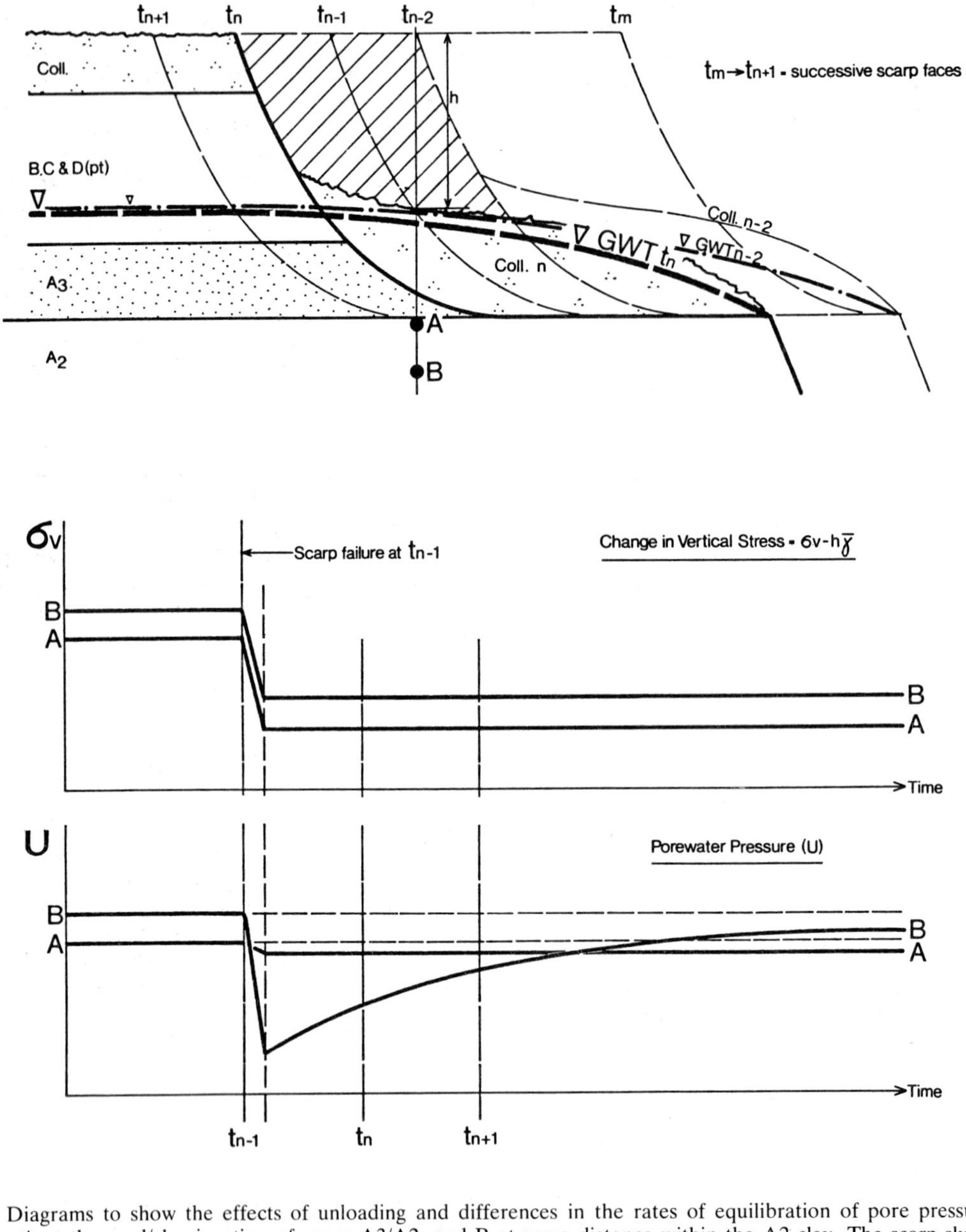

FIG. 7. Diagrams to show the effects of unloading and differences in the rates of equilibration of pore pressures at two positions: A at the sand/clay junction of zones A3/A2, and B at some distance within the A2 clay. The scarp slump at time t_{n-1} (forming the t_{n-1} scarp face) causes a marked reduction in vertical stress at points A and B. This results in a depression of the pore pressures at B but not at A where the pore pressure response is immediate. Stress and pore pressure changes at time t_n (forming the t_n scarp face) and stress changes other than vertical are ignored for the sake of clarity. When movement is renewed at time t_{n+1}, then shearing will be more likely at the sand/clay junction (A) than for planes at depths of up to several metres depth such as at B where equilibration is not complete.

(ii) Seepage Erosion and Piping

The efficacy of seepage erosion as a mechanism for initiating landsliding, where seepage gradients are increased by reservoirs or natural ponding of water, has been described by Terzaghi (1931). The mechanics of piping were further explained by Terzaghi & Peck (1948). Piping and soil movements in shallow drift deposits are clearly recorded by Masannat (1980) and Pierson (1983) among others. Seepage erosion, resulting in back-sapping, at deeper levels has been demonstrated by Hutchinson *et al.* (1981). Hutchinson (1982) has defined seepage erosion as the 2-dimensional version of the 3-dimensional process of piping and has produced a list of observed cases but for which good, quantitative information is lacking. While there is an association of possible seepage erosion with landsliding in many of these cases, the evidence for a causal relationship is purely circumstantial at present.

No positive information on the extent of seepage erosion or piping can be produced for the A3 Zone. The landsliding results in a lack of exposures except, as seen in Fig. 3, near the beach where wave erosion prohibits positive interpretation. Prominent vertical jointing is seen in Fig. 3 and it is probable therefore that the actual seepage through the A3 Zone is non-homogeneous with locally increased flows through the joints. The development of tension cracks and piping hollows within the colluvium may allow the removal of soil particles from the *in situ* A3 Zone itself but it is problematical as to the extent to which this occurs.

(iii) Equilibration Response Time

Unloading, due to the formation of a cut, leads to the depression of the pore water pressures in a clay slope (Bishop & Bjerrum 1960). The long time required for the equilibration of the depressed pore water pressures is an important factor in the delayed failure of some cutting slopes (Vaughan & Walbanke 1973). Similarly, recent landsliding results in unloading and the depression of pore water pressures as has been observed in cliff slopes by Hutchinson (1980) and Bromhead & Dixon (1984). A theoretical examination of an eroding natural slope suggests that substantial negative pore pressures are likely to persist for a long time (Koppula & Morgenstern 1984).

The influence of the delayed equilibration in the vicinity of a sand/clay junction such as the A3/A2 boundary is shown in Fig. 7. The pore pressures developing in response to the unloading due to periodic slumping of the scarp slope behind the colluvial bench are given for two points A and B. A is in the clay immediately below the sand/clay junction and B is at some depth within the clay. For clarity, only the effects of a vertical stress change from some previous scarp slump are shown. Recession of the cliff during the scarp slump (shown as taking place over a short period of time) changes the groundwater levels. Owing to the high permeability of the A3 Zone, the response at point A is immediate and at the end of the slumping process, the pore pressure there is equilibrated to the new groundwater level.

At point B, the pore water pressure is depressed by the unloading and equilibration to the new ground water level takes a long time. When renewed slumping occurs, the shear surface is more likely to pass through A where the pore pressures are controlled by the flow in the sand than through some point such as B where equilibration is not compete and the effective stress levels remain relatively high.

The scarp slump is induced by seaward movement of the colluvium on the bench, which allows a reduction of earth pressure in front of the scarp. The movement of the colluvium may be due to high pore pressures in the bench (e.g. from the high winter recharge) and/or from marine erosion at the toe. The movement of the bench colluvium and the subsequent scarp slump are dependent on the right combination of conditions: it is shown in Fig. 7 to occur at regular intervals only for the sake of clarity. It must be emphasised that, by this theory, it is not the equilibration of the depressed pore pressures which results in renewed scarp slumping: but that the slumping, occurring for other reasons, has its form controlled by the differences in the rate of equilibration. Shearing preferentially takes place at the sand/clay junction where there is no delayed equilibration and where response to slight fluctuations in water levels is immediate.

Discussion

Three possible causes for the development of landsliding at the A3/A2 junction have been outlined. The first two of these (pore pressure fluctuation and seepage erosion and piping) are mechanics frequently discussed in the literature although the true extent and effects of seepage erosion and piping at depth in natural slopes remains largely unknown. The third cause is, as far as the authors are aware, a mechanism not previously put forward. It is suggested here that all three mechanisms may be operative in the case of the 'A3 ' shear surface.

ACKNOWLEDGEMENTS: The authors would particularly like to thank Dr E.N. Bromhead, Messrs M.P. Chandler, B.J. Coles and N. Dixon for assistance with instrumentation at Naish Farm and Mr. A.I. Booth for assistance at Highcliffe Recreation Ground. They would also like to thank Mr. M.R. Cooper for helpful discussion and Mr. A. Brookes for drawing the diagrams.

References

BARTON, M.E., 1973. The degradation of the Barton Clay cliffs of Hampshire. *Quarterly Journal of Engineering Geology*, **6**, 423-440.

BARTON, M.E., 1977. Landsliding along bedding planes. *Bulletin of the International Association for Engineering Geology*, **16**, 5-7.

BARTON, M.E. & COLES, B.J., 1984. The characteristics and rates of the various slope degradation processes in the Barton Clay cliffs of Hampshire. *Quarterly Journal of Engineering Geology*, **17**, 117-136.

BARTON, M.E. & THOMSON, R.I., 1984. Studies of the water balance in a rapidly degrading soil cliff. *Proceedings of the 4th International Symposium on Landslides, Toronto, Canada*, **1**, 355-361.

BAYLEY, M.J., 1972. Discussion on Muir Wood, A.M. -Engineering aspects of coastal landslides. *Proceedings of the Institution of Civil Engineers (Pt. 2)* **53**, 405.

BISHOP, A.W. & BJERRUM, L., 1960. The relevance of the triaxial test to the solution of stability problems. *Res. Conf. Shear Strength Soils, Colorado, U.S.A.* 437-501.

BROMHEAD, E.N. & DIXON, N., 1984. Pore-water pressure observations in the coastal clay cliffs at the Isle of Sheppey, England. *Proceedings of the 4th International Symposium on Landslides, Toronto, Canada*, **1**, 385-390.

BURTON, E. St. J., 1933. Local geology and the stability of natural slopes. *Proceedings of the American Society of Civil Engineers, Journal of Soil Mechanics and Foundations Division, 93*, **SM4**, 437-446.

CRAIG, R.F., 1983. *Soil Mechanics* (3rd Ed.) Van Nostrand Reinhold (UK), Wokingham.

HENKEL, D.J., 1967. Local geology and the stability of natural slopes. *Proceedings of the American Society of Civil Engineers, Journal of Soil Mechanics and Foundations Division, 93*, **SM4**, 437-446.

HUTCHINSON, J.N., BROMHEAD, E.N. & LUPINI, J.F., 1980. Additional observations on the Folkestone Warren landslides. *Quarterly Journal of Engineering Geology*, **13**, 1-31.

HUTCHINSON, J.N., CHANDLER, M.P. & BROMHEAD, E.N. 1981. Cliff recession on the Isle of Wight S.W. Coast. *Proceedings of the 10th International Conference on Soil Mechanics and Foundations Engineering, Stockholm*, **3**, 429-434.

HUTCHINSON, J.N., 1982. Slope failures produced by seepage erosion in sands. *Landslides and Mudflows:* Reports of Alma-Ata International Seminar (UNESCO & UNEP), 1981 (Ed. by A. Sheko) 250-268. Centre of International Projects, GKNT, Moscow.

KOPPULA, S.D., & MORGENSTERN, N.R., 1984. Deficient pore pressures in an eroding soil mass. *Canadian Geotechnical Journal*, **21**, 277-288.

MAASLAND, M., 1959. Water table fluctuations induced by intermittent recharge. *Journal of Geophysical Research*, **64**, 549-559.

MASANNAT, Y.M., 1980. Development of piping erosion conditions in the Benson Area, Arizona, U.S.A. *Quarterly Journal of Engineering Geology*, **13**, 53-61.

MOCKRIDGE, R.G., 1983. Highcliffe cliffs - the maintenance of coastal slopes. *Proceedings of a Conference on Shoreline Protection, Southampton 1982*, 235-242. Telford, London.

PIERSON, T.C., 1983. Soil pipes and slope stability. *Quarterly Journal of Engineering Geology*, **16**, 1-11.

TERZAGHI, K., 1931. Earth slips and subsidences from underground erosion. *Engineering News-Record*, **107**, 90-92.

TERZAGHI, K. & PECK, R.B., 1948. *Soil Mechanics in Engineering practice (1st Edn)*. Wiley, New York.

VAUGHAN, P.W. & WALBANKE, H.J., 1983. Pore pressure changes and the delayed failure of cutting slopes in overconsolidated clay. *Geotechnique*, **23**, 531-539.

WARD, W.H., 1948. A coastal landslip. *Proceedings of the 2nd International Conference on Soil Mechanics and Foundation Engineering, Rotterdam*, **2**, 33-38.

M.E. BARTON & R.I. THOMSON, Department of Civil Engineering, University of Southampton, Highfield, Southampton, UK.

Appendix: Mathematic expressions for the drawdown curves

At *Highcliffe Recreation Ground* the three points fitted the parabolic equation of Kozeny for the phreatic surface in an earth dam (Craig 1983):

$$z^2 = 4x(A-x).$$

At Naish Farm the three points fitted the equation used by Maasland (1959) for groundwater flow over a horizontal impervious barrier:

$$z^n = B\,(L-x)x.$$

In these equations, z and x are the vertical and horizontal coordinates and A, B, L and n are parameters. No physical significance is attached to the use of either of these equations. The difference between the two sites may well result from variations in the permeability and other characteristics of the 'A3' bench colluvium.

Investigations into the role of groundwater in promoting floor heave in coal mine gateroads

P.A. Hart

ABSTRACT: The geological environment of two South Wales collieries at which site investigations were carried out is summarised. This is followed by a review of identified geological parameters and their effects on floor heave under the headings of discontinuities, mineralogy, and groundwater. Methods of underground site investigation, laboratory testing, and data treatment are summarised. The effect of an increase of groundwater content on mechanical parameter values was quantified for laboratory tested rock cores, and *in situ* interbedded strata sections. The extent to which groundwater affects floor heave was found to be controlled by the mineralogical and structural nature of the floor strata.

Introduction

One aspect determining the cost effectiveness of deep coal mines in the United Kingdom is the ability of longwall mining techniques to cope with anomalous, proximate geological situations. This is especially critical in our geologically complex coalfields. Detailed understanding of the effects of geological phenomena on mechanical behaviour of Coal Measures strata provides a sound basis for optimisation of tunnel layouts, support designs, and stability.

The aim of this work is to make progress towards quantifying relationships between Coal Measures geology, groundwater, and mechanical properties. The role of groundwater can then be better defined. In pursuance of this, the author was active in investigations at two collieries in South Wales coalfield from September 1979 to October 1982, forming part of continuous investigations carried out by Strata Mechanics Research Group, Cardiff.

Geological environment

Of significance to mining, coal seams in South Wales being autochthonous are usually underlain by a seatearth type rock of low strength which is often exacerbated by high water swelling capacity. Seams in South Wales coalfield are statigraphically sub-divided into Upper, Middle and Lower Coal Measures. Generally seams in the Upper Coal Measures and those is the upper part of the Middle Coal Measures are surrounded by predominantly arenaceous strata in comparison with current 'productive' seams in the lower part of the Middle Coal Measures, and the Lower Coal Measures which are generally more argillaceous and lithologically variable (George 1970). Thus mass strata permeability is considerably less in the 'productive' measures and the morphology and stratigraphic position of more permeable bodies is critical. These parameters and the fact that higher *in situ* stresses occur in the 'productive' measures due to greater burial depth, meant that increased and anomalous gateroad deformations were anticipated prior to and during investigations.

Betws Colliery, situated in the valley of the River Amman is approximately 50 km west of Taff Merthyr Colliery in the River Taff valley. At Betws 'Red Vein' anthracite in the Middle Coal Measures 137 m above the Cefn Coed Marine Band and 33–42 m below the Upper Cwmgorse Marine Band was worked from 'W' districts at approximately 600 m depth. The seam splits into three in the southwest of the Colliery, and thins towards the south-east. At Taff Merthyr the bituminous '6 Foot' seam in the Lower Coal Measures 22 m below the Amman Marine Band, was worked from 'B' districts at 450 to 600 m depth. At 15.5 m above the '7 Foot' seam lay the 'Yard' seam, and 6.2 m below lay the 'Amman Rider' seam.

Structurally, the Coalfield is a complex syncline, its trough axis striking approximately E-W, and is attributed to the Hercynian Orogeny. A simplified picture of fault trends is that the majority are normal faults, commonly displaying strikeslip, running NW and NNW, several compression faults running WNW-ESE, and three major NE-SW disturbances (George 1970; Owen & Weaver 1983). These sets often combine to form isolated mining blocks, and neutralise that part of the seams occurring within the weakened fault zones. Workings at both collieries were bounded by major faults, but joints and faults of small throw occurring within the designated working areas frequently had caused face and gateroad instability problems. Together with cleat it was clear that these discontinuities primarily controlled the transmission of groundwater between strata of mismatched permeability, and dictated the mechanical behaviour of the bridging strata.

From CRIPPS, J.C., BELL, F.G. & CULSHAW, M.G. (eds), 1986, *Groundwater in Engineering Geology,* Geological Society Engineering Geology Special Publication No. 3, pp. 115-126

Review of identified geological parameters and their effects on floor heave

It is of prime importance to review existing knowledge of the effects of discontinuity, mineralogy and groundwater occurrence to assess their roles in the context of this work.

Discontinuities

These are dominant in controlling mechanical parameter values (Platt 1957; Rowlands 1977; Briggs 1979), and in general, an increase in intensity of sedimentary discontinuities — stratification, lamination, graded bedding, cross stratification, ripple marks, and washouts — increases anisotropy and decreases mechanical parameter values (Peng 1978; Briggs 1979; Kulhawy 1980). In support of this it was reported that bearing capacity of underclay increased as the stratum thickness increased upto a maximum of 1.8 m (Rockaway 1979).

Post-depositional discontinuities are cleat, joints, faults (often slickensided) (Briggs 1979) and bedding plane shears (Salehy 1977; Jeremic 1980). Relative movement along these planes depends on surface roughness, length, orientation, tightness, infill material, and spacing. Generally conjugate sets and near vertical dipping discontinuities running near parallel to gateroads significantly destroy the integrity of and form unstable wedges in bridging beams in roof and floor. Thus floor heave can be caused by discontinuities inducing punching failure by pillars and/or supports, and by upthrusting of floor wedges due to differential lateral and vertical deformation of high clay beds beneath. Cleat in coal striking 20° with respect to gateroads with coal floor enhanced heave but otherwise was not significant (Holmes 1982). It was pointed out, however, that gateroads driven parallel to cleat and along seam strike were generally more stable (Afrouz 1975).

Cohesions in the range 0.014-0.035 MPa, obtained from twenty joint surfaces in dark grey mudstone taken from South Wales opencast sites were approximately ten times lower than those in siltier mudstones (Hassani *et al.* 1979). Sedimentary and tectonic discontinuities are frequently inter-related. For example, differential compaction faults are common in washouts, and bedding plane shears preferentially occur in mudstones and shales. Such occurrences emphasise the important role of mineralogy in determining mechanical behaviour.

Mineralogy

Coal Measures rock of South Wales predominantly comprises clays, quartz, carbonates and carbonaceous matter (Wilson 1965; Fayed 1970; Wislocki 1974; Rowlands 1977; Briggs 1979). Clay minerals present are illite and montmorillonite, often in mixed layer form, chlorite and kaolinite. Carbonaceous matter other than in coal is in the discrete form of fossil plant debris. Carbonates are those of calcium (calcite and aragonite), and iron (siderite) which is most abundant in British Coal Measures rock (Taylor & Spears 1970).

Where discontinuities are visibly absent mineralogical controls on mechanical parameter values have been identified. Direct relationships exist between mixed layer clay content and strain rate (White 1956; Reed 1972; Wood 1973; Kruszewski 1970); quartz and total carbonate content and uniaxial compressive strength, triaxial stress factor, Young's modulus, and rebound hardness (Price 1960; Taylor & Spears 1970; Barbour *et al.* 1980). It follows that inverse relationships exist between mixed layer clay content and strength parameter values (Briggs 1979; Barbour *et al.* 1979; Buffington *et al.* 1980); quartz and carbonate content, and strain rate.

The proportions of expandable clays present in Coal Measures rock directly influences water swelling ability. Montmorillonite and calcium illite are akin, and exhibit swelling due to their ability to adsorb water into their crystal lattices. Potassium illite, however, is non-swelling (Klosse 1979).

Groundwater

Typical permeabilities of mudstones and shales lie in the region of 1×10^{-8} to 5×10^{-12} mm/s measured in the laboratory, and 10^{-7} to 10^{-10} mm/s measured *in situ* (Goodman 1980). Increased *in situ* permeability is undoubtedly due to presence of discontinuities, emphasising their role as water conduits.

Groundwater geochemistry plays an important role in influencing the proportions of expandable calcium and potassium illites in argillaceous rocks (Klosse 1976). It is conceivable that groundwater charged with calcium ions percolating through fissures could significantly increase the proportion of expandable clays in proximate strata.

High natural water content of mudrocks containing mixed layer clays were responsible for low cohesion of the seam underclay and severe floor heave (Rockaway 1979). The effect of increasing the initial moisture content of mudstone from 7% to 17% had the effect of decreasing drained cohesion from 0.98 MPa to 0.19 MPa, angle of internal friction from 40° to 34° and deviator stress (Johnston 1980). Anisotropic deformation of mudrocks is exacerbated by influx of groundwater or increase in humidity. Influx increases intermolecular and pore water content, promoting early rock failure by weakening intermolecular bonds and grain boundary contacts (Murayama 1966; Van Eekhout 1976; Harper *et al.* 1979; Johnston 1980).

The laboratory time-strain response of different Coal Measures rocks in air dried and water saturated conditions was investigated (Afrouz & Harvey 1974). Applying the results to *in situ* behaviour, it was found that an increase of 1% water content in underclay produced 0.7% increase in floor heave.

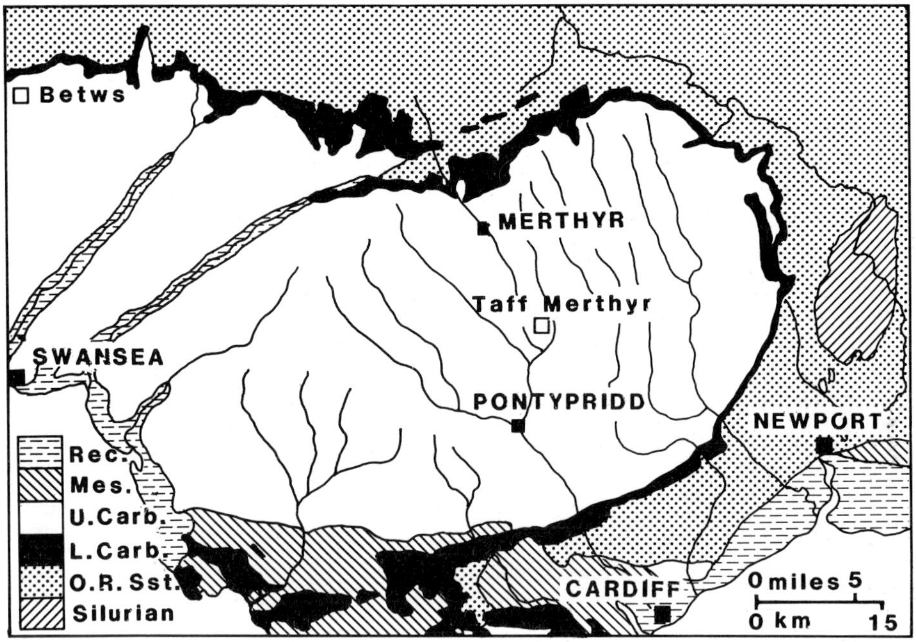

The South Wales Coalfield

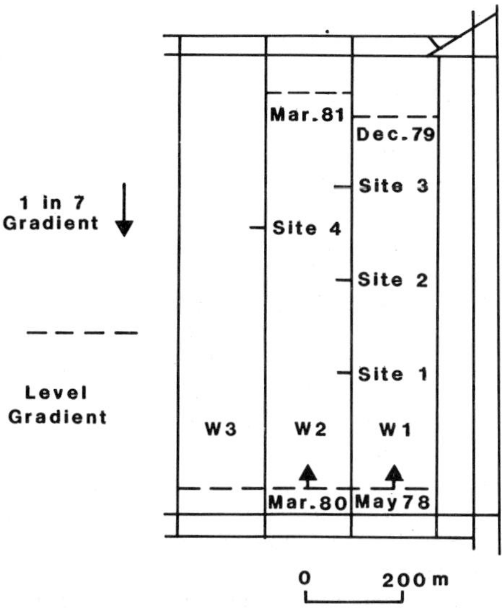

Layout of workings at 'Red Vein', Betws Colliery

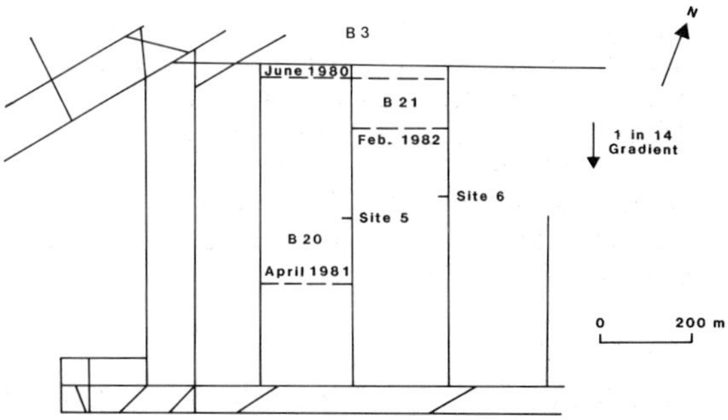

Layout of workings at '7 Foot' seam Taff Merthyr Colliery

FIG. 1. Investigation site locations

Investigation of geological parameters at 'Red Vein' and '7 Foot' seams

Investigation Methods

In situ investigations at longwall, retreating, rectangular profile gateroads, averaging a height of two metres and width of five metres, utilising pumped cement pack and roadside supports, and where rib pillars were eliminated, were carried out at both collieries. Investigation sites and locations are shown in Fig. 1.

The following, considered to be main influences upon long-term behaviour of a gateroad of given dimensions and at a give depth from surface, were regularly monitored in detail at each of the investigation sites, and on a less frequent and detailed basis along the whole gateroad length: geological features — dip of seam, cleat, fracturing, faulting, folding, lithological and mineralogical variations; groundwater conditions; efficiency of implementation, quality and performance of temporary and permanent supports; rate of face advance. This work deals only with influences on floor heave during Phase 1 deformation period, defined as occurring after gateroad creation but prior to effects caused by the front abutment pressure of the first face. Since temporary supports were essentially the same at each gateroad (rolled steel joists — RSJs — supported by hard and soft wood props), and permanent supports were not implemented, they shall not be further considered.

The full programme of *in situ* data collection techniques and procedure are shown in Fig. 2. Shortly after temporary support implementation geological mapping of seam, roof strata and exposed floor strata was carried out. Cored boreholes perpendicular to roof and floor strata to depths between eight and ten metres were complemented with channel samples collected from exposed underclay sections, and borehole flushings where cores could not be taken. Regular observations were made of water sources and quantities, run-off rates, standing water levels in boreholes and gateroads. Groundwater content was semi-quantified by putting the strata or rock cores into the following categories: air-dry (no visible water in gateroad, and cores tested when constant air-dry weight attained); saturated (standing water in gateroad, cores tested at constant saturated weight); and semi-saturated (falling somewhere in-between). No measurement of water pressure in proximate strata was possible in the investigations. The magnetic rod extensometer (Aziz 1978; Nicholas 1982), and the three dimensional rock displacement probe (Singh 1976; Smart *et al.* 1978) were used to monitor floor strata movement and provide a measure of *in situ* strain. Instrumentation implemented cored boreholes at each of the sites so that correlation of floor strata deformation with geological parameters was facilitated.

Fig. 3 shows laboratory data acquisition methods and procedure designed to produce a maximum amount of information from available rock samples. Lithological units, core quality and structure (Barton (Coordinator) 1978) were visually identified. Mineralogical content was determined using X-ray diffractometry (Briggs, 1979), and mechanical parameters - water swelling strain (Franklin (Coordinator) 1979), air-dry and water saturated elastic constants, uniaxial compressive strength, cohesion, internal friction angle, and triaxial stress factor were obtained using the unconfined compression test (Bienawski & Bernede (Coordinators) 1979), and the multiple failure state triaxial test (Vogley & Kovari 1978). The elastic modulus was taken at 50% elastic strain. Cohesion and internal friction angle were obtained using the straight line approximation. The triaxial stress factor was then calculated from these values.

Development of geological indexes

In order to assess the effect of groundwater in producing floor heave it was first necessary to quantify the individual contributions of structure and mineralogical content.

Demonstration of the effect of groundwater content on floor heave was aided by development of structural and mineralogical indexes that were readily evaluated and correlated with strata mechanics parameters.

Development of a discontinuity index

Rock Quality Designations (RQD) of AX diameter cores, taken from investigation sites, proved to be less useful than required for this work for the following reasons:

(1) Lamination in floor strata was highly variable often producing long runs where RQD was zero, thus making it impossible to distinguish one laminated section from another.

(2) Combined effect of lamination and cross-cutting fractures was not distinguished for RQD equal to zero.

(3) It was postulated that any relationship between RQD and *in situ* strain rate of floor strata would be of an inverse form, and would not comply with mineralogical and groundwater relationships being directly proportional to *in situ* strain rate.

An attainable, flexible core quality or discontinuity index was required that reflected geological structure, and that in combination with a mineralogical index could be directly correlated with *in situ* strain rate. Equation (1) represents the index, *Qs*, developed for use in laminated, jointed Coal Measures rock cores taken *in situ*, perpendicular to strata dip:

$$Qs = P_N + F_1 + ... + F_n \quad (1)$$

where *Qs* is the discontinuity index number per specified core diameter and length interval, P_N is the

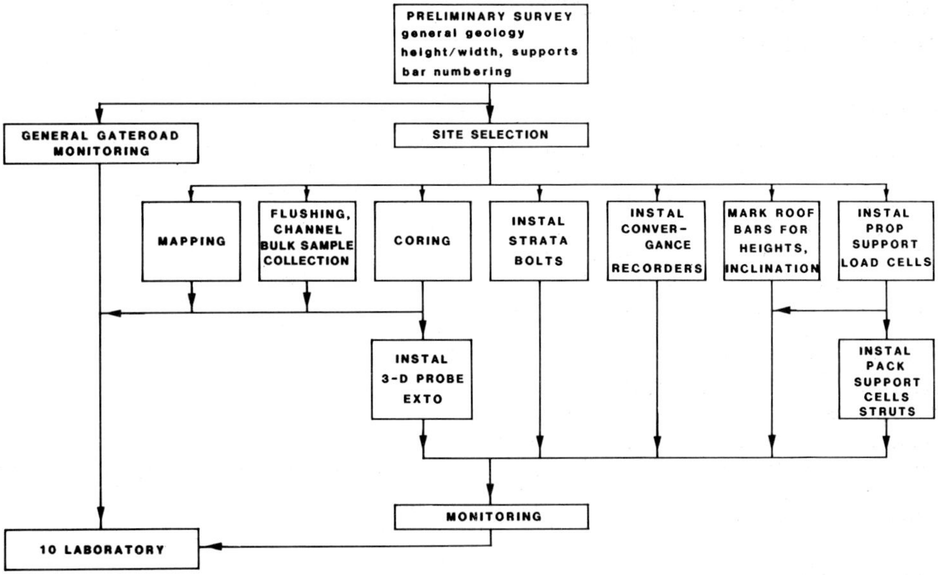

FIG. 2. Field investigation procedure.

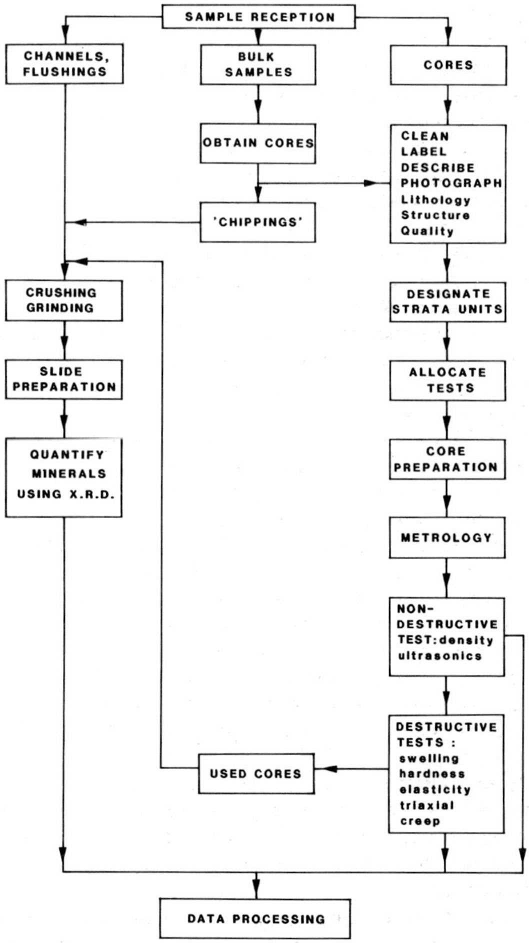

FIG. 3. Flow chart of laboratory data aquisition.

number of partings perpendicular to the core long axis, F_l is the length of influence of one inclined fracture, and F_n is the length of influence of an n th inclined fracture.

Maximum values of P_N, F_n, were governed by the AX core diameter and 0.5 m length interval adopted. The maximum possible 'density' of P_N was set at one parting plane per centimetre, facilitating visual evaluation. To comply with this F_n was also measured in whole centimetres. Evaluation of *Qs* was done upon arrival in the laboratory. All cores were subjected to approximately equal treatments (drilling and transportation methods) prior to laboratory scrutiny. The author recognises that in future investigations *Qs* should be evaluated on site.

Since intact cores were laboratory tested *Qs* was not utilised in this instance. Although the index, *Qs*, takes no account of the nature of the discontinuity surface, to a degree this is compensated for when it is incorporated with the mineralogical index, *XC*, given in Equation (2).

Development of a mineralogical index

To summarise, the following factors determined the form of the mineralogical index, *XC*:

(1) The index could be correlated with laboratory measured mechanical parameters and in combination with *Qs* could be directly correlated with *in situ* strain rate.

(2) Other than carbonaceous minerals, illite, kaolinite, quartz and siderite are the four significant minerals present in the South Wales Coal Measures (Rowlands 1977; Briggs 1979).

(3) An increase in clay mineral content produces strength parameter reduction and strain rate increase.

(4) Quartz and siderite essentially are strength promoting and strain rate reducing minerals.

(5) Inaccuracies in mineral content determination produced by using the X-ray diffraction technique were lessened by adopting the ratio form (Hart 1982).

The mineralogical index, *XC*, thus has the form:

$$XC = \frac{I + K}{Q + S} \qquad (2)$$

where I,K,Q,S are respectively crystalline illite, kaolinite, quartz and siderite contents in per cent. The form of *XC* conforms with *Qs* and groundwater category designation postulated as being directly proportional to *in situ* strain rate. It also proved adequate for correlation with laboratory determined strength parameters, postulated as being inversely related to *XC* in Equation (2).

Lithological descriptions of laboratory tested cores and flora strata are not presented, but as a guide some presentative lithologies and respective mineralogical analyses are presented in Table 1.

Although a wide variation of mineralogical contents was obtained for each type lithology, approximate relationships between lithology and *XC* were evident:

Lithology	*Range of XC*
Sandstone, siltstone	0.1 - 1.0
Silty mudstone, underclay	1.0 - 2.0
Mudstones, shales	> 2.0

Many samples had total crystalline contents less than 100%. The remainder was assumed to be taken up by amorphous minerals. Where rootlets were evident the majority of the amorphous mineral was probably carbonaceous in nature. The author has demonstrated the the degree to which lamination was developed in Coal Measures mudrocks was related to the ratio, *XA*, equal to the sum of illite, kaolinite and amorphous contents divided by the sum of quartz and siderite contents (Hart 1982). Index *XA* is not used here since it was found to be not widely applicable as compared with index, *XC*. However, lamination in mudrocks related to carbonaceous mineral content is accounted for later in the text by utilising *Qs* in the analysis of *in situ* strain.

TABLE 1. *Some mineralogical contents of type lithologies.*

Lithology	Crystalline mineral content (%)				Amorphous Content (%)	Index *XC*
	I	*K*	*Q*	*S*		
Underclay (no bedding)	45	10	35	3	7	1.45
Carbonaceous mudstone (with rootlets)	35	15	15	5	30	2.50
Uniform mudstone	42	15	18	5	20	2.49
Uniform silty mudstone	25	10	25	8	32	1.06
Uniform siltstone	1	9	40	10	41	0.18

Effect of water content on rock strength

Differences in strength parameter values resulting from laboratory tests on air dried and water saturated intact cores were quantified. Each strength parameter was correlated with index *XC*, so that 'dry' and 'saturated' filled curves were obtained. Utilising 'Minitab' (Pennsylvania State University 1981), linear regression analysis was used to curve fit the data.

Fig. 4 shows that reasonable correlation exists between swelling strain and index *XC*. The effect of saturation causes swelling strain which increases directly with relative increase in hydratable clay content. As a guide, a three metre section of strata having an average *XC* value of 2.0 could produce an increase in length perpendicular to bedding of 0.04 in less than one day. Action of such swelling would thus significantly reduce rock mass strength.

Values of 'dry' and 'saturated' strength parameters plotted against index *XC* together with fitted curves are shown in Figs 5, 6, 7, 8, 9. Fair to good correlation was attained with 'dry' and 'saturated' elastic modulus (E), internal angle friction (ϕ), and triaxial stress factor (K), 'saturated' uniaxial compressive strength (UCS) and cohesion (c). Poor correlation was identified between index *XC* and 'dry' uniaxial compressive strength, and cohesion. Possible respective explanations for this are that structural planes of weakness in the core were responsible for early failure, and in such cases index *XC* had no control, and that the straight line approximation of the Mohr envelope was not accurate enough. Relationships for these two strength parameters were included, however, to allow tentative assessments to be made. As expected the effect of saturation reduces strength parameter value for a given value of *XC*.

The reduction effect is quantified for each parameter in Fig. 10, where the per cent reduction in parameter value due to saturation is plotted against index *XC*. For a given value of index *XC* the reduction, R, in strength parameter value due to water saturation was calculated according to Equation (3):

$$R = \left(\frac{Y_d - Y_s}{Y_d} \right) \times 100\ \% \qquad (3)$$

Where Y_d, Y_s are the 'dry' and 'saturated' curve fitted strength parameter values respectively. A study of Fig. 10 reveals:-

(1) For value of index *XC* up to a maximum of about 2.0 there is a rapid reduction.
(2) For values of index *XC* between 2.0 and 7.0 there is either very little or no further reduction.
(3) minimum reductions occur where index *XC* is lowest and maximum reductions occur at high values of *XC*. There are summarised in Table 2.

Implications of these trends are that at *XC* values less than 2.0 the effect of an increase in hydratable clay content was critical, whereas above this value an increase in clay content had less dramatic effect. Hence Coal Measures rocks with *XC* values up to 2.0 (i.e. sandstone, siltstone, silty mudstone and underclay) were highly variably affected by groundwater, whereas those rocks with values greater than 2.0 (i.e. mudstone and shale) were less so.

It is clear that many more data points are required to increase confidence in all of these relationships. However, it is clear that water saturation of Coal Measures rocks had a significant weakening effect.

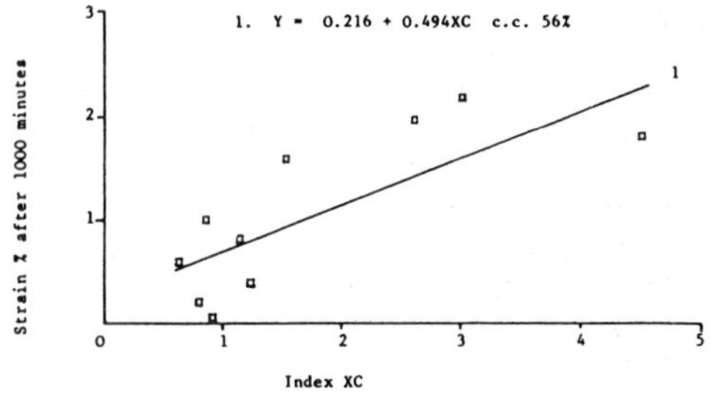

FIG. 4. Swelling strain per cent after 1000 minutes.

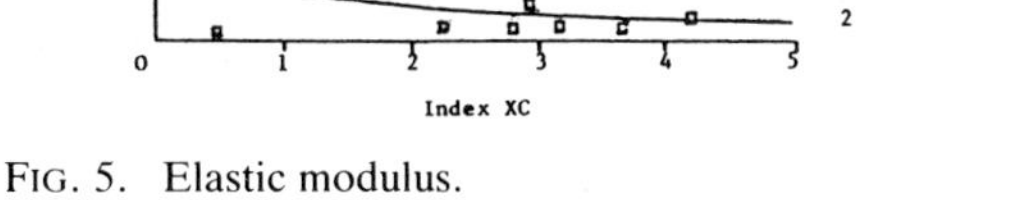

FIG. 5. Elastic modulus.

TABLE 2. *Minimum and maximum reductions in parameter values due to water saturation.*

Parameter	*XC*	minimum reduction (%)	*XC*	Maximum reduction (%)
E	0.35	54	7.0	88
UCS	0.25	38	7.0	69
K	0.25	45	7.0	63
ϕ	0.35	35	2.0	80
c	0.45	6	7.0	63

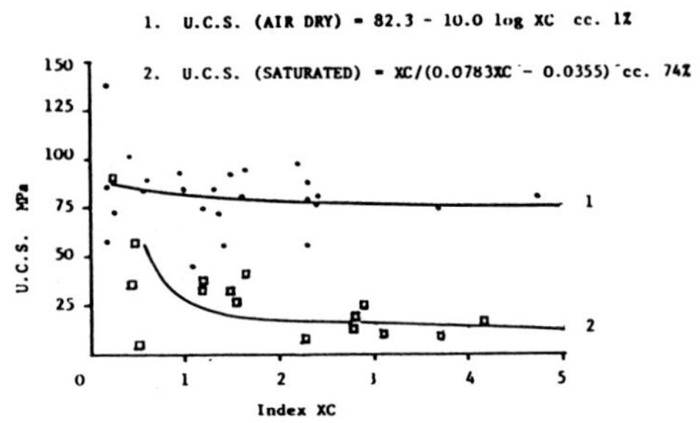

FIG. 6. Uniaxial compressive strength.

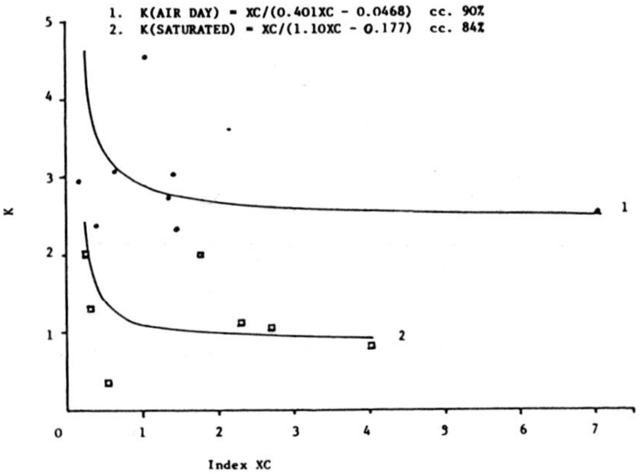

FIG. 7. Triaxial stress factor.

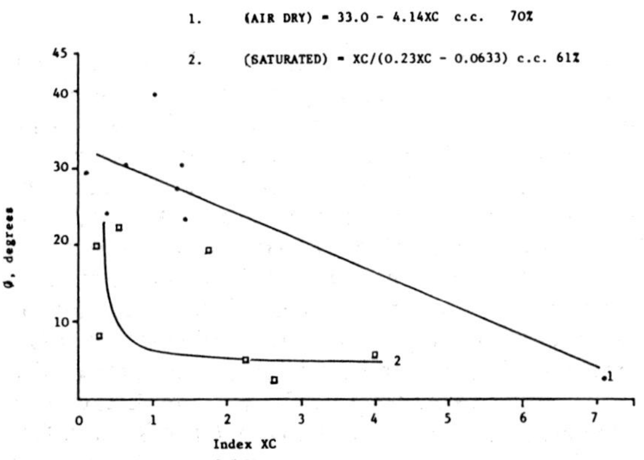

FIG. 8. Internal friction angle.

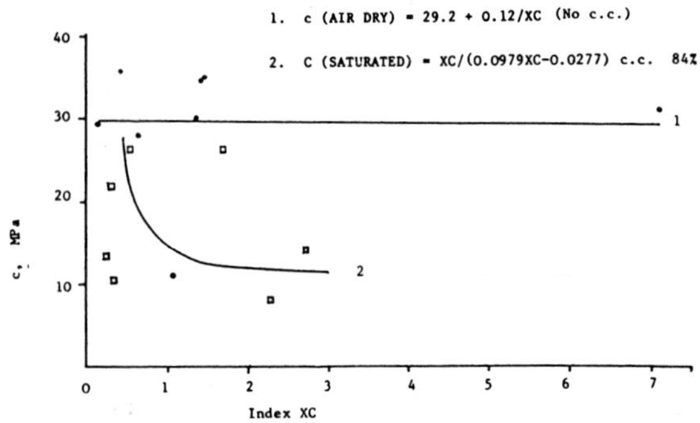

FIG. 9. Cohesion.

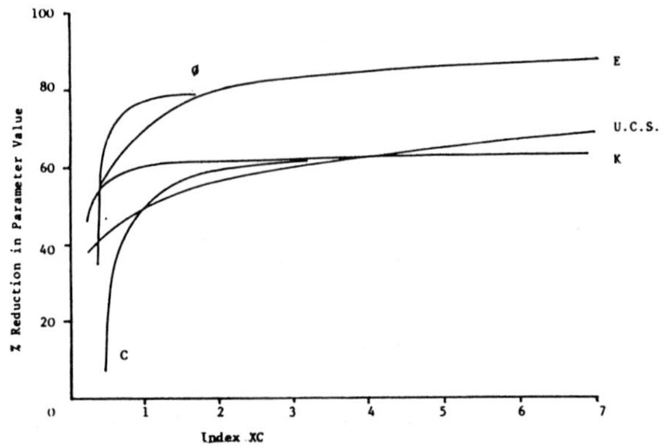

FIG. 10. Reduction in parameter values due to water saturation.

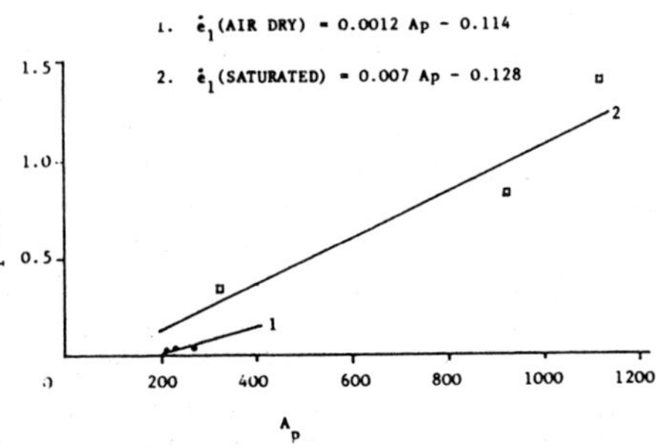

FIG. 13. Heave rate ($\dot{e}_1$) vs Ap.

1 B.W.1.1 $Y = 0.8236\,X^{1.6329}$ c.c 0.985
2 B.W.1.2 $Y = 0.4362\,X^{1.9706}$ c.c 0.977
3 B.W.1.3 $Y = 0.5409\,X^{1.9872}$ c.c 0.996
4 B.W.2.2 $Y = 0.6409\,X^{1.7219}$ c.c 0.998
5 T.B.1.1 $Y = 1.1685\,X^{1.1685}$ c.c 0.998
6 T.B.2.1 $Y = 1.5122\,X^{1.5122}$ c.c 0.989

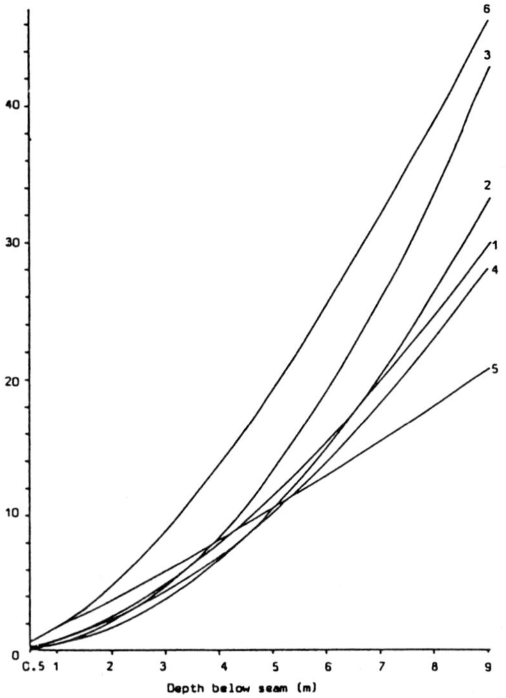

FIG. 11. Cumulative $I+K/Q+S$ ratio vs. depth 0.5m-9.0m.

1. B.W.1.1 $Y = 10.3813\,X^{1.3839}$ c.c. 0.994
2. B.W.2.2 $Y = 7.2787\,X^{1.6562}$ c.c. 0.995
5. T.B.1.1 $Y = 21{,}6809\,X^{1.3501}$ c.c. 0.993
6. T.B.2.1 $Y = 20{,}5429\,X^{1.2574}$ c.c. 0.966

N.B. Sites 3, 4 not available.
Aq values estimated from three surrounding sites.

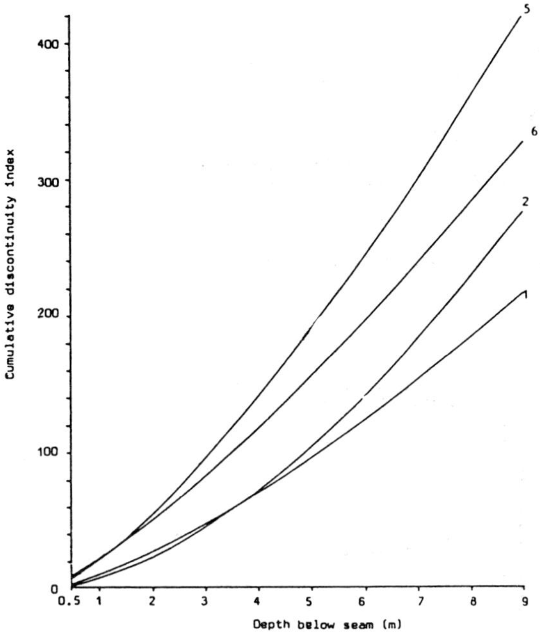

FIG. 12. Cumulative discontinuity index vs. depth 0.5-9.0 m.

Effect of groundwater on *in situ* strain rate

The aim here is to demonstrate the different effect of air dry and semi-saturated strata on *in situ* strain rate, utilising indexes *XC*, and *Qs*. Six investigation sites, 1 to 6 were available for analysis, sites 1 to 4 being at 'Red Vein' and sites 5 and 6 at '7 Foot' seam. Sites 1, 5, and 6 showed visible sources of groundwater continuously wetting the floor strata and hence were designated 'semi-saturated'. At sites 2, 3, and 4, there were no visible signs of groundwater, and were designated 'air-dry'.

In situ strain was calculated from instrumented borehole results. Movement of the top reference point near to the borehole collar, relative to the uppermost stationary reference point, was taken to be the increase in perpendicular length, l_i, in, millimetres of the strata section. The original length, l_o, was taken to be from the original position of the upper-most reference point down to the point where strata movement was not detected. All sites showed that the majority of strata movement in Phase 1 occurred in the top 3 m defining the value of l_o. *In situ* strain (e_l) and strain rate (e_l) during Phase 1 were calculated according to Equations (4) and (5):

$$e_l = \frac{l_l - l_o}{l_o} \tag{4}$$

$$\dot{e}_i = \frac{e_l}{t} \tag{5}$$

where t is duration time in days of Phase 1. At this stage it was pertinent to suggest that *in situ* strain in the 0 to 3 metre depth interval was in some way controlled by values of *Qs*, *XC*, and groundwater occurring in that interval, for a given gateroad dimension and *in situ* state of stress. Although the effect of water saturation on mechanical behaviour of cores of different rock type was clear, an assessment of this effect *in situ* was difficult since interaction of interbedded lithologies markedly influenced overall behaviour. To assess the effect of groundwater *in situ* it was first necessary to quantify the mass character of proximate floor strata in terms of indexes *XC* and *Qs*. In order to do this the following procedure was carried out for each site:

(1) So as to comply with index Qs, a weighted average of index XC was calculated, XC_1, for each 0.5 metre depth increment, according to Equation (6):

$$XC_1 = \frac{\Sigma(\text{Sample value of } XC \times \text{length of influence of sample})}{\Sigma(\text{length of influence})} \quad (6)$$

(2) The indexes XC_1 and Qs were accumulated per 0.5 metre interval starting from 0.5 metre depth below seam working downwards to a depth of 9.0 metres.
(3) Cumulative values of XC_1 and Qs were then plotted against depth in metres below seam at the mid-point of each depth increment. Points were then curve fitted, the results of which are shown in Fig. 11 and 12.

TABLE 3. *Values to show heave rate dependence on functions of mineralogy and discontinuity and groundwater condition.*

Site	Phase 1 heave rate ($\dot{e}_1$) mm/day	$Ax = \int_{0.5}^{3.0} (\Sigma XC)$	$Aq = \int_{0.5}^{3.0} (\Sigma Qs)$	Phase 1 groundwater condition	$Ap = Ax \times Aq$
1	0.33	5.59	58.92	Semi-saturated (SS)	329.5
2	0.01	3.82	55.23	Air-dry (D)	210.9
3	0.05	4.80	55.23	D	265.0
4	0.05	4.65	50.28	D	233.7
5	0.83	7.79	120.17	SS	935.5
6	1.39*	10.46	106.77	SS	1116.9

*Estimated since Phase 1 deformation was continuing.

Graphs of cumulative XC_1 (ΣXC_1) and cumulative Qs (ΣQs) versus depth below seam (d) characterise floor strata in such a way to allow the effect of interbedded lithologies on *in situ* strain to be quantified. For each site this was achieved by finding the areas Ax and Aq under their respective curves, XC_1 vs d and Qs vs d, in the 0.55 to 3.00 metre depth interval, and relating these areas to the relevant value of $\dot{e}_1$. Table 3 summarises these results, and also values for Ap equal to Ax multiplied by Aq. There being only three sites (data points) for each of the two groundwater conditions it was helpful to combine Ax and Aq and use the product as a single predictor in curve fitting $\dot{e}_1$ vs Ap values shown in Fig. 13.

A method of evaluating the effect of groundwater was to treat the air-dry and semi-saturated conditions as separate curves and quantify the difference between the two. With regard to Fig. 13 the following observations are apparent:

(1) The effect of 'semi-saturated' strata, as compared to 'air-dry' strata, was to increase heave rate considerably, the maximum increase achieved being a factor of ten.
(2) Heave rate generally increased directly with increase in values of mineralogical index and discontinuity index functions, and groundwater content.
(3) Many more data points are required to increase accuracy of and confidence in these relationships.

One factor that notably helped promote heave at site 6 was the occurrence of steeply dipping, tight and wet roof and floor strata fractures striking parallel and sub-parallel to the gateroad. In this case the Ap value quoted in Table 3 probably underestimates the heave rate. This is likely due to unaccounted for effects allowing groundwater to pervade into floor strata of high swelling capacity from a potential sandstone aquifer stratum occurring just above the immediate roof beam, and forming floor wedges unrestrained at their upper surface.

Conclusions

The following general conclusions are made:

(1) All other factors being equal, the effect the groundwater has on mechanical behaviour is directly controlled by the mineralogical and structural nature of Coal Measures rock.
(2) Where no effect of discontinuity was evident reduction of laboratory obtained strength parameter values due to water saturation was quantified aided by the mineralogical index, XC.
(3) Quantification of the increase in floor heave rate due to increase in groundwater content was attempted with some success, aided by use of mineralogical index and discontinuity index functions that, except for fracture

orientation, characterised rock masses at each of the sites. For a specific value of *Ap* the semi-saturated heave rate was increased by a maximum of a factor of ten compared to the air-dry heave rate.

(4) Heave in Phase 1 deformation period appeared to be initiated by the release of confining pressure as the gateroad was driven, allowing the water swelling effect of strata containing expansive clays to produce strain. The rate at which strain progressed in a stratum depended on its depth below the gateroad, its juxtaposition with respect to dissimilar strata, tightness, concentration, dip and strike of discontinuities, size and proximity of water sources linked to floor strata.

(5) Following the procedure presented, and using a more accurate method of measuring groundwater content, increased confidence would result by gathering of more data.

References

AFROUZ,A., HARVEY, J.M., 1974. Geology of rocks within the soft to medium range. *International Journal of Rock Mechanics and Mining Sciences* **11,** 281-290

AFROUZ, A., 1975. Floor behaviour along longwall roadways. *International Journal of Rock Mechanics and Mining Sciences,* **12,** 229-240

AZIZ, N.I., 1978. Development and application of extensometry techniques in rock mechanics investigations. *PhD. Thesis,* Univ. Wales, Cardiff.

BARTON, N. (Coordinator), 1978. Suggested methods for the quantitative description of discontinuities in rock masses. *International Journal of Rock Mechanics and Mining Science,* **15,** 319-368

BARBOUR, T.G., *et al.*, 1979. Relationship of mechanical index and mineralogic properties of Coal Measure rock. *Proceedings of the 20th US Symposium on Rock Mechanics,* June 4-6 Austin Texas, 189.

BIENAWSKI, Z.T., BERNEDE, M.J. (Coordinators), 1979. Suggested methods for determining the unconfined compressive strength and deformability of rock materials. *International Journal of Rock Mechanics and Mining Science,* **16,** 135

BRIGGS, D., 1979. An investigation of mineralogical and mechanical properties of Coal Measure rock. PhD Thesis, Univ. of Wales, Cardiff.

BUFFINGTON, D., STEPHENSON, R.W., ROCKAWAY, J.D., 1980. Residual strength of underclay. *Preprint of the 20th US Symposium on Rock Mechanics,* Univ. Missouri-Rolla.

FAYED, L., 1970. The clay minerals of some British Coal Measure rocks. *International Journal of Rock Mechanics and Mining Science* **7,** 253.

FRANKLIN, J.A., (Coordinator), 1979. Suggested methods for determining water content, porosity, density, absorption and related properties, and swelling and slake durability index properties. *International Journal of Rock Mechanicals and Mining Science,* **16,** No. 2, 141-156

GEORGE, T.N., 1970. *British Regional Geology, South Wales* (3rd Edn.). Natural Environment Research Council - Institute of Geological Sciences, London, HMSO.

GOODMAN, R.E., 1980. *Introduction to Rock Mechanics.* John Wiley and Sons, New York.

HARPER, T.R., APPEL, G., PENDLETON, M.W., SZYMANSKI, J.S. & TAYLOR, R.K., 1979. Swelling strain development in sedimentary rock in northern New York. *International Journal of Rock Mechanics and Mining Science,* **16,** 271-292

HART, P.A., 1982. A study of parameters influencing floor deformation in tunnels serving longwall coalfaces. *PhD Thesis,* Univ. Wales, Cardiff.

HASSANI, F.P., WHITTAKER, B.N., SCOBLE, M.J., 1979. Strength characteristics of rocks associated with opencast coal mining in the UK. *Proceedings of the 20th US Symposium of Rock Mechanics, Austin, Texas,* 347-355

HOLMES, P., 1982. Relations between geology and the stability of faces and roadways in the Barnsley seam. *Symposium on Strata mechanics,* Univ. Newcastle-upon-Tyne, 8-112

JEREMIC, M.L., 1980. Influence of shear deformation structure in coal on selecting methods of mining. *Rock Mechanics* **13,** 23-38

JOHNSTON, I., 1980. Properties of soft rock relevant to socketed pile design. *International Conference on Structural Foundations on Rock,* Sydney.

KLOSSE, 1967. In Kratzsch, H., 1983. *Mining subsidence engineering.* Springer-Verlag, Berlin,

KULHAWY, F., & GOODMAN, R.E., 1980. Design of foundations on discontinuous rock. *International Conference on Structural Foundations on Rock,* Sydney, 209.

KRUSZEWSKI, T., 1970. Influence of mineral composition of clay rocks on stability of slopes from lignite pits in Poland. *Proceedings of the 2nd International Congress of the International Association of Engineering Geologists,* Sao Paulo. V1.1-V1.6.

MURAYAMA, S., 1966. Swelling of mudstone due to sucking of water. *Proceedings of the 1st Congress of International Society of Rock Mechanics,* Lisbon, 495.

NICHOLAS, P.K., 1982. An investigation of yield zones surrounding coal mine excavations. *PhD Thesis,* Univ. Wales, Cardiff.

OWEN, T.R., WEAVER, J.D., 1983. The structure of the main South Wales coalfield and its margins. In: HANCOCK, P.L., (ed.). *The Variscan Fold Belt in the British Isles.* Adam Hilger Ltd, Bristol, 74-87.

PENG, S.S., 1978. *Coal Mine ground control.* John Wiley and Sons,

PLATT, J., 1957. Underclays of the South Wales coalfield and their influence on floor penetration. *PhD Thesis,* Univ. of Wales, Cardiff.

PRICE, N.J., 1960. The compressive strength of Coal Measure rocks. *Colliery Engineering,* July, 283.

REED, M, 1972. *MSc Thesis,* Univ. Newcastle-upon-Tyne. Data extracted from ROWLANDS.

ROCKAWAY, D.J., 1979. Investigation of the effects of weak floor conditions on the stability of coal pillars. U.S.B.M., O.F.R. (final), 15 July, 237.

ROWLANDS, N., 1977. An examination into mineralogical and mechanical properties of Coal Measures strata. *PhD Thesis,* Univ. of Wales, Cardiff.

SALEHY, M.R., MONEY, M.S. & DEARMAN, W.R., 1977. The occurrence and engineering properties of intraformational shears in Carboniferous rocks. *Proceedings of Rock Engineering,* Univ. of Newcastle-upon Tyne, 311.

SINGH, R.N., 1976. Measurement and analysis of strata deformation around mining excavations. *PhD Thesis,* Univ. Wales, Cardiff.

SMART, B.G.D., SINGH, R.N., ISAAC, A.K., 1978. A borehole instrumentation system for monitoring strata displacement in three dimensions. *International Journal of Rock Mechanics and Mineral Sciences,* **15,** 77-85.

TAYLOR, R.K., SPEARS, D.A., 1970. The breakdown of British Coal measure Rocks. *International Journal of Rock Mechanics and Mining Sciences,* **7,** 481-501.

VAN EEKHOUT, E.M., 1976. The mechanisms of strength reduction due to moisture in coal mine shales. *International Journal of Rock Mechanics and Mineral Sciences,* **13,** 61-67.

VOGLER, U.W., KOVARI, K., 1978. Suggested methods for determining the strength of rock materials in triaxial compression. *International Journal of Rock Mechanics and Mining Sciences,* **15,** No. 2, 47-51.

WHITE, A.W., 1956. Underclay squeezes in coal mines. *Transactions of the American Institute of Mining Engineers,* **205,** 1024-1028.

WILSON, M.J., 1965. The underclays of the South Wales Coalfield east of the Vale of neath. *The Mining Engineer,* April, 389-404.

WILSON, M.J., 1965. The origin and geological significance of the South Wales underclays. *Journal of Sedimenter Petrology,* **35,** No. 1, 91-99.

WISLOCKI, A.P., 1974. The mineralogical composition of the floor of the 6 Foot seams of Windsor Colliery, South Wales, *BSc Thesis,* Univ. Wales, Cardiff.

WOOD, P., 1973. *Msc Thesis*, Univ. Newcastle-upon-Tyne. Data extracted from ROWLANDS.

P.A. HART, Portsmouth Polytechnic, Department of Geology, Burnaby Building, Burnaby Road, Portsmouth PO1 3QL, UK.

The effect of groundwater flow on the strength and stability of silicate grouted soils

J.N. Shirlaw

ABSTRACT: A considerable volume of data on the strength of silicate grouted soils has been published in recent years. In particular there has been considerable discussion of the confined and unconfined strengths of grouted sands and the effect of creep on those strengths. Little has been said on the effect of groundwater seepage on the strength and residual permeability of the sand/silicate system.

Following observations on site into the effectiveness of silicate grouting, a small laboratory programme was set up to examine the effect of seepage on strength and residual permeability. The findings of the programme are summarised.

Introduction

Chemical grouts have been used to seal and consolidate sands for many years. The design of the zones to be grouted has so far been largely empirical. However computer and other techniques can now be used in analysis. Thus Tan & Clough (1980) have already used finite element analysis to produce specific design charts for treatment zones around tunnels.

In order to use numerical techniques, it is necessary to know what values to input for grouted sands. A considerable number of papers have been published on the behaviour of grouts, particularly silicate grouts, in recent years. One massive instance has been the three volume work by Baker *et al.* (1983). Considerable attention has also been paid to the creep behaviour of grouts, such as in the work of Borden *et al.* (1982).

One aspect of silicate grouting which has not been extensively documented is the effect of groundwater on the strength of the grouted sand mass. Although Hewlett & Hutchinson (1983) have published at least one paper dealing with the effects of groundwater flow in grouted sands, this was concerned mainly with dams, using high hydraulic gradients and measuring the long term dissolution of the grouts, and the resultant effect on permeability.

Grouting for tunnels and shafts presents an essentially different problem from that of dams. Hydraulic gradients through the treated mass are generally small, and the grouting is normally only required for the very short term between excavation and the introduction of effective support. For tunnels and shafts, the strength of the grouted sand mass is of equal interest to its permeability.

The problem

To study the effects of groundwater or silicate grouted sands, it is convenient to consider two aspects of the problem separately: grouting for tunnels and shafts is often carried out from the surface, before work commences. This allows tunnelling or sinking to proceed without intermittent delays to carry out grouting from within the tunnel or shaft. After grouting from the surface, it may take several months before any particular section of grouted sand is excavated. The effect on strength of this delay is of considerable interest.

Secondly, during excavation the difference in hydraulic head across the grouted zone will set up a flow, the size of which will depend on the residual permeability of the grouted sand. This flow could start before the face reaches any particular point (see Fig. 1). The flow may affect the strength of the sand/grout mass.

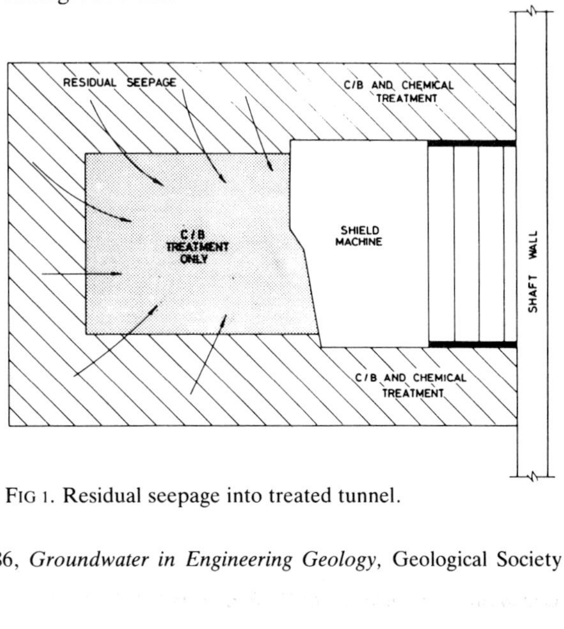

FIG 1. Residual seepage into treated tunnel.

From CRIPPS, J.C., BELL, F.G. & CULSHAW, M.G. (eds), 1986, *Groundwater in Engineering Geology*, Geological Society Engineering Geology Special Publication No. 3, pp.127-130

Sample preparation

In order to study the problems outlined above, the first stage was to construct a grout infusion apparatus. The apparatus allowed a column of saturated sand within a plastic tube to be grouted under a controlled pressure, and was similar to those described elsewhere (Rhone Poulenc).

Various factors have a significant effect on the unconfined compressive strength of the resulting sample. To get true comparability of results these factors have be standardised. The factors are extensively discussed by Baker *et al.* (1983), and it is sufficient to give the specification used in this study.

Size of sample: 40 mm diameter.
Length of plastic tube: 1 m.
Sand type: China river sand.
Dry Density of sand: 1.475 ± 2%. Mg/m^3.
Volume of grout used: 2 × theoretical.
Length of sample for compression test: 80 mm.
Rate of Strain: 1 mm/minute.
Grout: 250 silicate/55 Kg Organic reagent/water to make 1 m^3.
Silicate: 35-37° Baume, Molecular ratio 1:3.3.

Long-term strength testing

For long term strength testing a 0.15 mm – 0.425 mm sand was used. The prepared 1 m sample was cut into 80 mm lengths, stripped of the PVC tube, and immersed in a sand bath. A continuous, slow supply of water was provided at one end of the sand bath, with an outlet at the other end.

At intervals the samples were removed for unconfined compression testing.

Short-Term Stability Testing

Five samples were prepared using the following sand gradings: 0.15–0.212 mm, 0.212–0.3 mm, 0.3–0.425 mm, 0.425–0.6 mm and 0.6–1.18 mm. For 24 hours following infusion the grouted sand tube was left in the apparatus. The tube was then capped. After seven days two samples were cut. The first was removed from its PVC tube and the unconfined compressive strength measured.

The second sample was kept in its PVC tube, and placed in a simple constant head permeability apparatus. Tap water was allowed to flow through the sample at a hydraulic gradient of about 13. The volume and pH of the permeant were measured at regular intervals. After a set period (generally 7 days) the sample was removed from the PVC tube and the unconfined compressive strength measured.

Test result

The results of the long term testing are shown in Fig. 2. Fig.3 shows the results of standard (7-day cured)

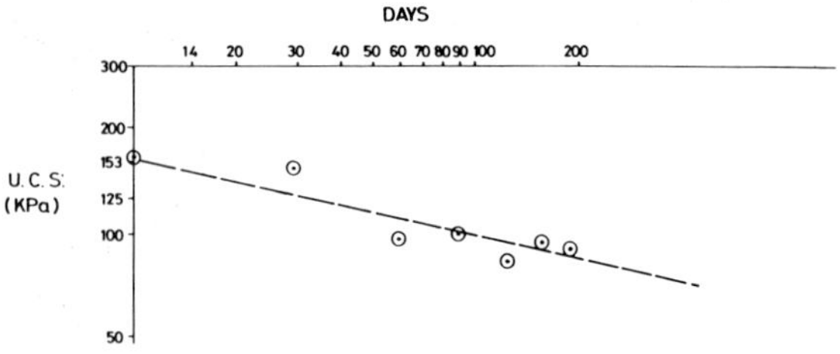

FIG. 2. Unconfined compressive strength (UCS)/time.

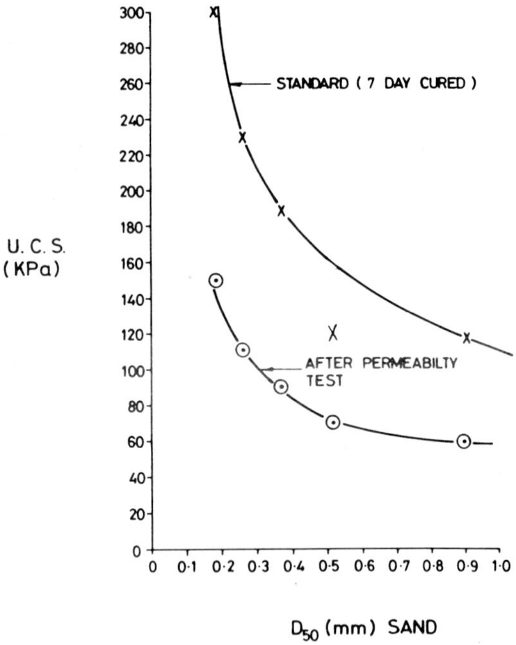

FIG. 3. Unconfined strength before/after permeability test.

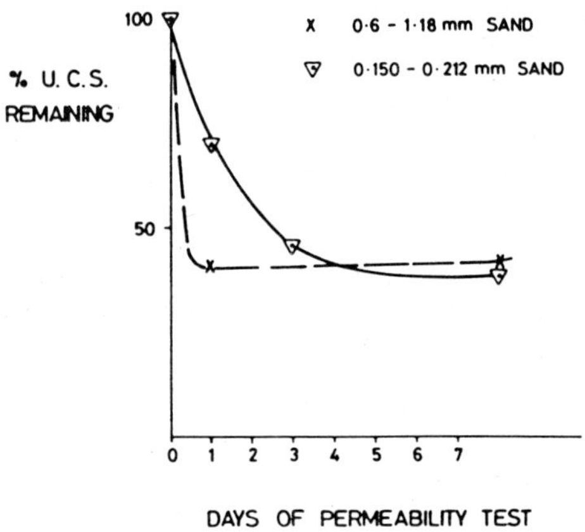

FIG. 4. Strength/number of days of permeability test.

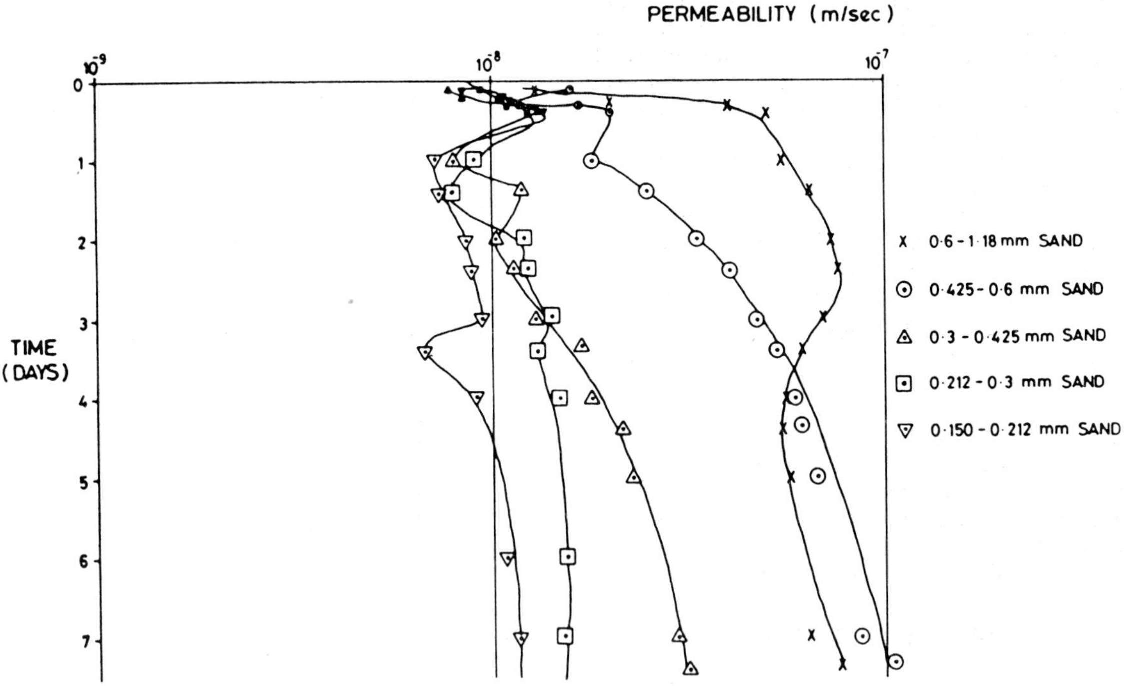

FIG. 5. Permeability/time.

result and the 7-day permeability test result. Strength tests were also carried out after 1 and 3 days of permeability test, and gave the results shown in Fig. 4. The change in measured permeability of the sand/gel sample during the seven day test is shown in Fig. 5.

Discussion

The long-term loss of strength shown in Fig. 2 is at variance to the increase in strength with time recorded by Clough *et al.* (1979), but this can be ascribed to the different method of curing.

Fig. 3 shows that there is a consistent loss of about 50% of strength following the 7-day permeability test. This percentage is independent of the grain size of the sand used. Fig. 4 implies that the loss occurs much more quickly in coarse sands than in fine sands. Fig. 4 also shows that the strength stabilises fairly rapidly, so this initial loss must be due to a different effect from that investigated by Hewlett & Hutchinson (1983), who looked at long term dissolution. Rhone Poulenc have suggested that the effect is due to unneutralised soda which reverses the silicate gel. As the unneutralised soda is flushed from the sample, so the effect ceases. Rhone Poulenc further suggest that in a column of grouted sand the grout undergoes successive dissolution and regellation. This may account for the fairly stable permeability over the 7-day test on grouted fine sand. However, Fig. 4 and 5 would suggest that in grouted coarse sand the unneutralised soda and dissolved get are flushed through the sample very quickly. The effect would probably have been slightly more pronounced if distilled water had been used as the permeant, as the mild acidity of chlorinated water would have tended to neutralise some of the otherwise unneutralised soda.

The study has identified a loss of strength in grouted sand samples both due to submergence in the long term and having a hydraulic gradient applied in the short term. In the time available it was not possible to check whether these phenomena were separate, or different aspects of the same underlying effect. Nor was it possible to check whether the creep behaviour is affected by the short term flow. It was recorded by Borden *et al.* (1983) that creep tests on grouted sands resulted in failure at between 30% and 50% of the 'quick' unconfined compression test result. Luong *et al.* (1976) have shown that this percentage is a function of the degree of neutralisation of the silicate. As the effect of the hydraulic gradient is to first remove the unneutralised soda, it is possible to speculate that the loss of 'quick loading' strength will be compensated to some extent by better creep behaviour.

Conclusions

This has been a very simple study, involving relatively few tests, only one grout mix, and very small samples. It is dangerous, therefore, to attempt to derive any universal conclusions on the behaviour of large grouted masses.

Nevertheless, the study does show that in choosing a suitable silicate grout, particularly in medium/coarse sands, the presence or absence of residual seepage must be taken into account. Thus for a tunnel constructed in grouted sands even a low compressed air pressure will reduce the necessary grout strength, both due to its support value and in the reduction of seepage and consequent effect on strength.

References

TAN, D.Y. & CLOUGH, G.W., 1980. Ground Control for Shallow Tunnels by Soil Grouting *Proceedings of the American Society of Civil Engineers, Journal of the Geotechnical Engineering Division,* 1037-1057.

BAKER, W.H., WALLER, M.H., HUCK, P.J. & KRIZEK, R.J., 1983 Design and Control of Chemical Grouting Report No. FHWA/RD-82/038 publ. U.S. Department of Commerce, National Technical Information Service.

BORDEN, R.H., KRIZEK, R.J. & BAKER, W.H., 1982. Creep Behaviour of Silicate Grouted Sand. *ASCE Proceedings of the Speciality Conference on Grouting in Geotechnical Engineering,* New Orleans, 450-469.

HEWLETT, P.C. & HUTCHINSON, M.T., 1983. Quantify Chemical Grout Performance and Potential Toxicity. *Proceedings of the 8th European Conference on Soil Mechanics & Foundation Engineering.* **1,** 361-366.

RHONE-POULENC. 'Hardener 600 Series' Handbook, Paris.

LUONG, M.P., GANDAIS, M. & ALLEMOND, P. 1977. Comportement Mechanique dessols lujectes aux produits Chimiques. *Annales de L'institut du Batiment et des Travaux Publics, Sols et Foundation* No. 145.

J.N. SHIRLAW (until recently with Hong Kong Mass Transit Railway), c/o Mass Rapid Transit Corporation, 25K Paterson Road, Singapore 0923.

Groundwater in Engineering Geology, London, 1986, pp. 131-138 Printed in England

Influence of water on engineering properties of weathered rocks

N. Turk & W.R. Dearman

ABSTRACT: The influence of groundwater on the engineering properties of rocks is well known. Because of its importance, rock mass classification methods include groundwater as one of the main classification parameters. Perhaps the best indication of the influence of groundwater on rock mass properties comes from the application of geophysical resistivity and seismic refraction techniques to investigate groundwater. When ground conditions change from an unsaturated to a saturated state, resistivity and velocity of the rock mass change markedly.

In nature, groundwater affects rock properties either by changes in pore water pressure or by chemical reaction. While increase in pore water pressure decreases the total normal stress in the rock mass, chemical reactions cause rocks to weather resulting in changes in physical and mechanical properties. Pore water pressure is best evaluated in the rock mass by *in situ* measurements. However, the properties of weathered rocks are very often best studied in the laboratory. Over the last decade the influence of weathering on the engineering properties of mainly igneous and sedimentary rocks, under dry and saturated conditions, has been investigated. A great deal of test data has been accumulated on the index and engineering properties of fresh and weathered rocks from the UK and abroad. Analysis of the available test data shows that while certain rock properties increase with weathering and saturation, others decrease.

In this paper changes with saturation in selected index and engineering properties of fresh and weathered porphyrites and andesites from the UK and Turkey are reported. Established relations between the dry to saturation index and engineering properties of weathered rocks will also be presented.

Introduction

The presence of water in rocks affects their physical properties, and in nature the degree of saturation will depend on position in the rock mass relative to the groundwater table. The position of the watertable, and moisture content in the overlying vadose zone both vary seasonally. Consequently the physical properties of rock material can be expected to fluctuate. In addition to this localised variability on the scale of an individual sample, within the rock mass itself there will be additional variation related to different degrees of weathering and alteration, and of fracturing.

In order to predict the engineering behaviour of structures built in or on a rock mass, the distribution of zones of equal rock mass quality and similar groundwater conditions should be known. Because of its importance, proposed rock mass classification methods include groundwater as one of the main classification parameters (Bieniawski 1973; Barton *et al.* 1974).

In nature, groundwater affects rock properties either by changes in pore water or by chemical reaction. While increase in pore pressure decreases the total normal stress in the rock mass, chemical reactions cause rocks to weather resulting in changes in physical and mechanical properties. Pore water pressure is best evaluated in the rock mass by *in situ* measurements. However, the properties of weathered rocks are best studied in the laboratory.

In the past research effort has been directed at the influence of water on the compressive strength of rocks. The seminal paper in this respect is that of Colback & Wiid (1965) who investigated the effect of a controlled degree of saturation of fresh rocks on their compressive strength. They demonstrated that as moisture content increases, compressive strength decreases in the fine-grained quatzitic shale associated with gold reefs of the Witwastersrand System in South Africa. This shale had a porosity of 0.28%. A parallel series of tests on a Coal Measures quartzitic sandstone, also from South Africa, showed the same trend. In this case porosity was 15%.

Before this time Kessler *et al.* (1940) had reported the effect of saturation on the compressive strength of a great number of American granites. Obert *et al.* (1946), Price (1960), Kjaernsli & Sande (1963), Zaruba (1965), and Salustowitz (1965) all report a decrease in compressive strength of igneous, metamorphic and sedimentary rocks on saturation with water. Results are frequently quoted for tests carried out at natural moisture content, air or oven dried, or saturated.

As far as can be judged from the descriptions, all of the rock types tested, mainly sedimentary, were fresh. Where description of some mineral constituents is mentioned, as or example in Kessler *et al.* (1940), no

From Cripps, J.C., Bell, F.G. & Culshaw, M.G., (eds), 1986, *Groundwater in Engineering Geology*, Geological Society Engineering Geology Special Publication No. 3, pp. 131-138

attempt has been made to quantify the degree of alteration in rocks that were commercially acceptable as building stone. The same comment applies to most of the work published subsequently, for example Parate (1973) on limestone, Ballivy *et al.* (1976) on sandstone, limestone and charnockite, Eeckhout (1976) on Coal Measures shales.

Over the last decade the influence of weathering on the engineering properties of mainly igneous and sedimentary rocks, under dry and saturated conditions, has been investigated by the authors. A great deal of test data has been accumulated on the index and engineering properties of fresh and weathered rocks from the UK and abroad. Analysis of the available test data shows that while the test values of certain rock properties increase with weathering and saturation, some others decrease.

In this paper changes in selected index and engineering properties of weathered rocks with saturation are reported.

Weathering of rocks

Weathering as a natural process may be either mechanical or chemical or a combination of the two. Mechanical weathering is dominated by physical breakdown with opening of discontinuities, the development of new discontinuities and the opening of mineral grain boundaries and cleavages. There is none of the decomposition that dominates chemical weathering which leads through stages of discolouration of the original fresh rock to decomposition of silicate minerals to clay minerals with attendant opening of grain boundaries. Chemical weathering leads to changes in colour, strength and porosity, but the original rock mineral texture is largely retained until the final stages when a residual soil is formed.

The rocks described in this paper have mainly been subjected to chemical weathering, or its analogue hydrothermal alteration.

Effect of water saturation on the properties of fresh rock materials

Parameters that are of importance in design in rock are density and porosity, strength and deformability. Strength may be unconfined compressive strength or point load index, and from triaxial testing c' and ϕ' may be determined. Deformability, as Young's modulus and Poisson's ratio, may be determined directly in the laboratory or estimated from other properties.

Mann & Fatt (1960) determined bulk compressibility, Young's modulus, and Poisson's ratio for three fresh sandstones in both oven dried and saturated cylinders. Bulk compressibility of wet sandstone was 10 to 30% greater than for dry. Young's modulus was 8 to 20% less for wet sandstone, and Poisson's ratio was 100% greater on one type of sandstone when wet, or about the same for wet samples of the others (Table 1).

In the same year, Price (1960) tested different sandstones at different moisture contents and showed that, for rocks with different porosities, the saturated strength was 45% of the completely dry unconfined compressive strength (Table 2).

For the effect of water on cohesion and friction angle (triaxial strength parameters) results for sandstones are provided by Coldback & Wiid (1965).

Effect of water saturation on the properties of weathered andesites

A range of andesites and porphyrites from Turkey and the UK has been examined in the laboratory using standard techniques (Brown 1981) as far as applicable. Laboratory specimens were 38 mm diameter and 76 mm long cylinders.

The UK samples were all from working or recently abandoned aggregate producing quarries, and the rock material was fresh. In contrast, andesite samples from Turkey showed a range of discolouration and alteration, and even the freshest samples had been weathered or altered to a noticeable degree. Combining the two groups of samples provided a range of weathering stages of rock material: fresh, slightly discoloured, and completely discoloured. As discolouration progressed the rocks became more altered to clay minerals but because of the very fine-grained nature of the groundmass conventional thin-section study was not appropriate and estimates of the type of alteration were assessed using XRD techniques. The resulting clay minerals were mixed-layer with montmorillonite. Increase in degree of alteration could be judged by colour and determination of unit weight and porosity.

Location of Samples

In the UK:

Ba.I	Bardon Hill Quarry, Leicestershire.
E.I	Enderby Warren Quarry, Leicestershire.
P.I	Pennycumquick, nr Burnmouth, Berwickshire.
H.I	Harden Quarry, Northumberland.
T.I	Threlkeld Quarry, Cumbria.
Bo.I	Bonawe Quarry, Argyllshire.
C.I	Cloburn Quarry, Lanarkshire.
A.I	Arenig Quarry, Bala, Wales.

In Turkey:

Bor.T1	Bornova, Izmir.
Bor.T2	Kestel Dam, Bergama.

TABLE 1. *Effect of water saturation on deformation properties of sandstones (Mann & Fatt 1960).*

	Young's modulus E (GPa)	Poisson's ratio ν	Bulk modulus K (GPa)
Boise — Dry	19.650 ± 0.000	0.264 ± 0.012	4.985 ± 0.365
No. of runs	4	4	7
Boise — Wet	16.685 ± 0.000	0.255 ± 0.013	5.978 ± 0.779
No. of runs	4	4	10
Berea — Dry	15.980 ± 0.138	0.242 ± 0.010	7.308 ± 0.703
No. of runs	8	8	8
Berea — Wet	18.340 ± 0.586	0.277 ± 0.016	7.860 ± 0.758
No. of runs	8	8	10
Bandera — Dry	12.686 ± 1.172	0.174 ± 0.021	5.764 ± 0.152
No. of runs	7	7	6
Bandera — Wet	10.273 ± 0.069	0.387 ± 0.007	7.274 ± 0.448
No. of runs	8	8	10

TABLE. 2. *Effect of moisture on compressive strength of sandstones (Price 1960).*

Type of Sandstone	Porosity	Strength		
		Oven dry MPa	Air dry	Saturated
Markham	16.0	190.98 (100%)	57%	-
Parkgate	10.0	164.09 (100%)	68%	45%
Pennant	2.5	268.98 (100%)	51%	45%
Darley Dale	19.5	80.67 (100%)	80%	45%

TABLE. 3. *Weathering grade in terms of porosity and density variations.*

Weathering grade	Porosity %	Density g/cm^3
Fresh	less than 3	more than 2.5
Slightly weathered	3 - 6	2.5 - 2.4
Moderately weathered	6 - 12	2.4 - 2.3
Highly weathered	more than 12	less than 2.3

Estimation of degree of weathering of the rock material

In the field, there is no difficulty in categorizing samples into a progressive weathering suite on the basis of qualitative, and to some extent subjective, parameters. Subsequent laboratory determinations of physical properties have confirmed the reality of the weathering sequence established in the field. The simplest diagnostic characteristics prove to be porosity and density, and an approximate classification of material weathering grades is given in Table 3.

Types of tests reported

The following standard tests were carried out on all samples in the oven dry and water saturated states:
Bulk density
Porosity
Ultrasonic velocity
Uniaxial compressive strength
Deformation properties (dry only)
Young's modulus
Poisson's ratio
Triaxial strength parameters
Cohesion (c)
Friction angle (ϕ)

Effect of saturation on individual properties

(a) Saturated density against dry density
A curvilinear relation has been obtained between saturation density (γ_s) and dry density (γ_d) for all rock types tested, both fresh from the UK and weathered from Turkey (Fig.1).

$$\gamma_s = 1.23\ \gamma_d^{\ 0.79} \quad (r=0.99;\ n=20)$$

(b) Saturated against dry ultrasonic velocity
The following curvilinear relation was obtained between saturation (V_s) and (V_d) dry ultrasonic velocity;

$$V_s = 3.0\ V_d^{\ 0.87} \quad (r=0.98;\ n=20)$$

and the saturation velocities are slightly higher than the equivalent dry velocities (Fig.2).

(c) Unconfined compressive strength.
There is a curvilinear relation throughout the weathering grades tested. Saturated strength (σ_s) is lower than the equivalent dry strength (σ_d). The relation is:

$$\sigma_s = 0.214\ \sigma_d^{\ 1.26} \quad (r=0.96;\ n=17)$$

and it will be noted that dry strengths are about 25% higher than the equivalent saturated strengths (Fig. 3).

Correlations between different properties and porosity

Physical and mechanical properties determined in the laboratory include porosity, density, ultrasonic velocity and unconfined compressive strength. Curves are presented for these parameters over the whole range of UK and Turkish rocks (Figs 4, 5, 6). Relations are best represented by negative exponential lines and power law type curves. If it is accepted that increase in porosity represents an increase in weathering 'grade', then density, sonic velocity and unconfined compressive strength all decrease systematically with increase in weathering grade.

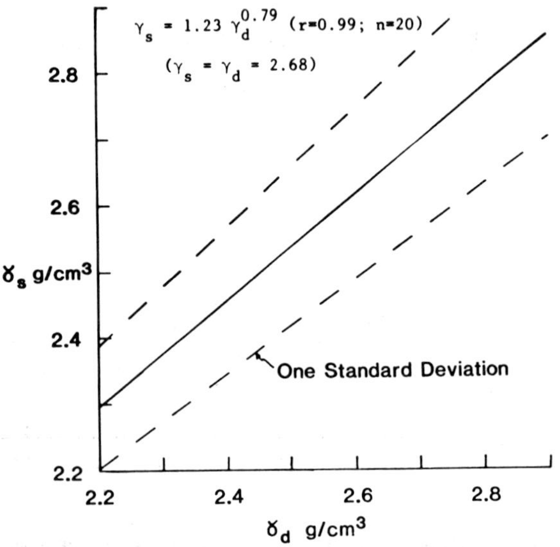

FIG. 1. Relation between dry and saturated densities of fresh porphyrites and weathered andesites.

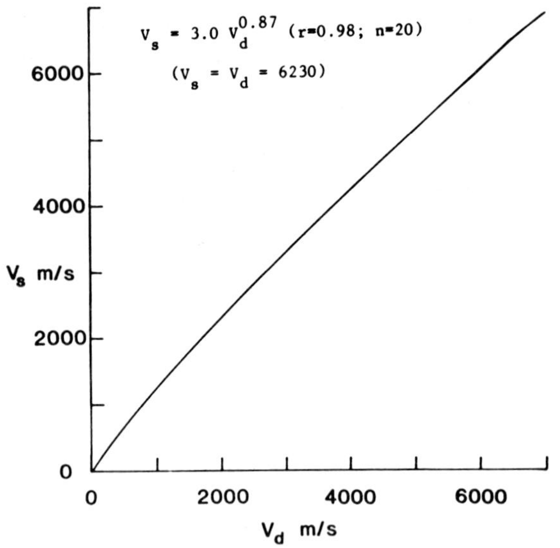

FIG. 2. Relation between dry and saturated ultrasonic velocities of fresh porphyrites and weathered andesites.

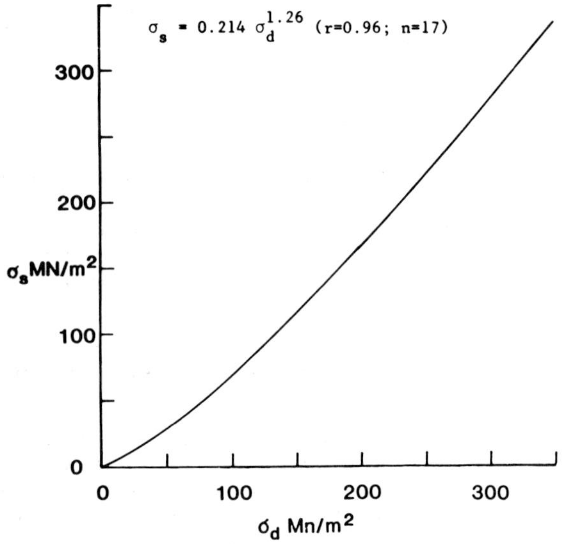

FIG. 3. Relation between dry and saturated uniaxial compressive strength of fresh porphyrites and weathered andesites.

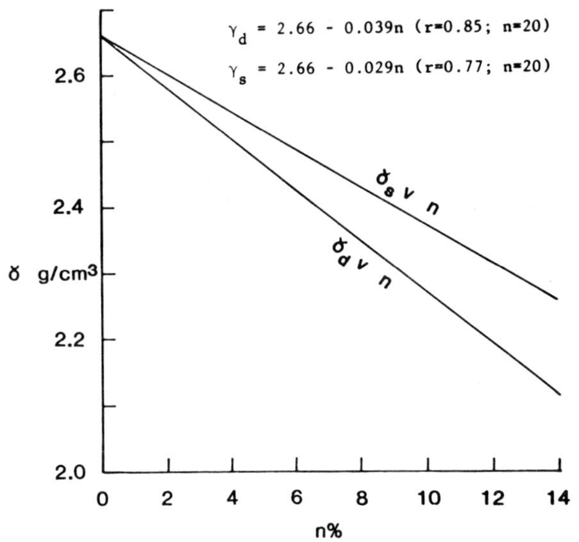

FIG. 4. Relations between dry and saturated densities and porosity of fresh porphyrites and weathered andesites.

It is also clear from the graph of porosity against density that, as expected, with increase in porosity there is a progressive increase in the difference between dry and saturated density (i.e. $\gamma_s - \gamma_d = n\%$).

For both sonic velocity and strength, for porosities greater than about 2% the differences between values for the dry and saturated states is constant. Dry strength is greater than saturated strength, and both decrease with increase in porosity (Fig. 6). In contrast, dry sonic velocity is less than saturated sonic velocity, but both decrease with increase in porosity (Fig. 5).

Other correlations between different properties and density

Another indicator of weathering 'grade' is change in density. In Fig. 7 strength is plotted against density, and the effect of saturation in decreasing strength is apparent over the whole density range. As densities increase, that is, the rock material becomes fresher, the two curves tend to converge, and the difference between dry and saturated strengths decreases and becomes very small.

Over the limited range of densities in the weathered rock suites tested, the results for sonic velocity are unexpected. Best fit curves for both saturated and dry determination are curves that cross (Fig. 8). Instead of the expected relation in which saturated sonic velocities are higher than dry velocities, this is found to apply only over part of the range with high density. At lower densities, the relation is inverted.

Plotting strength against sonic velocity (Fig. 9) gives a positive exponential power law fit. As strength decreases with increase in weathering 'grade', sonic velocity also decreases. Such a graph can be used to predict one property from another, which may be particularly useful since the sonic velocity test is non-destructive and may be carried out on non-standard sized test specimens.

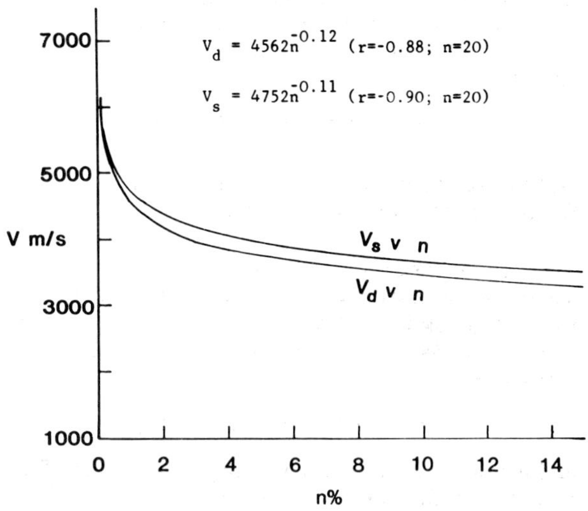

FIG. 5 Relations between dry and saturated ultrasonic velocities and porosity of fresh porphyrites and weathered andesites.

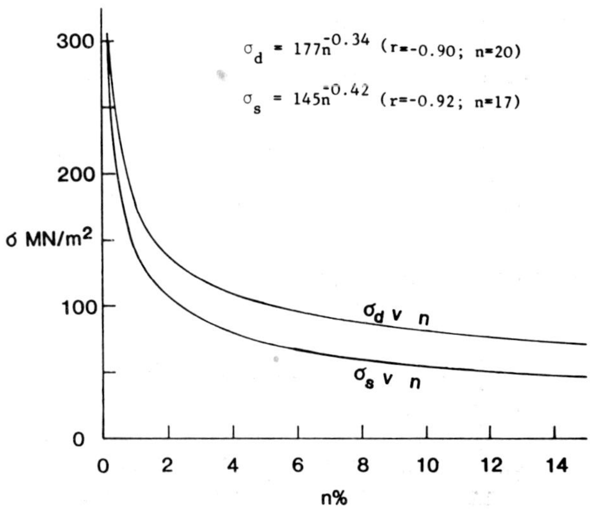

FIG. 6. Relations between dry and saturated uniaxial compressive strengths and porosity of fresh porphyrites and weathered andesites.

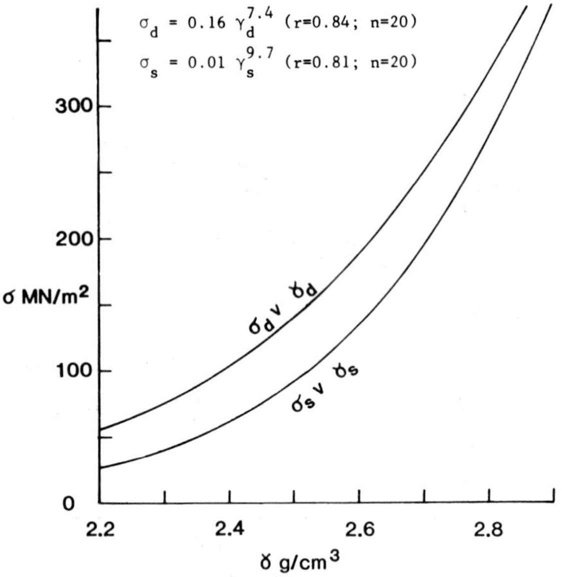

FIG. 7. Relations between dry and saturated uniaxial compressive strengths and corresponding densities of fresh porphyrites and weathered andesites.

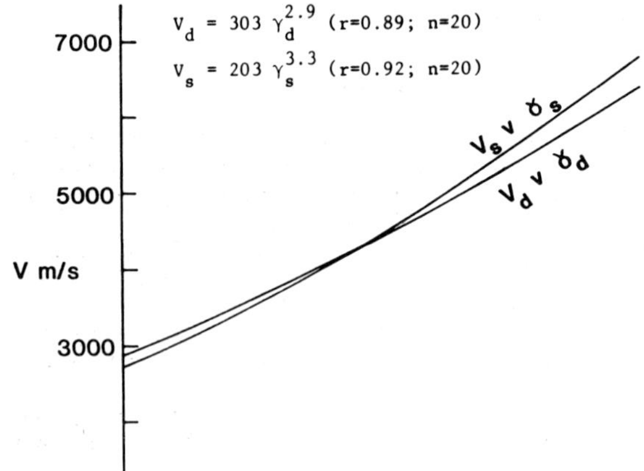

FIG. 8. Relations between dry and saturated ultrasonic velocities and corresponding densities for fresh porphyrites and weathered andesites.

Discussion and Conclusions

It is very well known that weathering generally affects the structure and behaviour or rocks. As the degree of weathering increases, rock become more porous and weaker. Additionally, it is also very well documented that the presence of water decreases the strength of rocks. However, in nature rocks are found under varying degrees of saturation. Therefore, it is important to know not only the weathering state of the rock mass but also its moisture content in order to predict the stability of structures and their design with confidence.

The influence of water on weathered rocks can be best appreciated by comparing properties of oven dry and saturated weathered rock specimens. In this paper, test results of fresh and chemically weathered andesite and related fine grained acidic rocks are given for ease of understanding. While most engineering tests could be carried out under both dry and saturated conditions, deformation properties were measured only in the dry condition.

Vutukuri *et al.* (1974) have given the following explanation for the effect of moisture and liquids on the strength of rocks, after reviewing the literature on both subjects. "Certain minerals decompose when they come in contact with liquids and are dissolved, creating more liquid-filled voids. The factor which contributes most in decreasing the strength of rocks seems to be the attack of liquids on crack tips, dissolving the material and increasing the stress of the apex, thereby helping in their propagation. It is also possible that liquids influence the surface energy of rocks and since creation of new surfaces during the process of fracturing is dependent upon the surface energy of the rock, its strength will depend upon the decrease or increase in the surface energy under the influence of the liquid. Liquids which wet the surfaces of the rock invariably decrease the surface energy of the rock and hence the strength".

Water exists in different forms in rocks and it is difficult to define the boundaries of each form. Fresh rocks tend to have adsorbed and free water, whereas weathered rock can have absorbed water in addition to adsorbed and free water. Since, the absorbed water is taken up within the crystal lattice of the clay minerals formed by weathering of the fresh rock minerals, this may result in expansion and in breaking up of the predominantly clay bearing rocks.

The importance of groundwater on the engineering properties of the rock mass has been recognised for a long time, but a detailed study has not been carried out. Recently suggested rock mass classification methods also include groundwater as one of the important classification parameters. (Bieniawski 1973 Barton *et al.* 1974). For example, Kendorski *et al.* (1983) give the following rating to different water conditions for rock mass classification purposes:

Water condition	*Rating*
Completely dry	15
Damp	10
Wet	7
Dripping	4
Flowing	0

It is interesting to note that such a grading of groundwater conditions in the rock mass does not give any indication of the effect of water on the rock mass properties as well as the groundwater pressure levels.

As there were not enough specimens of weathered andesite, extensive triaxial testing could not be carried out on these rocks. Thus concrete conclusions could not be reached about the triaxial testing of dry and saturated weathered andesites. However, available limited test results indicate that for fresh rocks, although saturation decreases cohesion, the friction angle remains the same, thus supporting the findings of Colback & Wiid (1965) and Parate (1973). The weathered rock specimens were tested under saturated conditions only and they showed that with increasing weathering not only cohesion but also the friction angle of rocks decreased.

Deformation properties of andesites were only tested on dry specimens due to laboratory testing difficulties. Even in the literature, there is very little information on the deformation properties of saturated rocks. However, the limited number of tests carried out on dry specimens of weathered andesites has shown that while Young's modulus decreased with increasing grade of weathering, Poisson's ratio increased.

The weathering grades of rocks can be best assessed by determining porosity and density of rock specimens. While and increase in weathering grade would increase the porosity it would also decrease the density. At zero porosity, dry and saturated densities should give the grain density of rock. The estimated grain density from Fig. 4 is 2.66 g/cm^3, whereas the mean of the measured grain density values is 2.70 g/cm^3. Such a discrepancy may have resulted from the presence of unconnected and unsaturated microcracks during porosity or density determination.

Relations presented in this paper cover a limited weathering range of andesites. Hopefully, in the long run, they will be extended to cover the whole weathering range.

The following conclusions can be reached from the analysis of test results on the influence of saturation on selected properties of fresh and weathered porphyrites and andesites:

(1) The graphs or relations presented depicting the interrelation between the dry and saturated state of rocks would enable a rock property to be estimated in one state from measurements in another state (Figs 1, 2, 3).

(2) An increase in porosity or a decrease in density due to weathering results in a decrease of not only strength but also ultrasonic velocity and other properties (Figs 6, 7, 8, 9). However, the decrease on saturation in strength or ultrasonic velocity with increase in porosity is not linear.
Above porosity of 1%, the difference between dry and saturated strengths and ultrasonic velocities seem to remain nearly constant (Figs 5, 6).

(3) The relation between the dry and saturated values of different properties enable one property to be estimated from another in the same or different state of saturation (Fig. 9).

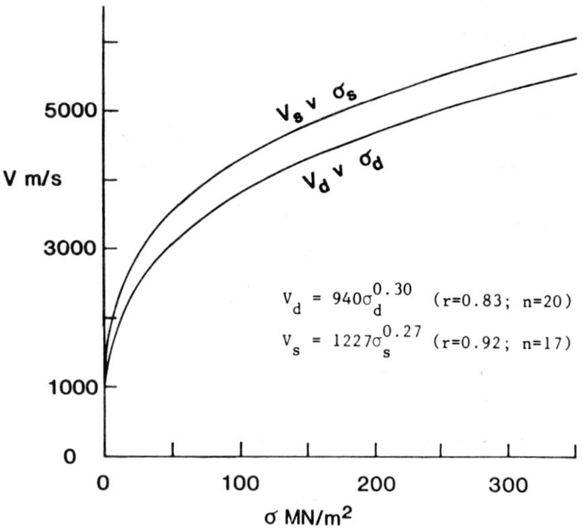

FIG. 9. Relations between dry and saturated ultrasonic velocities and corresponding uniaxial compressive strengths for fresh porphyrites and weathered andesites.

References

BALLIVY, G., LADANYI, B. & GILL, D.E. 1976. Effect of water saturation history on the strength of low porosity rocks. In '*Soil Specimen Preparation for Laboratory Testing*', ASTM.STP599 4 - 20

BARTON, N., LIEN, R. & LUNDE, I. 1974. Engineering classification of rock masses for the design of tunnel support. *Rock Mechanics*,**6**, 189-236.

BIENIAWSKI, Z.T. 1963. Engineering classification of jointed rock masses. *Transactions of the South African Institution Civil Engineers*. **15**, 335-344.

BROWN, E.T. 1981. *Rock Characterisation, Testing and Monitoring*. ISRM Suggested Methods. Pergamon Press, Oxford.

COLBACK, P.S.B. & WIID, B.L. 1965. The influence of moisture content on the compressive strength of rocks. *Proceedings of the 3rd Canadian, Rock Mechanics Symposium*, Toronto, 65-83.

KENDORSKI, F.S., CUMMINGS, R.A., BIENIAWSKI, Z.T. & SKINNER, E.H. 1983. A rock mass classification scheme for the planning of caving mine drift supports. *Proceedings of the 1983 Rapid Excavation and Tunneling Conference*, Sutcliffe, H. & Wilson, J.W. (Eds), SME-AIMMPE. New York. **1**, 191-233.

KESSLER, D.W., INSLEY, H. & SLIGHT, W.H. 1940. Physical mineralogical and durability studies on the building and monumental granites of the United States. *Journal of Research of the National Bureau of Standards (Research Paper PR 1320), 25*, 161-206.

KJAERNSLI, B. & SANDE, A. 1963. Compressibility of some coarse-grained materials. *Proceedings of the European Conference on Soil Mechanics and Foundations Engineering*, Wiesbaden, **1**, 245-251.

MANN, R.L. and FATT, I. 1960. Effect of pore fluids on the elastic properties of sandstone. *Geophysics*, **25**, 433-444.

OBERT, L., WINDES. S.L. & DUVALL, W.I. 1946. Standardized tests for determining the physical properties of mine rocks. *United States Bureau of Mines. R.I. 3891.*

PARATE, N.S. 1973. Influence of water on the strength of limestone. *Transactions of the Society of Mining Engineers. AIME.* **254,** 127-131.

PRICE, N.J. 1960. The compressive strength of Coal Measures rocks. *Colliery. Engineer,* **37,** 283-292.

SALUSTOWICZ, A. 1965. *Zarys Mechaniki gorotwotu,* Katowice, Wyawnictwo Slask.

VAN EECKHOUT, E.M. 1976. The mechanisms of strength reduction due to moisture in coal mine shales. *International Journal of Rock Mechanics, Mining, Sciences & Geomechanics. Abstracts,* **13,** 61-67.

VUTUKURI, V.S., LAMA, R.D. & SALUJA, S.S. 1984. *Handbook on Mechanical Properties of Rocks.* Trans. Tech. Publ. Clausthal, Germany.

ZARUBA, Q. (1965). Geology of the Orlik dam site. *Water Power,* **17,** 275-279.

N. TURK & W.R. DEARMAN, Department of Geotechnical Engineering, University of Newcastle upon Tyne NE1 7RU, UK.

Discussion Sessions 3 and 4

G. West (Transport and Road Research Laboratory) enquired whether the concept of soil suction might be sufficient to explain the loss in strength which occurs in specimens when their moisture content is increased.

M. de Freitas (Imperial College, London) suggested that although changes in suction might contribute to changes in the strength of weak rocks the effect is likely to be relatively small. Studies by the structural geologists do suggest that mineralogical effects also operate with an increase in the presence of water. *(See written contribution by Dr. G. West, this volume).*

G. West also suggested that a solution to the problem raised by Professor Dearman, viz that of attaining specific states of saturation, might be to investigate the use of sand columns with water tables at different levels for the wetter samples, pressure semi permeable membrane apparatus for samples required in the middle range and different hydroscopic salts for the dryer samples.

J. Pontin (Hydraulic Research Ltd.) felt that most of what had been said by Dr. de Freitas was not new. For example, what was so magical about the loss of strength with saturation? He had experienced this in Egypt where it was evident that the loss of strength of the soil in question could be attributed to the presence, at the grain contacts, of soluble salts which went into solution when wetted.

He would also like to comment on the process described by Dr. de Freitas as pneumatic failure. It seems most unlikely that high pressures are likely to develop in practice since although rapid infiltration may lead to this phenomenon in laboratory tests, lateral escape of air would occur in the field.

M. de Freitas. There is nothing magical about changes in strength occurring in the presence of water flow; it is a most natural response. However, we know almost nothing about the processes that produce this effect and that was why the phenomenon was included in the lecture: it is an area where much research remains to be completed. *(See written contributions by A.M. Podhalicz and D. Blythe this volume).*

The significant fact about pneumatic failure is that it need only occur once, over a limited area, for a cohesionless failure surface to be formed. Air pressures can be trapped below surface by local, impersistent hydrogeological boundaries such as might be created in water filled desiccation fissures.

J.K. Mitchell (University of California, Berkeley) suggested that slaking processes may be responsible for rapid strength loss observed in materials on saturation. Besides solutioning and slaking, this reduction could also be due to clay mineral hydration swelling at points of contact or dispersion of clay due to changes in the composition of electrolyte.

M. de Freitas agreed and thanked Professor Mitchell for his observation.

E.N. Bromhead (Kingston Polytechnic) referring to the process described by M. de Freitas as pneumatic failure observed that air bubbles have to be at higher pressure than the water due to curvature of meniscus surfaces. Flooding produces a seepage front which is drawn down by capillary suctions giving a negative pore pressure zone with an area of positive pore pressures above. The air pressure then increases due to confinement.

M. de Freitas added that during rapid infiltration experiments the response of the manometers in the sample is of interest. The pressure begins at atmospheric, goes negative as the capillary fringe passes by, then it increases to exceed atmospheric; once the wetting front passes by the pressure decreases to a negative value and then it increases to attain the hydrostatic level.

J.K. Mitchell drew attention to the similarity between the process envisaged and that of air slaking when dry fine grained materials are saturated.

W.R. Dearman (University of Newcastle) asked Dr de Freitas whether in his test to determine the effect of the degree of saturation on rock strength, the distribution of water within the specimens had been determined.

M. de Freitas stated that this had not been done, but a certain amount of water was added to specimens which were then allowed to mature. The results of the test suggest that this was satisfactory in as much that a reasonable series of relationships was observed between various parameters and presumed moisture content, but in any case most of the tests were performed on saturated samples. It has been observed that simple soaking is an inadequate method of saturation since even relatively permeable sandstones require the application of a vacuum, albeit small, to obtain full saturation.

Written contribution:

G. West (Transport and Road Research Laboratory)

I would like to congratulate Dr de Freitas on identifying promising areas of research in engineering

From Cripps, J.C., Bell, F.G. & Culshaw, M.G., (eds), 1986, *Groundwater in Engineering Geology,* Geological Society Engineering Geology Special Publication No. 3, pp. 139-142

hydrogeology, and I will comment on one of these, namely the marked increase in strength observed for certain rocks with decrease in moisture content at low moisture contents as reported by Priest & Selvakumar (1982). I suggest that this behaviour can be accounted for, at least in part, by *soil suction*. The study of soil suction was pioneered in the 1930s but the concept was not applied to soil mechanics until the 1950s in studies of soils beneath road and airfield pavements carried out by the Road Research Laboratory; this approach deserves wider recognition by the geotechnical community. It is a well known observation that a block of intact clay exhibits a marked increase in strength when it is dried, and this increase in strength with decreasing moisture content can be related to increase in soil suction (Croney & Coleman 1960). In a similar way, it is suggested, the increase in rock strength reported by Priest & Selvakumar may be due partly to an increase in suction.

Suction, which is the stress-free negative pore water pressure of a specimen of soil (or rock), is conventionally expressed on the pF scale. On this scale the logarithm to the base 10 of the suction in cm of water is the pF value; e.g. 100 cm water is pF 2. The use of this logarithmic scale, rather than a linear scale of moisture content, may have advantages when dealing with large changes in strength for small changes in moisture content, and it would be interesting to see Priest & Selvakumar's strength tests plotted against suction rather than moisture content. I have done this for their tests on Bunter Sandstone (Table 5.4.1(a) of their report) and the mean uniaxial compressive strength is shown plotted against the initial suction of the specimens in Fig. 1. It can be seen that from pF 0 to pF 5.5 the strength is little affected by suction (line AB), and this strength of about 37 MN/m^2 is probably due to cementing between the sand grains alone with little or no contribution from suction. However, from pF 5.5 to pF 7, the strength of the rock is markedly affected by suction (line BC) and it is suggested, therefore, that in this region suction is making a contribution to the strength over and above that of the intergranular cement.

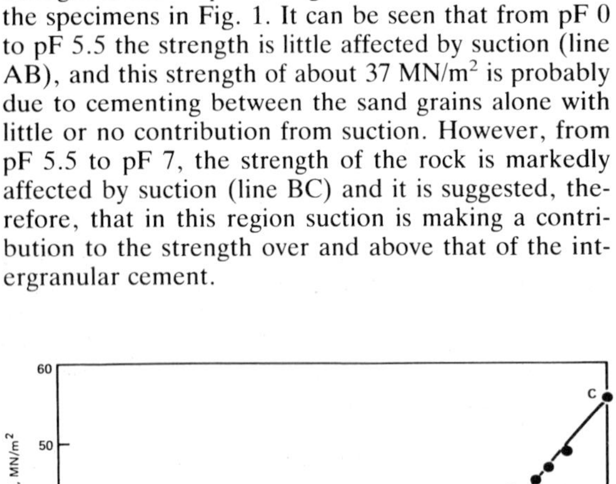

FIG. 1. Relationship between uniaxial compressive strength and initial suction for specimens of Bunter Sandstone (data from Priest & Selvakumar 1982)

In response to Professor DEARMAN's question as to how to prepare specimens of rock at moisture contents intermediate between saturation and air-dry such that they are in uniform moisture equilibrium, I can offer the following suggestions. The available methods that can be used, together with the appropriate pF range of each are:

Sand column over water table	pF 0 - 2
Suction plate	pF 0 - 3
Pressure membrane	pF 0 - 4.5
Relative humidity over various saturated salt solutions	pF 4.5 - 7

Details of some of these methods are given by Croney (1977). A sand column 1 m high was in fact used to prepare Bunter Sandstone specimens in some pilot tests carried out as a precursor to the work of Priest & Selvakumar (West 1979). For all methods, the specimens should be weighed daily so that the process of equilibration can be monitored, and uniform distribution of moisture throughout the specimen can be assisted by turning it over each time it is returned to the apparatus after weighing. With the first three methods it is essential to ensure good contact between the base of the specimen and the working surface of the apparatus.

ACKNOWLEDGEMENT: This contribution is published by permission of the Director, Transport and Road Research Laboratory.

References

CRONEY, D., 1977. *The design and performance of road pavements.* London (HMSO), Chapter 6.
CRONEY, D. & COLEMAN, J.D., 1960. Pore pressure and suction in soil *Proceedings Conference on Pore Pressure and Suction in Soils.* London, Butterworths, 31-37.
PRIEST, S.D. & SELVAKUMAR, S., 1982. The failure characteristics of selected British rocks. *Report to the Transport and Road Research Laboratory.* London, (Imperial College of Science and Technology), unpublished.
WEST, G., 1979. Strength properties of Bunter Sandstone. *Tunnels and Tunnelling,* **11** (7), 27-29.

Written contribution:

D. BLYTHE (National Coal Board, Opencast Executive)

During the discussion and question sessions following the presentation of papers, some of the problems regarding the laboratory saturation of rocks were mentioned. One of the problems in experimentation is not knowing the distribution of pore fluids in a partially saturated rock in which one may be carrying out strength tests.

Recently, some experiments were undertaken to examine the rate of saturation of freshly cored Coal Measures lithologies obtained from National Coal Board Opencast Executive exploration boreholes. Tests were restricted to mainly fine and medium grained sandstones together with some sandy siltstones. (Mudstones and silty mudstones have a tendency to slake during saturation.) Core samples were suspended in freshly prepared de-aired water after oven drying to constant weight. Immersed weights were recorded at intervals of time ranging from 5 seconds initially, to several days over a total period of a few months. When seemingly constant weights were obtained the samples were subjected to vacuum treatment to obtain maximum saturation. Saturation curves were then plotted by recalculating immersed weights as percentages of the total saturation value (saturated moisture content). Time was expressed as a logarithmic scale.

Two examples of the saturation curves obtained are shown in Fig. 1. The gradient of the initial part of the curve is influenced by grain size and shape as these two factors control the size and shape of voids. Gradient increases with increasing grain size. The degree of sorting and cementation will also be of significance. Clay minerals, with plate-like grains, will produce differing void shapes than quartz grains and therefore it is expected that the mineralogical composition of the specimen will be a controlling factor in the rate of fluid uptake of a specimen.

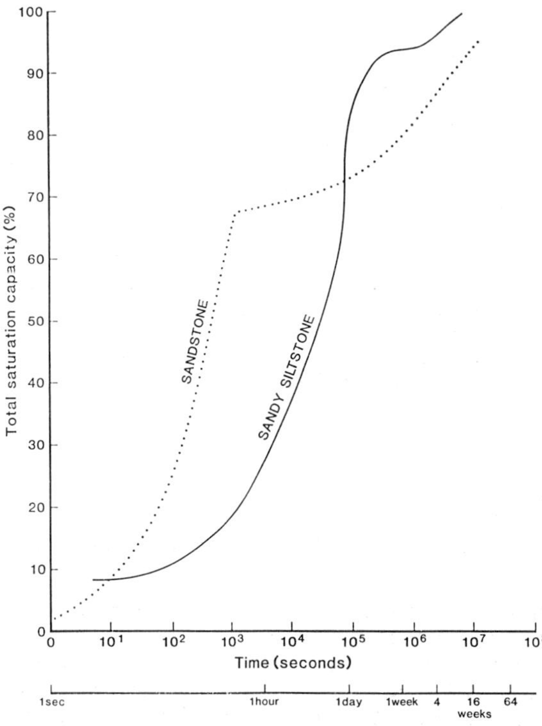

FIG. 1. Two examples of saturation curves found using de-aired water.

The points of inflection and duration of the 'plateau' region appear also to be governed by the pore size and shape. Coarser grained sandstones generally have inflection points after a shorter time interval and at a lower percentage of the total available void space than do finer grained or silty sandstones.

Determination of the actual physical process of displacing the air from the voids and replacing it with fluid is continuing with additional experimentation using different rocks and saturating fluids. It can be stated that the initial part of the curve corresponds to a rapid displacement of air from the rock as the saturating fluid invades the near surface pores. The rate of fluid uptake decreases with time as the more inaccessible pores are penetrated and the saturating fluid reaches the centre of the specimen. It is suggested that this corresponds to the point of inflection. Following this, there is a much slower rate of saturation during which it is thought that the remaining air in the rock (e.g. that trapped in pores with single openings) is gradually dissolved and displaced. This requires an exchange of air saturated fluid with other fluid in the rock pores and also an exchange of fluid within the rock and that in the immersion bath. Clay minerals (detrital and authigenic) will also absorb fluid into their crystal lattice. Solution of constituent minerals has been ruled out as a process affecting these results.

The manner of the saturation process and the time required without the application of a vacuum are quite striking and underline the importance of very carefully considered laboratory practices when determining effective porosity and saturated moisture contents of rocks.

Written contribution:

A.M. PODHALICZ (Sir William Halcrow and Ptns)

Various physical and chemical effects of action of groundwater on soil and rock materials have been described in papers in Session 3 also by Dr. M. de FREITAS. Dispersion of clay soils which may take place when there is no chemical equilibrium between soil and porewater is another possible phenomenon which should be considered.

Dispersion, or deflocculation, has been defined as a process where clay particles immersed in water are repelled from each other and pass into suspension even in practically still water with negligible erosive energy. Dispersive clays are sodium-rich and usually occur in hot semi-arid climates. The best documental occurrences are in south western United States, southern Africa and Australia. The fullest account of dispersion can be found in the proceedings of the 1976 ASTM-sponsored international symposium on the subject (Anon. 1977). Dispersion in undisturbed natural ground has been described by Goldsmith & Smith (1985), but most of the investigated cases of dispersion followed failure of water-retaining embankments which were built of dispersive soil, where impounding created disequilibrium between the embankment soil and the pore water. Investigations have shown that engineering causes such as inadequate compaction or omission of adequate provisions for drainage have contributed to failures. Sherrard (1984) gives a succinct summary of design measures which safeguard against failure due to dispersion.

Sir William Halcrow and Partners (1985) have carried out in 1984 investigations for an embankment dam at Lugoda in Tanzania which will impound 210 Mm^3 in a reservoir regulating the flow of the Ndembera river. The suite of tests on the completely weathered granite to be used as the embankment fill included pinhole tests (Head 1982) which are generally considered to be the most reliable indicator of dispersivity.

One of the samples from the borrow area has failed this test, a surprising find since literature search found no reference to residual soils produced by *in situ* weathering of igneous rocks being dispersive. Chemical analyses of the extracted pore water were then performed as a confirmatory test, and the sample which failed the pinhole test was found to have a 'percentage sodium[1] ratio of 64% which is high and indicative of dispersivity.

It was therefore accepted that some of the soils from the borrow area are potentially dispersive and the dam embankment was designed accordingly. The measures taken included provision of a 3 m wide vertical chimney drain with a D_{15} size of 0.25 mm and deepening of the cut-off trench in the foundations of the dam.

References

ANON, 1977. *Dispersive Clays, Related Piping and Erosion in Geotechnical Problems.* ASTM Special Technical Publication 623

GOLDSMITH, P.R. & SMITH, E.H., 1985. Tunnelling soils in South Auckland, New Zealand. *Engineering Geology* **22,** 1-11.

SHERRARD, T.L., 1984. Trends and debatable aspects in embankment dam engineering. *Water Power and Dam Construction* **36** (12), 26-32.

HALCROW, Sir William & Partners, 1985. Kapunga/Madibira Rice Project. *Detailed Engineering Study* (Unpublished Report)

HEAD, K.H., 1982. *Manual of Soil Laboratory Testing. 2,* Pentech Press, London: Plymouth.

[1] Percentage sodium ratio $= \left(\dfrac{Na}{Ca+Mg+Na+K} \right) \times 100\%$

SESSION 5

NATURAL CHEMISTRY OF GROUNDWATER AND CONTAMINATION

Hazardous waste containment

J.K. Mitchell

ABSTRACT: Containment systems that have been developed to isolate wastes from the soil and water that surround them in the ground include compacted clay barriers, slurry trench cutoff walls, geomembrane walls, concrete diaphragm walls, sheetpile walls, grout curtains, hydraulic barriers, compacted clay liners, and geomembrane liners. The choice in any case is dictated by whether waste from an existing facility or spill or from a new disposal site is to be contained and by the geological setting.

Properly compacted clays have properties that are well-suited for waste containment, as do flexible membrane liners. Compatibility between these materials and the waste must be considered in any case. Evidence is accumulating that the hydraulic conductivity of properly compacted clay is not significantly affected by exposure to most organic solvents and inorganic chemical wastes in dilute solution. Double liner designs that incorporate both compacted clay and flexible membrane liners as well as leachate collection systems have been proposed to satisfy the requirements of recent legislation in the USA.

The higher temperatures and long times associated with nuclear waste isolation by deep burial pose special problems of barrier stability. It appears that smectite clay has adequate stability to be used for repository buffer and seal material.

Introduction

Protection of the environment from the adverse consequences of hazardous and toxic wastes that are deliberately or accidentally disposed to land is one of the most important problems of the late twentieth century. Paramount among the concerns are the protection of groundwater quality and the prevention of toxic waste migration from one point to another.

When pollutants come into contact with the ground, several processes may ensue. Fig. 1, from Guswa *et al.* (1984), is a schematic representation of processes affecting organic pollutants in the soil/groundwater system. Many of these same processes could be important for inorganic pollutants and nuclear waste as well. Limited available data indicate that typical atmospheric releases of solvents from properly closed landfills are below established threshold limit values (TLV) (Environmental Institute for Waste Management Studies 1985), and this may be true also for the other waste types.

Consequently, isolation of the waste from contact with groundwater is the major concern. As barriers and seals of various types are employed as means for creating this isolation, a major related concern is the stability over long time periods of these barriers and seals in the chemical and temperature environments created by the waste materials.

Waste containment barrier systems are described in this paper. In addition, some consideration of waste-barrier compatibility are presented, and the problem of the establishment of threshold limits is described. Much of what is presented in this paper has been developed as a result of the author's participation as a member of the University of Alabama's Environmental Institute for Waste Management Studies, an interdisciplinary group of scientists and engineers from throughout the United States established to evaluate alternatives for waste management.

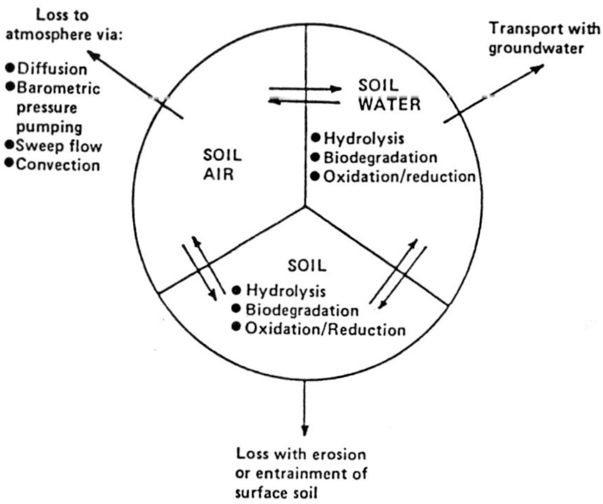

FIG. 1. Schematic representation of processes affecting organic pollutants in the soil groundwater system (from Guswa *et al.* 1984).

From CRIPPS, J.C., BELL, F.G. & CULSHAW, M.G., (eds), 1986, *Groundwater in Engineering Geology*, Geological Society Engineering Geology Special Publication No. 3, pp. 145-157.

Waste barrier systems

Introduction

Waste containment barrier systems can be divided into two major categories: (1) those used to protect against contaminant migration from existing disposal sites or spills, and (2) those for waste containment at new facilities. Choices among options in the first group are dictated by local conditions and the specifics of the required clean-up. Options in the second group are likely to be constrained by regulatory requirements.

Existing disposal facilities

Cutoff walls, with and without hydraulic barriers, and covers are the most commonly used methods for restricting waste movement from existing waste dumps and impoundments. Compacted clay, slurry trenches, grout curtains, geomembranes, sheetpiles, and concrete diaphragm walls have been used or proposed for use to provide vertical barriers against waste movement. Well-points and pumping may also be used to create an inward hydraulic gradient which opposes the outward flow of waste fluids and leachate. For these barriers to be effective there must be a stratum of low hydraulic conductivity materials, an aquitard, or other impervious bottom layer within attainable depth. The construction material should be resistant to; i.e. 'compatible' with, the waste over a long time period. That is, contact with the waste should cause no increase in permeability or decrease in durability of the cutoff material.

Compacted clay barriers consist of an excavated trench, backfilled with a compacted clay, to produce a wall of low hydraulic conductivity, as shown in Fig. 2. A compacted clay barrier provides a wide zone of low hydraulic conductivity, and it can be economic if there is a local borrow area of suitable clay and an aquitard at shallow depth. Construction costs increase rapidly for depths greater than a few meters and for construction below the water table.

Slurry trench cutoff walls (Fig. 3) consist of narrow (1 m or less), clay slurry filled trenches excavated into an aquitard using a backhoe or clamshell excavator. The trench is kept open during the excavation by the pressure exerted by the slurry. Soil–bentonite mixtures are often used for the slurry, and hydraulic conductivities as low as 1×10^{-8} cm/s can be achieved. An advantage of slurry trenches backfilled with soil-bentonite mixtures is that they can tolerate moderate deformations without cracking (Millet & Perez 1981).

Excavation of slurry trenches may be difficult or impossible if boulders or other subsurface obstructions are encountered. As inspection inside the trench is impossible, it may be difficult to insure continuity between the trench and the underlying aquitard.

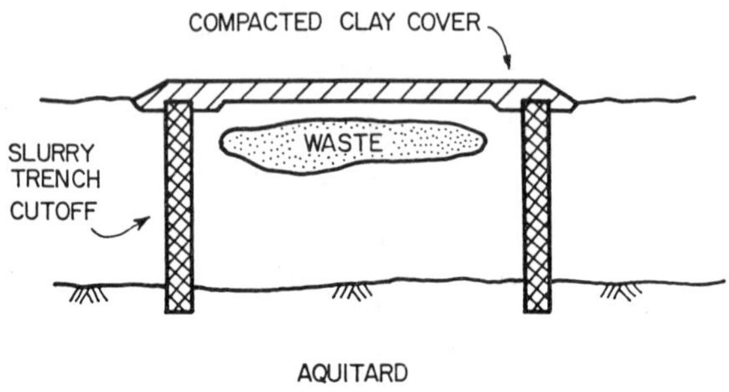

FIG. 3. Slurry trench cutoff wall with compacted clay cover.

Geomembrane walls are a modification of slurry trench technology (Fig. 4). The same method of construction is utilised, but U-shaped geomembranes are placed within the trench, and a water–sand mixture is pumped into the geomembrane so that its shape conforms to the walls of the trench. Monitoring wells can be installed in the sand interior for detection of leakage. The geomembrane material may be

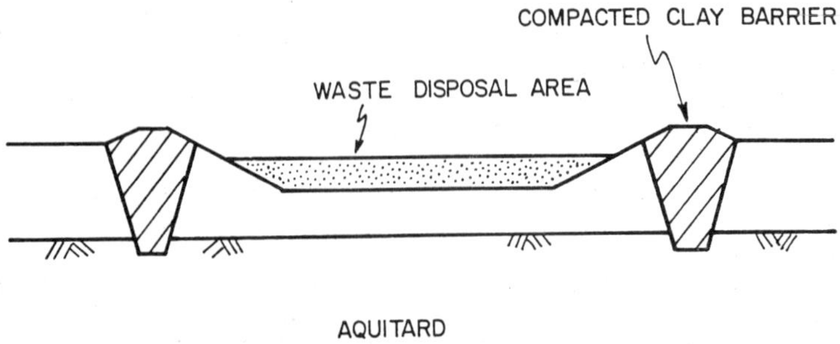

FIG. 2. Compacted clay barrier

chosen specifically to resist given chemicals, and repairs may be possible within the confines of the outside liner (Tallard 1984).

Diaphragm walls are constructed in the same manner as slurry walls, however, the clay slurry is replaced after excavation by pre-cast or cast-in-place concrete panels. Panels are connected using water-tight joint material or by the cement bentonite slurry used during excavation. The high cost of this type of cutoff restricts its use to situations where high structural stability is required (Kufs *et al.* 1983).

Sheetpile cutoff walls have been proposed for isolation of wastes in land disposal sites. Damage during driving and leakage through the interlocks are likely to adversely affect their watertightness, and dewatering wells may be required to insure effectiveness.

Grout curtains may be suitable in some situations. Cement, chemical, or bituminous grout can be used, with the choice dependent on the pore size of the material to be grouted and the leachate compatibility with the grout. Fine-grained soils are generally not suitable for grouting, because of the inability of either particulate or chemical grouts to penetrate interparticle voids. Three rows of grout holes may be required to achieve a sufficiently impervious barrier. The outer rows should be grouted first, followed by the inner row to seal remaining voids. Grouting may also be used to seal fractures in rock underlying a waste pile to reduce downward leakage of leachate. Unfortunately, total integrity of a grout curtain is both very difficult to obtain and to evaluate.

Barrier wall placement

Regional and local groundwater flow regimes must be considered when determining the location of cutoff walls. They may be placed upgradient, downgradient, or circumferentially around the waste, as shown schematically in Fig.5. Upgradient placement, Fig. 5b, is used to divert the groundwater away from the waste. This arrangement prevents contact between the waste and the barrier materials, thus eliminating the compatibility problem. Downgradient placement, Fig. 5c, directs the groundwater towards a well-point, and is a suitable arrangement if extraction and treatment are desired.

Hydraulic isolation

If a site is enclosed circumferentially, pumping from the inside will create an inward hydraulic gradient as shown in Fig. 6. An advantage of this system is that the barrier materials are exposed to an inward flow of fresh groundwater instead of leachate, thus minimizing compatibility problems between the waste and the barrier. It can be expensive, however, because of the need for continuous pumping and water treatment.

Hydraulic barriers can also be used without a physical cutoff barrier, however, the pumping requirements are much greater. Extraction wells are so located that the contaminant plume is directed towards them as withdrawal intake points. After withdrawal and treatment, the water is used to recharge the system. The concept, often called a pressure trough and ridge, is shown in Fig. 7.

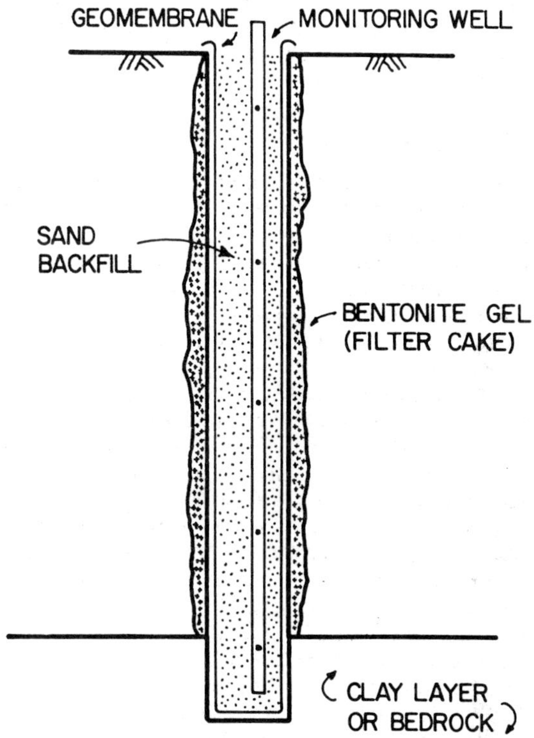

FIG. 4. Geomembrane cutoff wall (after Woodward-Clyde Consultants 1983).

New disposal facilities

Once a suitable site has been selected for a waste disposal facility, a properly designed and constructed liner and cover are the key considerations for the long term protection of the groundwater, surface water, and atmosphere above. It is important to note, however, that the initial site selection is critical. In a recent study, the Environmental Institute for Waste Management Studies (1985) concluded that the hydrogeological setting is paramount among criteria to determine the acceptability of sites proposed for disposal of wastes to land. It is also this area where the Institute found the greatest deficiencies in past practice.

Clay liners, geomembrane liners, leachate collection systems, and monitoring systems are the main components of most new waste storage and containment facilities.

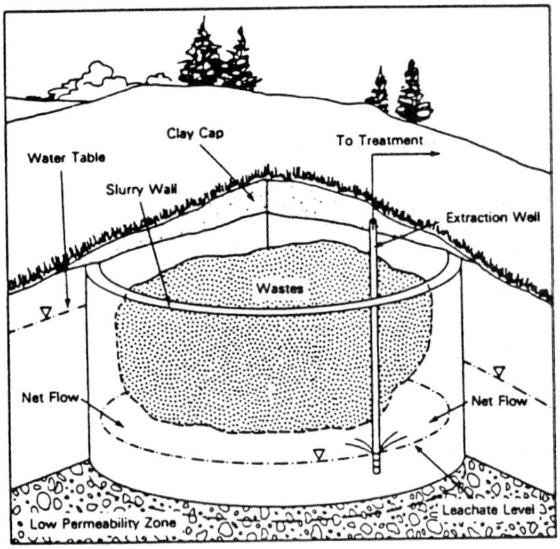

a. Circumferential placement

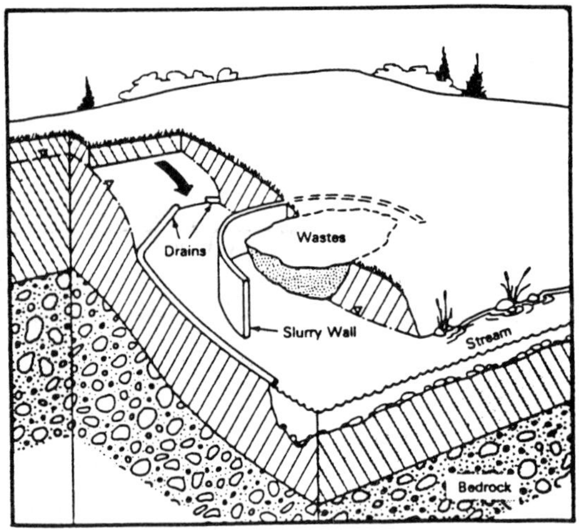

b. Upgradient placement

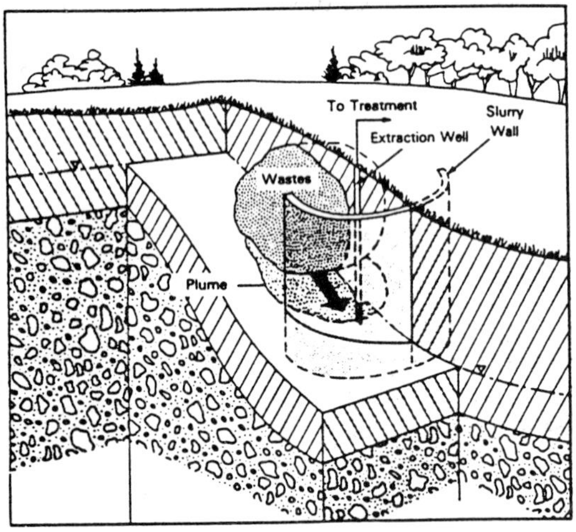

c. Downgradient placement

FIG. 5. Cut-away cross sections of different barrier wall orientations (from Spooner *et al.* 1983).

Clay liners are suitable for the containment of many wastes, because (1) clay is readily available in many areas, (2) low hydraulic conductivities (e.g., less than 1×10^{-7} cm/sec) can be obtained if proper water content and compaction procedures are used, (3) clay has adsorptive capacity for some wastes, (4) it is stable over a range of conditions, and (5) it may exhibit self-healing if cracks occur. Clay liners are constructed by compaction in lifts of about 150 m thickness. A sheepsfoot type roller is preferred in order to achieve maximum breakdown of soil clods. Compaction water contents slightly higher than required for maximum density will usually produce the minimum hydraulic conductivity. Minimum clay liner thicknesses are of the order of 0.6 m.

Liners can also be constructed using commercially processed bentonite by depositing a 25 to 50 mm thick layer of bentonite covered by a 200 to 300 mm layer of soil, or by compacting a 1:8 mixture of sand and bentonite. The main difficulty with thin bentonite layers is to insure uniform distribution of the bentonite without thin or bare spots. Success using a mixture requires thorough and uniform mixing of the sand and bentonite. Artificial clays (Jarrett & Marwood 1984) and clays treated to improve leachate resistance (Clem 1984) are also finding application for use as liners.

Care must be taken during the construction of clay liners to be sure that the material is free of contaminants, to avoid desiccation cracking during and after construction, and to insure that the material is placed at the specified water content and density. Even with good construction quality control, however, the hydraulic conductivity of field compacted clay liners may be substantially greater than that of laboratory compacted samples (Daniel 1984; Day & Daniel 1985).

A typical clay liner design for a hazardous waste landfill prior to 1984 is shown in Fig. 8. Recent

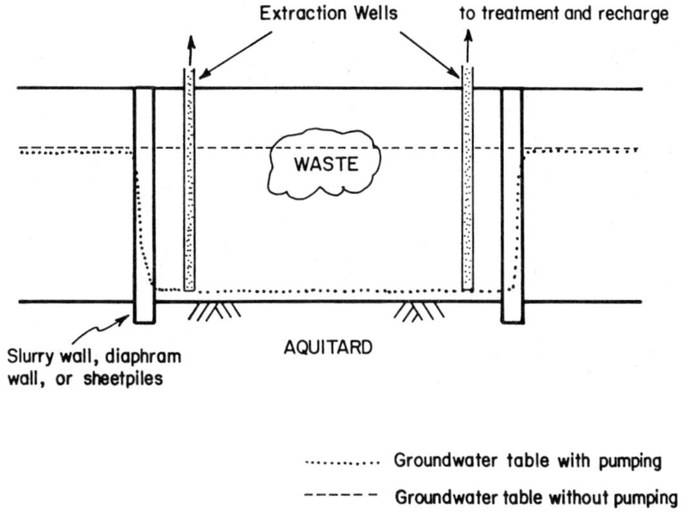

FIG. 6. Pumping from inside cutoff walls to create a hydraulic barrier.

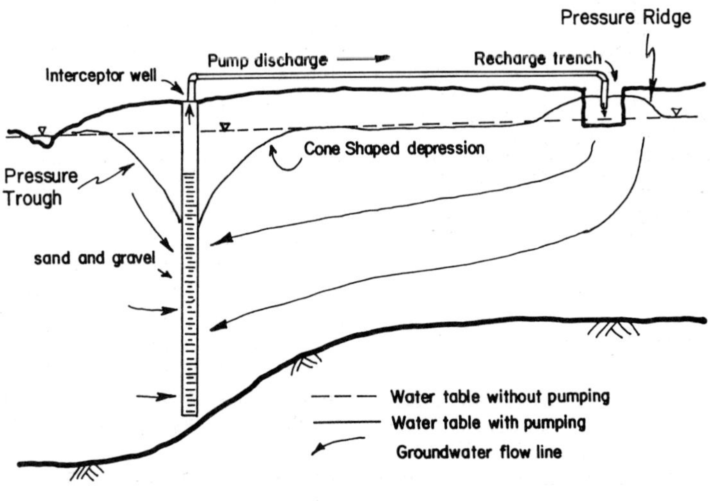

FIG. 7. Pressure ridge and trough method (from Woodward-Clyde Consultants 1983).

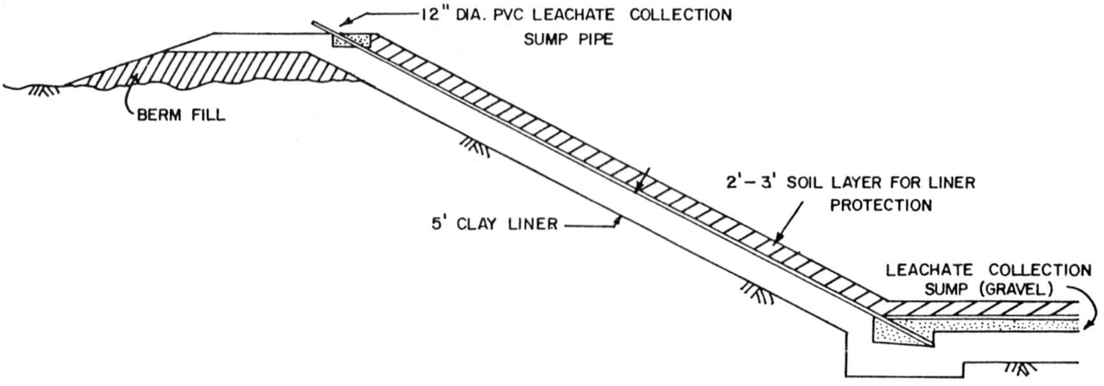

FIG. 8. Cross section of a clay-lined hazardous waste landfill constructed prior to the 1984 double liner requirements.

Federal legislation has resulted in minimum designs for both single and double liner systems, and these designs are described subsequently.

Flexible membrane liners (FML) composed of plastic or rubber sheets are now widely used because of their very low hydraulic conductivity (reported to be 1×10^{-11} cm/s or less). Much care is required for proper installation of geomembrane liners, and they must be protected from weather and vandalism prior to and during installation. Bonding and orientation of seams is an important factor in the subsequent performance of a geomembrane liner system. Many failures have been attributed to inadequate welding of seams (Ghassemi *et al.* 1983). After installation a soil cover is necessary to prevent damage from construction traffic and ultraviolet radiation.

Composite liners incorporate both clay blankets and geomembranes. Each can be used as a single standalone liner incorporated in a two liner system, or the liner can be a marriage of both, as in a smectite-treated membrane (Clem 1984). Other composite liners may include soil–cement mixtures, sprayed asphalt, and prefabricated asphalt. Soil–cement generally does not have a low enough hydraulic conductivity to act alone, and asphalt linings are not resistant to organic solvents.

Minimum designs

The most important recent legislation in the USA that relates to liner systems for new hazardous waste facilities is contained in the Hazardous and Solid Waste Amendments of 1984. In order to qualify for a permit, new or expanded permanent landfills and impoundments must have two liners, leachate collection systems above both liners, and a groundwater monitoring system. In addition the disposal of liquid wastes in landfills is banned so future disposal of liquid wastes will require their solidification by mixing with a suitable solid material, such as fly ash or cement kiln scraping.

The US Environmental Protection Agency has proposed two double liner designs in a draft technical guidance document (Second version, May 24, 1985). Schematic drawings of the proposed FML/composite and the FML/compacted soil double liner systems are shown in Figs 9a and 9b, respectively. These designs are intended to contain waste during operation of the landfill and for a 30 year post-closure monitoring period. They take into account that no clay or synthetic liner is completely impervious.

The FML/composite double liner system, shown in more detail in Fig. 10a, consists of, from top to bottom: (1) a granular or synthetic filter, (2) a granular leachate collection zone at least 300 mm thick, having a hydraulic conductivity greater than 1×10^{-2} cm/s, and containing internal drainage tiles, (3) a solvent resistant membrane at least 1 mm thick, (4) a secondary leachate collection system similar to above, having a recommended thickness of at least 300 mm, and a hydraulic conductivity greater than 1×10^{-2} cm/s, (5) a second geomembrane, and (6) a compacted clay layer at least 900 mm thick with a hydraulic conductivity less than 1×10^{-7} cm/s.

The FML/compacted soil liner design, Fig. 10b, is similar, except that the lower membrane is absent, and the compacted layer must be thick enough so that leachate can not pass completely through it during the operational and post closure monitoring periods. Thus the thickness of the compacted layer is dependent on the hydraulic conductivity of the compacted clay, which should not be less than 1×10^{-7} cm/s, and the leachate head. The latter is specified as a maximum of 300 mm, based on a properly functioning leachate collection system.

Single liner systems are allowed for certain landfills, surface impoundments, and waste piles. They must be designed to prevent waste migration to the adjacent soil, groundwater, or surface water during their active lives. Guidance for both compacted soil and FML single liner systems is being developed by the US EPA as shown in Fig. 11.

The new regulations apply to new landfills and impoundments that are located above the seasonal high water table. Thus a hydraulic barrier due to inward flow of groundwater would not appear to be a viable option under the new regulations.

Covers

Covers are required for the closure of hazardous waste sites. The purposes of the cover are to minimise the infiltration of water, thus minimising leachage generation, and to prevent escape of gases to the atmosphere. A cover must be impervious to water; be able to deform without cracking; be resistant to cycles of wetting and drying, and freezing and thawing, root growth, burrowing animals, and erosion; and be able to support vegetation (Daniel 1983). The cover should slope from a high point at the centre to

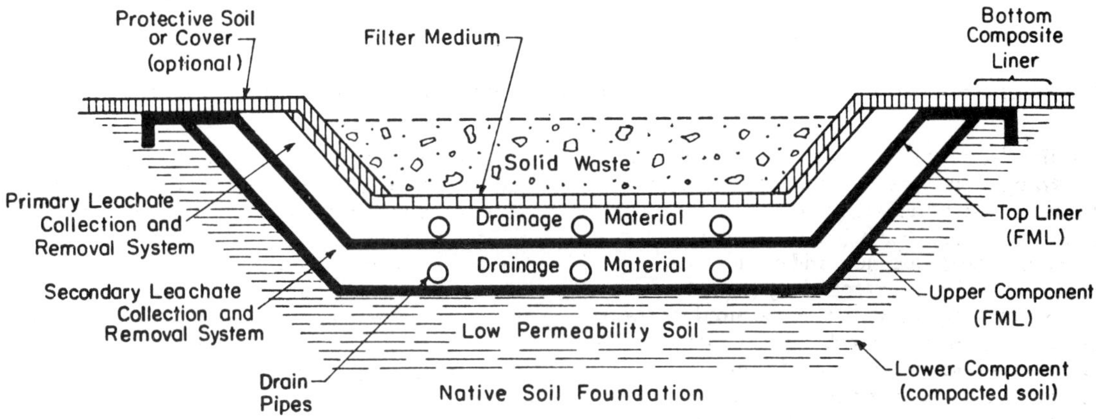

a. FML/composite double liner

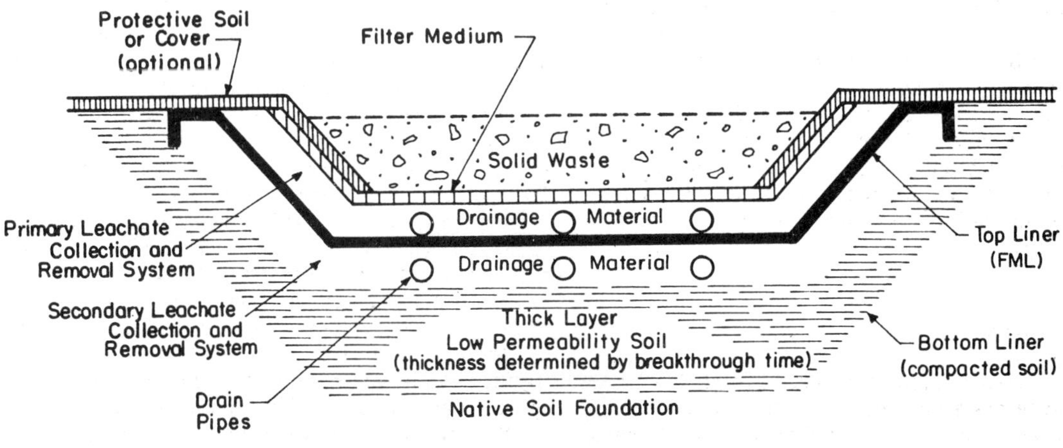

b. FML/compacted soil double liner

FIG. 9. Double liner systems proposed by the 1985 US EPA guidelines. The leachate collection layer is considered to also function as the geomembrane protection layer.

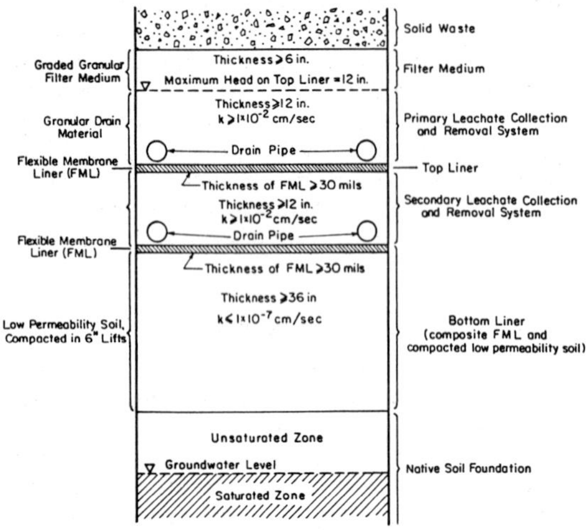

a. FML/composite

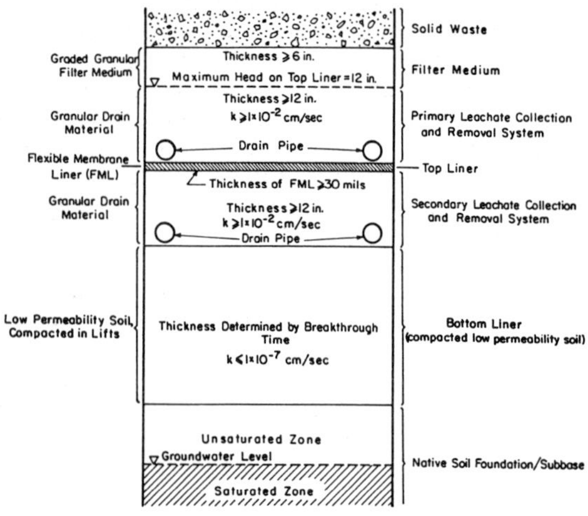

b. FML/compacted soil

FIG. 10. Schematic profiles of double liner systems.

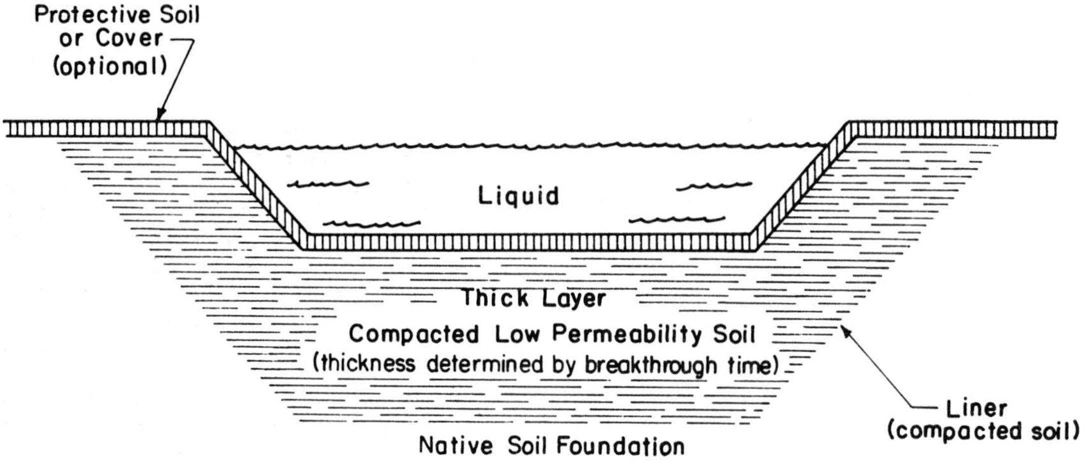

a. Compacted soil

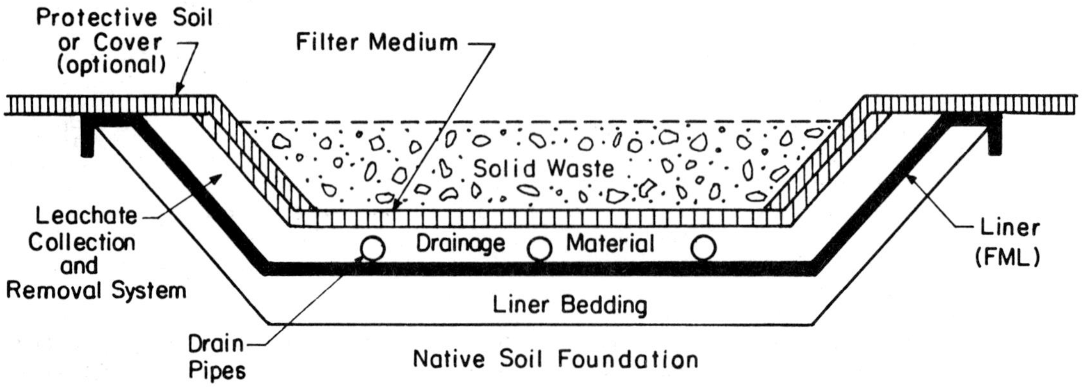

b. Flexible membrane

FIG. 11. Single liner systems for a landfill or surface impoundment peripheral diversion ditches that direct runoff away from the area. Covers can consist of compacted soil, geomembranes, or combinations of the two.

Composite covers containing both compacted clay and geomembranes, Fig. 12, combine the advantages and minimise the disadvantages of each material used separately. When a well-graded drainage layer is used above a clay layer, a geotextile material can be placed above the sand to act as a filter, with a geomembrane placed below it to act as a second barrier and to protect the clay from erosion. The upper geotextile must be covered by a layer of soil to protect it from ultraviolet radiation damage. If the landfill is too soft to allow adequate compaction of the clay, a lower geomembrane layer can serve as reinforcement, providing a firmer working surface and a stronger cover.

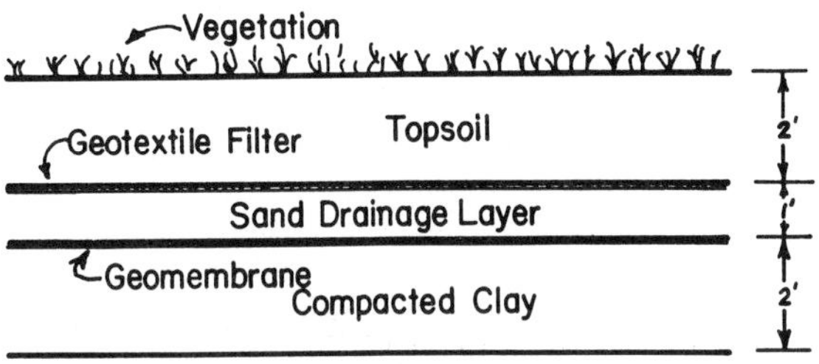

FIG. 12. Cross section of a geomembrane-compacted clay composite cover (from Waste Management, Inc. 1981).

Waste–barrier compatibility

Introduction

One of the issues of greatest concern in the underground disposal and long-term safe containment of wastes is whether or not prolonged contact between the barrier materials and the waste will adversely affect the properties and performance of the barrier. An incompatible waste–barrier system is considered to be one in which the properties of the barrier are sufficiently changed that it is unable to perform its design function over the full lifetime of the facility. In the case of hazardous and toxic chemical wastes, the chemical and physical interactions between the landfill leachate and liners or other barriers are of primary concern. They key question is whether or not these interactions will result in an increase in the hydraulic conductivity of the liner. In the case of nuclear waste, the chemical interactions may be of lesser importance; however, the effects of elevated temperatures and radiation must be considered.

Although current research on these questions is extensive, and information about specific interactions is being developed at a rapid rate, there remain many issues yet to be resolved before designers can be fully confident about the choices of materials and placement conditions to insure long-term compatibility. Important among the reasons for this are the inherent complexity of the materials and the reactions, the great range of chemicals and potential barrier materials to be considered, the difficulty of extrapolation of observations and results to periods of many years, and the great difficulty in reliable representation of actual field conditions by laboratory tests and analytical models. Some relevant conclusions are possible at this stage, however, and a few of them are given below.

Clay–chemical waste compatibility

Evidence in the literature is conflicting concerning the influences of chemicals on the hydraulic conductivity of compacted clay. Very high values; e.g., approaching 1×10^{-4} cm/sec, have been measured in some cases where solvents and liquid hydrocarbons dominate the liquid phase (Michaels & Lin 1954; Anderson & Brown 1981; Anderson *et al.* 1982; Brown *et al.* 1983). It is known that in some of these cases the high values are a consequence of direct mixing of the clay and hydrocarbon, thus giving a porous medium with a flocculated and open fabric that is quite unlike that of a clay compacted using water as the pore fluid. In others, testing anomalies may have been a factor.

Nonetheless, a review of the evidence shows that physico-chemical interactions can occur, and there are some data to show that exposure of clay to concentrated organic solvents can cause shrinking and cracking. If this does occur, then the hydraulic conductivity can increase gravely as a result of fracture flow. Inorganic chemicals can also influence the hydraulic conductivity as a result of internal swelling, shrinking, floccuation and deflocculation effects (Hardcastle & Mitchell 1974).

On the other hand, dilute solutions of organic solvents appear to have little adverse affect on hydraulic conductivity, and in some cases the conductivity may even be decreased (Daniel & Liljestrand 1984). Fernandez & Quigley (1985) present results that are quite consistent with the above. When a natural clay soil was mixed with different pure liquids and subjected to permeability tests using the same liquids, hydraulic conductivity values were obtained that increased from 5×10^{-9} to 1×10^{-4} cm/sec as the dielectric constant of the permeant decreased from 80 to 2. However, sequential permeation of compacted, water wet samples ($k \approx 10^{-8}$ cm/sec) showed no changes in hydraulic conductivity when permeated with water-insoluble hydrocarbons of low dielectric constant. These hydrophobic liquids displaced less than 10% of the pore water from samples at a void ratio of unity.

It was found by Fernandez & Quigley (1985), however, that permeation with water soluble alcohols led to substantial removal of pore water and increases in hydraulic conductivity of as much as a factor of ten. Subsequent permeation with liquid aromatics of very low dielectric constant resulted in 1000-fold increases in permeability even though only 30% of the pore space was occupied by the aromatic.

Taken collectively, the results of the various studies would seem to indicate that:

(1) Exposure of water-wet compacted clay to pure solvents may cause some physical changes that could cause an increase in the hydraulic conductivity.
(2) The hydraulic conductivity of compacted clay liners is not likely to be adversely affected by exposure to organics in dilute solution.
(3) Exposure to water-soluble solvents such as alcohol could not only lead to some increase in hydraulic conductivity, but also allow entry of organics of lower dielectric constant followed by large increases in hydraulic conductivity.
(4) Because the properties of high water content clay–water systems are more influenced by chemical effects than are those of low water content systems, the hydraulic conductivity of slurry wall barriers is likely to be more affected by exposure to leachate or chemical spills than that of compacted clay liners.

Geomembrane – waste compatibility

Although geomembranes have a very low hydraulic conductivity, they are susceptible to degradation in the presence of some chemicals or as a result of prolonged exposure to ultraviolet radiation. Tests by Haxo (1983) and Haxo & Nelson (1984) indicated that polymeric membranes are pervious to some organic molecules; e.g., acetone, xylene. Manufacturer's data compiled by Waste Management, Inc. (1983) are given in Table 1. It may be seen that hydrocarbons, especially halogenated hydrocarbons, can have severe effects on all five liner materials listed. The data indicate that high density polyethylene (HDPE) is the most resistant of the materials evaluated. Because of this resistance, HDPE liners, 1.5 to 2.5 mm in thickness are used in many chemical waste disposal facilities.

The long-term stability of smectites

The safe long-term containment of nuclear wastes by deep burial presents special problems owing to the very long times (10^5 to 10^6 years) that the groundwater must be protected and the higher temperatures associated with the radioactive decay. A variety of rock types have been considered for the location of nuclear waste repositories, and various natural and man-made materials have been proposed for use in borehole, tunnel, and shaft sealing and as a buffer mass around waste cannisters.

Smectite clays are among the prime candidates for this purpose because of their very low hydraulic conductivity, their swelling properties, their abundance, and their low cost. As smectites are naturally occurring earth materials and they represent near end products of weathering in many environments, they are expected to be rather stable over the long-term. It is known, however, that smectites can alter to illite. It is of interest, therefore, to enquire whether such alteration might occur during the life of a nuclear waste repository, and if so, whether this alteration could affect the physical properties of the clay so as to adversely impact its usefulness as a barrier material.

Two in-depth meetings have been held to consider the possible alteration of smectites if used for nuclear waste containment. The first was at the State University of New York at Buffalo in May 1982. The Proceedings of this meeting were published by the Swedish Nuclear Supply Company as SKBF/KBS Technical Report No. 83-03. The second was held in Washington, D.C. in December 1983, and the Proceedings were compiled by Duwayne M. Anderson, Texas A & M University, College Station, Texas, November 1984. An international group of earth scientists and engineers representing several relevant disciplines participated in the discussions. The major conclusions concerning smectite stability are summarised here.

The mechanism of conversion of smectite to illite involves three steps:

(1) Creation of a high lattice charge through removal of Si^{4+} and its precipitation in various forms adjacent to the clay lattice. The

TABLE 1. *Compatibility of geomembranes and various wastes (from Waste Management, Inc., 1983).*

WASTE	PVC	Hypalon	HDPE	XR-5	Neoprene
Inorganic Acids	0	M	0	0	M
Organic Acids	0	0	0	M	0
Inorganic Bases	0	0	0	0	0
Inorganic Salts	0	0	0	0	0
Alcohols	0	0	0	M	0
Hydrocarbons	S	S	M	S	S
Halogenated Hydrocarbons	S	S	S	S	S
Keytones	M	M	0	M	M
Detergents	M	M	0	0	M
Oils and Foods	S	M	0	0	M
Hydraulic Fluids	S	M	M	0	M

0 little or no effect; M moderate effect, S severe effect.

displaced silica is replaced by Al^{3+}, Mg^{2+}, etc. which are locally available from accessory minerals or the exchange ion complex.

(2) Influx of K^+ from the local solutioning of potassium bearing minerals, percolating groundwater, etc. to saturate the exchange capacity of the highly charged smectite intermediate.

(3) Collapse of the lattice to the typical non-expansive illite form.

The rate-limiting step in this conversion is the production of the highly charged smectite lattice by progressive substitution (step 1). The temperature must be high enough to allow the passage of ions out of and into the lattice. This can be relatively rapid at $T > 100°C$. It begins at about 60°C, but the rate is almost negligible at lower temperatures.

The results of geological and mineralogical studies, laboratory experiments, thermodynamic calculations, and some field studies led the participants of the second workshop to conclude the following relative to the use of smectities as barriers for radioactive waste isolation:

(1) Useful properties of smectites are:
   a. Mechanical buffering
   b. Hydraulic shielding that allows only diffusive transport
   c. Sorptive capacity for radionuclides. This may be unimportant because of limited capacity.

(2) The hydraulic barrier function is paramount, and integrity is required for 10^5 to 10^6 years.

(3) Whether or not cation exchange will adversely affect the smectite properties will depend on both the site and placement conditions.

(4) Reasonably conservative limiting temperatures are 100°C for clay or granite host rocks and 300°C in basalt.

(5) Provided the temperature is always less than 100°C, the hydraulic properties of a smectite buffer seal can be assured for 10^6 years.

(6) Techniques for placing and predicting the initial behaviour of both seals and buffers are well proven in crystalline rocks, but need further study in clay host materials.

(7) Smectites can be expected to partially collapse on a 10^5 to 10^6 year time scale; however, their overall function as a barrier should not be radically affected.

(8) Triotahedral smectites are likely to be more stable than dioctahedral smectites.

(9) The results of experimental and thermodynamic studies of clay mineral stability can be ambiguous.

(10) Mixtures of materials may be better for use as seals than pure clays.

(11) The following topics were identified as needing further study:
   a. The geotechnical properties of mixed clay backfills.
   b. The properties of stiff and plastic clay host rocks at depths of 300 to 500 m.
   c. The swelling, bonding, and pore pressure behaviour of clay seals in clay host rocks.
   d. The necessity for chemically tailored buffers.
   e. The role of micro-organisms in buffers, especially in sulphate reduction and corrosion.
   f. Quantitative values of properties required for seals.

Theshold limit concentrations

Given sufficiently long time and high enough gradients and waste concentrations, some contamination of the soil and groundwater beyond a waste containment barrier system is inevitable. It has become desirable, therefore, to begin development of a methodology for determining the maximum allowable concentrations of wastes at the point of disposal or discharge.

Conceptually, the problem is easily defined. A system consisting of a hazardous waste landfill or impoundment, a liner, an unsaturated zone above the water table, an aquifer with flowing groundwater, and a monitoring well is shown in Fig. 13. The monitoring well has been established as the point of compliance at which concentrations of different chemicals must always be less than some specified values, for example, drinking water standards. To be determined are the maximum allowable concentrations or total amounts of hazardous waste that may be placed in the disposal facility.

A number of difficulties in solving this problem are evident. To be considered are the following:

(1) The rate of leachate production in the waste, and the concentrations of chemicals in it.

(2) The leakage rate of leachate through the liner. This evaluation will require knowledge and/or assumptions concerning the hydraulic conductivity of the liner, leachate head and its variations with time, possible changes in liner properties with time, and possible absorption of pollutants by the liner.

(3) The rate of flow of leachate through the unsaturated zone above the water table. Partially saturated flow is a very complex problem, and existing predictive methods are severely limited by uncertainties about parameters.

(4) Contaminant transport through the groundwater. The advection-dispersion model may be well suited for this purpose. There are many numerical models now available. Unfortunately, their usefulness is questionable in many cases because of uncertainties about system geometry, the lack of information about diffusion and hydraulic conductivity coefficients, and incomplete knowledge of the many reactions and interactions that may occur during the transport process.

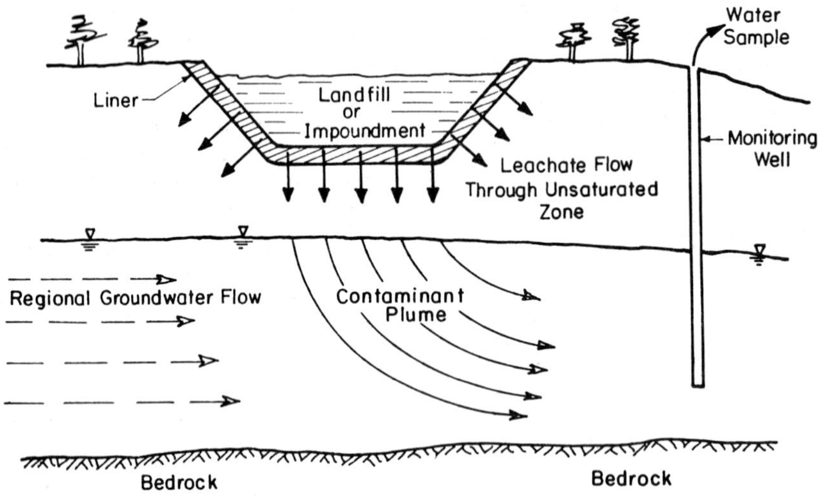

FIG. 13. Schematic diagram of a waste facility and groundwater system for illustration of the threshold concentration determination problem.

Nonetheless, a study is underway to develop a general methodology for establishing threshold concentrations that may be useful for obtaining approximate solutions to the problem outlined above for particular cases.

Conclusion

Protection of groundwater from contact with land-disposed toxic waste is perhaps the major environmental problem of the late twentieth century in the industrialised parts of the world. Containment systems that have been developed to isolate wastes from the soil and water that surround them in the ground include compacted clay barriers, slurry trench cutoff walls, geomembrane walls, diaphragm walls, sheetpile walls, grout curtains, hydraulic barriers, compacted clay liners, and geomembrane liners. The choice in any case is dictated by whether an existing waste disposal area or spill is to be contained or a new facility is to be constructed.

Initial site selection is critical to the successful performance of a waste disposal facility, and the hydrogeological setting is paramount among criteria for site acceptability. Unfortunately, this aspect has not been given adequate attention in the past.

Compacted clay barriers and liners are often well-suited for waste containment because of their low hydraulic conductivity, absorptive capacity for some chemicals, high stability and low cost. Proper placement water content and compaction procedures are essential for satisfactory performance. Compatibility between clay and waste must be considered in any case. Fortunately, evidence is now accumulating to indicate that the hydraulic conductivity of properly compacted clay is not significantly affected by exposure to most organic solvents and inorganic chemical wastes in dilute solution. Exposure to pure solvents may, however, cause interactions that yield increases in conductivity.

Flexible membrane liners, or geomembranes, composed of plastic or rubber sheets have wide application as waste containment liners because of their very low permeability. Great care is required in their installation to insure proper sealing of seams and to prevent damage from handling and equipment. Although resistant to many chemicals, some geomembranes are susceptible to degradation by organic solvents. The US Environmental Protection Agency now advocates the use of double liner systems. These 'belt and suspenders' designs incorporate two separate leachate collection and barrier systems. Both compacted clay and flexible membrane layers are used in these liners

Safe containment of nuclear waste after deep burial presents special problems owing to the long times (up to 10^6 years) that the groundwater must be protected and the higher temperatures associated with the radioactive decay. The evidence thus far is that smectite will be satisfactory for use as a buffer and hydraulic barrier seal in nuclear repositories, provided the temperature is always less than 100°C. Mineral mixtures may be more suitable than pure smectite.

A methodology is badly needed that will enable establishment of a maximum or threshold concentration of a specific waste in a disposal area such that the maximum allowable concentrations in the surrounding groundwater will not be exceeded under the most pessimistic leakage and groundwater flow conditions.

References

ANDERSON, D.C. & BROWN, K.W. 1981. Organic leachate effects on the permeability of clay liners. *Proceedings of the 7th Annual Research Symposium, Ft. Mitchell, KY,* EPA-600/9-81-002b, 199-130.

ANDERSON, D.C., BROWN, K.W., & GREEN, J.W. 1982. Effect of organic fluids on the permeability of clay soil liners. In SCHULTZ, D.W. (Ed) *Land Disposal of Hazardous Waste. Proceedings of the 8th Annual Research Symposium, Ft. Mitchell, KY,* EPA-600/9-82-002, pp. 179-190.

BROWN, K.W., GREEN, J.W., & THOMAS, J.C. 1983. The influence of selected organic liquids on the permeability of clay liners. In SCHULTZ, D.W. (ed) *Land Disposal Incineration and Treatment of Hazardous Waste. Proceedings of the 9th Annual Research Symposium, Ft. Mitchell, KY,* IPA-600-9-83-018, pp. 114-125.

CLEM, A.G., 1984. Modified bentonite sealant membrane. *International Symposium on Clay Barriers for Isolation of Toxic Chemical wastes, May 28-30, 1984, Stockholm, Sweden,* 158-163.

DANIEL, D.E., 1983. Shallow land burial of low-level radioactive waste. *Proceedings of American Society of Civil Engineers, Journal of Geotechnical Engineering,* **109,** GT. 1, 40-55.

DANIEL, D.E., 1984. Predicting hydraulic conductivity of clay liners. *Proceedings of American Society of Civil Engineers, Journal of Geotechnical Engineering,* **110,** GT. 2, 285-300.

DANIEL, D.E. & LILJESTRAND, H.M., 1984. *Effects on landfill leachates on natural liner systems.* Report to the Chemical Manufacturers Association, University of Texas at Austin, Department of Civil Engineering.

DAY, S.R. & DANIEL, D.E., 1985. Hydraulic conductivities of two prototype clay liners. *Proceedings of American Society of Civil Engineers, Journal of Geotechnical Engineering,* **111,** GT. 8, 957-970.

ENVIRONMENTAL INSTITUTE FOR WASTE MANAGEMENT STUDIES, 1985. *Disposal of solvents and solvent-contaminated wastes to land — a position paper.* (Preliminary draft, subject to revision), July 22, 1985, University of Alabama, Tuscaloosa, Alabama.

FERNANDEZ, F. & QUIGLEY, R.M., 1985. Hydraulic conductivity of natural clays permeated with simple liquid hydrocarbons. *Canadian Geotechnical Journal,* **22,** 205-214.

GHASSEMI, M., HARO, M., METZGER, J., POWERS, M., QUINLIVIN, S., SCINTO, L. & WHITE, H., 1983. Assessment of Technology for Constructing and Installing Cover and Bottom Liner Systems for Hazardous Waste Facilities, Vol. II. *EPA Contract No. 68-02-3174, Work Assignment No. 109.*

GUSWA, J.H., LYMAN, W.J., DONIGAN, A.S., JR., LO, T.Y.R. & SHANAHAN, E.W., 1984. *Groundwater contamination and emergency response guide,* Park Ridge, NJ, Noyes Data Corporation.

HARDCASTLE, J.H. & MITCHELL, J.K., 1974. Electrolyte concentration-permeability relationships in sodium illite-silt mixtures. *Clays and Clay Minerals,* **22,** 143-154.

HAXO, H.E., 1983. Analysis of fingerprinting of unexposed and exposed polymeric membrane liners. *Proceedings of the 9th Annual Research Symposium on Land Disposal of Hazardous Waste, Washington, D.C.,* EPA-600/9-83-018. 157-171.

HAXO, H.E. & NELSON, N.A., 1984. Permeability characteristics of flexible membrane liners measured in pinch tests. *Proceedings of the 10th Annual Research Symposium on Land Disposal of Hazardous Waste,* Ft. Mitchell, KY, EPA-600/9-84-007, 230-251.

JARRETT, P.J.D. & MARWOOD, E.W., 1984. The application of artificial clays in the control of toxic pollutants. *International Symposium on Clay Barriers for Isolation of Chemical Toxic Wastes, May 28-30, 1984, Stockholm,* Sweden, 74-82.

KUFS, C., WAGNER, K., ROGOSHEWSKI, P., KAPLAN, K. & REPA, E. 1983. Procedures and techniques for controlling the migration of leachate plumes. *Proceedings of the 9th Annual Research Symposium on Land Disposal of Hazardous Waste, Washington, D.C,* EPA-600/9-83-018, 87-113.

MICHAELS, A.S. & LIN, C.S. 1954. Permeability of kaolinite. *Industrial and Engineering Chemistry,* **46,** 1239-1246.

MILLET, R.A. & PEREZ, J.Y., 1981. Current USA practice: Slurry wall specifications. *Proceedings of American Society of Civil Engineers, Journal of the Geotechnical Engineering Division,* **107,** GT8, 1041-1056.

SPOONER, P.A., WETZEL, R.S., & GRUBE, W.E., 1983. Slurry trench construction of pollution migration cut-off walls. *Proceedings of the 9th Annual Research Symposium on Land Disposal of Hazardous Waste, Washington, D.C.,* EPA-600/9-83-018.

TALLARD, G., 1984. Slurry trenches for containing hazardous wastes. *Civil Engineering Magazine,* February, 41-45.

WASTE MANAGEMENT, INC. 1983. *Appendix: basis for tentative selection of HDPE liner material.* Internal File Document.

WOODWARD-CLYDE CONSULTANTS, 1983. Corrective action alternatives for goundwater contamination. Report prepared for the US Office of Technology Assessment, August.

J.K. MITCHELL, Department of Civil Engineering, University of California, Berkeley, CA 74720, USA

Trace organic compounds in groundwater: a survey of unpolluted groundwater from three major British aquifer systems

L. Clark & M.A.P. Kenrick

ABSTRACT: In view of the increased concern about the occurrence and persistence of organic chemicals in the environment, the Water Research Centre instigated a baseline survey of trace organics in groundwater from representative cross-sections of three major British aquifer systems.

A total of 32 public and private supply boreholes in selected areas of the Permo-Triassic Sandstone, Chalk, Lincolnshire Limestone and the Great Oolite aquifers have been sampled. The areas have been chosen with the aim of characterising the organic content of 'clean' groundwater from the confined and unconfined zones on each cross-section. The survey was designed to encompass groundwaters of varying ages, from recent recharge in outcrop areas through to ancient waters in deep confined sections. In this way, it was expected that any compounds of anthropogenic origin would be detected only in the younger groundwaters.

The total organic carbon content of the groundwaters sampled was generally very low, between 0.2 and 1.4 mg/l; low levels of aromatic halogenated hydrocarbons, which are generally taken to be contaminants, were found in many of the samples at levels above 0.01 μg/l; low levels of aromatic hydrocarbons were often detected. Although the number of compounds detected averaged 20 per sample (range: 7 to 52), few were present at concentrations greater than 0.1 μg/l. Over 100 different compounds were detected altogether but only 75 could be identified positively. Of these, nine occurred in more than half of the samples. There were no obvious differences in either compound types or levels between the confined and unconfined zones of any of the aquifers.

Uncontaminated groundwater examined in this survey may be typified as having a TOC level below 0.8 mg/l, and no more than four compounds (none of which are chlorinated hydrocarbons) present at levels above 0.1 μg/l.

Introduction

Groundwater may become contaminated by organic substances in a number of ways:

- leachate from landfill sites;
- agricultural and horticultural usage of organic chemicals;
- artificial aquifer recharge or effluent recharge;
- application of sewage sludge to land;
- atmospheric washout by rain or snow;
- infiltration of contaminated surface water; and
- natural processes such as animal or plant decomposition.

In the course of research into the effects of effluent recharge on groundwater quality in the major British aquifers (Baxter & Clark 1984), levels of total organic carbon in the groundwater around several sites were examined. More detailed studies on trace organic compounds showed that groundwater from control wells up gradient of the sites contained a range of compounds (Clark & Baxter 1982). Other studies on the identification of organic compounds in drinking waters (Fielding *et al.* 1981), have shown that groundwater may contain a wide variety of organic compounds.

Very little information exists on the identities and levels of organic compounds in uncontaminated groundwater. In view of the increased interest in organic pollution of groundwater throughout the world, the Water Research Centre considered it essential to make a survey of organic compounds in groundwaters considered to be uncontaminated when judged by the results of inorganic analyses. The results of such a survey would provide a baseline against which groundwater contamination may be defined.

The baseline survey was restricted to ten sites in each of the three major aquifer systems:

The Permo-Triassic Sandstones
The Jurassic Limestones
The Cretaceous Chalk

The groundwater samples were taken from active production boreholes and were analysed using a combination of gas chromatography and mass spectrometry (GC-MS).

From CRIPPS, J.C., BELL, F.G. & CULSHAW, M.G., (eds), 1986, *Groundwater in Engineering Geology,* Geological Society Engineering Geology Special Publication No. 3, pp. 159-167

Hydrogeology

The sites in each aquifer were chosen to lie along profiles extending from unconfined to confined parts of the aquifer so that changes in organic content as the groundwater moved through the aquifer could be adequately determined. The profiles selected (Fig. 1) are as follows:

Permo-Triassic : two closely-spaced profiles in the Nottinghamshire Sherwood Sandstone.

Jurassic : a profile in the Lincolnshire Limestone and one in the Great Oolite Series in the Cotswolds.

Cretaceous : one long profile in the Berkshire Chalk.

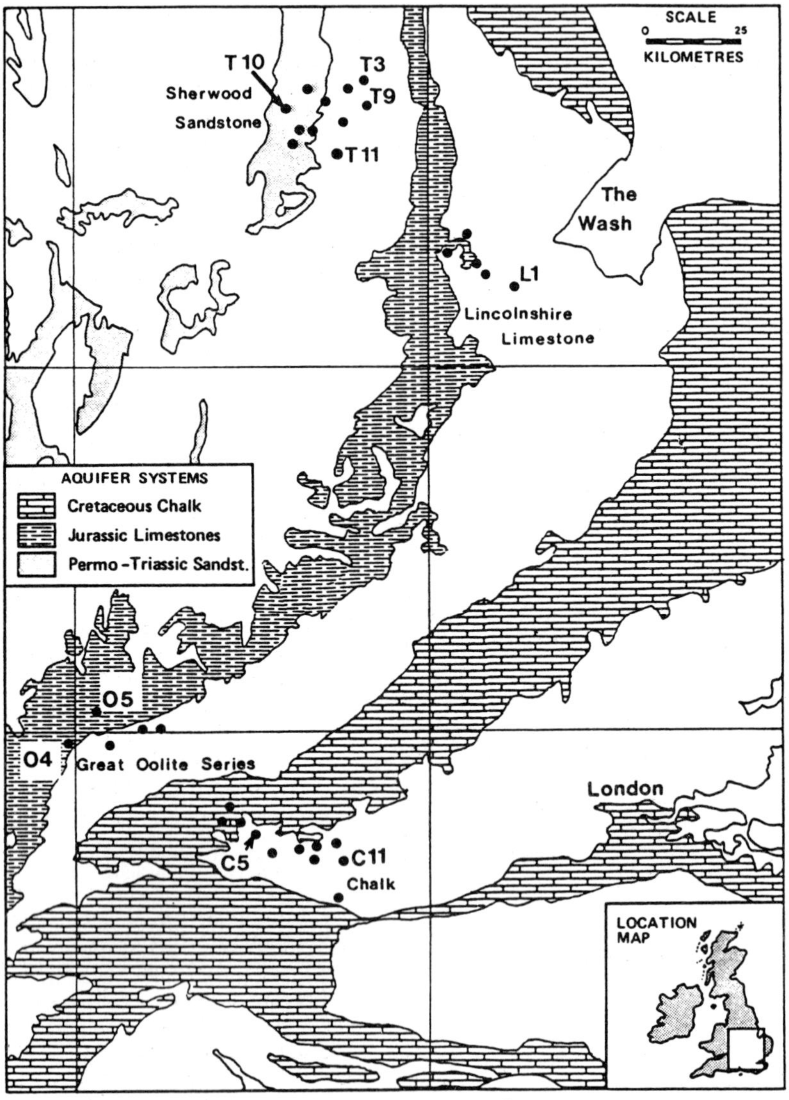

FIG. 1. Location of sampling profiles.

The criteria used to select these profiles were:

(1) the profiles had to extend from the unconfined outcrop areas to the deeply confined parts of the aquifer and include at least five operating boreholes;
(2) the boreholes had to have well-head sampling facilities situated so that chlorination or other treatment could not affect the sample quality;
(3) the inorganic chemistry of the groundwater had to have been studied in detail along each profile to provide a base for comparison with the organic results;
(4) the age of the groundwater should range from recent to many thousand years, that is, from the industrial age to the pre-industrial age;
(5) there should be no obvious groundwater pollution along the profiles; and
(6) the geological structure along each profile should be simple so that it would not interfere with the interpretation of analytical results.

The hydrogeology and inorganic hydrochemical evolution of the groundwater along each profile have been studied in great detail by previous workers. The references for these basic hydrogeological data are as follows:

Nottinghamshire Sherwood Sandstone — Edmunds *et al.* (1982)
Lincolnshire Limestone — Edmunds & Walton (1983)
Great Oolite Series — Morgan-Jones & Eggboro (1981).
Berkshire Chalk — Bath *et al.* (in preparation).

The profiles in all four aquifers show that there is a significant hydrochemical evolution in the groundwater as it passes from the unconfined to the confined zone of the aquifer. This transition is marked by a dramatic fall in the redox potential (Eh) of the water, and at, or close to, this Eh fall, those determinands characteristic of the oxidised, unconfined zone decrease in importance. Dissolved oxygen and nitrate fall close to zero while reduced iron and sulphur may appear in the groundwater.

The trace organic survey was designed to determine whether a quality evolution of organic compounds was present in the groundwater, comparable to those of their inorganic quality profiles.

Sampling and analysis

Sampling

The object when sampling groundwater for its organic content is to obtain a representative sample of water directly from the matrix of the aquifer under study. Special sampling apparatus (Pankow *et al.* 1984) could have been used, but would have been exhorbitantly expensive, so the present survey was based on existing water supply boreholes. Those boreholes with plastic used in their construction were avoided although, in most cases, the original plasticizers in the plastic would have been removed by leaching. The effects, if any, on the samples passing through the man-made components of a borehole (pump, rising main, taps) could not be assessed, but would be expected to be insignificant because of the age of the boreholes and the dilution available in their discharge.

Samples from each site were collected in three 5-litre glass bottles with ground glass stoppers. Two of the three 5-litre samples were needed to prepare organic extracts at two different pH levels; the third sample was a spare in case of mishaps.

In order to ensure that the water sampled was drawn freshly from the aquifer and had not stood for sometime in the borehole column, the borehole operators ran the borehole to discharge for some hours before sampling. Samples were taken, where possible, from sampling taps on the well-head pipework to avoid any influence of water treatment chemicals. The sample taps were flushed to waste for several minutes before sampling.

After rinsing each sample bottle twice with water from the borehole, the bottle was filled directly from the sample tap. Great care was taken to avoid contamination from condensate on taps and hands and the bottles were filled to the top to avoid any head-space volatile losses. Four further 125 ml samples were taken for total organic carbon (TOC) and total organo-halogen (AOX) analyses.

The samples from each site were taken directly by car to the laboratory to ensure that the organic extraction and concentration could be completed on the same day as the sampling.

Analysis

The analytical procedure for trace organics comprises:

(1) pH adjustment and addition of deuterated internal standards;
(2) extraction using a macroreticular styrene-divinyl benzene resin, XAD-2;
(3) elution of the adsorbed organic compounds from the resin using diethyl either followed by concentration of the diethyl ether extract to a volume suitable for examination by GC-MS;
(4) GS-MS examination of a portion of the concentrated extract to identify, where possible, the individual compounds present.

This procedure is discussed in some detail in Fielding *et al.* (1981), and Kenrick *et al.* (1985).

Gas chromatography is a routine method for the analysis of volatile organic compounds, and the use of a mass spectrometer as a detector for a gas chromatograph can provide sufficient information to enable a compound of interest to be identified. In theory, every organic compound has a unique mass spectrum which is reproducable, provided all mass spectrometric conditions are constant. In practice, however, the differences in the mass spectra of similar compounds may be very slight, and an unambiguous identification may not be possible for every compound detected.

The quantification of individual compounds also presents many problems in the analyses. The levels of the various compounds detected have been estimated from the response obtained for the various internal deuterated standards added to the samples prior to extraction. As is usual for general survey GC-MS analyses, certain assumptions have to be made concerning extraction efficiencies and GS-MS responses. This approach can give reasonable quantification (coefficients of variation <15% at the 0.1 μg/l level (James *et al.* 1980)) if the compound of interest is similar to one of the internal standards. However, for a totally unknown compound large error may occur. A minimum possible concentration can be calculated by assuming no losses (either by volatilisation or biodegradation) prior to extraction, 100% extraction efficiency, no losses during concentration of the extract, ideal behaviour on the GC column and excellent mass spectrometric response. Poor behaviour in any or all of these stages may result in this minimum possible amount being two orders of magnitude different from the actual amount.

The analyses for TOC were carried out on a Photochem Organic Carbon Analyser with a limit of detection of about 0.2 mg/1. The AOX analyses were carried out on a KIWA colorimetric instrument with a detection limit of about 10 μg/l.

Results

Relatively high compound concentrations were met in four of the initial 30 sites sampled. Three of these four sites were resampled and only the Cold Ash source in the Chalk (Fig. 1, C5) continued to show signs of contamination. The fourth site, Sheephouse Farm (Fig. 1, 05), in the Great Oolite Series, was not resampled because it contained such a complex suite of unusual compounds that its contamination was clear. Two extra boreholes were sampled (C11 and T11) to make up for the two contaminated ones. The Cold Ash site was also resampled four times to assess sampling repeatability and two sites in the Sherwood Sandstone (T9, T10) were resampled to five-monthly intervals to assess seasonal variability.

A total of 137 compounds were detected in the 32 samples of the general survey and of these 73 were positively identified (Table 1), 5 were tentatively identified, 19 were classified as isomers and 40 could not be identified. The majority of samples contained less than 30 compounds (Table 2), although some compounds could have been present below their detection limit.

The nine most commonly occurring compounds are listed in Table 3, together with their concentration data. Only three compounds (squalene, sulphur (S_8) and palmitic acid) were regularly detected at levels in excess of 0.1 μg/l. Toluene and the phthalates were generally in the range 0.05 to 0.1 μg/l, while the other compounds were usually below 0.03 μg/l.

Thirty-one compounds were detected at levels which occasionally exceeded 0.1 μg/l and nine of these were present at levels which exceeded 1 μg/l in some samples (Table 4).

TABLE 1.
Compounds identified, classified by functional groups.

1. Hydrocarbons	*3. Carboxylic acids and esters*
Hexene isomers (2)	Acetic acid
Benzene	Methyl benzoate
Cyclohexane	Hexanoic acid
Toluene	Octanoic acid
Heptadiene isomer	Lauric acid
o-Xylene	Diethyl phthalate
m-Xylene	Myristic acid
Ethylbenzene	Pentadecanoic acid
Styrene	Methyl palmitate
Nonane isomer	Dibutyl phthalate isomers (2)
C_3-Alkylbenzene isomers (8)	Palmitic acid
Decane isomer	Stearic acid
Limonene	Dioctyl adipate
C_4-Alkylbenzene isomers (2)	Dioctyl phthalate isomers (2)
Napthalene	bis-(2-Ethylhexyl) phthalate
1-Methylnaphthalene	bis-(2-Ethylhexyl) sebacate
2-Methylnaphthalene	
Dimethylnaphthalene	*4. Other oxygenated compounds*
Phenanthrene	
Fluoranthene	2,2-Dimethoxypropane
Pyrene	1,4-Dioxane
	Benzaldehyde
2. Halogenated hydrocarbons	2-Ethylhexanal (?)
	5-Methyl-5-hepten-2-one
cis-1,2-Dichloroethylene	Phenol
Chloroform	Xylenol isomer
1,1,1-Trichloroethane	iso-Propylphenol isomers (2)
Carbon tetrachloride	t-Butylcresol
Trichloroethylene	2,6-Di-t-butyl-p-benzoquinone
Chlorodibromomethane	Di-t-butylcresol isomer
Tetrachloroethylene	
1,1,2-Trichloroethane	*5. Miscellaneous*
Bromoform	
1-4-Dichlorobenzene	Hexyl cyanide (?)
$(CF_2\text{-}CFCl)_n$ (6)	Benzthiazole
Hexachlorobenzene	Heptyl cyanide (?)
	Octanamide (?)
	Decanamide (?)
	Sulphur, S_6, S_7 and S_8
	Diphenylamine
	Atrazine
	Pentachlorophenol
	iso-Propyldiphenylamine
	N-Butylbenzenesulphoneamide

(?) denotes tentative identification.

Of the 22 compounds detected at levels in the range 0.1 – 1 μg/l at at least one site, ten were detected at only one site — Sheephouse Farm. The remaining twelve are listed in Table 5.

TABLE 2. *Number of compounds detected at sites sampled.*

Sample No.	Sherwood Sandstone S	Berkshire Chalk C	Lincolnshire Limestone L	Great Oolite Series O
1	22	10	17	13
2	18	28	11	52
3	12	26	9	15
4	26	20	25	9
5	14	11	27	13
6	15	31		
7	24	7		
8	20	21		
9	16	31		
10	20	9		
11	39	20		
Average	21	19	18	20

TABLE 3. *Most commonly occurring compounds.*

	Concentration (μg/l)			
Compound name	Highest	Mean (1)	Average (2)	Occurrences
Cyclohexane	0.08	0.03	0.022	23
Toluene	0.2	0.046	0.027	19
m-Xylene	0.02	0.011	0.007	19
Ethylbenzene	0.07	0.016	0.014	28
Naphthalene	0.04	0.019	0.015	25
Squalene	1.0	0.26	0.20	25
Sulphur (S_8)*	2	0.4	0.22	18
Palmitic acid	5	0.78	0.51	21
Dioctyl phthalate	0.1	0.056	0.045	26

1. Mean level for samples in which the compound was detected.
2. Average level for all 32 sampled examined.
* Sulphur(S_6) and (S_7) were also present in some samples, but at much lower levels.

TABLE 4. *Identified compounds occasionally present at > 1 μg/l.*

	Concentration (μg/l)			
Compound name	Highest	Mean (1)	Average (2)	Occurrences
Chloroform	5*	2.6	0.16	2
bisa-(2-Ethylhexyl-sebacate)	1**	1	0.03	1
Myristic acid	1	0.26	0.10	12
Palmitic acid	5	0.78	0.51	21
Squalene	1	0.26	0.20	25
Sulphur (S_8)	2	0.40	0.22	18
1,1,1-Trichloroethane	4*	0.48	0.16	11
1,1,2-Trichloroethane	3*	3	0.09	1
Trichloroethylene	10*	1.9	0.36	6

1. Mean relates only to occurrences.
2. Average relates to all 32 sampling sites.
* Cold Ash.
** Sheephouse Farm.

TABLE 5. *Identified compounds occasionally present between 0.1 and 1 μg/l.*

Compound name	Concentration (μg/l) Highest	Mean (2)	Average (3)	Occurrences
Atrazine	0.2	0.2	<0.01	1
Bromoform	0.1	0.1	<0.01	2
Chlorodibromomethane	0.1	0.1	<0.01	4
1,4-Dichlorobenzene	0.1	0.03	0.01	14
cis-1,2-Dichloroethylene	0.5	0.5	0.02	1
Dioctyl phthalate	0.1	0.05	0.02	21
1,4-Dioxane	0.5	0.26	0.02	2
Hexanoic acid	0.2	0.2	<0.01	1
Pentachlorophenol	0.5	0.5	0.08	5
Phenol	0.1	0.1	<0.01	1
Sulphur (S_6)	0.2	0.05	0.01	7
Stearic acid	1.0	0.4	0.14	11
Tetrachloroethylene	0.5	0.27	0.04	5

1. Compounds detected only at Sheephouse Farm or Cold Ash are not included.
2. Mean for samples in which compound was detected.
3. Average for 32 samples.

TABLE 6. *Levels of halogenated hydrocarbons detected.*

Compound name	Concentration (μg/l) Cold Ash (1)	Max (2)	Mean (3)	Ave (4)	Occurrences
cis-1,2-Dichloroethylene	0.5		0.5	0.02	1
Chloroform	5	0.25	2.6	0.16	2
1,1,1-Trichloroethane	4	0.5	0.48	0.16	11
Carbon tetrachloride		0.1	0.05	<0.01	3
Trichloroethylene	10	1	1.9	0.36	6
1,1,2-Trichloroethane	3		3	0.09	1
Chlorodibromomethane	0.1	0.1	0.1	0.01	4
Tetrachloroethylene	0.5	0.5	0.27	0.04	5
Bromoform		0.1	0.1	<0.01	1
1,4-Dichlorobenzene	0.1	0.1	0.03	0.01	14

1. Cold Ash treated separately since it was significantly contaminated by halogenated compounds.
2. Maximum level at sites other than Cold Ash.
3. Mean relates only to the samples in which a particular compound was detected.
4. Average related to all 32 sampling sites.

The data in Tables 1 to 5 show that compounds which are most often detected are mainly hydrocarbons, while many of the compounds present at levels above 0.1 μg/l are halogenated compounds. Six hydrocarbons were regularly detected in this survey; cyclohexane, toluene, m-xylene, ethyl benzene, naphthalene and squalene. Halogenated compounds are of interest as they are normally assumed to be indicative of pollution and yet, in the present survey, several samples from each aquifer contained at least one halogenated compound. The highest levels detected, however, were in the slightly contaminated borehole at Cold Ash (Table 6).

Two compounds of specific interest are Atrazine, and Pentachlorophenol for both are man-made, the former a herbicide and the latter a wood preservative. Atrazine was detected at Kemble Junction in the Great Oolite (Fig. 1, O4) and later, in repeat sampling at Budby and Newton in the Sherwood Sandstone (Fig. 1, T9, T10). Pentachlorophenol was detected at six sites of which five were in the Sherwood Sandstone and one was in the Chalk. Two sites, Cottam and Newton (T3 and T9), are deeply confined and the presence of Atrazine and Pentachlorophenol is very difficult to explain as the water has been dated at several thousand years old (Edmunds *et al.* 1982).

The less precise measurements of organic content of groundwater, TOC and AOX, show that the overall levels of organics are low. AOX levels were below the detection limit of the method used (10 μg/l) and TOC levels were generally less than 1 mg/l (Table 7).

Repeat samples were taken at Budby (Fig. 1,T10) and Newton (T9) in June and November 1983 and April 1984 to see if there is a seasonal variation in quality. Slight variations between different samples were apparent, but no seasonal trend could be seen. Similarly, repeat samples at Cold Ash (C5), taken to test sample variations showed slight, but not significant, variations in the levels of compound detected.

Intra-aquifer trends

If any intra-aquifer trends exist, they are subtle. Various parameters which could be indicative of such trends are shown, with their corresponding values for the Sherwood Sandstone and Chalk, in Table 8. The low number of samples for the Lincolnshire Limestone and Great Oolite Series probably do not justify similar comparisons.

TABLE 7. *Results of TOC analyses.*

Sample No.	Sherwood Sandstone S	Chalk C	Lincolnshire Limestone L	Great Oolite Series O
1	0.3	0.2	0.7	0.35
2	0.45	0.35	0.2	0.85
3	0.8	0.4	0.3	0.55
4	0.4	0.55	0.85	0.2
5	0.2	0.4	1.4	0.65
6	0.25	0.2		
7	0.2	0.35		
8	0.2	0.2		
9	0.35	0.35		
10	0.25	0.35		
11	0.2	0.5		

TABLE 8. *Parameters possibly indicative of intra-aquifer trends.*

Aquifer zone	Sherwood Sandstone			Chalk	
	Unconfined (I)	Confined (II)	Confined (III)	Unconfined	Confined
TOC (mg/l)	0.43	0.22	0.27	0.38	0.34
Average number of compounds detected	18	20	25	21	19
Average number of detected >0.1 μg/l	5.6	4.7	3.0	5.3	5.0
Average number of halogenated compounds detected	1.8	1.3	0.7	2.8	1.4

Although there may be a reduction in, for example, the average number of halogenated compounds detected in the confined section of the aquifers compared to the unconfined sections, variations between individual sites preclude the use of all of the parameters listed to differentiate between confined and unconfined aquifers.

The occasional detection of trace amounts of halogenated compounds, such as pentachlorophenol and trichloroethylene in confined aquifers, where the groundwater is ancient, is very difficult to explain. If one accepts recent, anthropogenic origins for these compounds and a truly confined aquifer, then there must be local contamination via the borehole itself or surface contamination of sampling taps etc, or sporadic low-level contamination during sample handling and anlyasis. However, natural sources of halogenated compounds exist, usually relating to the marine environment, but these hardly provide a satisfactory explanation of the compounds encountered.

Conclusions

Thirty-two sites have been examined in total and of these, thirty were broadly similar in that usually less than thirty individual organic compounds were detected, with only five or less present at levels above 0.1 μg/l. Two sites, Cold Ash (C5) and Sheephouse Farm (O5), were markedly different, and are considered to to contaminated.

The inorganic constituents of groundwater exhibit marked changes along the profiles sampled in the initial survey with the result that the inorganic water quality in the unconfined zones of the aquifers is totally different from that in their confined zones. No such trends are apparent in the distribution of organic compounds through the aquifers.

Halogenated compounds were detected at 18 (56%) of the sites and at twelve of these sites the level of at least one halogenated compound was greater than 0.1 μg/l. Halogenated compounds have been found in almost all the boreholes in the confined zones or aquifers. Pentachlorophenol has been found at Newton in the Sherwood Sandstone (T9), for example, and bromoform at Spalding (L1). Bromoform is usually regarded as a byproduct of chlorination yet Spalding borehole, as far as in known, has never been chlorinated. The origin of these halogenated compounds is unknown but must be one of the following:

(1) Naturally occurring compounds unrelated to Man.
(2) Products of contamination of the aquifer or borehole.
(3) Products of contamination of the sample during sampling, transport or sample processing.

Halogenated compounds are often taken to be indicative of pollution because they are accepted to be anthropogenic. The second origin is possible, but from hydrogeological evidence, seems unlikely. If the halogenated compounds are derived from sample contamination during sampling, transport or analysis, then this fact underlines the problems of obtaining organically-uncontaminated water from boreholes, as the sites were selected for this survey on the basis of their suitability for sampling. However, the origins of these compounds are unknown and further investigation is needed to clarify the situation.

Total Organic Carbon levels (average 0.42 mg/l) were low and adsorbable organic halogen could not be detected (<10 μg/l). No obvious differences in the types or levels of compounds detected in samples from different aquifers were evident, and no obvious trends emerged when comparing confined and unconfined sections of any of the aquifers.

The ease with which the samples from Cold Ash and Sheephouse Farm could be identified as being contaminated suggest that the data base produced by the survey will be useful in defining uncontaminated groundwater, particularly when further analyses are added to broaden the data base. The water from the two contaminated sites is not grossly contaminated by other criteria.

ACKNOWLEDGEMENTS: This baseline survey is part of a major investigations by WRC into groundwater pollution by organic compounds. Many people are involved in this investigation which is under the management of C P Young. The GS-MS analyses have been done under the direction of Mr. M Fielding and Dr H A Jones. Thanks are also due to the Severn Trent, Anglia and Thames Water Authorities for their co-operation. The paper is published by permission of the Director of the Medmenham Laboratory of the Water Research Centre.

References

BATH, A.A., COOK, J.M., DARLING, W.G., EDMUNDS, W.M., KINNIBOROUGH, D.G., MILES, D.L. & MORGAN-JONES, M. Geochemical evolution of Chalk groundwater in the Berkshire Syncline. *British Geological Survey Report*, (in preparation)

BAXTER, K.M. & CLARK, L., 1984. Effluent recharge — the effects of effluent recharge on groundwater quality. *Technical Report, TR 199, Water Research Centre.*

CLARK, L & BAXTER, K.M., 1982. Organic micropollutants in effluent recharge to groundwater. *Journal of Water Science and Technology*, **14** (12)/B3, 15-30.

EDMUNDS, W.M., BATH, A.H., & MILES, D.L., 1982. Hydrochemical evolution of the East Midlands Triassic sandstone aquifer, England. *Geochimica et Cosmochimica Acta*, **46** (11), 2069-2081.

EDMUNDS, W.M. & WALTON, N.G.R., 1983. The Lincolnshire Limestone - Hydrogeochemical evolution over a ten-year period. In: W. Back and P.E. LaMoreaux (Guest Editors), V.T. Stringfield Symposium -Processes in Karst Hydrology. *Journal of Hydrology*, **61** (1/3), 201-211.

FIELDING, M., GIBSON, T.M., JAMES, H.A., MCLOUGHLIN, K. & STEEL, C.P., 1981. Organic micropollutants in drinking water. *Technical Report TR 159, Water Research Centre.*

JAMES, H.A., FIELDING, M., GIBSON, T.M., & STEEL, C.P., 1980. Quantitative aspects of the determination of organic micropollutants in raw and potable waters. In: *Advances in Mass Spectrometry*, **8**, Heyden, London, 429-435.

KENRICK, M.A.P., CLARK, L., BAXTER, K.M., FLEET, M., JAMES, H.A., GIBSON, T.M. & TURRELL, M.B., 1985. Trace Organics in British Aquifers. *Technical Report TR 223, Water Research Centre.*

MORGAN-JONES, M. & EGGBORO, M.D., 1981. The hydrogeochemistry of the Jurassic Limestones in Gloucestershire, England. *Quarterly Journal of Engineering Geology*, **14** (1), 21-27.

CLARK, L. & KENRICK, M.A.P. Water Research Centre, Henley Road, Medmenham, PO Box 16, Marlow, Bucks SL7 2HD, UK.

Groundwater in Engineering Geology, London, 1986, pp. 169-176 Printed in England

Application of isotope hydrology to groundwater problems in engineering geology

P.J. Leggo, C.D. Warren & D. Blakeway

ABSTRACT: Research on the natural variation of the isotopic species of hydrogen and oxygen has led to a firm understanding of the physical processes involved in the isotopic fractionation of these elements in water. Although much use has been made of artificially introduced chemical tracers to investigate flow patterns and characterise the groundwater, this method commonly suffers from bulk dilution and additional problems that limit its applicability to many investigations in engineering geology. Such limitations are avoided in the use of stable isotopic measurements as it is the variability in the isotope species of the water molecule itself that are used as the characterisation parameters providing a 'fingerprint' for each particular water sample.

In this paper background details of stable isotopic fractionation are summarised and two recent case histories presented to demonstrate its application to groundwater studies, on a small scale, commonly met in many civil engineering situations, namely investigation of possible leakage from a reservoir and identification of the source of water contained within a large landslip. By the seasonal variations in the isotopic values, trends in groundwater isotopic composition can be identified and interpreted within the often limited geological and hydrogeological constraints of the site so as to provide conclusions as to the origin of the groundwaters.

Introduction

In recent years increasing use has been made of the distribution of the various isotopic species occurring naturally in water, particularly where the introduction of chemical tracers suffers from the problem of bulk dilution (Siddle *et al.* 1984), or may be considered unacceptable if the reservoir or groundwater source is for consumer use. Interpretation of isotopic data can provide information on source of water, flow path, velocity, mechanism of recharge, chemical mixing and the age of the water. Isotopes commonly measured in water include the stable isotopes of oxygen, hydrogen and carbon, usually expressed as relative isotopic ratios $^{18}O/^{16}O$, D/H (D refers to deuterium or ^{2}H) and $^{13}C/^{12}C$ respectively, and the radioactive isotopes of tritium (^{3}H or T) and carbon (^{14}C).

Research on the natural variation in the isotopic composition of hydrogen and oxygen has led to an understanding of the isotopic fractionation of water that occurs during the meteoric cycle. In general the isotopic fractionation of hydrogen and oxygen occurs during every phase transformation and the meteoric cycle has been likened to a giant multi-stage distillation column (Epstein & Mayeda 1953). It is well established that fractionation patterns produced by evaporation and condensation of water in the lower atmosphere are preserved in precipitation and their accurate measurement provide parameters which can be used for the characterisation of water. Due to the higher vapour pressure of $H_2{}^{16}O$ relative to $HD^{16}O$ and $H_2{}^{18}O$, water vapour becomes enriched in the lighter molecule during evaporation and similarly during condensation the liquid phase becomes enriched in the heavy species. In this way repeated evaporation and condensation leads to progressively lighter isotopic values in the vapour phase.

The degree to which any particular rain is isotopically fractionated is a function of many variables but the main controlling influence of air temperature has been known for some years, (Dansgaard 1964). The isotopic composition of rain is also controlled, to various degrees, by such factors as: physiographic relief, latitude, inland distance and rainfall amount, and worldwide studies have shown that seasonal variations in air temperature can be directly correlated to systematic variations in isotopic composition. Thus rainfall at colder temperatures, higher altitudes and latitudes, and greater distance from the ocean will be characterised by isotopically light δD and $\delta^{18}O$ values (see below for terminology). The dependence of $\delta^{18}0$ values on altitude has been found to be from –0.13 to –0.3% per hundred metres (Evans, 1982). Such isotopic effects produce variations often orders of magnitude greater than analytical variance and can be used to characterise water from differing environments.

Isotope hydrology

In order to characterise the water so as to provide some clue to its source, one first has to measure the

From CRIPPS, J.C., BELL, F.G. & CULSHAW, M.G., (eds), 1986, *Groundwater in Engineering Geology*, Geological Society Engineering Geology Special Publication No. 3, pp. 169-176

concentration of the particular isotopic species in question. Because absolute measurement of isotopic ratios is analytically difficult relative isotope ratios are measured instead with the customary use of the symbol δ (delta) to designate the relative difference in concentration compared to an agreed International Standard. The δ value is thus a measure of the isotopic enrichment or depletion of a particular isotopic species being given in parts per thousand (units of permil ‰) and defined as:

$$\delta \left[\begin{array}{c} \text{Isotopic Species} \\ \text{D or } ^{18}\text{O} \end{array} \right] ‰ = \frac{(R_{\text{sample}} - R_{\text{standard}}) \times 1000}{R_{\text{standard}}}$$

where R is the isotopic ratio of the heavy to light species, D/H in the case of hydrogen and $^{18}O/^{16}O$ in the case of oxygen with the standard for these elements being Standard Mean Ocean Water (SMOW).

In all meteoric processes it has been found that a direct linear relationship exists between the isotopic fractionation of hydrogen and oxygen which is explained by the fact that equilibrium phase changes affect both δD and $\delta^{18}0$ sympathetically. Thus rainwater values, under equilibrium conditions, accord to a general equation commonly known as the 'world rainfall line' or 'global precipitation line':

$$D = 8\,\delta\,^{18}O + 10(‰) \text{ (Craig 1961).}$$

Whereas condensation is an equilibrium reaction, evaporation, because it results in irreversible kinetic effects, tends to bring about a deviation from the global precipitation line, see Fig. 1. Open systems such as lakes and reservoirs are subjected to repeated evaporation and condensation of their surface water with the result that samples taken from such sources are known to plot away from the global precipitation line. On the other hand rainwater entering the ground is generally isolated from such alteration effects and its isotopic character remains close to that formed at the time of precipitation. Hence δD and $\delta^{18}O$ values may 'fingerprint' water masses from difference sources.

In temperate climates the fractionation of hydrogen and oxygen during atmospheric precipitation closely follows a Rayleigh process at liquid — vapour equilibrium in which the isotopic composition of the precipitation depends mostly on temperature (Hoefs 1973). It is possible to relate values of $\delta^{18}O$ to mean precipitation temperatures (t) using the empirically derived equation:

$$\delta^{18}O = 0.69\,t - 13.6\ (‰).$$

By this means it is generally possible to distinguish summer and winter rain in that because the air temperature is highest during the summer months the rainfall during this period will be isotopically heavier than that falling during the winter provided that other factors such as rainfall amount and surface re-evaporation are not extreme enough to hide this seasonal effect.

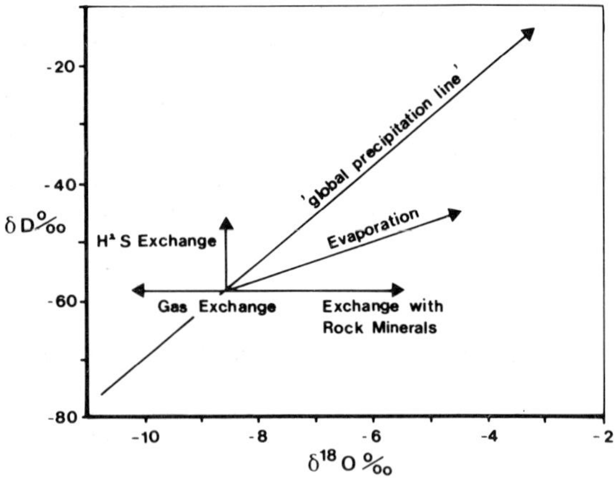

FIG. 1. Generalised $\delta^{18}O$ versus δD plot showing processes commonly responsible for deviation from the global precipitation line (modified from Toran 1982).

In the two case histories subsequently referred to seasonal trends in isotopic fractionisation patterns have been used to characterise rainfalls, surface water and groundwater so that clues can be given as to the likely origin of the groundwaters. The case histories concern:

FIG. 2. Case histories — location plan.

(1) *Moorgreen Reservoir, Nottinghamshire (Halcrow 1980)* where issues of water can be observed emerging near the downstream toe of the dam embankment and the question arises whether the seepage represents water leaking from the reservoir or is due to groundwater resurgence from the rocks beneath the reservoir site.

(2) *Taren Landslip, South Wales (Halcrow 1980)* where a perched water table exists in the landslip debris whose recharge could be directly from rainfall or could represent groundwater emerging from the surrounding Coal Measures bedrock.

The location of the two sites are shown on Fig. 2.

(1) Moorgreen Reservoir, Nottinghamshire

Moorgreen reservoir and associated embankment dam is situated 1.8 km north-east of Newthorpe Grange Estate, Eastwood, Nottinghamshire, some 10 kms north of Nottingham and is owned by the British Waterways Board. The reservoir extends approximately 1 km in a northerly direction from the dam embankment which is built on an outcrop of Middle Coal Measures strata. The dam crosses the valley on a north-west to south-east alignment and it has a crest length of some 230 m and a maximum height of 11 to 12 m. Its maximum width at the base is about 70 m.

The construction of the reservoir is presumed to have been completed soon after 1794 the source of the water being mainly the southward flowing Beauvale Brook and a small tributary stream, the Willey Spring, flowing into the north-west end of the reservoir. Both streams appear to be spring fed to some degree and to take surface run-off from the southern part of the catchment area.

The general relief of the catchment area is not great although shallow valleys do produce a relief of some 60 m, locally revealing the underlying geology shown on Fig. 3. The area was subjected to glaciation, which has left a fairly widespread but thin cover of drift deposits and occasional patches of till on the higher ground, although a thin mantle of weathered rock and alluvium is present in the valley floor. Bedrock cropping out at the dam site is composed of the Middle Coal Measures strata comprising sandstones, siltstones, mudstones, seatearths and coals which generally dip towards the north-east at about 3° to 4°. The Middle Coal Measures are unconformably overlain further east by the Lower Permian Marl and Magnesian Limestone which dip at a shallow angle eastwards. The Lower Permian Marl acts as an aquiclude to the Magnesian Limestone and numerous springs emerge at the contact although many of these disappear during the summer months suggesting that the aquifer has only a limited capacity as compared to the underlying Coal Measures whose springs discharge throughout the year and appear to represent a separate groundwater system.

Moorgreen reservoir is subject to Statutory Inspections under the Reservoir (Safety Provisions) Act 1930 and in his report on the inspection of 3rd March, 1977, the Inspecting Engineer, N.J. Cochrane, (Sir William Halcrow and Partners) indicated the need to ascertain the extent to which water seepages downstream of the dam embankment originated from water leaking from the reservoir, or from groundwater issuing as springs. The possibility that the water seepages at the downstream toe could represent leakage from the reservoir is further advanced by the fact that the dam and underlying Coal Measures rocks have been subjected to some 4.8 m subsidence over the past 110 years with the total seam extraction under the dam head being some 8.3 m. The alternating strains which would have developed in the strata due to the subsidence has greatly damaged the strata and opened up pre-existing joints. Minor faulting also appears to have affected the strata beneath the dam on its eastern side.

In order to investigate the source of the issues emerging at the downstream toe of the dam, isotopic analyses were carried out on fifteen samples, six of which were taken from the reservoir itself and four from the issues being studied (one each from the sandstones on either side of the spillway, one from the ochreous sinkhole, and one from the spring issuing from the gravel stream bed close to the north-west corner of the dam). The remaining samples were of groundwater taken from springs within the catchment and from a well on the western side of the catchment. The location of the sample sites relative to the reservoir are briefly described in Table 1 and illustrated on Fig. 3.

Samples were taken in July 1979, the 'summer' samples, and also in February 1980, the 'winter' samples. Owing to the fact that air temperature is highest and evaporation greatest during the summer months, the rainfall during this period will be isotopically heavier than that falling during the winter, as long as other factors, for instance, relief, are not so extreme that this seasonal effect is hidden. In the case of Moorgreen Reservoir catchment with its low relief and small extent, it is to be expected that fractionation effects, other than temperature, will be negligible.

The results are plotted on the delta diagrams shown in Fig. 4. The samples taken in July 1979 showed that the reservoir water exhibited far less isotopic variability than the resurgent groundwater. Data obtained on five reservoir water samples define a small area on the delta diagram, and none of the groundwater samples have isotopic parameters which plot inside this field (Fig. 4a). However, a linear relationship exists between the isotopic fractionation of hydrogen and oxygen for the data obtained from groundwater samples collected close to the dam embankment. Because the reservoir data field lies on this 'groundwater' trend line, there still exists a

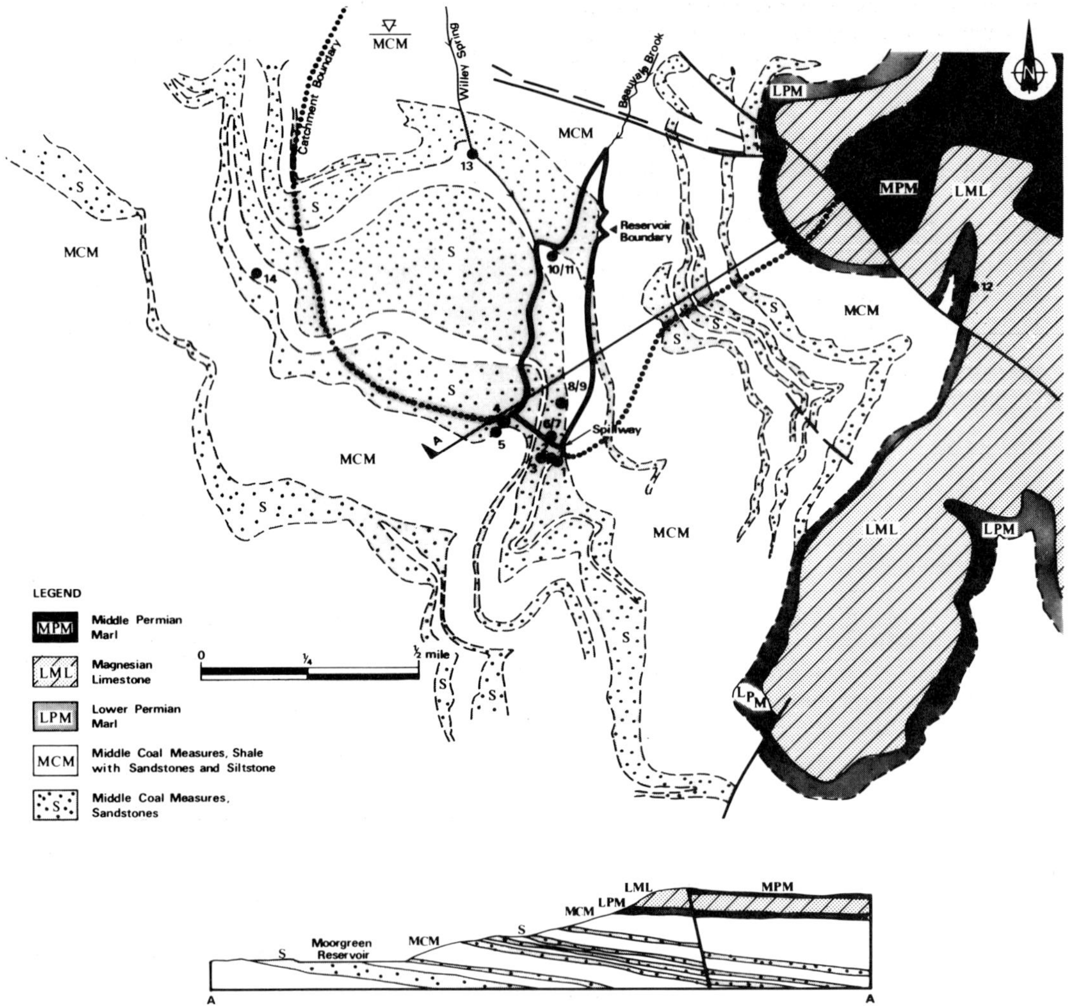

FIG. 3. Geology of the reservoir area and location of sample sites.

TABLE 1. *Sample description.*

Sample no.	Location
1	Spring Water from MCM sandstones from bottom of spillway left-hand side facing downstream.
2	Spring water from MCM sandstones from bottom of spillway right-hand side facing downstream.
3	Spring water from MCM sandstones 10 m below dam embankment adjacent to the bottom of spillway. Water feeding from ochreous sinkhole.
4	Spring water from foot of embankment at the north-west corner of the dam embankment issuing from gravel stream bed.
5	Spring water from near surface outcrop of MCM sandstones in hillside field approximately at the 90 m contour, 55 m south-west of sample 4.
6	Reservoir water just below surface 14 m, north of spillway.
7	Reservoir water near bottom 14 m, north of spillway.
8	Reservoir water just below surface 185 m north of spillway.
9	Reservoir water near bottom, 185 m north of spillway.
10	Reservoir water just below surface, 685 m north of spillway.
11	Reservoir water near bottom, north of spillway.
12	Spring water from Robin Hood's Well, junction of Magnesian Limestone with Lower Permian Marl.
13	Fast flowing surface stream 500 m due west of the northern end of reservoir.
14	Well water from Willeywood Farm.
15.	(i) Spring water from Marl/Limestone contact, hillside site 780 m south of Beauvale Manor Farm.
	(ii) Spring water below tree roots 460 m east of Beauvale Manor Farm, near contact of Limestone Marl.

possibility that some mixing is taking place between reservoir water and resurgent groundwater issuing from the Middle Coal Measures bedrock at the downstream toe of the dam. If this were the case, groundwater samples whose isotopic parameters plot nearest the reservoir water field (e.g. samples 2 and 3) would represent the worst cases of leakage and those plotting furthest away would contain the smallest component of reservoir water. In this respect it is interesting to note that the distant groundwater samples issuing from the Magnesian Limestone (i.e. samples 12 and 15) and from Willey Spring and Willeywood Farm well (i.e. samples 13 and 14) appear to be quite distinct from the groundwater samples emerging at the downstream toe.

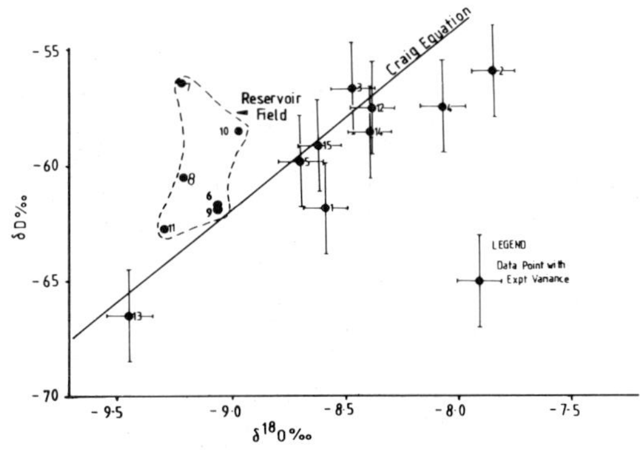

(b)

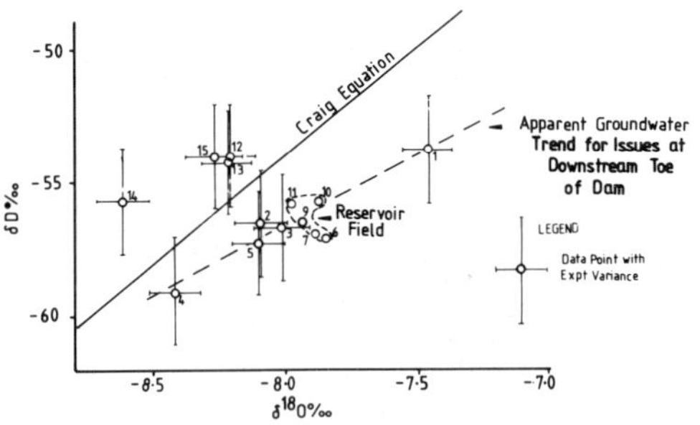

(a)

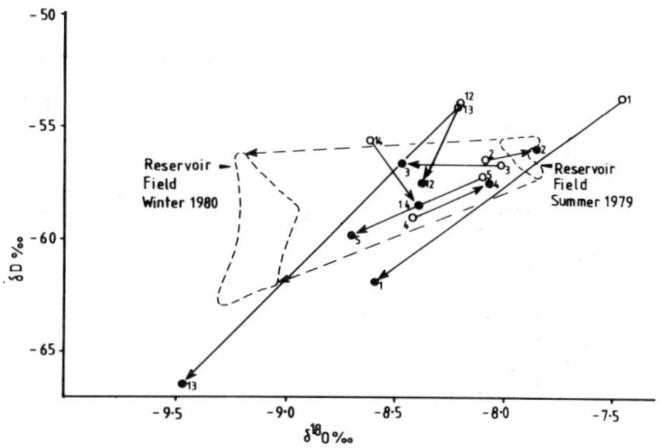

(c)

FIG. 4. Delta diagrams: (a) Summer data; (b) Winter data; (c) isotope movement picture.

A second set of samples taken in February 1980 was taken to substantiate this 'mixing' hypothesis, on the assumption that the seasonal isotopic variations might enable a further distinction to be made between reservoir water and groundwater. As winter rain is depleted in the heavy isotopic species, the surface water should register this effect within a short time, but the resurgent groundwater will follow a behavioural pattern controlled by the transmissibility of the strata through which the water is passing. The reservoir water should therefore show a high degree of isotopic homogeneity, while at the same time showing the bulk winter depletion.

For the 'winter' samples, the reservoir data define a discrete data field as did the 'summer' samples although the field has a greater area (Fig. 4b). The isotope plot has moved in the direction hypothesised, with the winter samples being depleted in the heavy isotopic species compared to the summer samples, although since none of the groundwater data fall within the reservoir data field, they can still be distinguished. The spread of the winter reservoir data is thought to be due to the far greater volume of water passing through the reservoir system.

As shown by the isotope movement picture (Fig. 4 c) the resurgent spring water appears entirely random as compared to the movement of the reservoir data and it is this fact that leads to the conclusion that there is little if any relationship between the two. Clearly at Moorgreen the reservoir water has a very uniform, single well defined isotopic composition which can be distinguished from the groundwater in both summer and winter. This alone could be used as an argument to show that reservoir leakage is unlikely but when the seasonal movement picture is taken into consideration a firmer conclusion can be reached supporting the view of non-hydraulic connection between the reservoir water and issues emerging at the downstream toe of the dam.

If a high degree of correlation had existed between the two systems, then they would be in close hydraulic contact and reservoir leakage would have been indicated. However, this is certainly not the case based on the isotope movement picture with the scatter of the groundwater samples being indicative of control by bedrock transmissivity to a degree that suggests a wide range of hydrogeological properties operating in a system or systems, isolated from the reservoir. This conclusion is supported by correlations between rainfall, reservoir water level, vee notch flows and temperature measurements at the issue locations.

(2) Taren Landslip, South Wales

The A470 Taff Vale Trunk Road extends northwards from Cardiff in the valley of the river Taff which cuts through the broad synclinal basin of the South Wales coalfield. Stages I-III were completed in 1972 and Stage IV representing the section from Abercynon to Pentrebach is nearing completion, including the crossing of a large late to post-glacial translational rock slide south of Aberfan known as the Taren Landslip (Fig. 5).

The consultants to the Welsh Office for the project are Rendel Palmer and Tritton, but in 1979 and 1980 Sir William Halcrow and Partners carried out detailed studies in connection with hydrology and hydrogeology of the Taren Landslip and advised the Consultants concerning the design of remedial drainage measures.

The Taren Landslip covers an area of some 30 hectares and lies on a steep hillside west of the River Taff with most of the land lying between the 120 m and 340 m contours. It is one of a number of late to post-glacial landslips which exist in South Wales and has resulted from translational rock slides at several levels on weak layers in the Pennant Measures with movements in an easterly direction away from the Kilkenny Fault located near the backscar. A continuous slip surface consisting of three connected slip zones has been identified in the main landslip, the Upper and Central Zones overlying bedrock whilst the Lower Zone overlies glacial till and alluvium, Figs 6 and 7. The northern areas of the Landslip, known as the North-West Block and the Northern Wing, are less disturbed than the main landslip.

FIG. 5. View northwards showing the A470 road berm being constructed across the toe of the slip.

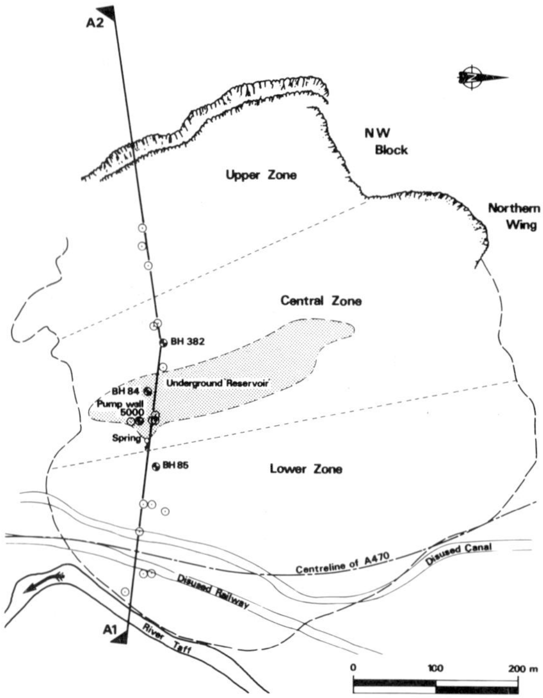

FIG. 6. General plan of landslip.

Investigations carried out prior to construction indicated the presence of a major perched water table and substantial underground 'reservoir' existing above the slip base towards the middle and southern end of this Central Zone. Overflow from the Central Zone 'reservoir' appeared to be from one spring located near its south-eastern side, although other minor issues were also observed following periods of prolonged rainfall.

As part of the hydrogeological studies to ascertain the most beneficial drainage design for the area, which in turn was part of the overall stabilisation measures for the slip itself, measurements of isotopic fractionation on rainwater and groundwater samples were made to provide clues as to the origin of water contained within the landslip material in relation to that in the underlying Coal Measures bedrock. In the time available during the particular phase of investigation it was only possible to measure relatively few sites over a limited period of two and a half months, i.e. water samples taken in April 1979 and June 1979. This time limitation places a restriction on the 'isotope movement picture' but nevertheless a systematic variation emerges that contributed valuable information and aided the Engineer involved with the problem of drainage of the Central Zone.

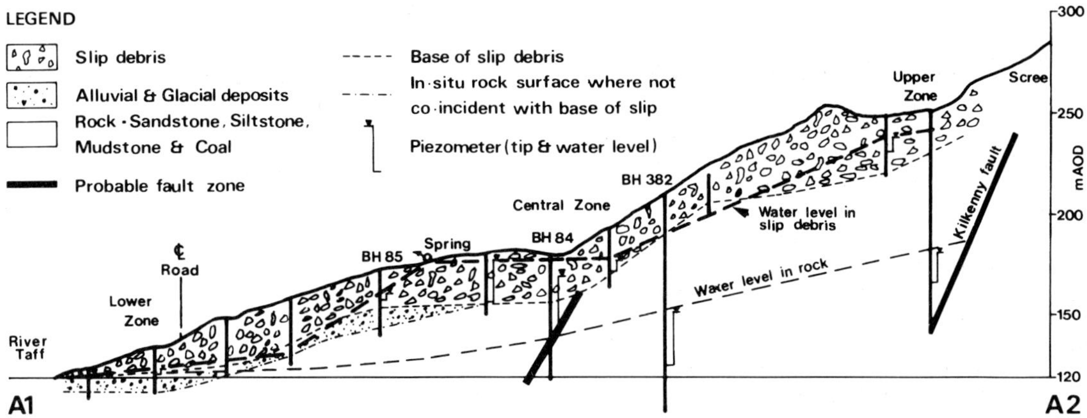

FIG. 7. Geological cross-section and sample locations (for location of section line A1-A2 see Fig. 6).

Water samples were analysed for four locations, namely, for borehole piezometers 382 and 84 sited in sandstone bedrock beneath the landslip (84 being traversed and affected by a major fault), from borehole piezometer 85 located in landslip material and from the spring issuing out of the slip debris at the south-eastern side of the underground 'reservoir'. The location of the sample sites are shown on Figs 6 and 7. In addition, electrical conductivity of the samples was measured and a sample of rainfall was taken for purposes of comparison with the groundwater samples.

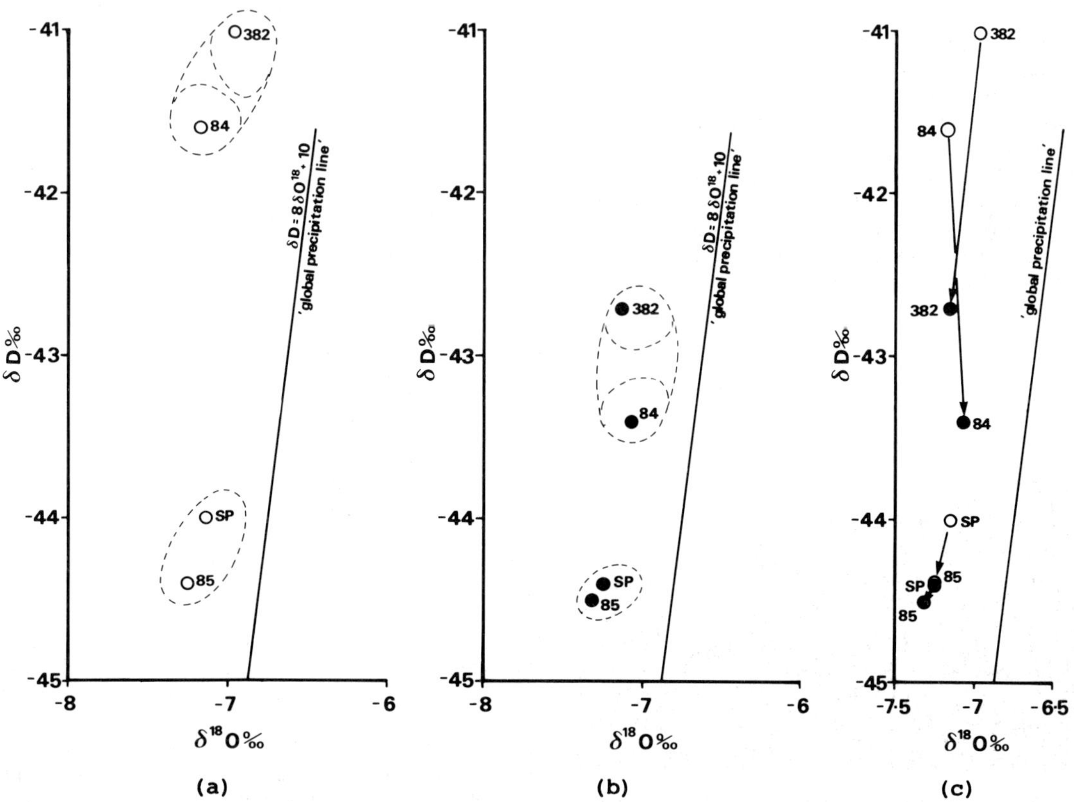

FIG. 8. Delta diagrams: (a) April data; (b) June data; (c) isotope movement picture.

The results reveal firstly that the rain sampled on 12th April, 1979 gave δ D ‰ and δ ^{18}O ‰ values (i.e. -29.7 ‰δD and -4.86‰ $\delta^{18}O$) close to the accepted global precipitation line. There is, therefore, little to indicate the presence of unusual and non-typical precipitation factors. Secondly, when the groundwater samples are plotted on delta diagrams (Fig. 8) two (possibly three) distinct groupings are apparent: the spring (SP) and sample 85 for landslip debris comprising one and the bedrock samples 382 and 84 the other(s). Both sets of data (i.e. those taken in April and June of 1979) give the same results suggesting that the water in the underlying Coal Measures rocks is not in total hydraulic continuity and freely mixing with that found in the landslip debris. Pumping tests have, however, indicated a certain degree of hydraulic continuity between borehole piezometer 84 sited in faulted sandstone bedrock and the overlying landslip debris and this might explain the difference in values obtained for the two samples of strata water. Conductivity measurements demonstrate the differentiation more clearly as follows:

Rain	25	μmho
Spring	60	μmho
Borehole 85	85	μmho
Borehole 84	120	μmho
Borehole 382	190	μmho

The 'isotope movement picture' illustrated in Fig. 8 c is interesting in that all the data shows the same systematic trend. In each case, samples collected in June 1979 are more depleted in the heavy isotopic species than those collected in April, particularly the samples taken from bedrock, which is the reverse of the expected trend towards isotopically heavier rainfall in the summer months. In order to study this effect and to obtain the 'hold-up' time and relative transmissivities of both bedrock and landslip debris, it would have been necessary to record air temperatures and rainfalls at the site and collect a great many more samples, both of rainfall and groundwaters over a much longer period.

ACKNOWLEDGEMENTS: The authors wish to thank Sir William Halcrow and Partners for the facilities used in the preparation of this paper and the assistance of colleagues. They also acknowledge the kind permission of the British Waterways Board (North), Mr N.J. Cochrane (Sir William Halcrow and Partners), the Welsh Office and their consultants, Rendel Palmer and Tritton for the information presented. Laboratory analyses of the Moorgreen sample were carried out at AERE, Harwell, under the direction of Mr R.L. Otlet and the analysis of the Taren samples were supervised by Dr J.A Miller of Cambridge University.

References

CRAIG, H., 1961. Isotope variations in meteoric waters. *Science,* **133,** 1702.

DANSGAARD, W., 1964 Stable isotopes in precipitation. *Tellus,* **16,** 436-468.

EPSTEIN, S. & MAYEDA, T.K., 1953. Variations of the $^{18}O/^{16}O$ ratio in natural waters. *Geochimica et Cosmochimica Acta,* **4,** 213.

EVANS, C.V., 1982. Harwell involvement in Isotope Hydrology. *International Symposium on Applications and Technology of ionising radiations, Riyadh, Saudi Arabia,* 12-17th March, 1982.

EVANS, G.V., OTLET, R.L., DOWNING, R.A., MONKHOUSE, R.A. & RAE, G., 1979. Some problems in the interpretation of isotope measurements in United Kingdom aquifers. *Proceedings of the International Symposium of IAEA., Neuherberg,* **2,** 679-708.

HALCROW, SIR WILLIAM & PARTNERS, 1980. Chapter B6 Hydrology and hydrogeology. Taff Vale Trunk Road A470 Stage 4, Abercynon – Pentrebach. Interpretative Report on Taren Landslip. *Unpublished Report by Rendel, Palmer and Tritton for Welsh Office.*

HALCROW, SIR WILLIAM & PARTNERS, 1980. Moorgreen Reservoir : Report of Site Investigation of the Dumbles, August and October 1979 and September 1980, and of the isotopic investigation of water samples from the Reservoir and Reservoir Area. *Unpublished report for British Waterways Board (North).*

HOEFS, J., 1973. *Stable Isotope Geochemistry.* Springer-Verlag, New York.

SIDDLE, H.J., JONES, D.B. & WARREN, C.D., 1984. The use of tracer test to assess groundwater flows in site investigations. *Engineering Group of the Geological Society;* 20th Regional Meeting (Guildford), September 1984.

TORAN, L., 1982. Isotopes in Groundwater Investigations. *Groundwater* **20,** No. 6.

P.J. LEGGO, Cambridge House, 21 High Street, Linton, Cambridge, CB1 6HS, UK.
C.D. WARREN & D. BLAKEWAY, Sir William Halcrow & Partners, Vineyard House, 44 Brook Green, London W6 7BY, UK.

Groundwater geochemistry and mass transfer in the East Yorkshire Chalk

J.I. Pitman

ABSTRACT: The chemical composition of groundwater from the East Yorkshire Chalk exhibits a systematic change in chemistry as it moves downdip from the recharge area of the Wolds to the Holderness lowland. Three distinctive hydrogeochemical zones, coincident with the unconfined, semi-confined artesian aquifer and fully confined aquifer of Holderness exist. Chemical composition in the unconfined aquifer is dominated by carbonate equilibria at a closed system PCO_2 of -2.09 with calcite and one mole % Mg-calcite. (Ca^{2+}=2.3 mmol/l; Mg^{2+}=0.2 mmol/l; HCO_3^-=4.08 mmol/l; ΣCO_2=4.48 mmol/l; pH=7.44). In the semi-confined artesian flow aquifer, ingress of soil CO_2 and limited reduction of dissolved O_2 increase ΣCO_2 to approximately 6.4 mmol/l, whilst Ca^{2+} increases to 3.69 mmol/l at a PCO_2 of -1.76, and pH of 7.17. In the fully confined aquifer, reduction of O_2, NO_3, SO_2^{-4} occurs, with increased ΣCO_2 to 6.86 mmol/l, HCO_3^- to 6.13 mmol/l, and PCO_2 to −1.87. Cation exchange of Ca^{2+} for Na^+ also occurs. Mass transfer calculations show that the major reactions in the aquifer take place in the soil zone, and within the semi-confined zone, with low Mg-calcite undergoing solution whilst calcite is precipitated. Throughout the aquifer, groundwaters are saturated with calcite, after allowing for the effects of ion pairing and complex formation.

Introduction

The Yorkshire Wolds is a gently dipping limestone aquifer, dominated by fissure flow in the saturated zone, and by intergranular flow in the unsaturated zone. The outcrop area of unconfined Chalk is 822 km^2, and is typical of the 17,200 km of the English Cretaceous Chalk outcrop.

The chemical composition of groundwater in the East Yorkshire Chalk shows a systematic change as groundwater moves from the recharge areas downdip eastward and southward towards Holderness.

There are few studies of the groundwater chemistry of the Chalk, apart from those of Casey (1969) and Casey & Newton (1973), who described the chemistry of chalk springs and rivers in Hampshire and Dorset, and the British Geological Survey (Edmunds *et al.* 1973; Bath & Edmunds 1981). Pitman (1978a) described in detail the groundwater geochemistry of a small outlier of chalk at Givendale, East Yorkshire, and showed that simple carbonate geochemical models could be used to predict the chemical composition of chalk groundwaters, and hence the rate of solute transfer for a given water flux through the system.

The objective of this paper is threefold. Firstly, it describes the chemical variation observed throughout the aquifer. Secondly, the major chemical processes that control this variation are identified and, thirdly, the hydrogeological significance of those processes is assessed.

Location and geology

Lithology and stratigraphy

The total thickness of the Yorkshire Chalk exceeds 420 m, and consists of four distinctive lithostratigraphical units (Wood & Smith 1978; Kent 1980). These are in ascending order, the Ferriby, Welton, Burnham and Flamborough Chalk Formations.

The Ferriby Chalk corresponds to the Lower Chalk of southern England and consists of 18 to 36 m of flintless, marly, soft chalk, relatively high in insoluble residue (Jeans 1968; Pitman 1978b). The Welton Chalk forms 150 m of massively bedded flinty chalk, with well developed fissures. This is succeeded by the flinty, thinly bedded Burnham Chalk (130 to 150 m thick), and the flintless, well bedded Flamborough Formation.

Physical Properties of the Chalk

Chalks are fine grained micrites, characterised by calcitic particles in the size range 0.5 μm to 4 μm, and 10 μm to 100 μm (Hancock 1975). The finer fraction are coccoliths (Mimran 1978) within the protective framework of larger polygonal intergrown calcite crystals.

Yorkshire Chalks have densities in the range 2.00 to 2.56 g/cm^3 (Mimran 1978), corresponding to porosities of 8 to 25%. Measurements by Bird (1976) on borehole samples from the Welton and Burnham

From CRIPPS, J.C., BELL, F.G. & CULSHAW, M.G., (eds), 1986, *Groundwater in Engineering Geology,* Geological Society Engineering Geology Special Publication No. 3, pp. 177-185

Chalks ranged from 14 to 25%, with outcrop samples in the range 20 to 26%. Outcrop samples from the upper part of the Burnham Chalk, and the Flamborough Chalk ranged between 26 and 38%.

The small void size associated with these small porosities results in the massive chalk having very low permeabilities, generally in the order of 0.0002 m/day (Foster & Milton 1974,1976). Other physical properties of chalk have been reviewed by Scholle (1977), Hancock (1975), and Pitman (1984). It should be noted that the Yorkshire Chalks differ from the more southerly outcrops in having a much higher density and lower porosity.

Mineralogy

Calcite and low magnesium calcite, (mole % Mg of 1%) are the dominant minerals in the aquifer, constituting over 98% of the aquifer material. The insoluble residue of the chalks are dominated by clays, particularly smectites and illite; apatite; quartz; and trace amount of feldspar, limonite, pyrites, hornblende, rutile and zircon (Jeans 1968;Pitman 1978b). All of these minerals will influence the final water composition to some degree, although it is reasonable to assume that only the most abundant and fast reacting minerals, such as calcite, will dominate the water chemistry (Garrels 1976).

Hydrogeology

The Yorkshire Chalk is an important regional aquifer, with a large, as yet untapped potential for water supply to the Humberside region (Foster & Milton 1976). Springs at Driffield, and wells and bores in the confined aquifer are used to supply water to industry, agriculture and domestic use. Over extraction of groundwater around Hull has resulted in the intrusion of saline water from the River Humber (Foster *et al.* 1976). Effective recharge of the Chalk aquifer averages 300 mm, usually between November and March, with average rainfall over the Wolds varying between 700 to 900 mm.

The Chalk aquifer can be subdivided into three hydrogeological units:

(1) The unconfined aquifer of the Chalk Wolds
(2) The semi-confined artesian flow zone of the Chalk dip slope
(3) The fully confined aquifer beneath the boulder clays of Holderness (Fig. 1).

The unconfined Chalk aquifer

This aquifer has been intensively investigated by the BGS as part of their investigations into nitrate pollution of the Chalk groundwater (Foster & Crease 1974; Lawrence *et al.* 1983), saline water intrusion (Foster *et al.* 1976) along the Humber, and large scale development of groundwater resources.

Detailed groundwater investigations around Etton have shown that the aquifer approximates to a layered, moderate transmissivity (*T* averages 1200 m/d) low specific yield (0.005) water table system, dominated by fissure flow. There are two distinct zones of fissure development, with the upper, 7 m thick zone, just below the water table giving the greatest flow. A lower zone, 8 m thick, was also observed 35 m below the water table. Pumping tests showed that the aquifer was isotropic.

Similar zones of high permeability have recently been observed in the Chalk of the Thames Basin (Connorton & Reed 1978), and of Hampshire (Headworth *et al.* 1982). The best explanation of

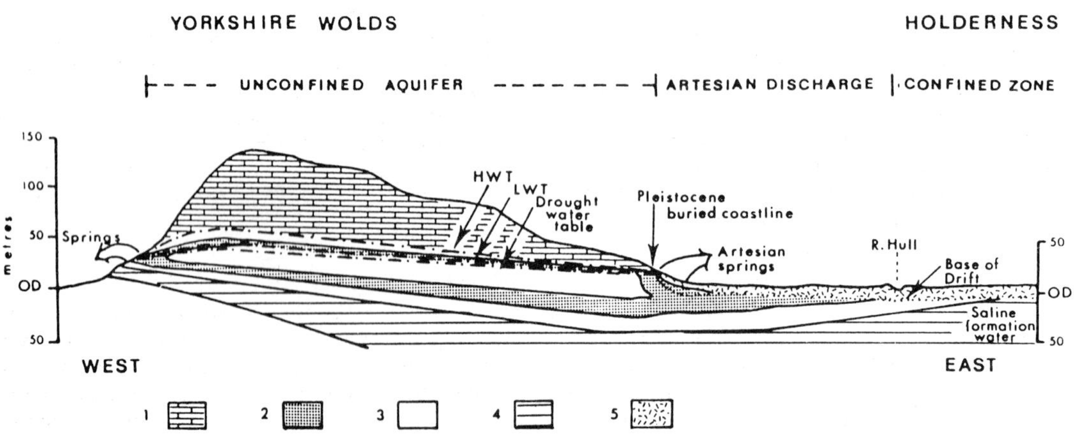

FIG. 1. Hydrogeological cross section of the Yorkshire Chalk. 1=unsaturated zone of the chalk wolds, 2=saturated aquifer zones of major permeability identified by Foster & Milton (1974), 3=zones of minor permeability, 4=zones of very low permeability, and 5, the confining Pleistocene tills of Holderness.

the values of T and S observed is fissure flow through fissures of average opening 0.5 to 1 mm, with a density of 10 per m. Actual flow velocities measured at Etton varied between 13 to 50 m/d. Little flow was detected in the aquifer below −30 m OD (Foster & Milton 1974). Flow through the unsaturated aquifer, as determined by tritium profiles, approximated to plug flow conditions through the intergranular pore space at velocities of 0.88 to 1.4 m/yr (Foster *et al.* 1982; Lawrence *et al.* 1983).

Natural groundwater discharge occurs through springs along the foot of the escarpment, along a line coincident with the buried cliff line in Holderness, and as an extensive area of artesian flow between the cliff line and the River Hull.

The confined Chalk aquifer

Data on flow in the confined aquifer is meagre, although large public pumped well supplies occur in this zone north of Beverly. Tritium counts in this zone show pre-bomb levels.

The semi-confined artesian flow aquifer

The area of artesian flow averages 170 km^2, with numerous kelds and springs being common, particularly in the zone west of the River Hull and the edge of the buried cliff line at about 35 m OD.

Groundwater sampling and analyses

Details of the procedures used to sample rainfall, soil water and spring waters are described by Pitman (1978a). Groundwater samples from the unconfined aquifer were collected with a battery operated stainless steel 850 ml well sampler, lowered to the appropriate depth by cable. Temperature, conductivity, pH, dissolved oxygen and HCO_3^- were determined at the well head. Field filtered (0.45 μm) samples were then acidified and analysed for cations by AAS, whilst anions were determined on unacidified samples. Samples from the confined aquifer were sampled at the well head, after the pump had been running for 30 minutes, using the same procedures as above. Soil gases were sampled at the Givendale site (Pitman 1978a). Kelds and springs in the artesian flow zone were also sampled.

Data Analysis

Free ion activities and molalities were calculated from all water sample analyses in which the ion charge balance was better than 5%, using the computer programme described by Pitman (1978), and after allowing for the effects of ion pairing and complex formation. For the $CaCO_3$–H_2O–CO_2 system, the equilibrium constants given by Plummer & Busenberg (1982) were utilised; for the non-carbonate species, the values of the equilibrium constants used were taken from Parkhurst, *et al.* (1980). Sample saturation indexes were calculated from the ratio IAP/K_t, where IAP is the water sample ion activity product of the free ion, and K_t the mineral equilibrium constant at the field temperature t. Sample PCO_2 was calculated, from the field measured HCO_3^- and pH.

Results

The chemical composition of selected waters along an east-west transect of the unconfined, semi-confined and fully confined Chalk aquifer is shown in Fig. 2.

The data above clearly show that the chalk aquifer hydrogeology is reflected in the aquifer hydrogeochemistry, corresponding to the unconfined, semi-confined and fully confined aquifer previously described. Systematic increases in Mg^{2+}, Na^+, Cl^-, and K^+ downdip are evident, as are decreases in pH and PCO_2. Values of log SI_c (calcite), however, remain almost constant. Reduction of NO_3^- and SO_4^{2-} in the confined aquifer also occurs.

Unconfined aquifer chemistry

Statistics for 22 well samples collected in October are given in Table 1. The most significant feature of this data is the remarkable constancy of the chemical composition, for both the measured and calculated parameters. All waters were saturated with calcite at a PCO_2 of -2.02. Seasonal variation of the parameters are within the range observed in Table 1.

In an attempt to establish whether the spatial variation reflects the variation of water chemistry with depth in the unconfined aquifer, depth samples were taken from Etton 4E, for both high (March) water table positions, and for low (October) water table positions. The extent to which vertical mixing influences the data shown in Fig. 3 is not known.

Water samples from the fissure zones, particularly between -5 to -15 m, are saturated with calcite at a PCO_2 of -2.0, whereas the zones between the major fissures is slightly supersaturated with calcite. Profiles obtained during October were similar, as were samples collected from Etton 3E, 244 m WSW of 4E. Pumping tests have shown that between 70 to 90% of the pumped water comes from the middle fissure zone (Foster & Milton, 1974).

Artesian zone chemistry

Data for samples collected from artesian springs and flowing wells are given in Table 2. Because they represent natural groundwater discharge, their chemical properties will indicate the extent to which solution is active in this part of the aquifer. Considerable variation exists in the carbonate chemistry parameters, with several samples being very undersaturated with calcite. Very high PCO_2 values were also encountered within this zone.

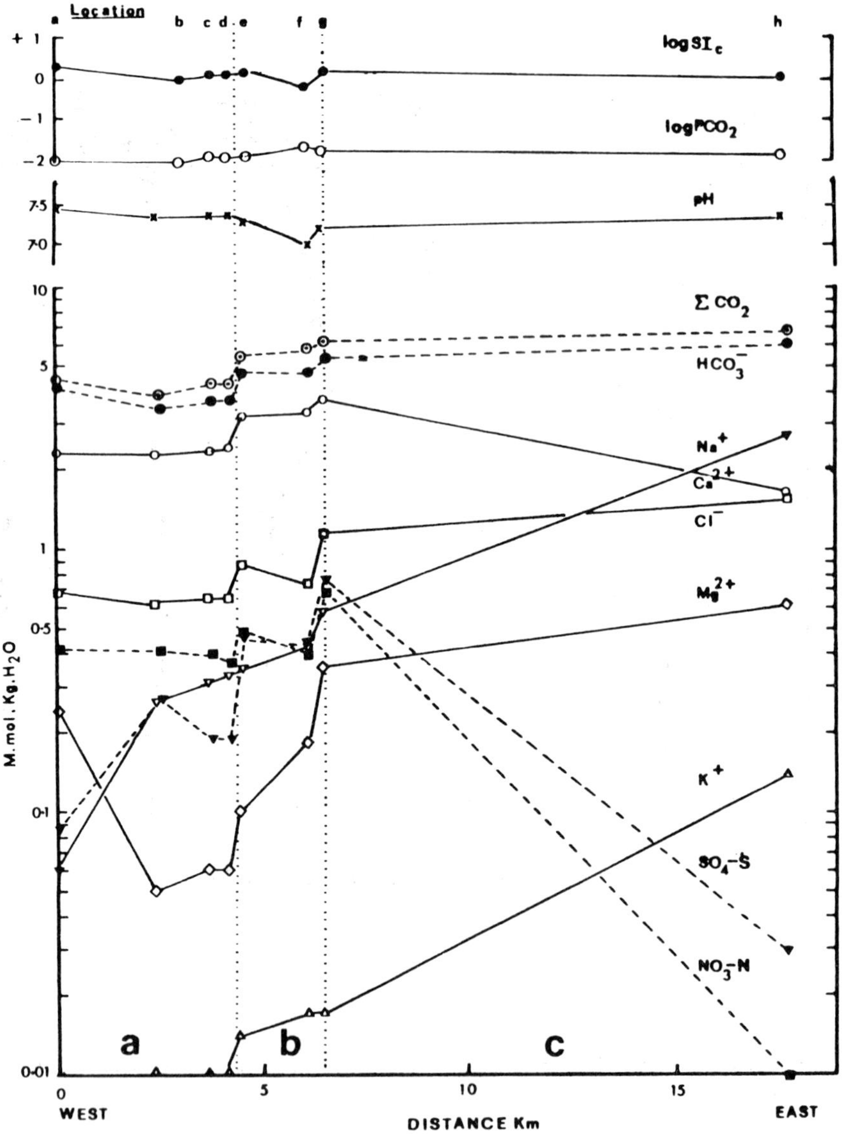

FIG. 2. Observed changes in groundwater chemistry along the transect shown in Fig. 1. Large a,b and c refer to the unconfined, artesian and confined aquifer units respectively, whilst the locations are a=Arras, b=Etton Wold House, c=Etton 2, d=Etton 4E, e=Etton Rectory, f=Woodhouse Fm, g=Bygott and h=Connygarth.

TABLE 1. *Mean composition of 22 unconfined aquifer samples, October, 1977 (mg/l).*

	Ca^{2+}	Mg^{2+}	Na^+	K^+	HCO_3^-	NO_3-N	SO_4-S	pH	PCO_2	Log SI_c	T°C
Means	92.9	1.3	6.1	0.6	234.5	5.2	21.9	7.31	-2.02	-0.02	11.3
SD	5.7	0.3	1.2	0.0	17.8	0.8	5.4	0.08	0.09	0.08	0.6

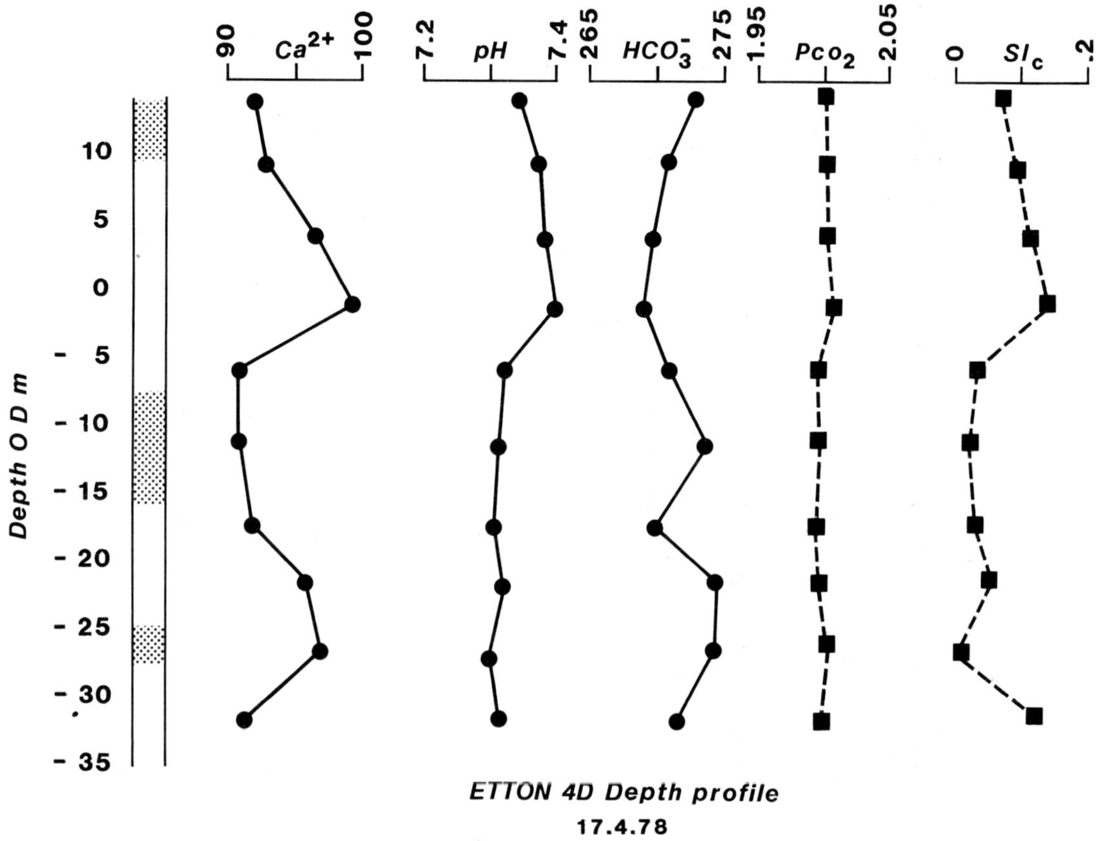

FIG. 3. Depth profile of groundwater chemical composition at Etton 4E. The shaded zones are the zones of maximum fissure development (Ca^{2+} in mg/l).

TABLE 2.
Range of chemical composition of groundwater samples from the artesian flow zone of Holderness, for 17 samples, in mg/l.

	Ca^{2+}	Mg^{2+}	Na^+	K^+	HCO_3^-	NO_3-N	SO_4-S	pH	PCO_2	Log SI_c	T°C
Max	120.0	8.0	3.9	10.3	316.9	6.7	35.5	7.48	-1.42	-0.3	13.6
Min	85.0	0.9	0.2	5.6	212.8	0.1	4.5	6.82	-2.12	+0.29	10.4
Mean	97.8	3.4	1.4	7.4	278.3	3.8	11.3	7.17	-1.88	-0.06	11.6
SD	11.5	1.5	0.9	2.5	32.9	1.9	9.1	0.17	0.19	0.13	0.9

Confined aquifer

The confined aquifer to the east of the artesian flow zone is marked by an oxidation-reduction barrier, characterised by the progressive disappearance of dissolved oxygen, nitrate, and sulphate along the flow path, whilst alkalinity, total carbon, sodium and iron steadily increase. This sequence (Fig. 2) is typical of that predicted by theory (Champ *et al.* 1979) and observed in other English limestone aquifers (Ineson & Downing 1963; Edmunds 1973) under reducing conditions. Additionally, cation exchange of (Ca^{2+}.... $+Mg^{2+}$) for 2 ($Na^{+}+K^{+}$) also occurs.

Identifying and modelling reaction paths

Three approaches were used to identify chemical reactions within the East Yorkshire Chalk.

(1) Observations of seasonal changes in springwater chemistry, combined with mineralogical studies of the Chalk (Pitman 1978a,b) and soil gas composition were utilised to graphically identify simple carbonate chemistry evolution models.

(2) Given observed changes in chemistry along the flow path, and know mineralogy, mass balance reactions were derived that account for the changes in anion and cation concentrations measured.

(3) Deriving stoichiometric coefficients of the mass balance reactions which account for the observed changes in anion and cation concentrations.

Carbonate chemistry

Major differences in water chemistry of carbonate aquifers are best examined in terms of the mode of carbonate evolution, with CO_2 being the major variable (Hendy 1971; Langmuir 1971; Pitman 1978a). Three modes of evolution are possible:

(1) Open system

$$mHCO_3^-=(K_1BPCO_2)/(10^{-pH}.HCO_3^-) \qquad (1)$$

(2) Closed system

$$mHCO_3\text{-}Constant/(2.10^{-pH}+K_1) \qquad (2)$$

(3) Equilibrium with calcite

$$mHCO_3^-=0.5\,(pH\text{-}log(2Kc/K_2)+log\,(Ca^{2+}+HCO_3^-) \qquad (3)$$

In a detailed study of springwater chemistry of a small chalk outlier at Givendale, Pitman (1978a) showed that shallow groundwaters evolved under open system conditions, whereas deeper groundwaters were closed to soil CO_2. It was concluded that most deep groundwaters evolve under closed system conditions.

Fig. 4 is a plot of all the groundwaters samples from the confined and unconfined aquifer, plotted in terms of the sample pH and HCO_3. Groundwaters appear to essentially evolve under equilibrium conditions with respect to calcite, in response to changing PCO_2 and temperature, according to the reaction

$$CO_2 + CaCO_3+H_2O = Ca^{2+} + 2HCO_3^- \qquad (4)$$

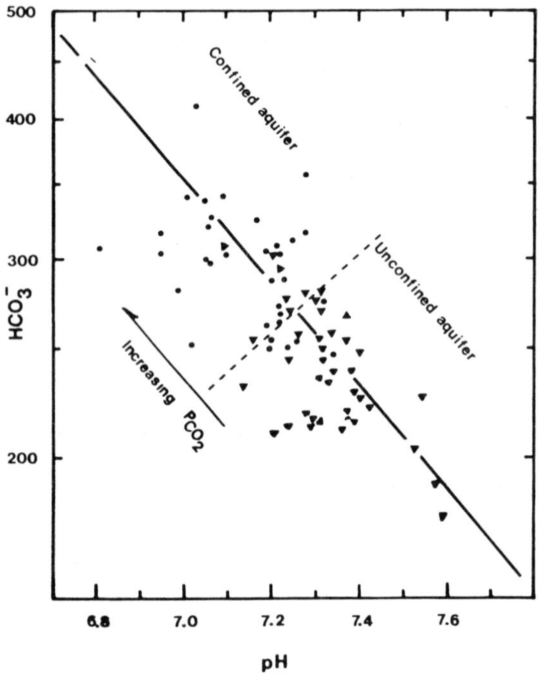

FIG. 4. Distribution of unconfined (triangles) and confined (dots) groundwater samples about the calcite saturation line (solid diagonal) for 9°C.

In an attempt to further refine the stoichiometric coefficients, and also to account for the changes observed in Mg^{2+} and Na^{+} and K^{+}, further reactions are required. Table 3 summarises the water chemistry changes along the flow path from recharge water to the fully confined aquifer of Holderness. This data will form the basis for the application of mass balance reactions to the Chalk aquifer.

Chemical mass transfer

Plummer & Back (1980) have shown that if the changes in water composition between any two points in a groundwater flow system can be determined, and is only due to *in situ* chemical reactions, then the net chemical change can be quantified by mass balance relations of the form:

Initial solution composition + 'Reactant phase'=
Final solution composition + 'Product phase'.

TABLE 3. *Observed changes in solute concentrations, mmol/kg H_2O.*

	Ca^{2+}	Mg^{2+}	Na^+	HCO_3^-	CO_2	pH	$logSI_c$	PCO_2	Type
Rain	0.02	0.00	0.11	0.07	0.02	4.83	--	–3.40	Recharge
Arras	2.29	0.04	0.06	4.06	4.49	7.40	+0.11	–2.09	Unconfined
Etton	2.35	0.06	0.33	3.82	4.46	7.33	+0.02	–2.05	Semi-confined
Bygot	3.69	0.35	0.58	5.34	6.24	7.17	+0.03	–1.76	Artesian
Connygarth	1.69	0.62	2.71	6.13	6.86	7.34	+0.02	–1.87	Confined

The 'reactant phase' and 'product phase' are constituents that enter or leave the aqueous phase during chemical reaction (Plummer *et al.* 1983). If the chemical composition of the initial and final waters are known, the geochemical reactions defined, and the solid phases present known, a mass balance can be described between the reactants and products, using N equations of the form

$$\sum_{j=1}^{\phi} a_j \beta_{c,j} = \Delta Mc_{c=1,N} \qquad (5)$$

where ϕ = total number of minerals and gases; N = minimum number of constituents needed to define the composition of the chosen minerals and gases; a_j = stoichiometric coefficients of the mineral or gas in mol/kg; $\beta_{c,j}$ = stoichiometric coefficient of the cth constituent in the jth mineral or gas, and ΔMc the change (mol/kg H_2O) of the Cth constituent in the aqueous phase along the reaction path.

In this paper the programme BALANCE (Parkhurst *et al.* 1982) was used to determine the reaction coefficients. Plummer *et al.* (1983) have described the difficulties associated with this method, as well as indicating its power to answer questions on plausible phases and reactions in aquifer systems.

The phases considered in deriving mass balance reactions in the chalk aquifer are calcite, low-Mg-calcite, CO_2(soil), CO_2 (from CH_2O), and ion exchange of the form

$$Na_2X + (Ca^{2+}+Mg^{2+}) = 2Na^+ + (Ca^{2+}+Mg^{2+})X. \qquad (6)$$

Evolution of Arras water

If the rainfall of Table 3 is added to the soil with a PCO_2 of -2.02 (average PCO_2 for a soil at 10°C; see Pitman 1978a), and reaction takes place via reaction (1) and (4), with CO_2 gas, calcite, Mg-calcite as phases, then

$$\text{Rain} + 1.96CO_2 + 4.80\,(Ca_{0.99}\,Mg_{0.01})CO_{3(s)}$$

$$= \Delta M_{CaT} + \Delta M_{MgT} + \Delta M_{CO_{2T}}$$

$$= \text{Arras water} + 2.27CaCO_{3(s)}$$

implying that the dissolution of the Chalk involves solution of the Mg-calcite and simultaneous precipitation of calcite. Recently Dever *et al.* (1983) have shown that calcite deposition in the vadose zone and soils is common in chalk terrains, and necessary to explain the observed ^{13}C found by them. It has also been shown (Pitman 1978a; 1985) that over 90% of the initial solution takes place within the first metre of percolation of the vadose zone.

Evolution of Etton water

Between Arras and Etton 4E, (4.2 km), only a small change in solution chemistry is evident. Na^+ increases by 0.27 mmol, so that mass transfer reactions including cation exchange are required:

$$\text{Arras} + 2(Ca^{2+}_{0.99}\,Mg^{2+}_{0.01})CO_3 + 0.135\ \text{Ion exchange}$$

$$= \text{Etton} + 1.775\,CaCO_{3(s)} + 0.455\,CO_2 \qquad (8)$$

implying that the dissolution of 1 mol% Mg-calcite and simultaneous precipitation of calcite occur, with loss of CO_2 gas. Additionally, ion exchange of Ca^{2+} for $2Na^+$ explains the increase of Na observed. Mimran (1978) has shown that secondary calcite deposition is needed to explain the $\delta^{18}O$ = –8‰ found in these chalks; additionally he has shown that the lower porosity of the Yorkshire Chalks is also due to secondary calcite precipitation.

Evolution of artesian zone water

Between Etton and Bygott, 2.3 km downdip, the water chemistry undergoes a significant change (Table 3). Additionally, SO_4^{2-} and NO_3^- and Cl^- all increase (Fig.2), whilst dissolved O_2 decreases to 8.8 mg/l.

Clearly the large increase of anions and cations observed here is due to the very shallow groundwater, and the very leaky confining tills. The effect here is complex, with fertilizers probably contributing most of the SO_4^{2-} and Cl^-, whilst the aquifer becomes open to soil CO_2, and CO_2 from the reaction $CH_2O + O_2 = CO_2 + H_2O$. The mass transfer between Etton and Bygott is given by the relation

$$\text{Etton} + 29\,(Ca_{0.99}\,Mg_{0.01})CO_3 + 0.295CO_2$$

$$+ 0.07CH_2O + 0.125CaXNa$$

$$= \text{Bygott} + 27.25\,CaCO_{3(s)} \qquad (9)$$

with solution of Mg-calcite (to supply the Mg) and simultaneous precipitation of calcite, the net transfer into solution being 1.34 mmoles Ca^{2+}. This zone of the aquifer is thus characterised by significant solutional activity, mainly due to the influx of CO_2, 80% from soil gases, 20% from reduction of O_2 by CH_2O.

Evolution in the confined aquifer

Between Bygott and Connygarth, 11.3 km downdip, concentrations of O_2, NO_3^-, SO_4^{2-}, decrease and disappear, whilst Cl^-, Na^+, HCO_3^-, Fe^{2+}, and K^+ all increase. Because the chemical analyses did not include all dissolved ions, particularly HS^-, Mn^{2+} and CH_4, it was not possible to determined an electron balance necessary to solve the mass balance equations under redox conditions. If, however, the chemistry of Ca^{2+}, Mg^{2+} and Na^+ is followed, then the dominant reactions are ion exchange of Na^+ for Ca^{2+}, and extensive precipitation of calcite, resulting in cementation of the aquifer.

Discussion

The hydrogeochemical data presented here show that the changes in groundwater composition for the unconfined and semi-confined aquifers can be described by relatively simple mass balance relationships involving calcite, low Mg-calcite, soil CO_2 gas, and dissolved organic matter (CH_2O), with ion exchange of Na^+ for Ca^{2+}. However, these mass balance relationships are not unique, as they depend upon the phases chosen. Carbon isotopic data is required to clarify some of the conclusions reached.

However, what does emerge is that the zones of higher permeability reported by Foster & Milton (1974) are formed under present day conditions, although this does not mean to say that the processes were not more active in earlier time. In particular, the models tested here suggest that the area of highest K and T will occur wherever there is an ingress of CO_2, either from the soil, or from dissolved O_2 and CH_2O reduction. Because these areas are also characterised by shallow groundwater, pollution from fertilizers is likely to be rapidly transported into the aquifer because of the enhanced permeability development.

References

BATH, A.H. & EDMUNDS, W.M., 1981. Identification of connate water in interstitial solutions of chalk sediments. *Geochimica et Cosmochimica Acta.*, **45**, 1449-1461

BIRD, M.J., 1976. Core analyses results from the East Yorkshire Chalk. *IGS Rept., 76/3, HMSO, London,* 58-62.

CASEY, H., 1969. The chemical composition of some S. English chalk streams and its relation to discharge. *Association of River Authorities. Year Book,* 100-113

CASEY, H. & NEWTON, P.R.V., 1973 The chemical composition and flow of the River Frome and its tributaries. *Freshwater Biology,* **3,** 317-333

CHAMP, D.R., CULENS, J. & JACKSON, R.E. 1979. Oxidation-reduction sequences in groundwater flow systems. *Canadian Journal of Earth Science,* **16,** 12-23.

CLAYTON, C.R.I., 1983. The influence of diagensis on some index properties of chalk in England. *Geotechnique,* **33,** 225-241

CONNORTON, B.J. & REED, R.N., 1978. A numerical model for the prediction of long-term yield in an unconfined aquifer. *Quarterly Journal of Engineering Geology,* **11,** 127-139.

DEVER, L., DURAND, R., FONTES, J.CH., & VACHER, P., 1983. Etude pedogenetique et isotopique des neoformations de cacite dans un sol sur craie. *Geochimica et Cosmochimica Acta,* **47,** 2079-2090.

EDMUNDS, W.M., 1973. Trace element variation across an oxidation-reduction barrier in a limestone aquifer. *Proceedings of the Symposium on Hydrogeochemistry and Biogeochemistry, Tokyo,* 1972. (Ed.E.Ingerson). Clarke, Washington, DC, 500-526.

EDMUNDS, W.M., LOVERLOCK, P.E.R., & GRAY, D.A., 1973. Interstitial water chemistry and aquifer properties in the Upper and Middle Chalk of Berkshire. *Journal of Hydrology,* **19,** 21-31.

FOSTER, S.S.D. & CREASE, R.E., 1974. Nitrate pollution of chalk groundwater in East Yorkshire - a hydrological appraisal. *Journal of the Institution of Water Engineers,* **28,** 178-194.

FOSTER, S.S.D. & MILTON, V.A., 1974. The permeability and storage of an unconfined chalk aquifer. *Hydrological Science Bulletin,* **18,** 485-500.

FOSTER, S.S.D. & MILTON, V.A., 1976. *Hydrological basis for the large scale development of groundwater storage capacity in the East Yorkshire Chalk.* IGS Rept. 76/4, HMSO, London.

GARRELS, R.M., 1976 A survey of low temperature water-mineral relations. *Interpret. Environmental Isotope Hydrochemistry Data Groundwater Hydrology Proceedings Advisory Group Meeting* 1975, 65-84.

HANCOCK, J.M., 1975. The petrology of the Chalk. *Proceedings of the Geologists Association,* **86,** 499-535

HEADWORTH, H.H., KEATING, T. & PACKMAN, M.J. 1982. Evidence for a shallow highly-permeable zone in the chalk of Hampshire, U.K. *Journal of Hydrology,* **55,** 93-112

HENDY, C.H. 1971. The isotopic chemistry of spelotherms-1. *Geochimica et Cosmochimica Acta,* **35,** 801-824.

INESON, J. & DOWNING, R.A. 1963. Changes in the chemistry of groundwaters in the Chalk passing beneath argillaceous strata. *Bulletin of Geological Survey of Great Britain,* **20,** 176-92

JEANS, C.V., 1968. The origin of the montmorillonite of the European Chalk with special reference to the Lower Chalk of England. *Clay Minerals,* **7,** 311-330.

KENT, P.E., 1980. *Eastern England from the Tyne to the Wash.* 2nd ed. IGS, HMSO, London.

LANGMUIR, D. 1971. The geochemistry of some carbonate groundwaters in central Pennsylvania. *Geochimica et Cosmochimica Acta,* **35,** 1023-1045.

LAWRENCE, T., FOSTER, S.S.D. & IZZARD, W., 1983. Nitrate pollution in East Yorkshire revisited. *Journal of the Institute of Water Engineers and Scientists,* **38,** 408-420.

MIMRAN, Y., 1978. The induration of the Yorkshire and Irish Chalks. *Sedimentary Geology,* **20,** 141-164.

PITMAN, J.I., 1978a. Carbonate chemistry of groundwater from Chalk, Givendale, East Yorkshire. *Geochimica et Cosmochimica Acta,* **42,** 1885-1897.

PITMAN, J.I., 1978b. Chemistry and mineralogy of some Lower and Middle Chalks from Givendale, East Yorkshire. *Clay Minerals,* **13,** 93-100.

PITMAN, J.I., 1984. Thermal diffusivity of Upper Chalk from Hampshire, England. *Engineering Geology,* **20,** 207-218

PARKHURST, D.L., PLUMMER, L.N. & THORSTENSON, D.C. 1982. BALANCE, a computer programme for calculating chemical mass transfer from geochemical reactions in groundwater. *US Geological Surveyed Water Research Investigation,* 82-14, pp 33. NTIS Technical Report. Pb 82-255902, Springfield, VA 22161.

PARKHURST, D.L., THORSTENSON, D.C. & PLUMMER, L.N. 1980. PHREEQE-a computer program for geochemical calculations. *US Geological Survey Water Research Investigation,* 80-96, NTIS Techical Report PB81-167801, Springfield, VA 22161.

PLUMMER, L.N. & BACK.W., 1980. The mass balance approach: its application to interpreting the chemical evolution of hydrologic systems. *American Journal of Science,* **280,** 130-142.

PLUMMER, L.N. & BUSENBERG, E., 1982. The solubilities of calcite, aragonite and vaterite in CO_2-H_2O solutions between 0-90°C, and an evaluation of the aqueous model for the system $CaCO_3$-CO_2-H_2O. *Geochimica et Cosmochimica Acta.* **46,** 1011-1044.

PLUMMER, L.N., PARKHURST, D.L. & THORSTENSON, D.C. 1983. Development of reaction models for groundwater systems. *Geochimica et Cosmochimica Acta.,* **47,** 665-686

SCHOOLE, P.A., 1977. Chalk diagensis and its relation to petroleum exploration. *American Association of Petroleum Geologists Bulletin,* **61,** 982-1009.

WOOD, C.J. & SMITH, E.G. 1978. Lithostratigraphical classification of the Chalk of North Humberside and Lincolnshire. *Proceedings of the Yorkshire Geological Society,* **42,** 263-287.

J.I. PITMAN, Department of Geography, King's College, University of London, London, UK

Hydrogeological investigation of a proposed waste disposal site at Nant Y Gwyddon, South Wales

R. Tonkin

ABSTRACT : The paper describes an investigation into the feasibility of siting a large landfill site on an extensive hilltop plateau, in an area receiving annual rainfall in excess of 2000 mm. The underlying rocks mainly comprise jointed sandstones. Water resources are located on and adjacent to the area and are fed by streams rising within it. The method of site operation required (i.e. containment or dispersal) could be determined only be means of detailed investigation capable of quantifying pollution risks. A groundwater tracing test was performed in conjunction with a surface water survey. These are described and the results and interpretation presented. The proposed method of site development is summarised.

Introduction

The Rhondda Borough encompasses the valleys of the Rhondda Fawr and its tributary the Rhondda Fach. The valley sides almost everywhere rise through a succession of crags, benches and glacially oversteepened slopes to the interfluves which are typically over 300 m above sea level. The population of some 80,000 resides in the valley floors where civic and social amenities are sited and the communication network of road, rail, and formerly canal, routes compete for space with rivers and housing.

The Rhondda Borough Council, as Waste Disposal Authority for its area, is required to make arrangements for the disposal of 'controlled' wastes arising in the borough. A need for new landfill capacity was indicated in its Waste Disposal Plan of May 1980 (Rhondda Borough Council 1980). The process of site selection, undertaken by the Council in 1980 and early 1981, identified a small number of potentially suitable landfill sites of which that occupying the upland catchment of the 'Nant y Gwyddon' stream was most favoured. This paper describes the investigation into the site's feasibility undertaken by Sir William Halcrow and Partners from August 1981 to April 1984.

Location

The proposed landfill lies within a large, steepsided almost flat-topped spur of high ground some 1.2 km wide and known as Llwynypia Mountain (Fig. 1). It forms an easterly extension of a moorland plateau, which separates the Rhondda Fawr and Ogwr Fawr Valleys, with elevations up to 530 m OD. Llwynypia Mountain, some 260 ha in extent, is bound to the east by the River Rhondda Fawr, which passes not less than 600 m from the site, and to the south and north by the tributary valleys of Clydach Vale and of the Nant y Gelli. The waste tip will occupy some 24 ha of a shallow hollow between 295 and 360 m OD.

Meteorology

The average annual rainfall was estimated at 2100 mm based on observations at the Penrhys Cemetery rain gauge (No. 490566) 3 km NE, elevation 287 m OD. Variations of between 77 and 132% of the average might be expected from inspection of long-term annual records for a selection of stations in Glamorgan.

Estimates of potential evapotranspiration (PE), (Doorenbos & Pruitt 1977), derives from data recorded at climatological stations 11 and 28 km south-east and compensated for elevation differences, indicate a monthly variation from 5 mm (December) to 86 mm (June). Comparison of mean monthly rainfall and PE values showed there would usually be sufficient total rainfall for full PE to be achieved. The mean monthly effective rainfall would range from 42 mm (June) to 237 mm (December); the mean annual total being 1620 mm.

Geology

The site is underlain by the Rhondda Beds of the Upper Coal Measures (Woodland & Evans 1964). The strata proved in boreholes comprise thick developments of medium and fine grained sandstone interbedded with laterally variable, usually thinner, argillaceous strata containing coals and associated seatearths. Vertical and inclined planar joints were common being usually open and clean, although some calcite was present locally. Core was often fresh but

From Cripps, J.C., Bell, F.G. & Culshaw, M.G., (eds), 1986, *Groundwater in Engineering Geology*, Geological Society Engineering Geology Special Publication No. 3, pp. 187-196.

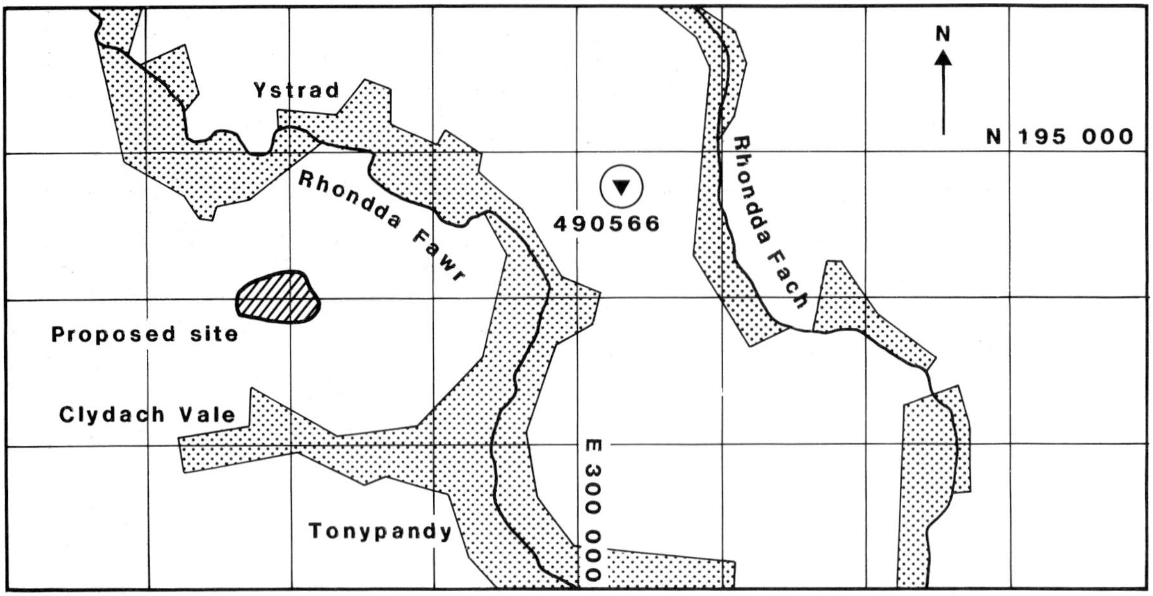

FIG. 1. Location of site. (Based upon the 1970 Ordnance Survey 1:50,000 map with permission of Her Majesty's Stationery Office, Crown copyright reserved)

otherwise partial discolouration indicated downward percolation of groundwater. Average fracture spacing was 75 to 125 mm. Seatearths exhibited little or no evidence of jointing or weathering which accords with their usual role as aquicludes; albeit possibly leaky where affected by mining. Strata exposed in disused quarries showed three intersecting very widely spaced vertical joint sets with a maximum aperture of 100 mm.

The superficial deposits include 0.6 to 1.0 m of head deposits, 0.1 to 0.2 m of peaty topsoil and impersistent patches of thin clayey plateau drift (discontinuous with glacial fill in the valleys). The total thickness proved in the 7 boreholes and 14 trial pits ranged from 0.2 to 2.2 m. Laboratory tests revealed a minor clay fraction (mean 7 and maximum 12%), with substantial sand and gravel fractions (mean 43 and 31 % respectively). Decomposing sandstone lithorelics are almost universally present as angular cobbles. Constant head laboratory tests yielded a mean permeability value of 3×10^{-8}m/s which corresponds to a silt.

The site lies at the western end of the Abercynon Syncline where there is a regional dip east of some 4° but varying in direction and amount. The Cymmer and Dinas Faults trending SE to NW with throws of 32 and 74 km respectively create a graben some 1.3 m wide within which the site lies; other minor faults are also present (Fig. 2).

Mining

Eleven seams were worked beneath and within influencing distance of the site by deep mining methods and a further four seams were worked locally from their crops between the mid 1800s and 1962. A possible arithmetical total thickness varying from nil to 12 m was extracted beneath Llwynypia Mountain. The area of lesser extraction correspond to:

(1) pillars of unworked coal coincident with colliery boundaries (providing security against trespass and water ingress)

(2) disturbed ground within seams adjacent to the major faults.

Zones of extension (i.e. tensional strain) associated with edges of workings are known to cause dilation of joints which in turn can significantly modify the hydrogeological regime to provide greatly increased potential for transmission of groundwater (Daughton *et al.* 1976). A study of the patterns of extraction shown on abandonment plans indicated that zones of extension lay between the proposed landfill site and much of the hillside surrounding it.

Hydrogeology

The first stage of the investigation, involving desk study and field mapping, reported that two major abstraction points for public water supply (together

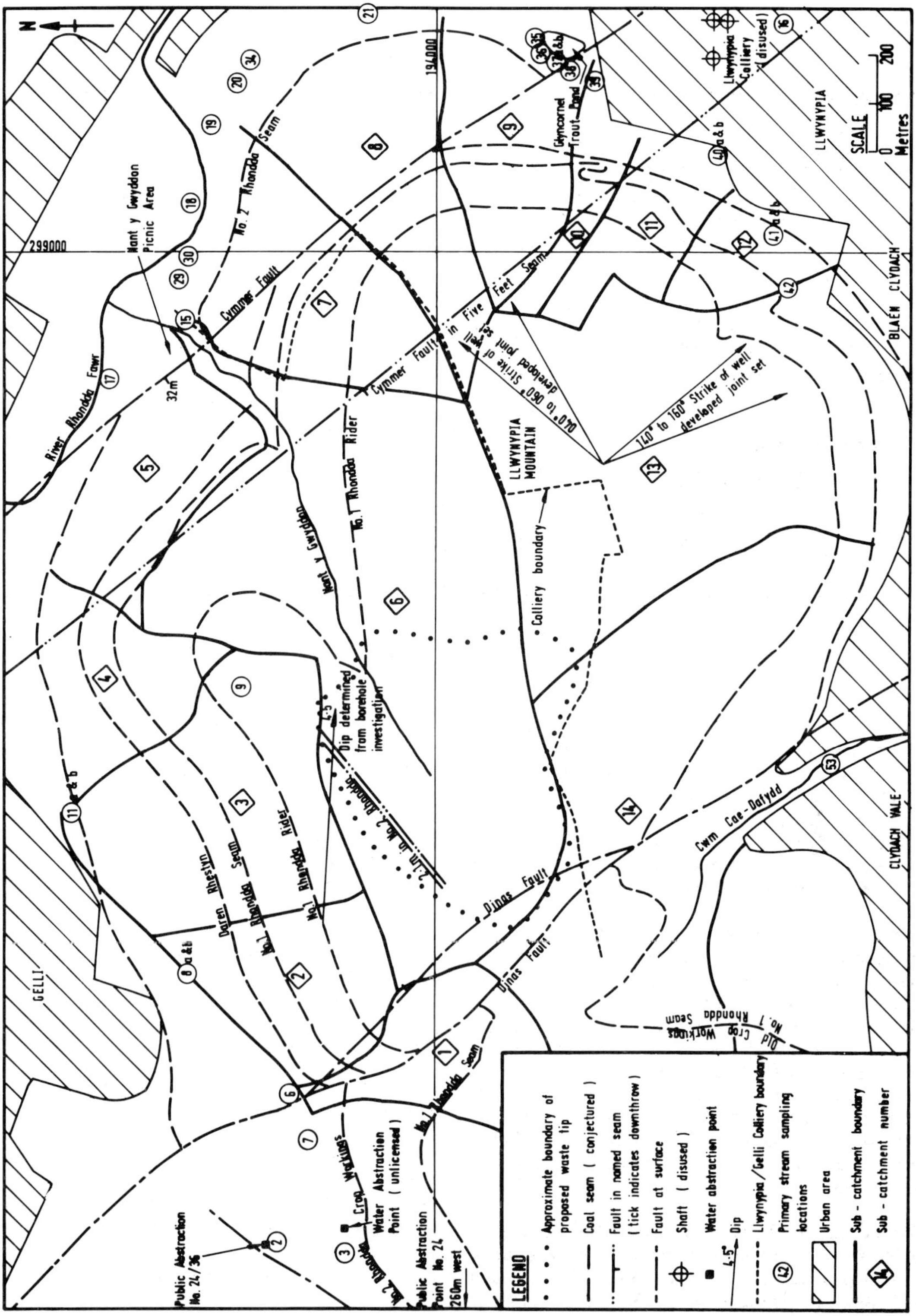

FIG. 2. Geological and hydrological plan. (Based on the 1948 Ordnance Survey 6 inch map with permission of Her Majesty's Stationery Office, Crown copyright reserved)

licensed for 1.66×10^6 m^3 pa.), an unlicensed domestic abstraction, a recently developed picnic and nature trail amenity area centered on the Nant y Gwyddon, the Glyncornel Trout Pond and stock watering facilities were potentially at risk from leachate migration. It also confirmed the presence of issues and spring lines along the recorded crops of various seams on the north-east, east and south-east facing slopes. Preliminary estimates of leachate columns and characteristics indicated that leachate entering the groundwater system might contaminate some local water resources to an unacceptable degree suggesting that the site should operate on 'concentrate and contain' principles (Department of the Environment 1984). The estimated costs would greatly exceed those incurred in operating a 'disperse and attenuate' site. However, the groundwater system's potential for dispersion, attenuation and dilution could not be quantified without further investigation which, while costly, was considered a worthwhile investment. The objectives of the detailed study may be summarised as:-

(1) To quantify issues in relation to their catchments; as an indication of the extent to which mining induced zones of extension might have perforated aquicludes thereby intercepting groundwater and transmitting it to deeper seams without local crops.
(2) To establish the velocity of groundwater flows to the crops of aquicludes underlying the site and thereby the role of fissure flow in groundwater movement and the potential for leachate in groundwater.
(3) To quantify the extent to which flows emerging from beneath the site were augmented by the large upland groundwater catchment extending west beyond the Dinas Fault.
(4) To evaluate the capacity of the soil cover for reducing infiltration and providing potential for natural containment.

Seven boreholes were sunk by rotary drilling methods to prove the No. 1 Rhondda Rider at depths of 20 to 67 m along the crop of which most flows emerge. One borehole was taken to 87 m to prove the No. 1 Rhondda Seam and to check its capacity as an aquiclude and that of the aquicludes above it. The vertical succession encountered in boreholes, is presented in Fig. 3. Nine piezometers were installed. Variation in piezometric head provides an indication of the relative competence of an aquiclude and it is interesting to note that the No. 1 Rhondda Rider was able to sustain a head of 19.75 m in piezometer NG2a. Piezometric response to rainfall indicated a delay of about 7 days even in piezometer 3a at 68 m depth and lying beneath a number of argillaceous units, which behaved as leaky aquicludes. Contouring of the phreatic surface above the No. 1 Rhondda Rider produced a mean value for the hydraulic gradient of 1 in 18 east beneath the centre of the proposed landfill varying to south-east on its eastern margin. This confirms conclusions drawn from field mapping and demonstrates that, after allowance for upward drag along the Dinas Fault shown by spot heights recorded on mine plans, there would be little capacity for spillage west across the fault and towards the water supply points.

Piezometric levels were high during the exceptionally wet four month period of October 1982 to January 1983. Rain in late February re-established these levels which were sustained by a further unusually wet March to late June period. It was not until mid-August that piezometric levels fell back to those of early February. The 25 week period from 23rd February to 10th August was used for a water balance calculation for Llwynypia Mountain as a zero net change in groundwater storage in the aquifers could be assumed.

Hydrology

Llwynypia Mountain was divided into 14 surface subcatchments by reference to contoured topographic plans (ref. Fig. 2). Drainage from the hill top and sides is provided by eighteen watercourses with outfalls to the bordering valleys. A multiplicity of flows feed them from a number of seepage lines. Flow gauging points were established at or below the lowest confluences ('primary points') and above confluences progressively further upstream ('secondary' and 'tertiary' points etc.). Gauging was undertaken by vee-notch weir, flow meter or by bucket and stopwatch. In addition, the National Coal Board was then regulating water levels beneath this part of the Rhondda Fawr by pumping from the No. 4 shaft of the disused Glamorgan Colliery and the Welsh Water Authority recorded daily discharges.

The relative areas of each sub-catchment together with the proportions of all flows off the mountain that each conveys is illustrated in Fig. 4. Subcatchment 9, containing Glyncornel Trout Pond, corresponds to only 4.9% of the total area but receives 66.4% of flows. Catchment 6, in which the Nant y Gwyddon amenity area lies, while representing 22.7% of area received 16.4% of flow. It was inferred that a significant groundwater contribution enhances flows in the disproportionately wetter sub-catchment 9 and, to a lesser extent, sub-catchments 6, 10, and 12. The water balance calculation referred to above was based on an estimated total rainfall of 612 mm and calculated actual evaporation of 239 mm producing an effective rainfall of some 373 mm over the catchment. Cumulative flows off the site were 1.154×10^6 m^3 implying a net groundwater enhancement of 0.186×10^6 m^3, equivalent to 17% of the effective rainfall on the surface catchment over the period. The extensive and high sandstone catchment west of the Dinas Fault is assumed to constitute the source of the inflow. Large flows may in fact traverse the fault and pass beneath much of Llwynypia Mountain, though interception and downward percolation along zones of extension may prevent their emergence at rockhead.

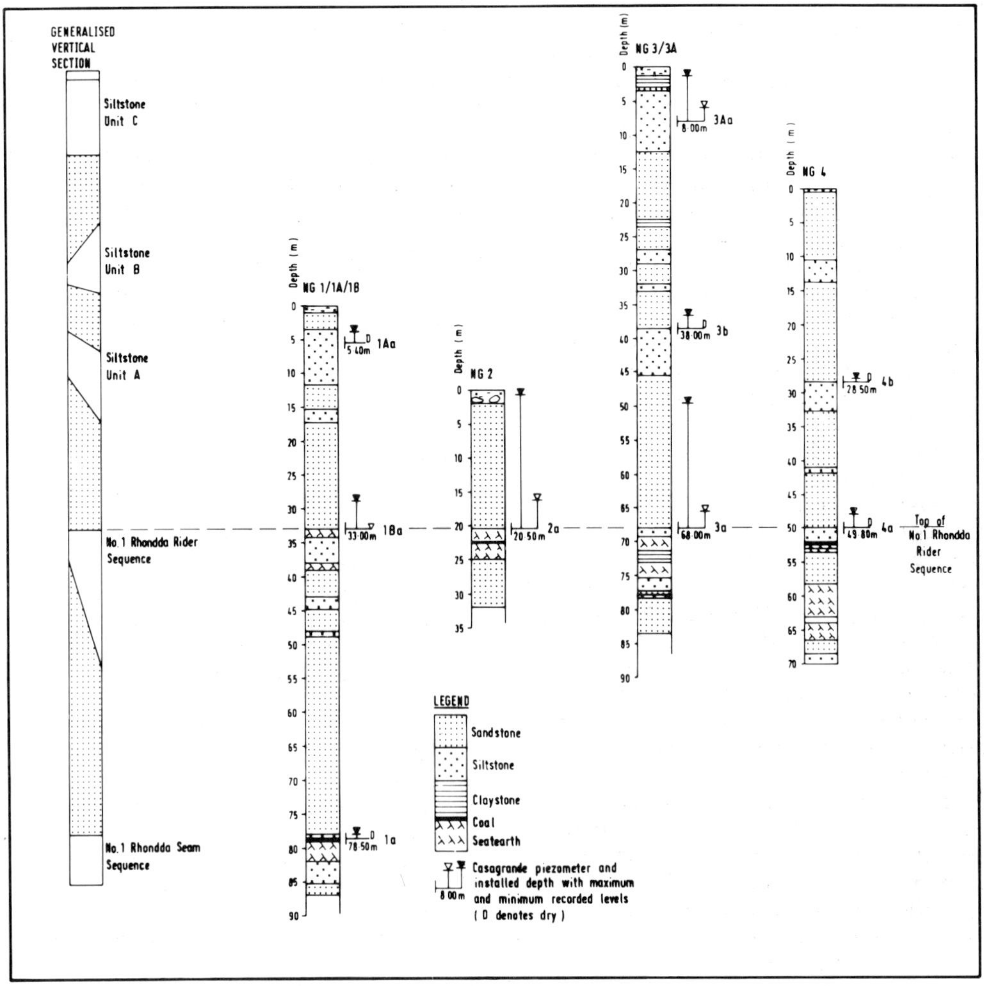

FIG. 3. Succession proved in boreholes, and piezometric variation.

Annual flows from December 1982 to December 1983 based on stream gauging totalled 3.89×10^6 m^3. Rainfall over this period was 114% of the annual average. Estimates of flows in average and exceptionally wet and dry years were obtained from these values.

Groundwater tracing

To quantify groundwater parameters a tracer test was undertaken using water labelled with tritium, as it can be measured with great sensitivity, has an adequately long half-life and can be wholly relied upon to follow groundwater in all circumstances and without the problems of absorption which might invalidate surveys based on other tracers (Siddle *et al.* 1984). Four injection points at mutually distant locations were specified to be representative of most of the extensive landfill area. They included locations from which migrating leachate could most adversely affect local water resources.

370 GBq of tritiated water were injected by the Geophysical Tracers Group of AERE Harwell on 23rd February 1983. Each injection point comprised two 1 m deep pits excavated in the soil cover and lined with metal cylinders fabricated from 200 l drums. Covers prevented windblown dispersal. A surface

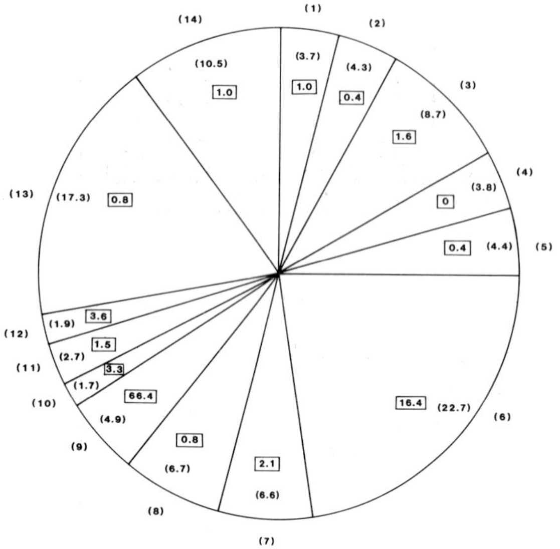

FIG. 4. Catchment area and flows.

infiltration method was adopted in order that groundwater movement through interflow should remain possible. Injection into boreholes in rock would have prejudiced the test by precluding movement above rockhead. Injection into shallow boreholes in soil was estimated to be an unrealistically lengthy operation. An initial concentration of 1.5×10^5 Bq/ml (i.e. 1×10^7 disintegrations per minute per gram, dpm/g) was obtained by diluting each phial of tracer in a polythene tub with 200 l of freshwater from a bowser before draining it into the pits. The tub and pits were then flushed with a further 200 l of freshwater. This brought the concentration down to a value that would ensure no member of the public could ingest more than the permitted I.C.R.P. annual limit of intake if tracer should enter the public water supply as a result of the test. The method of execution was approved by the then 'Environmental Protection Unit' of the Welsh Office and the conditions set by their technical advisors at the Radiochemical Inspectorate.

It was recognised that every possible emergence of tritium needed to be sampled and gauged to quantify:

(1) the total proportion of tritium, reappearing on the surface and
(2) the relative quantities being transmitted in each sub-catchment.

The proportions of leachate likely to emerge in the various sub-catchments could then be inferred and concentrations of contaminants estimated from flow data.

The cost of tritium analyses can form the greatest component of a tritium tracer test. It was reduced in two ways.

(1) by testing only primary point samples on a regular basis
(2) by combining aliquots of primary samples into composite samples until they became positive.

Full suites of samples were taken and retained to enable 'backtracking' to identify positive issues retrospectively. Piezometers were sampled with balers kept in each piezometer to avoid cross-contamination.

Tracer appeared first in piezometer 1Ba when sampled on 4th March. It was first detected on the surface in the Nant y Gwyddon at at flows into Glyncornel Lake when sampled on 15th March 1983. Transit times yielded values of permeability around 5×10^{-4} m/s indicating a high permeability typical of a fissured formation. Substantial proportions of the injected tracer were recovered over a period exceeding 300 days. Concentrations were generally low with a maximum value of 79 dpm/g at a surface issue, which corresponded with a dilution of 1.3×10^5 relative to the injected concentration. A high degree of mixing was indicated.

Observations over the test period demonstrated a clear response of flows to antecedent rainfall, for example in the June to August drought (Fig. 5). Rain in September and October ended the drought and substantial increases in tracer recovery followed, exceeding the initial May peak in the case of sub-catchment 6 (Nant y Gwyddon). The pattern in sub-catchment 9 (Trout Pond) was similar but less pronounced. It is noteworthy that had the test been halted in early August, when the tracer recovery rates had declined to 12 and 1% of the initial peaks, very misleading conclusions could have been drawn. Presentation of cumulative recovery vs. cumulative flows can provide a more accurate model which may be expected to demonstrate an asymptotal-like progression towards the eventual total recovery value (Fig 6.). The test was halted when concentrations had substantially declined and it was feasible to extrapolate eventual recoveries.

Tracer was only detected in sub-catchments 6, 7, 9, 10 and 11 and in the estimated proportions tabulated below. It was inferred that the portion unaccounted for was transmitted along vertical joints, principally in zones of extension, to deeper strata without local crops.

TABLE 1. *Measured and inferred tracer recovery.*

Flow-path	Percentage tracer
Sub-catchment 6	20
Sub-catchment 9	55
Sub-catchment 7,10,11	1
Deep strata (inferred)	24
Total	100

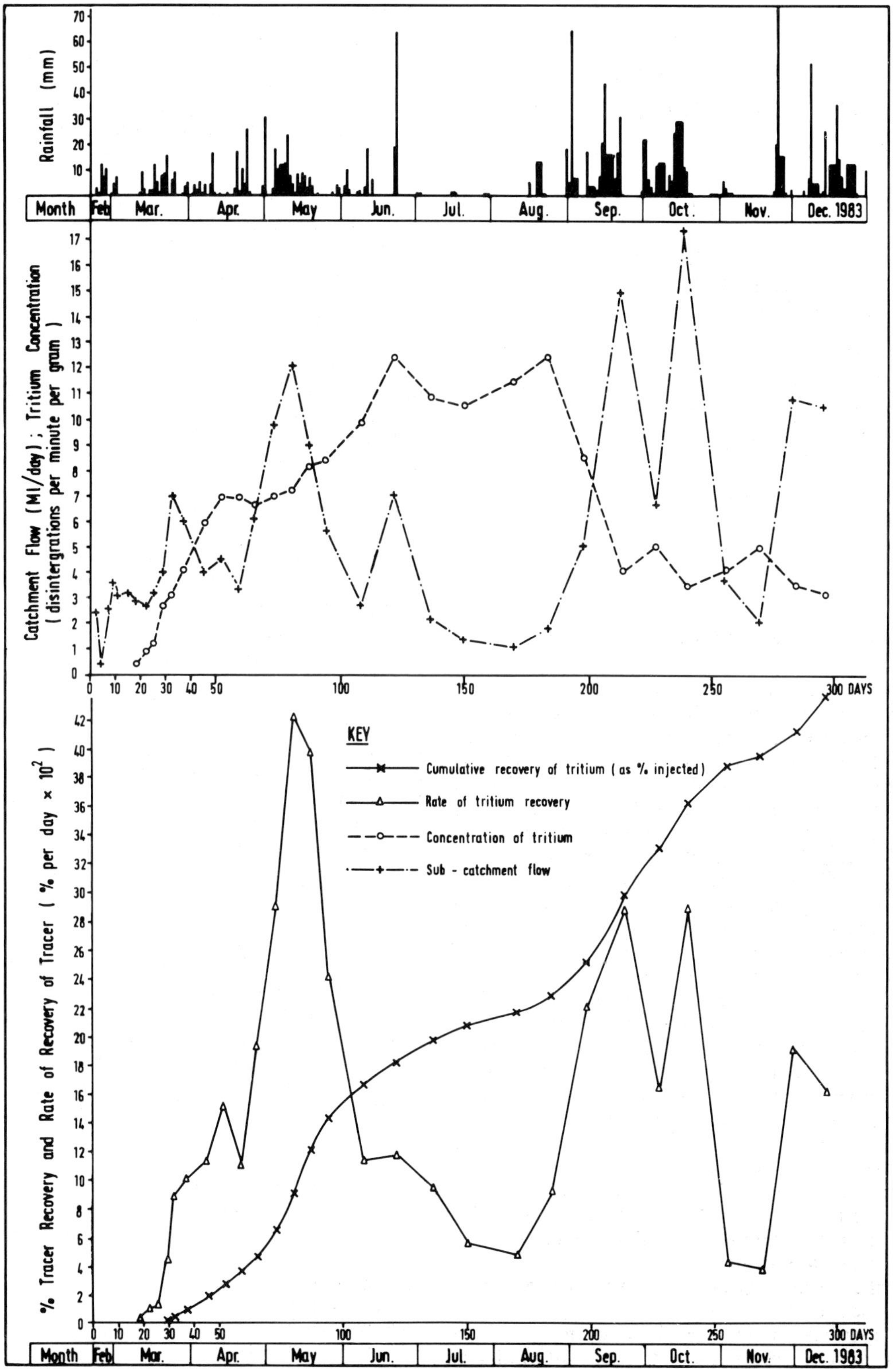

FIG. 5. Daily rainfall, stream flows and tracer recovery in sub-catchment 9.

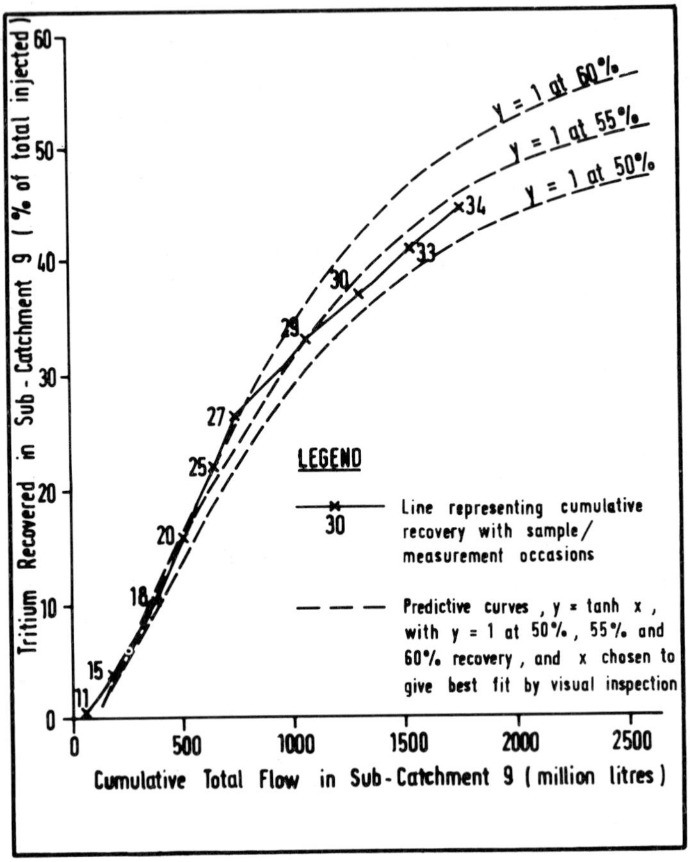

FIG. 6. Cumulative recovery vs. cumulative flow.

It is interesting to note that little tracer emerged in sub-catchments 7 and 8, which lie between 6 and 9, although the latter together comprise 13% of the whole area. This is reflected also in the relative flows transmitted by them (2.1 and 0.8% respectively). It was observed towards the end of the drought that whereas flows to sub-catchment 9 (Trout Pond) fell to about 9% of the May peak, the decline in sub-catchment 6 (Nant y Gwyddon) was to 1.6% of that in May. These results suggest that the hydraulic gradient to sub-catchment 9 is greater than that to sub-catchment 6 which greatly exceeds that to intervening sub-catchments. This may be explained by postulating a local anticlinal flexure diverting flows to north and south but unequally when the phreatic surface is critically depressed such that groundwater ceases to spill north over its crest.

Site development

Estimates of the loads of major pollutants likely to be leached by rainwater infiltrating the proposed landfill (DoE 1961, Robinson & Maris 1974; DoE 1984) and to percolate into the groundwater system were converted into probable mean concentrations at each of the water resources deemed at risk by reference to:

(1) estimated annual flows off the whole catchment for average and exceptional years,
(2) proportions emerging in sub-catchments 6, 7, 9, 10 and 11, and
(3) proportions of leachate like to emerge in each sub-catchment.

In the case of the Glyncornel Trout Pond this revealed that ammoniacal nitrogen might vary between 2.1 and 3.7 mg/l in exceptionally wet and dry years respectively. The mean value of 2.8 mg/l was in excess of the value which trout can tolerate. Value of BODs (Biochemical Oxygen Demand) were similarly estimated at unacceptably high levels. It was concluded that the site should operate on the concentrate and contain principle.

Piezometric response to rainfall demonstrated that the soil cover provided little obstacle to infiltration and would be unable to contain a head of leachate within the waste tip. The soil's relative paucity and variable but coarse composition were considered not to provide a feasible medium for bentonite admixture. Commercial quantities of a suitable, naturally-occurring clay were not locally available. As a consequence it was recommended that a manufactured liner be used and HDPE based products were considered suitable.

Disposal and treatment options considered (County Surveyors' Society, 1984) included combinations of:

(1) on-site aerobic treatment,
(2) spray irrigation,
(3) disposal to the River Rhondda Fawr,
(4) disposal directly to disused mine working,
(5) disposal to the Rhondda Fawr Trunk Sewer (which outfalls directly to the Bristol Channel),
(6) tankering.

It is beyond the scope of this paper to do more than summarise the arrangements eventually selected following a series of meetings with the Welsh Water Authority and Rhondda Borough Council. In outline they provide for:

(a) leachate collection by gravity flow in pipes laid above the liner.
(b) discharge to a holding reservoir (this is required as a result of undercapacity of the combined foul and storm water trunk sewer which is surcharged by heavy rain),
(c) regulated discharge by pipe to the trunk sewer,
(d) regulated discharge to the River Rhondda Fawr if, after extended heavy rain, the leachate holding reservoir approaches capacity (by which time flows in the River Rhondda Fawr would be enhanced by run-off from its essentially 'flashy' catchment),
(e) control involving level sensors in the sewer and river and automatic valve control at the reservoir and the outfall.

Access

The only existing vehicular access from the lower valley sides is via unmetalled, circuitous track through an extensive Forestry Commission plantation to the south. A new access road was therefore considered necessary. As 102 landslips have so far been reported in the Rhondda Valleys (Conway *et al.* 1980) and this appears to understate the total by some 60 per cent (Sir William Halcrow and Partners, 1985), it was essential that the route alignment was investigated in relation to areas of active instability and potential reactivation of dormant areas. This investigation formed part of the feasibility study but is not described here.

The site in currently expected to become operational in 1986.

Conclusions

The paper set out to describe a hydrogeological investigation of a proposed landfill site in an area receiving high rainfall and in which numerous water resources have been developed. It revealed that the tip site is underlain by a thin soil cover which readily permits infiltration of rainwater. Recharge of underlying aquifers is rapid and emergence of groundwater along the crops of aquicludes is sustained by variable but occasionally large heads of water providing a mean hydraulic gradient of 1 in 18. Despite a fissure flow mode of groundwater transmission, considerable mixing and dilution potential is indicated. A net inflow of groundwater to aquifers underlying the site appears derived from an extensive upland catchment beyond a fault zone to the west. Adoption of an analysis of tracer recovery based on comparison with cumulative flows facilitates interpretation of results during drought conditions. Loss of groundwater through downward percolation to deep strata is significant but would not carry away sufficient leachate to enable the site to function without risk of pollution except by containment principles. To protect established fishing and other social amenities, site lining with leachate disposal mainly to trunk sewer are proposed.

ACKNOWLEDGEMENTS: The Author wishes to thank the Partners of Sir William Halcrow and Partners for their assistance in the preparation of the paper and records his indebtedness to colleagues for their help in the accumulation and analysis of data, to Mr E.W. Clipsham of AERE, Harwell, and to Rhondda Borough Council for permission to publish these data. The views expressed are those of the Author.

References

CONWAY, B.W., FORSTER, A., NORTHMORE, K.J. & BARCLAY, W. J., 1980. South Wales Coalfield Landslip Survey. *Report No. EG 80/4 of the Institute of Geological Sciences, Special Surveys Division Engineering Geology Unit,* **2** 1-131.

COUNTY SURVEYORS' SOCIETY, Committee No. 4, Special Activity Group No. 6, 1984, Report No. 4/1, The treatment of leachate from landfill sites, 1-69.

DAUGHTON, G., NOAKE, J.S. & SIDDLE, H.J., 1976, Some hydrogeological aspects of hillsides in South Wales. *Proceedings of a Conference Rock Engineering, Newcastle upon Tyne,* 423-439.

DEPARTMENT OF ENVIRONMENT, 1961, Pollution of water tipped refuse, Report of the technical committee on the experimental disposal of house refuse in wet and dry pits. DoE London.

DEPARTMENT OF ENVIRONMENT, Landfill Practices Review Group, 1984, The selection of landfill sites, WLR Technical Note Series No. 64, DoE, 1-72.

DOORENBOS, J. & PRUITT, W.O., 1977, Crop Water requirements, F.A.O. *Irrigation and Drainage Paper No. 24, Food and Agriculture Organisation, Rome,* 1-144.

HALCROW, SIR WILLIAM AND PARTNERS, 1985, Interim statement on stage 1 assessment of landslip potential : South Wales, *Dept. of Environment Contract ref. PECD/7/1/149-63/84; Private communication.*

RHONDDA BOROUGH COUNCIL (Local Plans Unit), Waste Disposal Plan, Report of Survey, 1980.

ROBINSON, H.D., & MARIS, P.J., 1974, Leachate from domestic waste : Generation, composition and treatment. A review. Water Reserve Council Technical Report TR108.

SIDDLE, H.J., JONES, D.B. & WARREN, C.D,, 1984, The use of tracer techniques to assess groundwater flows in site investigations, *Proceedings of Engineering Group of the Geological Society, Site Investigation Practise : Assessing BS 5930,* Univ. of Surrey pp419-431.

WOODLAND, A.W. & EVANS, W.B., 1964, The Geology of the South Wales Coalfield, Part IV, Pontypridd and Maestef, H.M.S.O., London.

R. TONKIN, Sir William Halcrow and Partners, 25 Windsor Place, Cardiff CF1 3BZ, UK.

Site investigations techniques used to assess the likely hydrological impact of a proposed landfill site

J.H. Whittle & A.C.S. Swanson

ABSTRACT: The construction of a landfill site is proposed at Dalinlongart, near Dunoon, Argyllshire. The site forms an unspoilt and secluded area within the hills and glens of the Cowal Peninsula on the Clyde Estuary and lies to the north of the nearby Highland Boundary Fault.

A development of this type has to satisfy the rigid constraints on the emission of pollutants and in this respect it was necessary to assess the likely impact of such a development on the existing hydrological conditions.

To obtain information for the design and subsequent operation of the landfill site, a site investigation was carried out involving a variety of techniques, both within the site and its environs. The merits and limitations of the differing investigation methods are described and discussed with respect to providing the required hydrogeological data in the context of this site.

This paper illustrates the benefits of a desk study, in particular the aerial photograph interpretation, in assessing the hydrogeology of the site area and in the economic planning of the subsequent ground investigation. The advantages of mapping techniques and deep trial pits to expose the structure of soil and rock strata are outlined together with the requirement to carry out water quality analyses before, during and after the construction of the landfill site. Boreholes were required not only to confirm the ground conditions indicated by other works, but also to provide permeability data and install standpipes and piezometers for the continued monitoring of groundwater conditions.

Introduction

The construction of a landfill site is proposed at Dalinlongart, near Dunoon, Argyllshire, Scotland. The site lies within an area of high relief, 2 km to the west of the Holy Loch within the foothills of the Western Highlands. The surrounding area comprises hills rising to heights in the order to 300 m and incised glens with associated water courses discharging into the Holy Loch at sea level. The proposed site is located within a gently sloping topographic hollow at a general level of 82 m AOD on the north western slopes of Dalinlongart Hill which rises to 196 m AOD. The area is presently covered by a forestry plantation of small conifer trees and a peat bog.

The proposed landfill site occupies an area of about 6.5 hectares and is intended to cater for domestic refuse from a scattered rural populations. The development is required to comply with the relevant legislation and regulations of the licensing river purification board and to avoid undesirable aqueous emissions (or leachates) from the landfill which could contaminate water courses in the vicinity of the site.

In order to satisfy these requirements and so minimise the impact of the landfill site on its environs, correct design, construction, operation and final closure/reinstatement is necessary. The hydrogeology of the site area and its environs required to be established, and accordingly a site investigation was carried out. The extent of the investigatory works included for future monitoring to enable the continual assessment of the landfill development. Many existing landfill sites exhibit significant discharges of leachate, which can cause serious pollution. This is frequently due to poor planning, design, construction and site operation. In the Dalinlongart area, the existing Glen Saul landfill site has been operational for some years, and some contamination of a nearby water course has occurred due to direct discharges of leachate. The site investigation was extended to include this area and so provide a comparison with the proposed landfill. This paper discusses the extent of these site investigation works.

Landfill design

In general, the suitability of prospective landfill sites requires to be established by a team of professionals including geotechnical and civil engineers and chemists and each requires to understand aspects of the selection and future control of landfill sites. Therefore, in order to provide suitable geological expertise, the following general comments on the development of landfill sites are worthy of consideration.

Leachate generation

To achieve the lowest quantities of leachate generation at a landfill site it is necessary to define the

From Cripps, J.C., Bell, F.G. & Culshaw, M.G., (eds), 1986, *Groundwater in Engineering Geology,* Geological Society Engineering Geology Spcial Publication No. 3, pp. 197-205.

hydrological regime, including an appraisal of meteorological data. Prospective sites require to be engineered to ensure that ground and surface waters have minimal impact on the site operations and leachate generation. Domestic waste in saturated conditions produces large quantities of leachate.

In order to minimise leachate generation, water entering the landfill site should be intercepted and taken away from the site by a suitable drainage system. The position of regional and perched water tables need to be established together with surface water runoff and groundwater inflows.

For sites such as Dalinlongart, and UK sites in general, precipitation infiltrating the waste is a major water source and during the tipping operations low permeability intermediate cover materials are required to minimise rainfall percolation. Upon completion, the accepted restoration design is to minimise long term rainfall percolation to restrict post-operational leachate generation by the final capping and contouring. To form such cover/capping layers can involve significant quantities of selected cohesive materials and the identification of suitable materials within the site or its vicinity can result in considerable cost savings. The operation of the landfill by the tipping of waste into cells (i.e. small contained areas) can minimise the surface of waste exposed to rainfall.

Leachate treatment

Most landfill sites require to incorporate a leachate collection system in order to carry out treatment processes which can significantly reduce leachate concentrations. One of, or a combination of, the following methods are generally adopted:

(a) Spray irrigation; leachate is collected and sprayed either over a section of the site or a selected area adjacent to the site. This creates the leachate and can be combined with recirculation techniques. The hydrogeology of those selected areas requires to be established to ensure satisfactory attenuation of the leachate.

(b) Leachate recirculation; involving the direct pumping of leachate onto selected areas underlain by suitable ground conditions to attenuate the leachate. This method is generally most effective in unsaturated ground conditions of medium intergranular permeability. In relatively dry landfill conditions, recirculation of leachate through the waste can be utilised to reduce the leachate volume although the concentration of the leachate issuing from the refuse can increase.

(c) Storage of leachate in lagoons; the leachate is collected in lagoons for other treatment processes such as aeration and chemical treatment and including its removal off site. The local availability of inert low permeability materials can facilitate the construction of lagoons, otherwise suitable construction materials will require to be imported.

Leachate emissions

With respect to the control of leachate emissions and pollution two main types of landfill site exist, namely:

(a) Containment sites are those sites which are hydrologically secure, where leachate is prevented from infiltrating the underlying soil/rock strata or groundwater by impermeable barriers. Low permeability ground conditions will favour this type of site, otherwise the construction of impermeable barriers is required. The volume of liquid within the site requires to be monitored as the leachate content may increase in time resulting in the refuse becoming saturated which could give rise to surface discharge or potential overflow beyond the barriers.

(b) Dilution/Dispersion sites; leachate is permitted to discharge from the site such that natural dilution, attenuation and dispersion in the surrounding hydrological regime reduces leachate concentration to acceptable levels. A regime of moderate intergranular permeability conditions can be appropriate for a landfill of this type. Dilution/dispersion can also be achieved in suitable surface or groundwater flows.

In the selection of sites, it should be appreciated that high secondary permeability conditions can result in a rapid rate of leachate transport along discrete pathways with little dilution or dispersive effects to reduce leachate concentrations. For a landfill in such hydrogeological conditions measures would require to be introduced to prevent high leachate concentrations entering these pathways of preferential flow. It is frequently landfills in these latter conditions which are more problematic, complex and costly to maintain.

Domestic refuse can take several decades to decompose to an essentially inert material, and leachate control systems may be required to function for a considerable time after tipping operations have ceased. The control system should include for the provision of ground and surface water quality monitoring facilities and incorporate suitably located standpipes in boreholes and selected sampling of watercourses.

General considerations

Information on the ground conditions should also be obtained for other elements of the landfill development which are relevant to a civil engineering project where substantial earthmoving and drainage work are involved. At Dalinlongart these included stability of temporary excavations in both soil and rock, rock mass characteristics including ease of excavation and the design of access roads.

Initial site appraisal

An initial site walkover revealed the proposed area of landfill to take the form of a topographic hollow, enclosed to the north and west by small rocky knolls, to the south by a low ridge and to the east by Dalinlongart Hill (Figs. 1 and 2). The site itself slopes very gently down to the north west and is currently covered by soft saturated peat with occasional areas of standing water, and a vegetation cover of rough grass, reeds, heather and scattered small conifers.

A network of deep furrows and open drainage ditches cover the site, associated with a forestry drainage scheme. These open drains discharge into the watercourse at the north western corner of the site, which joins the Little Eachaig River to the north at a level of some 60 m below the site. A further watercourse lies close to the eastern site boundary, and also discharges into the Little Eachaig River, which in turn flows into Holy Loch at sea level.

The higher ground around the site has a covering of peat resting on variable glacial soils with outcrops of schist and psammite bedrock.

The area is subject to a high rainfall of about 2500 mm/year. The main elements of water storage are the peat deposits and to a lesser extent the small ponds scattered within the site. The underlying glacial tills were expected to have a low permeability, although more rapid groundwater flows are possible in coarser morainic debris. The localised deposits of fluvial sands and gravels are likely to be the only highly permeable deposits present. The metamorphic bedrock may be expected to have a low permeability unless frequent and open discontinuities permit significant fissure flow.

Preliminary site assessment

It was concluded at this stage that the proposed landfill development would require the removal of free water, drainage of the peat together with the construction of a perimeter drainage system to minimise water inflows from the surrounding high ground. It was anticipated that the refuse would be tipped into cells. The soils initially excavated to form the cells could be stockpiled and utilised for cover and/or reinstatement of the site depending on their extent and suitability. It was expected that bedrock would be present at relatively shallow depths and this would favour a containment design concept for the landfill, which could be further enhanced if bedrock reflected the topography and formed a shallow basin across the site area.

Site investigation procedure

In order to determine the general ground conditions and hydrological regime and so provide the required geotechnical data for design, construction, operation and final closure or reinstatement a site investigation was carried out comprising the following elements:

Desk study

A desk study was carried out as outlined below, to supplement and confirm the initial site appraisal and to plan the required fieldwork

(a) Topographical maps — 1:25 000 and 1:10 000 scale Ordnance Survey sheets were used to identify the general surface drainage pattern and for the calculation of catchment areas contributing water to the site or its vicinity. Larger scale maps were not readily available due to the remote location of the site.

(b) Meteorological data — precipitation and temperature data were examined in order to estimate the residual rainfall at the site and across nearby catchment areas contributing water to the site area. Estimates of surface runoff flows, evaporation losses and infiltration assessments were made from this data. The Meteorological Office provided data for three locations within this region, however, these were close to the sea level, and therefore an assessment of the altitude effects of the site was required. Data covering a prolonged period of time was examined in order to take account of climatic variations and storm effect.

(c) Geological records — the 1:10560 scale geological sheet Argyllshire 173 indicates the superficial deposits to comprise peat overlying glacial till over the southern and eastern sections of the site. Elsewhere, the peat is indicated to be lying directly on bedrock. Exposures of glacial till are variously described as 'red sandy drift', 'boulder clay' and 'gravels'. Similar soil conditions are indicated across the upland areas surrounding the site, with locally thin superficial cover and small rock outcrops. Granular fluvial deposits are shown on the narrow valley floors.

The underlying bedrock is shown to comprise metamorphic strata of the Upper Dalradian Beinn Buela Grits. These generally consist of quartz psammites (locally massive and coarse grained) together with quartzose and micaceous schists. The geological structure is complex; the general strike of both bedding and cleavage trending north east - south west. Some faulting is conjectured to exist close to the site and minor Tertiary dolerite dykes are also present in the area.

The Memoirs of the Geological Survey confirm the general or geological conditions to be as indicated on the geological sheet. The general composition of the Beinn Beula Grits comprises the interbedded strata of the following types, which frequently include quartz veins:

quartz psammites
quartz muscovite schists
chlorite schists.

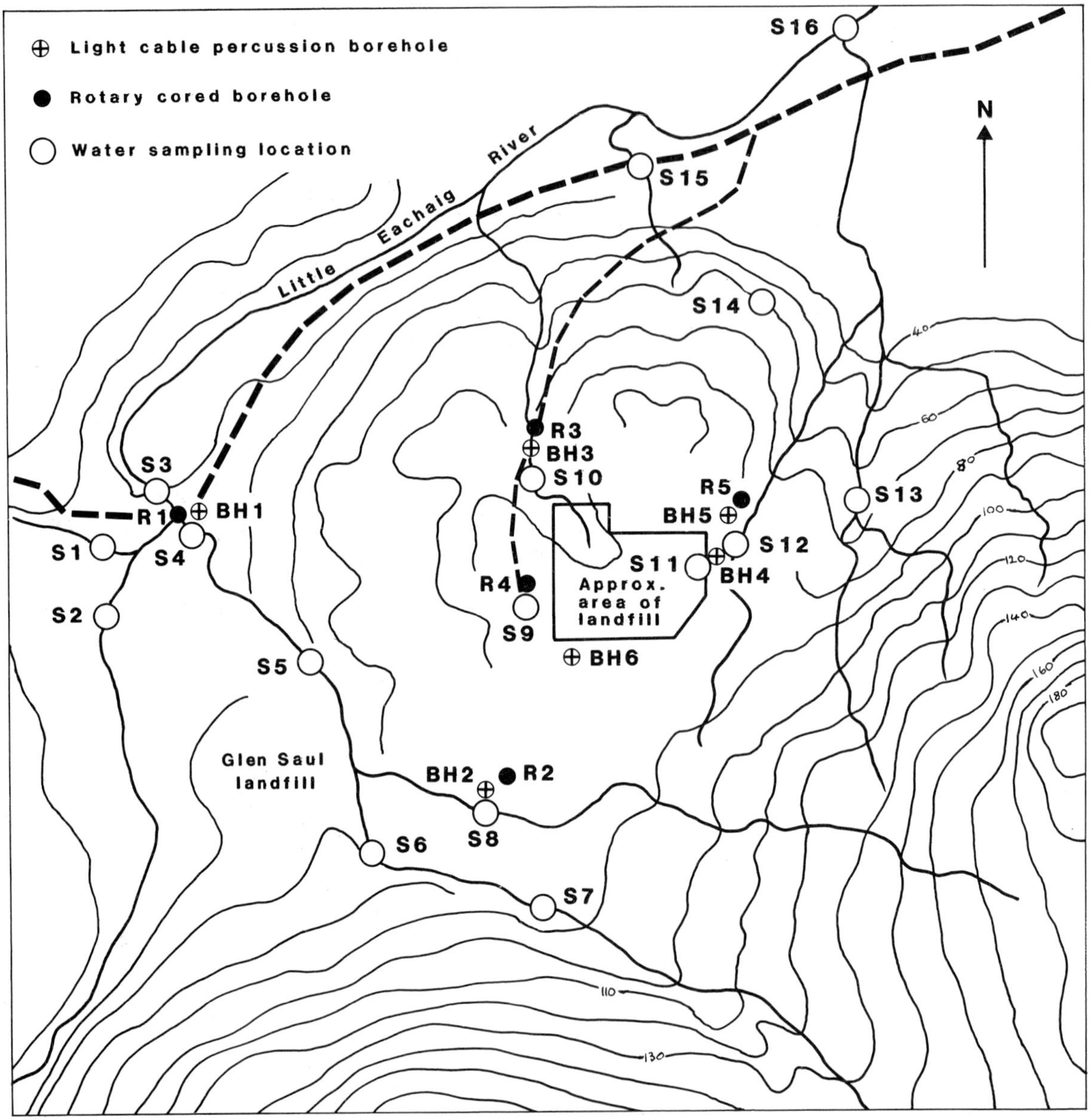

FIG. 1. Plan of site and environs showing borehole and water sample locations.

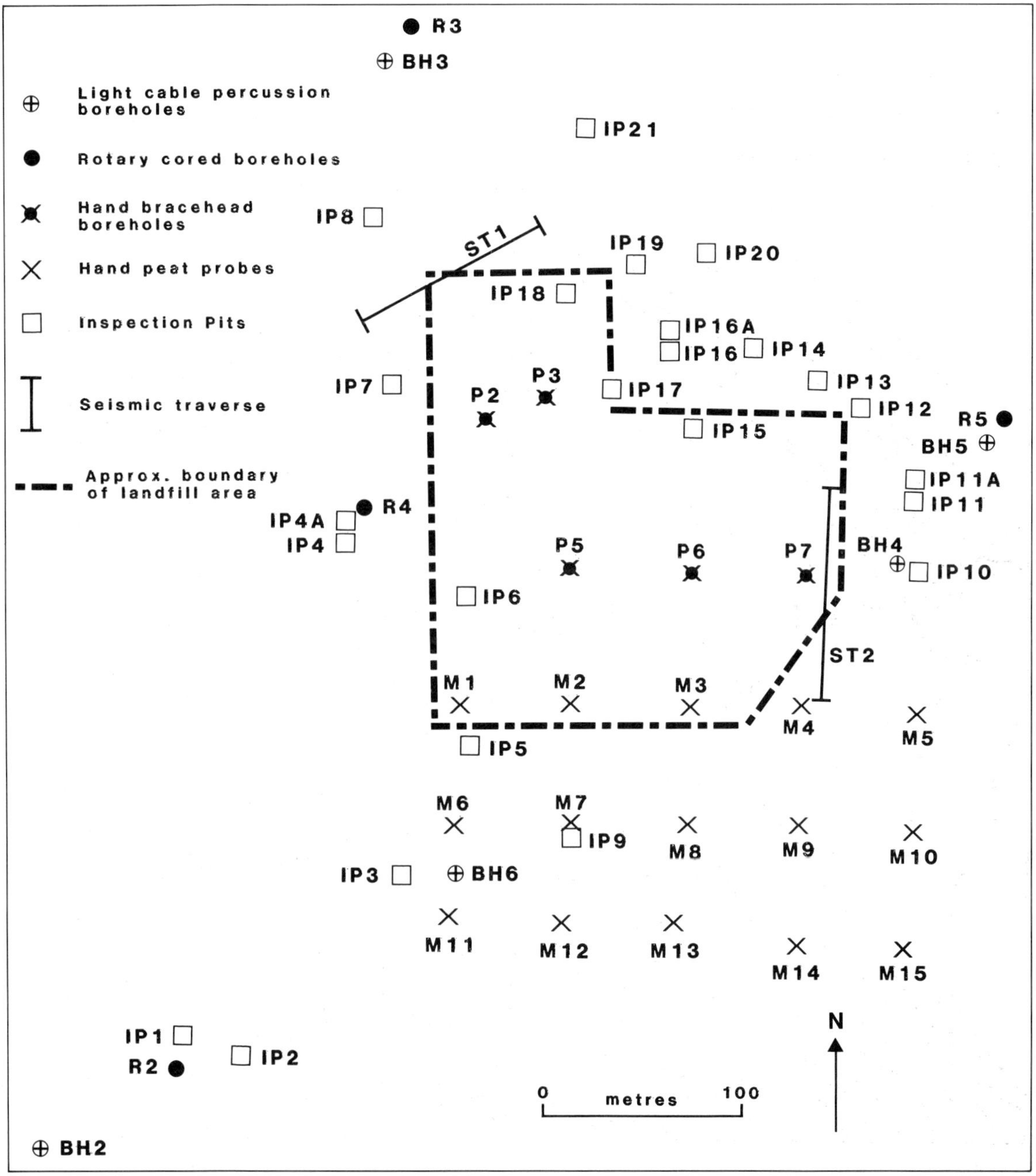

FIG. 2. Site plan showing locations of site investigation works.

Enquiries to the British Geological Survey confirmed that the above general geological conditions complied with their records of this area.

(d) Aerial photographs — vertical colour 1:10 000 scale aerial photographs were examined stereoscopically in order to assess the general geological and geomorphological features in the vicinity of the site.

The photographs were taken in May 1975 before any large trees were present in the region and the detailed drainage pattern provided by the deep ploughing and ditching works, associated with the forestry planting, were visible in detail. This confirmed the surface drainage within the site to discharge principally into the water course at the north-west corner of the site, with some minor ditches located close to the north-east corner of the site. Areas of waterlogged ground and small areas of standing water, were visible, correlating with those currently visible on site; some of these appeared to be slightly larger than at present.

The vegetation was assessed to comprise predominantly heather, rough grasses, bracken, reeds in waterlogged areas and small conifers — possibly newly planted. The varieties and distribution of vegetation gave an indication of the drainage characteristics of the superficial deposits. Relatively free draining ground conditions were indicated along the southern boundary of the site becoming progressively wetter and more waterlogged towards the northern end of the site, where areas of standing water were also located.

The distribution of soil types was assessed to be generally as shown on the geological sheet. From the vegetation changes referred to above, the thickness of peat appeared to increase northwards across the site. A low rounded ridge of relatively free draining soils along the southern margin of the site was identified as a possible morainic deposit.

Only limited information concerning the underlying bedrock structure was apparent due to the thickness of superficial deposits across the site area. However, rock exposures and areas of apparently thin drift cover on the higher ground to the north-east and west of the site suggest that bedrock may reflect the basin form of the topography, although there was limited evidence of the bedrock level beneath the thicker drift along the southern margin of the site. A visible major lineation coincided with the fault conjectured on the geological sheet, although no evidence of any relative displacement across this feature was apparent from the aerial photographs.

Field mapping

Geological mapping of the site and its environs was carried out by examining the soil and rock exposures in the several stream sections, forestry tracks and along a recently excavated road cutting to the north of the site. Features identified during the desk study were also examined and plotted on the field maps. A visual assessment was made of the permeability characteristics of the soil and rock exposures, by carefully identifying the extent and condition of discontinuities, zones of weathering and locations of groundwater seepages. In order to assess any preferential flow paths along fissures, the field data was analysed by stereographic plots.

The field mapping also included a physical land survey of the site and its immediate vicinity, together with possible routes for future access to the landfill development. Contoured site plans were prepared at a scale of 1:1000. This survey was required in order to obtain accurate level and positional data of features across the site and also to facilitate the plotting of accurate geological cross sections.

The initial surface water sampling was carried out in conjunction with the field mapping. The purpose of the sampling programme and associated chemical analyses was to establish the existing water quality and general chemical composition of the surface water. Watercourses and ponds of standing water were sampled in areas potentially vulnerable to contamination by leachate emission from the landfill, and also in the vicinity of the existing Glen Saul landfill site. The sampling locations also took into account the variations in water appearance across the area, namely, clear water in the larger watercourses, peaty water within the site area, widespread ferric deposits, local areas of algae in ponds and white suspensions and thick ferric deposits near the Glen Saul landfill. The sampling was commenced at an early stage in order to provide a data base for future comparison with samples collected as part of the site investigation fieldwork and subsequent long term monitoring of the landfill site. The parameters required from the chemical testing are listed below:

pH
BOD (Biological Oxygen Demand);
COD (Chemical Oxygen Demand);
TOC (Total Organic Carbon);
ammoniacal nitrogen;
conductivity;
chloride;
sulphate;
nitrate;
iron.

This combination of tests was selected to identify any leachate emissions at a future date, and to assess the following properties of the surface water as a measure of current condition and quality:

dissolved oxygen levels, from BOD to COD;
organic load, from BOD and TOC;
organic load, from COD, major anions and conductivity;
potential for ferric precipitates, from iron content;
acidity from pH value.

Ground investigation

The ground investigation works were carried out in order to confirm and supplement the ground conditions data as indicated by the desk study and geological mapping, and to determine the sub-surface conditions within the site and adjacent areas, particularly along potential routes of leachate outflow from the site. The ground investigation also included *in situ* tests, sampling for laboratory tests and the provision of standpipes and piezometers for the future monitoring of groundwater. The ground investigation works were carried out and recorded in accordance with BS 5930 (1981), and comprised the following:

(a) Inspection pits - 23 pits were excavated to assess the nature, condition and continuous vertical sequence of strata underlying the site and its environs. In particular, the soil mass structure was examined and any groundwater inflows recorded. Whenever possible, the underlying bedrock was exposed in order to establish the effects of surface weathering. The inspection pits were also utilised for soil and groundwater sampling.

Selected inspection pits were secured and left open for several months so as to observe long-term groundwater fluctuations, and any structural features of the soils exposed which can become more readily visible on superficial weathering. Mass permeability tests were also intended, but it proved difficult to prevent significant inflows of water into the excavation from the saturated surface cover of peat. A qualitative permeability assessment of the glacial till and bedrock was however possible, in that the excavations remained full of water for a considerable time, even in pits located on the summit of a small ridge, indicating the soils and bedrock to have low permeabilities.

The presence of soft peat across much of the site restricted the locations of the inspection pits to the margins of the proposed landfill area. An additional series of inspection pits were excavated across part of the knoll to the north of the proposed landfill area in order to assess if this area could provide suitable capping/cover materials.

(b) Light cable percussion boreholes — these were sunk to supplement the information obtained from the inspection pits, and to install standpipes for permeability testing and future monitoring and sampling of groundwater.

The boreholes were located as follows to confirm specific aspects of the ground conditions:

Borehole 1 — close to the existing landfill at Glen Saul to check for ground pollution.

Boreholes 2, 6 — in the area of thick morainic soils to investigate strata below depths attainable by the inspection pits.

Borehole 3 — close to the watercourse which exits from the north-west corner of the site and at the position of the conjectural fault indicated by the desk study.

Boreholes 4, 5 — between the proposed landfill area and watercourse located beyond the eastern site boundary.

Soil samples were obtained for identification and classification tests. In order to estimate the flow rates of groundwater through the soil mass *in situ* falling/rising head permeability tests were carried out and comparisons made with the estimate of permeability based on partial size data.

(c) Rotary cored boreholes — to determine the nature and condition of the underlying rock mass, rotary cored boreholes were sunk alongside the light cable percussion boreholes. The majority of the boreholes were sunk specifically using a compressed air flush, to enable accurate observations of any groundwater seepages. However, the drilling progress was found to be slow due to the presence of very strong and thick psammite horizons, and subsequently water flush was used to improve the drilling rate.

In situ single packer water injection tests were carried out in order to determine the permeability characteristics of the rock mass. These tests provided satisfactory results when considered in conjunction with possible errors inherent with this type of test (Bliss 1984). Recognition was given in the interpretation of the results to water leakage past the packer due to irregularities in the surface on the borehole caused by drilling through the very strong psammite horizons.

Standpipes were also installed for the future monitoring of groundwater within bedrock.

(d) Peat probes — probing using Mackintosh probe equipment and hand bracehead boring enabled the rapid assessment of peat thickness in areas inaccessible to the drilling and excavating plant. Some localised areas of peat were unsafe to walk on.

(e) Seismic survey — two seismic refraction traverses were carried out to determine the location of rockhead in potentially critical areas within the soft peat area. Single channel enhancement hammer seismic equipment was used. The ends of the traverses were located close to other boreholes or inspection pits to assist in the interpretation of results. One end of the traverse was also located outside the area of soft peat to minimise the absorption, scattering and non-elastic effects of the peat. The data was interpreted by the Plus-Minus method (Hagedoorn 1959).

(f) Groundwater sampling — the borehole standpipes were utilised for groundwater sampling as part of the water sampling and analysis programme previously described.

Summary of results

The ground conditions indicated by the site investigation may be summarised as follows:

(a) Soft fibrous PEAT to a general depth of 1 to 2 m increasing to 3 m in the central portion of the site.

(b) Stiff clayey sandy SILT with gravel and medium dense silty SAND and angular gravel with cobbles and boulders representing glacial till and morainic soils respectively. Variable permeability values of 10^{-8} to 10^{-4} m/s were recorded.

(c) The underlying bedrock was found to be in the general form of a basin or trough, with the lowest point at the north west corner of the site. Very strong faintly weathered chlorite and quartz mica SCHIST locally psammitic was revealed with tight and impersistant discontinuities with no indication of significant fissure flow. Low rock mass permeabilities are indicated, in the range 10^{-4} to 10^{-7} m/s, the higher values reflecting weathered strata close to rockhead.

The results of the water sampling and analysis have been classified. The larger river watercourses in the area to be Class 1 quality, namely the Little Eachaig River and Glen Kin Burn. The small watercourses and ponds close to the site correspond to Class 2 and 3 conditions due to peat carried in suspension increasing BOD levels and natural ferric discolouration. Groundwater samples also comply with Class 3 conditions. The watercourse adjacent to the existing landfill at Glen Saul is currently in a Class 3 to 4 condition with a thick bed load of ferric deposits.

General landfill design

The ground conditions revealed by the site investigation confirmed the initial hydrogeological assessment and design concept for the site. A containment landfill type was considered appropriate given the existence of a low permeability bedrock basin, with drainage of peat within the site, and perimeter drainage to prevent water inflows. The glacial till soils were considered suitable as providing cover materials and as a recirculation medium for a spray irrigation leachate treatment scheme.

The borehole standpipes and surface water monitoring points will be used for the continued collection of water samples in order to assess the hydrological development of the site, tipping procedures and the effectiveness of leachate treatment processes.

Conclusions

(1) In order to obtain geotechnical information required for the design of the proposed landfill site at Dalinlongart, and assess its likely hydrological impact, a site investigation was required to define the hydrogeological regime and general ground conditions, both within the site and its environs.

(2) Appreciation of the design options available to landfill engineers assisted the planning the control of the investigation.

(3) A site walkover enabled a preliminary assessment of the prevailing hydrogeological conditions and a possible outline of a suitable landfill scheme to be established.

The desk study provided additional and detailed information, and identified areas requiring further examination by geological mapping, inspection pits and boreholes.

The ground investigation comprised a variety of techniques, reflecting the ground conditions, particularly the presence of soft peat which constrained access to heavy plant and also the data required, namely, strata location, classification and *in situ* characteristics — particularly permeability values. The information obtained from each exploratory method was complementary and contributed to the overall hydrogeological model of the site and its environs.

(4) The surface and groundwater sampling programme was an essential component of the investigation. This has provided monitoring facilities and obtained data indicating the current water qualities present in the area, for future comparison during and following the landfill development.

(5) Landfill developments are now controlled by strict planning and legal requirements, particularly with regard to any emissions of leachate.

In order to design an effective landfill this paper has illustrated that extensive site investigation works are required to define the hydrogeological conditions and to assess the hydrological impact on the environment.

ACKNOWLEDGEMENTS: The authors would like to thank Argyll and Bute District Council for their kind permission to publish this case study. Leachate management advice provided for this development by Water Research Centre Environment Ltd is acknowledged. Thanks are also given to staff at Thorburn Associates for their assistance.

References

BARBER, C & MANIS, P.J., 1984 Recirculation of leachate as a landfill management option: benefits and operational problems. *Quarterly Journal of Engineering Geology,* **17,** 19-29.

BLISS, J.C. & RUSHTON, K.R., 1984. The reliability of packer tests for estimating the hydraulic conductivity of aquifers. *Quarterly Journal of Engineering Geology,* **17,** 18-91.

British Standard 5930, 1981 *Site Investigations* BSI, London.

CRAWFORD, J.F. & SMITH, P.G., 1985 *Landfill Technology* Butterworths, London.

GUNN, W. *et al.,* 1897 The Geology of Cowal. *Memoirs of the Geological Survey of Scotland,* HMSO, Edinburgh.

HAGEDOORN, J.C., 1959 The Plus Minus method of interpreting seismic refraction sections. *Geophysical Prospecting.* **7,** 158-182.

WELTMAN, A.J. & HEAD, J.M., 1983 Site Investigation Manuail, *CIRIA, Special Publications* 25. Property Services Agency, DoE, London.

J.H. WHITTLE, Applied Geology Ltd., Camden House, St Johns, Kenilworth, Warwicks CV8 1FB, UK. (formerly at Thorburn Associates).
A.C.S. SWANSON, Thorburn Associates, 145 West Regent Street, Glasgow G2 4SA, UK.

Discussion Session 5

E. JAMES (Hatfield Polytechnic) asked Professor J.K. MITCHELL whether account had been taken of the possibility of significant climatic changes in the 10^6 years period envisaged for nuclear waste containment. Mr JAMES also referred to the difficulty of installing a flexible impermeable membrane in a slurry filled trench.

J.K. MITCHELL (University of California, Berkeley) replied that since disposal also involves deep burial (at hundreds of metres below the surface) the impact of climatic changes would not be too important. A membrane can be installed using a rack and frame device.

D.W. COX (Polytechnic of Central London) asked Professor MITCHELL how high horizontal permeability compared with vertical permeability can be taken into account in compacted clay liners.

J.K. MITCHELL stated that most specifications required interlayer scarification by the use of sheep's foot compactor at a moisture content wet of optimum so giving a mingled surface. Hence a homogeneous material is produced.

D. LERNER (University of Birmingham) asked Professor MITCHELL to be more specific as to what was meant by 'hazardous waste'. Does this include domestic refuse for which the requirement for containment may not be so stringently applied.

J.K. MITCHELL answered that hazardous waste includes ignitable, corrosive, toxic and radioactive materials of which the toxic ones are volumetrically more important.

R.J. HILL (A. Monk & Co. plc) asked Mr C. WARREN to explain what was meant by the meteoric cycle.

C. WARREN (Sir William Halcrow & Pts) replied that it referred to the normal natural cycle of rainfall and evaporation. Infiltration is a function of latitude, relief and air temperature and is predicted from these and precipitation records.

L. CLARK (Water Research Centre) asked Mr J.H. WHITTLE for clarification regarding the method used (this volume, p. 203) to determine the dissolved oxygen levels.

J.H. WHITTLE (Thorburn Associates) explained that the values obtained from Biological Oxygen Demand (BOD) and Chemical Oxygen Demand (COD) are not precise measures of dissolved oxygen. A visual assessment of watercourses has been made and compared with measured BOD and COD values at other sites to provide an indication of oxygenation state.

S. TONKIN (Sir William Halcrow & Pts) responded to a question put by Mr J. PONTIN by stating that the tracer was injected at four points within the proposed disposal area and collection points were established in each sub-catchment area. Mr TONKIN also referred to the legal requirements relating to the use of tritium tracers in the United Kingdom. He stated that the practice is governed by the International Commission of Radiological Protection Units for tritium injection. Approval for the use of this method is controlled by the Radiochemical Inspectorate in London; in this case, they advised the Welsh Office who gave permission.

From CRIPPS, J.C., BELL, F.G. & CULSHAW, M.G., (eds), 1986, *Groundwater in Engineering Geology,* Geological Society Engineering Geology Special Publication No. 3, p. 207.

SESSION 6

ENGINEERING INVESTIGATION FOR GROUNDWATER QUALITY

Pollution investigation of a Triassic sandstone aquifer: the role of mineralogy

D. A. Spears

ABSTRACT: Soil and sediment samples below an effluent spreading site were analysed for mineralogy, whole rock major and trace element geochemistry and porewater chemistry. In spite of the high surface loading of trace metals, sediments are contaminated to depths of only 1 m. Sediment contamination at depth in the unsaturated zone was not detected. Pore water concentrations do decrease with depth, but this is attributed to dilution rather than reaction. Trace metal porewater concentrations at depth approach the 'natural' background levels for Permo-Triassic sandstones. Dissolution of minerals is predictably limited, although iron oxyhydroxide and clay minerals are distributed to give large effective surface areas and probably are responsible for trace metal absorption and exchange, and the attainment of low solute concentrations.

Introduction

The Permo-Triassic sandstones form an extremely important aquifer in the United Kingdom. Extraction and recharge take place, however, in areas of high land utilisation which leads to an anthropogenic input into the system. This input may arise from restricted area sites, such as landfill operations, or from larger sites where domestic/industrial and agricultural effluent has been spread. On an even larger scale are inputs arising from the use of fertilisers in modern agricultural practice. Reaction between infiltrating pore water and sediment may reduce or, ideally, eliminate the contaminant *en route* to the aquifer. The unsaturated zone may therefore be an important buffer which protects the aquifer. The movement of pore fluids through the unsaturated zone has therefore attracted considerable interest in recent years. The problem of the increasing nitrate concentrations in groundwater has necessitated comprehensive surveys of the recharge into major aquifers. Many specific sites have also been investigated in the United Kingdom by a number of workers, which although of post-completion nature do provide predictive information for future sites.

Changes in pore water quality from the surface down into the aquifer, that is, through the unsaturated zone, are studied by extracting pore waters from borehole cores. The extraction method adopted by most United Kingdom workers is that of centrifuge extraction. Reaction between sediment and pore fluid also involves either loss or gain from the sediment itself and therefore complimentary geochemical analyses of the whole rock may be used in conjunction with the pore water analyses. Mineralogical data is also essential if reactions with the sediment are to be understood and predicted. Without mineralogical data equilibrium calculations are of little value. It is this approach of pore water and sediment analysis that is illustrated in this paper drawing on data obtained in conjunction with the Water Research Centre from sites overlying Permo-Triassic sandstones and one site in particular, at Whittington Hall, where sewage effluent has been spread for a period of close to 100 years.

Mass balance and reaction-path simulation

The approach of combined whole rock and pore water analyses is shown diagrammatically on Fig. 1. Changes in pore water and sediment compositions should balance if the total volume of fluid (per unit cross-sectional area) is known. In practice, this latter quantity may be unknown and the geochemical data may be used to calculate volumes. This is what could be referred to as the total mass balance. Although the total mass balance may be a difficult goal to achieve nevertheless pore water and sediment analyses compliment one another with respect to time. Thus changing pore water compositions down the infiltration pathway are the result of current reactions. How long this takes is a function of the rate of infiltration and the depth of the unsaturated zone. The sediment on the other hand, is not transitory and the changes are preserved, thereby providing a longer term record of infiltration reactions.

The mass balance method described by Plummer *et al.* (1983) differs from the approach described above in that input and output concentrations into a system are known and the differences are balanced by postulating changes in the reactant and product phases, that is, the information on the minerals is derived from the solutions. In an elegant example described by Garrels & Mackenzie (1967) the compositional difference between snow water and

From CRIPPS, J.C., BELL, F.G. & CULSHAW, M.G., (eds), 1986, *Groundwater in Engineering Geology*, Geological Society Engineering Geology Special Publication No. 3, pp. 211-218

groundwater emerging from granite in the Sierra Nevada was attributed to a series of mineral reactions. The correct choice of product phases was based on the known thermodynamic stability in the pore fluid composition. The considerable refinement introduced by Plummer *et al.* (1983) permits computer calculation of speciation and stability of possible mineral reactions. Reaction in an aquifer between two wells was attributed to incongruent dissolution of dolomite driven irreversibly by gypsum dissolution and accompanied by calcite precipitation, oxidation of organic matter and sulphate reduction. Reaction-path simulation (Plummer *et al.* 1983) represents another step in which the initial water composition only is known and the final concentration is predicted following a series of hypothetical reactions.

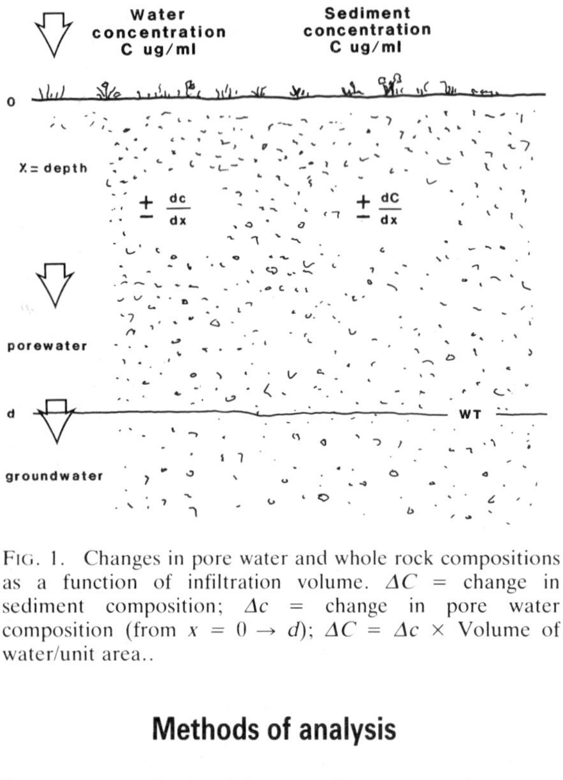

FIG. 1. Changes in pore water and whole rock compositions as a function of infiltration volume. ΔC = change in sediment composition; Δc = change in pore water composition (from $x = 0 \rightarrow d$); $\Delta C = \Delta c \times$ Volume of water/unit area..

Methods of analysis

Pore waters obtained by centrifuge extraction were analysed for metals by atomic absorption spectrometry (AA, Perkin-Elmer 460) operating in either the flame or furnace mode. The latter provides detection limits of the order of 1–2 μg/l for a number of elements of interest, which coupled with the small volume requirements makes the AA an ideal instrument for pore water analysis. Even so our efficiency of porewater extraction was relatively low for the sandstones from the unsaturated zone (≈5%), which meant that we were unable to routinely determine major ions. However, these analyses were made by WRC on the Whittington samples using the same or similar methods to those described in this volume by Stewart & Hitchman. These authors stress the need for the on-site measurement of some groundwater parameters and acidification for solution preservation. The same also applies to pore waters but in order to extract pore waters on site a mobile laboratory with a high speed centrifuge is required. These facilities were not available during the Whittington investigation and samples were therefore frozen on site and later extracted in the laboratory. The cores were frozen to prevent bacterial reduction which was found to be satisfactory by WRC for their nitrate work. The influence of freezing on trace metal concentrations was investigated in the Sheffield laboratories using Permo-Triassic sandstones from a site near to Sheffield which was sufficiently close to extract pore waters with the minimum of delay.

Borehole samples were crushed in a Tema disc mill and further ground by hand in an agate mortar. Major and trace elements in whole rock samples were analysed by X-ray fluorescence spectrometry (XRF) using the lithium tetraborate fusion method of Norrish & Hutton (1969) for the major elements, excluding Na, and pressed powders for the trace elements. We now determine Na by XRF but at the time of the analyses a separate determination involving $HF-HClO_4$ digestion and measurement by AA was necessary. Currently both major and trace elements may be determined on the same fused disc, which takes approximately 10 minutes per sample for the major and rather longer for the traces. Accurate elemental analyses may therefore be obtained in minutes on modern instruments compared with the rather longer time required for the then so called, rapid wet methods, based mainly on colorimetric determinations.

Identification of the minerals present in the sandstones was achieved by standard petrogaphical thin sections and by X-ray diffraction. Whole rock samples were run as back-filled mounts and separated clay fractions as smears on glass slides. Quantitative data can be obtained by XRD for the minerals present in sandstones, including, with care, the clay minerals (Brindley & Brown 1980), but in this work XRD was used for identification and quantification was based on the chemical analyses.

Finally, differential dissolution methods were used to identify the association of metals with specific reactive phases present in the sediment. Solutions were prepared using water, sodium acetate, acetic acid and sodium citrate (Ross 1980) and measured by AA. Such associations should also emerge in a statistical analysis of the whole rock geochemistry.

Sample location and site history

Samples were obtained from four boreholes drilled by WRC as part of an investigation of groundwater contamination below an effluent spreading site situated at Whittington Hall (NGR, SO 859826), near Stourbridge. At this site primary sewage effluent of domestic and industrial origin has been spread for close to 100 years on agricultural land overlying the Permo-Triassic aquifer. The sandstones belong to the Wildmoor Sandstone Formation, which occur stratigraphically above the Kidderminster Formation (formerly Bunter Pebble Beds). Soil samples were obtained from the depth interval 0–1 m by hand augering, whereas the borehole samples covered the depth interval 1–86 m. Control samples were also taken adjacent to the effluent spreading areas of both soil and sandstone.

Petrography and mineralogy

The grain size of the sandstones ranges from medium to very fine grained. The sandstones were deposited from water on alluvial fans grading into an alluvial plain. Such organic matter as there was at the time of deposition was destroyed by oxidation and therefore the potential for reduction during diagenesis was severely restricted. The sandstones therefore contain free Fe_2O_3 (as opposed to Fe^{3+} in mineral structures) which imparts the characteristic red colour to the sediments. The depositional environment was also one in which chemical weathering was restricted. The sandstones are sub-arkosic with a feldspar content of about 20% with K-feldspar much more abundant than plagioclase (K-feldspar averaged 19% by point counting and plagioclase 2% Spears 1983). Clay minerals, although present in amounts not exceeding 1–2%, are nevertheless important because of their reactive nature and accessibility to percolating solutions. The clay minerals present include smectite, illite, chlorite and kaolinite which therefore provide a significant cation exchange capacity. Chlorite in near-surface samples did demonstrate some features which could be due to infiltration and weathering. Calcite is also present in some of the sandstones, and its presence or absence is depth related. Below the irrigation site calcite was recorded at depths greater than 46 m, whereas in a control borehole off the site calcite first appears at 7 m. If present the calcite is uniformly distributed with a mean value of 4.5 ± 1.8% (Spears 1983) representing cement, with occasional higher values due to carbonate clasts.

With the exception of calcite and a minor change in the chlorite these are not depth related changes attributable to infiltration, although a variation in the total amount of clay was noted which proved to be an original, grain size related feature. Infiltration could be eliminated as a cause because the trend with depth was seen in reverse in one of the boreholes. Feldspar variation as a function of weathering was sought in vain.

Mineral stability and rates of reaction

The minerals of sandstones may be classified as either resistate, hydrolyzate or precipitate. Quartz and feldspar are resistates representing a detrital fraction which survives the weathering process in one or more sedimentary cycles. The hydrolyzate fraction is also dominantly detrital and consists mainly of the clay minerals produced during weathering reactions. To these are added clay minerals formed during burial diagenesis in what is still a relatively low-temperature environment. These minerals are either stable or metastable in the present weathering environment. In the latter case the rates of reaction are sufficiently slow that the minerals appear stable. These minerals in sandstones and also mudrocks, which simply represent an increase in the proportion of the hydrolyzate fraction, have therefore previously been through the weathering mill and it is not surprising that changes are not detected in what is on a geological time scale a relatively short period, that is 100 years of enhanced infiltration. In an earlier investigation of infiltration into the Permo-Triassic sandstone aquifer in the Vale of York (Spears & Reeves 1975) a small variation in the whole rock K_2O percentage was detected and in part this could represent some loss of K-feldspar but stripping of K from illite is also a possibility.

Although whole rock changes are relatively minor, weathering of feldspars and in particular the presence of etch pits may be demonstrated by SEM (Berner 1981). The rate of reaction is therefore controlled by surface reactions and not by solute transport. Berner (1981) calculated that if feldspar dissolution were transport controlled, grains of 200 μm radius would survive for about 14 months in soils. The survival of feldspars in the present sandstones is ample demonstration of the slowness of reaction. This means in the context of groundwaters that the input from resistate minerals will be negligible in the short term (10^{-1} - 10^{1} yr) but important in the long term (> 10^{3} yr). This will also apply to the input from structural changes of the clay minerals. Saline groundwaters at depth do not necessarily represent trapped seawater, but may well be meteoric water which is sufficiently old (and static) to have achieved equilibrium with the minerals present.

The other group of minerals, the precipitate minerals, are important in the short-term. At the Whittington site absence of calcite near the surface is attributed to dissolution. Such decalcification is common in soils and rocks, provided that concentrations do not exceed equilibrium values. (This is achieved by the system having sufficient throughput, which is a function of climate). Rates of reaction are sufficiently rapid for there to be a weathering front below which the infiltrating pore solutions are in equilibrium with calcite and are therefore unreactive. Unlike the weathering of the silicates the rates of reaction owe more to solute transport from the reacting surface and the reduction of surface area leads to rounded surfaces. At Dunsforth in the Vale of York (Spears & Reeves 1975) the depth of decalcification is 7.3 m, with an original calcite concentration of 3.6%. Knowing the pore water concentrations, it is possible to calculate the flow rate and thus the aquifer recharge (Spears & Reeves 1975). At the Whittington site, in areas of natural infiltration (control boreholes), concentrations and depth of calcite loss are very similar to the Dunsforth figures which suggests that so too is the total volume which has passed through the system.

In the work by Plummer *et al.* (1983) cited earlier, the minerals controlling the groundwater composition are all precipitate minerals including, in addition to carbonates, sulphates (gypsum) and sulphides (FeS_2). Sulphides are not involved in the Whittington reactions because in red-bed sandstones there is little possibility of production of both Fe^{2+} and reduced sulphur species and even less of both production and survival or sulphides. The problems caused by pyrite weathering are sufficiently well documented in both foundation instability and acid-mine drainage to appreciate that reactions are rapid in the short term in oxygenated high Eh soils and groundwaters. Although sulphides are absent in the Whittington Permo-Triassic sandstones pyrite was recorded in the glacial sediments overlying this aquifer in the Vale of York. The resulting sulphate contributes to the groundwater, whereas the Fe_2O_3 remains immobile in the unsaturated zone.

Although structural changes involving the clay minerals are slow this does not apply to the exchange reactions. Similarly the adsorption of trace metals onto iron and manganese oxyhydroxides is rapid. Experimental work has shown that there is an initial rapid reaction (order of minutes) attributed to surface bonding followed by much slower reaction (order of days) possibly involving solid state diffusion (Tessier *et al.* 1985). The mean Fe_2O_3% in the Whittington sediments is 1.58 ± 0.60% and MnO is 0.06%, which coupled with the clay minerals represents a very effective substrate for rapid reaction.

In the above discussion the emphasis is on the mineral reaction and the supply of solutes into the aquifer. Mineral stability and reaction rates are also an important consideration in the physical behaviour of rocks during weathering but not considered, however, is reduction of interparticle forces in mudrocks.

In the Whittington sandstones the loss of calcite is a partial loss of cement and the rock is physically weak and friable. The distinction drawn in this work between soil and sandstone owes more to method of collection (auger v. borehole) than the physical properties.

Major ions in the Whittington pore waters and groundwaters

The natural evolution of pore waters with the minerals in the sandstones is somewhat disrupted at the Whittington site. In the absence of an anthropogenic input the resistate minerals would provide a source for the slow transfer of ions to the groundwater. Concentrations would be low and probably the rainwater contribution could be recognised. The effluent spreading has, however, added major ions to the system in quantity.

Representative analyses from the Whittington site are given in Table 1. Concentrations decrease from the unsaturated zone into the saturated zone below the site. In the saturated zone concentrations decrease with distance from the site although zones of preferential flow have been identified. The improvement in water quality is attributed by Baxter (1985) to dilution based on ratios of ions against conservative Cl. Solutes derived from the effluent are thought to be progressively mixed with lower concentration water of meteoric origin. The effluent solutes are present in sufficiently high concentrations that additions from the minerals are overshadowed. The dissolution of calcite is certainly one reaction responsible for the addition of ions to the pore waters, but even in areas where calcite still survives its contribution to the pore waters is not readily recognisable. Thus in the Whittington sediments the ionic strength of the pore waters does not increase in the direction of infiltration, as would be predicted from mineral dissolution reactions if these were important. The lack of change in the whole rock geochemistry demonstrates that dissolution reactions are not important, with the exception of calcite. The decrease in pore water concentrations with depth is attributed to dilution on the evidence of pore water chemistry and also from the lack of either observed precipitation reactions from the whole rock geochemistry or theoretically possible reactions which could account for the pore water changes.

Metal retention in the Whittington sediments

The average metal composition of the effluent spread on the Whittington site is shown in Table 2a. Concentrations for the metals are in the order Zn>Pb>Cu>Ni>Cr>Cd. A significant fraction of this metal loading is present associated with particulate matter (Table 2b) rather than in solution.

Representative soil analyses are shown in Table 3 for depths of 0.1 m and 0.87 m. Many of the elements show an order of magnitude decrease over this depth interval. Analyses of intervening samples and other profiles demonstrate an exponential decrease in concentrations with depth such that background concentrations are achieved at approximately 1 m depth. The soils at Whittington developed from the underlying sandstones and an insight into background concentrations may also be achieved by comparing soil and mean sandstone analyses (Table 3), which again demonstrates negligible contamination at depths close to 1 m. Elements concentrated in the soil as a result of effluent additions are Zn, Pb, Cu, Cr, Ni and Sr, in that order which is very much in line with

TABLE 1: *Major ions in pore waters and groundwater, Whittington, mg/l.*

	Ca	Mg	Na	K	Nitrate	Nitrite	Cl	SO_4
Unsaturated zone								
1-10 m, $n = 10$	179	34	83	35	84	0.6	72[1]	158[1]
polluted	(50)	(20)	(42)	(8)	(55)	(0.3)		
Unsaturated zone	170	31	105	28	47	0.5	148[1]	
11-26 m, $n = 15$	(46)	(6)	(8)	(5)	(28)	(0.6)		
polluted								
Saturated zone	119	14	60	8	29	<0.1	115	140
30-83 m, $n = 24$	(15)	(9)	(16)	(1)	(8)		(30)	(28)
Regional groundwater [2]	65	5	12	2	9	<0.01	26	43

[1] Number of samples, $n \leq 5$
[2] Baxter, 1985

the effluent composition (Table 2). The elements Zr, Y and V were weakly concentrated in the soil compared with bedrock, but exponential concentration trends characteristic of an anthropogenic element were absent. These three elements probably reside in the resistate heavy minerals and thus reflect a minor concentration effect during soil formation.

The sediment analyses demonstrate retention of anthropogenic elements in the upper part of the unsaturated zone. This could be due entirely to a filtration effect with the solutes passing through the system in the pore waters. However, the order of metal concentrations in the soil is very similar to that in the effluent which does suggest that not only has the particulate matter been retained but the solutes also. This question is resolved by the pore water analyses.

Identification of the metal input into the unsaturated zone is recognised both by decreasing concentrations along the infiltration pathway and also by comparison with background concentrations. The mean analyses quoted in Table 3 are based on 42 sandstone samples which are themselves not uniform. The background variation may therefore be reduced if the mineralogical controls are recognised. In these sandstones the metals are associated with the clay minerals and the oxyhydroxide material which both correlate positively. Using the Al_2O_3 concentration as a measure of the metal host it is possible to reduce the background variation or rather the predicted background concentrations with a greatly reduced error margin. For example, in the case of Ni the background concentration is 7 ± 3 μg/g and for a normal distribution 95% of samples fall in the interval $\pm\ 2\sigma$. Although this is sensitive an improvement is achieved by utilising the Ni v Al_2O_3 correlation. Thus the equation for this relationship is Ni = 2.23 Al_2O_3 –5.95% and the 95% confidence limit on the intercept is $\pm$ 2.94 (Spears 1983; Table III) which provides a measure of the dispersion about the equation. Expressed simply, the variation of an element in a sediment is reduced if account is taken of the association of that element with a specific measurable fraction of the sediment. A number of elements correlate with Al_2O_3 in the Whittington sediments, expressing a clay association, but only in the case of Pb were the potentially contaminated samples found to differ from the unpolluted. Other elements, including Ni were not enriched at depth in the unsaturated zone.

In order to check the above conclusions, and in particular the exceptional Pb behaviour, sequential leaching was undertaken. The idea is that specific fractions of the sediment are dissolved in turn and the solutions analysed for metal association. Included in these fractions are water soluble ions, exchangeable cations, and carbonate and oxide associated cations. The analysis of reactive phases failed to reveal metal contamination at depth in either the unsaturated or the saturated zone and the Pb concentrations were not significantly different between contaminated and uncontaminated samples. The difference in Pb noted in the whole rock data is therefore thought to be original and by chance coincides with the distribution of potentially polluted samples.

The geochemistry of the Whittington soils and sandstones therefore reveals significant contamination of only the upper metre of the system. Metals are not moving through this zone to react deeper in the system and one explanation is that equilibrium values are quickly achieved within the soil and are not disturbed deeper in the unsaturated zone by possible reactions involving major phases.

Trace metal concentrations in the pore waters

As noted in an earlier section the borehole cores from the Whittington site were frozen on site to preserve the pore water chemistry. This procedure is known to be effective for the nitrate system but the influence of freezing on the trace metals had not been ascertained. This was done in the Sheffield laboratories using

TABLE 2: *Metal composition of Whittington effluent.*

(a) Mean metal concentrations, µg/l, in the final Whittington effluent for the period 1975-77 (Baxter 1985).
(b) Particulate and solution composition of Whittington effluent — analysed by J.R. Newman, University of Sheffield. Particulate is that fraction retained on a 0.45 µm filter. Concentrations in µg/l.

(a)

	Cr	Cu	Ni	Zu	Cd[1]	Pb[1]	Mn	Fe
n = 26	35	95	48	578	11	104	N/A	N/A
σn	23	34	29	275	9	61		

[1]n = 17. N/A = not analysed.

(b)

	Cr	Cu	Ni	Zn	Cd	Pb	Mn	Fe
Solution	34	-	67	33	27	7	174	5
Particulate	62	174	55	690	13	195	185	5770

TABLE 3: *Representative soil and mean sandstone concentrations, Whittington, µg/g.*

	Cr	Cu	Ni	Zn	Pb	V	Rb	Sr	Y	Zr	Mn
Soil 0.1 m	650	275	52	2220	339	46	72	112	18	250	360
Soil 0.87 m	40	24	9	235	52	31	86	63	25	280	240
Mean sandstone	33	8	7	28	33	22	81	58	16	185	460
Standard Deviation	11	3	3	14	18	6	17	12	7	93	390

TABLE 4: *Comparison of pore water compositions before and after freezing. Significance of correlation, correlation coefficient and slope of reduced major axis regression equation shown.*

	Before v. after freezing												
	Ca	Mg	K	Na	Fe	Mn	Cr	Cd	Co	Cu	Ni	Pb	Zn
Correlation coefficient	0.97	0.99	0.88	0.97	0.20	0.86	0.72	0.76	0.06	0.58	0.97	0.64	0.82
Significance at 99% level	✓	✓	✓	✓	×	✓	✓	✓	×	✓	✓	✓	✓
Slope	0.80	1.12	0.60	0.88	0.16	0.41	1.19	0.76	0.07	1.01	0.64	0.92	0.82

borehole samples of Permo-Triassic sandstones obtained from a local site, sufficiently close to shuttle unfrozen cores into the laboratory for pore fluid extraction. Pore waters were also extracted from a sample split which was frozen on site. The pore water chemistries are compared in Table 4. Most elements correlate significantly and with slopes of the regression equations close to 1.0, thus demonstrating the validity of freezing. Notable exceptions are Fe, Mn and Co. The latter is present in very low concentrations close to, or below, its detection limit of 2 µg/l which accounts for the lack of correlation. The changes in Fe and Mn are fundamentally more serious in that a change in Eh could be indicated, but even so other elements are not affected, which indirectly suggests that changes in oxidation states are not responsible for metal immobility. The freezing experiments demonstrate that meaningful concentration data may be obtained for pore waters, although trends are more reliable than absolute concentrations.

The variation in metal concentrations in pore waters through the unsaturated zone and into the aquifer are illustrated in Table 5. Concentrations decrease from the upper 10 m of the unsaturated zone compared with the next 10 m which are comparable with values obtained in the saturated zone. The latter are grouped because a depth grouping has less meaning than in the unsaturated zone.

The saturated zone values are also compared with 'natural' groundwaters (Table 5) and Cu, Cd and Ni would appear to be enriched beneath the Whittington site, but other elements, including Cr, Pb, Zn and Co, are at natural background levels. Only in the case of Cu do concentrations exceed WHO limits (Table 9; Forstner & Wittman 1981). Nevertheless, high metal loadings at the surface do not lead to metal enriched pore waters. It should be stressed that concentrations in the pore waters are in the parts per billion range.

Although pore waters decrease in metal concentration below 10 m, there is no evidence of metal enrichment in the sediment at depths as great as 10 m. If pore water concentrations decrease without reaction as was noted for the major ions, the same cause, namely dilution, may be invoked to explain the fall in concentrations.

Concentrations in solution are comparable with those reported by Tessier *et al.* (1985) in oxic sediments, where general agreement with laboratory experiments on trace metal absorption onto iron oxyhydroxides was noted. Trace metal concentrations were also below those predicted by equilibrium solubility models. In the present oxic samples iron oxyhydroxides probably control the trace metal concentrations. Adsorption is pH dependent and lack of precise pH pore water measurements in this work prevents calculation of the average equilibrium constant for the reactions. However, complications will arise in the effluent system due to the formation of both inorganic and organic metal complexes which may either increase or decrease adsorption (Tessier *et al.* 1985). Nevertheless a notable feature of the Whittington soils and sediments is the presence of this reactive oxyhydroxide material which because it is distributed as grain coatings has a high effective surface area. This material is available for reaction, not only in adsorption reactions, but also in oxidation reactions involving organic matter. The lack of iron reduction and loss, as revealed from the whole rock geochemistry, demonstrates that the oxygen demand within the pore waters must be catered for by dissolved oxygen, which in turn is a function of the permeability. Thus the sand grains not only provide an efficient substrate for adsorption reactions but also maintain an oxic system through the high permeability.

Conclusions

A geochemical and mineralogical investigation of the sediments and pore waters below an effluent spreading site has reached the following conclusions:

1. There is limited mineral dissolution with infiltration except for calcite.
2. The limited dissolution agrees with predicted mineral reactivity and high major ion concentrations in near-surface pore waters are derived externally rather than internally.
3. High metal loadings onto the surface present in both solution and particulate matter, have resulted in exponential depth decreases in whole rock concentrations which achieve background concentrations at depths of approximately 1 m.
4. Lack of metal pollution at greater depths (> 1 m) in the unsaturated zone was confirmed by selective dissolution experiments to isolate reactive phases and by reducing the statistical variance attributable to specific minerals.
5. There is a decrease in trace metal pore water concentrations at depths > 10 m, which together with a decrease in major ion concentrations is attributed to dilution. Concentrations approach natural background levels in spite of the surface additions.
6. The use of frozen cores to preserve pore water chemistries does not in general produce irreversible changes in the composition of the pore waters.
7. Iron oxyhydroxide material is present as grain coatings presenting a high effective surface area in a permeable substrate and adsorption equilibrium with trace metals may be responsible for the low concentrations at depth.
8. Analysis of sediments and pore waters provides a double-edged approach to the study of groundwater evolution.

TABLE 5.

Mean metal concentrations in Whittington pore water, concentrations expressed in μg/l, standard deviation in brackets.

Samples	Cu	Cr	Pb	Cd	Zn	Co	Ni	Mn	Fe
Unsaturated zone:									
2-10 m, n=5	445(230)	<2	105(120)	10.5(0.5)	1270(495)	<2	40(28)	380(280)	60(55)
Unsaturated zone:									
12-20 m, n=5	170(14)	<2	9(2)	6.5(3)	505(180)	<2	20(8)	545(360)	42(20)
Saturated zone:									
31.0-82.7 m, n=26	135(50)	4	7(7)	5(3)	270(130)	<2	16(20)	135(220)	970(1400)
Natural Permo-Triassic groundwater									
median [1]	9.2	7.1	2.7	0.5	107	1.4	3	-	150
minimum-maximum	3-30	2-18	1-6	0.5-0.8	8-600	1-5	1-3		10-500

[1]Table 7, Edmonds and Morgan-Jones, 1976.

ACKNOWLEDGEMENTS: The work described in this paper could not have been undertaken without the resources, financial assistance and co-operation of the Water Research Centre. Many individuals have contributed, but in particular I would like to thank Lewis Clark and Chris Young. The author takes, however, full responsibility for the views expressed in this paper.

References

BAXTER, K.M., 1985. The effects of discharging a primary sewage effluent on the Triassic Sandstone Aquifer at a site in the English West Midlands. First International Conference on Groundwater Quality Research, Rice University, Houston, Texas, October 1981. In: Ward, C.H., Giger, W. & McCarthy, P.L. (eds) *Groundwater Quality*, John Wiley & Sons, New York.

BERNER, R.A., 1981. Kinetics of weathering and diagenesis, in 'Kinetics of geochemical processes', Lasaga, A.C. and Kirkpatrick, R.J., (eds), *Reviews in Mineralogy*, Vol. 8, Mineralogical Society of America, 111-134.

BRINDLEY, G.W. & BROWN, G. 1980. Crystal structures of clay minerals and their X-ray identification. *Minerological Society Monograph* No 5, Mineralogical Society of London.

EDMUNDS, W.M. & MORGAN-JONES, M., 1976. Geochemistry of groundwaters in British Triassic sandstones: The Wolverhampton-East Shropshire area. *Quarterly Journal of Engineering Geology* **9,** 73-101.

FORSTNER, U. & WITTMANN, G.T.W., 1981. *Metal pollution in the aquatic environment.* Sprinter-Verlag, Berlin, Heidelberg, New York.

GARRELS, R.M. &. MACKENZIE, F.T., 1967. Origin of chemical compositions of some springs and lakes: In Stumm, W. (ed), *'Equilibrium concepts in natural water systems'*, Advanced Chemical Series, **67,** 222-242.

NORRISH, K. & HUTTON, J.T., 1969. An accurate X-ray spectrographic method for the analysis of a wide range of geological samples. *Geochimica et Cosmochimica Acta,* **33,** 431-453.

PLUMMER, L.N., PARKHURST, D.L. & THORSTENSON, D.C., 1983. Development of reaction models for groundwater systems. *Geochimica et Cosmochimica Acta,* **47,** 665-686.

ROSS, C.A.M., 1980. Experimental assessment of pollutant migration in the unsaturated zone of the Lower Greensand. *Quarterly Journal of Engineering Geology* **13,** 177-187.

SPEARS, D.A. &. REEVES, M.J., 1975. The infiltration rate into an aquifer determined from the dissolution of carbonate. *Geological Magazine,* **112,** 585-591.

SPEARS, D.A., 1983. Geochemistry and mineralogy of Triassic sandstones and implications for groundwater composition. *Minerological Magazine,* **47,** 183-190.

STUMM, W. & MORGAN J.J., 1970. *Aquatic chemistry: an introduction emphasising chemical equilibria in natural waters.* Wiley-Interscience, New York.

TESSIER, A., RAPIN, F. & CARIGNAN, R., 1985. Trace metals in oxic lake sediments: possible adsorption onto iron oxyhydroxides. *Geochimica et Cosmochimica Acta.,* **49,** 183-194.

D.A. SPEARS, Department of Geology, University of Sheffield, Sheffield, S1 3JD, UK.

Groundwater in Engineering Geology, *London*, 1986, pp. 219-223 Printed in England

A technique for groundwater quality monitoring using resistivity measurements made through a modified plastic well-screen

P. Jackson, J. Hallam, D. Allen, S. Hassett & M.G. Culshaw

ABSTRACT: A composite well-screen has been developed for resistivity measurements using high permeability 'terram' fabric wrapped around plastic well-screen containing 1.5 mm slots. Laboratory and *in situ* tests indicate that this screen presents a relatively small impedance to the flow of water and electricity while excluding fines that may be present.

Experiments with the well-screen were made *in situ* after emplacement in an unconsolidated, dense, fine coastal sand body. Results are presented of electrical resistivity measurements made through the well-screen, using a focussed resistivity tool, which indicate that changes in groundwater conductivity could be monitored as changes in formation resistivity. The groundwater inside the well-screen has been logged but suffers from the possibility of vertical mixing whereas measurements through the casing do not.

The well-screen also offers the possibility of taking water samples in unconsolidated sediments using a simple, very low pressure, double packer system. The effect of the well-screen on the resistivity measurements was found to be small after comparison with a self-bored resistivity log where the sand has been in direct contact with the formation.

Introduction

Electrical resistivity measurements in boreholes have been extensively developed over the past 50 years (Doll 1951; Schlumberger 1969) in response to the needs of hydrocarbon exploration. A massive volume of literature exists on the borehole techniques and attendant interpretation schemes which are used to assess the properties of porous reservoir rocks *in situ*. The classic work of Archie (1942, 1950) is applicable to most porous media and provides the link between the resistivities of the pore fluid and the formation. This work infers that a knowledge of formation resistivity can be used to assess pore fluid resistivity if the formation is clay-free and of know porosity and particle shape (Archie 1942; Atkins & Smith 1961; Jackson *et al.* 1979; Worthington 1982).

The electrical resistivity of porous media is measured in terms of a formation factor (*FF*) which was introduced by Archie (1942):

$$FF = \frac{\text{Formation resistivity}}{\text{Pore fluid resistivity}} = n^{-m} \qquad (1)$$

where n is the fractional porosity and m is a constant for a particular formation and is controlled by particle shape (Atkins & Smith 1961; Jackson *et al.* 1979). If the formation contains clays, then the matrix will conduct and the measured *FF* will be too small unless the pore fluid resistivity is very low (Wyllie & Southwick 1954; Worthington 1982).

The electrical resistivity method can therefore be used as an indicator of pore fluid resistivity and is particularly suited to monitoring formations where the pore fluid resistivity may change over a period of time, perhaps due to the seepage of pollutants (Mansfield Adams 1984).

Resistivity measurements in porous reservoir rocks are made routinely by borehole logging methods. However, unconsolidated sediments, or rocks that require casing, may be difficult or impossible to log. Metal casing would short circuit the electric current flow while plastic casing would prevent current from reaching the formation. Inductively coupled resistivity methods exist but are very insensitive at the resistivities expected for formations containing groundwater.

There is some evidence in the literature of resistivity measurements made through slotted casing (Chapalier 1982); if this method proves to be feasible then properties outside the borehole could be assessed and would be unaffected by mixing within the casing (Mansfield Adams 1984). The major problem in this approach is to devise a well-screen that has the following attributes:

1. A very large open area in order to produce an acceptably small increase in measured resistivity (say 10-20%)
2. A hole size small enough to act as a barrier to the individual grains of the formation.
3. Sufficient strength to ensure the hole stays open.

From Cripps, J.C., Bell, F.G. & Culshaw, M.G., (eds), 1986, *Groundwater in Engineering Geology*, Geological Society Engineering Geology Special Publication No. 3, pp. 219-223

Well-screen development

In situ trials

Field trials were undertaken utilising a borehole drilled in medium to fine running sand and lined with perforated plastic pipe normally used for land drainage. Resistivity measurements were made using a focussed borehole tool, ('laterolog 7'; Doll 1951; Cambefort 1957; Jackson 1981) with the aim of forcing current to flow through the slotted casing rather than up and down the borehole. The casing had a small number of relatively small slots 25 mm long and 0.6 mm wide giving an open area of 0.32%.

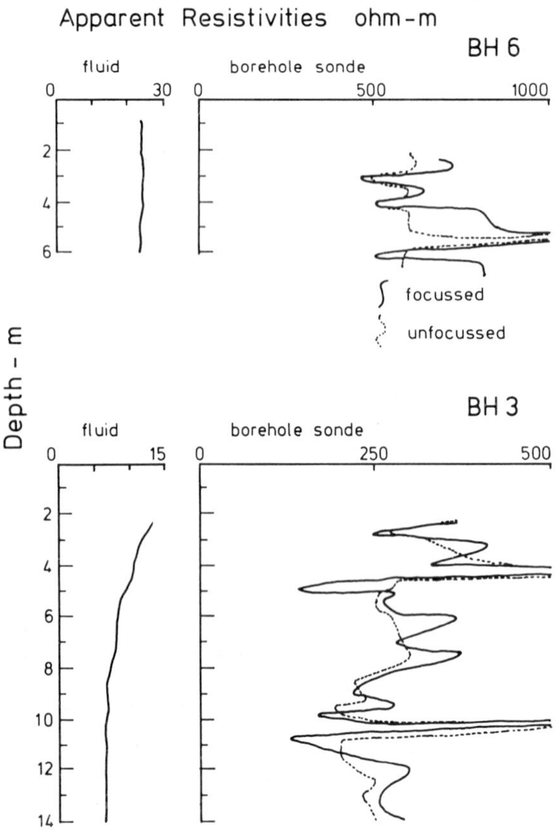

FIG. 1. Resistivity measurements in slotted casing (0.6 mm wide and 0.32% open area) emplaced in a saturated coastal sand. In BH6 the casing is free of drilling mud and full of groundwater; BH3 contains drilling mud which had settled out over a period of a year but still reduced the fluid resistivity to 6 ohm-m as opposed to 24 ohm-m in BH6. The majority of the measured electrical resistance is caused by the flow through the small area of slots.

The slotted casing proved to be unsuitable in almost every respect. Resistivity measurements inside the casing are shown in Fig. 1 where the high peaks are due to the unperforated joins between sections. The bulk resistivity of the sand was known from surface soundings to be 100 ± 5 ohm-m between depths of 1 and 25 m, so the values in Fig. 1 are very high, showing the casing to have a large effect. Boreholes BH6 and BH3 are lined with similar slotted casing but filled with different resistivity fluids, due to drilling mud remaining in BH3, and illustrate the high resistance of slots and its dependence on the resistivity of the fluid in the casing. The measured values are far in excess of the 100 ohm-m expected from the resistivity soundings. The extra resistance is due to current flowing through the fluid filled slots which are of known dimensions; given the resistivity of the fluid in the slots the resistance of the slotted casing could be calculated for the focussed case where the electric current flow should be radial. The fluid resistivity was logged and the resistance of the screen calculated. The fluid resistivity within the casing was reduced by the addition of a small amount of common salt to check the above hypothesis, and the logging repeated. The resistance of the casing was not consistent in the two cases; it was also noted that sand was entering through the slots. The conduction path through the slots was being blocked by bridging of sand grains, which changed in time, due to the action of the logging. The resistance afforded by a particular slot was unknown and changing with time. It was not possible to distinguish between the formation resistance and the casing resistance.

The effects of mixing due to logging activities is illustrated in Fig. 2 which demonstrates the benefits of assessing the pore fluid in the formation rather than logging the resistivity of the fluid within slotted casing.

Although these experiments gave a series of negative results they provided the basis of the design of a screen that would prove more satisfactory.

Model tank experiments

Experiments were undertaken to identify a well-screen which would provide minimum impedance to radial current flow, and yet completely exclude fine sand grains (0.2 mm). These two conditions were achieved using a slotted well-screen of large open area and a 'terram' fabric having high fluid permeability and yet very small pore size. Slotted well-screens were obtained commercially and various fabrics were wrapped around and tested using a scaled down version of the focussed borehole tool (Fig. 3). Well-screen with 2 mm wide slots and an open area of 12% caused an increase in resistance of 50% when immersed in water of conductivity 0.5 ohm-m. A 260% increase as produced by a similar screen having 3% open area and 0.2 mm wide slots. The 2 mm slots were chosen as being suitable; it was noted that for the full scale focussed borehole tool the percentage increase due to the casing would be close to 50%/*FF* because most of the measured resistance apertains to the formation outside the borehole; the casing effect will be proportional to the fluid resistivity if the individual grains of the formation are excluded from the slots. This would typically result in a 10-15% increase in measured resistance for a saturated clean sand having *FF* = 4.

A large number of fabrics were tested in order to find one with small resistance to electric current flow and to eliminate ones with a large number of small holes which may become blocked by individual grains. Fig. 3 illustrates such a fabric where the holes of a cross-weave of nylon ('terram' 280) have become blocked by air bubbles. The composite 'terram' fabrics 12M12 and 20M20 have very small resistance to electric current flow; they have a very loose cross-weave of nylon on the outside backed with a filter similar to fine glass-fibre mat. The nylon weave provides strength while the filter provides a barrier to small particles while allowing electric current to flow almost unimpeded. The permeability of the 12M12 (to water) was high, giving a pressure drop of 0.1 m for a flow of 30 l/s over an area of 1 sq m, which is consistent with its low electrical resistance when saturated.

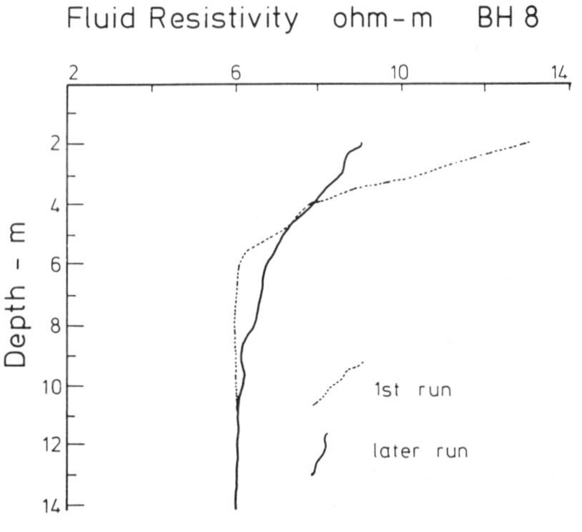

FIG. 2. The mixing of fluid inside slotted casing due to the movement of logging tools. The stratification is due to groundwater (25 – 30 ohm-m) entering the slotted casing, over a period of 1 year, which contained low resistivity fluid due to the presence of residual conductive drilling mud.

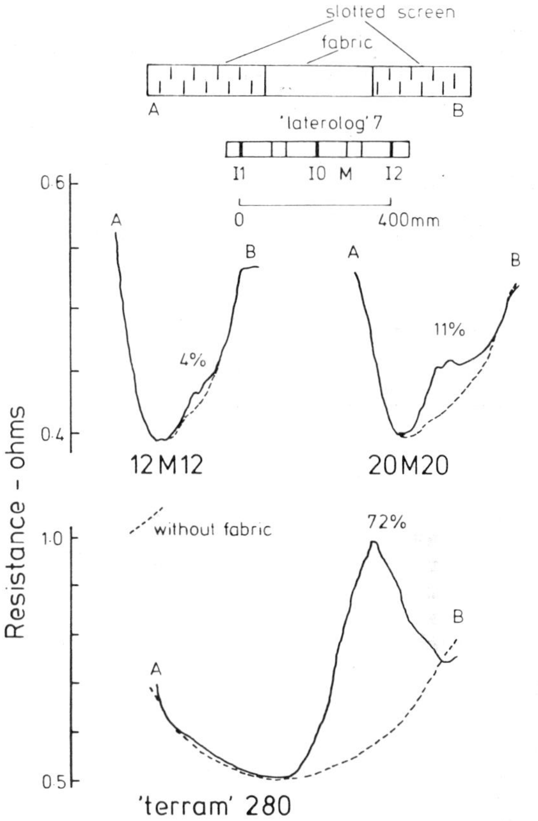

FIG. 3. Electrical resistance measurements made inside a length of slotted well-screen (2 mm slots 12% open area) showing the effects of three different 'terram' fabrics which were used as an outside wrap. A 'laterolog 7' array was used with potential measurements made at M. The screen was fixed in a model tank filled with water of resistivity 0.5 ohm-m. The model laterolog probe was pulled automatically through the wellscreen from A to B and its position was determined using a shaft encoder connected to a chart recorder.

In situ testing

A test site in sand

The same sand site was used for *in situ* testing of the composite well-screen. The sand had a mean size of 0.2 mm and was known from previous drilling and sampling to be dense. A large number of resistivity soundings were made over an area of 0.1 sq km and a typical example, close to the location of the drillholes, is shown in Fig. 4. Sand of resistivity 100 ± ohm-m extends from the water table (0.1-1 m seasonal) to a depth of 25 m where it overlies boulder clay, which in turn overlies high resistivity basement rocks. The resistivity is consistent with a dense sand saturated with groundwater of 20–27 ohm-m.

The location for the well-screen was drilled using a self-bored resistivity device, which contained a modified 'laterolog 7', in order to obtain a resistivity profile of the sand body which is unaffected by casing. The resistivity sonde was an integral part of the drilling equipment, and in direct contact with the formation. The self-bored results are shown in Fig. 5 and are in good agreement with the resistivity soundings. The resulting borehole was cased with steel to 15 m and remained full of the drilling mud used in the self-boring process. The drilling mud was replaced with fresh water and the fabric-covered well-screen (3 m sections) lowered down inside the steel casing which was then withdrawn leaving the fabric well-screen in the ground to a depth of 15 m. A small amount of clean water was pumped into the hole to disperse any remaining mud, and then water was pumped out for one hour at a rate of 1.5 l/s. Initially the pore water was discoloured, after 5 minutes it became clear. A constant head permeability test was carried out at the start of pumping; the head required to produce the 1.5 l/s flow was 0.965 m. The application of the standard Horslev formula produced a permeability of 1×10^{-4} m/s which was in good agreement to that calculated from the grain size data.

The borehole was left to stabilise for a year, in order that the sand would be packed tightly against the well-screen, before resistivity logging and water sampling was attempted through the well-screen. After stabilisation the fluid inside the composite screen was logged for temperature and electrical conductivity; the conductivity results were converted to resistivities, and are shown in Fig. 5. A flexible, low-pressure double packer system was used to extract pore fluid from the formation and to conduct permeability tests. The pore fluid from the packer testing had resistivities in the range 24-29 ohm-m which is slightly lower than from the fluid logging inside the casing.

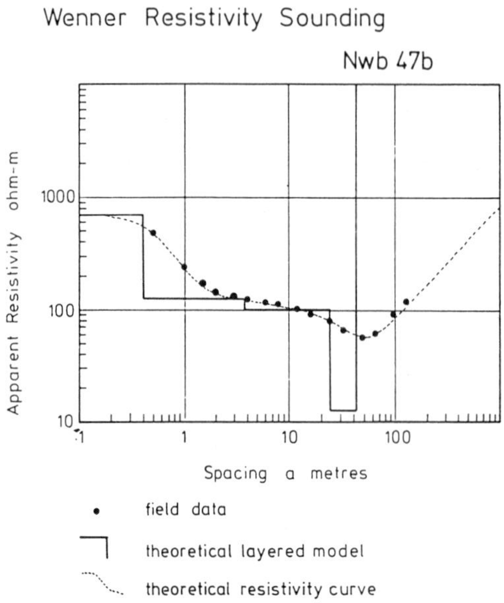

FIG. 4. A resistivity sounding at a sand test site showing sand of resistivity 100 ohm-m to a depth of 25 m.

Intervals of the well-screen were tested for permeability using the double packer system and the values were in the range 2.3×10^{-5} to 1.3×10^{-4} m/s. These values are similar to previous draw down tests and suggest that there was not a disturbed zone of high permeability next to the well-screen.

A focused resistivity log was made inside the slotted casing and is shown in Fig. 5 for comparison with the self-bored results. These resistivity results are approximately 10% greater than the self-bored results and the surface sounding data, showing the effect of the casing to be small. Some of the general trends shown by the self-bored data are seen in the log using the borehole sonde but are complicated by the large peaks at the casing joints.

The effect of the casing is small and changes in the resistivity of the formation could be seen and would not be confused with changes in slot resistance.

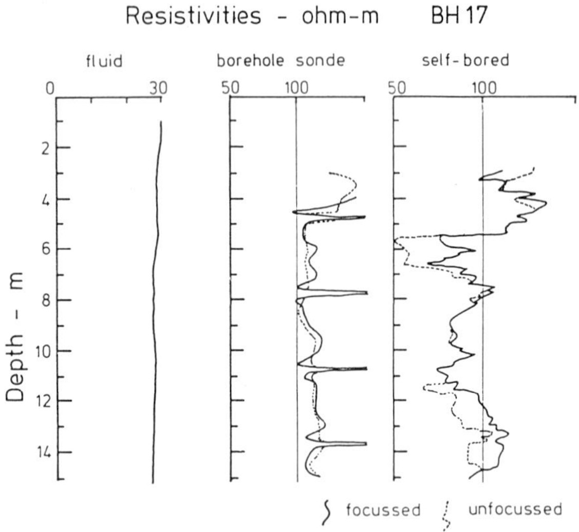

FIG. 5. Test results at the sand site (Fig. 4) showing self-bored results and a resistivity log through the composite well-screen which was placed in the hole produced by the self-boring. A comparison of the sets of results shows the well-screen increases the values by approximately 10%.

Conclusions

Resistivity measurements made through slotted well-screen, having 12% open area and wrapped with 'terram' 12M12 fabric, will be increased by approximately 20% compared with measurements made in direct contact with the sand using a self-bored resistivity sonde, and as such are suitable for monitoring changes in pore fluid resistivity because it is sensitive to changes outside the well-screen and will therefore not be affected by mixing of the water inside. If natural mixing does not occur, then the results presented show that the action of logging will itself mix the water inside the screen and so destroy the resistivity distribution.

The resistivity measurements inside the well-screen exhibit large peaks and small troughs when the electrodes pass through the unperforated joints between individual 3m sections. The peaks and troughs tend to obscure the detailed resistivity structure that is seen in the self-bored results, and may be reduced in vertical extent using higher resolution electrode arrays.

The self-bored results have greater resolution than the borehole measurements because the self-bored sonde is approximately one third of the total length of the borehole sonde and it is in direct contact with, ideally, undisturbed formation where there will be no averaging effect of the well-screen and the fluid in the borehole. The borehole sonde was designed to have a greater depth of investigation than the self-bored sonde specifically to reduce the dependence on the extra resistance caused by the presence of the well-screen.

The effect of the well-screen is small enough to allow changes in the resistivity of the adjacent formation, caused by changes of pore fluid resistivity due to the seepage of pollutants, to be assessed.

ACKNOWLEDGEMENTS: We wish to thank Dr G. J. Mayhead and Mr J.A. Carter of the North Wales Forestry Commission and Dr G. Howells and Mr A. Bennett of the Nature Conservancy Council, Bangor, Gwynedd, for providing the site and assistance generously afforded.

The contribution of D. Mitchell and J. Paulson to the fieldwork and to some discussions is gratefully acknowledged.

This paper is published with the permission of the Director of the British Geological Survey (NERC).

References

ARCHIE, G.E., 1942. The electrical resistivity log as an aid in determining some reservoir characteristics. *Transactions of American Institute of Mining and Metallurgical Engineers,* **146,** 54-62.

ARCHIE, G.E., 1950. Introduction to petrophysics of reservoir rocks. *Bulletin of the American Association of Petroleum Geologists,* **34,** 943-961.

ATKINS, E.R. and SMITH, G.H., 1961. The significance of particle shape in formation factor-porosity relationships. *Journal of Petroleum Technology,* **13** 285-291.

CHAPELLIER, D. 1981. Logs in plastic cased wells. *Seventh European Logging Symposium Transactions* Paper 28, Society of Professional Well Log Analysts, Paris, France.

CAMBEFORT, H., 1957. La mesure in situ de la porosity des sables (Measurement in situ of the porosity of sand). *Proceedings of the 4th International Conference on Soil Mechanics and Foundations Engineering, London,* **1,** 213-215.

DOLL, H.G., 1951. The Laterolog: A new resistivity logging method with electrodes using an automatic focussing system. *Transactions of the American Institute of Mining and Metallurgical Engineers,* **192,** 305-316.

JACKSON, P.D., TAYLOR SMITH, D. and STANFORD, P.N., 1978. Resistivity-porosity-particle shape relationships for marine sands. *Geophysics,* **43,** 1250-1268.

JACKSON, P.D., 1981. Focussed electrical resistivity arrays: some theoretical and practical experiments. *Geophysical Prospecting,* **29,** 601-626.

MANSFIELD ADAMS, W., 1984. Geophysical monitoring of underground hydrocarbon leaks and spills: Two case histories. Paper 343, *EAEG 46th Annual Conference,* London, 1984.

SCHLUMBERGER, 1969. *Log interpretation principles.* Schlumberger Ltd., 277 Park Ave., New York, N.Y. 10017

WORTHINGTON, P.F., 1982. The influence of shale effects upon the electrical resistivity of reservoir rocks. *Geophysical Prospecting,* **30,** 673-677.

WYLLIE, M.R.J. and SOUTHWICK, P.F., 1954. An investigation of the S.P. and resistivity phenomenon in dirty sands. *Journal of Petroleum Technology,* **6** (Feb), 44-57.

P. JACKSON, J. HALLAM, S. HASSETT & M.G. CULSHAW, British Geological Survey, Keyworth, Nottingham NG12 5GG, UK.

D. ALLEN, British Geological Survey, Crowmarsh Grifford, Wallingford OX10 8BB, UK.

Borehole sampling techniques and field analysis of groundwater in landfill pollution studies

A. Stuart & S. P. Hitchman

ABSTRACT: A borehole drilled at a landfill for site investigation and/or monitoring purposes will generally intercept a vertical interval within an aquifer which may have a wide range of permeabilities and so the entry of a pollutant into a borehole column will not be uniform. Drilling methods and the need to avoid the introduction of contaminants to groundwater mean that hydraulic testing may be restricted, particularly in unconsolidated formations. Techniques that do not add water should be used in preference to injection or tracer tests. Sampling and testing during borehole drilling are necessary to define the groundwater flow system within the borehole, to enable an effective completion to be made. In unconsolidated formations, silt and clay horizons may restrict vertical groundwater movement with the result that pollutants become stratified and move in discrete zones within the aquifer. Therefore taking a bulk or average water sample from a borehole completed to monitor full aquifer thickness will usually result in dilution and mixing of the contaminant with cleaner groundwater. Formation disturbance caused by percussion drilling can also cause vertical groundwater flow and mixing outside the borehole casing making the isolation of discrete zones difficult.

Groundwater samples obtained during site investigation drilling or from routine sampling visits will be subject to temperature and pressure change. As a sample is depressurised, variables such as pH, Eh, dissolved gas content and the speciation of certain constituents can be altered. Quantification of these changes and their minimisation means that strict procedures should be adhered to in sample handling and analysis on site.

This paper, covering landfill research work funded mainly by the Department of the Environment and carried out by the Fluid Processes Research Group of the British Geological Survey describes techniques and equipment used to overcome some of the problems cited, to obtain representative groundwater samples.

Introduction

Domestic and industrial wastes disposed of into shallow land burial sites in humid climates will generally produce a leachate which may migrate through underlying strata and enter the groundwater flow system. It is important that the likely impact of this leachate on groundwater quality is fully assessed prior to disposal taking place, and that groundwater is adequately monitored during the often beyond the operational life of the landfill, to confirm the predicted performance of the site. The detail of monitoring required for each site will vary and depend, among other things, on the hydrogeological complexity and the degree of risk the landfill is thought to present.

The hydrogeological assessment of the site will first involve consideration of existing background data (local water well quality and levels, borehole records etc) followed by drilling additional boreholes to provide more detailed hydrogeological information for inclusion in a predictive model. Whenever possible, initial site investigation boreholes should be retained for monitoring purposes, so that natural variations in groundwater levels and chemistry may be obtained prior to waste disposal for comparison with future groundwater data, and so that the actual impact of the landfill can be compared with that predicted. Significant departures between predicted and observed events may lead to a refinement of a predictive model, more detailed investigation of the hydrogeology, modification of the disposal operation, or, in the extreme case, interception of the pollution plume with a wellpoint system to prevent further contaminant migration.

This report describes techniques for sampling and hydraulic testing during site investigation drilling, and illustrates various types of borehole completion and sampling equipment that enable representative groundwater samples to be obtained from a variety of aquifer types. Details are given for on-site analysis of these samples,which could be carried out as soon as possible after collection, and for chemical preservation for later laboratory analysis.

Since a borehole will generally intercept a vertical interval within an aquifer which may have a wide range of permeabilities, the entry of a contaminant into a borehole column will not be uniform so precise sampling is required to identify the contributing zone. A borehole drilled in a fissured aquifer such as a limestone or sandstone may only intercept one or two

From CRIPPS, J.C., BELL, F.G. & CULSHAW, M.G., (eds), 1986, *Groundwater in Engineering Geology,* Geological Society Engineering Geology Special Publication No. 3, pp. 225-246

fissures that transport groundwater and these must be identified and the flow system within the borehole defined before accurate groundwater sampling is possible. In unconsolidated formations such as sands, silts and gravels, one would expect groundwater flow to be intergranular and therefore more uniform and predictable than fissure flow. However, silt and clay horizons within the geological sequence will restrict vertical groundwater movement with the result that pollutants stratify and move in discrete zones within the aquifer. Therefore taking a bulk or average water sample from a borehole will usually result in dilution and mixing of the contaminant and give non-reproducible results. Open boreholes will allow vertical groundwater flow and mixing in multi-aquifer systems, and even under less dynamic conditions, diffusion is possible up or down the borehole column.

Zonation of a contaminant may also occur as a result of density contrasts between leachate and groundwater. Density contrasts may effectively separate certain immiscible components of the leachate so that the lighter fraction floats at the water table while heavier fractions sink to the base of the aquifer, and each fraction experiences different environments in which to react or degrade, for example, aerobic conditions may prevail at the water table, while anaerobic reducing conditions may exist at greater depths.

Therefore representative sampling, during drilling is vital since this will define the extent and distribution of contaminants within the aquifer before the construction of boreholes provides vertical high permeability routes for groundwater movement and pollutant mixing. Similarly, it is necessary to use appropriate equipment to yield representative groundwater samples for completed boreholes.

Site investigation — drilling, formation sampling and hydraulic testing

Drilling

Boreholes are drilled to prove the geological succession, to allow *in situ* measurements of hydrogeological parameters and access by logging tools, to provide water samples for the evaluation of groundwater chemistry and rock samples for evaluation of the formation mineralogy and physical properties. The drilling technique used is dictated by the formation geology and the type of sampling required during drilling. A percussion-type rig is most suitable for drilling in unconsolidated formations such as silts, clays, sands and gravels, and for drilling through the landfill itself where precise depth samples are required. In consolidated formations such as limestones, cemented sandstones or siltstones, rotary drilling techniques are more suitable. Air is preferable to water, mud, or foam as a flushing medium since there is least risk of geochemical contamination of groundwater samples. However, air flush can oxidise groundwater, and the return of contaminated water at some pressure to the surface may present a hazard. It may be necessary to use two rigs in some investigations, a percussion rig to penetrate the landfill, and a rotary rig to extend the borehole into the underlying formation. Harrison (1976) describes drilling and sampling methods used by the British Geological Survey (BGS) as part of a research programme (Department of the Environment 1978) investigating the behaviour of hazardous wastes in landfill sites, and similar work is reported by Barber *et al.* (1976).

Formation sampling during drilling

Precise sampling of the unsaturated zone using percussion methods is achieved by the recovery of steel tubes, 100 mm in diameter, and 0.45 mm long, known as U 100s, which are driven into the formation. These can be taken at regular intervals, or continuously. They will normally contain no free water, but interstitial water can be be recovered from the samples in the laboratory by high speed centrifugation (Edmunds & Bath 1976) and a profile of pore water composition obtained. This technique is also suitable for pore water extraction from cores obtained from rotary drilling in both unsaturated and saturated zones of consolidated formations, but these samples may be partly aerated by the drilling fluid. Subsamples can be dried to determine moisture content and grain size distribution and subjected to a comprehensive extraction programme for chemical characterisation at this stage. During rotary drilling through the saturated zone, groundwater samples may be recovered from the air/water return, or by bailing from the borehole once the drill stem is withdrawn.

As an alternative, during percussion drilling, bulk disturbed samples can be recovered from the drilling tool (shell, bailer, clay-cutter etc.). This may be necessary where the unconsolidated formation is saturated, and samples will not stay in the U 100 tube when withdrawn. Bulk disturbed 'slurry' samples can be taken from the shell or bailer and transferred to wide-mouthed 4 litre capacity polyethylene bottles with urea lids. Temperature, pH, and dissolved oxygen can be measured immediately, but depending on the coarseness of the solids, it may be necessary to allow the sediment to settle out before determining Eh and conductivity; total alkalinity should always be measured on a filtered sample.

Although the drilling process may have greatly disturbed the sample every effort should be make to minimise contact with the atmosphere during this period (i.e. by completely filling the bottle, keeping the lid on when not in use and, agitating the sample as little as possible) in order to discourage further chemical changes (see below). Once on-site determinations have been performed, the supernatant may be drawn off for preparation for more detailed analysis (see below). A syringe, rinsed at least twice with the sample, may be useful for this; alternatively, the liquid may be siphoned off using a clean plastic tube — the first few millilitres should be discarded.

All recovered samples may be aseptically subsampled and stored at 4°C for microbiological examination. Solid samples which are to be returned to the main laboratory for further processing must be protected from change as much as possible by sealing from the atmosphere and by keeping cool (4°C). Samples not in capped bottles or U 100 tubes can be isolated in heat-sealed polythene sheaths or double wrapped in sealable polythene bags, taking care to expel the air before closing.

If the geology and groundwater flow regime for a site have already been well defined, or a borehole is required for the measurement of background water quality or simply groundwater level monitoring formation samples may not be required. In this case, drilling rates can be acce erated since U 100 samples and cores would not be recovered. Alternatively, a flow-through sampling technique using a coring device driven with a hydraulic hammer can be used where samples are required, but where no instrumentation or completion is contemplated or where minimal surface disturbance is necessary. This equipment has been used for the extraction of small diameter (20 mm approx) core samples from unconsolidated formations at depths up to 10 m. A sample tube is driven to the required depth, with the formation able to pass through and out the top of the sampler during drilling. This method causes little disturbance to the sample or the formation, and the small diameter means that the core is not lost when the sampler is recovered, even from saturated deposits.

Hydraulic testing during drilling

In situ determination of formation properties is important for adequate hydrogeological assessment of sites. During percussion drilling in unconsolidated formations, access to the formation is limited since temporary drill casing must be installed as the borehole progresses to prevent collapse. However, standard falling head, constant head, or recovery tests can be carried out to measure permeability of the formation at the bottom of the casing at regular intervals during drilling (Hvorslev 1951). The drilling technique will cause disturbance of the formation, and so results obtained by these methods will probably be overestimated. Another disadvantage of such testing is that subsequent chemical sampling may be invalidated once water has been injected, unless formation water, previously pumped from the borehole is used. It is therefore preferable to use methods that withdraw water from the borehole, such as slug tests, or single well pump testing.

Hydraulic measurements in consolidated formations are simplified since they are more likely to remain open without the use of temporary casing. Slug (Cooper *et al.* 1967), falling head and constant head injection tests can be carried out progressively in the bottom section of the borehole (usually 2 or 3 m), isolated by a single inflatable packer. Alternatively, to avoid sample contamination and interruption of drilling, the borehole can be drilled to its required depth, and the tests carried out using a straddle packer arrangement.

Laboratory determinations of permeability bulk density, grain density, and porosity can be carried out on recovered U 100 and core samples using techniques described by the American Petroleum Institute (1960). U 100 samples will be subject to drilling disturbance, while permeameter measurements on core samples will provide measurements of intergranular permeability only, and underestimate and bulk formation value, which will comprise intergranular and fracture/fissure components.

Borehole completion

After drilling and sampling, a borehole may be completed to provide a long-term groundwater monitoring point. Completions will vary, depending on geology, aquifer complexity, and sampling requirements, but can be broadly classified into three types:

(1) open borehole;
(2) lined, screened, borehole;
(3) enclosed instrumented borehole.

Open borehole

A borehole drilled in a consolidated formation such as a cemented sandstone or siltstone may remain open without the installation of a permanent lining, but there is always a risk that the borehole walls will collapse. Since boreholes around landfill sites are invariably required to provide a long-term monitoring point, it is suggested that a lining should always be used.

Lined, screened borehole

A lined borehole, screened below the water table has the main advantage of accessibility to the full thickness of the aquifer penetrated by the borehole. In addition to groundwater sampling, water level measurements, geophysical logging and hydraulic testing (e.g. pumping tests) for hydrogeological parameters can be carried out. However, they will also allow vertical flow to take place and mixing of groundwater if several horizons are intercepted by the borehole. Boreholes penetrating unconsolidated formations (including the landfill itself) require the installation of casing to prevent collapse, A permanent monitoring point in its simplest form will comprise casing perforated beneath the water table possibly with a gravel pack for wall support and to prevent the ingress of fine particulate matter which will silt up the hole (Figure 1a). If one or more perched water tables exist, multiple piezometers can be installed at different levels in one borehole, efficiently sealed from each other (Figure 1b). In practice, the emplacement of seals (using a cement/bentonite mix) is very difficult and may allow vertical leakage if not fully effective. To reduce this problem, several boreholes can be drilled close together, each one completed and screened to monitor a different zone of the aquifer.

Improvements in well screen design have been made in recent years, particularly with the introduction of a synthetic filter fabric with a choice of mesh sizes (Fig. 1c). This had advantages over basic borehole completions as described earlier, viz.

(a) the filter fabric of the screen can be of a mesh size to suit the application, thus minimising clogging and well silting;
(b) gravel packing, with the associated problems it may cause by jamming drill casing and well screen, is unnecessary;
(c) screen lengths and spacings can be manufactured to suit particular requirements;
(d) well efficiency is improved and head losses reduced.

However, this design encourages groundwater flow within the screen, providing a vertical high permeability corridor. For this reason, results from hydraulic testing or groundwater sampling which involve the addition or withdrawal of water from discrete zones within the borehole, will be unreliable. Preferable to one continuous screen through the aquifer, the incorporation of several separate screened sections of this type in one borehole means that several zones can be isolated and each of these sampled and tested. Using inflatable packers to straddle the screened section, constant and falling head injection (Hvorslev, 1951) and point dilution tests (Halevy *et al.* 1967) can be carried out to measure permeability and groundwater velocity respectively. Interpretation of results from these tests carried out in boreholes in unconsolidated formations should take into account the presence of a high permeability disturbed zone around the borehole (Drost *et al.* 1968). This will be created by:

(a) percussion drilling in the saturated zone;
(b) installation of a smaller diameter screen than that at which the borehole is drilled, allowing the formation to collapse once drill casing is removed.

If suitable low permeability horizons exist, external casing packers may be effective in isolating discrete zones within an aquifer. These serve as an alternative to bentonite seals in multiple aquifer systems. There is a current requirement for development of an adjustable inflatable sleeve packer which can be located in borehole casing prior to installation. Inflation using grout will form an effective seal by keying with adjacent low permeability horizons. Neoprene rubber sleeves can be used to similar effect, to prevent confined groundwater invading the annulus of boreholes completed in a shallow glacial sand aquifer (Fig. 2). The neoprene sleeve is inflated by compressed gas to abut the clay, and grout then pumped in to form a permanent seal.

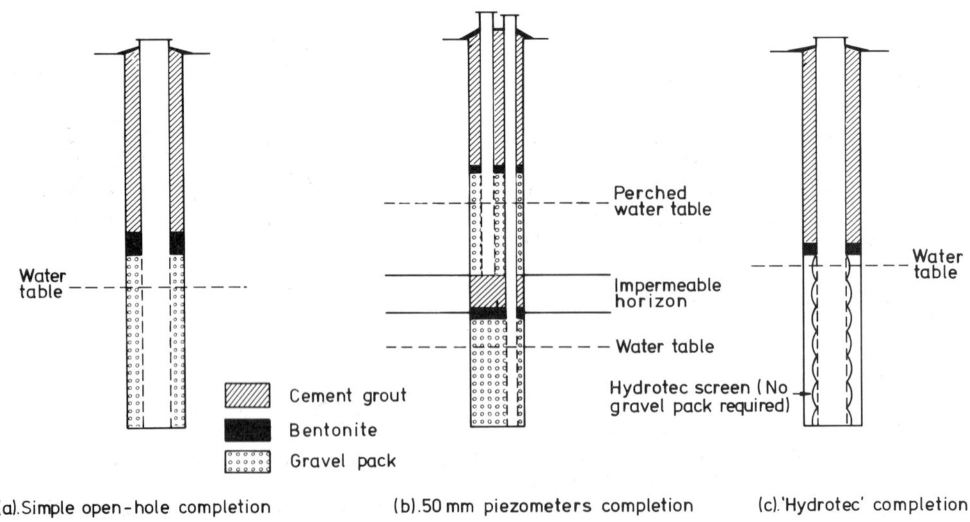

FIG. 1. Cased open-hole completions in unconsolidated formations.

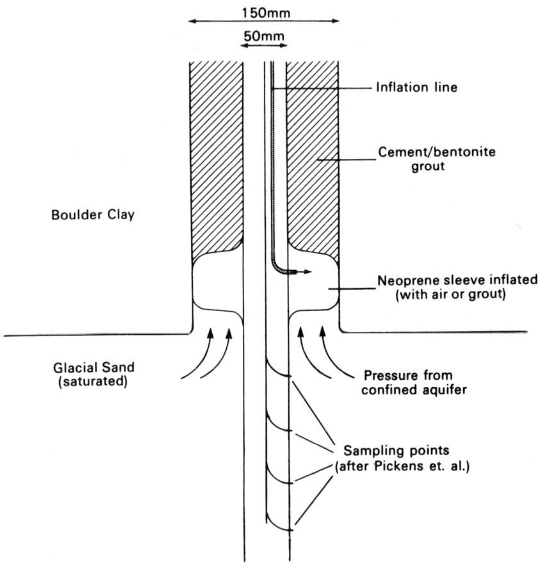

FIG. 2. Sampling installation using external casing packer.

Permanent external packers have been incorporated into a multiple peizometer system suitable for hard rock applications by a Canadian company. With this system there is practically no limit to the number of piezometer ports that can be installed in a single drillhole (Fig. 3), since no wires or tubes are required to join each port to the surface. The packers are inflated using water, bentonite slurry, or cement grout, from within the casing through a one-way valve. Samples are taken through valves installed in casing couplings. Valves can be of two types: it can be a sliding valve which can remain open or closed or it can be a one-way check valve which remains closed due to external water pressures until it is opened from the inside (Patton 1979). The sliding valve type is operated in conjunction with internal packers on either side of the valve to seal off a portion of the inside of the piezometer casing, while the one-way check valve is actuated by a special tool which locates the valve, seals it from the water inside the piezometer casing, and then opens the valve and takes a water sample. The principal advantages of using this system is that repeated tests and sampling can be carried out on the same zones of the aquifer.

Instrumented boreholes

Rather than a series of open boreholes as the end product for a site investigation, it may be necessary for sampling devices to be permanently installed in the borehole. This may be because a borehole would cause an obstruction, or possibly where instrumentation will provide a cheaper acceptable completion than permanent casing. Several instruments can usually be installed in one borehole, allowing sampling of several horizons within an aquifer. This enclosed system will improve the reliability of chemical data by reducing oxidation and aeration of groundwater, which may occur in open boreholes. Instrumented boreholes will provide samples from precise depths, but vertical coverage of the aquifer will be incomplete. Malfunction of installed samples will be impossible to rectify. Details of permanent sampling installations are given in the following section.

Groundwater sampling techniques in boreholes

A variety of methods are available for the extraction of groundwater, and these will depend on the type of borehole completion, borehole diameter etc.

Discrete samplers

Described below are some of the installations possible in the instrumented boreholes, described in the previous section.

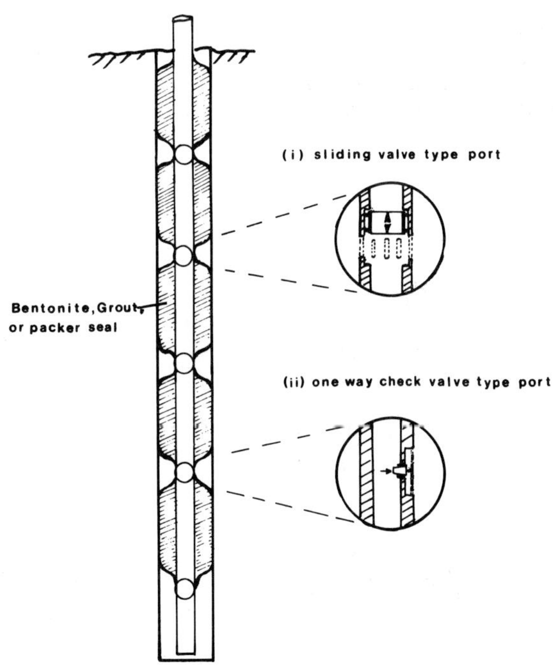

FIG. 3. Westbay multiple piezometers.

Water Research Centre (WRC) in situ samplers

These can be installed at several depths within a borehole, and sealed in place to prevent vertical groundwater flow (Barber *et al.* 1976). Samples are obtained by pressurising the sampler with compressed gas passed down one tube which seals the sample

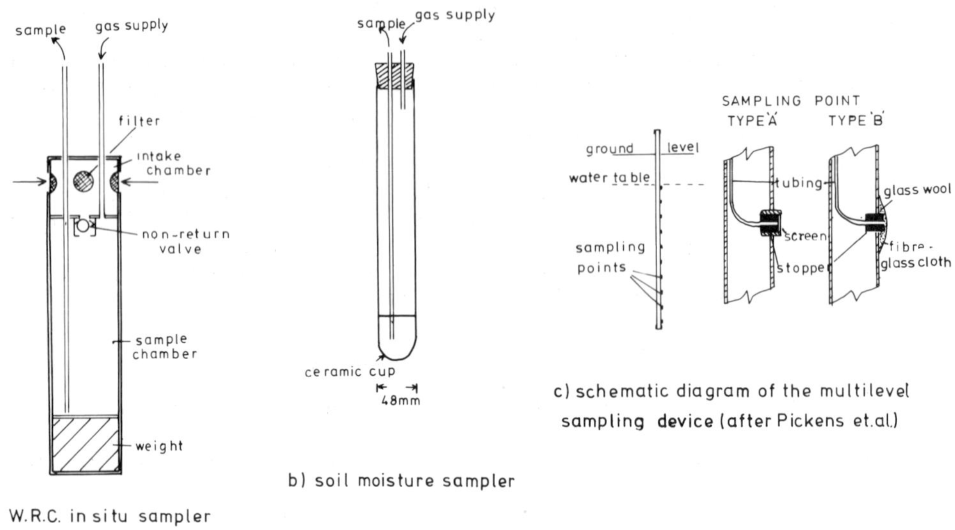

FIG. 4. Permanent sampling installations.

chamber and allows water to be expelled via the other tube to the surface (Fig 4a). The sampler should be evacuated at least once before a sample is taken, to allow groundwater from the surrounding formation to enter the chamber. A shortcoming of this type of sampler is that the stainless steel mesh intake filter allows fine grained saturated sediments into the sample chamber which in time can cause it to clog.

Soil moisture samplers

These are operated in similar fashion to WRC *in situ* samplers but are of a much simpler design. They are in fact manufactured for unsaturated zone sampling, and consist of a 1 bar entry valve porous ceramic cup cemented to a plastic pipe (Fig 4b). The sampler is supplied with a stopper and single rubber tube which is used for evacuation of air to encourage water entry into the cup. Once water has accumulated, a vacuum is applied to the tube to extract a sample. It has been found, however, that if the single tube stopper is replaced by a 2 tube system, compressed gas can be used to force a sample to the surface from depths up to 10 m below water level. The sampler should be flushed once before a sample is recovered for analysis. The fine pore size of the ceramic cup means that the water sample is pre-filtered. Several of these devices can be installed in one borehole with each isolated from the others to prevent vertical leakage.

Multilevel sampler

This device can be used for the collection of groundwater samples at many levels from a single borehole installation (Fig 4c). Pickens *et al.* (1978), used a bundle of polypropylene tubes contained inside a PVC pipe which is installed in the aquifer. Simple construction means that the frequency and spacing of sampling points can be chosen on site to suit requirements. Each tube protrudes through the wall of the pipe to form a sampling point. The end of each tube is protected by either; (a) a circular piece of 100-mesh stainless steel screen set at one end of a short length of small diameter PVC tubing (type A); or (b) a glass wool covering wrapped with fibreglass cloth (type B). The authors report that both types of sampling points have been used successfully, and pointed out that the device is best suited to unconsolidated deposits where piezometric levels are close enough to ground level to allow sample extraction by suction.

Collection of large volume samples may take some time, and this is exacerbated by the incorporation of in-line filters which may be used as an early stage preparation of the sample for field analysis or preservation.

Pumps for 50 mm diameter boreholes

Electric submersible pumps suitable for use in 50 mm diameter boreholes are extensively available in the United States. In this country, however, reliance is often made on adaptations of other equipment, such as peristalitc pumps, vacuum pumps and small 'caravan type' 12 volt DC water pumps for extracting bulk groundwater samples, cleaning out the borehole or inducing flow into the hole from the surrounding formation to provide a representative sample. They are most suitable for sampling piezometer installations in which only a small zone of the aquifer is screened, and they are limited to shallow applications, normally extracting samples from boreholes where the water level is less than 10 m below ground level. Peristaltic pumps will only exert a partial vacuum and consequently can only be used to approximately 6 m, but are simple to operate by lowering a tube from the pump down the borehole (Fig. 5a). A continuous sample will be supplied, but at a rate of less than 1 litre per minute. Vacuum pumps require a similar single tube installation (Fig. 5b), and will draw water to the surface and into a collection vessel. If on site analyses are required, instruments such as conductivity probes, specific ion electrodes, or dissolved oxygen cells can be incorporated in line, together with filters if required. Small (38 mm o.d.) submersible pumps which operate using 12 volt DC are also available. While one of these will only pump to a head of about 4 m a throughflow design enables them to be used in series which together with the incorporation of a non-return valve, improves overall performance (Fig. 5c).

Several types of gas driven pumps have also been used in attempts to obtain accurate samples. Buss & Bandt (1981) have recovered water samples from depths in excess of 30 m at a rate of approximately 3 litres per minute (Fig. 5a). The sample chamber is alternatively pressurised and depressurised, forcing water past the check valve to the surface. BGS have used a similar pump in conjunction with an adjustable packer system (Fig. 8).

A modification of this design uses an internal flexible bladder to prevent direct contact between sample and the gas (Fig. 6b). When inflated, the bladder displaces water from the sampler to provide a pulsed sample. Riha (1980) describes a similar instrument (Fig. 6c) in which compressed air actuates a flexible membrane which squeezes water from the sampler. A refinement is the incorporation of simple rubber packers at top and bottom which help to ensure that water is taken from a restricted and isolated zone of the aquifer.

To eliminate contamination of water samples, Tomson *et al.* (1980) have developed a nitrogen powered continuous delivery pump made entirely from glass and Teflon for the measurement of trace level organic compounds (Fig. 7). Wells with water levels as deep as 38 m below ground level have been sampled successfully with this system, and while the all-Pyrex-glass components are not as rugged as metal pumps, the authors report only one breakage during field use.

Sampling methods in larger boreholes

Of the samplers so far described, only that of Riha incorporates single rubber packers in an attempt to limit the size of the sampling zone in boreholes which have an extensive screen length. Inflatable borehole packers have been used in the oil industry for many years to isolate one or more zones in open hole or casing, to facilitate testing, acidizing, hydraulic fracturing etc., generally in boreholes between 75 and 300 mm in diameter. For these applications they need to be rugged, able to withstand high pressures, and of high quality, and are consequently expensive. However, for monitoring and sampling of groundwater in shallow boreholes in unconsolidated aquifers, there is a requirement for a more portable packer system that can be used in lined boreholes of 100 mm diameter and greater, possibly in conjunction with some of the small diameter pumps described earlier.

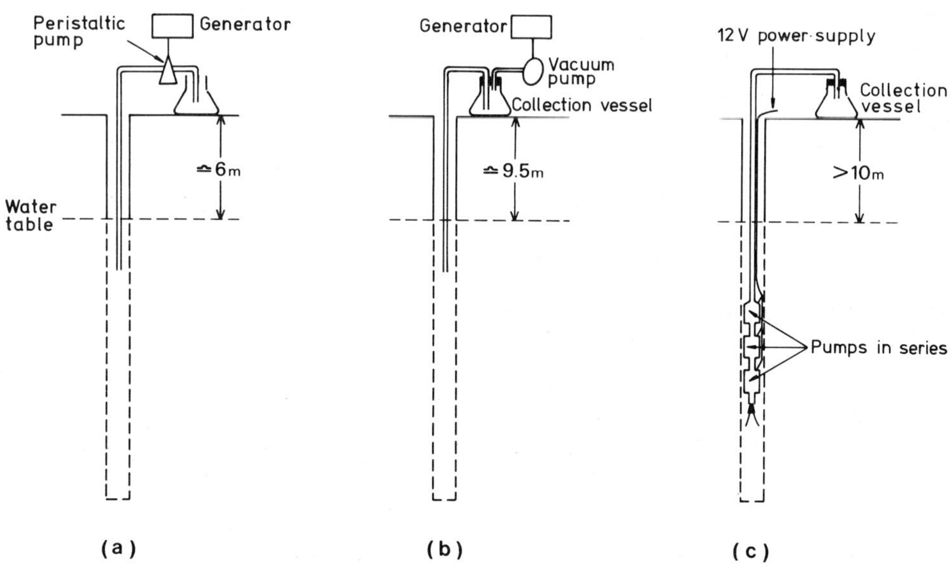

FIG. 5. Adaptations of pumping systems for 50 mm diameter boreholes.

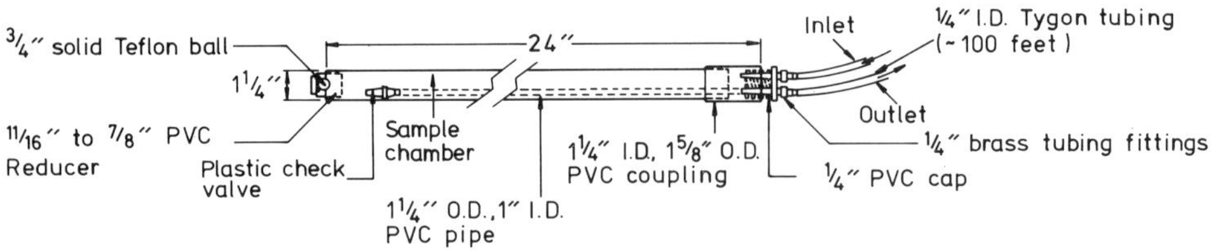

FIG. 6(a). Air-driven pump for 2 in I.D. wells (after Buss & Bandt 1981).

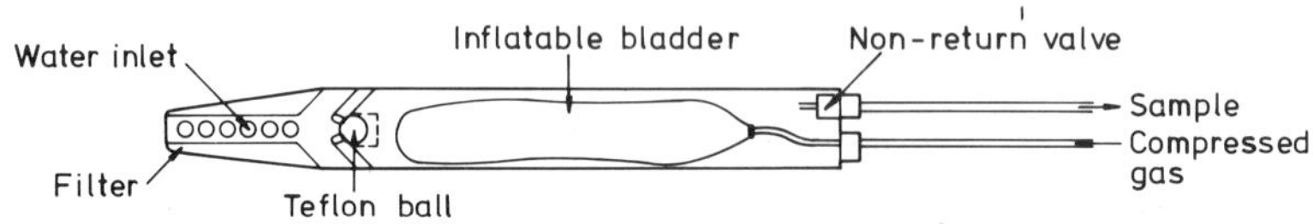

FIG. 6(b) Schematic diagram of bladder pump manufactured by Industrial and Environmental Analysts inc., Vermont, USA.

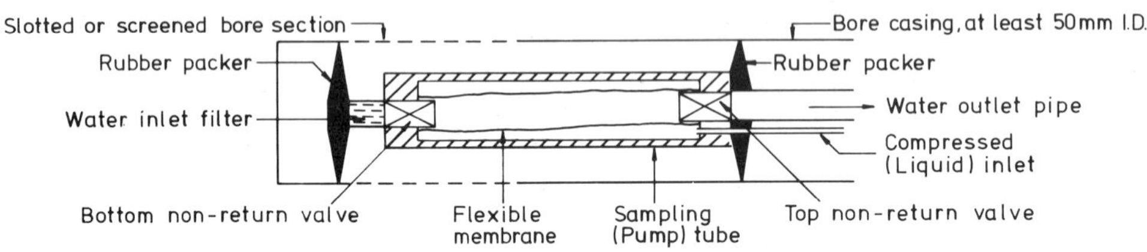

FIG. 6(c) Multi-stage sampler (after Riha 1980).

FIG. 6. Samplers.

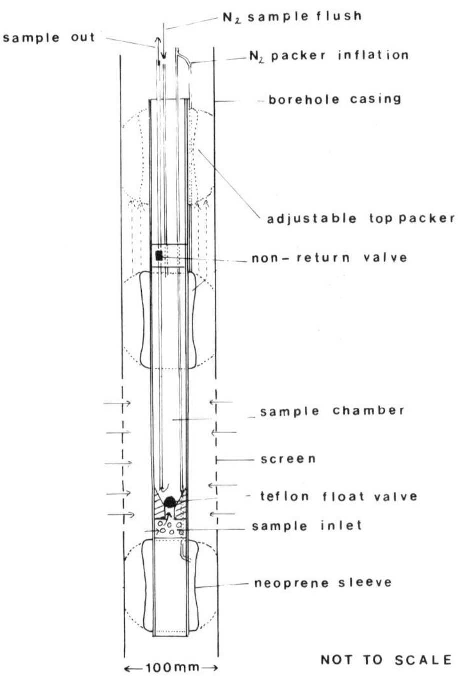

FIG. 8. Sampler used by BGS.

A lightweight packer system based on a 50 mm diameter access tube and neoprene rubber inflatable packers can be used to isolate sampling zones in 100 and 150 mm diameter screened boreholes. The access tube to the surface allows installation of a small diameter pump for sample discharge and collection. The incorporation of several screened sections using the synthetic screen material, described above, means that several zones can be isolated and sampled in one borehole (Fig. 9a). This arrangement would be of benefit in layered aquifers or where density stratification is taking place in polluted groundwater. It will also allow hydraulic testing (injection test, point dilutions) to be carried out at several discrete horizons within the aquifer to produce permeability and groundwater velocity profiles. A sophistication is the use of a four packer system incorporating three pumps. The pumps, used at the correct rate produce horizontal flow to the borehole and allow the central zone to be sampled accurately (Fig. 9b), reducing the risk of vertical groundwater flow in the borehole disturbed zone.

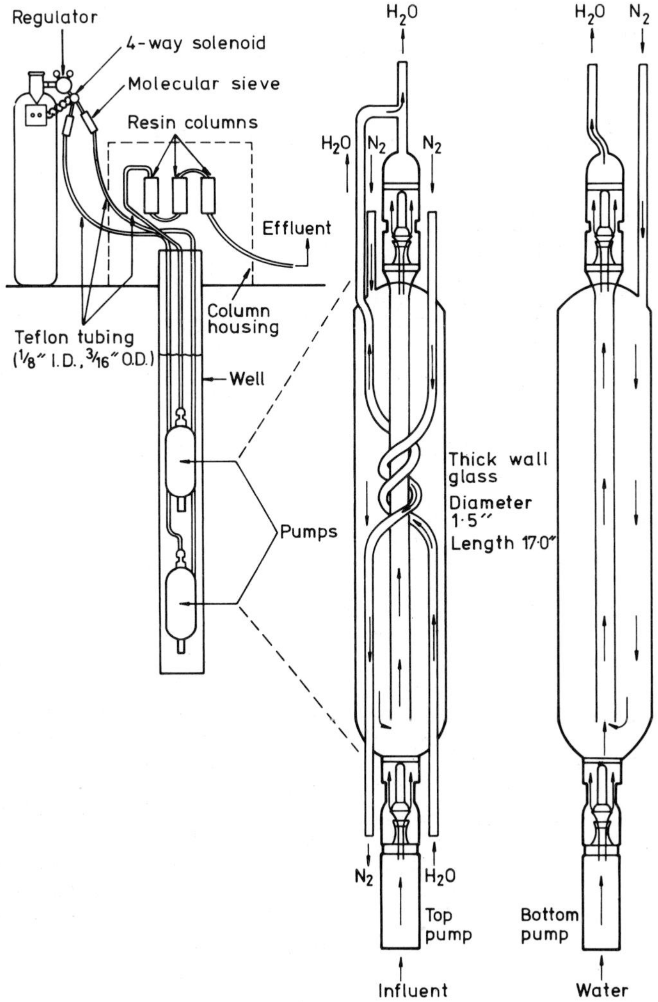

FIG. 7. Continuous discharge/high lift pump arrangement (after Pettyjohn *et al.* 1981).

This system, but simplified by the incorporation of only one pump, has been used in Denmark by Andersen (1979). Fig. 10 shows ports in the sampler wall positioned between packers allowing groundwater entry at three points. Water from the upper and lower pumped intervals is removed to the surface by an electric submersible pump, while water from the sampling zone will flow into an inflatable sampling vessel via a check valve. Compressed inert gas (eg. nitrogen) is used to force the sample to the surface. Once the sampling vessel is empty, pressure is released and water will again flow into the vessel. The procedure is repeated to remove the volume of the packered interval, before a representative sample is taken.

Even the installation of a single packer in a borehole means that water can be extracted from either the top or bottom zones of the aquifer and large scale mixing, as a result of pumping is prevented.

Electric submersible pumps for boreholes 100 mm diameter

An extensive range of electrically powered submersible pumps is available in Britain. These are generally designed for high discharge rates for water supply or pumping tests, and may not be suitable for groundwater sampling.

On-site groundwater chemistry determinations

Once a groundwater sample has been recovered strict procedures should be followed in its handling. Removed from depth in an aquifer, the sample will undergo temperature and pressure changes. As it is depressurised, variables such as pH, Eh, dissolved gas content and the speciation of certain constituents can be altered to varying degrees. Atmospheric contact is another potential source of alteration resulting in pH, Eh and speciation changes, and therefore contact with the atmosphere should be minimised by filling sample bottles, and keeping them tightly stoppered.

Because of physical and chemical changes as a sample is collected, speed is essential for on-site work. Temperature, pH, Eh, and dissolved oxygen must be determined on each sample at the borehole, as soon as it is produced. Care must also be taken to ensure that the samples obtained are representative of the target zones; usually this means taking a number of samples until key parameters are steady. However, groundwater may be drawn from a predominant flowpath and hydraulic factors should also be considered.

Prior to preservation, temperature, pH, Eh, conductivity and total alkalinity, as well as any other relevant determinations should be measured. Temperature, pH and Eh measurements are duplicated in the laboratory to provide an indication of the extent of the chemical changes that may have occurred between collection and analysis.

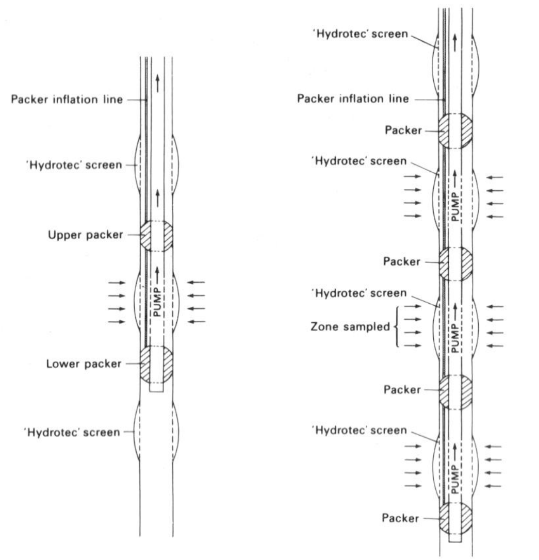

(a) isolation of one zone for sampling within a borehole

(b) 4 - packer system for sampling central zone of three

FIG. 9. Lightweight packer installations for discrete zone sampling

When a continuous sample flow is produced, the use of a flow-through cell (Fig. 11) allows measurements to be made on samples that have not been in atmospheric contact. Flow-through cells are available which hold a number of electrodes or probes, so that temperature, pH, Eh and conductivity can be measured in the same cell. It should be remembered that Eh and conductivity must be determined on a 'stationary' sample, to avoid any contribution due to streaming potential (Wood 1976).

The common techniques of field analysis are included in the Appendix of this report, and additional information is given by Cook *et al.* (1979), Wood (1976), and the appropriate instruction manuals.

Temperature

Physical and chemical properties vary with temperature, therefore, the temperature at which a sample was analysed, together with the original groundwater temperature, must be measured to enable correct interpretation of results.

A measured property without its corresponding temperature is meaningless.

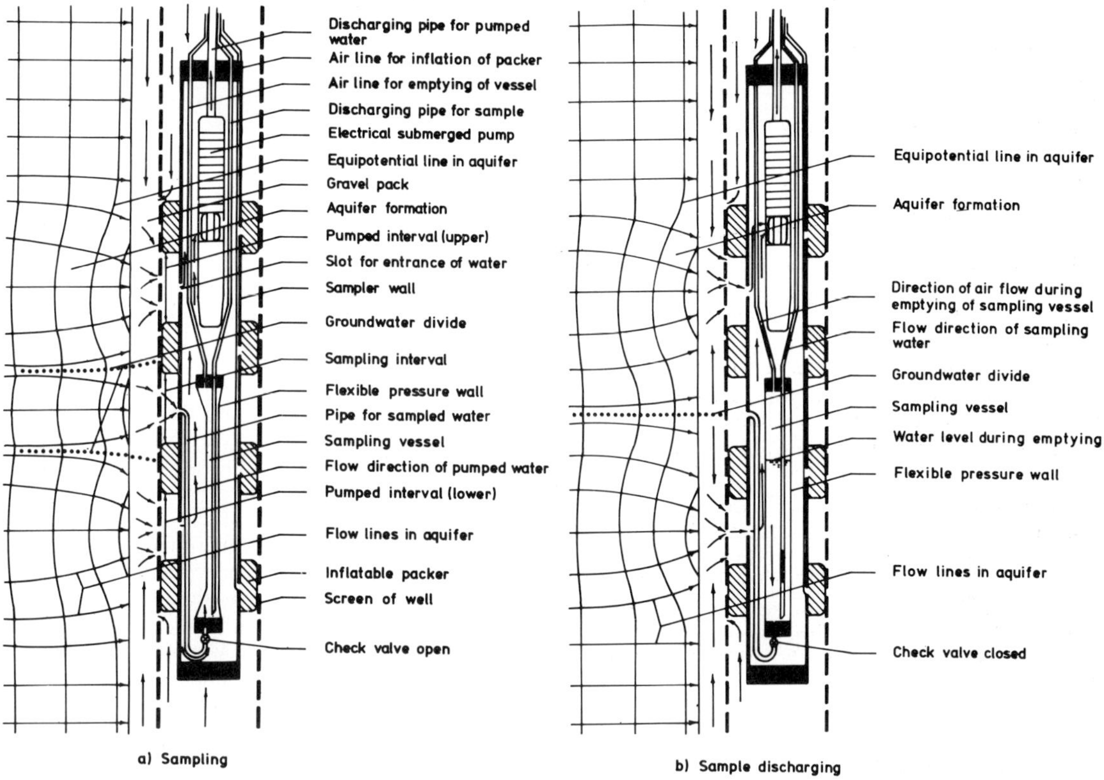

FIG. 10. Semi-automatic level-accurate groundwater sampler (after Andersen 1979).

pH

pH is a function of the activity of hydrogen ions in a solution: pH = $-\log [H^+]$.

An acid is defined in the Bronsted-Lowry classification as a proton donor, and a base as a proton acceptor:

$$\underset{\text{acid}}{HB} \rightleftarrows \underset{\text{proton}}{H^+} + \underset{\text{base}}{B^-}.$$

For the dissociation of water:

$$H_2O \rightleftarrows H^+ + OH^-$$

the equilibrium constant, K_w, at one atmosphere and 25°C is:

$$1.008 \times 10^{-14} \approx 1 \times 10^{-14}$$

$$\frac{[H^+]\,[OH^-]}{[H_2O]} = K_w = 10^{-14}$$

By convention the activity of water is unity in this very dilute solution and at neutrality $[H^+] = [OH^-] = 10^{-7}$, hence at one atmosphere and 25°C neutral pH = 7.00. At higher temperatures K_w increases and the neutral pH value decreases at 100°C $K_w = 10^{-2.32}$ and neutral pH = 6.2.

The presence of anions that form strongly dissociated acids in solution (e.g. Cl^-, SO_4^{2-}) will decrease the pH, and those that form weakly dissociated acids in solutions (e.g. NH_2^-, HCO_3^-), increase it. In any groundwater the interactions of the various acid-base equilibria present will be extremely complex; thus the pH value represents a composite measurement that is a reference for the interpretation of other data.

Several factors can influence the pH of a sample once removed from its normal environment. On contact with air, for example, carbon dioxide dissolved in the sample equilibrates to the partial pressure of atmospheric carbon dioxide, which disturbs the carbon dioxide - bicarbonate - carbonate buffer system and changes the pH (see below). Contact with air can also oxidise ions in solution — again, a pH change may result.

Eh (Oxidation-reduction potential, Redox potential)

Redox potential is a measure of the oxidising or reducing capacity of a sample: the greater the negative value of Eh (in mV) the more reducing the sample. (An oxidising substance (oxidant) is an electron acceptor, which is itself reduced as it oxidises a second substance.)

A reducing substance (reductant) is an electron donor, which is itself oxidised as it reduces a second substance; so:

$$A + B^- \rightarrow A^- + B$$

where A oxidises B^-, and is itself reduced. This introduces an important aspect of redox reactions: that for every oxidation reaction there is a corresponding reduction reaction, and vice versa. The above equation may now be seen to consist of two half-reactions:

$A + e^- \rightarrow A^-$	reduction
$B^- \rightarrow B + e^-$	oxidation
$A + B^- \rightarrow A^- + B$	redox reaction

Redox reactions are often the result of the transfer of electron carrying groups rather than free-electrons; this may be represented as:

$$W + X \rightarrow Y + Z$$

or, as half reactions:

$W + e^- \rightarrow Y$	reduction
$X \rightarrow Z + e^-$	oxidation
$W + X \rightarrow Y + Z$	redox reaction

As each redox reaction has an associated redox potential, the redox potential of a sample represents a combination of the individual potentials of each redox equilibrium present. (A parallel may be drawn here with the composite nature of the pH measurement).

The redox potential is related to the standard redox potential, E° by the Nernst equation:

$$\mathrm{Eh} = E^\circ + \frac{RT}{nF} \ln \frac{(\text{oxidised species})}{(\text{reduced species})}$$

where:

R = the gas constant, 8.314 J K^{-1} mol^{-1}

T = the solution temperature in degrees Kelvin, K

n = the number of moles of electrons transferred

F = the Faraday constant (the charge on one mole of electrons), 96,487 coulombs mol^{-1}

(oxidised species) = the product of the activities of the oxidised species

(reduced species) = the product of the activities of the reduced species

and

E° is relative to the standard hydrogen electrode (SHE) as zero.

Under standard conditions the oxidised and reduced species are both at unit activity, at 25°C (298K) and at one atmosphere pressure.

Elements in groundwater can form different compounds depending on whether they are under oxidising or reducing conditions. As an example: in a reducing environment iron exists mainly as organic complexes and sulphides; however, as conditions become more oxidising, iron silicates, oxides and carbonates begin to predominate, with an accompanying change in iron solubility (Stumm & Morgan 1970).

As for pH, contact with air may alter the Eh reading due to oxidation of the sample.

Dissolved oxygen

The dissolved oxygen content of a groundwater is an indicator of the amounts of oxidisable substrates present, for example organic pollutants. It may be expressed either as percentage saturation or as parts per million. The ultimate source of dissolved oxygen is the atmosphere, but it can remain in solution for long periods in the absence of reducing material.

Obviously when measuring dissolved oxygen there must be no contact with the atmosphere. A slow-through cell is the ideal way of excluding air, but often there is insufficient groundwater flow for this to be feasible. In this case the inlet and outlet pipes of the cell should be sealed with rubber stoppers, the cell filled to the brim and the lid/electrode attachment screwed down before the measurement is taken.

Electrical conductivity

Conductivity K (usually measured in microsiemens per centimetre (μS cm^{-1})) is equivalent to specific conductance, that is, the capacity of a 1 cm cube of a substance at 25°C to conduct an electrical current. H is thus an indication of ionic concentration, and, by extension, of total dissolved solids (TDS), although, there is no exact relationship between conductivity and TDS, an approximation is given by:

$$K\mathrm{A} = \mathrm{TDS}.$$

In most natural waters the value of constant A is between 0.55 and 0.75 (Hem 1970), but in polluted groundwaters, where the ionic composition is more complex, the relationship between K and TDS is more variable. As a general rule the higher the conductivity, the greater the concentration of dissolved solids; conductivity can therefore be used as a rapidly determined broad indicator of groundwater quality.

Total Alkalinity

Total alkalinity, measured by titrating with a standard acid, represents the acid-neutralising capacity of a sample (Stumm & Morgan 1970). Because of the abundance of carbonate minerals and the availability of carbon dioxide (from the atmosphere or microbial activity), the carbonic acid equilibria system (Eq.1) dominates alkalinity in most groundwaters.

$$CO_2(g) \rightleftarrows CO_2^*(aq) \underset{\updownarrow}{} (H_2CO_3) \rightleftarrows HCO_3^- + H^+ \rightleftarrows CO_3^{2-} + 2H^+ \quad (1)$$

$$H^+ + OH^- \rightleftarrows H_2O\ . \quad (2)$$

Titration with a standard acid takes the reaction shown in equation 1 completely to the left (H_2CO_3), and is taken to represent the sum of the initial concentrations of hydroxide, carbonate, and bicarbonate; i.e.:

$$\text{total alkalinity} = [OH^-] + 2[CO_3^{2-}] + [HCO_3^-] - [H^+] \quad (3)$$

Total alkalinity can also include contributions from other anions that take part in hydrolysis, for example silicates, borates, ammonia, organic bases,sulphides and phosphates.

Consideration of Equations 1 and 2 shows that there are potentially three equivalence points during the acid titration of total alkalinity: at approximately pH 10.6, corresponding to the alkalinity due to the hydroxide ion (OH^-); at approximately pH 8.3, corresponding to that of the carbonate ion (CO_3^{2-}); and at approximately pH 4.5, corresponding to that of the bicarbonate ion (HCO_3^-). As groundwater is commonly in the range pH 6 to pH 8, the contribution to total alkalinity by OH^- and CO_3^{2-} is negligible, and alkalinity due to the bicarbonate ion is the only capacity actually measured; for this reason total alkalinity is often (inaccurately) referred to as bicarbonate concentration.

Similarly, a total acidity titration can be performed, using strong standard base. The equivalence points are the same as for a total alkalinity titration, this time corresponding to the acidity due to H^+ (pH 4.5), CO_2^* (aq) ($=H_2CO_3$), at pH 8.3 and HCO_3^- (total acidity) at pH 10.6.

The total alkalinity or total acidity curve is shown graphically in Fig. 12, with x, y and z marking the equivalence, or end, points.

As with total alkalinity (Eq.3) the various acid- and base-neutralising capacities in Fig. 12 may be expressed algebraically (Table 1).

The carbonic acid/carbonate equilibrium system acts as a buffer, effective in the range pH 4.5 - 8.2. Bringing the sample into contact with the atmosphere may affect this equilibrium, as the concentration of hydrated carbon dioxide ($CO_2^*(aq)$) adjusts to the atmospheric partial pressure of carbon dioxide. It is important to appreciate that although this equalization process can change the pH, the total alkalinity remains unaltered, i.e. alkalinity is a conservative property of groundwater with respect to the concentration of carbon dioxide. This is not immediately obvious, and warrants some discussion.

$$CO_2(g) \rightleftarrows CO_2^*(aq) \rightleftarrows HCO_3^- + H^+ \rightleftarrows CO_3^{2-} + 2H^+ \quad (1)$$
$$\updownarrow\ (H_2CO_3)$$

and

$$\text{total alkalinity} = [HCO_3^-] + 2[CO_3^{2-}] + [OH^-] - [H^+] \quad (3)$$

Reconsider Equations 1 and 3: should CO_2 be dissolved from the atmosphere, the equilibrium in equation 1 shifts to the right; the resulting increase in $[HCO_3^-]$ and $[CO_3^{2-}]$ is balanced by a corresponding increase in $[H^+]$, so that there is no change in titratiable alkalinity even though the pH has decreased and the dissolved carbon dioxide concentration has increased.

TABLE 1. *Acid- and base-neutralising capacities of the carbonic acid system.*

		End point	Equation
Total alkalinity	$= [HCO_3^-] + 2[CO_3^{2-}] + [OH^-] - [H^+]$	x	3
Carbonate (CO_3^{2-}) alkalinity	$= 2[CO_3^{2-}] + [OH^-] - 2[CO_2^*(aq)] - [H^+]$	y	4
Hydroxide (OH^-) alkalinity	$= [OH^-] - [HCO_3^-] - 2[CO_2^*(aq)] - [H^+]$	z	5
Total acidity	$= [H^+] + 2[CO_2^*(aq)] + [HCO_3^-] - [OH^-]$	z	6
Dissolved $CO_2(CO_2^*(aq))$ acidity	$= [H^+] + 2[CO_2^*(aq)] - [CO_3^{2-}] - [OH^-]$	y	7
Hydrogen ion (H^+)	$= [H^+] - [HCO_3^-] - 2[CO_3^{2-}] - [OH^-]$	x	8

Conversely, if the sample degasses, losing CO_2, the equilibrium shifts to the left and $[HCO_3^-]$, $[CO_3^{2-}]$ and $[H^+]$ decrease. The $[H^+]$ decrease is accompanied by a corresponding $[OH^-]$ increase ($H_2O \rightleftarrows H^+ + OH^-$, Eq.2); again, the titratable alkalinity is unchanged, although the pH has increased and carbon dioxide concentration has decreased. This emphasises the importance of measuring pH immediately after sample collection and again before titration.

Total alkalinity should be measured less than 24 h after collection, to minimise microbial changes.

Preparation of samples for detailed analyses

As it is impossible to perform detailed analyses in the field, it will be necessary to take samples back to the laboratory for more thorough characterisation. Unless treated, the composition of the samples may change with time, and analyses will be unrepresentative.

Filtration

Even a clear-looking groundwater sample may contain suspended particulate matter, which, if not removed, will influence analytical results. During storage unfiltered solution composition may be changed by reaction such as ion exchange and dissolution of particulate and colloidal materials; in addition, acidifying an unfiltered sample may dissolve any solids present.

Filtration apparatus is shown in Fig. 13. The nitrogen pressure system (A) is to be preferred, as it minimises contact with the atmosphere; the suction apparatus is, however, rather more convenient. Two filters are used, the first a coarse glass fibre pre-filter, and the second of cellulose acetate with a 0.45 μm pore-size. Colloidal material less than 0.45 μm in diameter may occur in groundwater, but smaller pore-size filters make the filtration process too lengthy to be practical.

The first few millilitres of filtrate are used to rinse out a clean 500 ml and clean 60 ml reagent bottles (60 ml is sufficient sample volume for analysis by many analytical laboratories but some may require different volumes and should be consulted before embarking on a field sampling programme). The filtrate should be collected in the large bottle and decanted into the 60 ml bottles for treatment. This ensures homogeneity, since the composition of filtrate may change as filtration proceeds, due to the gradual clogging of the filter progressively excluding smaller colloids.

Note that if silica is to be an analyte, polythene bottles should be used, contact with glass minimised, and suitable blanks processed.

Preservation

The common preservation techniques are outlined in Table 2, and their effects are shown in Table 3.

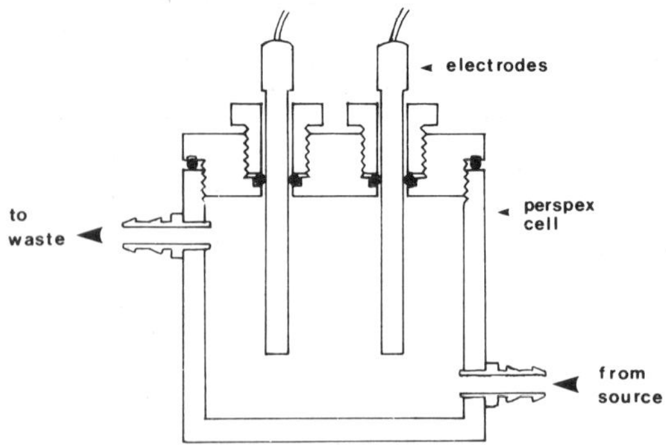

FIG. 11. Flow-though cell based on that of Wood (1976).

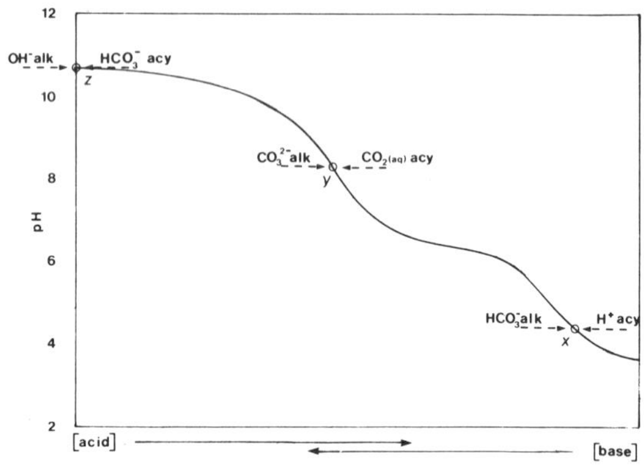

FIG. 12. Equivalence points in the titration of total alkalinity or total acidity.

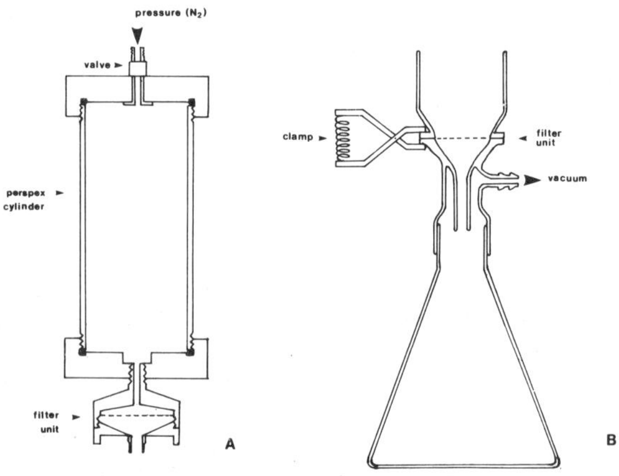

FIG. 13. Groundwater filtration apparatus.

TABLE 2. *Summary of preservation techniques (USEPA, 1974).*

Preservative	Filter 0.45 μm	Analytes	Max. storage	Container	BGS label	Volume /ml
HCl, pH≤2	yes	Ca, Mg, Na, K, Fe, Mn, Ni, Zn, NO_3^-, SO_4^{2-}	6 m	Glass or polythene	C	60
H_2SO_4, pH≤2, 4°C	yes	TOC, Total PO_4^{3-}, NH_4^+, Cl^-	24 h	Glass	NP	60
NaOH, pH≥12, 4°C	no	Total CN^-	24 h	Glass or polythene	CN	60
ZnOAc, ≈ 1g dm^{-3}, 4°C	no	Total S^{2-}	24 h	Glass	S	60
H_3PO_4, pH≤4 + $CuSO_4$, ≈ 1g/dm^{-3}, 4°C	yes	Phenols	24 h	Glass	P	60
4°C	yes	SiO_2, Al	6 m	Plastic	SA	60

1. Check the pH after adding acid/alkali — samples with similar pH values do not necessarily require similar amounts of reagent to reach the desired pH; the governing factor is the pH buffering capacity, measured in the total alkalinity titration.
2. Storage times are the 'lowest common denominators'; consult reference for the stabilities of individual analytes.
3. As an additional precaution for preserving cyanides, CN^-, dark bottles may be used.
4. Volumes are those required by the EMSc Division of AERE Harwell; other analytical laboratories may have different requirements.

TABLE 3. *Summary of the effect of preservatives.*

Preservative	Effects
Acid (HCl,HNO_3, H_2SO_4,H_3PO_4)	Metals solvent - prevents precipitation and adsorption to bottle surfaces; bacterial inhibitor; forms salts with organic bases, thus stabilising them.
NaOH	Forms salts and stabilises organic compounds
$CuSO_4$	Bacterial inhibitor
ZnOAc	Precipitates sulphide
Refrigeration, 4°C	Bacterial inhibitor; slows chemical and physical reactions

Cations in untreated samples may precipitate as hydroxides or form complexes; they also tend to be adsorbed onto the surface of glass, plastic, and quartz containers (Robertson 1968) — for this reason it is good practice to add the acid to a few millilitres of filtrate before pouring in the whole sample, as it is easier to prevent adsorption than to reverse it.

Micro-organisms can affect composition in several ways — by changing the nitrate - nitrate - ammonia balance, reducing sulphate to sulphide, decreasing the dissolved oxygen content, altering the biochemical oxygen demand (BOD), and metabolising organo-phosphorus and easily degraded organic compounds.

Properly treated, the above effects can be halted or decelerated. (Ogawa *et al.* 1981; German Chemists Assoc. 1981; Batley & Gardner 1977; Maienthal & Becker 1976; Philbert & Traversy 1973).

Summary

The preceding sections on analysis and treatment may be usefully summarised as a flow chart (Fig. 14).

Detailed laboratory analysis

The methods generally used for the analysis of samples collected by the Fluid Processes Research Group of BGS are described in Tables 4, 5 and 6. The concentrations of metals are determined by Atomic Absorption Spectrophotometry and of other constituents by Autoanalyser and other methods.

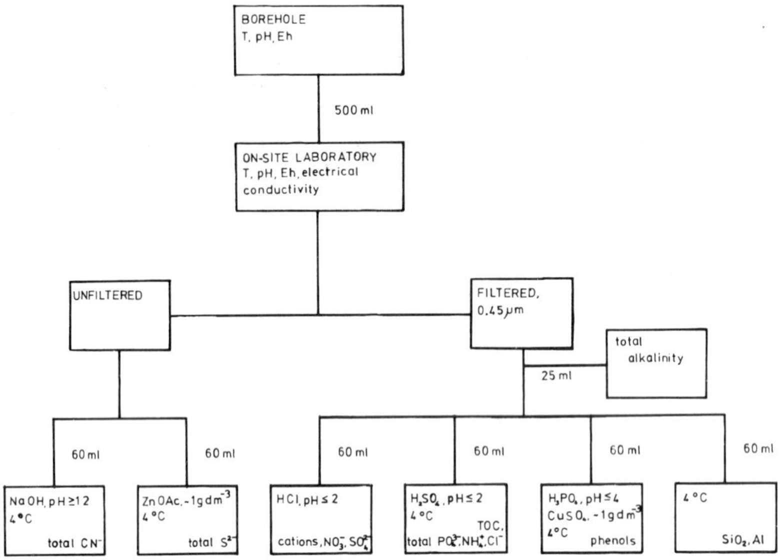

FIG. 14. On-site sample treatment.

TABLE 4. *Detailed analysis of groundwater — Autoanalyser methods.*

Analyte	Detection limit/ppm	Method	Reference
SO_4^{2-}	5	Turbidometric*	Based on various 'Technicon' methods, similar to Ogner G & Haugen, A (1977). Automatic determination of sulphate in water samples and soil extracts containing large amounts of humic substances. *Analyst,* **102,** 453-457.
SO_4^{2-}	0.1	Dimethyl sulphonazo	Based on: Rejnders, H. F., van Staden, J.J., and Griepink, B. 1979. Photometry in a continuous flow system. *Fresenius Zeitschrift fur Analytische Chemie,* **295,** 122-124. See Also: Rejnders, H. F., van Staden, J. J., and Griepink, B. 1980. Flow through determination of sulphate in water using different methods: A comparison. *Fresenius z. Anal. Chem.* **300,** 273-276.
Total PO_4^{3-}	0.1	Colorimetric (Molybdenum blue)	Based on: Czech, F. W., Hrycyshyn, T. B., 'Technicon' Symposium 1966. *Automation in Analytical Chemistry,* **I,** 163-1967, Medied Inc., N.Y.C.
Cl^-	0.01	Colormetric (Mercury thiocyanate)	Based on: Zall, D. M. and Garner, M. Q. 1956. Photometric determination of chlorides in water. *Analytical Chemistry,* **28,** 1665-1668.
Total CN^-	0.001	Colormetric	Based on: 'Technicon' Industrial Method 315-74W.
NO_3^-	0.001	Volumetric	Based on: Henriksen, A. and Selmer-Olsen, A. R. 1970. Automatic methods for determining nitrate and nitrite in water and soil extracts. *Analyst,* **95,** 514-518.
Phenols	0.001		Temple, W., The quantitative determination of phenolic compounds in aqueous solution. Part 1, Phenol equivalents of phenolic compounds using continuous flow analysis techniques. AERE M 3035.

* By far the most common method of determining sulphate concentration.

TABLE 5. *Detailed analysis of groundwater — Atomic absorption spectrophotometry methods.*

Analyte	Detection limit/ppm	Interferences	Notes
Ca	0.01	Si, Al, PO_4^{3-}, SO_4^{2-}	Add La or Sr at 0.1 –1% to combat interferences
Mg	0.001	Si, Al	Add La at 0.1 – 1% to combat interferences
K	0.01		K and Na are partially ionized in the flame, but the effects can be largely overcome by adding a large concentration (1000-2000 ppm) of other alkali salts to standards and samples
Na	0.001		
Fe	0.02	HNO_3, Ni	A very lean flame is used to overcome the reduction in sensitivity caused by HNO_3 and Ni
Mn	0.01		
Ni	0.02		
Zn	0.001		

TABLE 6. *Detailed analysis of groundwater — other methods.*

Analyte	Detection limit/ppm	Method	Reference
TOC	50	Flame ionisation	Cropper, F. R., Heinekey, D. M. and Westwell, A. 1967. The determination of total organic matter (carbon content) in aqueous media. *Analyst,* **92,** 436-449.
Total S^{2-}	20	Spectrophotometry	American Public Health Association (1976). Standard methods for the examination of water and waste water (14th ed). (Washington: American Public Health Association).
NH_3		Spectrophotometry (nesslererization or phenate methods); Titration (acidimetric)	
SiO_2		Spectrophotometry (UV)	

Conclusions

Any groundwater monitoring and sampling scheme must take several factors into consideration before borehole completions are finalised. These include geology, depth to groundwater, aquifer thickness, and nature of the contaminants. In few instances will the provision of information on rock type, water levels, aquifer properties and bulk groundwater chemistry from a simple borehole array be sufficient. In the case of established or sensitively sited landfills it is likely that specialised installations and sampling equipment will be necessary to define groundwater quality and trends therein accurately.

Several borehole completions and sampling techniques have been described, suitable for a wide range of applications. Once groundwater samples have been obtained, they must be dealt with rapidly; either by taking on-site measurements or by preservation for later detailed laboratory analysis, or both. It is pointless taking precise, representative groundwater samples if their subsequent handling and analysis are not similarly well controlled. Equally, detailed analysis of samples from poorly completed or unidentified zones within boreholes will provide questionable data.

Only a limited amount of equipment suitable for sampling is commercially available in Britain, and there is a need for a lightweight system for routine discrete zone sampling within boreholes. Only by careful sampling and on-site analysis and treatment of samples will it be possible to predict and evaluate the effects of waste disposal on the environment accurately. This will in turn lead to a more effective and economic management of waste disposal practices involving the minimum of groundwater pollution.

Groundwater in Engineering Geology, London, 1986

References

AMERICAN PETROLEUM INSTITUTE, 1960. Recommended practice for core analysis procedure. *American Petroleum Institute* (AP1). Report. 40. API, Dallas Texas.

ANDERSEN, L. J., 1979. A semi-automatic level-accurate groundwater sampler. *Danmarks Geologiske Undersogelse* Arbog Kobenhaven, 165-171.

BARBER, C., NAYLOR, J. A. and MARIS, P. J. 1976. Techniques for investigating groundwater pollution at landfill sites. Paper presented at Groundwater Quality, Measurement, Prediction and Protection Conference, organised by Water Research Centre at Reading University.

BATLEY, G. E. and GARDNER, D. 1977. Sampling and storage of natural waters for trace metal analysis. *Water Research*, **11**, 745-756.

BUSS, D. F. and BANDT, K. E., 1981. An all-teflon bailer and an air-driven pump for evacuating small-diameter groundwater wells. *Groundwater,* **19**, 429-431.

COOK, J. M., EDMUNDS, W. M., and MILES, D. L., 1979. Water sampling and field chemical measurements for overseas hydrogeological surveys. *Report of the Institute of Geological Sciences*, WD/ST/79/11.

COOPER, H. H., BREDEHOEFT, J. D., and PAPODOPULOS, I. S., 1967. Response of a finite-diameter well to an instantaneous charge of water. *Water Resources Research*, **3**, 263-269.

DEPARTMENT OF THE ENVIRONMENT, 1978. Co-operative programme of research on the behaviour of hazardous wastes in landfill sites: final report of review committee HMSO London.

DROST, W., KLOTZ., D., KOCH, A., MOSER, H. NEUMAIER, F. and RAUERT, W., 1968. Point Dilution Methods of Investigating Ground water Flow by means of Radioisotopes. *Water Resources Research,* **4**, No. 1.

EDMUNDS, W. M. and BATH, A. H., 1976. Centrifuge Extrection and chemical analysis of interstitial waters. *Environmental Science and Technology,* **10**, 467-479.

GERMAN CHEMISTS ASSOCIATION, 1981. Preservation of water samples. *Water Research*, **15**, 233-241.

HALEVY, E., MOSER, H., ZELLHOFER, O. and ZUBER, A., 1967. Borehole dilution techniques: a critical review. *IAEA Proceedings Series, Isotopes in Hydrology, Vienna*, 531-564.

HARRISON, I. B., 1976. Construction of investigatory boreholes in landfill sites. *Surveyor*, 25th June, 22-24.

HEM, J. D., 1970. Study and interpretation of the chemical characteristics of natural water. *US Geological Survey Water-Supply Paper 1473.*

HVORSLEV, M. J., 1951. Time lag and soil permeability in groundwater observations. Bulletin 326, *US Waterway Experimental Station, Vicksburg, Miss.*

MAIENTHALL, E. J. and BECKER, D. A., 1976. A survey on current literature on sampling, sample handing, for environmental materials and long term storage. *Interface,* **5**, No. 4, 49-62.

OGAWA, I., JUNK, G. A. and SVEC , H. J., 1981. Degradation of aromatic compounds in groundwater and methods of sample preservation. *Talanta*, **28,** 725-729.

OWENS, J. W., GLADNEY, E. S. and PURTYMUN, W. D., 1980. Modification of trace element concentrations in natural waters by various field sampling techniques. *Analytical Letters,* **13,** 253-260.

PATTON, F. D., 1979. Groundwater instrumentation for mining projects. *Proceedings of the First International Mine Drainage Symposium,* May 20-23, 1979, Denver, Colorado.

PETTIJOHN, W. A., DUNLAP, W. J., COSBY, R. and KEELEY, J. W., 1981. Sampling Groundwater for Organic Contaminants. *Groundwater,* **19**, 180-189.

PHILBERT, F. J. and TRAVERSEY, W. J., 1973. Methods of sample treatment and analysis of Great Lakes water and precipitation samples. *Proceedings of 16th Conference on Great Lakes Research,* 294-308. International Association for Great Lakes Research.

PICKENS, J. F., CHERRY, J. A., GRISAK, G. E., MARRITT, W. F. and RISTO, B. A., 1978. A multilevel device for groundwater sampling and piezometric monitoring. *Groundwater,* **15,** 322-327.

RIHA, M. 1980. Groundwater sampling and testing. *National Water Well Association of Australia, Conference Cowes.*

ROBERTSON, D. E., 1968. The adsorption of trace elements in sea water on various container surfaces. *Anal. Chim. Acta,* **42,** 533-536.

STUMM, W. and MORGAN, J. J., 1970. *Aquatic Chemistry: an Introduction Emphasizing Chemical Equilibria in Natural Waters.* New York: Wiley Interscience.

TOMSON, M. B., HUTCHINS, S., KING, J. M. and WARD, C. H., 1980. A nitrogen powered continuous delivery, all-glass-Teflon pumping system for groundwater sampling from below 10 metres. *Groundwater,* **18,** 444-446.

UNITED STATES ENVIRONMENTAL PROTECTION AGENCY 1974. Methods for chemical Analysis of Water and Waste Water, Washington DC.

WOOD, W. W. 1976. Techniques of water resources investigations of the United States Geological Survey. Guidelines for collection and field analysis of groundwater samples for selected unstable constituents. US Dept. of the Interior. Washington DC.

A. STUART & S. P. HITCHMAN, British Geological Survey, Keyworth, Nottingham NG12 6GG, UK.

Appendix: Methods of on-site measurements

pH

Apparatus and reagents
Combinations pH electrode (Orion 91-05); pH meter (Orion 407A); buffer solutions — pH 4, 7, 9.2; distilled water; tissues.

a. The pH electrode bulb should be kept moist during storage with pH 7 buffer solution or distilled water. New electrodes and those that have been allowed to dry out must be soaked in pH 7 buffer overnight before use. Store in pH 7 buffer solution or distilled water between determinations.
b. Buffer solutions are conveniently prepared from ready-formulated powder or tablets, available commercially. The pH values of such solutions according to temperature are defined in tables supplied by the manufacturer; those for BDH buffer powders are shown in Table 7.
c. Turn the meter to 'OFF' when not actually measuring.
d. The buffer solutions and samples should be at approximately the same temperature, small variations can be compensated for with the temperature knob.
e. Check calibration periodically during use.

TABLE 7. *Changes in the pH of BDH buffer solutions with temperature.*

T/°C	pH of standard buffer solutions		
0	4.00	7.11	9.49
10	4.00	7.06	9.35
20	4.00	7.01	9.24
25	4.01	7.00	9.19
30	4.02	6.98	9.14
40	4.04	6.97	9.07
50	4.06	6.97	9.01

Measurement
The procedure outlined is appropriate to the Orion 407A meter.

1. Remove protective electrode cap and check that there are no air bubbles in or near the electrode bulb; bubbles may be removed by shaking the electrode downwards. Connect electrode to meter.
2. Rinse electrode with distilled water, remove excess water by shaking and dabbing with a tissue.
3. Place the electrode in pH 7 buffer solution and switch on meter.
4. Adjust pH reading to the exact buffer value with the calibration control.
5. Repeat step 2 and place electrode in pH buffer solution.
6. Adjust pH reading to the exact buffer value using the temperature compensator knob. Turn the slope indicator to the solution temperature and read off the percentage slope; a slope of less than 90° indicates faulty electrode or buffer solution.
7. Repeat step 2 and check pH 7 calibration.
8. Repeat step 2 and place electrode in sample; record the pH value.

Redox

Apparatus and Reagents
Combination redox electrode (Orion 96-78, 97-78); electrode filling solution (Orion 90-00-01); meter (Orion 407A); ZoBell's solution:
1.408g potassium ferrocyanide ($K_4Fe(CN)_6$. $3H_20$), 1.098g potassium ferrocyanide ($K_3Fe(CN)_6$), 7.46g potassium chloride (KCl), made up to 1l with distilled water (ZoBell's solution is stable for several weeks if kept in a dark bottle); distilled water; tissues.

a. Take care — ZoBell's solution is poisonous.
b. Place electrode in distilled water between determinations. For storage the filling solution should be drained and the electrode flushed with distilled water.
c. The 90-00-01 filling solution is suitable for samples of total ionic strength less than 0.2 M. Filling solution 90-00-11, 4 M KCl saturated with AgCl, is used for samples of total ionic strength greater than 0.2 M. Top up filling solution as necessary during use.
d. The redox electrode has an associated reference electrode potential, given for various temperatures in Table 8. The sample redox potential relative to the standard hydrogen electrode is given by:

sample Eh = absolute mV reading
+ reference electrode potential.

TABLE 8. *Reference electrode potentials at various temperatures.*

T/°C	Reference electrode potential/mV	
	90-00-01	90-00-11
0	257	
10	251	
20	244	
25	241	199
30	238	

e. Turn meter to 'OFF' when not actually measuring.

Measurement
The procedure outlined is appropriate to the Orion 407A meter.

1. Remove protective electrode cap, flush through and fill electrode with appropriate filling solution; connect electrode to meter.
2. Check electrode operation by placing it in ZoBell's solution; the meter should read around 210 mV at 10°C and 190 mV at 20°C.
3. Rinse electrode with distilled water and remove excess water by gently dabbing with a tissue.
4. Place electrode in sample; wait for reading to stabilize and record.

Dissolved oxygen

Apparatus and reagents
Dissolved oxygen electrode (pHOX IC.600 electrode with 660 head); dissolved oxygen meter (pHOX 62tE); 2% sodium sulphite solution; 100% air saturated distilled water; distilled water.

a. 100% air saturated water can be prepared by shaking a 1 l narrow mouthed bottle half full of distilled water for 15-20 minutes. Table 9 shows the dissolved oxygen concentration of air-saturated water at sea-level at various temperatures, and Table 10 the correction factors to be applied at various altitudes.
b. Avoid touching the membrane; periodically check for damage, replace when necessary and check that the output reaches zero by conditioning the electrode in 2% sodium sulphite solution.

TABLE 9. *Saturation values for dissolved oxygen in pure water at various temperatures.*

Temp/°C	$[O_2]$/ppm	Temp/°C	$[O_2]$/ppm
0	14.63	21	8.91
1	14.23	22	8.74
2	13.84	23	8.57
3	13.46	24	8.42
4	13.11	25	8.26
5	12.77	26	8.12
6	12.45	27	7.97
7	12.13	28	7.84
8	11.84	29	7.70
9	11.55	30	7.57
10	11.28	31	7.45
11	11.02	32	7.33
12	10.77	33	7.21
13	10.53	34	7.09
14	10.29	35	6.98
15	10.07	36	6.88
16	9.86	37	6.77
17	9.65	38	6.67
18	9.46	39	6.57
19	9.27	40	6.47
20	9.08		

TABLE 10. *Oxygen saturated water — correction factors for altitude.*

Altitude/m	Correction factor	Altitude/m	Correction factor
0	1.00	1300	1.17
100	1.01	1400	1.19
200	1.03	1500	1.20
300	1.04	1600	1.22
400	1.05	1700	1.24
500	1.06	1800	1.25
600	1.08	1900	1.26
700	1.09	2000	1.28
800	1.11	2100	1.30
900	1.12	2200	1.31
1000	1.13	2300	1.33
1100	1.15	2400	1.34
1200	1.16	2500	1.36

c. Leave in air between determinations. Store in distilled water overnight, or in 2% sodium sulphite solution for long periods.
d. Because a small amount of oxygen is consumed from the surrounding liquid during use, the electrode must be gently agitated (unless a flowing sample is used) to ensure stable readings.
e. Turn meter off when not in use.

Measurement

1. Remove electrode cap and rinse off storage solution, connect electrode to meter.
2. Switch meter on and depress 'ZERO OXYGEN' button, if necessary set reading to zero with the zero adjustment knob.
3. Depress 'READ TEMP' button, push 'SET 30°C' button and adjust if necessary; if 30°C cannot be set the battery must be renewed.
4. Immerse electrode in 100% air saturated distilled water and select 'X2' range; when the reading is stable set to 100% on the scale 'X1' range for final adjustment.
5. Switch to 'X2 range and immerse electrode — sample, allow reading to stabilise and if necessary select different range, record reading.

Electrical conductivity

Apparatus and reagents
Conductivity probe (pHOX systems); conductivity meter (pHOX systems series 52); conductivity standard (potassium chloride, KCl, 7.436 5 g l^{-1}); cleaning solution (2% sulphuric acid, H_2SO_4); distilled water; tissues.

a. Store probe in distilled water or cleaning solution between determinations. Ensure that cleaning solution is thoroughly rinsed off before measuring.

b. Check calibration periodically, either as at step 2 below, or using the standard solution, details of which are given in Table 11.

TABLE 11. *Conductivity standard readings at various temperatures.*

T/°C	Conductivity (μS)
0	7.138 x 10^3
18	11.167 x 10^3
25	12.856 x 10^3

c. With the meter in the 'TEMP. COMP. IN' mode, readings are automatically referred to 25°C.

d. Turn meter to 'OFF' when not actually measuring.

Measurement

1. Connect probe to meter.
2. Calibrate the meter by selecting any range with the function switch, and 'TEST' on the Test/Read knob. A correctly calibrated instrument will read full scale to 10, adjustable by turning a small screw under the scale. If it is impossible to set the needle to 10, the batteries need to be renewed.
3. Rinse probe with distilled water, shake off excess water and wipe with a tissue.
4. Place probe in sample to a depth of at least a 12.5 mm above the hole in the guard tube.
5. Allow about 60 s for the reading to stabilize, then with the Test/Read knob on 'READ' and the 'TEMP/ COMP. IN' position selected, rotate the range switch until a reading is obtained; record conductivity.

Total alkalinity

Apparatus and reagents

Pipette (25 ml); pipette filler; titration containers (100 ml weighing bottles); burette (50 ml) with stand and clamp OR 'Hach' digital titrator; standard acid solution (0.05 M H_2SO_4) OR 'Hach' standard acid cartridge; indicator solution (bromocresol green) OR pH measuring equipment; white tile.

a. Samples should be filtered to exclude suspended carbonates.

b. 25 ml is a convenient sample volume, more or less may be used according to conditions.

c. A standard acid solution is conveniently prepared from a commercial concentrated solution (e.g. 'Convol' by BDH), following the manufacturer's instructions carefully.

Measurement

1. Rinse and fill the burette with standard acid.
2. Transfer an aliquot (25 ml) of sample to a weighing bottle and add a few drops of indicator.
3. Record the initial burette reading.
4. Carefully add acid to the sample, swirling continuously.
5. As the indicator starts to turn from blue to pale yellow, add the acid much more slowly.
6. When the colour change is complete, record the final burette reading and calculate the total volume of acid added.

An alternative method dispenses with the indicator, using instead pH measurement. The pH is measured after each addition of acid, the acid being added in smaller amounts around pH 4.5. A plot of pH against total volume of acid added shows a point of inflexion, corresponding to the end-point of the titration. The indicator method is, however, more convenient, and with practise can be both accurate and precise.

Probably the easiest method of determining total alkalinity is to use a 'Hach' digital titrator. A standard acid cartridge is fitted to the 'gun', the reading zeroed, and acid added to the sample by turning a knob on the 'gun'. Either indicator or pH can be used to gauge the end-point. The digital titrator gives a reading that is directly related to total alkalinity.

A digital titrator and indicator solution are recommended for field use.

Calculations

Titration using a burette

$$H_2SO_4 + 2HCO_3^- \rightarrow SO_4^{2-} + 2H_2CO_3$$

Hence

$$1 \text{ mole } H_2SO_4 \equiv 2 \text{ moles } HCO_3^-$$

$$\Rightarrow 2(\text{molarity of } H_2SO_4 \times \text{vol. } H_2SO_4)$$

$$(\text{molarity of } HCO_3 \times \text{vol.} HCO_3)$$

Or

$$2M_a\, V_a = M_b\, V_b$$

$$\Rightarrow M_b = \frac{2M_a V_a}{V_b} = [HCO_3^-] \quad \text{total alkalinity.}$$

Molecular weight of HCO_3 is 61 g

$$\text{total alkalinity} = \frac{61 \times 2M_a V_a}{V_b} \text{ gl}^{-1}.$$

For 0.05 M H_2SO_4 and a 25 ml sample volume this becomes:

$$\text{total alkalinity} = \frac{61 \times 2 \times 0.05 \times V_a}{25}$$

$$= 0.244\, V_a \text{gl}^{-1}, \text{ with } V_a \text{ measured in millilitres.}$$

Titration using a 'Hach' digital titrator

The relationship between the digital readout and total alkalinity is defined on the label of the acid cartridge used.

Using a 1.589 ± 0.006N (= 0.796 ± 0.003M) H_2SO_4 cartridge and a 100 ml sample, the readout is the total alkalinity measured in mgl^{-1} (ppm) of $CaCO_3$. to convert to mgl^{-1} of HCO_3 this value is divided by 0.8202; for a 25 ml sample, therefore,

$$\text{total alkalinity} = \text{readout} \times \frac{4}{0.8202}$$

$$= \text{readout} \times 4.88\ mgl^{-1} \text{ of } HCO_3^-.$$

Economical collection of reliable groundwater information

J.S. Sutton & C.D. Eldred

ABSTRACT: Many site investigations fail to provide reliable information on site groundwater levels and quality yet such information is frequently crucial for successful design and construction. In many cases instrumentation and long term monitoring are either unjustifiable or impracticable. However, adherence to a series of simple rules ensuring accurate recording of water levels and flush returns with reference to borehole and casing depth, throughout the drilling or boring programme, can ensure that reliable information is obtained. Similarly simple and cheap equipment can ensure that general parameters defining water quality (pH and EC) and unstable parameters such as dissolved gases, hardness and alkalinity can be reliably determined, although care is required to ensure that samples are representative particularly if no trial excavations are made.

The use of this type of information and the requirement for it where it has not been properly collected are illustrated with reference to investigations for road construction, an opencast coal mine and a major dewatering operation.

Introduction

In the course of many small to medium sized site investigations sufficient date to provide a reliable basis for assessment of groundwater conditions is available for collection at little or no additional cost. Yet, in many cases, little or no effort is made to collect such data in a systematic manner and this is in turn reflected in a poor understanding of the groundwater conditions on site. For large investigations where resources may be available for piezometer installation and long term monitoring a much more complete picture is obtained if the location and levels at which piezometers are installed take account of information derived from systematic water level recording during drilling.

The value of the systematic recording of water levels during drilling or boring operations has long been recognised and is discussed at some length by Knill (1971). It is not, however, referred to in more recent standard publications such as BS 5930 (1981). Most standard daily drilling record sheets provide a section for routine recording of water levels which more often than not is left incomplete. Routine recording of features such as electrical conductivity of water which provide a reliable index of water quality is even less frequently carried out.

Data Collection

Water levels

Groundwater level data recorded during drilling can only be reliably interpreted when viewed together with records of borehole depth, casing depth and whether or not water has been added to assist drilling progress. Even where water is regularly added to the hole or continuously used as a flushing medium useful information can be inferred from borehole water levels.

There can be few, if any, boreholes no matter how shallow, that reach completion without at least one tea or meal break and most holes continue through a long succession of interruptions. Each one of these interruptions provides an opportunity for assessment of groundwater simply by careful recording of the water level in the hole and the time when work on the hole stops and starts. All that is required for this operation is a standard well dipping tape and sufficient care to ensure that measurements in any one hole refer to the same reference point.

Interpretation of the results of these observations must be carried out in the light of the overall subsurface condition revealed by the investigation but is based primarily on very simple rules:

(i) Consistently falling water levels over each break suggest that the water table has not yet been reached or that during hole advancing water levels are maintained above the water table.

(ii) Consistently rising water levels indicate that holes are being advanced below the water table.

(iii) Should rising or falling levels stabilise within the same general range across the site a good indication of the water table level is available.

(iv) Variation with depth of borehole or casing may suggest a pattern of confined or partially confined water levels.

(v) The rate of change of water level gives a good measure of permeability and an indication of rate of water movement into any subsequent excavations.

From CRIPPS, J.C., BELL, F.G. & CULSHAW, M.G., (eds), 1986, *Groundwater in Engineering Geology*, Geological Society Engineering Geology Special Publication No. 3, pp. 247-250

If this information is not recorded during drilling it is irretrievably lost and unlike sample description cannot be even partially recovered at a later date.

Flush returns

Accurate recording of flush returns and casing levels during water flush rotary drilling are also a source of information on rock fracturing and permeability. Sudden losses indicating major fractures are perhaps the most obvious. Equally valuable information can be obtained from indications of recovery of lost flush returns with advancing of casing or irregular fluctuations of returns during drilling in variable strata. Because flush returns are an integral part of efficient drill operation this information is often well recorded by drilling operators. It could, however, be more widely used in interpretation of site conditions.

Water quality

Water quality data is frequently difficult if not impossible to obtain reliably from boreholes and drillholes where it is common practice to add water during hole advancing. In these conditions elaborate pumping or prolonged bailing of the hole may be necessary to ensure that uncontaminated samples of groundwater are obtained. However, where groundwater is relatively shallow a simple pit provides an extremely simple and cheap means of providing a simple assessment of water quality. In general terms the electrical conductivity measured in micromhos per cm is equivalent to about two thirds of the total dissolved solids expressed as milligrammes per litre.

The value of the conductivity gives a preliminary indication of the origin of the groundwater and can be compared with values measured from local surface water sources, it also gives an indication as to whether or not potential problems such as corrosion or encrustation merit more detailed study. Other parameters such as pH can be readily measured with simple equipment and analysis for specific features such as chloride content, hardness or dissolved gases can be performed on site with equipment costing no more than £20 to £30. Frequently these parameters can only be reliably measured on site, unless elaborate precautions are taken for stabilising samples during transport and storage prior to laboratory analysis.

Regular monitoring of the electrical conductivity of flush returns in bailer water during drilling can also provide valuable information in the level of groundwater entry into a borehole.

Examples

Road construction

At the site of a proposed road diversion 8 to 10 m of glacial till covers Old Red Sandstone bedrock on the sides of steeply incised overdeepened valley infilled with dense fluvioglacial sands and gravels (Fig. 1). Water levels encountered during drilling of two of the boreholes are summarised in Table 1.

A more complete record would include levels at either side of meal and other breaks but even the data given in Table 1 indicates a locally perched watertable in the glacial till on the valley sides with a deep water level in the underlying rock. This water level in the bedrock is artesian in the valley bottom with a slightly lower water level close to soil in the alluvial and fluvioglacial deposits.

Opencast development

At a proposed opencast site reliable estimates of volumes of groundwater which could enter the pit during working were required in order to relate discharge volumes to the capacity of nearby culverts. The site is cut by a major fault and contains two small areas of abandoned underground workings. Both the fault and the workings could be potential sources of large volumes of groundwater in otherwise a generally low permeability rock mass.

The site was investigated by boreholes, primarily for coal reserve calculations, at about 50 m intervals and the main source of water level data was the morning and evening levels recorded on the driller's logs. This data showed a general groundwater level close to the ground surface falling to 30 to 40 m below the surface in the areas of abandoned workings.

This pattern was interpreted as indicating drainage through the more fractured rocks immediately surrounding the workings to a level below the proposed opencast pit. The line of the fault was identified from stratigraphic interpretation. Water levels in boreholes on either side of this line also indicated a water table close to the ground surface. It was therefore concluded that the fault zone was not an area of increased rock mass permeabiltity.

While piezometric data might have been preferred it was possible, on the basis of water levels systematically recorded during drilling, to conclude that major groundwater flows into the excavation close to either the fault zone or the abandoned workings were unlikely.

Site dewatering

At a major coastal construction site about 50 m of dense permeable sands overlie impermeable deposits. The groundwater level is slightly above sea level, 3 to 4 m below the ground surface. Foundation excavations may be up to 15 m deep and could be open for 2 to 3 years. Detailed geotechnical investigations including four pumping tests, have been carried out on site. In connection with the dewatering the question of groundwater quality is important both in the evaluation of the possible extent of saline intrusion into the aquifer and in establishing the extent of corrosion or encrustation of well screens and pipework that might take place. These questions cannot be answered as no reliable water representative of the aquifer was collected.

As a result of this omission the cost of individual dewatering wells has undoubtedly been increased and no baseline data is available for rigourous evaluation of possible changes in water quality.

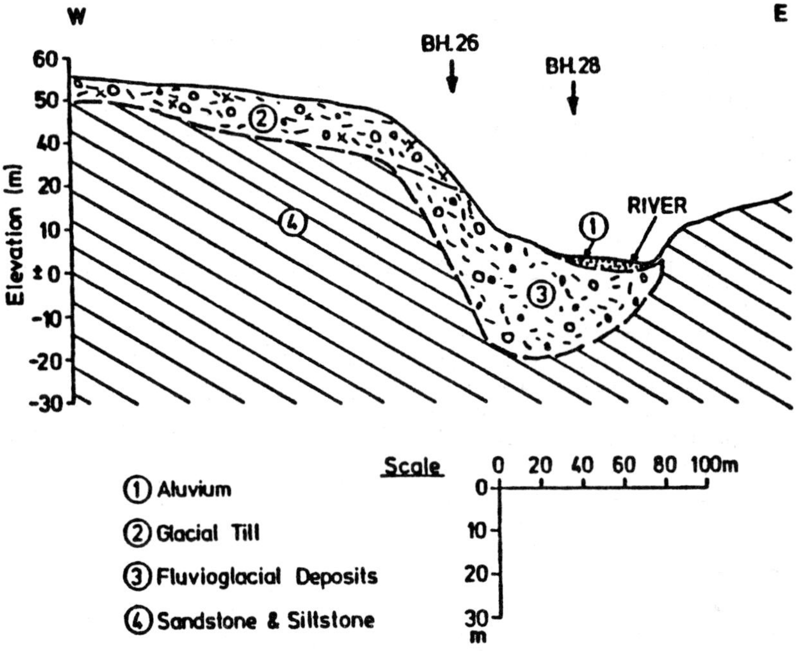

FIG. 1. Summary geological cross-section.

TABLE 1. *Summary of water levels encountered in boreholes.*

	Date	Time	Depth of hole (m)	Depth of casing (m)	Depth of water (m)
B.H. 28	9.12	18.00	8.0	6.5	6.5
(rock head 15.2 m)	10.12	08.30	8.0	6.5	0.6
		17.30	9.7	9.7	0.8
	11.12	09.00	9.7	9.7	1.1
		14.00	15.00	12.7	1.6
	12.12	09.00	15.00	12.7	artesian
		16.00	17.9	12.7	artesian
B.H. 26	14.1	17.00	1.5	open hole	0.5
(rock head 12 m)	15.1	08.30	1.5	open hole	0.5
		17.30	6.7	open hole	dry
	16.1	08.30	6.7	open hole	dry
		17.30	14.6	6.7	13.8
	17.1	09.30	14.6	6.7	13.6

Conclusions

Questions concerning site groundwater conditions can often be reliably answered from systematic collection of groundwater data available in the course of routine site investigation. Frequently such data is not systematically recorded resulting in misinterpretation or poor understanding of the local hydrogeology. This can in turn result in either expensive precautionary designs or perhaps difficulties during construction giving rise to later claims. No matter how small a site investigation may be, every effort must be made to ensure systematic collection of groundwater levels encountered during drilling accompanied by at least the measurement of simple water quality parameters such as electrical conductivity.

References

B.S. 5930, 1981. *Code of Practice for Site Investigations.* British Standards Institution, London.

KNILL, J.L., 1971. Assessment of reservoir feasibility. *Quarterly Journal of Engineering Geology,* **4,** 355-373.

J.S. SUTTON & C.D. ELDRED, Sir Alexander Gibb & Partners, Earley House, London Road, Earley, Reading RG6 1BL, UK.

Dynamic sampling of groundwater for dispersion and corrosion investigations

D.C. Watkins

ABSTRACT: Principles of sampling groundwater for the determination of its *in situ* inorganic and organic composition are described in the context of purpose and accuracy required. Two examples of techniques developed and used by Hydraulics Research are presented. Sampling of groundwater in a natural gradient regime containing artificial contamination by a tracer species is described. Special pneumatic piezometers were installed and employed to produce micro samples. Methods of installing, commissioning and monitoring these for extended periods are explained. Dynamic sampling of groundwater in a laminated glacial lake formation is also illustrated. Data were needed to define the corrosion and incrustation potential of the native groundwater on, for example, well screens. A method was devised which enabled a rapid assessment of relevant parameters using hand-portable equipment. Dynamic measurement of EC, pH, Eh and O_2 were accomplished. Samples were also taken for off-site chemical and bacteriological analyses.

Micro sampling as part of a dispersion experiment

Background

An experiment was to be conducted to determine the dispersion coefficients of a low-permeability sand aquifer. Tracer was to be introduced into the aquifer and regular sampling of the groundwater at various locations down-gradient of the injection point carried out to enable the tracer migration to be monitored.

The test site encompassed an area of 0.12 ha. The stratigraphic sequence below the site is shown in Table 1 and the flow characteristics of the aquifer are given in Table 2.

TABLE 1. *Stratigraphy of test area.*

Depth (m)	Strata
0 – 2.25	Semi-permeable sandy CLAY aquitard
2.25 – 4.0	Low permeability SAND aquifer (semi-confined)
>4.0	Impermeable indurated CLAY aquiclude

TABLE 2. *Flow characteristics of test aquifer.*

	Permeability		Mean hydraulic gradient	mean pore velocity
	m/s	m/day		m/day
max	1×10^{-4}	10	0.01	0.10
min	2×10^{-5}	2	0.01	0.02

These low pore water velocities inhibit the volume of water that may be withdrawn for sampling purposes. If excessive water was drawn from the aquifer the abstraction would cause an interference to the natural flow regime which would detract from the authenticity of the experiment. This effect is emphasised by the fact that the aquifer was in a variably cemented condition and flow occurred mainly through discrete horizontal bands. Removal of water from a thin permeable band would exaggerate the horizontal effect due to abstraction in comparison to that for a homogeneous aquifer.

Sample-point piezometers

The method used to obtain the groundwater samples entailed the installation of Cambridge Drive-in Tip piezometers within the aquifer. These piezometers were cone tipped with a maximum diameter of 33 mm. They contained a porous plastic element 300 mm in length and 20 mm diameter. A stud nut and olive connector was fitted to the piezometer to accommodate 3.8 mm I.D. pneumatic piezometer tubing as an alternative to a rigid standpipe.

Method of installation

A 36 mm diameter bore was made to 3.15 m depth using a Marlow Flow-Through probe. The probe rods were then removed retrieving a sample of the aquifer in the process. The pneumatic tubing attached to the piezometer was threaded through a 3.5 m long 25 mm I.D. rigid tube which located over the piezometer top as a mandrel. The end of the piezometer tubing was then clamped at the top of the tube and the assembly

From CRIPPS, J.C., BELL, F.G. & CULSHAW, M.G., (eds), 1986, *Groundwater in Engineering Geology*, Geological Society Engineering Geology Special Publication No. 3, pp. 251-254

pushed into the pre-bored hole by hand. The piezometer tubing was then released and the mandrel removed leaving the piezometer tip in place within the aquifer and the piezometer tubing leading to the surface.

A small quantity of washed sand was placed into the annulus above the piezometer tip and allowed to settle through the water column. This preceded bentonite pellets which sealed the bore to the surface as shown in Fig. 1.

Water was then drawn through the sample point in the manner described below to expel any air trapped in the system.

Method of sampling

Groundwater was drawn from the piezometer tip and through the piezometer tubing to the surface by use of a hand-operated vacuum pump. The sample was collected in a vacuum vessel as shown in Fig. 1. To ensure that the sample collected was representative of the mobile groundwater at that location it was necessary to remove and discard the possibly static water occupying the sample point system (flushing). As the aquifer is semi-confined, use of small bore tubing had a distinct advantage over a standpipe because of the reduction in the volume of static water that required flushing before a dynamic sample could be obtained direct from the aquifer around the piezometer tip. Flushing of the system also had the effect of flushing the sampling equipment at the surface in preparation for receiving the real sample and avoiding contamination from previous samples.

The volume of water required for the flush was calculated to be 110 ml. The minimum volume of water sample required for an accurate determination of tracer concentration was 50 ml which means a total abstraction per sample of 160 ml and this volume was considered negligible in terms of the experiment. However, care was still taken not to sample excessively.

Results

A total of 30 sample points were installed by the method described above. Only one installation proved unsuccessful and this is considered to be due to a gravel pocket within the aquitard encountered by the pre-bored hole and which slumped back slightly preventing the piezometer tip from being pushed into position. The piezometer tip was recovered and successfully reinstated in an offset position.

It was also possible to dip the piezometers to record the piezometric level in each. This was accomplished by using a specially designed electrical contact probe which fitted inside the 3.8 mm pneumatic piezometer tubing.

The experiment ran for 18 months and more than 1000 groundwater samples were taken.

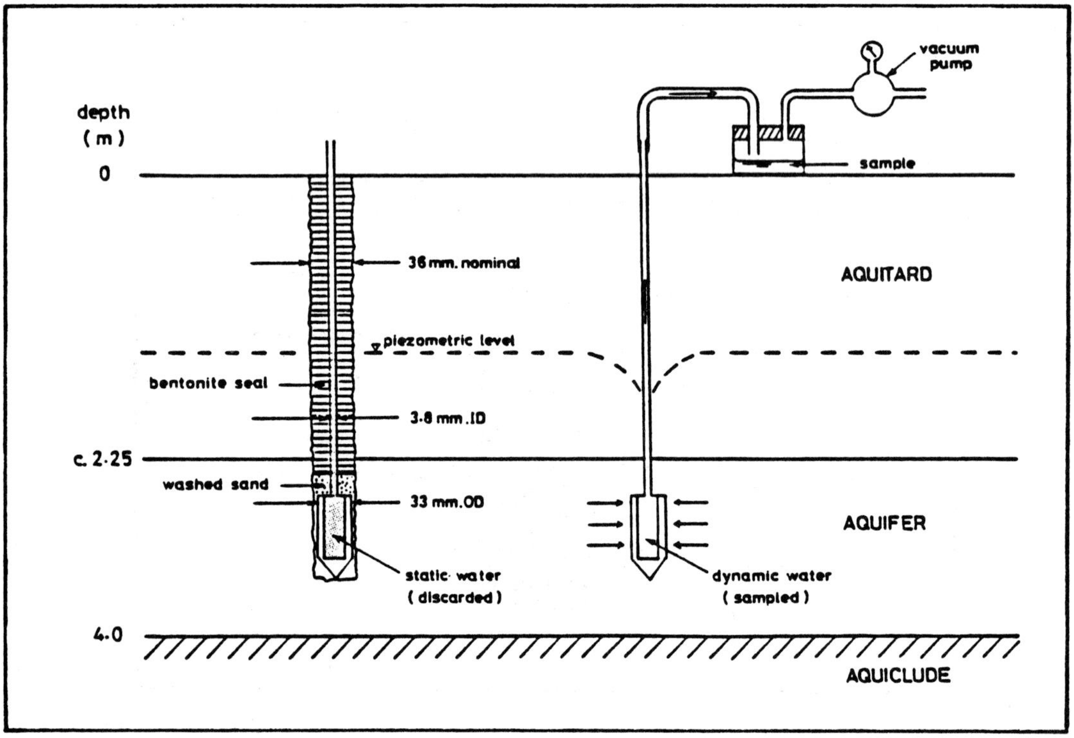

FIG. 1. Diagrammatic section through sample point piezometers.

Conclusions

The sampling method described above has been used successfully to produce plots of tracer breakthrough and tracer peaks during a dispersion experiment. Samples have been obtained which contain up to 20,000 times the background level of tracer without contamination to adjacent samples. This verified that the technique offered discrete and accurate sampling whilst causing minimum disturbance to the natural groundwater regime.

Dynamic sampling to assess corrosion/incrustation potential of groundwater

Background

A large dewatering exercise to reduce pore pressures in a formation of laminated silts and clays (lake deposits) was to be studied. Water quality data was needed to assess the effect of long term pumping on well screens. It was invisaged that a full scale pumping test would eventually be carried out as part of a permeability testing exercise and water quality tests would be made on discharges from these. However, a preliminary water quality test was considered applicable prior to the main site investigation.

A borehole from a previous site investigation in which the laminated lake deposits were screened was used to provide the groundwater for the test. The area under investigation was located in a tidal estuary with ground level at 0 m OD and was therefore inundated by the tide much of the time. At this stage in the investigation the test did not warrant the construction of a platform above high tide level and so a rapid test procedure using portable equipment during low tide was required.

Method

Certain parameters have to be measured in a dynamic condition in order to obtain values representative of the *in situ* state. The equipment used for dynamic sampling is listed in Table 3.

TABLE 3. *Dynamic determinations and measuring equipment.*

Parameter	Measuring equipment
Electrical conductivity, EC	pHOX 52
Hydrogen ion activity, pH	Philips PW9411
Redox potential, Eh	Philips PW9411
Dissolved oxygen, O_2 (aqueous)	pHOX 62T
Temperature, T	pHOX 62T

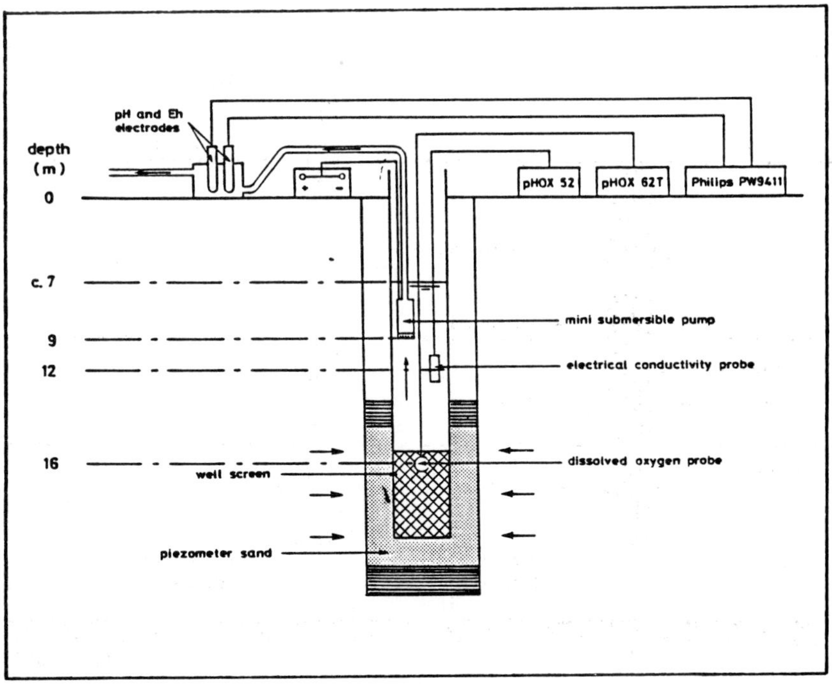

FIG. 2. Diagrammatic section through borehole.

The equipment was set up in the borehole as shown in Fig 2. The pH and Eh electrodes were set in a flow through cell at the ground surface. The discharge from the well was produced by a pump powered by a 12 volt car battery. The pump used was a mini submersible type which was 40 mm diameter, 150 mm length, lightweight, cheap and manufactured as a general purpose pump for small boats. The pump proved ideal for this application.

The EC probe was positioned 1 m below the level of the pump. The O_2 probe was positioned below this and opposite the well screen in order to intercept the incoming groundwater at the earliest possible point to obtain the most relevant values. The dissolved oxygen meter also incorporated a temperature sensor so that the temperature of the water could be ascertained before being drawn through the pump.

The borehole used had been standing sealed for two years prior to the test. It was therefore plumbed to ensure that it had not silted up. Once this was confirmed it was pumped out using the mini-pump. A steady drawdown of 7 m was achieved after 30 minutes pumping and after 90 minutes it was calculated that the whole of the original water column had been removed. The borehole was then allowed to recover and the test was carried out during the next low tide period.

During the test three stages of discharge were produced as shown in Fig 3. The water level in the well was maintained above the pump inlet to ensure that air remained excluded from the flow through cell. Stage 1 pumped from 0 to 17 minutes from the start of the test. The well was then allowed to recover partially from 17 to 38 minutes and then Stage 2 pumped from 38 to 50 minutes. Pumping then recommenced between 65 and 70 minutes from the start of the test in order to collect water samples for off-site analyses. Eh and pH measurements were made during Stages 1 and 2. Measurements of EC, O_2 and T were made throughout the test as the probes were below the pump level and therefore continuously in the water flow.

Three bulk samples were collected at the end of the test (Stage 3). Sample 1 was collected in a 2 litre bottle for determination of dissolved constituents Na, K, Al, total Fe, Mn, Cu, Cl, F, SO_4, NO_3 and SiO_2. This sample was acidified soon after collection to avoid any deposition of metals on the bottle surface. Sample 2 was collected in a 1 litre bottle for determination of total hardness and total alkalinity. This sample was obviously not acidified. Sample 3 was collected in two 2 litre bottles for microbiological analysis. these bottles had been sterilized with chlorine prior to the test. All sample bottles and their tops were washed thoroughly in the discharge before sampling and were filled to the brim with the sample to exclude air.

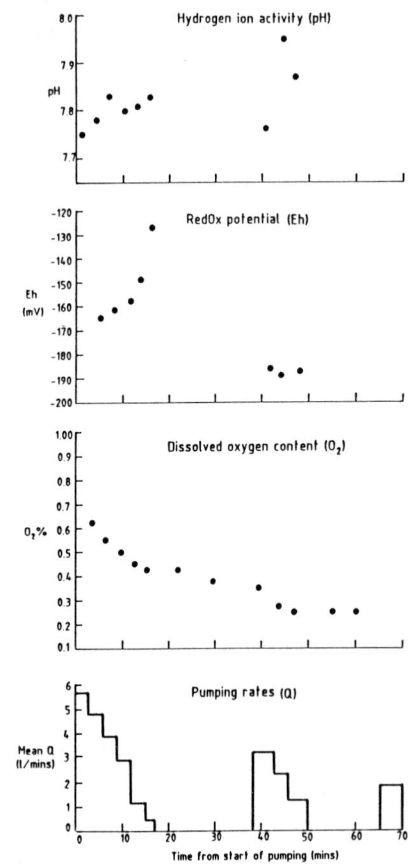

FIG. 3. Results of dynamic sampling

Results

The pH, Eh and O_2 measurements are shown in Fig 3. EC and T have been omitted because they remained constant at 1.2 mS/cm and 10°C throughout the test. Dissolved O_2 content is shown as % saturation which was later converted to μg/ml making the appropriate corrections for temperature and salinity.

Conclusions

The results of the dynamic sampling exercise were used in conjunction with the chemical and bacterialogical analyses of the bulk water samples to assess *in situ* conditions of water quality and thus to predict the corrosion/incrustation potential of the groundwater with regard to long term pumping.

It must be stressed that this was a small scale preliminary test. Any prolonged pumping as part of a dewatering exercise would be likely to draw water from a wide area and may include leakage from other facies present. Water quality measurements during a full-scale pumping test would be necessary in order to achieve truly representative results.

D.C. WATKINS, Groundwater Section, Hydraulics Research Ltd, Wallingford, Oxon OX10 8BA, UK.

Discussion Session 6

A. CAMPBELL (Ove Arup & Pts) asked Dr SPEARS about the effects of drilling techniques on contamination of samples for mineralogical and pore water studies.

D.A. SPEARS (University of Sheffield) stressed the need to take great care with drilling and sampling. In the case of the research described in the paper, the sampling was carried out by the Water Research Centre who are well experienced with work of this type.

L. CLARK (Water Research Centre) added that 100 mm diameter percussion core samples are taken using two 100 mm i.d. open drive sampler barrels (U100 barrels) in sequence so that driving can continue beyond the length of the first barrel. Hence samples are taken from at least 20 cm below the previous base of the hole. The core is retained in its U100 liner, sealed with endcaps and either frozen or used immediately. In consolidated formations rotary drilling with air flush is employed, but even using mylar liner, contamination can occur. However, consistent results have been obtained by splitting the core and sampling only the central part.

G. WEST (Transport and Road Research Laboratory) asked Dr SPEARS why samples are frozen.

D.A. SPEARS stated that freezing inhibits microbiological decomposition and reduction in samples. Ideally pore waters should be extracted on site and acidified to prevent chemical changes. However, this requires a mobile laboratory with a high speed centrifuge and even then core production may exceed the rate at which it can be handled in the laboratory. The use of some method of core preservation may therefore prove inevitable.

J. PONTIN (Hydraulic Research Centre) in answer to a question regarding the peaks in iodine traces which were presented in his paper, explained that they are a consequence of high rainfall events and the movement of water at different velocities through a system of high and low permeability zones. Sampling from piezometers illustrates the coalescing and divergent nature of the flow and smearing of the contamination during longitudinal dispersion.

From CRIPPS, J.C., BELL, F.G. & CULSHAW, M.G., (eds), 1986, *Groundwater in Engineering Geology,* Geological Society Engineering Geology Special Publication No. 3, p. 255

SESSION 7

INVESTIGATIONS FOR GROUNDWATER QUANTITY AND FLOW

Instrumentation in groundwater

J.A. Charles

ABSTRACT: The problems posed by the flow of groundwater are many and diverse. Instrumentation can only be usefully applied when the nature of the problems that may be encountered in a particular situation are appreciated. Instrumentation may be employed to monitor the rate of flow of groundwater, to measure the seepage velocity, to measure permeability and hydraulic gradient, to detect flow and leakage and to monitor the effect of flow (e.g. settlement). Instrumentation should be reliable and great care should be taken during installation and operation as incorrect measurements may be much worse than no measurements.

Introduction

At the 15th International Congress on Large Dams, held in Lausanne in June 1985, there was a session on 'Dam and Foundation Monitoring'. The General Reporter of this session, G. Post, remarked that "It must be accepted that the information from instrumentation is limited ... one only finds what one is looking for". Instrumentation must be related to the problems of the particular situation under investigation and unless we have a clear idea of the nature of these problems it is most unlikely that anything useful will be achieved by instrumentation.

An illustration of the varied nature of the problems that can be caused by the flow of water is given by the history of an earth dam with a central clay core.

(i) During the first impounding of the reservoir, the fill upstream of the core will become saturated as the reservoir water level rises. If the fill was not well compacted during construction, it may suffer a reduction in volume (collapse compression) on saturation causing an undesirable settlement of the upstream slope.

(ii) If subsequently the reservoir water level is drawn down rapidly, the upstream slope could become unstable. The stability of the upstream slope against shear failure during drawdown will depend partly on the permeability of the fill and the magnitude of excess pore pressures remaining in the fill as the reservoir empties.

(iii) With the reservoir full, leakage of water under the embankment could cause erosion of the foundations.

(iv) With the full reservoir head acting at the core/upstream fill interface hydraulic fracture of the core could occur and erosion of the embankment commence.

(v) If the spillway capacity is inadequate a large storm could cause the reservoir to overtop the embankment and erode the downstream slope.

(vi) In certain circumstances chemical reactions could occur between the reservoir and the embankment fill and cause pollution problems.

In geotechnical engineering there is often a concentration on problems concerned with the stress-strain properties of soils. The flow of water is important in this type of problem because of its influence on effective stress. Nevertheless of the 6 types of problems listed above only item (ii) is of this character. Clearly different types of instrumentation would be needed to deal with these diverse problems. In considering suitable instrumentation one important distinction needs to be kept in mind. De Mello (1977) distinguished between problems that depend on average values and those which, like the failure of a chain in tension, depend on extreme values. Leakage and internal erosion are cases where extreme rather than average values are the controlling factors.

Basic relationships for groundwater flow

The slow flow of a fluid through a porous medium is governed by Darcy's Law:

$$q = kiA \qquad (1)$$

where q is the volume of water seeping through an area A per unit of time (rate of flow), i is the loss of hydraulic head per unit length in the direction of flow (hydraulic gradient), k is the coefficient of permeability or hydraulic conductivity.

From CRIPPS, J.C., BELL, F.G. & CULSHAW, M.G., (eds), 1986, *Groundwater in Engineering Geology,* Geological Society Engineering Geology Special Publication No. 3, pp. 259-262

The apparent velocity of flow (v) is the average rate of flow across unit area of soil, that is:

$$v = \frac{q}{A} \quad (2)$$

The actual seepage velocity (v_s) is related to v by:

$$v_s = \frac{v}{n} \quad (3)$$

where n is the porosity of the ground.

These simple relationships provide a means of categorising the different types of measurement that can be made with suitable instrumentation:

(i) measurements of rate of flow (q),
(ii) measurement of seepage velocity (v_s)
(iii) measurement of permeability (k) and hydraulic gradient (i),
(iv) detection of flow or leakage,
(v) measurement of effects of flow e.g. settlement.

Measurement of rate of flow

This is often feasible when the groundwater level has been either lowered or raised by human activities. When the water level is being lowered by pumping the discharge from the pumps can readily be measured. Where the water level has been raised by a dam, suitably designed drainage measures within the dam can make it possible to control the seepage and leakage through the dam, identify the location and source of leakage, and monitor the rate of flow. Siddle (1985) described the construction of a drainage gallery, a reinforced concrete box culvert was constructed, in which a steel vee notch weir and autographic recorder were installed to provide a continuous record of flows.

Measurement of seepage velocity

Flow velocity can be measured using various types of tracer. Siddle *et al.* (1984) presented a comprehensive summary of the use of tracer techniques; natural isotopes and artifical tracers such as coloured dye-stuffs, particulate tracers, stable chemical compounds and radioactive isotopes were described. This type of measurement may be of particular importance in problems involving erosion as flow velocity may itself be of major significance quite apart from the possibility of using it to calculate a volumetric rate of flow. Tracer techniques are often used to determine a direction of flow and source of leakage. It should be remembered that in ground of low permeability, the time required for tracers to move even short distances will be very long.

Measurement of permeability and hydraulic gradient

The great range of permeabilities found in soils and rocks is of considerable significance when instrumentation for the measurement of permeability is considered. In soils permeability might vary from one m/s in a clean gravel to 10^{-10} m/s in an unfissured clay. As a soil profile can include soils with widely different permeability in close proximity to each other, it is important to define whether it is the local permeability of a particular soil that needs to be measured or whether the average bulk permeability over a wide area and large depth of ground is required. It has been recognised that testing soft alluvial soils in small oedometers to determine their consolidation properties can lead to large over-estimates of the time required for the soils to consolidate in the field. This is probably due to the presence of drainage paths in the field which are not present in the small clayey samples tested in the small oedometer. The problem can be overcome either by testing larger and more representative samples in the laboratory or by carrying out field permeability tests in piezometers which measure the permeability of a much larger volume of soil. Pumping tests, in which the groundwater level is drawn down some distance from the well, go a stage further and give information about the bulk permeability of a very large volume of soil around the well. The Code of Practice for Site Investigation (BS 5930: 1981) gives details of borehole permeability tests, packer or Lugeon tests and large scale pumping tests.

In his general report on the session on 'Foundation Treatment for Control of Seepage' at the 25th International Congress on Large Dams, A. Bozovik remarked that "The review of the submitted papers demonstrates that the water adsorption or Lugeon test remains the main engineering tool in assessing the permeability of dam foundations and in evaluating the achieved efficiency of the grouting treatment". Criticisms of this test are well known and improved procedures have been proposed (Pearson & Money 1977). Walthall & Campbell (1985) considered that the most reliable results are obtained using outflow tests and recommended tests at very low excess pressures (around 1 m head). They quoted the modelling work of Bliss & Rushton (1984) as supporting the view that the controlling influences on the behaviour of packer tests extend for only 10 m into the surrounding aquifer.

The installation of a piezometer makes it possible to monitor pore water pressure and, with certain types of piezometer, also to measure permeability and hydraulic fracture pressure. A piezometer consists of a filter and a pressure measuring device. Porous tips which function as filters come in different shapes and sizes and the ceramic filters can be either high or low air entry. The tips may be pushed into direct contact with the ground or installed in a sand cell.

Piezometers are often classified as hydraulic, pneumatic or electrical depending on the nature of the pressure measuring device. This can be misleading as hydraulic piezometers are now frequently used with electrical transducers. The significant distinction concerns the location of the transducer; whether it is attached to the tip and permanently buried in the ground or whether the water pressure is transmitted from the tip to a remote and accessible location where the transducer is housed. The advantages of the latter system (the twin tube hydrualic piezometer) are obvious. When necessary deaired water can be circulated around the tubing and the piezometer tip can be deaired. Permeability tests can be carried out. The measuring device can be recalibrated or replaced if it malfunctions. However, there can be problems in cold climates where the water in the connecting tubes may freeze and where very rapid changes in pore pressure have to be monitored (i.e. under dynamic loading conditions) the response time of the hydraulic piezometer although small may not be short enough and an electrical piezometer may be required.

Twin tube hydraulic piezometers are not suitable for installation in deep boreholes because of the difference in elevation between the tip and the measuring point at ground level. Pneumatic or electrical piezometers can be used but very often the simple Casagrande standpipe piezometer, with the tip in a sand cell sealed into the borehole with a plug of bentonite pellets, is the best solution. This type of instrumentation makes it possible to carry out permeability tests. A major limitation of the Casagrande standpipe piezometer for pore pressure measurement is the slow response time in fine grained soils (over a week in a low permeability soil). Electrical transducers can be installed in standpipes to facilitate automatic logging of water pressures. The use of an inflatable packer above the transducer eliminates the long response time and there is still the facility for removing the transducer and packer and carrying out permeability tests in the standpipe when required.

It is difficult to install more than two standpipe piezometers in one borehole and be confident that the tips are reliably sealed into the borehole. A system has been introduced recently which it is claimed overcomes these difficulties (MP system of Westbay Instruments). It consists of casing, couplings and packers permanently installed in a borehole. Valved measurement port couplings are located wherever pressure measurements are required. Permeability tests can also be carried out. A probe is lowered into the casing and locates the measurement ports.

The measurement of permeability by tests in boreholes and in hydraulic and standpipe piezometers involves the application of a hydraulic pressure different from that in the ground. The flow caused by this differential pressure is measured and the permeability is calculated. A crude test can be carried out in the bottom of a borehole but preferably the test is performed in a piezometer that has been installed and sealed into the borehole. When a positive excess pressure is applied the test is called a falling head or inflow test, or where the pressure is decreased it is termed a rising head or outflow test. The pressure may be kept constant (constant head test) or allowed to come back into equilibrium with the surrounding ground (variable head test).

It has been mentioned that before installing instrumentation for the measurement of permeability it is essential to define whether a local permeability or an average bulk permeability is required. When a piezometer is used to monitor pore water pressure it will be essential that it measures a very localised pressure. For example if a one metre deep clay layer has free draining soils above and below it, a standpipe piezometer installed in a 2 m long sand cell will not give useful measurements of pore pressure within the clay layer. The 'piezometer' is in reality acting as a drain.

Detection of flow or leakage

Concentrated leakage through or under an embankment built to retain water may cause erosion that imperils the safety of the embankment. In such situations detection of leakage (i.e. recognition and location of it) is more important than precise measurement of the rate of flow. If seepage flows are collected in a drainage system, a sudden increase in flow unrelated to reservoir level or rainfall could indicate the onset of such leakage. An increase in pore pressures recorded by piezometers in the fill downstream of a central core could also give an indication of this type of problem if the piezometers happened to be very close to the leakage path. However, many old earth dams have no proper drainage system and no piezometers. Visual inspection can locate damp patches and tracers may help in the detection of the source of the flow.

Infrared thermography is currently being examined by the Building Research Establishment to assess its usefulness in locating wet areas on the downstream slope of earth dams and in identifying the source of the flow of water. The method depends on wet areas being at a different temperature from the surrounding ground which has not been affected by leakage. With areas of particular interest identified, probes can be inserted into the fill to measure subsurface temperatures. If the temperature of the wet area can be related to reservoir temperature, this provides evidence about the source of the leakage. A number of old earth dams in the Pennines have been surveyed using this method. The technique is based on the fact that all objects above absolute zero emit infrared energy in the form of electromagnetic radiation. The infrared camera detects and measures the invisible thermal radiation by converting it into an electrical signal for display on the screen of a cathode ray tube to form a visible image known as a thermogram. The thermal image may be recorded using conventional photographic equipment or it can be recorded by an analogue video recorder for subsequent thermal analysis.

Measurement of effects of flow

Field measurements have shown that a rising ground water level can cause settlement due to collapse compression in an opencast mining backfill (Charles *et al.* 1984). Instrumentation was required that would make it possible to measure settlement at different depths within opencast mining backfill. This was essential to demonstrate that surface settlement was due to compression occurring at the elevation of the rising water level. The BRE magnet extensometer system (Marsland & Quarterman 1974) was successfully used in this situation. There are two basic components; a ring magnet which acts as a marker and a reed switch which is used as a sensor. Each magnet is anchored to the sides of the borehole by a strong spring. The reed switch is lowered down a central rigid access tube which is isolated from the ground by an outer helically reinforced flexible tube. The extensometer is lowered down a borehole on a nylon rope attached to a heavy base weight. Each magnet spring is held back by a nylon cord which can be cut by a pneumatic cutter when it is at the correct depth. The base weight contains a porous stone attached to the end of the central access tube so that it can be used to monitor the water level in the backfill as well as for settlement measurements.

Concluding remarks

Post (1985) commented that one only finds what one is looking for. Equally it can be added that one only gets what one pays for. Not only must instrumentation be designed to measure those quantities and properties that are relevant to the problems of the particular situation under investigation, it must also be purchased in an appropriate manner. It is essential to use high quality equipment and both installation and measurements must be carefully supervised. Failure to do this can be serious as in many situations it would be preferable to have no measurements rather than have incorrect measurements.

ACKNOWLEDGEMENT: This paper is published by permission of the Director of the Building Research Establishment.

References

BLISS, J.C. & RUSHTON, K.R., 1984. The reliability of packer tests for estimating the hydraulic conductivity of aquifers. *Quarterly Journal of Engineering Geology,* **17,** 81-91.

BOZOVIK, A., 1985. General report on foundation treatment for control of seepage. *Proceedings of 15th International Congress on Large Dams, Lausanne,* **3,** pp 1467-1583.

BRITISH STANDARDS INSTITUTION, 1981. Code of Practice for Site Investigations (BS:5930), London.

CHARLES, J.A., HUGHES, D.B. & BURFORD, D., 1984. The effect of a rise of water table on the settlement of backfill at Horsley restored opencast coal mining site, 1973-1983. *Proceedings of 3rd International Conference on Ground Movements and Structures,* UWIST, Cardiff, 423-442.

MARSLAND, A. & QUARTERMAN, R., 1974. Further development of multipoint magnetic extensometers for use in highly compressible ground. *Geotechnique,* **24,** 429-433.

de MELLO, V.F.B., 1977. Seventeenth Rankine Lecture: Reflections on design decisions of practical significance to embankment dams. *Geotechnique* **27**, 279-355.

PEARSON, R. & MONEY, M.S., 1977. Improvements in the Lugeon or packer permeability test. *Quarterly Journal of Engineering Geology,* **10,** 221-239.

POST, G., 1985. General report on dams and foundation monitoring. *Proceedings of 15th international Congress on Large Dams, Lausanne,* **1,** 1623-1727.

SIDDLE, H.J., JONES, D.B. & WARREN, C.D., 1984. The use of tracer techniques to assess groundwater flows in site investigations. *Site Investigation Practice: Assessing BS:5930, 20th Regional Conference of the Engineering Group of the Geological Society,* University of Surrey, 419-431.

SIDDLE, H.J., 1986. Groundwater control by drainage gallery at Aberfan, South Wales. *Groundwater in Engineering Geology, 21st Annual Conference of the Engineering Group of the Geological Society,* University of Sheffield, this volume.

WALTHALL, S. & CAMPBELL, J.E., 1986. The measurement, interpretation, and use of permeability values with specific reference to fissured aquifers. *Groundwater in Engineering Geology, 21st Annual Conference of the Engineering Group of the Geological Society,* University of Sheffield, this volume.

J. A. CHARLES, Geotechnics Division, Building Research Establishment, Garston, Watford, UK.

Crown Copyright. Reproduced by permission of the controller, HMSO.

Rapid permeability testing by the pressurisation method

J.M.A. Pontin & M.A. French

ABSTRACT: Design and use of equipment developed to determine permeability and compressibility of saturated ground is described. The basis of the method is to produce compressible flow of groundwater by pressurising special boreholes with sealed screened zones.

The development of the principles and theory of the method from existing techniques is described. Analysis of data is illustrated and the advantages of the method discussed.

Introduction

As part of a major site investigation the hydraulic and geotechnical characteristics of a series of laminated clays and silts were to be examined. The site was in an inter-tidal area, and the restrictions on access, combined with a fluctuating response of pore pressures to the tide, demanded a rapid means of testing the permeability of those soils.

Conventional falling head tests were found to be inaccurate for the reasons described below and so a new test and method of analysis was required in order to achieve reliable results. This test procedure was used at the site in July 1982, November 1984 and March 1985.

FIG. 1. Test site with equipment set up.

From CRIPPS, J.C., BELL, F.G. & CULSHAW, M.G., (eds), 1986, *Groundwater in Engineering Geology*, Geological Society Engineering Geology Special Publication No. 3, pp. 263-272

Theory of permeability tests in pressurised boreholes

Incompressible soil theory

The conventional variable head test uses the theory, due to Hvorslev (1951) for an incompressible rigid soil. By conservation, the volume of water entering the soil in time Δt is equal to the change in volume of the water column in the borehole. This rate of flow at any time is assumed to be proportional to the product of the head differential and the soil permeability. These two equations allow q (see Appendix for notation) to be eliminated:

$$q\Delta t = \Delta V_w$$

$$q = FK(h_w - h_\infty) = (FK/\gamma_w)(P_w - P_\infty). \quad (1)$$

If the change in volume is expressed in terms of h_w then the differential equation can be integrated directly.

In a borehole open to the atmosphere this volume is that change in the water column produced during Δt:

$$\Delta V_w = \pi r_c^2 \Delta h_w = (\pi r^2_c/\gamma_w)\Delta p_w. \quad (2)$$

The change in head, or pressure, toward the equilibrium state at 'infinite' time can then be expressed by

$$H_w/H_o = (h_w - h_\infty)(h_o - h_\infty) = \exp - (t - t_o)/T \quad (3)$$

or

$$P_w/P_o = (P_w - p_\infty)/(P_o - p_\infty) = \exp - (t - t_o)/T$$

where T is the 'basic time lag' defined by

$$T = \pi r^2_c/FK$$

$$T = \gamma_w V/FK \quad (4)$$

for head and pressure respectively. We have introduced two factors, defined physically by

$$F = q/KH_w = \gamma_w q/KP_w \quad (5)$$

$$V = q\Delta t/\Delta p_w = q\Delta t/\gamma_w \Delta h_w \quad (6)$$

which are the form or shape factor F, the rate of flow per unit head difference into soil of unit permeability and the volume factor V, the volume of flow per unit change in pressure in the borehole. The latter is independent of permeability and is expressed by

$$V = \pi r_c^2/\gamma_w. \quad (7)$$

The form factor is related in a complex manner to the dimensions of the borehole and boundaries of the soil into which it is inserted. (see British Standard Code 5930: 1981, for example).

From Equation (3) the head ratio is reduced to about 37% of the initial value after a time equal to T from the start of the fall in head, t_o. The value of T is proportional to the square of the radius of the falling column and therefore the duration of the test is reduced considerably by smaller radii. For example a borehole with uniform soil flush with the bottom of the casing, diameter D, has a basic time lag approximately equal to

$$T = \pi D^2/4 \times 2.75DK = \pi D/11K. \quad (8)$$

For soil permeability 10^{-6}m s^{-1} and D equal to 200 mm, typical of the laminated clay situation, T is equal to about 16 hour, if the normal casing test is made. In 1 hour the head ratio would fall only to 94% of the initial value. This is obviously impractical in general and is the reason why the constant head test is normally used for soils of lower permeability.

From Equation (4) the diameter of a standpipe extension to the borehole that would produce a specified head ratio of 0.37 after 15 minutes (a 'rapid test') after the initial (Maximum) value is in fact about 25 mm with a permeability of 10^{-6}m s^{-1} in an isotropic soil.

For the laminated clay series the permeability was undoubtedly anisotropic with the horizontal value expected to be more than 10 times the vertical. When this factor is introduced the time lag equation is modified to:

$$T = \pi D/11K_m$$

where

$$K_m = \sqrt{K_v K_h} \quad (9)$$

and the standpipe diameter needed for a 'rapid' test reduced to less than 15 mm.

A new approach was therefore developed by recognising that the same effect can be produced by making ΔV_w very small in the basic Equation (2). This can be achieved if the volume entering the soil is derived by water expanding from a fixed volume, V_w, which is pressurised above the ambient groundwater pressure in the borehole.

The definition of volume compressibility for water,

$$C_w = (\Delta V_w/V_w)\Delta p_w \quad (10)$$

is substituted into the basic Equations (1) with the result that the basic time lag is now expressed by

$$T = C_w V_w \gamma_w/FK \quad (11)$$

if the expansion and contraction of the volume containing the pressurised water are assumed to be negligible. With the numerical values above and a borehole depth, d, of 20 m the 0.37 head ratio would be achieved after a time

$$T = \pi\, d\, D\, C_w \gamma_w / 11 K_m. \qquad (12)$$

with

$$C_w = 4.65 \times 10^{-10}\ \mathrm{m^2\ N^{-1}}$$

and

$$\gamma_w = 9920\ \mathrm{N\ m^{-3}}$$

(for sea water), this is evaluated to be 16.6 seconds. The advantage of using a pressurised system is apparent on the basis of this principle applied to an incompressible soil.

Compressible soil theory

In site investigation work the accepted method of *in situ* permeability testing of compressible strata is by the constant head method applied through hydraulic piezometers. The theory is due to Gibson (1963, 1966) but is limited to short filter sections in reasonably isotropic soils.

For cylindrical piezometers, or boreholes with high length to diameter ratio, in well-stratified soils, whose horizontal permeability exceeds the vertical by more than 5, another theory can be used. This is the finite-well theory developed quite recently in groundwater hydraulics, hitherto neglected in soil mechanics practice. The theory was developed in a series of papers by Papadopulos *et al.* (1967, 1967, 1973) in the US Geological Survey.

On the basis of this theory the head ratio varies after an instantaneous change in the water level of a finite well, radius r_w, according to

$$H_w/H_o = F_1(\alpha, \beta) \qquad (13)$$

where

$$\alpha = r_w^2 \gamma_w m_v l / r_c^2, \ \beta = K_h l t / r_c^2$$

and F_1 is an infinite integral which is tabulated by Papadopulos & Cooper (1967). The other symbols are defined in the Notation. The 'effective' or virtual radius of the well is not necessarily equal to the physical bored radius, r_w, or the radius of casing wherein the fall of water level occurs.

The solution was developed with the following assumptions:

(a) the flow from the borehole is radial and sensibly uniform from the filter length, l;

(b) the stratum is extensive (infinite) and uniform;

(c) the inertia of the water column is negligible.

In the groundwater and oil industries tests of this kind are called 'slug tests'. It is well known that the method is reliable for finding an estimate of the mean horizontal permeability in the vicinity of the borehole; actually this is the 'borehole transmissivity' equal to $K_h l$. The reliability of the estimate of storativity and vertical compressibility, m_v, is not so certain according to Papadopulos *et al.* (1973).

An extension of this test for the pressurised borehole condition was suggested very recently by Bredehoeft & Papadopulos (1980), to be used in testing strata of very low permeability at great depth. The motive was to reduce the duration of the test period.

With the assumptions above the formal solution for instantaneous pressurisation is now expressed by

$$H_w/H_o = F_2(\alpha, \beta) \qquad (14)$$

where now

$$\alpha = r_w^2\, m_v l / V_w C_w, \quad \beta = \pi K_h l t / V_w C_w \gamma_w$$

and the function F_2 is identical to F_1 with these parameters.

We can note that the pressurised solution is obtained from the open borehole form by making the substitution

$$\pi r_c^2 = V_w C_w \gamma_w \qquad (15)$$

which was developed above for the incompressible soil theory.

The permeability and compressibility are found from a 'type curve' comparison with measured data which is described in detail by Bredenhoeft & Papadopulos (1980).

However, in the testing of the laminated clays, the application of Equation (14) to test data produced unrealistic results for these coefficients. The reason was that the mathematical assumptions used to derive this equation were not satisfied physically in the test operation.

The most important departure appeared to be the failure to achieve a satisfactory approximation to instantaneous pressurisation, i.e. the period of rising pressure to the maximum value was too long. During this period water was being injected into the soil in relatively large quantity compared to the zero value assumed by the mathematical description of the process. To compensate for this effect we have developed a new approach for a well of finite diameter. This will now be described, and called the 'HR method'.

Theory of the HR method

During the injection period, t_o, the rate of discharge into the borehole and adjacent soil is kept constant. The head inside the borehole is then described by

$$H_w = (Q/4\pi K_h l)\, F(u_w, \alpha) \qquad (16)$$

where

$$u_w = r_w^2/4c_v t, \quad \alpha = r_w^2 m_v l/\, V_w C_s \gamma_w$$

and c_v is the coefficient of consolidation for horizontal radial flow from the well. Equation (16) is a 'pressurised' form of the solution presented in (14), with the substitution, Equation (15). In addition, account has been taken of the non-rigidity of the borehole system, so that C_w has been replaced by C_s, the 'system compressibility'. C_s can be calculated by considering the borehole design and material properties.

The residual head difference after injection has ended is expressed by superimposing an image well in the normal manner

$$H_w' = H_w(t) - H_w(t')$$

$$= (Q/4\,\pi K_h 1)\,[F(u_w, \alpha) - F(u_w', \alpha)] \qquad (17)$$

where

$$u_w' = r_w^2/4c_v t', \quad t' = t - t_o$$

This equation can be simplified in two stages to provide a very convenient form for data analysis. It has been established from tabulation that if

$$t > 250\ \text{'}r_c^{2}\text{'}/K_h l = t_c$$

then

$$F(u_w, \alpha) \approx W(u_w)$$

where $W(u_w)$ is the well (Theis) function for 'non-leaky' groundwater flow.

In the inequality we have used

$$\text{'}r_c^{2}\text{'} = V_w C_w \gamma_w/\pi$$

for the pressurised version. Also, if

$$u_w < 0.05 \text{ then}$$

$$W(u_w) \approx -\gamma - \ln u_w.$$

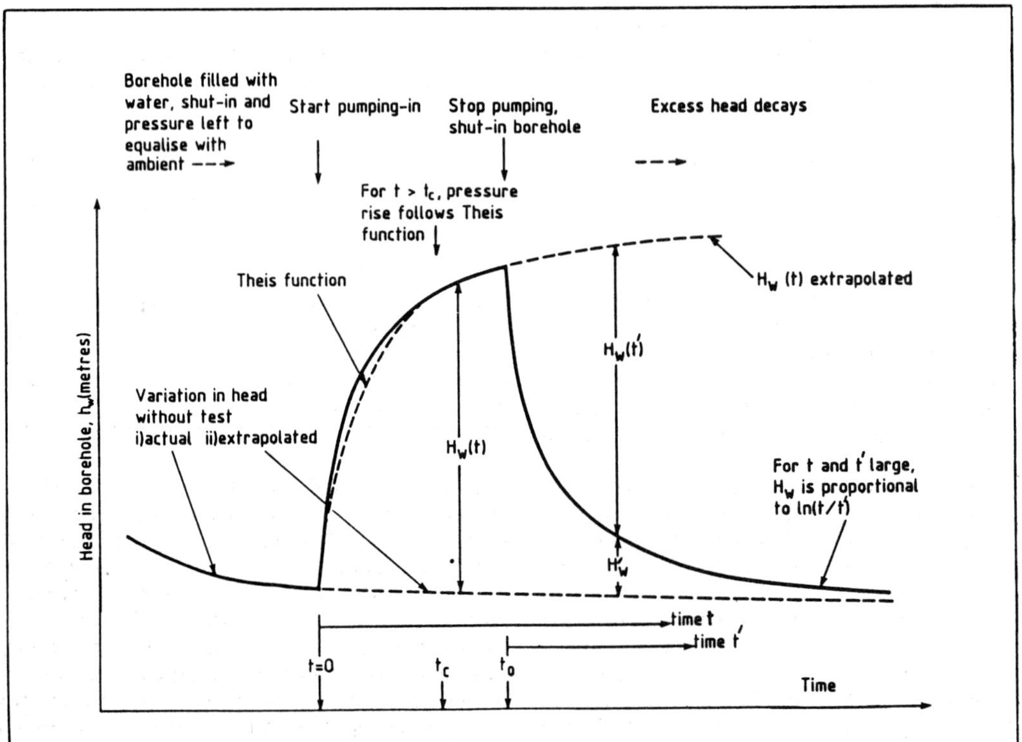

FIG. 2. Head variation in borehole.

When these approximations, valid if t and t' are sufficiently large, are substituted into Equation (17) a very simple equation is obtained, viz:

$$H_w' = (Q/4\pi K_h l) \ln (t/t') \tag{18}$$

Thus when the residual pressurised head is plotted against log (t/t') the data should ultimately fall on a straight line through the origin. The slope of this line provides the estimate of K_h.

The compressibility can also be found, not directly from this line, but in the following way from the pressurising period.

During pressurisation, for $t > t_c$,

$$H_w = (Q/4\pi K_h l)\ W(u_w) \tag{19}$$

At the end of the injection period $t = t_o$ the value of W can therefore be calculated and the value of the argument, u_w, found from tables of the function. The compressibility is then computed from

$$c_v = r_w^2 / 4u_{w,o}\ t_o$$

$$m_v = K_h/\gamma_w\ c_v$$

where the nominal well radius is used. If the soil permeability is altered by constructing the filter section the effective radius will be either more or less than r_w. A first approximation is to assume that the degree of alteration is small. The permeability value is not dependent on which radius is used.

Borehole design and construction

The possibility of performing pressurisation tests in normal borehole casing, examining the soil over a short, uncased length at the bottom of the hole, was considered but rejected for the following reasons:

(a) cleaning the bottom of a borehole is a notoriously difficult operation in clay strata.

(b) ordinary casing joints are liable to leak

(c) it could not be guaranteed that the laminated clay would not squeeze into the uncased section

(d) the permeability determined for the tested volume would be influenced by vertical flow from the hole and not horizontal flow alone.

For these reasons, special boreholes were designed incorporating well-screen and filter sand, and using careful drilling techniques. Details of dimensions are given in Table 1 and the design of the pressurisation borehole used in the November 1984 tests is illustrated in Fig. 3. To drill the hole, casing was advanced to the top of the strata to be tested. Boring was then continued to the bottom of the hole by the reverse circulation method using a 350 mm drill bit.

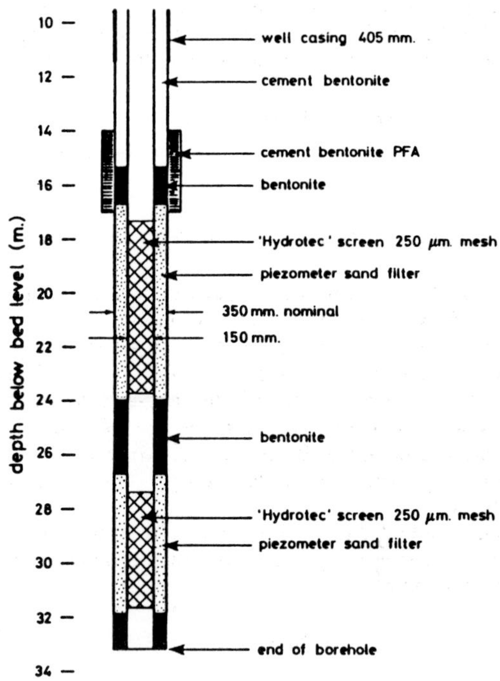

FIG. 3. Construction of test borehole.

TABLE 1. *Construction details of pressurisation boreholes.*

Borehole		1982	Nov 1984
Diameter of boring (mm)		200	350
Depth of boring (mm)		25.0	33.2
Diameter of screen (mm)		150	150
Upper filter	Depth to top (m)	14.0	16.7
	length (m)	7.25	7.2
Lower filter	Depth to top (m)	-	26.7
	length (m)	-	5.1

Well screen, 150 mm Hydrotec screen with a 250 µm fabric filter, was installed in the hole. As Fig. 3 illustrates, the 1984 well was designed for testing 2 different strata. The two screened sections were surrounded by piezometer sand filters which were placed by tremmie pipe and with sodium tripolyphosphate added to the flush water to disperse clay caked to the face of the exposed deposits. After completing tests on the whole length of the well, it was grouted up to cover the lower screened section and more tests were carried out on the upper screened length alone. In this way, it was possible to determine the permeability of both of the tested strata. During the 1982 tests, only one screened section was used.

The top of the well string, with a screw thread, was left standing about a metre above bed level so as to permit work to be carried out at low tide.

FIG. 4. Borehole sealed by 'top hat'.

Equipment

The equipment used is illustrated schematically in Figs 1,4,5,6 & 7.

The most important condition assumed in the analysis, which the equipment was designed to satisfy is that the flow rate into the well during the pressurisation stage should be constant. In addition it is clearly important to perform several tests on each section and so a means is required of achieving various flow rates each of which is, nevertheless, constant during a test.

As a check on the pressurisation method, provision was made for carrying out a conventional falling head test and a short-duration pumping test at the same time.

Electricity was provided at 110 Volts by a 2.5kVA Model 25P Knylor generator driven by a Petter AB1 4.5HP diesel engine. This was used to power the pump, a Mono GH Pump coupled to a Var Spe variable speed unit, the flowmeters and lighting for night work.

For the pressurisation stage, water was pumped from a supply (two 250 l capacity barrels) through a strainer and in-line filter, then via a 'circuit board' to the well, which was capped by a 'top-hat' and was therefore sealed.

The circuit-board was found to be a convenient way of mounting various pieces of equipment which are required for accurate control and measurement. These are:

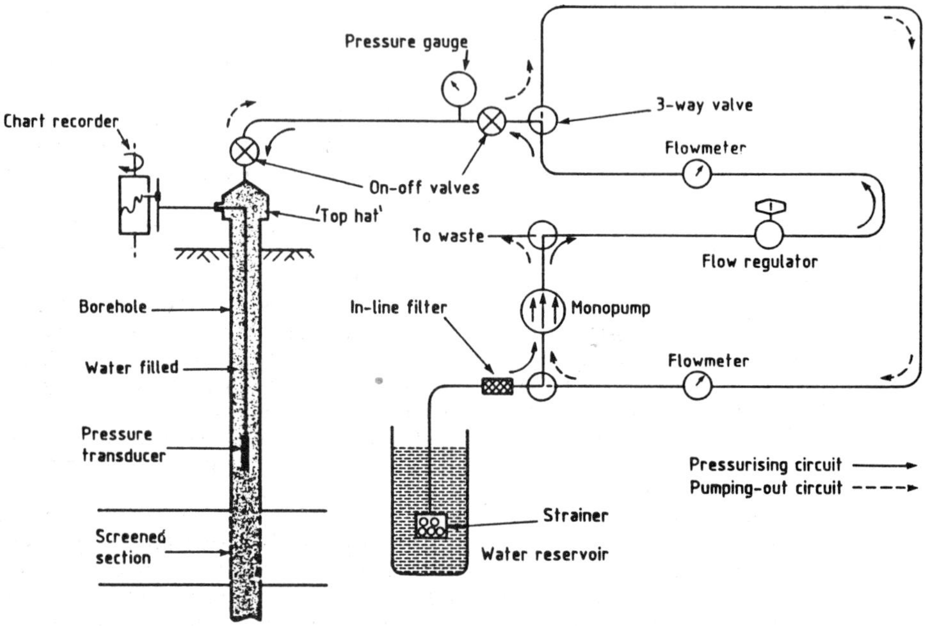

FIG. 5. Schematic diagram of equipment.

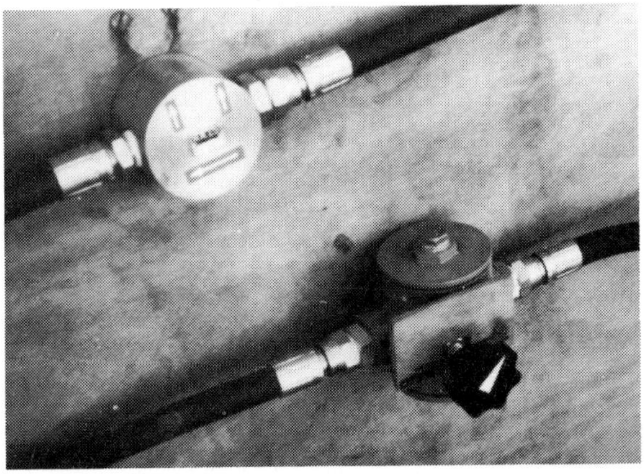

FIG. 6. Flow regulator and flow meter.

FIG. 7. Chart recorder.

(a) A Flostat Type V automatic flow regulator, which performed well in maintaining a steady rate of water injection into the borehole against the varying well pressure.
(b) Two Litre Metre MM flowmeters, one for the pressurising circuit and one for pumping-out. These incorporate an orifice plate which causes a small flow to be diverted from the mainstream. This flow is measured by a Pelton Wheel and so indicates the total discharge, which is shown on a digital display. The flow range chosen was 1–65 l/min, displayed to the nearest 0.1 l/min.
(c) A pressure gauge, for approximate measurement of borehole pressure.

These were connected by Hi-flex high pressure hose. Three-way control valves, quick release couplings and on-off valves were also used. Since the pumped water was brackish, all metal parts were of stainless-steel to avoid corrosion.

The borehole pressure was measured accurately by a rapid response strain-gauge pressure transducer supported by a cable at a fixed depth in the borehole. The pressure was recorded by a multi-speed pen recorder and displayed also in digital form.

Test method

Access to the site was achieved by loading equipment into a boat and floating out on the ebbing tide. The boat settled out at low tide and provided a clean working area.

The well was first flushed out by pumping water down to the bottom and allowing it to overflow for some 20 minutes. In the most recent tests (March 1985) some development was carried out by alternate pumping and recharging of the well. This was done a day before testing commenced, in order to allow full recovery.

A falling head test was performed and then the series of pressurisation tests as follows:

(a) Pump water in to borehole and allow overflow until air is removed. Adjust flow by varying pump speed and flow regulator setting until desired flow rate is achieved.
(b) Shut-in borehole and leave until pressure equalises with ambient.
(c) Pump in water at a constant rate.
(d) Stop pumping and shut-in borehole. Observe head decay.
(e) Repeat (c) and (d) 2 or 3 times once head has decayed to near ambient, varying the injection rate.

This was followed by a short-duration (about 1 hour) pumping-out test where the well was pumped to the list of the pump suction and the quasi-steady state flow thus produced was measured.

During the pressurisation tests, it was important to observe certain conditions, namely the time limits described in the theory and also the avoidance of hydraulic fracture.

This latter phenomenon was reported by Bjerrum *et al.* (1972) to occur in clay soils during constant head permeability tests using small drive-in piezometers. Experiment and analysis suggest that this will occur if the applied excess pressure, Δu exceeds the effective horizontal or vertical stress of the soil. In that case, cracking of the soil may occur leading to a marked increase in local permeability.

Hydraulic fracture was avoided by calculating an upper limit on the applied excess pressure and not exceeding this value.

A test suite such as that described above, comprising a falling head test, 3 or 4 pressurisation tests and an approximately hour long pumping out test was completed over one low tide period of about 5 hours. The work required two operators, partly due to the heavy equipment and party to the need for haste because of the tide.

TABLE 2. *Summary of pressurisation tests.*

Borehole	1982		1984 both sections		1984 upper section		
Test	1	2	1	2	1	2	3
Pressurising period (minutes)	1.5	1.3	4.0	4.1	8.5	8.5	9.5
Declining period (minutes)	21.0	17.0	12.0	10.0	12.0	10.9	11.5
Max. excess head above ambient (m)	9.7	9.9	6.30	7.75	4.29	5.76	7.15
Excess head at end of test (m)	0.17	0.17	0.35	0.09	0.20	0.70	0.98
Injection rate (1/min)	20	20	3.75	5.35	1.9	2.75	3.85

Analysis and results

Theory predicts that values of H_w plotted against log (t/t') should ultimately fall on a straight line and Fig. 8 is an example of how this was indeed so in all of the tests. It is clear however that the values of H_w needed

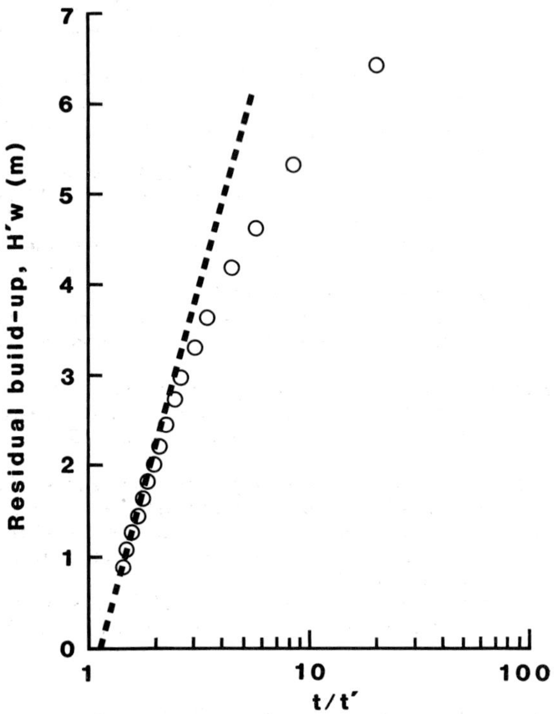

FIG. 8. Graph of H_w against log (t/t').

for plotting are not simply the same as the reading, h, produced by the transducer. This is because the reference head in the well is not constant due partly to the fluctuating response of the ambient pore water pressure to the tide, and primarily to the effect of head decay from any previous tests.

Referring to Fig. 2, it can be seen that the reference head measured before the test must therefore be extrapolated in order to provide a value of H_w.

Results were calculated as described in the theory and are quoted here for the 1982 tests and for the upper section of the 1984 tests. These are of interest because the same soil stratum, the laminated clays, was tested in both cases and because a later major pumping-test provides a comparison for the values obtained. Results for the more recent tests have not been published.

Table 3 summarises the results. The value obtained for the coefficient of permeability, K_h is remarkably consistent both between successive tests at different flow rates on the same borehole, and between the 1982 and 1984 boreholes.

These values, however, are approximately 1/8.5 of that derived from the later full-scale pumping test. It should be noted that the well used for that test was of more sophisticated design than the pressurisation wells, and that it had been extensively developed.

The value calculated for compressibility, m_v, is, as mentioned, sensitive to the effective well radius used in calculations. Thus Table 3 gives a range of values for m_v, based on the two extremes of using the bored radius of the hole and the radius of the well screen. For this reason it is felt that the test will predict compressibility to within a factor of 10, and this is indeed the evidence of the results in Table 3.

TABLE 3. *Comparison of results from various test methods.*

Borehole		1982		1984 upper section		
Test		1	2	1	2	3
K_h (10^7m/s)	pressurisation test	3.0	3.0	2.7	3.1	3.5
	falling head test*	2.5			-	
	major pumping test	-			26	
m_v (m^2/MN)	pressurisation test	0.20–0.36	0.17–0.31	0.02–0.11	0.022–0.12	0.025–0.14
	falling head test*	0.0012			-	
	major pumping test	-			0.034	

*by Gibson (compressible) method.

Conclusion

The HR pressurisation test has been used to predict permeability to within an order of magnitude of that determined by a large scale pumping test. This has been achieved using a well which was far cheaper than the pumping well and which, significantly, was not developed. Had the pressurisation well been developed (as in the more recent tests) it would clearly have yielded a value for permeability closer to that of the main well.

Compressibility was also predicted to within a factor of 10 and it is felt that this margin of error may be typical for values of compressibility measured by this test.

The HR method has the advantage over a full-scale pumping test of speed and low cost. Its superiority over conventional falling head tests lies partly in the construction of a special borehole, allowing clear definition of the section being tested and reducing problems of smearing etc, but above all in the consistency between test method and analysis.

Results obtained from tests to date have produced reasonable estimates of permeability and compressibility (and hence diffusivity). More trials would lead to still more understanding of the reliability of the method.

References

BISHOP, A.W. & AL-DHAHIR, Z.A., 1970. Some comparisons between laboratory tests, *in situ* tests and full-scale performance with special reference to soil permeability and coefficient of consolidation. *Proceedings of the Conference on In Situ Investigations in Soils and Rocks,* London, 251.

BJERRUM, L. NASH, J.K.T.L., KENNARD, R.M. & GIBSON, R.E., 1972. Hydraulic fracturing in field permeability testing. *Geotechnique,* **22**, 319.

BRAND, E.W. & PREMCHITT, J. 1980. Shape factors of cylindrical piezometers. *Geotechnique,* **32**, 203.

BRAND, E. W. & PREMCHITT, J. 1982. Response characteristics of cylindrical piezometers. *Geotechnique,* **32**, 203.

BREDEHOEFT, J.D. & PAPADOPULOS, S.S. 1980. A method for determining the hydraulic properties of tight formations. *Water Resources Research* **16**(1), 223.

BRITISH STANDARDS INSTITUTION 1981. *Code of practice for site investigations* BS5930.

COOPER, H.H., BREDEHOEFT, J.D. & PAPADOPULOS, S.S., 1967. Response of a finite diameter well to an instantaneous charge of water. *Water Resource Research,* **3**, 263.

DOMENICO, P.A. & MIFFLIN, M.D., 1965. Water from low permeability sediments and land subsidence. *Water Resource Research,* **1**, 563.

GIBSON, R.E., 1963. An analysis of system flexibility and its effect on time-lag in pore water pressure measurements. *Geotechnique,* **13**, 1.

GIBSON, R.E., 1966. A note on the constant head test to measure soil permeability in situ. *Geotechnique* **16**, 256.

HVORSLEV, M.J., 1951. Time lag and soil permeability in groundwater observations. *Waterways Experimental Station US Corps of Engineers* Vicksburg, Miss. Bull No 36.

NEUZIL, C.E. 1982. On conducting the modified 'slug' test in tight formations. *Water Resource Research* **18**(2), 439.

PAPADOPULOS, S.S., BREDEHOEFT, J.D. & COOPER, H.H., 1973. On the analysis of 'slug test' data. *Water Resource Research* **9**, 1987.

PAPADOPULOS, S.S. & COOPER, H.H. 1967. Drawdown in a well of large diameter, *Water Resource Research* **3**, 241.

J.M.A. PONTIN & M.A. FRENCH Groundwater Section, Hydraulics Research Limited, Wallingford, Oxfordshire OX10 8BA.

Appendix: Notation

c_v	$= K_h/\gamma_w m_v$, coefficient of consolidation for horizontal flow
C_s	system compressibility
C_w	compressibility of groundwater
d	depth of borehole casing
D	depth of borehole filter, diameter of borehole
F	form or shape factor for piezometers
F, F_1, F_2	finite well functions defined in references
h_w, h_∞	$= z + p_w/\gamma_w$, hydraulic head in well or piezometer at some or infinite time after start of test
H_o, H_w	excess head in well or piezometer above ambient at infinite time, at the start or during equalisation
K, K_h, K_v	coefficient of permeability, horizontal or vertical
l	length of borehole filter
m_v	$= \Delta e/(1+e)\Delta p'$, coefficient of volume compressibility
P_w, P_∞	pressure in well or piezometer at some or infinite time
P_o, P_w	excess pressure in well or piezometer above ambient at infinite time, at the start or during equalisation
q	rate of flow from well or piezometer
Q	injection or discharge rate from well
r_c	radius of borehole casing or standpipe
r_w	radius of borehole
t, t'	time from start of injection, time from end of injection
t_o	injection
T	basic time lag (Hvorslev)
u_w	a parameter defined in text
V	volume factor of piezometer
V_w	volume of water in well or piezometer
$W(u)$	well function for non-leaky aquifers (Theis)
α, β	parameters of the finite well functions, defined in test
γ_w	unit weight of groundwater

Groundwater in Engineering Geology, London, 1986, pp. 273-278 Printed in England

The measurement, interpretation and use of permeability values with specific reference to fissured aquifers.

S. Walthall & J.E. Campbell

ABSTRACT: The measurement of permeability, or more correctly, hydraulic conductivity of an aquifer is of fundamental importance in the understanding of water movement within it and to the prediction of the behaviour of engineering works. A wide range of techniques is available to the hydrogeologist and engineering geologist, and includes: laboratory measurement, using both gas and water permeameters; pumping test, with and without observation boreholes; inflow and outflow packer tests; and numerical and analogue aquifer models. Each technique is described, and its practical problems and relative merits discussed. It is important to realise that some of the techniques measure the hydraulic properties of the borehole while others measure the properties of the surrounding aquifer. Care is needed to choose the appropriate methods for different problems. A brief case history is presented explaining the use of these techniques for the dewatering of a large civil engineering structure.

Introduction

The Geologist or Engineer faced with the problem of determining hydraulic properties has a wide choice of techniques available to him. Each technique has strict limitations, not all of which are readily obvious, and some of which arise from the method of carrying out the measurement. The techniques fall into four categories, each having its origin in a different branch of science. The first is the Petroleum Geologist's laboratory technique, where the properties of small rock samples are measured in the laboratory using a permeameter. The second technique is the Hydrogeologist's pumping test using purpose-constructed abstraction and observation boreholes to test a small section of aquifer in the field. Thirdly, the Engineer has developed individual well testing in the form of the packer test, used in grouting as the Lugeon test, and in oil exploration as the drill stem test. Finally, the mathematicians have devised numerical techniques for modelling aquifer behaviour both on a regional scale and locally around a borehole.

Permeability measurements

The slow flow of a fluid through a porous medium is governed by Darcy's Law (Darcy, 1856) which in its original form is

$$Q = \frac{-K \, \mathrm{d}h \, A}{\mathrm{d}L} \qquad (1)$$

Q is the rate of flow,

$\frac{\mathrm{d}h}{\mathrm{d}L}$ is the hydraulic gradient, and

A is the cross sectional area.

The coefficient of proportionality K is a function not only of the porous medium but also of the fluid and is known as the hydraulic conductivity. It is usually expressed in units of metres per day (m/day) or in centimetres per second (cm/s).

It is possible to rewrite Darcy's law in the form

$$Q = \frac{k\sigma g}{\mu} \frac{\mathrm{d}h}{\mathrm{d}L} A \qquad (2)$$

where
σ is the fluid density
g is the acceleration of gravity,
μ is the fluid viscosity, and
k is the specific or intrinsic permeability.
k is independent of the fluid, and depends only on the properties of the porous medium. It is usually expressed in Darcys or millidarcys (md) and the relationship between K and k is

$$K = \frac{k\sigma g}{\mu} \qquad (3)$$

For water at 20°C, 1 Darcy is equivalent to 0.8346 m/day, while for water at 10°C, 1 Darcy = 0.639 m/day.

In fissured rock the classical Darcian approach is not strictly applicable (Sharp & Maini 1972) but for conditions where flow within the fissures is laminar, a Darcian expression can be used, namely:

From CRIPPS, J.C., BELL, F.G. & CULSHAW, M.G., (eds), 1986, *Groundwater in Engineering Geology,* Geological Society Engineering Geology Special Publication No. 3, pp. 273-278

$$q = K_J i \quad (4)$$

q is the flow-rate per unit width of fissure
i is the effective hydraulic gradient within the fissure
K_J is the proportionality constant for the fissure under a given geometrical condition and for a given fluid.

The term transmissivity (T) is defined as the hydraulic conductivity (K) multiplied by the saturated thickness of the aquifer. T values are usually expressed in metres squared per day (m^2/d).

Laboratory measurements

Gas

Gas permeameters are used extensively in core analysis laboratories, and the American Petroleum Institute has published (1952, 1960) two recommended practices dealing with their use. The method is quick and allows a large number of samples to be tested with moderate precision. Some correction may be necessary to allow for gas slippage (Klinkenberg 1941) in low permeability material. In additional it is important to establish the validity of using gas as the permeant if reaction is likely between the porous medium and a permeating fluid in the field.

Water

Two types of permeameter are available for determining hydraulic conductivity, the constant head permeameter and the falling head permeameter. The apparatus is described by Todd (1980) and, although simple to use, involves greater preparation time than the gas permeameter in that samples must be thoroughly saturated before testing.

Several workers (Lovelock 1977; Monicard 1980; Campbell 1982) have reported lower intrinsic permeability values when testing with water than with gas. This discrepancy can result when a reaction takes place between the porous medium and the test fluid (Smith & Stallman 1954). However, it may result from experimental errors inherent in water permeability testing. These include:

(1) migration of fine material during testing (Bodman & Harradine 1938, Younger & Lim 1972),
(2) the application of unrealistic hydraulic gradients to samples, particularly when using a falling head permeameter (Younger & Lim 1972),
(3) the presence of entrapped air within the sample (Christiansen 1944; Pillsbury & Appleman 1945; Lovelock 1977),
(4) the clogging of pores by micro-organisms (Klinkenberg 1941; Allison 1947; Gupta & Swartzendruber 1962).

These errors can combine to produce a discrepancy of one or two orders of magnitude between gas and water results, particularly in low permeability rocks. A further problem with laboratory testing in general is that of obtaining samples which are representative of field conditions.

Field measurements

Laboratory measurements provide values of intergranular, or primary, permeability. Features of secondary permeability, such as joints and fissures, however, may significantly affect the flow in an aquifer. It is therefore important to measure hydraulic conductivities in the field to take account of secondary permeability.

Pumping Tests

These have been carried out by Hydrogeologists and Water Engineers for many years and the techniques are well established. The field techniques, installation and instrumentation are described in the British Standard Code of Practice for Test Pumping Water Wells (BS 6316: 1983). The classical methods of analysis using curve fitting techniques are reviewed and described in detail by Kruseman & De Ridder (1970) who draw attention to the need to use the appropriate method of analysis for different aquifer conditions. Pumping tests fall into two categories, those without observation boreholes where all measurements are taken in the abstraction borehole, and those with observation boreholes to monitor the behaviour of the surrounding aquifer.

Pumping tests without observation boreholes

The common tests are step drawdown tests, packer tests, slug tests and variations on these including drill stem tests, all of which examine the pressure versus flow relationship. It is vital during any of these tests that the well face is clean and that any clogging effect caused by drilling debris, chemical precipitate, smearing or organic slime is removed. The effect of leaving a borehole unused for 1 year after cleaning was to halve its effective permeability, and hence its yield. Similarly, the yield of an abstraction borehole at Kenyon Junction, near Warrington, was reduced by an order of magnitude after being out of commission for a period of years.

The *step drawdown* test is used to determine the well characteristics and to predict the maximum yield of the borehole. If the tests are carried out carefully and a minimum of 5 different steps are used, it is also possible to determine the aquifer properties in the immediate vicinity of the well. Reviews of the methods of analysis have been presented by Clark (1977), Brereton (1979) and Kaergaard (1982). The test is economical both in time and effort, since it can be completed within a day, including the installation and removal of equipment from the borehole. Slug tests may also be used for a rapid assessment of the behaviour of a well (Black 1985).

In order to examine the distribution of permeability with depth it is necessary to use *packer testing*. A section of the borehole is isolated using expanding packers and fluid movement is induced, as either inflow or outflow. The pressure/flow relationship is monitored and from this the permeability of the section can be determined. The test has two basic forms. The first is the drill stem test used in the oil industry which has been modified for use in civil engineering (Daw 1984). The second is the Lugeon test which was originally designed for measuring the permeability of dam foundations and has been adapted for hydrogeological use (Pearson & Money 1977; Brassington & Walthall 1985). The analysis of packer tests assumes steady state conditions and it is therefore easy to calculate the permeability value from the pressure/flow ratio for a given borehole size (Anon. 1974).

The most reliable results are obtained using outflow tests. Injection tests have several disadvantages, the most important being the problem of the test section becoming clogged (Brassington & Walthall 1985). Packer tests can be used to examine in detail the nature of flow into a borehole and the influence of fissures. For this, testing must be carried out at very low excess pressures (around 1 head), where flow is linear and Darcian. At higher pressures fissure flow rapidly becomes non-linear as the flow rate in the fissures increases.

Analogue models have been constructed to simulate packer tests and to examine the validity of the test and the influence of high and low permeability zones (Bliss & Rushton 1984). The results indicate that packer tests can give a very detailed picture of the aquifer in the vicinity of the borehole. However, great care is needed during testing and the best results are obtained using outflow testing techniques at very low excess pressures.

Pumping test with observation boreholes

A conventional pumping test involves constant rate abstraction from a borehole, and water level monitoring in small diameter wells at various distances from the abstraction borehole. Recent work by Rushton & Howard (1982) and Walthall & Ingram (1984) has highlighted the unreliability of open observation boreholes. Such a borehole can act as a large vertical fissure and its presence can induce flow in a layered aquifer. The resultant water level measured in the borehole is not necessarily related to any particular part of the aquifer. Readings from open observation boreholes should therefore be treated with caution. To overcome this problem piezometers should be installed in the borehole to prevent vertical flow. Normally, analysis is carried out by fitting type curves to the field data, but the assumptions on which the type curves were produced are not always met. Features such as stratification in the aquifer, the presence of fissures, part confined/unconfined conditions and radial variations in aquifer properties can be represented individually. However, there are combinations of these conditions that occur in practice for which no analytical solutions are available. A numerical method of analysis which is described by Rushton & Redshaw (1979) can overcome this problem. Many conditions can be included in a single numerical solution. The method involves changing the aquifer parameters in successive model runs until a match is obtained between modelled drawdowns and field measurements.

Numerical modelling

As mentioned above numerical models can be used to analyse pumping tests, and to reproduce the behaviour of an aquifer in the vicinity of an abstraction well. The aquifer parameters are varied until a match is obtained between observed and modelled drawdowns. On the same principle a regional groundwater model can be used to investigate aquifer parameters on a regional scale. A finite difference model, as described by Rushton & Redshaw (1979), is commonly employed, but it must be realised that a model is only as good as the data used to construct it. The groundwater balance within the aquifer system must be examined in detail to evaluate inflows and outflows. Inflow data include recharge estimates (rainfall minus evaporation), intrusion from estuaries and rivers, and flows across the boundaries of the model from adjacent strata. Outflows include abstractions from boreholes, spring flows and losses to rivers. Long-term records, preferably greater than 10 years, of water levels in observation boreholes throughout the aquifer are also required to highlight any changes in the groundwater balance. Provided sufficient detailed data are available to calibrate the model it can be used to ascertain the range and any real variation of transmissivity values with the aquifer.

Result of testing

The techniques described above were used to examine the hydraulic properties of the Permo-Triassic sandstone aquifer of the Lower Mersey Basin, during a joint study between North West Water and the University of Birmingham. The results of testing at two sites, namely, Kenyon Junction and Padgate, both near Warrington, are presented in Table 1 and are discussed to illustrate how the various tests produce different results because they measure different parameters. These results have been presented by the various authors of the joint study in other papers, which are referred to below.

Laboratory tests

At Kenyon Junction, samples of core were taken at 0.5 m intervals to the full depth of the borehole (102 m).

The average horizontal permeability to gas was 928 millidarcys equivalent to a hydraulic conductivity of 0.59 m/day for water at 10°C. Individual measurements were multiplied by the sampling interval and summed to give cumulative transmissivity values. These are shown on Fig. 1, plotted at intervals of 3 metres down the borehole, and indicate an intergranular transmissivity (T_i) of 36 m²/day. At Padgate a value of 0.48 m/day was measured for the hydraulic conductivity of the aquifer. When multiplied by the saturated thickness of aquifer penetrated by the borehole, this gives a T value of 93 m²/day.

Packer testing

Injection packer testing at Kenyon Junction revealed zones of high hydraulic conductivity, which coincide with fissures in the borehole, and much lower hydraulic conductivities in the unfissured zones. The cumulative transmissivity of the borehole with depth was determined from the packer tests by summing the transmissivity value of each test zone. This is shown on Fig. 1, and indicates a total transmissivity (T_t) of 165 m²/day. It can be seen that above 65 m depth the T_i and T_t values are very similar, but a large difference develops when the fissured sections are encountered below 65 m. At Padgate similar tests revealed a total transmissivity of 185 m²/day. (Brassington & Walthall 1985).

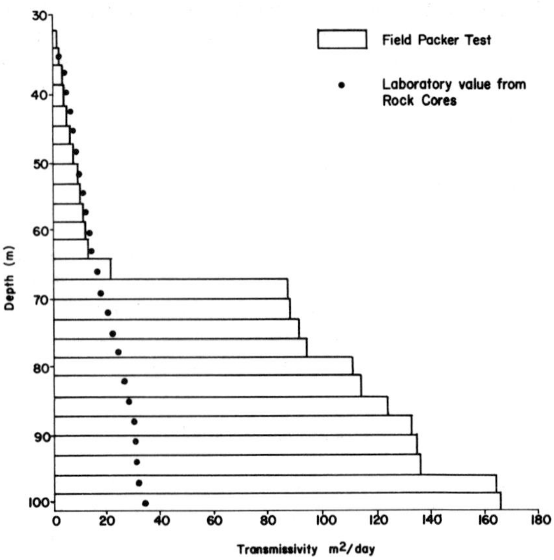

FIG. 1. Cumulative transmissivity at Kenyon Junction.

Pumping tests

Step drawdown tests at Kenyon Junction and Padgate indicated transmissivity values of 250 m²/day and 300 m²/day respectively. Aquifer pumping tests at Kenyon Junction using the boreholes as an open observation

TABLE 1.
Transmissivity and hydraulic conductivity values for laboratory packer and pumping tests, and regional modelling.

	Kenyon Junction.		Padgate	
Method of Testing	T(m²/day)	K(m/day)	T(m²/day)	K(m/day)
Laboratory Tests	36	0.51	93	0.48
Injection Packer Tests	165**	2.36	185**	0.95
Step-Drawdown Tests	250	3.57	300*	2.68
Aquifer Pumping Test with open observation borehole	50 – 150	0.7 – 2.1	-	
Aquifer Pumping Test with piezometers.	85	1.2	-	-
Regional Flow Modelling.	300	1.5	300	1.5

1. Kenyon Junction saturated aquifer thickness 70 m.
2. Padgate saturated aquifer thickness 190 m, except for * where thickness 110 m.
** Tests affected by clogging and high injection pressure. See Brassington & Walthall (1985).

well yielded transmissivity values in the range 50–150 m^2/day. Further tests with piezometers installed in the boreholes, when analysed by numerical modelling techniques, gave transmissivity values of 85 m^2/day. (Rushton & Howard 1982).

Regional flow modelling

In the model used to simulate groundwater flows in the Lower Mersey Basin aquifer a regional transmissivity value of 300 m^2/day has been found to provide the best simulation (Anon. 1981). Assuming an effective thickness of 200 for the aquifer, a hydraulic conductivity value of 1.5 m/day is indicated.

Discussion of results

Laboratory measurements on core samples provide values of intergranular permeability. In a fissured aquifer therefore they are almost certain to underestimate the hydraulic conductivity of the aquifer. This is borne out by the results from the Lower Mersey basin where the laboratory value of 0.5 m/day is much lower than the 'true' hydraulic conductivity value between 1.2 and 1.5 m/day derived from aquifer pumping tests and flow modelling.

Packer testing clearly demonstrates the significance of fissures in controlling groundwater flow into boreholes, a feature discussed by Rushton & Howard (1982), Bliss & Rushton (1984), and Brassington & Walthall (1985). As Fig. 1 shows, at Kenyon Junction the measured field transmissivity increases dramatically between 64 m and 70 m as fissures are encountered. If the boreholes had been drilled to 64 m a much lower transmissivity value would have been obtained than at 70 m. It must be realised that packer tests in a borehole measure the hydraulic properties only of that particular borehole. Analogue modelling work by Bliss & Rushton (1984) supports this view by demonstrating that the controlling influences on the behaviour of packer tests extend for only 10 m into the surrounding aquifer. Packer testing results, therefore, should not be used as a measure of the overall permeability of the aquifer.

Similarly, step-drawdown tests measure well performance, but without the detail provided by packer testing. Their results are more dependent on fissures encountered in an individual borehole, than the permeability of the surrounding aquifer. The hydraulic conductivities obtained at Kenyon Junction (3.57 m/day) and Padgate (2.68 m/day) are considerably higher than the 'true' value of 1.2 – 1.5 m/day.

To obtain information on the behaviour of an aquifer during a pumping test, observation wells must be monitored. However, an open observation borehole can give anomalous results due to vertical flow occurring in the borehole itself and disturbing the groundwater regime. To overcome this problem piezometers should be installed in the borehole to prevent vertical flow. Measurements in such piezometers are regarded as the best means of monitoring aquifer behaviour, and determining the hydraulic properties of the aquifer.

Case History

It was proposed to construct a large shaft 35 m in diameter and 30 m deep in Permo-Triassic sandstones. In order to provide design data for the shaft construction a number of boreholes were drilled and tested to determine the hydraulic properties of the sandstones. Two boreholes were drilled to 40 m depth within the perimeter of the proposed shaft, and a further eight observation boreholes in the surrounding area. Piezometers were installed in the observation wells with filter sand and bentonite seals to provide on average 6 m long response zones. The testing procedure involved:

(i) packer testing in the two boreholes within the perimeter of the shaft.
(ii) step-drawdown tests in the same two boreholes.
(iii) a 48 hour constant rate pumping test with abstraction from one borehole within the shaft and monitoring at all the other sites.

Using the results of the packer tests and step-drawdown tests, plots of Quantity Pumped versus Drawdown were prepared to indicate the specific yield of the holes. This was used to design the depth of the dewatering wells and to define the performance requirements of the dewatering system.

The analysis of the constant rate test was carried out on a computer using the radial flow described by Rushton & Redshaw (1979). Values of permeability and storativity for the aquifer were obtained which provided a good match between field and modelled results in the observation wells. These values enable predictions of aquifer behaviour to be made, and thus the rate of pumping, necessary to achieve the desired dewatering, can be defined. Using a computer program by King (1984) the drawdown distribution within the aquifer was examined for various patterns of dewatering wells. The optimum number of wells and their location was defined, together with a predicted drawdown distribution.

Summary

The quantity of water to be abstracted to produce the required drawdown was derived from the regional aquifer parameters which were determined from the constant rate test and local knowledge of the aquifer, including a regional model. The number and location of the abstraction points were designed using the regional aquifer parameters in King's (1984) simple model. The depths and diameters of the wells were designed using the value of specific yield obtained from the packer tests and step-drawdown tests. The specific yield is dependent on the local aquifers characteristics and the hydraulics of the well itself (well losses), which can be obtained only from packer tests and step-drawdown tests.

ACKNOWLEDGEMENTS: The authors are grateful to the Director of Planning and the General Manager of Rivers Division, North West Water Authority, for permission to publish this paper. We wish to acknowledge the work by our colleagues in the Authority and staff from the University of Birmingham during the joint study. any opinions expressed are the authors' own and do not necessarily reflect those of the Water Authority.

References

ALLISON, L.E., 1947. Effect of microorganisms on permeability of soil under prolonged submergence. *Soil Science,* **63,** 439-450.

ANON., 1984. *Earth Manual.* 2nd Edition. Dept. of the Interior, Bureau of Reclamation, Washington, D.C.

ANON., 1981. *Saline Groundwater Investigation Phase 1 -Lower Mersey Basin.* Final Report to the North West Water Authority, University of Birmingham.

ANON., 1983. *Code of Practice for Test Pumping Water Wells.* BS 6316:1983. British Standards Institution, London.

A.P.I., 1952. *Recommended Practice for Determining Permeability of Porous Media.* American Petroleum Institute, Dallas, Texas. 3rd Edition.

A.P.I., 1960. *Recommended Practice for Core analysis procedure.* American Petroleum Institute, Dallas, Texas.

BLACK, J.H., 1985. The interpretation of slug tests in fissured rocks. *Quarterly Journal of Engineering Geology,* **18**, 161-171.

BLISS, J.C. & RUSHTON, K.R., 1984. The reliability of packer tests for estimating the hydraulic conductivity of aquifers. *Quarterly Journal of Engineering Geology,* **17**, 81-91.

BODMAN GB. & HARRADINE, E.F., 1983. Mean effective pore size and clay migration during water percolation in soils. *Proceedings of the Soil Science Society of America.*

BRASSINGTON F.C., & WALTHALL, S., 1985. Field techniques using borehole packers in hydrogeological investigations. *Quarterly Journal of Engineering Geology,* **18**, 181-193.

BRERETON, N.R., 1979. *Step Drawdown Pumping Tests for the Determination of Aquifer and Borehole Characteristics.* WRC. Report TR103. Medmenham.

CAMPBELL, J.E., 1982. *Permeability characteristics of the Permo-Triassic Sandstones of the Lower Mersey Basin.* MSc Thesis, University of Birmingham.

CHRISTIANSEN, J.E., 1944. Effect of entrapped air upon the permeability of soils. *Soil Science,* **58**, 355-365.

CLARK, L., 1977. The analysis and planning of step drawdown tests. *Quarterly Journal of Engineering Geology,* **10**, 125-144.

DARCY, H.M 1856. *Les Fontaines Publiques de la Ville de Dijon.* Victor Dalmont, Paris.

DAW, G.P., 1984. Application of aquifer testing to deep shaft investigations. *Quarterly Journal of Engineering Geology,* **17**, 367-386.

GUPTA R.P. & SWARTZENDRUBER D., 1962, Flow-associated eeduction in the hydraulic conductivity of quartz sand. *Proceedings of the Soil Science Society of America,* 6-10.

KAERGAARD, H., 1982. The Step Drawdown Test and Non-Darcian Flow; A critical review of theory, methods and practice. *Nordic Hydrogeology,* **13**, 247-256.

KING, J.M., 1984. Computing drawdown distributions using microcomputers. *Groundwater*, 212, 780 - 784.

KLINKENBERG, L.J., 1941. The Permeability of Porous media to liquids and gases. *Drilling and Production Practice*, 200-213.

KRUSEMAN G.P. & DE RIDDER N.A., 1979. *Analysis and Evaluation of Pumping Test Data.* International Institute for Land Reclamation and Improvement, Wageningen, The Netherlands.

LOVELOCK P.E.R., 1977. Aquifer Properties of Permo-Triassic Sandstones in the United Kingdom. *Bulletin of the Geological Survey of Great Britain* No. 56

MONICARD, R.P., 1980. *Properties of Reservoir Rocks: Core Analysis.* Editions Technic, Paris.

PEARSON R. & MONEY, M.S., 1977. Improvements in the Lugeon or packer permeability test. *Quarterly Journal of Engineering Geology*, **10**, 221-40.

PILLSBURY A.F. & APPLEMAN, D., 1945. Factors in permeability changes of soils and inert granular material. *Soil Science*, **59**, 115-123.

RUSHTON, K.R. & HOWARD, K.W.F., 1982. The unreliability of open observation boreholes in unconfined aquifer pumping tests. *Goundwater,* **20,** 546-550.

RUSHTON K.R. & REDSHAW S.C., 1979. *Seepage and Groundwater Flow.* Wiley, Chichester.

SHARP J.C., & MAINI, Y.N.T., 1972. Fundamental considerations on the hydraulic characteristics of joints in rocks. *Proceedings of the Symposium on Percolation through Fissured Rock, Stuttgart.*

SMITH, O. & STALLMAN R.W., 1954. Measurement of Permeabilities in Groundwater Investigations. *American Society for Testing Materials* Technical Publication No. 163, 98-122.

TODD D.K., 1980. *Groundwater hydrology*, 2nd. ed. Wiley, New York.

WALTHALL S. & INGRAM J.A., 1984. The investigation of Aquifer parameters using multiple piezometers. *Groundwater*, **22**, 25-30.

YOUNGER, J.S. & LIM C.I., 1972. An investigation into the flow behaviour through compacted saturated fine-grained soils with regard to fines content and over a range of applied hydraulic gradients. In: *Fundamentals of Transport Phenomena in Porous Media.* International Association for Hydraulic Research. Elsevier, Amsterdam.

S. WALTHALL, North West Water, Design & ConstructionUnit (East) Oakland House, Old Trafford, Manchester M16 0QF,UK.
J.E. CAMPBELL, North West Water, Rivers Division, New Town House, Warrington, WA1 2QD, UK.

Design and application of a borehole flowmeter

J.L.Wild

ABSTRACT: A borehole flowmeter has been developed to measure low flow rates in NX (76 mm) boreholes. The instrument can be used to determine the standing water level and permeability of a test section of the borehole.

The borehole flowmeter is a short, lightweight wireline packer which is inflated pneumatically and contains a turbine flowmeter transmitter unit. The useful flow range of the standard instrument is approximately 0.7 cm^3/s to 70 cm^3/s. The unit was originally designed to measure small axial flow rates in an array of boreholes which was thought to intersect a series of perched aquifers.

Experimentation showed that an increased rate of flow could be induced through the unit by piping water into the top of the borehole, the flow rate being proportional to the excess head of water above the unit. Using this principle, a form of *in situ* falling head permeability test was devised. The corrected head of water at a zero flow rate was found to correspond to the standing water level in a borehole, and the linear relationship between applied head and flow rate was used to calculate the permeability of the section of borehole below the unit. The results are compared with those of packer tests carried out in the same test sections.

The method has several advantages over packer testing in many conditions. The wireline equipment is light and can be operated by one man. The permeability test results are repeatable and are completed on average within one hour, though a satisfactory relationship between flow and head is usually established within ten minutes in a moderately permeable formation.

Introduction

The borehole flowmeter was originally designed to confirm the presence of flow cells in a closely spaced array of drillholes. The array, in the Old Red Sandstone near Newcastleton, Roxburghshire, was thought to intersect a dipping series of perched aquifers. The natural groundwater regime was modified by the array, and axial flow was thought to be occurring in the holes. An instrument was required to measure the flow at any point in the hole. This would have to be sensitive to low rates of flow, and sufficiently rugged for field use in uncased drillholes.

Design of the borehole flowmeter

Axial flowmeters have been described before (e.g. Tornaghi 1979; US Army 1954; Tate *et al.* 1970) which use an open turbine and photocell arrangement. These are easily damaged, and lack the required sensitivity. The measurement of a low flow rate is made easier if the velocity is increased by confining the flow to a pipe within a packer. The aim was to build a flowmeter into a packer which was small enough for one-man wireline operation. A suitable range of flow measuring equipment is manufactured by Litremeter Ltd., a versatile system of transmitters and remote display units. The transmitter units contain a robust turbine, and resin encapsulated electronics which can survive accidental submergence. Three interchangeable transmitters have been used, with a total range of measurement of 0.3 to 250 cm^3/s.

The diameter of the transmitter as supplied is too large to fit in a standard pneumatic packer, and a special packer based on a short section of thin-walled aluminium pipe was constructed (Fig.1). Two packer ends heavy enough to overcome the buoyancy of the packer were machined from mild steel, and the packer had a thin walled inflatable rubber sealing section with a nominal length of 50 cm. Since the transmitter units are uni-directional, duplication of the internal pipework was necessary in order to measure flows both up and down a drillhole. External rubber plugs on the packer ends were used to reverse the direction of measurement. The 4 kg package was suspended from a graduated stainless steel wireline with a four-core electrical cable and an airline. The packer was inflated by footpump, and the cable from the flowmeter transmitter was connected to a digital meter.

Method of use

The packer was lowered down the borehole to the required depth and inflated. The inflation pressure required was little more than hydrostatic, since the packer had little differential pressure to withstand. The annulus between the packer and an NX hole is

From Cripps, J.C., Bell, F.G. & Culshaw, M.G., (eds), 1986, *Groundwater in Engineering Geology,* Geological Society Engineering Geology Special Publication No. 3, pp. 279-281

small making over-inflation difficult despite the relatively thin sealing material. A submersible pressure transducer measured the head of water above the unit, and this was displayed on a pen recorder.

The borehole flowmeter was put to three uses:
(i) groundwater flow measurement;
(ii) detection of permeable zones;
(iii) permeability testing.

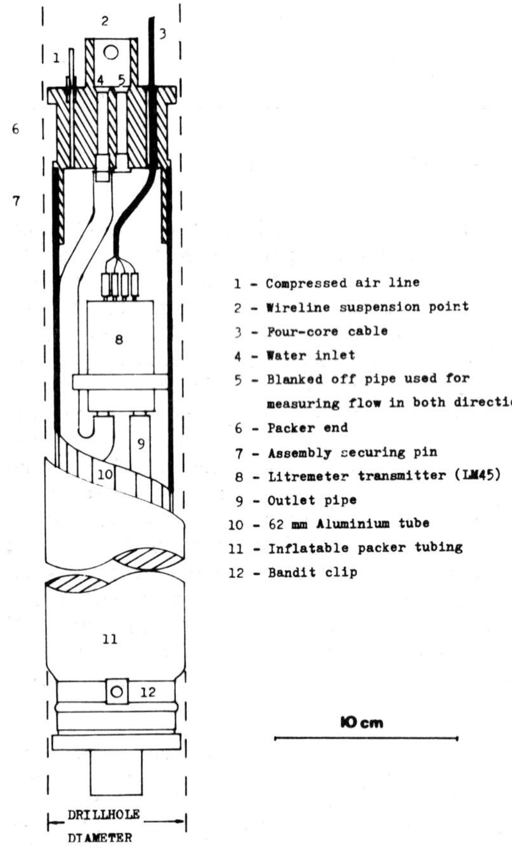

FIG.1. Borehole flowmeter showing internal detail.

Groundwater flow measurement

To carry out these tests, the flowmeter was lowered and inflated in stages down the water-filled sections of a drillhole. At each stage, a flow rate was recorded. The lowest flow which can be measured accurately by the Litremeter system is 0.17 cm^3/s and rates of up to 8 cm^3/s were detected. To find the direction of flow, the profile of measurements was repeated with the direction of flow through the turbine reversed. The direction of flow is that in which the higher reading is recorded.

Detection of permeable zones

An artificial flow down the drillhole can be induced by pouring water into the top of the hole. If a small constant excess head is maintained, the flow at a given depth quickly approaches a constant value. A profile of flow measurements down the hole can efficiently locate zones of higher permeability. Bypass of the packer is indicated by a section of hole in which the flow rate is lower than the sections above and below. The location of packer bypass and permeable sections can assist in optimisation of a subsequent double packer test programme and interpretation of the results.

Permeability testing

In a test in which a flow rate through the packer is induced, the rate of flow is proportional to the excess head in the test section. This forms the basis for the variable head flowmeter permeability test. A submersible pressure transducer was lowered to the packer for monitoring the head of water, and a constant flow of water was introduced into the top of the hole (Fig. 2). After a few minutes, the flow rate is adjusted to give a small excess head. After approximately 30 minutes the excess head approached an equilibrium value and the flow of water from the surface was stopped. Measurements of induced flow through the packer and head of water were taken as the excess head in the borehole decreased. Since the head falls quickly at first, the majority of the data are gathered in the first few minutes of a test. The head was then corrected for losses through the packer and a graph of Q (flow) against H (head) plotted. The correction would not be necessary if an internal transducer was used to measure the excess pressure in the test section.

In the majority of the tests, the relationship between Q and H was found to be linear (Fig. 3). The head at the corrected zero flow intercept corresponds to the standing water level. In some cases the results show a repeatable sharp increase in the ratio of Q to H when the excess head passed a particular depth. This was due to bypass of the packer in the early stages of the test, which ceased when the water level fell below the bypassing fracture.

Comparison of the results with packer tests

A programme of single and double packer tests was carried out in the drillhole array, some using an instrumented double packer which incorporated a transducer for the measurement of test section pressure. The majority of the values of hydraulic conductivity calculated from the results of the borehole flowmeter and packer tests are very similar, except in one 30 m drillhole.

To a depth of 14 m the drillhole intersects a series of conductive bedding planes and fractures, and below this depth the sandstone is relatively intact. The standing water level is approximately 12.5 m. Below 14 m the single packer and flowmeter tests give very similar values of *Q/H* and a calculated hydraulic conductivity of 10^{-8} m/s. In the upper section the packer results give values which are two to seven times greater than the flowmeter derived values of approximately 10^{-5} to 10^{-6} m/s. It is thought that this discrepancy, which increases towards the surface, is the result of fissure flow in the unsaturated rock mass. In the flowmeter tests, the water used to create an excess head saturated a large volume of rock along connecting discontinuities, and the flow from the test section was essentially radial. During packer tests water passing into the fractures near the top of the test section travels upwards and outwards, and this results in a larger flow for a given applied head. The results of double packer tests confirm the high conductivity of these near surface discontinuities, and in long higher pressure packer tests the measured hydraulic conductivity was found to decrease substantially with time, to values comparable with those derived from the flowmeter tests.

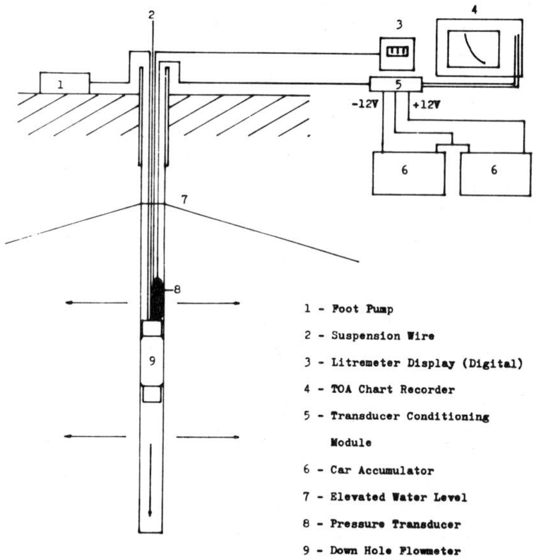

FIG. 2. Diagram of equipment used for permeability testing.

Conclusions

The borehole flowmeter is a versatile tool with several uses including permeability testing. The equipment is easily operated by one man, and provides an alternative to packer tests in many situations. A double system with integral pressure transducers would have wider application, and could be used for both inflow and outflow testing.

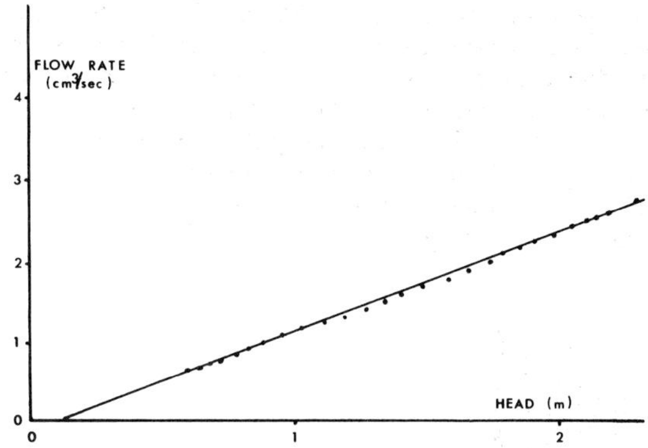

FIG. 3. Graph of a typical test result.

The main practical advantages over a packer are that head losses in drill pipes are eliminated, and testing at applied heads lower than ground level is possible, avoiding turbulent flow and fracture dilation. Leakage of test section water past the flowmeter packer is less likely, since the differential pressure across the flowmeter packer is low. End effects are minimal at the top of the test section as water is free to enter the rock mass from the borehole above the packer. In a double system, water could be piped through the test section to beneath the lower packer to reduce end effects at the base of the test section.

The hydraulic conductivity is calculated from a large number of observations, and should be more reliable than values calculated from a packer test based on a few incremental head stages. An X-Y plotter could be used to display the progress of a test, facilitating real time calculation of hydraulic conductivity in the field.

ACKNOWLEDGEMENTS: The financial support for this research and for the construction of the drillhole array was provided by the NERC. I wish to thank Dr M.S. Money for help, guidance and encouragement during this research, and Mr A. Blythe for assistance in the construction of the equipment.

References

LITREMETER LTD. Unit 9, Park St., Industrial Estate, Aylesbury, Bucks HP20 1ET.

TATE, T.K. ROBERTSON, A.S. & GRAY, D.A. 1970. The hydrogeological investigation of fissure-flow by borehole logging techniques. *Quarterly Journal of Engineering Geology,* **2.** 195-216.

TORNAGHI, R. 1979. Experimental criteria for the design and control of grouting in sandy-gravelly soils. *Tunnelling 1979,* Institution of Mining and Metallurgy, London.

U.S. DEPT. OF ARMY, CORPS OF ENGINEERS. 1954. Waterways experimental station relief well flow meter. *Misc. paper 5-83,* Vicksburg, Miss.

J.L. WILD, Engineering Geology Unit, Department of Geology, University of Newcastle Upon Tyne, Newcastle Upon Tyne, U.K.
Presently at Britoil plc, 150, St. Vincent Street, Glasgow, G2 5LJ, UK.

Results of an experimental programme of *in situ* permeability testing in rock

J.L. Wild & M.S. Money

ABSTRACT: In an earlier pilot programme of tests in shallow drillholes (Pearson & Money, Quarterly Journal of Engineering Geology, 1977) it was shown that significant improvements could be made at low cost to packer testing in rock, which lead to greater reliability in test performance and interpretation. It was also demonstrated that non-equilibrium flow occurred in some tests which, if not recognised, could lead to anomalies in test results.

A second programme of tests has been completed under commercial conditions using significantly deeper arrays of 76 mm drillholes in two rock masses, one in the Old Red Sandstone, and the other a set of basalt flows. Single packer tests were carried out as drilling proceeded using a standard mechanically-expanded packer with continuous autographic recording of flows and of input pressures. Some of the holes were also tested after drilling with a prototype double packer system instrumented with transducers to measure pressure within, above and below the test section. Results of the two methods are compared and individual test behaviour has been examined and it has been found that results were affected by changes in groundwater level produced by drilling. In any test it is unwise to assume that the effective piezometric level in the rock mass at the time and position of a test is necessarily the same as that observed in a completed drillhole.

The tests have produced further practical evidence of the need to consider the scale of tests and observations in relation to the inhomogeneity of the rock mass under investigation. At these test sites there were marked variations in standing water levels, and in measured permeabilities in adjacent closely spaced drillholes.

Introduction

The packer or Lugeon test is a well established technique for estimating the mass permeability of rocks in investigations for dams and underground works. In recent years the test has also been used to an increasing extent in hydrogeological investigations, particularly where aquifer behaviour is, or is suspected to be dominated by fissure flow. The principles of the test are well known; water is injected at a series of constant pressures into a section of drillhole sealed off by expandable packers and measurements of flow and pressure used to calculate permeability. Test results often exhibit anomalies and a variety of explanations have been given in the extensive literature to explain various aspects of test behaviour.

In an earlier pilot programme of tests (Pearson & Money 1977) it was shown that significant low cost improvements could be made to test equipment and techniques which lead to greater reliability in test performance and interpretation. It was also shown that non-equilibrium flow occurred in some tests which if not recognised could lead to anomalies in test results and interpretation.

Testing in this earlier programme had been confined to shallow depths due to the limitations of the drilling equipment and the resources then available. Provision of a grant by the Natural Environment Research Council made it possible to drill holes to greater depths and to improve instrumentation of the tests.

It was intended to continue the investigation with the following programme:

(1) Drill in 3 rock types to greater depths, (up to 30 m) to form arrays of drillholes. Single packer testing to be done under 'commercial' conditions.

(2) Carry out double packer tests to compare with single packer tests.

(3) Use the arrays of holes for longer term pumping-in and pumping-out tests, using each hole in turn as the pumped well and the others as observation wells.

(4) Carry out other forms of permeability tests such as slug tests.

The first and second of these objectives was broadly achieved, although there was insufficient time to complete the double packer testing. The existence of low permeabilities and variable water levels made it impossible to complete objective 3 although drawdown tests using a syphon technique were successfully carried out at two sites. The development of a borehole flowmeter to investigate the anomalous water levers, however, proved to be a useful testing tool (Wild

From Cripps, J.C., Bell, F.G. & Culshaw, M.G., (eds), 1986, *Groundwater in Engineering Geology*, Geological Society Engineering Geology Special Publication No. 3, pp. 283-293

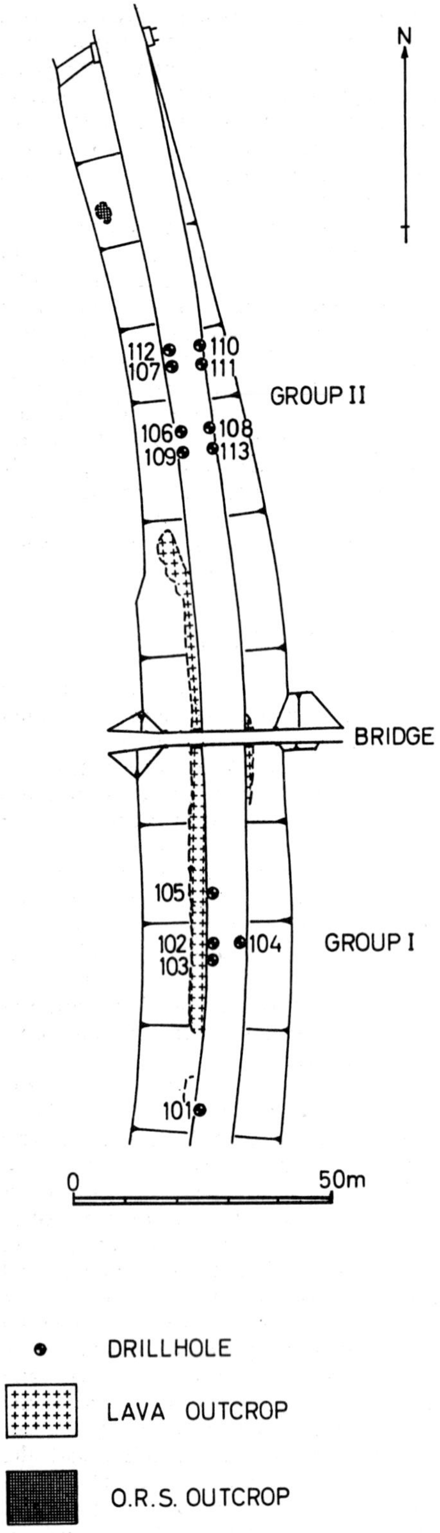

FIG. 1. Layout of main test site.

1985). Slug tests were performed but proved difficult to interpret as the recent analysis of Black (1985) has showed.

Test sites

It was not possible to drill at the sites previously used due to difficulties with access for the rig but the work was carried out in the same general area of Roxburghshire and in two of the same formations, the Upper Old Red Sandstone and Silurian greywackes with, in addition tests in Lower Carboniferous basalts.

The site in which most of the work was carried out is in a cutting on the disused Hawick to Carlisle railway line between the former Steele Road and Riccarton Junction Stations. The cutting was constructed c. 1860, probably by hand excavation of overburden and by blasting with black powder so that it is unlikely that the rock mass has been significantly affected except by unloading. The maximum depth of the cutting is approximately 15 m and the width of the track bed is approximately 8 m. It was necessary to keep the track open for forestry traffic and the drillholes are therefore sited along the edges of the track. The railway was closed in 1971 and about 1 m of ballast was left *in situ*. The original level of rock head was some 1 to 4 m above track level.

The cutting contains exposures of the Upper Old Red Sandstone and of the (supposed) Lower Carboniferous Birrenswark Lavas and drillholes were sunk into both these formations.

Drilling was also carried out at two other sites, both shallow quarries in Silurian greywackes. It proved impossible, however, to maintain stable uncased holes and although some permeability tests were carried out they were all at shallow depths and are not reported on here in detail.

Drilling

Drilling was carried out with a rotary rig and all the holes were cored in rock with a double-tube swivel barrel using water flush. Mylar lining was used in some holes but in general was found to be unnecessary as good core recovery was obtained throughout. Drilling in each hole commenced at 84 mm diameter but as soon as casing had been drilled through the ballast and seated in sound rock, generally at depth of 2 m to 4 m, the hole was completed at 76 mm diameter taking TNX core. In order to keep the maximum length of each hole available for testing the contractor was not permitted to reduce the hole size or to install further casing. Only one hole had to be abandoned at less than the planned depth due to caving. These holes have remained open for five years.

Most of the holes were drilled to 20 m depth but one in the sandstone was taken to 30 m and two were limited to 8 m.

The layout of the drillholes in the railway cutting is shown in Fig. 1. The first four holes, Group I, were drilled 20 m deep in the Birrenswark Lavas, a second group of eight holes, Group II, were drilled into the Sandstone, the four southerly holes penetrating the Lava – Sandstone contact.

Summary logs of these twelve drillholes are shown in Figs 2 and 3 on isometric projections.

Geology

The Upper Old Red Sandstone is poorly exposed in the cutting but there are good exposures in an adjacent stream section. Detailed examination of the cores shows that the sequence of lithologies is similar to that described by Leeder (1973) with much of the succession formed by cyclic deposition of intraclastic conglomerates, cross-bedded and ripple marked sandstones and calcretes. Discontinuities seen in the cores are mainly bedding or cross-bedding planes with some vertical joints and a few more fractured zones suggested minor faulting.

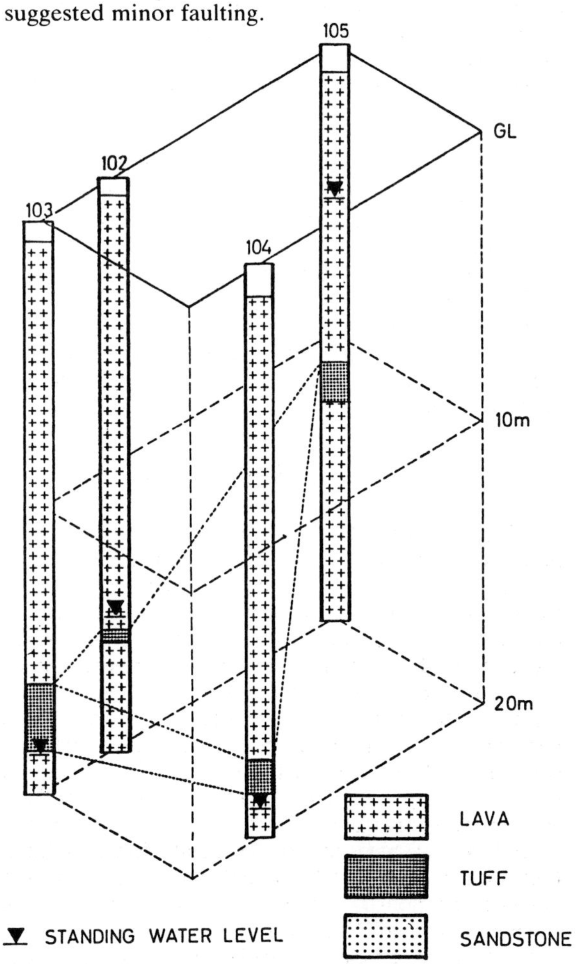

FIG. 2. Isometric projection of Group I drillholes.

Problems in correlating the closely spaced Group II holes also suggest the presence of faulting which is not revealed by exposures in the cutting. It appears from local mapping that the Lava/ORS contact dips at 25° to 30° to the south-east, but the drillholes show that basalt flows interdigitate with sandstones, and either that the contact is faulted or that the flows wedge out very rapidly.

The Birrenswark Lavas crop out in the cutting and were encountered in the Group I holes and in four of the Group II holes. The lavas are microporphyritic basalts of Dalmeny type (Lumsden *et al.* 1967). Most outcrops in the area are heavily weathered and even apparently fresh material from cores has an altered groundmass, the rock nevertheless proved extremely hard to core. Most of the flows are amygdaloidal and when weathered are vesicular. Thin tuffaceous sandstones separate some flows. The rock is generally closely jointed near the surface with the joint frequency increasing with depth, although as measurements of RQD show (Fig. 4) the increase is not regular or even similar in adjacent holes.

Packer testing equipment

Packer testing of the holes was carried out in two stages; during drilling 3 m lengths of hole were tested with single packers, in the following field season, some nine months later, two holes were tested with double packers; some long sections of the same holes were also tested with the same equipment used in a single packer mode.

The equipment used in the initial tests was supplied by the contractor and the packer consisted of a series of corrugated rubber sleeves that were expanded by the weight of the rods and, if necessary, by additional force from the drill chuck. The effective length of the seal was 0.5 m which although rather short appeared to be effective. No provision was made for monitoring test section pressures directly.

The double packer equipment was based on the instrument developed by Pearson & Money (1977) but with important modifications. It consisted of two inflatable 'ductube' packers, each 1.5 m in length separated by 3 m of perforated pipe. The instrument was fitted with three transducers to measure pressures within the test section, below the lower packer and above the upper packer. The transducers were all housed in a sealed container inside the lower packer and connected to the appropriate section of hole by piezometer tubes. Power supplies and signals were conducted by a multicore cable. Testing and development of this equipment took a considerable time, the chief practical difficulty being that of keeping the water, the air for inflation and the electrical signals separate from each other. The choice of a screened multicore cable was dictated by economy and it was not suitable for this application, although the equipment eventually worked satisfactorily. Any future equipment of this type

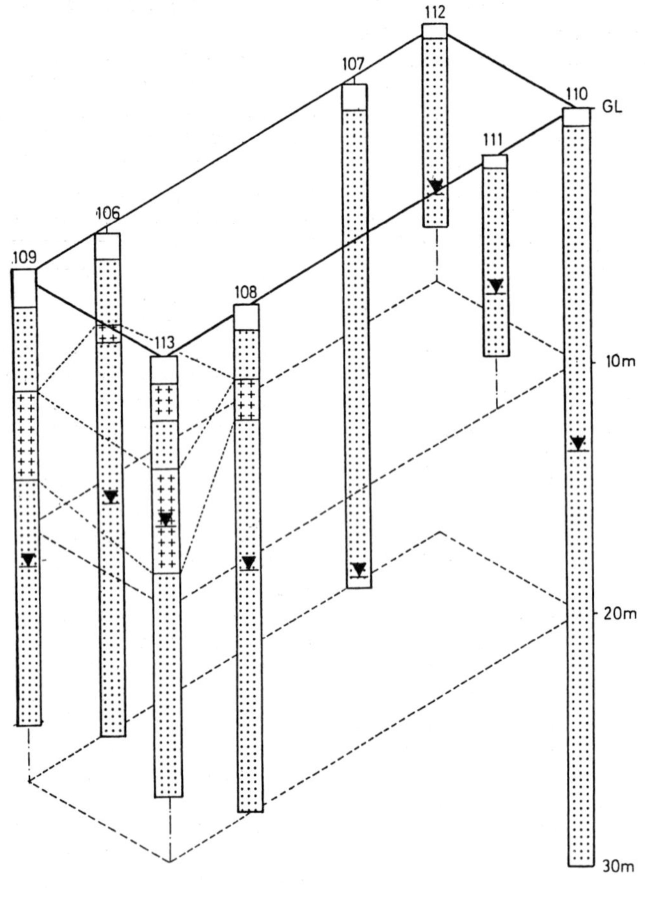

FIG. 3. Isometric projections of Group II drillholes. For key see Fig. 2.

would be designed around a proper underwater cable and connector system. Fully waterproof connectors are however large in relation to equipment that will fit inside a 76 mm drillhole and leave little room for air and water inputs.

The water supply, as in the previous series of tests was obtained from a series of constant head tanks consisting of 200 l drums arranged at various heights on the sides of the cutting. These were fed by gravity from an adjacent stream, the water being passed through a larger set of tanks to allow settlement of suspended material and to act as a reservoir. 200 mesh (75 μm) filters were fitted on the input side of all measuring equipment. Pumps were used in a few tests where large flows were required.

Flow rate and pressure were monitored continuously in all tests with a variety of instruments, all installed, operated and read by the investigators and their field assistants. During drilling the contractor was only required to install and remove the packers and connect them to the measuring system.

Flows were measured with 'litremeter' turbine and orifice-turbine flow meters covering the range 0.02 to 100 l/min and in the later stages of the programme these could be linked to a chart recorder. Pressures were measured with bourdon test gauges, a mercury manometer and pressure transducers. A screw hydraulic ram salvaged from triaxial equipment enabled pressure checks and calibration to be carried out on site. A differential pressure gauge connected across the complete measuring network provided useful indication of blockages or excessive head loss in flowmeters. All the equipment was mounted in a small converted caravan which also housed battery power supplies and served as a site cabin and workshop for minor repairs.

Groundwater level

The depth of the cutting, its general dampness and the presence of nearby streams all encouraged the view that the water table would be close to the surface. Little information was gathered during drilling since the use of water flush and frequent packer testing in daily 12 hour shifts allowed little time for equilibration. The sandstone holes were drilled very rapidly and few observations were possible. The result of the packer tests showed that the effective levels at the time of testing were generally within a few metres of the surface, regardless of depth. Once drilling had been competed, however, water levels declined to a very different depths in adjacent holes. These levels as shown diagrammatically on isometric projection in Figs 2 and 3.

In the Group I basalt holes, three of the levels are near the base of the hole at depths of 15 to 19 m and these are consistent with the high permeability attributed to a single fracture in 104 and loss of flush water with drilling in 103. They are less consistent with the low permeability measured at the base of 102 and bear no relation to the much higher level of 5.4 m depth in 105.

In the Group II mixed basalt and sandstone holes, the levels are reasonably consistent but differ by more than might be expected from the measured values of permeability and the proximity of the holes.

In the Group II holes which are all in sandstone the two deeper holes have similar levels whilst the two shallow holes have much higher levels.

It is known that the levels fluctuate seasonally and that they drop significantly in a dry summer. It has not been possible to monitor the levels on a regular basis due to the remoteness of the site.

No general pattern of groundwater movement can be deduced from these levels and if single holes rather than groups of holes had been drilled these variations would not have been apparent. It is difficult to propose a simple model to explain these variations but the following factors may be significant:

(a) It is possible that the holes intersect discontinuous fracture systems, each with its own flow regime and piezometric levels.

(b) Lithological variations in the sandstone sequence and vesicular margins to the lava flows may form discrete aquifers, each of which may also function independently.

(c) There may be general underdrainage of the site to the ORS/Silurian unconformity which probably underlies the Group II holes at no great depth.

(d) The drillholes themselves may be acting as drains and producing anomalous levels.

Observations during pumping-in tests showed connections between some of the holes in the sandstones but more detailed observations, including identification of levels of inflow and outflow are required before any more specific conclusions can be drawn.

Single packer test results

The total test time for each 3 m section of hole was kept to approximately one hour to limit rig standing time and, in accordance with the recommendations of Pearson & Money (1977), only 3 or 4 ascending pressure stages were used. Continuous monitoring of flow and of input pressure made it possible to retrieve useful results from almost all of the tests and to calculate a value of permeability. In a few tests the capacity of the flowmeters was exceeded and only a lower bound value could be determined.

The results were calculated on the basis of Darcy's law of permeability, using a standard shape factor, but are shown for convenience in 'Lugeons' using the common conversion factor 1 Lugeon = 1×10^{-7} m/s. the results are summarised in graphical form on Figs 4 and 5.

The logarithmic scale necessary to plot the results disguises the wide range of values and also variation of results between adjacent drillholes.

Tests in two of the Group I holes in basalt clearly indicate a decline in permeability with increasing depth, the other two do not, although none of the holes is more than 12 m apart.

The tests in the Group II holes are almost entirely within the sandstones yet here too there are substantial differences between results in adjacent holes at similar depth.

It appears, therefore, that the testing of 3 m long sections in these rock types does not yield representative values of the rock mass permeability. The section is too short to intersect sufficient discontinuities and the result, in a single section is profoundly influenced by a few of the wider joints or bedding planes.

The calculation of permeability from a packer test is based on the plotting or calculation of the flow/head ratio, the pressure head being the excess pressure applied to the test section above a known or assumed static water level. This static water level is often taken to be the ultimate standing water level in the hole on completion or after equilibration. Experience in these tests showed, however, that this assumption can lead to serious errors and misinterpretation of the results.

Since the values of flow rate (Q) and applied pressure head (δAH) are known with some confidence from good instrumentation, the ratio Q/H and hence permeability can be calculated from any pair of test stages.

This is illustrated in Fig. 6 where it is clear that in this test $Q = 0$ when $H = 6.2$ m. The *effective* standing water level in the hole at the time of test is therefore 6.2 m which is 0.2 m below ground level, while the long term standing water level is 15.2 m below ground level. This could be another result of the anomalous water levels described above but it

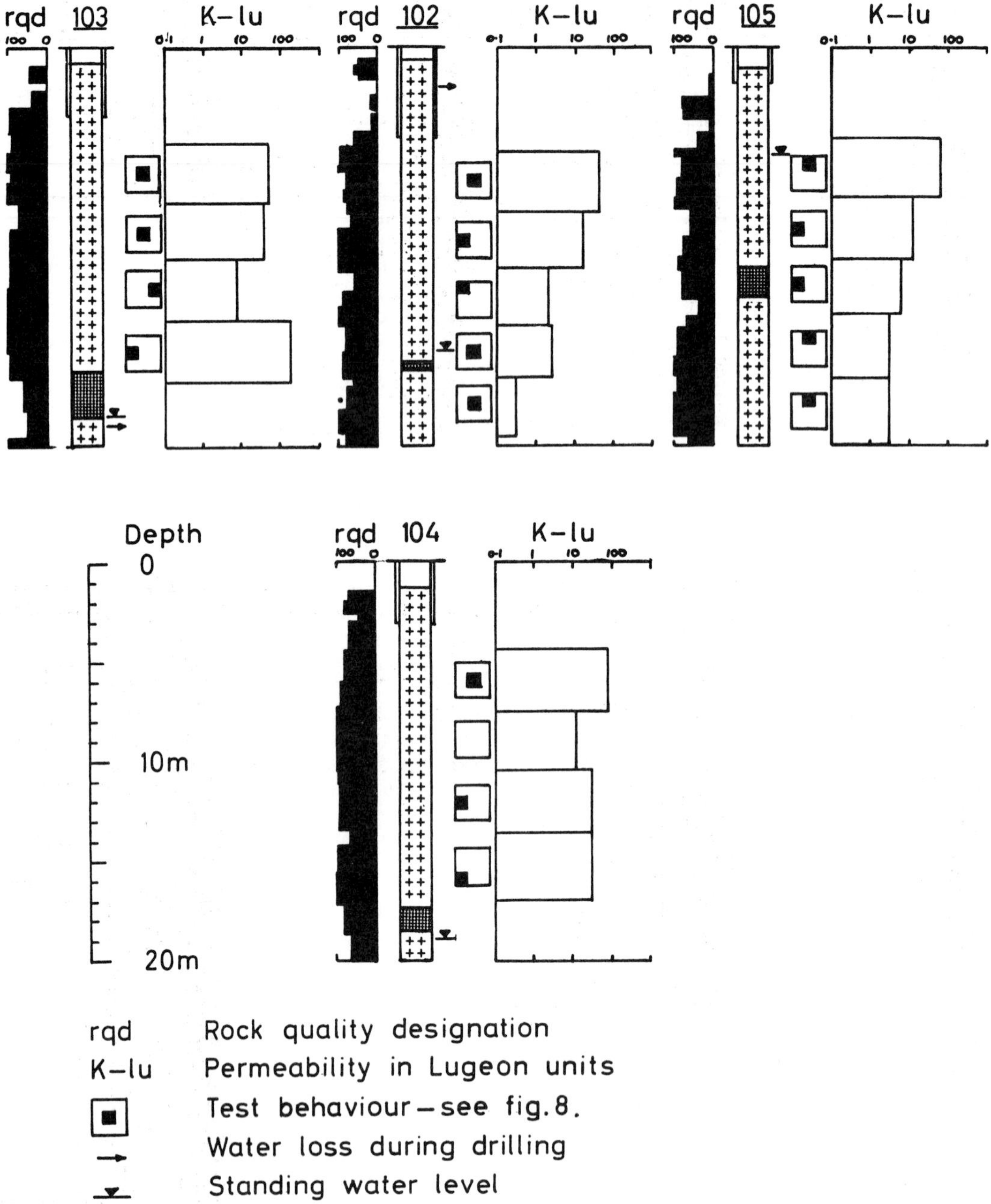

FIG. 4. Single packer test results in Group I drillholes. Legend for lithologies as Fig. 2.

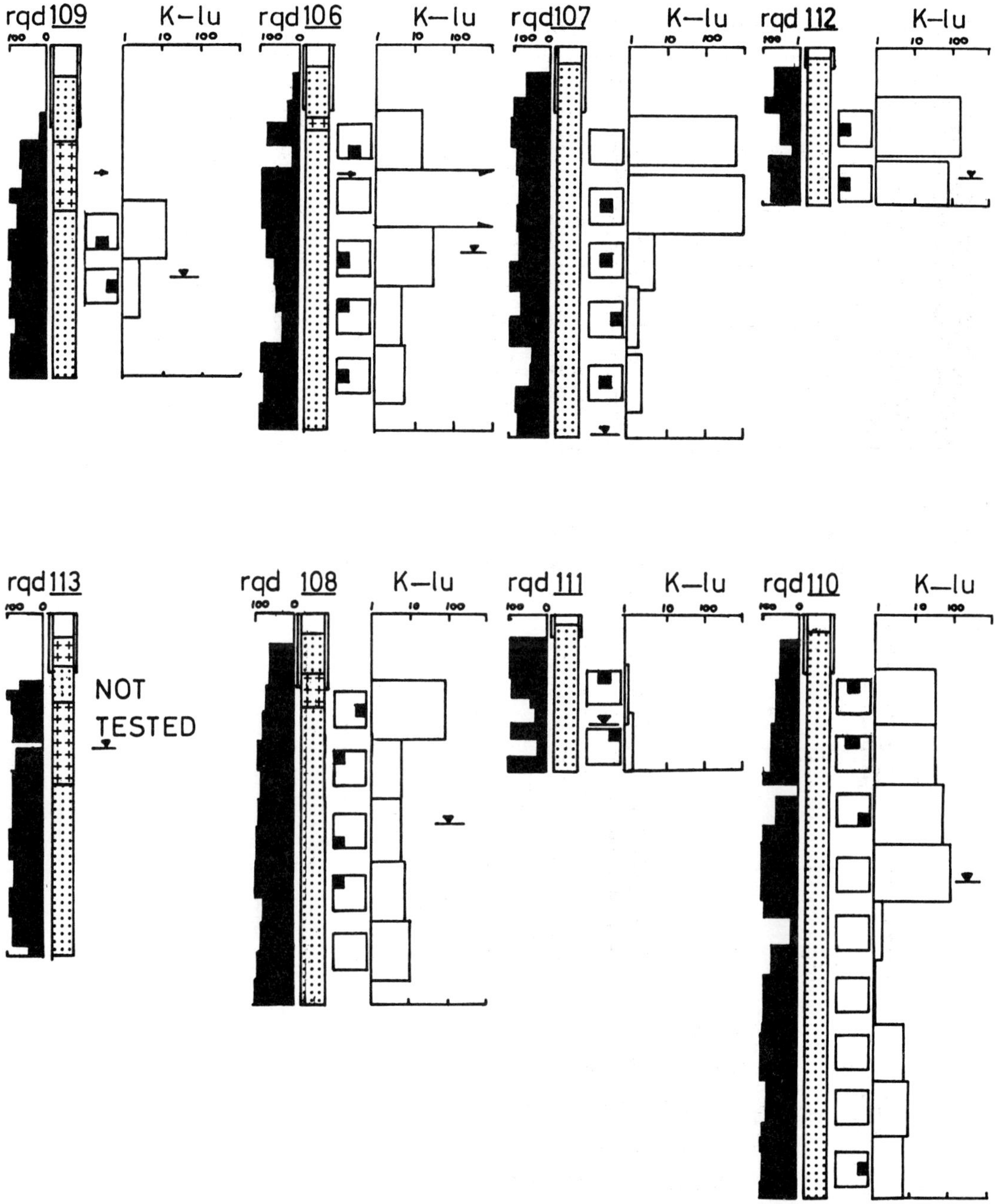

FIG. 5. Single packer tests is Group II drillholes. Symbols as Fig. 4.

seems more likely that it is due to the artificial raising of water level in the drillhole by prolonged drilling with water flush and successive packer testing. This effect was not detected in the earlier test programme in which drilling and testing were not continuous and there was ample time for equilibration.

The use of incorrect standing water levels will not only lead to erroneous results but to false assumptions about the shape of the *Q* vs *H* curve. This point is discussed further below.

Double packer tests

The programme of double packer tests was restricted by the difficulties in development already described but a series of 12 successful tests was carried out in two of the Group II drillholes. A further 12 tests were carried out with the double packer equipment and its instrumentation used in single packer mode to test longer sections. In most tests both ascending and descending pressure stages were used, in some cases 6 to 8 in number.

The results of these tests in hole are shown graphically in Fig. 7 and it will be seen that in general there is good agreement between the first stage single packer and the later tests. There seems no reasons to mistrust the lower seal with pneumatic packers of this type and the double packer has considerable advantages in terms of flexibility of testing, especially if it can be handled without the need for expensive rig standing time. The measurement of pressures down the hole greatly improves the reliability of the test and the confidence with which the results can be interpreted and as a generalisation the double packer tests were 'better' than the single packer tests.

Fig. 7 also shows the results of some long-section tests carried out with the borehole flowmeter (Wild 1985) and these also agree well with the packer test results.

Test behaviour

The results of packer tests are usually presented as a statement of the calculated value of permeability,

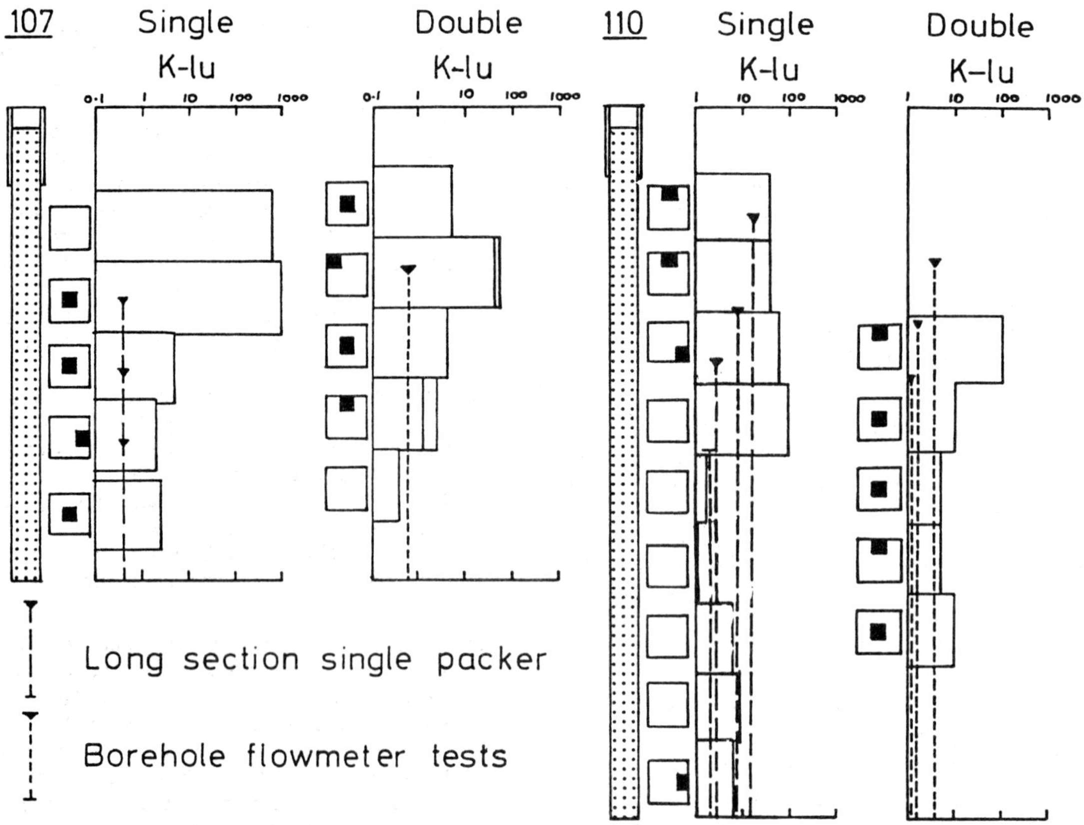

FIG. 6. Comparison of single packer, double packer and flowmeter tests in drillholes 107 and 110.

sometimes accompanied by a plot of Q against H. Pearson & Money (1977) showed the value of plotting transient values of Q and H and of plotting the ratio Q/H against time in those tests which show marked non-equilibrium behaviour. Houlsby (1976) has classified test behaviour into five groups and has given guidelines for the selection of appropriate lugeon values. His groupings, however, are based solely on the Q vs H plot and do not take into account behaviour during individual test stages.

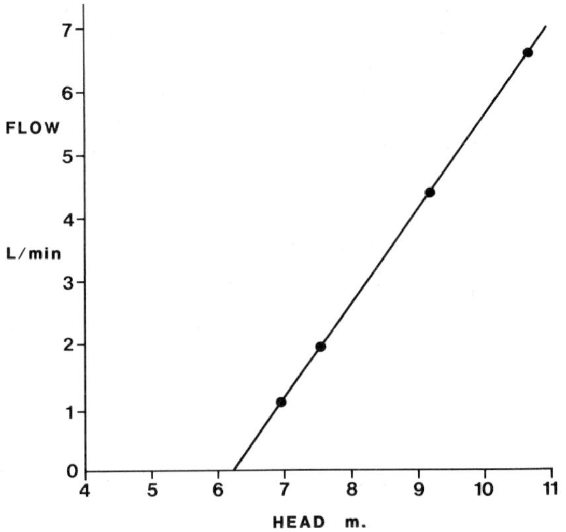

FIG. 7. Single packer test result showing effect of piezometric level raised by water flush. The head is measured relative to the mid-point of the 3 m long test section.

An alternative classification is shown in Fig. 8 in which test results are assigned to a simple matrix. There are three classes of behaviour during test stages at constant head:

Row
A the flow falls continuously
B the flow is steady
C the flow rises.

In most tests it is commonly found that there is a significant drop in flow in the first five minutes after the application of pressure and the assignment of a test to classes A or C is based on a 10% change in flow rate in the 15 minutes following the initial 5 minute period. An example of changing flow and pressure in a single test stage is shown in Fig. 9.

The test can also be classified in terms of the Q vs H plot as shown on Fig. 8.

Column
1 Q/H declines with increasing head
2 Q/H is constant in all stages
3 Q/H increases with increasing head.

The definition of linearity of a plot with few points and an uncertain origin is somewhat arbitrary, however, it was possible to assign a class to most tests with some confidence.

These test characteristics can then be combined into the 9-division classification which is shown graphically in Fig. 8. A few tests do not fit any of these categories and form a 'miscellaneous' tenth class.

The numbers of single packer tests in each group are shown separately in Figs 8a and 8b and the percentage of each test type for all the hole are shown in Fig. 8c.

Although all test types are possible and did occur, the majority of tests, 62%, fall into the four classes A1, A2, B1, B2. Of the remainder it is considered that the 15% of tests in Column 3 almost all suffered from packer leakage mainly at the highest pressure. The 15% of 'miscellaneous' tests showed variable in of flow rate with time but nearly all were of Column 1 type with combinations of A and C behaviour.

Discussion

The distribution of the various types of test is shown by a miniature representation of the matrix alongside the results on Figs 4, 5, and 7. It appears that there is no particular relationship between test behaviour, position and depth except that:

(i) There appears to be some tendency for successive tests in a hole to be of the same type, rather than for the types to occur randomly.
(ii) Type A (flow falling) behaviour tends to be more common near the surface and this is consistent with Pearson & Money's view (1977) that this phenomenon is due to storage capacity in the rock and thus to the raising of groundwater level around the hole during the test. It should also be noted that this type of behaviour was again most marked in the few tests carried out at shallow depths in the greywackes.
(iii) The few type A.1 tests all gave values of permeability less than 10 lugeons. This may be fortuitous but is possibly associated with the incidence of a few narrow fractures.
(iv) The characteristics of the ascending pressure stages of the double packer tests are also shown on Fig. 7 but little can be deduced, partly because of the 'miscellaneous' nature of the drillhole 110 tests.

Most of the double packer tests and the single packer tests performed with the fully instrumented packer showed type 2, that is, linear Q vs H plots and as already noted are considered to be reliable tests. Apart from the circumstances listed above under (ii) and (iii), there seems little evidence for correlating test behaviour with fundamental hydraulic properties of the rock masses. In most cases variations in test behaviour are probably a function of test procedure and water levels in the hole prior to testing.

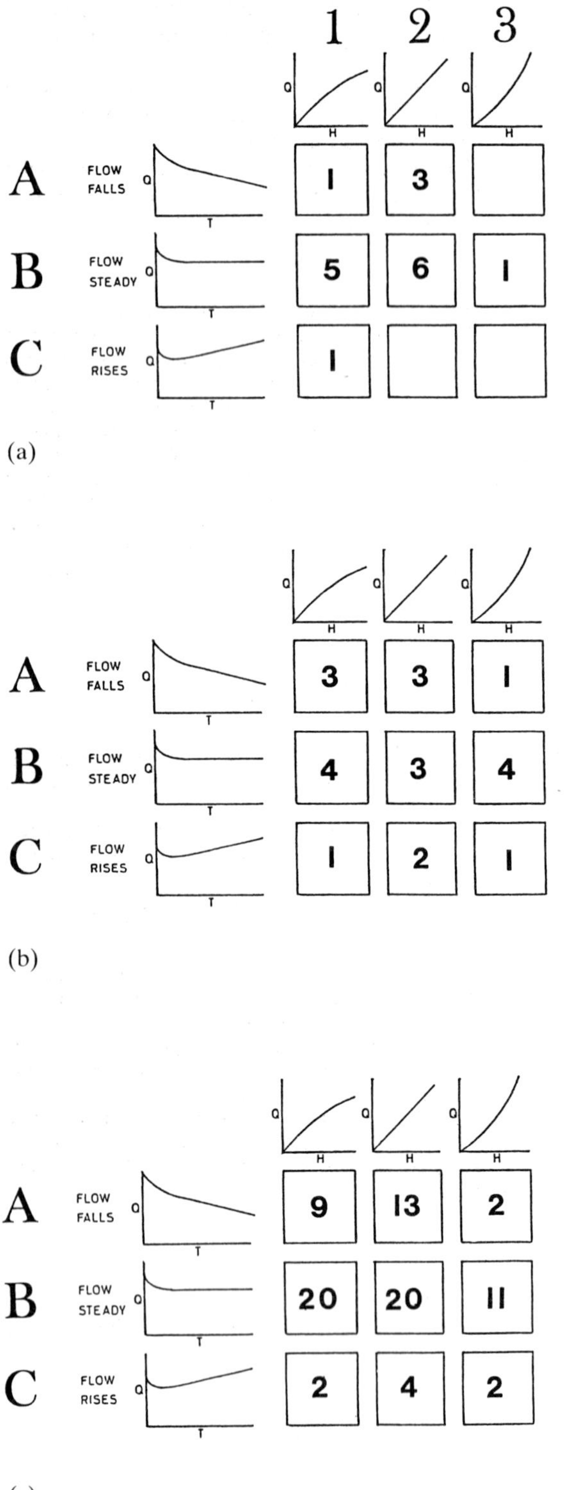

FIG. 8. Test behaviour in (a) Group I drillholes, (b) Group II drillholes, (c) Group I and II drillholes.

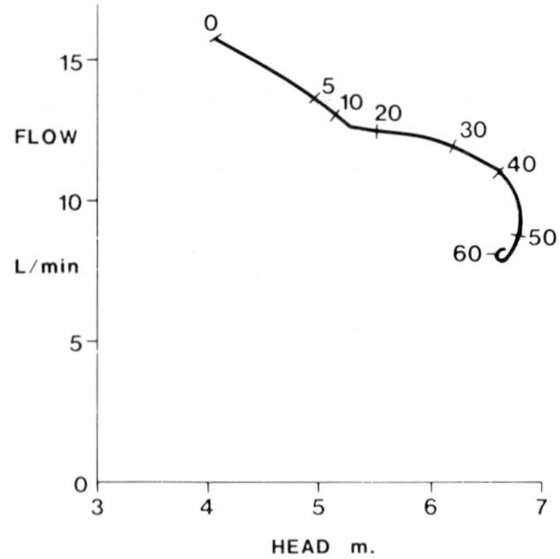

FIG. 9. Variable flow and head in a single test stage. Numbers on the curve show minutes after start of test.

Attention has already been drawn to the effect of drill flush water in raising groundwater levels in the test section and this will be particularly significant in zones of very low permeability. Two other factors may also be important in such tests:

(i) very small quantities of water are injected into the zone in a test of normal duration so that near-hole conditions are likely to be particularly important.

(ii) removal of the drill string and insertion of the packers may raise or lower the water level in the hole by a significant amount due to displacement.

It has already been suggested that in some near-surface tests the groundwater level around the test section is raised during the test. At any stage in such a test the true applied head ΔH will always be less than the applied head calculated on the basis of a static water level. This will therefore tend to produce results of both Type 1 and Type A, and thus especially Type 1.A. As a corollary if a water level artificially raised by drilling were to decline during a test then the calculated head would be too small and results would be of Types 3 and Type C. It should also be noted that such changes would also produce 'hysteris' loops in the Q vs H plot, as pressures are reduced. Pearson & Money (1977) demonstrated evidence of 'back-pressure' effects in some test. In the present programme of tests reducing pressures were used only in the second series of tests with the instrumented packer. Some of these showed hysteresis loops, others did not, but there appears to be no systematic variation in this further aspect of test behaviour.

Future developments

It has again been shown that the reliability of packer testing can be improved by the use of constant head tanks, down-the-hole pressure measurement and continuous monitoring and useful results can be retrieved from most tests. The interpretation of 'spot' flow and pressure measurements will present problems, however, and it requires at least 15 minutes at each pressure stage to produce a single reliable point on the Q vs H plot. It is also difficult to make reliable installations of pressure transducers in small diameter packer systems at low cost.

Two other methods have been used in which continuous Q vs H plots are obtained and the static water level is determined from the plot.

One method utilises the borehole flow meter to which reference has already been made (Wild 1985), and which has been shown to produce results comparable with packer tests.

A second method utilises conventional packers suspended on drill casing instead of drill rod. The casing is topped up with water above the static level and the test may be carried out in either constant or falling head mode, using a single submersible pressure transducer to record head differences. Tests have recently been carried out to depths of 100 m in Coal Measures using this technique.

Conclusions

1. It has again been shown that good instrumentation, continuous monitoring and the use of constant head tanks make it possible to obtain repeatable results from packer tests.
2. Variable flows were found in the first 5 minutes of most test stages and at least a further 10 minutes of continuous monitoring is required to obtain consistent readings.
3. Water levels in the rock mass may be significantly altered by the use of water flush in drilling and the static water levels in a test may not be the same as the ultimate standing water level in the completed drillhole.
4. Single packer, double packer and borehole flow meter tests gave comparable values of permeability.
5. The drilling of multiple closely-spaced holes revealed wide variations in standing water levels and measured permeabilities which would not have been apparent from isolated holes in a standard site investigation.
6. The 3 m long test sections appeared to be too short in relation to the inhomogeneity of the rock masses.
7. Non-equilibrium flows do not occur in some test but it appears that this effect may be most common near the surface.
8. Various types of test behaviour have been classified but there is no clear relationship with rock properties at these sites.

ACKNOWLEDGEMENTS: This work was supported by the Natural Environment Research Council who provided a grant for drilling and instrumentation and a research studentship for JLW. The authors are grateful to Mr D.R. Dowson and Mr and Mrs J.R. Harris for permission to use the test sites, to Mr D. Woolfenden of Economic Forestry (Scotland) Ltd. for his cooperation and to the late Mr Andrew Douglas for provision of accommodation at Saughtree. The authors wish to thank Mr A. Blyth and the late Mr W Proctor for help with equipment and Mr G.D. Barton, Miss S.E. James and Mr M. Townsley for invaluable assistance in the field. The drilling was carried out by Geotechnical Engineering (Northern) Ltd.

References

BLACK, J.H. 1985. The interpretation of slug tests in fissures rocks. *Quarterly Journal of Engineering Geology,* **18,** 161-71.

HOULSBY, A.C. 1976. Routine interpretation of the Lugeon water test. *Quarterly Journal of Engineering Geology,* **9,** 803-14.

LEEDER, M. 1973. Sedimentology and paleogeography of the Upper Old Red Sandstone in the Scottish Borders Basin. *Scottish Journal of Geology* **9**(2), 117-44.

LUMSDEN, G.I. et al. 1967. *Geology of the Neighbourhood of Langholm. Memoirs of the Geological Survey of Scotland.* HMSO. Edinburgh.

PEARSON, R. & MONEY. M.S., 1977. Improvements in the Lugeon or packer permeability test. *Quarterly Journal of Engineering Geology,* **10,** 221-40.

WILD, J.L. 1985. Design and application of a borehole flowmeter. *Proceedings of Conference on Groundwater in Engineering Geology.* University of Sheffield, (this volume)

J.W. WILD, & M.S. MONEY, Engineering Geology Unit, Department of Geology, University of Newcastle upon Tyne, Newcastle upon Tyne, UK. (J.W.W. presently at Britoil plc, 150 St. Vincent Street, Glasgow G2 5LJ, UK).

Discussion Session 7

J.S. SUTTON (Sir Alexander Gibb & Pts) queried Dr MONEY's dismissal of the Lugeon test in which although only a limited volume of rock in the immediate vicinity of the test length is tested, the conditions are comparable to those developed in curtain grouting. Hence the test results are usually comparable with grout take and illustrate the more general principle that the applicability of any test is greatest when its scale is closest to the scale of the engineering problem being analysed. Often external constraints preclude the use of appropriately scaled tests and it is perhaps a major justification for the use of specialist engineering geologists to relate small scale test results to actual conditions.

M.S. MONEY (University of Newcastle-upon-Tyne) stated that the Lugeon test may well be adequate for assessing grouting criteria, but frequently the test is used to indicate mass rock quality. As such it is not better than RQD. There is a particular problem with using high pressure permeability determinations. In a deep hole or a low water table the addition of the static head to the pressure applied at the surface produces points on the flow head plot which are a long way from the origin. Extrapolation to lower flows and pressures gives unreliable values of the permeability. Tests should be carried out at the lowest possible pressures, and these should be measured in the test section.

S. WALTHALL (North West Water Authority) added that well inflow tests are unreliable since clogging of pores occurs; even in the laboratory tests let alone those done in the field. He also pointed out that inflating the packers alters the aquifer conditions and the results should be interpreted in terms of the water pressure measured in the isolated test section. It is apparent that the effects of testing can be transmitted over large distances. In tests at Warington noticeable changes occurred at a distance of 90 m. Finally, although storage is one of the most important aspects of civil engineering groundwater control, no easy, yet adequate, method of determination is available.

A. FLETCHER (Orisis Cesco Ltd.) asked Mr WALTHALL for clarification regarding his methods of field measurement compared with conventional packer testing.

S. WALTHALL explained that the data were obtained using low heads under controlled conditions with experienced staff. The data were used solely to determine the well parameters, the aquifer parameters having been found from conventional pumping tests and mathematical aquifer models.

M. NOEL (University of Sheffield) asked Mr FRENCH or Mr PONTIN a question about the measurement of *in situ* permeability. The method is analogous to measuring thermal conductivity by monitoring the decay of heat pulse. In the latter, applying a time delay to the origin time of the heat pulse has been found to improve the fit of decay of the pulse to the measurement of temperature compared with theoretical values.

J. PONTIN (Hydraulics Research Ltd.) pointed out that the test does not rely on an instantaneous pressure pulse. A period of constant discharge is used to increase the pressure within a sealed, water-filled and air-free borehole, hence permeability and compressibility can be determined.

C. WARREN (Sir William Halcrow & Pts) asked Dr MONEY to clarify points regarding water levels in closely spaced boreholes. It was implied that there was no hydraulic connection between the holes but would pumping tests in the aquifer give differing results for each hole?

M.S. MONEY explained that low water levels and the small diameter of the holes had ruled out conventional pumping tests. There were interconnections between some of the holes as shown by a rise of water level in holes adjacent to those being tested, but in a few cases the levels remained high for some time after the completion of the test.

J. PONTIN suggested that this effect may be due to trapped air, and Dr MONEY agreed that this was a possible explanation.

J.K. MITCHELL (University of California, Berkeley) referred to a Speciality Conference of the American Society of Civil Engineers that included *in situ* permeability testing. It is only possible to measure permeability to within one order of magnitude or so. Fortunately, in most cases this is adequate but in some instances it is more critical. Even in a relatively simple geological condition (uniform clean sand) with a well defined material, permeability values ranging between 0.1 and 40 $\times$ 10^{-4} cm/s were obtained from laboratory and field tests and correlations. Measurements of post-construction seepage confirmed a permeability of about 5 to 10 $\times$ 10^{-4} cm/s.

From CRIPPS, J.C., BELL, F.G. & CULSHAW, M.G., (eds), 1986, *Groundwater in Engineering Geology,* Geological Society Engineering Geology Special Publication No. 3, p. 295

SESSION 8

PREDICTION OF GROUNDWATER CONDITIONS

A model for predicting groundwater level response to meteorological changes

M. E. Barton & R. I. Thomson

ABSTRACT: A simple model, based on the overall soil water balance, has been developed for the prediction of groundwater levels in response to meteorological changes. The model consists of two parts as follows:

(i) A soil moisture deficit model which uses daily rainfall and potential evapo-transpiration data to estimate the effective rainfall entering the ground beyond the influence of evapo-transpiration.

(ii) A model relating the effective rainfall to the water table and predicting the groundwater level response.

Rainfall, potential evaporation and water level records are required for at least two years in order to perform a split record test. The data from the first year are used for calibrating and from the second year for testing the model.

The model has been applied to a cliff top site adjacent to the degrading coastal cliffs at Naish Farm, near Highcliffe, Hampshire. The necessary data has been collected over the two year period from October 1982 to October 1984. The groundwater level data used is for the upper permeable layer of Plateau Gravel. The Plateau Gravel acts as a groundwater store, supplying seepage both down into the underlying Barton Clay and, under winter conditions, outwards to the cliff face. The meteorological data were obtained by setting up a local weather station. The results show a good model fit to the observed data.

The model is suitable for general application. It is considered that it may prove of particular value for areas with active landslides as a means of understanding the incidence and rates of slope movements.

Introduction

The aim of the model is to predict groundwater level fluctuations caused by meteorological variations. The relationship between meteorological conditions and ground water level is complex. It has been simplified in this model by using a water balance approach.

Theory of the model

The model consists of two separate water balances. The first is for the upper part of the profile and calculates the effective rainfall percolating down below the influence of evaporation. It is:

$$\mathrm{ER} = \mathrm{P} - \mathrm{RUNOFF} - \mathrm{AE} - \Delta\mathrm{S} \quad (1)$$

where P is rainfall; RUNOFF is surface runoff; AE is actual evaporation; ER is effective rainfall; ΔS is the increase in soil moisture storage.

The effective rainfall percolates down the profile to become recharge to the groundwater table where a second water balance is applied:

$$\mathrm{RECHARGE} = \mathrm{DRAINAGE} + \Delta\mathrm{W} \quad (2)$$

where DRAINAGE is the loss of water to groundwater flow; and ΔW is the change in water storage which is achieved by a change in the water table level.

As the water balances involve volumes of water they need to be solved over a set period of time. This is called the *time step of calculations*.

Calculation of effective rainfall

Effective rainfall is calculated by solving Equation 1. RUNOFF is approximated as a constant percentage of rainfall.

AE and ΔS are estimated by using a soil moisture deficit (SMD) model. This model assumes that when the SMD is satisfied, there is no further change in soil moisture storage ($\Delta S = 0$). It has been assumed by some authors (e.g. Penman 1949) that no effective rainfall occurs while there is an SMD although a number of researchers have reported this not to be the case (e.g. Kitching *et al.* 1977). Smith *et al.* (1970) introduced a direct effective rainfall component where a proportion of the infiltrated rainfall bypasses the SMD model to go directly as effective rainfall. Rushton & Ward (1979) examined a number of different possible direct effective rainfall models. The need for a direct effective rainfall component may

From Cripps, J.C., Bell, F.G. & Culshaw, M.G., (eds), 1986, *Groundwater in Engineering Geology,* Geological Society Engineering Geology Special Publication No. 3, pp. 299-311

component may be due to the presence of a system of cracks (e.g. tension cracks), or a reflection on the limitation of the SMD concept for representing the soil moisture fluxes within the soil matrix. The model here includes a direct effective rainfall component (ER_1) in addition to the component available when SMD = 0, (ER_2). The value of ER_1 is taken as a proportion (A) of the infiltration.

When water is plentiful it is freely evaporated to the atmosphere at a rate which is determined by meteorological factors. Under these conditions, the evaporation is *flux controlled* and its rate is referred to as the potential evaporation (PE). When water is not plentiful, i.e. as the soil dries out, the evaporation rate may fall below PE and is *profile controlled,* i.e. control is by the maximum rate at which moisture is transferred to the evaporation surface. The two conditions may be defined as:

Flux Control : $PE \leqslant P - RUNOFF$; $AE = PE$,
Profile Control : $PE > P - RUNOFF$; $AE \leqslant PE$.

The evaporation rate, AE, is a function of PE and SMD. The function used here is the same as that used by Lloyd *et al.* (1966) and is shown in Fig. 1.

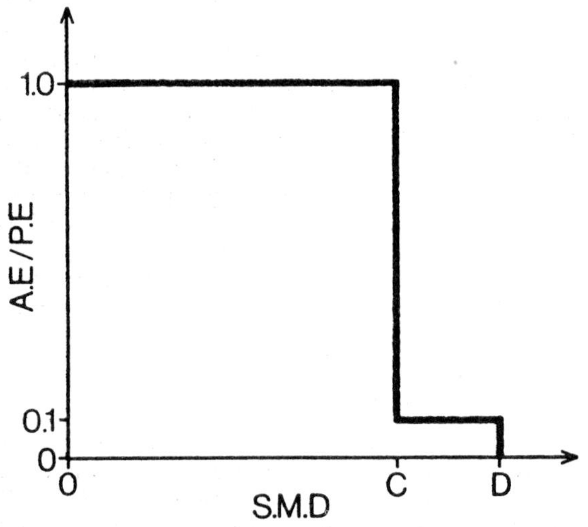

FIG. 1. Relationship between AE and PE and SMD used in the model.

The calculation of effective rainfall is summarised in Fig. 2. The term DETENTION is a model concept introduced for the rainfall less the surface runoff. Physically DETENTION may be considered to consist of interception storage on vegetation and ponded storage at the soil surface. When the evaporation rate under flux control has been satisfied any excess water infiltrates the soil surface. INSTORE is the amount by which SMD is reduced and ER_2 is the excess after SMD has been reduced to zero. A mathematical summary of the model to calculate effective rainfall is given in Appendix A.

Calculation of recharge

Effective rainfall entering the top of the unsaturated zone eventually displaces a similar quantity at the bottom as recharge although the two distributions will not be the same.

Headworth (1972) noted there was a delay between an effective rainfall and a groundwater response (and therefore recharge) which he termed the 'response interval' and Harper (1975) termed as the 'time lag'. In the model it is termed the DELAY time and is taken to be a whole number of time steps.

The unsaturated zone is considered to act like a leaky reservoir with input ER (displaced in time by DELAY) and output recharge. The concept of representing hydrological processes by reservoirs has been used by others (e.g. Nash 1960). The outflow response from a single reservoir having received an instantaneous unit input is to decay exponentially as described by:

$$q = \frac{1}{K} \cdot e^{-t/K}, \quad (3)$$

where t is the time since the input, and K is the decay constant with units of time. The unit of time is equal to the length of one time step. The model uses this equation and assumes the effective rainfall to occur instantaneously at the start of the time step to which it refers. Assuming proportionality and superposition and including the delay time Equation 3 gives the total recharge for the nth step (see Appendix B) as:

$$RECHARGE_n = e^{-1/K} \cdot RECHARGE_{n-1} + (1-e^{-1/K}).ER_{n-DELAY} \cdot \quad (4)$$

This is an iterative equation which must be started by assuming an initial value of recharge well before the start of the calibration period so that there are enough iterations to make the error due to the estimation of the initial recharge negligible. The number of iterations necessary before the start of water level prediction depends upon the effective rainfall distribution and the value of K. A value in excess of $5K$ iterations should be more than adequate.

Calculation of a change in water level

A change in water level is calculated by solving Equation 2. However, to do this two other relationships are needed. The first relates a change in storage to a change in groundwater level. This is termed the specific yield, SY, which may vary with water level but is assumed to be constant for a given time step. Equation 2 is re-written as:

$$RECHARGE = DRAINAGE + (h_1 = h_0).SY \quad (5)$$

where h_0 and h_1 are the water levels at the start and end of the time step.

The second relationship relates the drainage to the storage. If drainage and storage are transformed by dividing by SY they become the instantaneous rate of fall of water level due to drainage only and water level respectively. This relationship was derived by Thomson (1921, 1931) and Headworth (1972) for locations in chalk. The drainage relationship was derived by the method used by Headworth (1972) and is described below.

The water level record is examined for periods of falling water levels. The mean fall in water level per time step between readings is determined for the entire range of recorded water levels. Each value of the rate of fall in water level (groundwater recession) is then plotted against the mean water level between readings. Ignoring the experimental errors, the plot must inevitably give a wide scatter of points due to the continued recharge lessening the recession rate. However, for any given water level, the groundwater recession rate due to drainage alone is the maximum rate at which the water level will fall. An envelope curve drawn through the highest values of groundwater recession rate gives the requires relationship between water level and ground water recession due to drainage alone. The curve is approximated by a series of straight lines (Headworth 1972) each of the form:

$$h = h_m - \frac{1}{\text{RECESS}} \cdot \frac{dh}{dt} \qquad (6)$$

where h is the water level, h_m is the 'minimum' water level, and RECESS is a recession constant with units of 1/time.

The instantaneous drainage flow is:

$$q = -\text{SY}.\ \frac{dh}{dt}. \qquad (7)$$

Rearranging equation 6 in terms of dh/dt and inserting in Equation 7 gives:

$$q = \text{SY.RECESS.}(h - h_m)\ . \qquad (8)$$

This shows that if the water level varies linearly over the time step then so does the instantaneous drainage flow. The total drainage for the time step is:

$$\text{DRAINAGE} = \left(\frac{q_1 + q_0}{2}\right) . T \qquad (9)$$

where q_0 and q_1 are the instantaneous drainage flows for the start and end of the time step; and T is the length of the time step and equals one by definition. Substituting Equation 8 in Equation 9 gives:

$$\text{DRAINAGE} = {}^1/_2\ .\ \text{SY}\ .\ \text{RECESS}\ .\ (h_1 + h_0 - 2h_m)\ . \qquad (10)$$

Substituting Equation 10 in Equation 5 and rearranging gives:

$$h_1 = \frac{\dfrac{\text{RECHARGE}}{\text{SY}} + h_0\left(1 - \dfrac{\text{RECESS}}{2}\right) + \text{RECESS}\ .\ h_m\ .}{\left(1 + \dfrac{\text{RECESS}}{2}\right)} \qquad (11)$$

This is an iterative equation which requires h_0 to be initialised. The values of SY, RECESS and h_m for each iteration are taken to apply to the water level h_0. The time step may be of any length provided the assumptions of a linear variation in water level and a constant SY hold. The units of h_1, h_0, h_m and RECHARGE should be the same. A schematic view of the whole model is given in Fig. 2.

Testing of the model

The model has been tested on water level data collected near the actively degrading coastal cliff slopes formed in the outcrop of the Barton Clay at Christchurch Bay, Hampshire (Fig. 3). The engineering geology (Barton 1973), degradational processes (Barton & Coles 1984), hydrogeology and piezometric and meteorological observations (Barton & Thomson 1984) have been described elsewhere. The water level data used is for the drift deposit (Plateau Gravel) overlying the Barton Clay on the cliff top and were not affected by cliff top slumping. Meteorological measurements were obtained by setting up a weather station (Fig. 3), supplemented with data from Hurn Airport. Potential evaporation for the grass covered cliff top was estimated using Penman's Method (Penman 1963).

Time step for calculations

The water level data indicate that a time step of one day is necessary to satisfy the assumption of a linear variation in water level. Howard & Lloyd (1979) presented evidence that the length of time step also affects the calculation of effective rainfall. They found a significant difference between one day and ten day time steps. A one day time step has been used to test the model.

Split record test

At least one year's data is needed to calibrate the model. It is not always possible to obtain unique estimates for the parameters (Gupta & Sorooshian 1983). Therefore, a second year's data is used to test the calibrated solution to ensure that the fit is still adequate. This is called a split record test and has been used here with two years' data (October 1982 to October 1984).

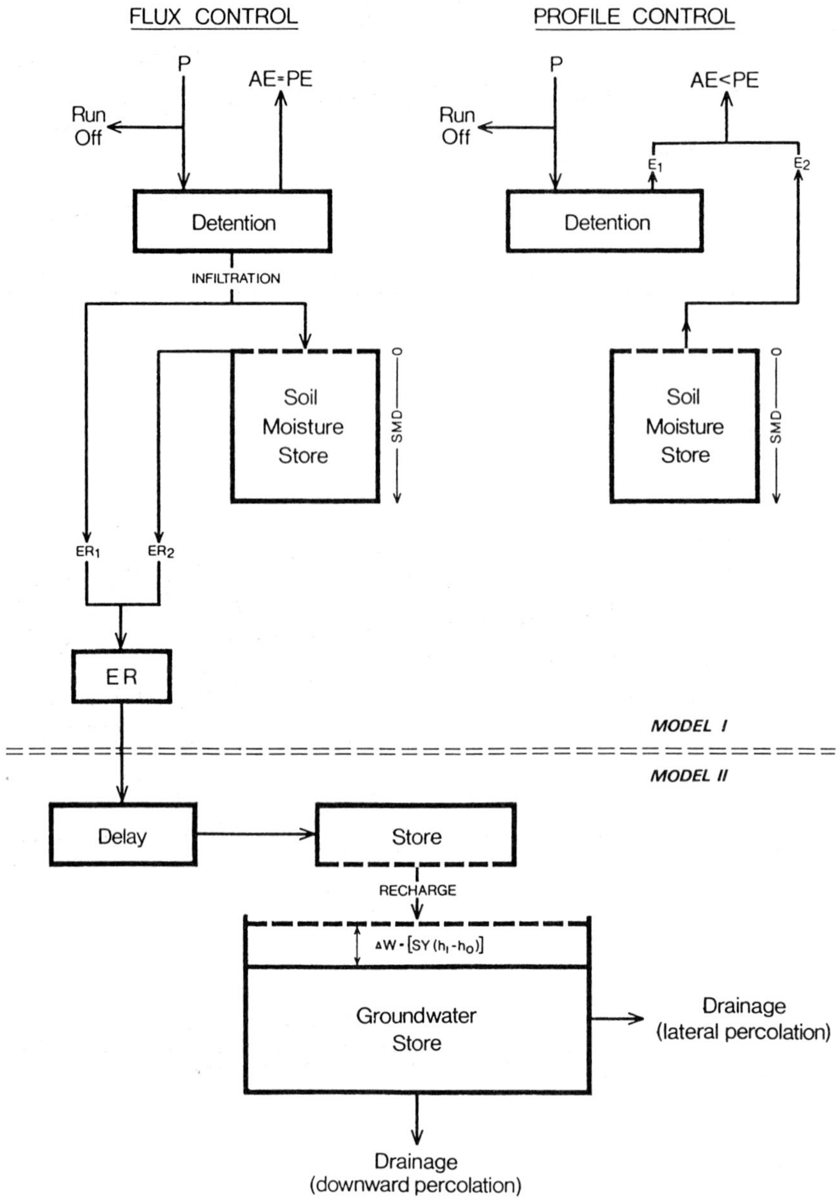

FIG. 2. Schematic view of two-part model for predicting groundwater level response to meteorological changes.

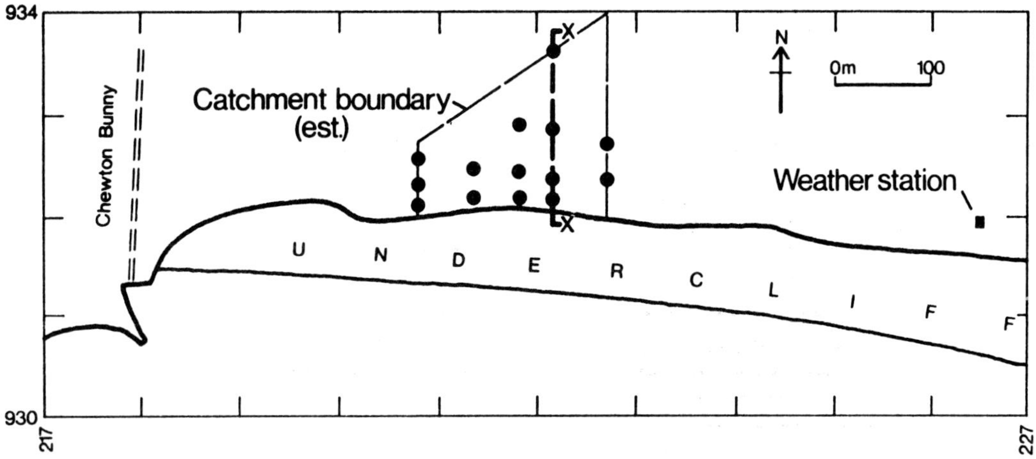

(a)

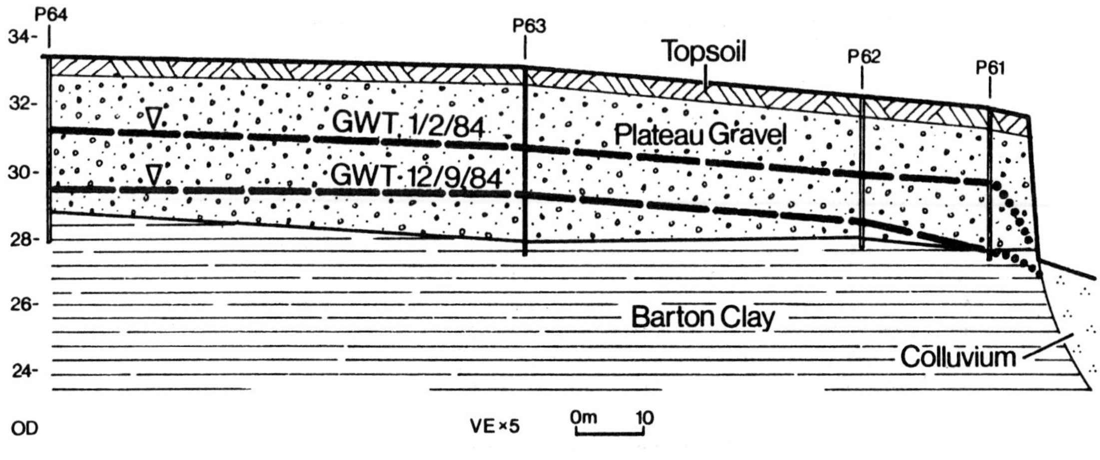

(b)

FIG. 3 (a) Location of test area showing weather station, standpipe/piezometers in the Plateau Gravel and the estimated catchment boundary for seepage through the gravel to the undercliff. (b) Section XX through the standpipe/piezometers for which the model has been tested (results presented in Figs 4 & 5 for P64). All piezometers shown are open to the Plateau Gravel.

Performance criterion

Model performance was assessed objectively using a performance criterion. The one used here (CORREL) is similar to that (U_7) used by Diskin & Simon (1977) in their comparison study for a number of performance criteria:

$$\text{CORREL} = \frac{C1 - C2}{C1} = 1 - U_7 \qquad (12)$$

where

$$C1 = \Sigma(\text{obs}_i - \overline{\text{obs}})^2 \quad \text{and} \quad C2 = \Sigma(\text{obs}_i - \text{pred}_i)^2$$

obs_i are the observed water level data with mean obs
pred_i are the predicted water levels corresponding to the observed water levels.

Model calibration

Model calibration is made easier by some prior estimation of parameter values. Surface runoff has not been observed to occur and is therefore taken to be zero. The value of the parameter *D* (Fig. 1) was made large so as to make it redundant. Thus the number of model parameters to be optimised was reduced by two. Penman (1949) gives a value of *C* (Fig. 1) of 100 mm for grass and 25 mm for bare ground. Thompson *et al.* (1981) reduce the "available water" by an arbitrary 25% for permeable soil. The soil at Highcliffe is permeable and the grass cover variable, being very good in places and almost bare in others. The parameter *C* is a measure of the available water so that its value is expected to vary between 20 and 75 mm depending on the grass cover. For the water level test data the grass cover was good which was reflected by an optimised value of *C* of 65 mm. The value of ER_1 is obtained using a proportion (*A*) of the infiltration ranging from 0.25 to 0.34 with a value of 0.33 for the test data. The parameter DELAY can be estimated from rainfall and observed water level data. The optimised values of DELAY and *K* for the test data were 1 day and 2.3 days respectively. The value of SY was assumed to be constant for all water levels with a value of 0.065 for the test data. An initial estimate of the drainage relationship was made using the method in Headworth (1972) as previously described. Fig. 4 shows the scatter of points obtained. The error bars show the estimated error in water level observations (±1 cm). The errors may be due to frequent adverse weather conditions and the use of two alternating observers. Thus the use of an envelope curve was only used as an initial estimation. The final drainage relationship was found by successive estimations until the optimised model adequately predicted the water level recession.

Results

Fig. 4 shows the final drainage relationship of the optimised model. Fig. 5 shows the observed and predicted water levels and the distributions of rainfall, effective rainfall, recharge, and drainage. The large increase in drainage flow at high water levels is due to seepage at the cliff face. The fairly constant drainage flow throughout the year is due to leakage into the Barton Clay (between the profile section and the cliff edge). The model performance over the two year period is very good with a value of CORREL of 0.975.

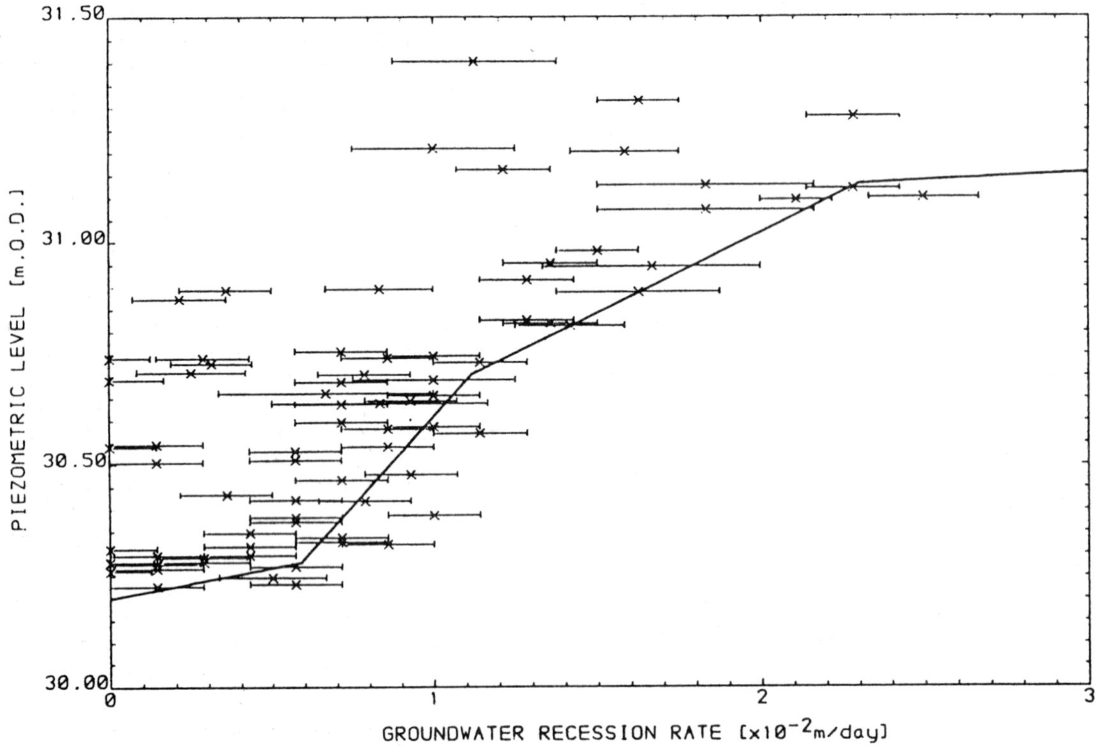

FIG. 4. Relationship between water level and groundwater recession due to drainage.

Discussion

Precision

There will always be some discrepancy between the observed and predicted outputs. These will be due to a combination of approximations and assumptions made in the model, the size of the time step, errors in the input data, errors in the observed output data and the choice of performance criterion used (Diskin & Simon 1977).

Ambiguity (in the pattern of water level fluctuation between readings) and errors in the output data can be reduced by more frequent observations. Errors in the input data may be observational, due to the method of measurement, or due to the non-representativeness of the measurement. Rodda (1967) found that standard rain gauges underestimate rainfall by varying amounts depending on the wind speed. Rainfall varies spatially so that the rain gauge should be sited as near as possible to the water level measurement site. Howard & Lloyd (1979) examined the sensitivity of the Penman equation to errors in the measurement of the relevant meteorological variables and in the estimation of parameters in the equation. Chidley & Pike (1970) tabulate parameter values established by various researchers in different parts of the world. There are a number of methods of estimating potential evaporation, each of which gives a different estimate (Sangrey *et al.* 1984). Kitching *et al.* (1977) measured different recharges for two sites close to each other. They attributed the difference to one site being more sheltered so that evaporation was less.

The runoff proportion can vary with rainfall intensity and surface moisture conditions. The assumption of runoff being a constant percentage of rainfall is an approximation. A number of alternative direct effective rainfall models are also possible (Rushton & Ward 1979). The method of estimating runoff and direct effective rainfall in the model can easily be altered if desired as can the choice of the SMD model. Calder *et al.* (1983) and Alley (1984) compared different SMD models. Effective rainfall and recharge could, alternatively, be measured directly by lysimeter (e.g. Kitching *et al.* 1977, and Keating 1984), or from profile moisture content and potential measurements (Freeze & Banner 1970) although this would require a large amount of field work.

Similar models

Other models for predicting groundwater level fluctuations have been produced by Sangrey *et al.* (1984) and Canuti *et al.* (1984). The models investigated by Alley (1984) were developed to model stream flows but could be adapted to estimate groundwater level fluctuation.

The model of Sangrey *et al.* (1984) is of particular interest in being developed for general landslide studies (*vide infra*). It differs from the model presented here. Sangrey *et al.* calculated effective rainfall as the value of rainfall minus potential evaporation when positive (else zero) using a time step of one month. They did not include surface runoff or surface soil moisture storage effects. Their equation for the calculation of recharge was the same as here (Equation 4) except that they did not allow for a delay in groundwater response. They used a simple linear relationship between water level and recharge whereas here, a water balance has been used so that Equation 11 is a function of the water level at the start of the time step as well as the recharge.

Canuti *et al.* (1984) also use a linear correlation of water level but with the logarithm of a term they call the antecedent precipitation index (API) which is analogous to recharge. It is a measure of past rainfall. However, the API did not account for evaporation and so gave high summer readings.

Groundwater pressures

The model has been developed and shown to simulate fluctuations in the level of the groundwater table. The model may also be used to predict groundwater pressures below the water table which are influenced by the position of the water table. The specific yield, SY, would have to be redefined as a relationship between the recharge (drainage) and the resultant increase (decrease) in piezometric level if there were no drainage (recharge). This is evidenced by data for some piezometric pressures in the Barton Clay at the test site which show a similar, but more damped, response to meteorological influences than does the water table in the Plateau Gravel. Harper (1975) presented evidence that suggested that piezometric pressures in confined aquifers may also be predicted from records of rainfall (although observations need to be adjusted to account for the effect of fluctuations in atmospheric pressure). He found that the plot of piezometric pressure versus recession rate was non-linear but this could be approximated by a series of straight lines as in the test example presented here. The data Harper presents would need a time step of considerably less than one day. This would require the use of a continuously recording rainfall gauge. The diurnal variation of PE should be estimated although a daily average could be sufficiently accurate.

Thus the model can be used to simulate the piezometric pressure of any point below the water table provided it is only influenced by meteorological conditions and stable drainage conditions. Other effects such as tidal changes, atmospheric pressure changes, piezometric time lag, or land movement have not been included. Due consideration for these effects would have to be made before applying the model.

Application of the model in geotechnical engineering

While the model has some general similarity with models used in the field of hydrology, such as the regional water balance models (discussed by Alley

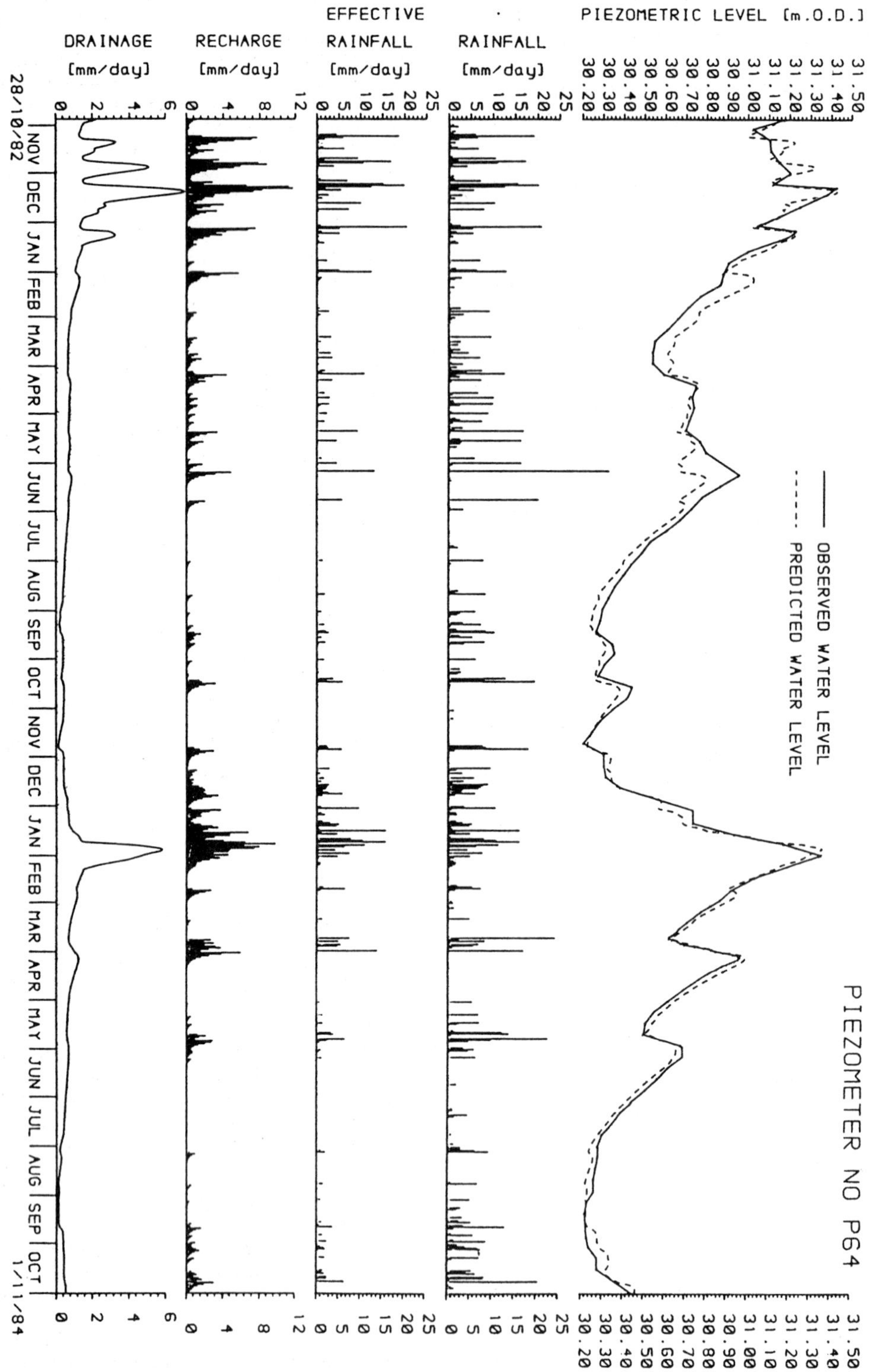

FIG. 5. Observed and predicted water levels for Piezometer No. P64. Also shown are the distributions of rainfall, effective rainfall, recharge and drainage.

1984) and recharge calculations for aquifers (discussed by Freeze & Banner 1970; Howard & Lloyd 1979; Keating 1984; Rushton & Ward 1979 and others) it has been formulated particularly for use in Geotechnical Engineering with special reference to landslide studies. Sangrey (1982) and Sällfors & Svenson (1984) have drawn attention to the comparative neglect of studies of the temporal variations of water tables and pore pressures despite their crucial role in slope stability. Much work is currently in progress to remedy this deficiency (see, for instance, Bertini *et al.* 1984; Brand *et al.* 1984; Canuti *et al.* 1984; Greenway *et al.* 1984 and Sangrey *et al.* 1984).

Three particular applications of the model presented here can be summarised as follows:

1. Prediction of groundwater levels resulting from meteorological changes. Owing to the delayed response of groundwater levels to rainfall events, the advance warning given by the meteorological observations could prove an advantage over direct piezometric readings (Harper 1975; Barton & Thomson 1984).
2. Prediction of maximum likely, future ground water levels. Given the continuity of the recharge/drainage conditions, then the meteorological records could be used to find the probability (as expected recurrence interval, or return period) of any given water level.
3. Prediction of water levels backwards in time from suitable meteorological records to improve the accuracy of back analysis of stability. Back analysis is an important means of determining soil parameters in stability studies (Chandler & Skempton 1974) and the accuracy of the result could be improved if the recharge/drainage conditions of the calibrated site were analogous to those that existed at the position of the failure plane prior to the failure.

ACKNOWLEDGEMENTS: The authors are grateful to Mr C. W. Salter of Naish Estates (Hants) Ltd., for permission to work in the study area. They wish to record their especial thanks to Dr E. N. Bromhead and Messrs M. Chandler, B. J. Coles, M. R. Cooper and N. Dixon for their assistance with the instrumentation and to Mr A Brookes for technical assistance and drawing the illustrations.

References

ALLEY, W. M., 1984. On the treatment of evaporation, soil moisture accounting, and aquifer recharge in monthly water balance models. *Water Resources Research* **20,** 1137-1149.

ANDERSON, M. G. & POPE, R. G., 1984. The incorporation of soil water physics models into geotechnical studies of landslide behaviour. *Proceedings of the 4th International Symposium on Landslides*, Toronto, Canada, 1, 349-353.

BARTON, M. E., 1973. The degradation of the Barton Clay cliffs in Hampshire. *Quarterly Journal of Engineering Geology*, **6**, 423-440.

BARTON, M. E. & COLES, B. J., 1984. The characteristics and rates of the various slope degradation processes in the Barton Clay cliffs of Hampshire. *Quarterly Journal of Engineering Geology*, **17**, 117-136.

BARTON, M. E. & THOMSON, R. I., 1984. Studies of the Water Balance in a rapidly degrading soil cliff. *Proceedings of the 4th International Symposium on Landslides*, Toronto, Canada 1, 355-361.

BERTINI, T., CUGUSI, F., D'ELIA, B. & ROSSI-DORIA, M., 1984. Climactic conditions and slow movements of colluvial covers in Central Italy. *Proceedings of the 4th International Symposium on Landslides*, Toronto, Canada, **1**, 367-376.

BRAND, E. W., PREMCHITT, J & PHILLIPSON, H. B., 1984. Relationship between rainfall and landslides in Hong Kong. *Proceedings of the 4th International Symposium on Landslides*, Toronto, Canada, **1**, 377-384.

CALDER, I. R., HARDING, R. J. & ROSIER, P. T. W., 1983. An objective assessment of soil moisture deficit models. *Journal of Hydrology*, **60**, 329-355.

CANUTI, P., FOCARDI, P., GARZONIO, C. A., RODOLFI, G. & ZANCHI, C., 1984. Analysis of the dynamic of a mass movement on silty clayey lacustrine deposits in North-Central Italy (Mugello, Tuscany). *Proceedings of the 4th International Symposium on Landslides*, Toronto, Canada, **1**, 391-397.

CHANDLER, R. J. & SKEMPTON, A. W., 1974. The design of permanent cutting slopes in stiff fissured clays. *Geotechnique,* **24,** 457-466.

CHIDLEY, T. R. E. & PIKE, J. G., 1970. A generalised computer program for the solution of the Penman Equation for evapotranspiration. *Journal of Hydrology,* **10,** 75-89.

DISKIN, M. H. & SIMON, E., 1977. A procedure for the selection of objective functions for hydrologic simulation models. *Journal of Hydrology,* **34,** 129-149.

FREEZE, R. A. & BANNER, J., 1970. The mechanism of natural ground water recharge and discharge: 2. Laboratory column experiments and field measurements. *Water Resources Research,* **6,** 138-155.

GREENWAY, D. R., ANDERSON, M. G. & BRIAN-BOYS, K. C., 1984. The influence of vegetation on slope stability in Hong Kong. *Proceedings of the 4th International Symposium on Landslides, Toronto, Canada,* **1,** 399-404.

GUPTA, V. K. & SOROOSHIAN, S., 1983. Uniqueness and observatibilty of conceptual rainfall - runoff model parameters: the percolation process examined. *Water Resources Research,* **19,** 269-276.

HARPER, T. R., 1975. The transient groundwater pressure response to rainfall and prediction of rock slope instability. *International Journal of Rock Mechanics and Mining Science & Geomechanics Abstracts,* **12,** 175-179.

HEADWORTH, H. G., 1972. The analysis of natural ground water level fluctuations in the Chalk of Hampshire. *Journal of the Institution of Water Engineering,* **26,** 107-124.

HOWARD, K. W. F. & LLOYD, J. W., 1979. The sensisitivity of parameters in the Penman evaporation equations and direct recharge balance. *Journal of Hydrology,* **41,** 329-344.

KEATING, T., 1984. Recharge into a shingle beach. *Journal of Hydrology,* **72,** 187-194.

KITCHING, R., SHEARER, T. R. & SHEDLOCK, S. L., 1977. Recharge to Bunter Sandstone determined from lysimeters, *Journal of Hydrology,* **33,** 207-232.

LLOYD, J. W., DRENNAN, D. S. H. & BENNELL, B. M. U., 1966. A groundwater recharge study in north eastern Jordan. *Proceedings of the Institution of Civil Engineers,* **35**, 615-631.

NASH, J. E., 1960. A unit hydrograph study, with particular reference to British catchments. *Proceedings of the Institution of Civil Engineers,* **17**, 249-282.

PENMAN, H. L., 1949. The dependence of transpiration on weather and soil conditions. *Journal of Soil Science,* **1**, 74-89.

PENMAN, H. L., 1963. Vegetation and Hydrology. Technical Communication No. 53, Comm. Bur. Soils, *Comm, Agric. Bur.,* Farnham Royal, England.

RODDA, J. C., 1967. The Systematic error in rainfall measurement. *Journal of the Institution of Water Engineers,* **21**, 173-177.

RUSHTON, K. R. & WARD, C., 1979. The estimation of groundwater recharge. *Journal of Hydrology,* **41**, 345-361.

SÄLLFORS, G. B. & SVENSSON, C., 1984. Prediction of maximum groundwater pressure. *Proceedings of the 4th International Symposium on Landslides,* Toronto, Canada, **1**, 437-440.

SANGREY, D. A., 1982. Evaluation of landslide properties; in 'Application of Walls to Landslide Control Problems'. (Ed. by R. B. Reeves), *American Society of Civil Engineers.*

SANGREY, D. A., HARROP-WILLIAMS, K. O. & KLIEBER, J. A., 1984. Predicting groundwater response to precipitation. *Proceedings of ASCE, Journal of Geotechnical Engineering Div.* **110**, 957-975.

SMITH, D. B., WEARN, P. L., RICHARDS, H. J. & ROWE, P. C., 1970. Water movement in the insaturated zone of high and low permeability strata by measuring natural tritium. *In: Symposium of the use of isotopes in hydrology*, IAEA., Vienna, 73-87.

THOMPSON, N., BARRIE, I. A. & AYLES, M., 1981. The Meteorological Office Rainfall and Evaporation Calculation System: MOREECS Hydrological Mem. No. 45, Met 08 (Hydrometeorological Services). **HMSO**.

THOMSON, D. H., 1921. Hydrological conditions in the Chalk at Compton, W. Sussex. *Transactions of Institution of Water Engineers*, **26**, 228-251.

THOMSON, D. H., 1931. Hydrological conditions in the Chalk at Compton, W. Sussex. (Second Paper). *Transactions of the Institution of Water Engineers,* **36**, 176-195.

M. E. BARTON & R. I. THOMSON, Department of Civil Engineering, University of Southampton, Southampton, Hampshire, SO9 5NH, UK.

Appendix A:
Mathematical summary of the calculation of effective rainfall (Model 1, Fig. 2)

DETENTION = P–RUNOFF

(i) For flux control, i.e. DETENTION $\geqslant$ PE

AE = PE
INFILTRATION = DETENTION – PE
ER_1 = A.INFILTRATION, where A in a parameter
INSTORE = INFILTRATION – ER_1

If INSTORE $\leqslant SMD_{old}$

then $SMD_{new} = SMD_{old}$ – INSTORE

and $ER_2 = 0$

$ER = ER_1 + ER_2$ = A.(P–RUNOFF–PE)

If INSTORE $> SMD_{old}$

then $SMD_{new} = 0$

and ER_2 = INSTORE – SMD_{old}

$ER = ER_1 + ER_2$ = P–RUNOFF–PE–SMD_{old}

(ii) For profile control, i.e. DETENTION < PE

E_1 = DETENTION
INFILTRATION = 0
ER = 0
INSTORE = 0

If $PE - E_1 \leqslant C - SMD_{old}$

then $SMD_{new} = SMD_{old} + (PE–E_1)$

If $PE - E_1 > C - SMD_{old}$ and $SMD_{old} < C$

then $SMD'_{new} = C + 0.1[(PE–E_1)-(C–SMD_{old})]$

If $SMD_{old} \geqslant C$

then $SMD'_{new} = SMD_{old} + 0.1(PE–E_1)$

If $SMD'_{new} \geqslant D$

then $SMD_{new} = D$

If $SMD'_{new} \leqslant D$

then $SMD_{new} = SMD'_{new}$

$E_2 = \Delta S = SMD_{new} - SMD_{old}$

$AE = E_1 + E_2$ = P–RUNOFF+SMD_{new} –SMD_{old}

SMD_{old} and SMD_{new} are the SMDs at the start and end of the time step.

Appendix B: Calculation of the total recharge for the nth time step (model 2, Fig. 2)

The outflow response from a single reservoir having received an instantaneous unit input is:

$$q = \frac{1}{K} \cdot e^{-t/K} \tag{B1}$$

where q is the instantaneous outflow at time t after the input, and K is a decay constant. Assuming proportionality:

$$q = \frac{\mathrm{ER}}{K} \cdot e^{-t/K} \tag{B2}$$

where q is now the instantaneous recharge and ER is the instantaneous input. ER is the effective rainfall displaced in time by DELAY time steps (see Fig. 2). The effective rainfall is assumed in this analysis to occur instantaneously at the start of the time step to which it refers. Therefore, the instantaneous input in Equation B2 which occurs at time $t = 0$ is the effective rainfall calculated for the time step $t = -\mathrm{DELAY}$ to $t = {}^{-}\mathrm{DELAY}+1$. Integrating equation B2 gives the total output for any one time step as:

$$Q = \int_{t}^{t+1} \frac{\mathrm{ER}}{K} \cdot e^{-t/K} \cdot dt = -\mathrm{ER}[e^{-t/K}]_{t}^{t+1} = \mathrm{ER} \cdot e^{-t/K} \cdot (1 - e^{-1/K}) . \tag{B3}$$

Let ${}_jQ_i$ be the output for the ith time step since the jth input.

Let ${}_{j\text{-DELAY}}\mathrm{ER}$ be the jth input. Equation B3 gives the recharge due to the jth input as:

$$\text{for } t = 0 \quad {}_jQ_1 = {}_{j\text{-DELAY}}\mathrm{ER} \cdot (1 - e^{-1/K})$$

$$\text{for } t = 1 \quad {}_jQ_2 = {}_{j\text{-DELAY}}\mathrm{ER} \cdot e^{-1/K} \cdot (1 - e^{-1/K}) = e^{-1/K} \cdot {}_jQ_1$$

$$\text{for } t = i \quad {}_jQ_{i+1} = {}_{j\text{-DELAY}}\mathrm{ER} \cdot e^{-i/K} \cdot (1 - e^{-1/K}) = e^{-1/K} \cdot {}_jQ_i$$

$$\text{for } t = i+1 \quad {}_jQ_{i+2} = {}_{j\text{-DELAY}}\mathrm{ER} \cdot e^{-(i+1)/K} \cdot (1 - e^{-1/K}) = e^{-1/K} \cdot {}_jQ_{i+1} .$$

$$\left.\begin{array}{l} \text{In general, for } i = 1,\ {}_jQ_i = {}_{j\text{-DELAY}}\mathrm{ER} \cdot (1 - e^{-1/K}), \\ \text{for } i > 2,\ {}_jQ_i = e^{-1/K} \cdot {}_jQ_{i-1} . \end{array}\right\} \tag{B4}$$

Assuming superposition, the total groundwater recharge for the nth time step is the sum of the ground water recharges for that time step due to all the previous individual inputs of effective rainfall. This is given by:

$$\mathrm{RECHARGE}_n = \sum_{j=1}^{n} {}_jQ_{n-j+1} . \tag{B5}$$

Substituting Equations B4 for terms in equation B5 gives:

$$\mathrm{RECHARGE}_1 = {}_{1\text{-DELAY}}\mathrm{ER} \cdot (1 - e^{-1/K})$$

$$\mathrm{RECHARGE}_2 = e^{-1/K} \cdot {}_1Q_1 + {}_{2\text{-DELAY}}\mathrm{ER} \cdot (1 - e^{-1/K})$$

$$\text{RECHARGE}_3 = e^{-1/K} \,.\, {}_1Q_2 + e^{-1/K} \,.\, {}_2Q_1 + {}_{3\text{-DELAY}}\text{ER} \,.\, (1 - e^{-1/K})$$

$$\text{RECHARGE}_n = e^{-1/K} \,.\, {}_1Q_{n-1} + e^{-1/K} \,.\, {}_2Q_{n-2} + e^{-1/K} \,.\, {}_3Q_{n-3} + \ldots$$

$$\ldots + e^{-1/K} \,.\, {}_jQ_{n-j} + \ldots + e^{-1/K} \,.\, {}_{n-1}Q_1 + {}_{n\text{-DELAY}}\text{ER} \,.\, (1 - e^{-1/K})$$

$$= e^{-1/K} \,.\, \left(\sum_{j=1}^{n-1} {}_jQ_{n-j} \right) + {}_{n\text{-DELAY}}\text{ER} \,.\, (1 - e^{-1/K})$$

$$= e^{-1/K} \,.\, \text{RECHARGE}_{n-1} + {}_{n\text{-DELAY}}\text{ER} \,.\, (1 - e^{-1/K}) \,.\, \text{(B6) (= 4 in paper)}$$

Thus the groundwater recharge for any given time step is the weighted average of the effective rainfall of DELAY time steps before and the previous time step's recharge. The relative weighting is determined by the parameter, K. The model uses Equation B6 for the calculation of recharge, with $\text{RECHARGE}_{n-1} = 0$ for $n = 1$ and at the start of water level prediction $n \geqslant 5K$.

Groundwater in Engineering Geology, London, 1986, pp. 313-320 Printed in England

Groundwater modelling by microcomputer: applications to dam and reservoir slope stability

E. N. Bromhead, A. L. Havouzari, M. Brice & B. H. Rofe

ABSTRACT: The paper describes the use of computer seepage modelling in three earth dams. In the first case, it is used principally as a design tool; in the second, as an aid to interpreting site investigation data and observations; in the third to investigate the effect of reservoir raising on downstream valley side slopes.

Comments are made on the use of the techniques in practice, and typical results are given.

Introduction

The prediction of groundwater levels is an integral part of slope stability analysis and design. It can include localised investigations, or what would normally be considered to be 'engineering seepage analysis' usually carried out on a cross section through the slope; and it can encompass more generalised groundwater models. In the case of the latter, water flow is again taken to occur in one plane only, but this time it is a plan of the area of the seepage investigations. Three dimensional analyses are rarely carried out.

This paper describes some examples of both approaches, and their application in practice. The calculations referred to, make use of the finite element method. Initial work has been done on a VAX minicomputer at Kingston Polytechnic, and later work on Apricot personal computers which although smaller and slower, are quite able to do this work.

Theoretical background

The theory of seepage has its origins in classical hydro-dynamics as the flow of an ideal fluid. Expressed as potential flow (potential = ϕ) in an isotropic region, the general continuity equation (see Fig. 1):

$$\frac{\partial Vx}{\partial x} + \frac{\partial Vy}{\partial y} + \frac{\partial Vz}{\partial z} = 0 \qquad (1)$$

gives the conventional Laplace equation:

$$\frac{\partial^2\phi}{\partial x^2} + \frac{\partial^2\phi}{\partial y^2} + \frac{\partial^2\phi}{\partial z^2} = 0 \qquad (2)$$

for potential flow in an istropic, homogeneous seepage domain when Darcy's law:

$$Vn = -Kn \, \frac{\partial\phi}{\partial n} \qquad (3)$$

is incorporated. Strictly, in the more general case,

$$\frac{\partial}{\partial x} Kx \frac{\partial\phi}{\partial x} + \frac{\partial}{\partial y} Ky \frac{\partial\phi}{\partial y} + \frac{\partial}{\partial z} Kz \frac{\partial\phi}{\partial z} = 0, \qquad (4)$$

allows for arbitrary variation of the permeability coefficients in both space and direction.

Finite element solution to this equation subject to conventional boundary conditions in two and three dimensions is trivial, and the basic techniques are at least 20 years old (Zienkiewicz & Cheung 1965). Extensions to the theory permit time dependent behaviour (drawdown, compressibility, etc.) to be included.

The solution to this equation is by means of fairly complex mathematics (e.g. Bromhead 1984) which is only possible in simple seepage domains; by finite difference numerical methods, or by flow net sketching. Although it is possible to tackle complex *shaped* problems with the latter two, non homogeneous permeability presents the analyst with major problems.

The finite element method, in contrast, takes an energy approach, leading to the following integral:

$$I = \int_{\text{vol of seepage domain}} Kx\left(\frac{\partial\phi}{\partial x}\right)^2 + Ky\left(\frac{\partial\phi}{\partial y}\right)^2 + Kz\left(\frac{\partial\phi}{\partial z}\right)^2 \text{d vol.} \qquad (5)$$

which must be a minimum for a valid solution to the seepage pattern. Subdivision of the seepage domain into *elements* of finite size enables the integration to be carried out simply, and the potentials evaluated at

From CRIPPS, J.C., BELL, F.G. & CULSHAW, M.G., (eds), 1986, *Groundwater in Engineering Geology,* Geological Society Engineering Geology Special Publication No. 3, pp.

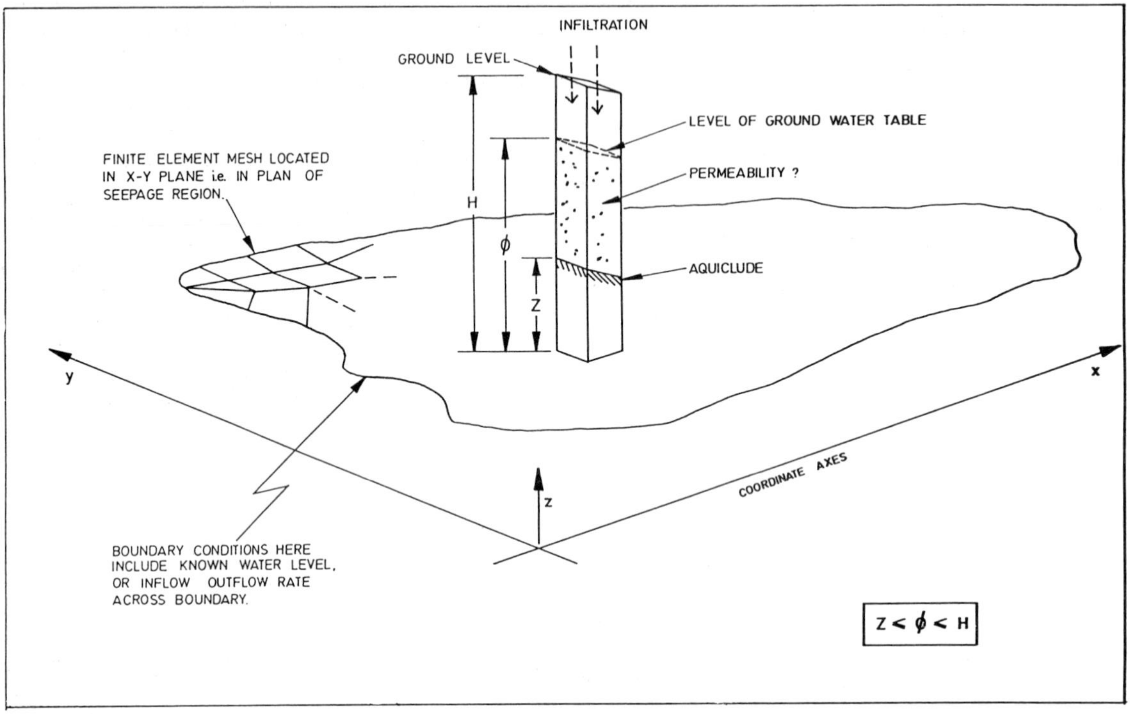

FIG. 1. Theoretical model.

key points or *nodes* where the elements 'connect'. The background to this is widely understood in structural and continuum mechanics (e.g. Zienkiewicz 1972) and will not be further pursued here. However, any constitutive law for seepage cast in a form similar to Equation 4 can be interpreted into an equivalent to Equation 5, without recourse to the energy equations which are sometimes difficult to handle.

Ordinarily, seepage in a plane section through a slope, such as an earth dam, is subject to a number of boundary conditions. These may be physical boundaries to flow, and/or hydraulic boundaries. The software utilised for these analyses allows the following to be used:

- impermeable,
- equipotential,
- seepage surface,
- phreatic line (located via Taylor & Brown's method, 1966),
- surface of known infiltration rate.

Computer graphics techniques are used both to check data for correctness as well as to interpret the results.

For the regional groundwater flow models where hydraulic gradients are low, vertical sections are comparatively less useful, (although Freeze & Witherspoon (1969) use them successfully), and a plan type model is used. The theory of this follows. Although apparently complex, it is quite simple a numerically based extension to Dupuit's theory, taking into account flow in both plan directions.

If Equation 4 is recast into two dimensions and the aquifer thickness is included, then the Equation becomes:

$$\frac{\partial}{\partial x}\, Kxt\, \frac{\partial \phi}{\partial x} + \frac{\partial}{\partial y}\, Kyt\, \frac{\partial \phi}{\partial y} = 0\,. \qquad (6)$$

For *confined* flow, ϕ is then the elevation of the piezometric surface and t can, of course, vary with x and y in an arbitrary fashion (provided that this is approximately represented by the finite elements and mesh used). For *unconfined* flow, t is given by:

$$t = \phi - z \qquad (7)$$

where z is the elevation of impermeable bedrock. This cannot easily be solved directly but a method of successive trials, analogous to the procedure for dealing with effective stress dependent permeability (Bromhead & Vaughan, 1980) can be used. An index of the convergence of this procedure is:

$$\iint \phi \, dxdv \qquad (8)$$

evaluated over the whole plan area of the mesh. Convergence problems can be encountered using this simple iterative procedure, but they are more prevalent where higher order (i.e. more complex) finite elements are used.

Time dependent analyses, incorporating storativity effects either due to aquifer compressibility or depletion can be incorporated into the finite element method without difficulty as can infiltration from rainfall or leaky aquicludes. The final case record incorporates all of these refinements in an analysis of regional groundwater flow in the east abutment of Grimwith Dam.

Methodology

In the following sections there are three case histories, each of which illustrates a different *philosophy* behind the application of seepage analyses.

Existing software was used for the first two, but for the third, new techniques were evolved concurrently with the analysis stages. There is no doubt that the former stage of affairs is preferable, since this speeds up the process and reduces error, but provided the constitutive laws can be defined and/or re-cast in the appropriate form, even the latter can be made practicable.

Seepage analysis as an engineering design tool: Kiteta Dam

A detailed review of the design calculations for this 34 m high dam to be constructed 100 km south-east of Nairobi is given by Pugh & Bromhead (1985). Although a detailed site investigation was carried out, laboratory testing was by necessity restricted and although design parameters could be obtained, these would be subject to quite large variations. Particularly in the case of the seepage pattern, this would be greatly affected by such variations, and the resulting changes in pore water pressures would have a major impact on stability. It was therefore essential to not only evolve a design which was insensitive to the likely variations in material properties but also be economic in its use of materials. This was only possible by undertaking systematic parametric analyses of the extreme case.

An example of the major variation in seepage pattern, caused by only a small change in permeability in a transition zone, is shown on Fig. 2.

Using a preliminary design section it was found that there was little head loss through the rock foundations of the dam so that there were high uplift pressures under the downstream shoulder. The parametric studies also showed that a grout curtain on its own would not provide sufficient control of foundation pressures and needed a drain on the downstream side, or an upstream impermeable blanket. Simple stability calculations showed that the latter would have an adverse impact on stability and was not proceeded with (upstream blankets in steep shouldered dams are a stability risk).

Further 'fine tuning' of the sections made it possible to reduce the width of the core, although at the cost of ensuring tighter construction control at the downstream transitions and filters. Nevertheless, in view of the shortage of suitable core material on site this was a desirable change, and led to the maximum utilisation of spoil from spillway excavations.

In order to capitalise on the fine detail from the seepage analyses, the transfer of pore water pressure data to the stability analysis programs was automated.

The main advantage of the finite element analyses of seepage in this case was that design assumptions could be fully tested in a simple, economic way. Ultimately, this led to the evolution of a design cross section which was relatively insensitive to the exact nature of the various materials involved in the construction of each zone. Obviously, there will be cases where this cannot be arranged, and a tighter site control on those areas must be arranged. It is beneficial to be aware of these at the outset. Studies of seepage in which the governing parameters are systematically varied is an ideal method of acquiring this knowledge, and the finite element method a highly effective tool.

Mitchell's House Dam: interpretation of site investigation findings

Mitchell's House Reservoirs are located 2.5 km east of Accrington, Lancs. The two reservoirs were constructed separately in the latter half of the nineteenth century. They are separated by a common central embankment, but are retained by dissimilar embankment dams. Although analyses of sections through both dams were made, the principles involved, if not the sections, were the same, and only one is illustrated here (section 2, Fig. 3).

A detailed site investigation of the dam was undertaken in 1983 because the reservoirs have had a history of leakage, both through the embankments (which were raised in 1958) and through their foundations into the underlying Coal Measures rock.

The Coal Measures hereabouts have been worked, and one particular seam, 70 m below dam foundation level, which although worked and thought possibly to have collapsed, causing distress to the dams, was found to be still open and the surrounding rock mass intact. The uppermost 2 m or so of rock is extensively

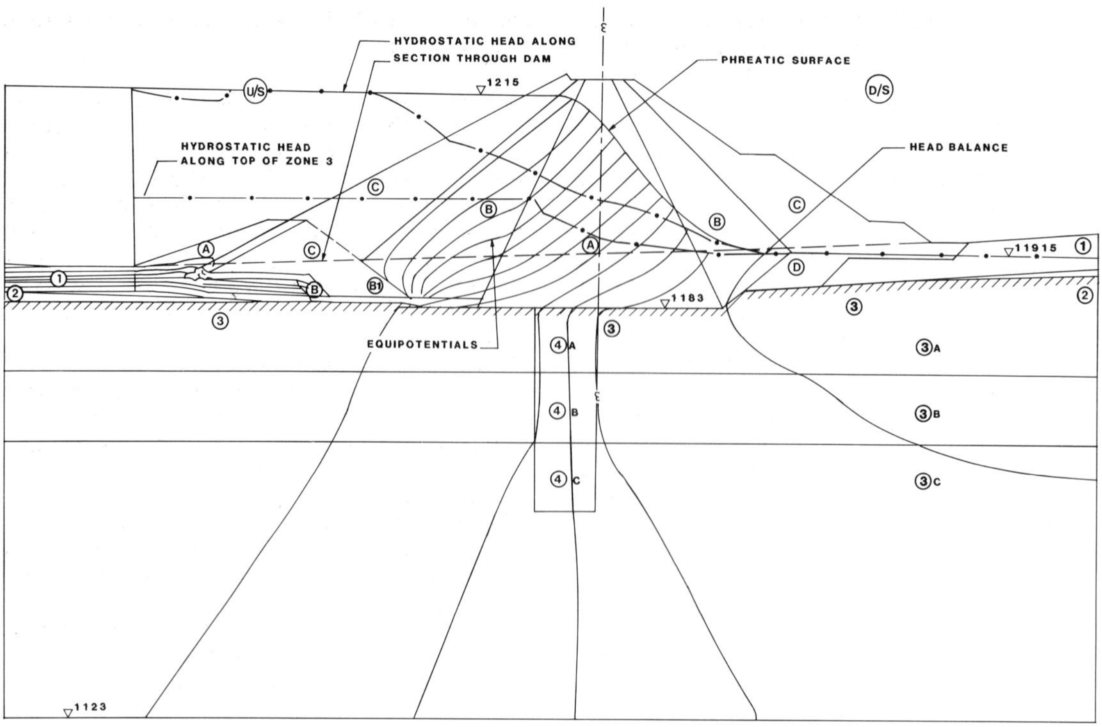

(a) Parametric study no. 1 (min k in transition (3))

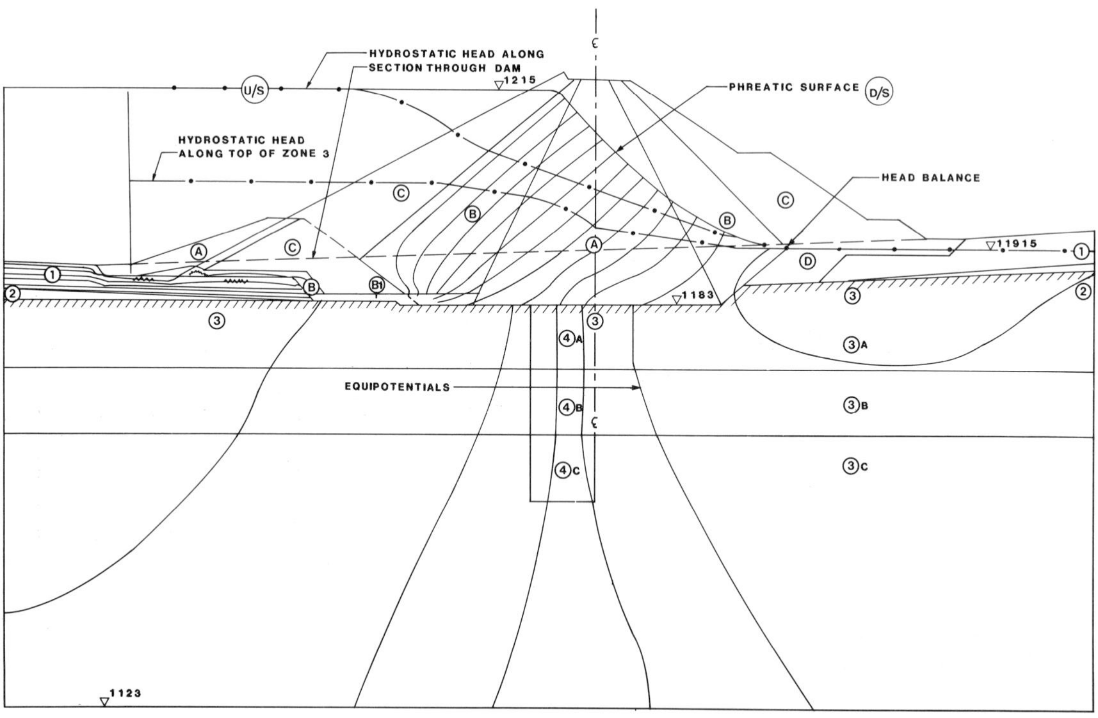

(b) Parametric study No. 2 (max k in core (A))

FIG. 2. Kiteta Dam.

weathered, resulting in a high bulk permeability, beneath this the rock is fresh and the fractures tight.

Piezometric observations, Fig. 3, showed low pore water pressures in the dam, and finite element analyses were undertaken to elicit a convincing explanation for these. They appeared not to be concordant with observations of seepage occurring from the downstream side of the dam.

A finite element mesh broadly similar to that used for the Kiteta Dam (above) was used. Even from the outset, the measured pore water pressures could only be approximated by the analyses if underdrainage into the weathered rock horizon played a significant role in the flow pattern. Furthermore, despite the drill holes which had been logged in detail, the internal structure of the dam was still fairly poorly understood, and its zonation for seepage analysis more uncertain than was desirable. In the event, and following much trial and error, an internal subdivision of the dam and foundation was made, matching the observed piezometric levels quite well. Since this drew the phreatic surface well down below the downstream shoulder of the dam and its toe, it indicated that the observed seepages were unconnected with the main seepage pattern in the dam. Quite possibly they are a secondary flow, perched on the contact between the fill used for raising and the original dam, though the possibility also exists that discrete flow paths have been established through the core.

Grimwith East Hillside — impact of reservoir raising on surrounding areas

Impounding of the new Grimwith Reservoir is nearly complete. The reservoir lies on a series of interbedded limestones, sandstones and shales of the Yoredale Series of the Carboniferous. These strata have a dip of approximately 8° NW in the eastern sides of the valley downstream of the dam. In the past, slope failures have been experienced, partly influenced by this unfavourable dip, in these slopes. Slides here were stabilised during construction of the embankment dam.

However, the uppermost shale horizon, termed Shale H, and its overlying sandstone, Sandstone I, were thought to be susceptible to destabilisation if ground water levels were affected by the latter stages of impounding. The following analyses were undertaken to check this possibility.

The steps in the analysis were as follows:

1. Develop a finite element model which reflected the geological structure of the East Hillside.
2. Analyse this, and by varying (a) the inputs (infiltration, levels of water table at watersheds etc.) and (b) the basic parameters of the model (e.g. permeability – transmissivity) obtain a match between predicted and known piezometric elevations (CALIBRATION PHASE)

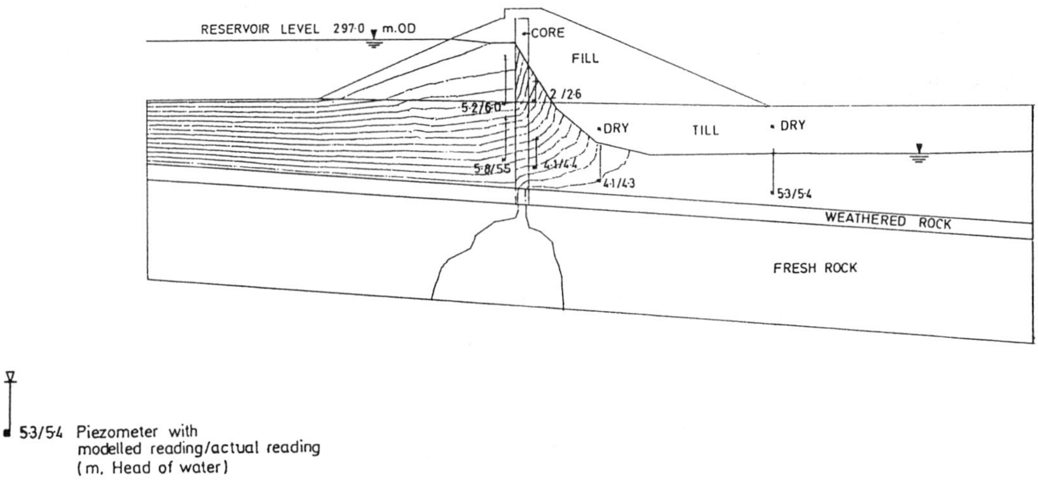

FIG. 3. Mitchell's House Dam (Section 2) showing equipotentials at 1 m intervals.

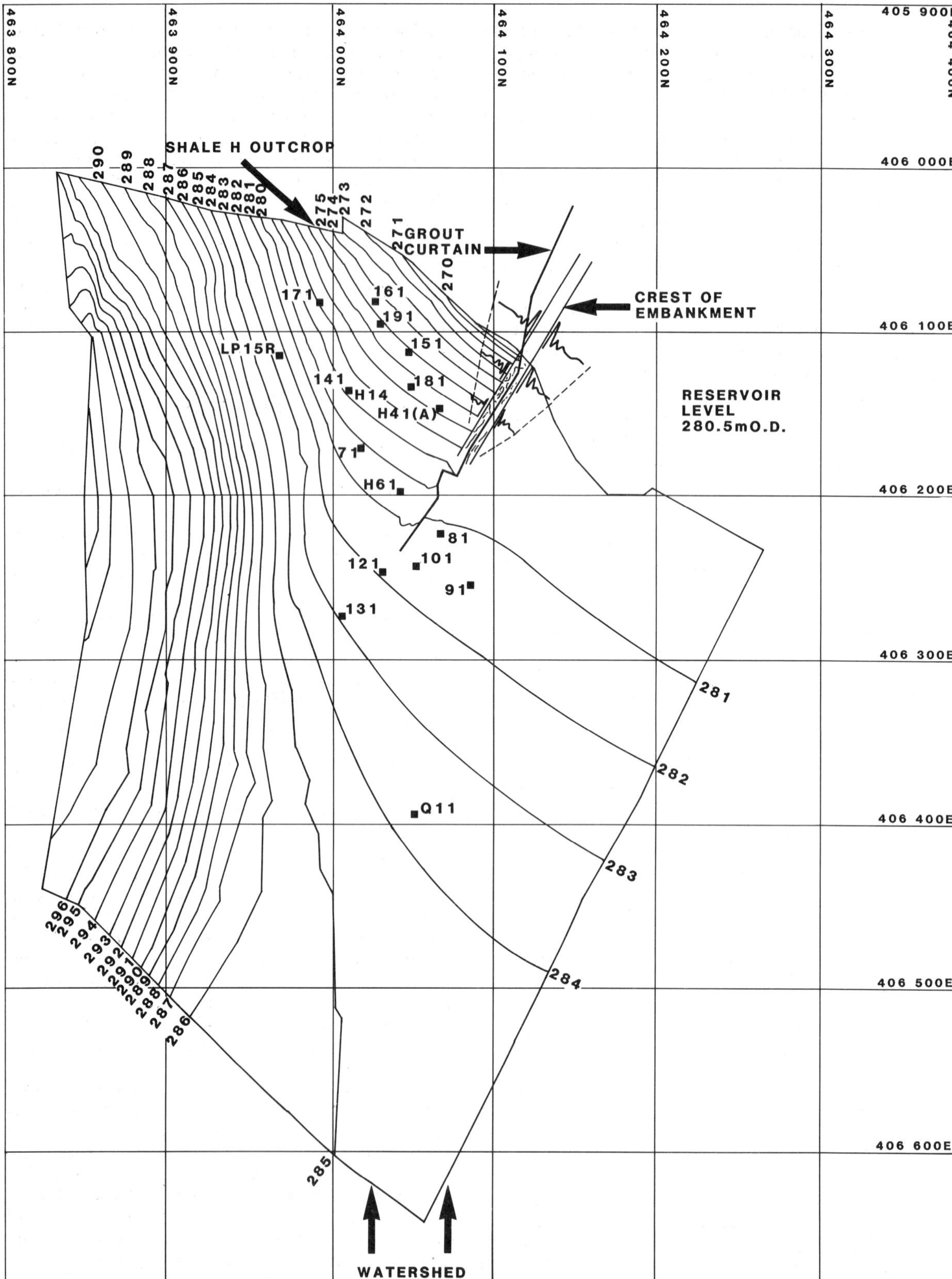

FIG. 4. Grimwith East Hillside — equipotentials (m.OD).

3. Apply the raised reservoir level, keeping the other factors constant to discover changes in the water table elevation in critical zone(s). (ANALYSIS PHASE)
4. Incorporate the changed piezometric levels in the stability analyses for the critical sections.
5. Take appropriate action.

A large model plan area was selected to minimise the influence of bad choices in 2(b) above, and parabolic isoparametric elements were used to reduce data preparation. Neither of these was a particularly good choice, since the large mesh size not only increased solution times, but also introduced areas where the Shale H aquiclude should project above the regional water table like an island. This posed some numerical problems solved ultimately by setting a bottom level for the movement of the water table in the iterative solution and moving the boundaries. The element choice was particularly unfortunate, since these elements can suffer localised internal reversals of the seepage direction when used in simple iterative schemes, and so oscillate about the correct solution. Where computer costs are not at a premium as was the case, the answer is to allow an excessive number of iterations and to average (say) the last ten.

A typical plot of contours of groundwater level in the East Hillside is shown in Fig. 4. The influence of dam cut-off, and grout zones in the hillside, are clearly seen in the concentration of head drop in that region. A perfect match with the piezometric data was not obtained during the calibration phase, but the results (Table 1) are thought to be adequate.

TABLE 1. *Comparison between piezometer field readings and model prediction.*

Piezometer	Range of piezometric levels corresponding to a reservoir level of 280 - 281 m OD	
	Field readings (m OD)	Model prediction (m OD)
H6I	280.1 - 282.4	280.4
7I	280.1 - 284.4	280.1
8I	279.5 - 284.4	281.2
9I	279.6 - 284.5	281.5
10I	279.5 - 284.4	281.6
12I	280.6 - 284.7	281.9
13I	283.2 - 284.8	283
* 14I	278.5 - 278.9	278.8
* 15I	274.2 - 274.3	275.1
* 16I	274.3 - 274.4	274.6
* 17I	278.3 - 278.5	278.1
* LP15R(I)	283.6 - 284.6	282.6
* H4I(A)	276.9 - 279.2	276.9
* 18I	274.1	276.5
* 19I	274.6	275.2

* These piezometers were installed more recently than the others, the number of readings corresponding to a reservoir level within the specified range were therefore fewer, resulting in a tighter range.

Subsequently, the analysis using the raised level was undertaken, and plots similar to Fig. 4, but showing the *differences* between the two cases were automatically produced. As a test on sensitivity, analyses incorporating nominal extra infiltration were made, and from these, the likely change in piezometric heads in the problem area was obtained.

These likely changes were then added to the initially observed heads to obtain the probably head after reservoir raising for use in the stability analysis of non-circular sliding surfaces largely along slip zones within Shale H.

The calibration could have been performed more easily if investigatory drillholes distant from the dam had been made, and if the distribution of piezometric observations had been more widespread.

Discussion

The problems of seepage analysis using finite elements are severalfold, and include:

1. Identification, classification and zonation of materials in the seepage domain.
2. Measurement or estimation of permeability (this can be *relative* permeability).
3. Identification, location and modelling of boundary conditions.
4. Subdivision of the overall model into elements and solution.
5. Interpretation of the results.

In all of these, interactive graphical techniques are essential, and convenient access to computing facilities vital. Both terminal systems, connected to mini computers, and personal computers (e.g. Apricot, Sirius, IBM PC etc.) are capable of performing these tasks provided that the basic software is available. However, the computer is the least of the problems in 1–5 above.

In each of the three case histories, the use of seepage modelling is for a different purpose. For Kiteta, for instance, although used primarily as an analysis of a conceptual design the lessons learnt from each analysis were incorporated in the next evolutionary stage of the design process, until ultimately an economic design was achieved. This was not done by seepage analysis alone, but was interspersed with stability studies, without which the upstream blanket might have been proceeded with at a cost of stability.

In the second case example, the analyses showed conclusively that the main seepage domain was strongly influenced by underdrainage. It confirmed the correct functioning of the piezometers since their readings were self-consistent and demonstrated that the downstream seepage originated in discrete flows.

Finally, the Grimwith Dam East Hillside model showed how reservoir levels could influence the stability of the downstream valley slopes. At present, further work to investigate possible remedial measures, is in hand.

In all three cases, however, the use of these analytical techniques have forced a discipline on their users: identify and model the problem fully; interpret the results correctly; and take the appropriate actions. The computer merely reduces the tedium of the calculations, it cannot replace the trained engineering geologists and geotechnical engineers whose proper role lies in performing these functions.

ACKNOWLEDGEMENTS: Ms A. L. Havouzari is supported by a SERC/DTI Teaching Company Award. The authors gratefully acknowledge the permission of Rofe, Kennard & Lapworth, Consulting Engineers to publish the data in this paper, and the help of colleagues at their respective institutions.

References

BODEN, J. B. & CHARLES, J. A., 1984. The safety of old embankment dams in the United Kingdom — some geotechnical aspects. *Municipal Engineer,* **111,** 46-60.

BROMHEAD, E. N., 1984. An analytical solution to the problem of seepage into counterfort drains. *Canadian Geotechnical Journal,* **21,** 657-662.

BROMHEAD, E. N. & VAUGHAN, P. R., 1980. Solutions for seepage in soils with an effective stress dependent permeability. *Proceedings of the 1st International Conference on Numerical Methods for Non-Linear Problems.* Pineridge Press, 567-578.

FREEZE, R. A. & WITHERSPOON, P. A., 1966, 1967, 1968. Theoretical analysis of regional groundwater flow (three papers). *Water Resources Research,* **2,** 641-656; **3,** 623-634; **4,** 581-590.

PUGH, R. S. & BROMHEAD, E. N., 1985. Design of seepage control measures for an embankment dam using the finite element method. *Proceedings of the 15th International Conference on Large Dams,* Lausanne, 1167-1183.

TAYLOR, R. L. & BROWN, C.B., 1967. Darcy flow solutions with a free surface. *Proceedings of ASCE, Journal of the Hydrology Division* **93,** 25 - 33.

ZIENKIEWICZ, O. C., 1972. *The Finite Element Method in Engineering Mechanics.* Wiley.

ZIENKIEWICZ, O. C. MAYER, P. & CHEUNG, 1966. Solution of anisotropic seepage by finite elements. *Proceedings of ASCE, Journal of the Engineering Mechanics Division,* **92,** 111-142.

E. N. BROMHEAD, Reader in Geotechnical Engineering, Kingston Polytechnic, Canbury Park Road, Kingston-upon-Thames KT2 6LA, UK.
A. L. HAVOUZARI, M. BRICE & B. H. ROFE, Rofe, Kennard & Lapworth, Raffety House, 2-4 Sutton Court Road, Sutton, Surrey SM1 4SS, UK.

The prediction of groundwater levels using computer based mathematical models

G.A. Hurley

ABSTRACT: The prediction of water level from rainfall events in landslip areas is important as maximum levels are required to determine minimum factors of safety for the slip. Also, rates of slip movement are likely to be linked to water level and an accurate prediction of water level fluctuation may lead to the modelling of slip movement.

Using real data, this paper presents two mathematically based models which have been used to predict water levels in a South Wales Landslip. The first approach uses time series and the second a conceptional idea. The models are compared and their applicability to other sites discussed.

1. Introduction

The Blaina landslips, situated near Abertillery, are two of the largest inland landslips in Great Britain. The Bournville slip, which was used for this study, is approximately 420 m long and 450 m wide. Its exact date of origin is not recorded, but it is believed to be about 100 years old. The toe of the slip is moving at a rate of 2–3 m per year. A number of autographic piezometers have been installed within the landslip and all show a similar response to rainfall. Those piezometers with the longest periods of record were used for the modelling process and the results for piezometer 10, situated in the centre of the slip, are presented.

The Bournville slip consists of a large boulder field and a number of earthflows at the toe. It is believed that there are topographic, geological and geotechnical characteristics which make the hillside susceptible to landslipping.

Geology

Solid geology

The backscar of the landslip passes through the Hughes and Brithdir Beds. The junction between these beds is defined by the Cefn Glas Seam. Both units consist predominantly of strong Pennant Sandstone. All beds dip at an angle of approximately 4 degrees, towards the south-west and with the topographic slope. The Cefn Glas Seam consists of between 5 and 10 m of claystones, with thin coal seams and associated seatearths. Behind the backscar two arrays of massive fissures exist to an unknown depth. The most notable set of fissures trend north–south. The other is north-west to south-east and parallel to faults. In 1973 both sets of fissures were noted to have been reactivated.

The Brithdir seam occurs at the base of the Brithdir Beds and the seam has been worked to exhaustion.

The underlying Rhondda and Llynfi Beds are entirely concealed by the landslip and superficial deposits. The Rhondda Beds consist of an alternating sequence of sandstones, conglomerates and quartzites, interbedded with layers of argillaceous strata, which are typically coloured red, purple and green. They have previously been given the informal name 'Red Beds' or 'Deri Beds'. The argillaceous layers soften to clays of low shear strength.

The Llynfi Beds generally comprise of claystones, siltstones and seatearths. Sandstones rarely exceed 5 m in thickness. The junction with the overlying Rhondda Beds is defined by the horizon of the No. 2 Rhondda Seam, but in the eastern part of the coalfield the seam is often absent.

Superficial deposits

Superficial deposits consist of the weathering products of the locally cropping strata. Typically they consist of soft to stiff grey, brown and red silty clays with gravel sized fragments of siltstone and some large clasts of sandstone.

The earthflows forming the downhill part of the landslip consist of displaced superficial deposits and are approximately 6 m thick. These landslip deposits are underlain by 4–6 m of superficial deposits. The landslip deposits and superficial deposits gradually thin out as one moves uphill.

Groundwater modelling

In both modelling exercises the longest continuous periods of piezometric record were used to 'fit' the models. The 'goodness of fit' criteria used to compare the predicted and actual water levels was the 'R-squared' value where:

From Cripps, J.C., Bell, F.G. & Culshaw, M.G., (eds), 1986, *Groundwater in Engineering Geology,* Geological Society Engineering Geology Special Publication No. 3, pp. 321-325

$$R^2 = \frac{(Y_a - Y)^2 - (Y_a - Y_p)^2}{(Y_a - Y)^2} \quad (1)$$

and

Y_a = actual water level (m),

Y = average water level (m),

Y_p = predicted water level (m).

Once the optimal variables and parameters had been determined the entire length of piezometric level was simulated using only values of effective rainfall and the initial piezometric level.

Time series modelling

Through a process of trial and error it was found that the optimal time series model contains three independent variables, lagged piezometric level, rainfall, and lagged rainfall. Its general form can be represented by the following notation:

$$H_t = \alpha H_{t-1} + \beta R_t + \alpha R_t \gamma R_{t-1} + C, \quad (2)$$

where

H_t = piezometer level on day t,

H_{t-1} = piezometer level on day $t-1$

α,β,γ,C = time independent parameters,

R_t = effective rainfall on day t,

R_{t-1} = effective rainfall on day $t-1$.

The parameters were estimated using the instrumental variable algorithm. This was done using a program developed by Dr M.B. Beck (Department of Civil Engineering, Imperial College, London). The program carries out a process similar to the kalman filter, essentially a recursive least squares calculation (Young 1974).

The model fit is given in Fig. 1 and the simulation of the entire piezometric record, for piezometer 10, given in Fig. 2.

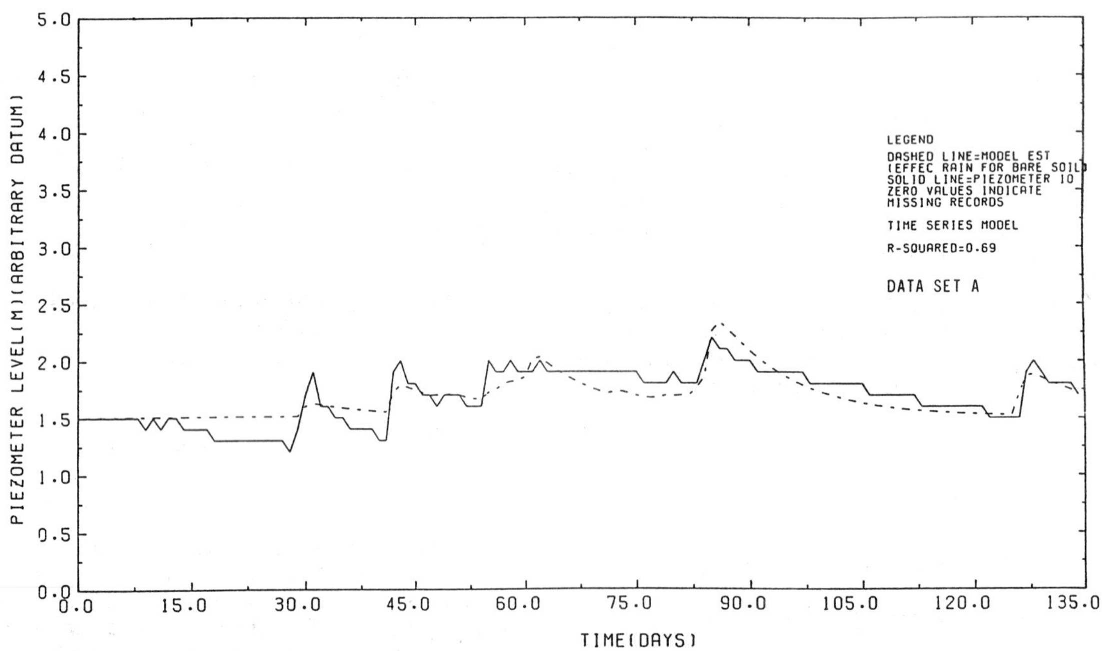

FIG. 1. Piezometer Level (m) vs. Time (days) (Dates 22.7.83–3.12.83).

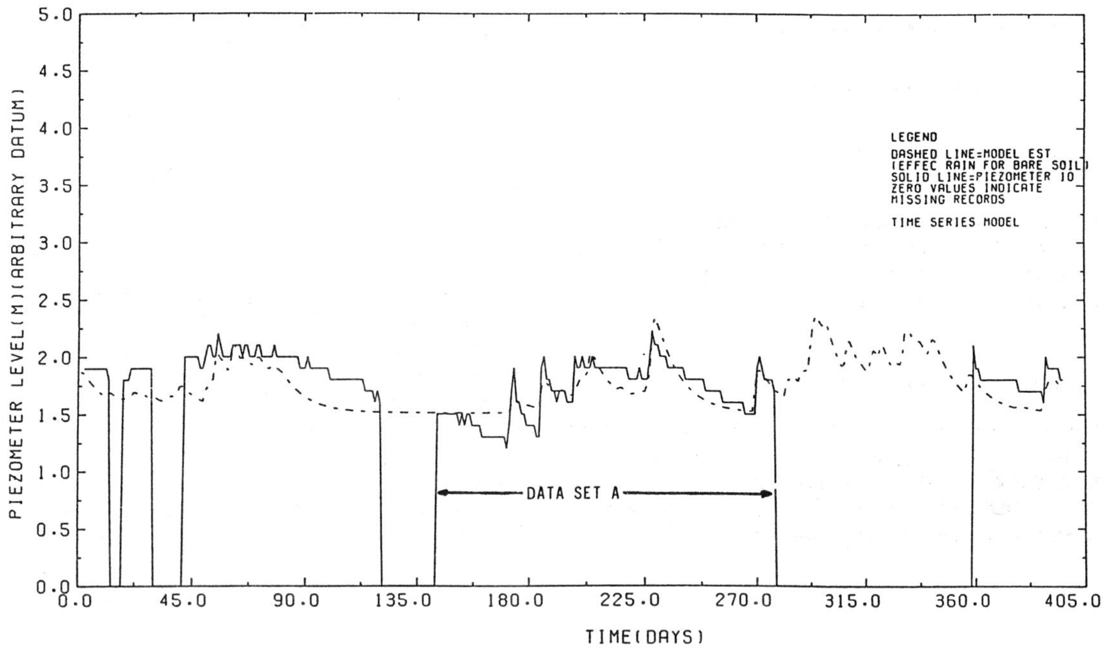

FIG. 2. Piezometer Level (m) vs. Time (days) (Dates 1.3.83–31.3.84).

Conceptional modelling

Of the conceptional models available, a simple storage model is closest to reality under conditions where rainfall intensities are generally less than infiltration capacities; it is also most useful under conditions where rainfall is frequent. For these reasons a simple storage model was chosen to model the groundwater response to rainfall.

The model used in this analysis was derived from work done by Kirkby (1975). The model behaves like a leaking bottle. The rate of drainage of leakage from the store is represented by the equation:

$$Q_t = Q_o \, e^{St/C} \qquad (3)$$

where

Q_t is the rate of outflow at time t,

e is the base of natural logarithms,

S_t is the storage at time t,

Q_o and C are model parameters.

A mass balance equation expresses the link between input, output and change in storage:

$$\mathrm{HER} - Q = \frac{\mathrm{d}s}{\mathrm{d}t} \quad (\text{for HER} > 0), \qquad (4)$$

where

HER = hydrologically effective rainfall (mm/day),

Q = drainage (mm/day),

$\frac{\mathrm{d}s}{\mathrm{d}t}$ = change in storage (mm/day).

The model is therefore a combination of two storages and is shown diagramatically in Fig. 3.

To satisfy continuity the value of effective rainfall must be positive. For this reason when the value of effective rainfall is zero the model assumes that a very small amount of effective rain has fallen, in this case 0.001 mm.

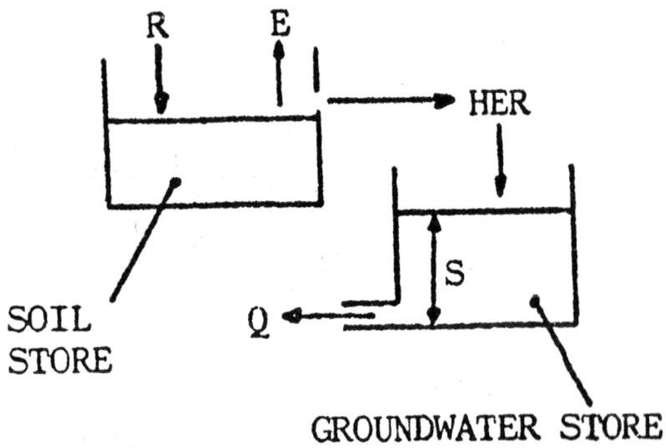

(HER = hydrological effective rainfall)

FIG. 3 The Storage Model

Equations 3 and 4 may be combined in the following way:

Let HER = i

then

$$i - Q = \frac{ds}{dt} \text{ and } Q_t = Q_o \, e^{S_t/C}.$$

Therefore

$$S_t = C \log_e (Q_t/Q_o).$$

Substituting for S and integrating gives:

$$Q_{t+1} = \frac{i}{(1 - e^{-i/c}) + \frac{i}{Qt}(e^{-i/c})} \quad (5)$$

Using Equations (3) and (5) the piezometric levels may be simulated, given an initial storage value, effective rainfall readings and initial estimates for the parameters Q_o and C. The model was fitted using the longest continuous data set (A). A program was written to calculate, sequentially, simulated piezometric values. These simulated values were compared with the recorded values using the R-squared value as the 'goodness-to-fit' criteria.

After each calculation the initial parameters, Q_o and C, were automatically changed by a set amount. Once the optimal parameters were found using data set (A) the whole length of the piezometric record was simulated using these values. The model fit and simulation of the entire piezometric record are given on Figs 4 and 5 respectively.

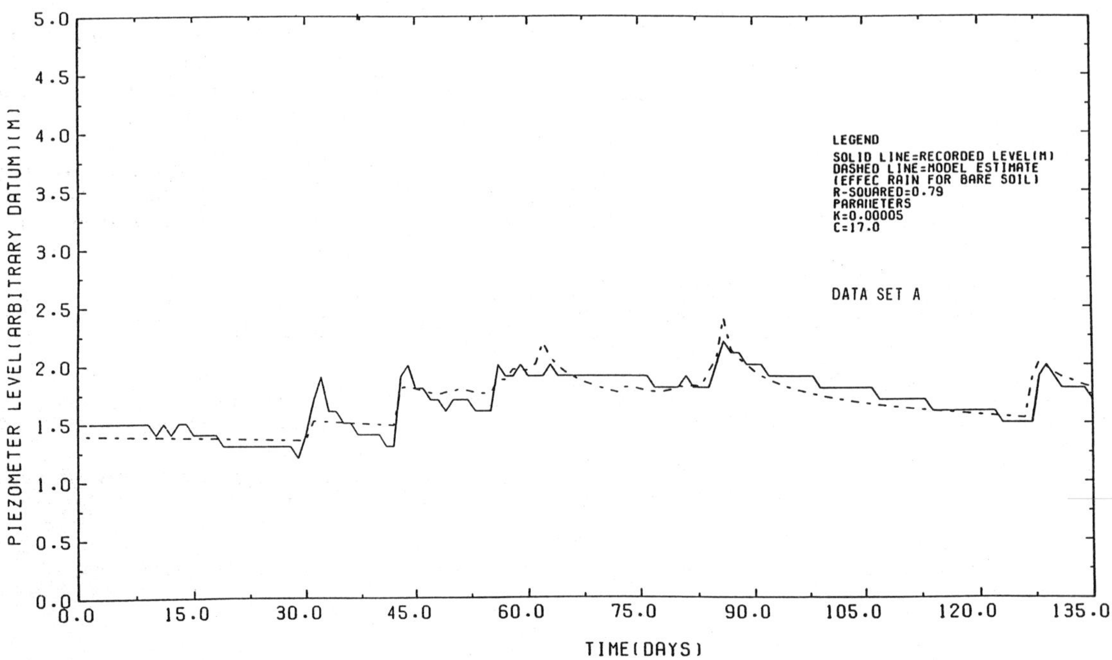

FIG. 4. Piezometer level (arbitrary datum) (m) vs. time (days) (dates 22.7.83–3.12.83).

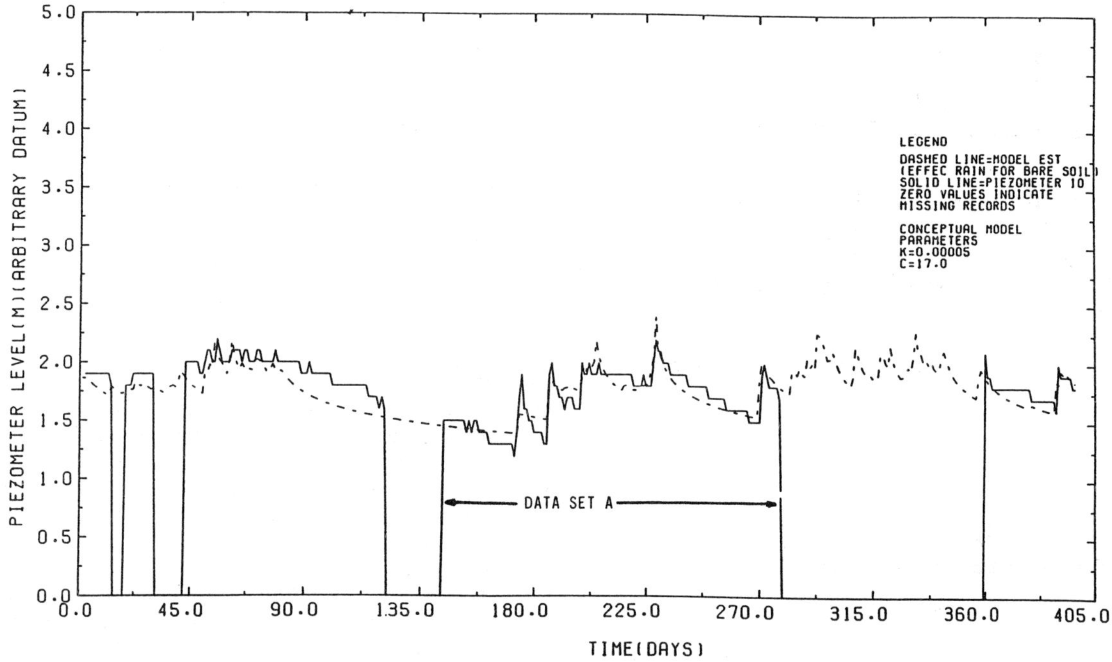

FIG. 5. Piezometer level (m) vs. time (days) (dates 1.3.83–31.3.84).

Model comparisons and their applicability to other sites

Table 1 compares the performance of the two models.

TABLE 1. *Model performance comparison.*

Model type	*R*-squared value
Time series	0.69
Storage	0.78

The conceptional model gives a higher *R*-squared value than the time series model and appears to simulate the entire length of the piezometric record slightly better. For the conceptional model the error between the simulated and recorded readings is generally less an 0.20 m. It should also be noted that the recorded readings are only accurate to ± 0.10 m.

The applicability of the storage model was discussed above. As the time series model is more of a 'black-box' approach, its applicability to other sites is probably greater than the storage model. Its results, however, may be less accurate than a conceptional model.

The parameters for both models are site specific. Therefore, the applicability of both models to other sites would also depend upon the amount of data available to 'calibrate' the model.

ACKNOWLEDGEMENTS: The author would like to thank Sir William Halcrow and Partners and Gwent Council for providing the piezometer data and background information for the site.

References

KIRKBY, M.J. (1975). *Hydrograph modelling strategies in Process on Physical and Human Geography.* Peel, P.F. et al. Heinmann, London.

YOUNG, P. (1974). Recursive approaches to time series analysis, *Bull, I.M.A.*, **10.**

G.A. HURLEY. Geotechnical Engineering (Southern) Limited, Rock House, Lower Tuffley Lane, Gloucester GL2 6DT, UK.

Predicting piezometric levels in steep slopes

D.N. Lerner

ABSTRACT: Three methods of predicting piezometric levels in steep slopes are presented and illustrated for a case history in Hong Kong. Statistical correlation of levels with various measures of rainfall showed some promise as a method, but poor choice of an objective function biased the models away from the events of most importance. A groundwater flow model of a cross section through the slopes had too coarse a discretization to be accurate enough, but helped develop an understanding of the slopes' hydrogeology. Empirical methods based on classifying piezometers and extrapolating and regionalising storm and seasonal responses were finally adopted.

Introduction

The Mid-Levels are an urbanised and densely populated area of Hong Kong island (Fig. 1). Slopes are naturally steep and have been modified by cut and fill in the construction of tower blocks and roads (Fig. 2). There have been many serious landslips on steep slopes in Hong Kong. In his excellent review, Lumb (1975) reported that there were 285 slips in the disaster of June 1972 alone. One of these slips was at Po Shan in the Mid-Levels; it killed 67 people (Anon 1972).

In 1979, the Hong Kong Government imposed a temporary restriction on building development in the Mid-Levels. This gave time for a study of the geotechnical factors that should govern future development in the area. Lumb (1975) has shown how closely failures were correlated with heavy rainfall in Hong Kong, and clearly surface and groundwater hydrology were important parts of the geotechnical study. This paper concentrates on the methods used to predict piezometric levels. Details of all the hydrological studies can be found in GCO (1982), together with descriptions of geology and soil properties. Much of the rest of the study remains unpublished.

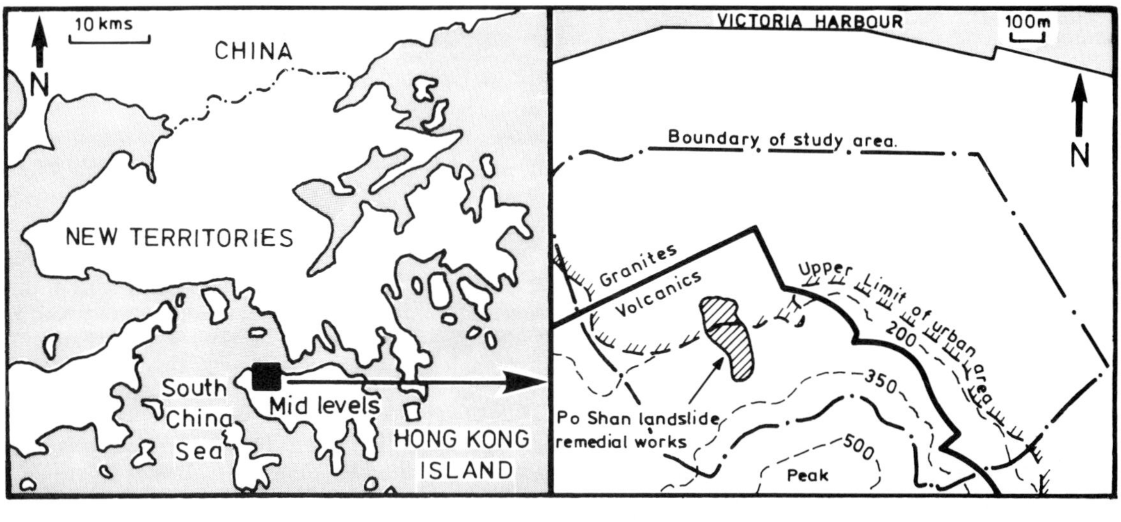

FIG.1. (a) Location of Mid-Levels, Hong Kong, (b) some features of study area.

From CRIPPS, J.C., BELL, F.G. & CULSHAW, M.G., (eds), 1986, *Groundwater in Engineering Geology,* Geological Society Engineering Geology Special Publication No. 3, pp. 327-333

FIG.2. Upper limit of urban area and natural slopes above the Mid-Levels. Horizontal bands show position of remedial works for Po Shan Landslide.

Groundwater in the Mid-Levels

Hydrogeology

Hong Kong island consists of volcanic rocks with a granite intrusion. The contact runs roughly east–west through the study area (Fig. 1) with granite in the lower slopes.

The rocks have weathered to a three layer hydrogeological system:

(a) Bedrock of low permeability.

(b) Decomposed *in situ* rock; thickness ranges from zero on upper slopes to 90 m with considerable variations. Permeabilities are 7–10 × 10^{-6} m/s (granite) and about 4 × 10^{-6} m/s (volcanics).

(c) Colluvium, derived mostly from volcanics on the higher slopes; thickness varies from zero on both upper and lower slopes to 5–20 m in mid-slope. Permeability varies with age and nature, from >1000 × 10^{-6} m/s in a boulder field down to 5 × 10^{-6} m/s in the oldest material.

A schematic cross-section is shown in Fig. 3. The colluvium and decomposed rock layers are the 'aquifers' of interest where the important groundwater flows and piezometric responses are found. The main groundwater body is continuous across the bedrock–decomposed rock contact.

Recharge

Recharge to the two aquifers of interest is from three sources: rainfall on the upper slopes, from bedrock, and leakage from services. Above the urban area, colluvium and decomposed rock are exposed, and a proportion of the rain falling here infiltrates into the aquifers. Direct recharge from rainfall is small in the urban area where there is little ground not covered by roads, concrete, buildings or chunam, a protective covering.

During the course of groundwater modelling studies (see below), rainfall recharge was found to be too low to sustain observed piezometric levels in the dry season. A possible source of extra recharge is slow from bedrock, with rainfall infiltrating into exposed

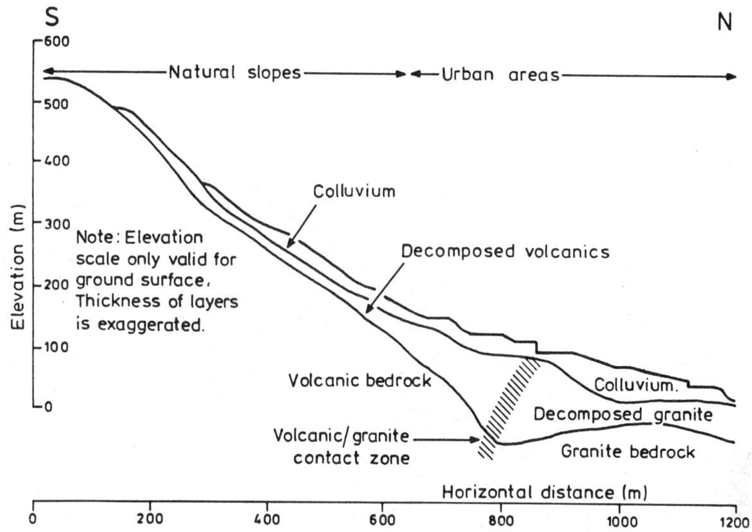

FIG. 3. Schematic cross-section through the Mid-Levels.

bedrock on the highest slopes, and flowing up to the aquifers lower down the slopes. Leach & Herbert (1982) suggested that this source is sufficient to maintain observed piezometric levels, but this would require infiltration to be greater than rainfall because the area of exposed bedrock is much smaller than the area where extra recharge to the principal aquifers occurs. Nevertheless, some piezometer nests in the study area do show upward flow from bedrock and there is clearly some contribution from this source.

Leakage from services (water mains, salt water mains, sewers and storm drains) is the third source of recharge in the Mid-Levels. The possibility has been recognised in the past (Anon 1979) and leaking services were at least partially implicated in the Po Shan landslide in 1972 (Anon 1972). Lerner (1985) reviewed the piezometric and chemical evidence for leakage being an important recharge source both in the Mid-Levels and elsewhere. He calculates it may be as high as 2000 mm/yr in full urbanised parts of the Mid-Levels.

Piezometric fluctuations

The 400 piezometers installed in the Mid-Levels showed a wide variety of responses to rainfall. Some rise gradually throughout the wet season, some rise and fall in response to one day's rain and intermediate degrees of response are also seen.

Some piezometer nests show that a perched water table forms on the top of decomposed rock, with a main water table below. Some areas of perched water table are permanent, for example, sustained by leaking services. Other areas are transient, appearing in response to significant rainfalls.

Other watery factors

Several other groundwater factors have important influences on piezometric levels and stability in steep slopes, but detailed discussions are beyond the scope of this paper. These include:

(a) Pore water pressures above the water table. Measurements in the Mid-Levels showed that suctions dropped to zero and some positive pressures developed in response to moderate storms, thereby reducing stability (GCO 1982).

(b) Pipes, that is continuous voids up to 200 mm in diameter formed by internal erosion. Pierson (1983) has presented circumstantial evidence that pipes cause slope failures in some circumstances. Pipes were found in the Mid-Levels, particularly in the upper slopes.

(c) Drainage, earth retaining structures and other man-made features will affect groundwater flows both adversely and beneficially in different places.

Predicting piezometric levels

Return period of predicted levels

Recommended practice in Hong Kong was to consider the piezometric levels resulting from storms with 10 and 1000 year return periods (GCO 1979). With piezometric records of only 6–18 months, extrapolating to 1000 years was not reasonable, and only 10 year storms were considered.

Levels of a particular return period do not necessarily result from a rainstorm of the same return period. Piezometric levels may be affected by:

(a) storm duration,
(b) temporal distribution of rainfall during the storm,
(c) antecedent soil and groundwater conditions,
(d) total rainfall.

Probability analysis of a long sequence of piezometric levels would give the most direct and best estimates of levels of a given return period. However, this option was not available for the Mid-Levels because of the short records, which included one drier than average year. An attempt was made to transform a long rainfall record into equivalent piezometric levels using statistical methods but was not completely successful (see below). In the end, piezometric levels resulting from 10 year storms were adopted, accepting the possible errors involved.

Leaking water services

In some areas, leaking services have a greater effect on piezometric levels than natural recharge. There is potential for leaks throughout the urban area, which could cause piezometric levels much higher than those in response to 10 year rainstorms. The following example should illustrate the important of leaking services.

At one piezometer group in the urban area, a perched water table 12 m thick has been present throughout the period of observation; there is no source of natural recharge in the area. The steady leakage causing this has not been detected in routine surveys or has not been enough to warrant repair. This piezometer also shows a storm response which would be about 5 m for the 10 year event. Thus leaking services can easily raise piezometric levels by 17 m at this site, possibly by more at other sites.

Leaks, their timing, location and size, are poorly documented and piezometric levels resulting from them cannot be given return periods. However, because leaks can cause very large rises in piezometric levels, the position and possible effect of all water services must be taken into account in stability analyses.

Methods used

Three methods were used to analyse piezometric levels:

(a) Statistical correlation of levels and rainfall. The method showed some promise, although did not produce many results for this study.
(b) Finite difference models of vertical two-dimensional sections. These were useful aids in understanding the groundwater system but were not accurate enough to provide estimates.
(c) Empirical extrapolation of observed responses provided the results used in stability analyses.

Statistical models

Forty-eight piezometers were selected for analysis, representing most geologies and response types. Depth to piezometric level was regressed by least squares on a variety of measures of rainfall, such as cumulative rainfall for various durations, lagged cumulative rainfall, antecedent precipitation indices (APIs) and lagged API. In almost all cases an API variable explained most variance. Thirty-five of the final regression equations had a correlation coefficient of 0.84 or better (HIP, 1980).

In the second phase of the modelling, rainfall records for 1949–80 were put through the 35 models. Annual maximum levels were taken and probability analyses made. Levels of various return periods could then be estimated.

Review of statistical models

Only 14 of the regression models passed a detailed subjective review based on a visual inspection of observed and predicted levels for the short period of field data.

A further review looked at the predictions for 1966 and 1973, two notable wet years. Of course no piezometer records went back that far, and the comparison was of return periods of rainfall and predicted piezometer levels.

As discussed above, return periods of rainfall and piezometric levels do not always correspond. But by looking at a range of rainfall durations, some matching would be expected if the piezometers critical storm duration is included.

The regression models suggest that most piezometric levels in 1966 had less than 5 year return periods, but rainfall probabilities were typically 1 in 15 to 1 in 30 years. For 1973, predicted return periods ranged from 2.5 to 210 years, with half over 100 years. However, rainfall return periods are only over 10 years for very long events of 3–6 months.

The models were seen to be biased towards long duration events. This is probably due to the least squares method of fitting the equations; the few data points associated with short storms are heavily outweighed by the greater number of points between events and due to long events. Lerner (1975) and Diskin & Simon (1977) discuss the biases in using different objective functions in fitting such models.

Overall the regression models were not good enough for the Mid-Levels study. However, they did show promise, partly because they were quick and easy to use. Similar approaches should be considered where longer records are available.

Groundwater flow models

A second method of predicting piezometric levels is to solve the time variant equation of groundwater flow and head. Both finite difference and finite element methods are suitable, although finite differences are easier when developing a new model. Two models were built for the Mid-Levels. One has been described by Leach & Herbert (1982); the other is briefly outlined below. However, the main emphasis of this section will be the difficulty of constructing an accurate model of such slopes.

Model description

Groundwater flow in the Mid-Levels is down the topographic gradient. Therefore flow could reasonably be modelled in a vertical cross section down the gradient. The model contains the colluvium and decomposed rock layers, and can accept interruptions in the colluvium layer. One boundary was a fixed head representing Victoria Harbour, the other an impermeable boundary above the highest point where groundwater had been observed in the decomposed rock layer (Fig.4). A backward difference, successive over-relaxation algorithm was used for the numerical solution. Time steps were 15 minutes during storms, increasing gradually to two days in times of no rain.

Numerical stability

Both models suffered from numerical stability problems. These arose from (a) difficulties in modelling aquifers changing from dry to water table to full or confined conditions, and (b) from their coarse discretization, that is, their subdivision in space and time.

Accuracy

Mature reflection suggests that the discretization was too coarse to allow accurate predictions of piezometric levels. Rushton & Tomlinson (1977) defined a dimensionless time $t'=tT/L^2S$), where t = time, T = transmissivity, L = representative dimension of model and S = storage coefficient.

They showed by numerical experiment that reasonable accuracy is only obtained for $t'>2$. For typical Mid-Levels conditions, this means that accurate predictions will only be available 10 days or more after the end of a recharge event. This is of little use for modelling storm responses! However, Rushton & Tomlinson showed that finite difference models do maintain an accurate water balance at all times.

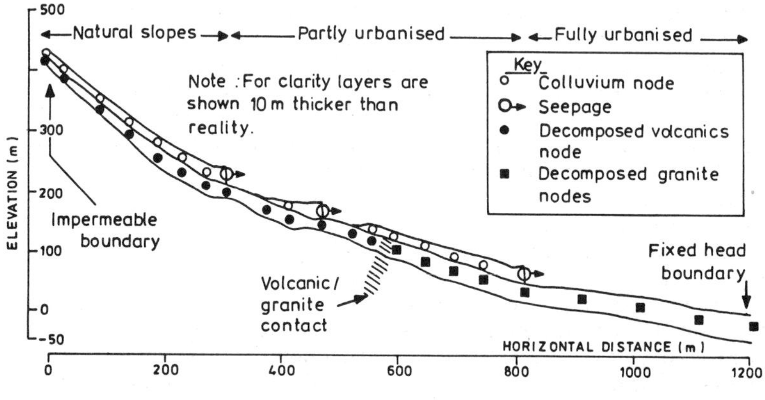

FIG.4. Finite difference model of groundwater flow in the Mid-Levels.

Another way of looking at the accuracy of models is to consider the accuracy of the field measurements used to construct and calibrate them. Four sets of field data are needed:

(a) Piezometric levels. Accurate measurements are possible, but groundwater flows are controlled by gradients, not absolute levels, and small errors in levels will give larger errors in gradients. Even with 400 piezometers in 1.5 km^2 of the Mid-Levels, gradients are not well defined when average slopes are 20°.

(b) Geometry of aquifers. Similar considerations apply.

(c) Aquifer properties (e.g. permeability). Properties measured on a small scale, such as a field permeability test, are not the same as those controlling large scale flow in an aquifer and there is no easy link between the two sets of values (Lerner *et al.* 1984).

(d) Recharge. Recharge cannot be measured in the field, only estimated from rainfall, evapotranspiration and runoff, all of which are difficult to measure accurately on steep slopes.

The brief discussion above suggests that *none* of the data sets are well defined, and groundwater modelling becomes an art of adjusting uncertain field estimates to form a consistent description of aquifers, flows and heads. High accuracy cannot be expected, especially where rapid changes occur in space and time. The main benefit of modelling the Mid-Levels was that it forced us to develop conceptual models of the groundwater hydrology which were numerically consistent. These lead directly to the acceptance of recharge from bedrock and to seek evidence of leaking services.

Empirical extrapolation

Neither of the above methods gave detailed predictions for the whole study area. Final predictions were made by empirical methods based on detailed inspection of observed responses. The piezometers were subdivided into groups of similar responses and a method of extrapolation was devised for each group. The methods used for the western half of the study area are described here; the empirical methods used in the east are described by GCO (1982).

Piezometers were separated into main and perched water table groups. Within each group, piezometers were classified as responsing to storms or seasonable events.

In all cases, the 10 year level was predicted to be the sum of the response to a 10 year event, and a typical, pre-event base level. This approach has several advantages, including:

(a) Responses were found to show regional patterns, allowing extrapolation into areas without piezometers.

(b) Base levels can be estimated for any site of interest, either by drilling or from topographic contours and structural geology.

Base level

For seasonally responding piezometers, the 'typical base level' was the seasonal low in early 1980, because there was not enough data to estimate an average seasonal low. For storm responses, the base level was the average level in the 1979 wet season, ignoring storm rises (but see following paragraph on perched water tables). Thus the base level reflected 1979 conditions in all cases. 1979 was wetter than average, but inspection of the few long piezometer records suggested that high levels show much more variation than lows and averages. That is, minor adjustments to the base level were less important than predicting responses.

Perched water tables only showed storm responses. Where the perched water table was permanent, the average 1979 level was used as a base level. For transient cases, the top of the decomposed rock was adopted as the base level, as this was the horizon on which perching was thought to occur.

Response to design rainfall

Seasonal rainfalls for 1979 were about the same as 10 year events. Many piezometers did not have complete records for 1979, and so 1979 responses were extrapolated from 1980, using data from those piezometers which had both sets of data.

Extrapolation of storm responses was much more difficult, requiring consideration of storm duration.

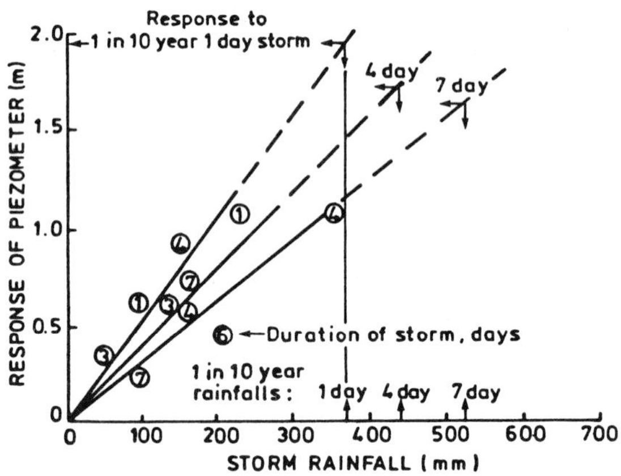

FIG. 5. Extrapolation of storm response, taking account of storm duration.

The method adopted is related to rainfall amount and storm duration. Such plots were used to find the critical duration of 10 year storm, that is, the duration giving the largest response. Fig.5 is for a piezometer with particularly good records; nevertheless there is barely sufficient data to make a reasonable extrapolation.

Conclusions

Three methods of predicting piezometric levels in steep slopes have been presented, illustrated with a case study in the Mid-Levels, Hong Kong. All three have pros and cons.

Statistical, time series, analysis of piezometric levels and rainfall is a valid approach. Details of statistical technique can significantly affect results, even though they may not be obvious to the non-specialist. In the case cited, choice of a least squares method biased the models towards long duration events.

Groundwater models are always useful. They require that a conceptual model of the hydrogeology is developed, and they may help to ensure that there is numerical consistency between the various data sets. The models of cross-sections through the Mid-Levels had very coarse discretizations, and so were too inaccurate to make quantitative predictions of piezometric levels.

Empirical methods, based on classifying, extrapolating and regionalising observed responses, were finally adopted. Although not best estimates in any statistical sense they have the advantages of being easily understood, using all the data and allowing regionalisation of results.

The response of water tables to rainfall is not the only groundwater factor affecting slope stability in urban areas. In particular, leakage from water carrying services can cause very large rises in piezometric levels.

ACKNOWLEDGEMENTS: The Mid-Levels study was carried out by Binnie & Partners (Hong Kong) and Ove Arup & Partners Hong Kong Ltd., for the Geotechnical Control Office of the Government of Hong Kong. I am grateful to Binnie & Partners and the Geotechnical Control Office for permission to prepare and present this paper although the opinions expressed are mine and do not necessarily reflect official policy.

References

ANON, 1972. *Final report of the Commission of Inquiry into the Rainstorm Disasters 1972.* Government Printer, Hong Kong.

ANON, 1979. *Geotechnical manual for slopes (draft).* Geotechnical Control Office, Public Works Department, Hong Kong.

DISKIN, M.H. & SIMON, E., 1977. A procedure for the selection of objective functions for hydrologic simulation models. *Journal of Hydrology,* **34,** 129-150.

GCO (Geotechnical Control Office), 1982. *Mid-Levels Study: Report on Geology, Hydrology and Soil Properties. Government Printer, Hong Kong.* (Published version of reports [prepared by Binnie & Partners (Hong Kong) and Ove Arup & Partners Hong Kong Ltd.)

HIP (Halcrow International Partnership), 1980. *Statistical correlation of groundwater response, Final Report. Mid-Levels Study, Part 2A.* Geotechnical Control Office Government of Hong Kong (unpublished).

LEACH, B. & HERBERT, R., 1982. The genesis of a numerical model for the study of the hydrogeology of a steep hillside in Hong Long. *Quarterly Journal of Engineering Geology* **15,** 243-259.

LERNER, D.N.., 1975. Development of a conceptual model for a chalk catchment in East Anglia. *MSc Thesis,* Imperial College, London.

LERNER, D.N., 1985. Leaking pipes recharge groundwater. Submitted to *Ground Water.*

LERNER, D.N., LLOYD, J.W.. & WALSH, D.M.., 1984. Computer modelling of alluvial aquifers for water resources studies, Proceedings of 4th Congress, Asian-Pacific Division, IAHR, Chang Mai, Sept. 1984, 1553-1565.

LUMB, P., 1975. Slope failures in Hong Kong. *Quarterly Journal of Engineering Geology.,* **8,** 31-65.

PIERSON, T.C., 1983. Soil pipes and slope stability. *Quarterly Journal of Engineering Geology,* **16,** 1-11.

RUSHTON, K.R. & TOMLINSON, L.M., 1977. Permissible mesh spacing in aquifer problems solved by finite differences. *Journal of Hydrology,* **34,** 63-76.

D.N.LERNER, Hydrogeology Section, Department of Geological Sciences, University of Birmingham, P.O. Box 363, Birmingham B15 2TT, UK. (Formerly with Binnie & Partners, London)

Geophysical prediction of fluid flow in sediments

M.A. Lovell

ABSTRACT: The remote assessment of fluid flow by geophysical means relies largely on empirical formulae relating particular geophysical measurements to the permeability for a specific geological situation. This approach has been proved successful by other workers in the field for delineating and quantifying the nature of aquifers. Recently, however, the proposals to store radioactive waste in deep-sea sediments, together with the increasing interest in hydrocarbon formation processes and basin development, have provided further impetus to the study of geophysical–permeability interactions.

A series of laboratory experiments involving a wide range of natural and artificial sediments has enabled the relationships between the geophysical characteristics and the fluid flow to be investigated. The results show very clearly the interrelationship between permeability and the geophysical parameters, electrical resistivity and compressional wave velocity, and the capability of predicting the one by measuring the other. In particular, extensive measurements of electrical formation factor and permeability in clean sands show the dominance of particle shape, and hence pore shape, in defining the interaction of these energy transfer processes.

In defining a dependence on pore shape the variation of the results effectively identifies differences in the structure of the sediments and hence different distributions of pore geometry. This may in turn be related to depositional environments, and, in extending the results to sedimentary rocks, to the diagenetic and lithification processes involved. Each of these fabrics, however, tends to be anisotropic, and this is borne out by the geophysical measurements. Since the geophysical characteristics are closely linked with permeability then it follows that permeability anisotropy may be reasonably expected also. While this directional dependence may be minimal in surficial marine sediments, it may be considerable at depth in the sediment column, particularly as consolidation progresses. The identification and measurement of anisotropy should help in defining geophysical-fluid flow relationships, especially for fluid flow predictions by remote measurements.

Introduction

The flow of fluids through the ground is currently of considerable interest to geoscientists and engineers involved in aquifer and hydrocarbon reservoir studies, civil engineering projects, radioactive waste storage, basin dewatering research and geothermal energy projects. In each of these programmes fluid flow concerns movement through large volumes of rock and depends on the solid structure and, moreover, the distribution of interconnecting voids, cracks and fissures. Geophysical techniques are currently employed to evaluate, quantitatively, many physical properties of the ground together with their lateral and vertical variations. Any extension of these geophysical methods to the evaluation of permeability would thus be most welcome.

This paper is concerned with the empirical use of electrical resistivity and seismic propagation in evaluating the permeability of sediments. Such techniques have been used before and some examples are detailed below. However, electrical resistivity – permeability results for unconsolidated sediments in the laboratory provide a better understanding of the physical constraints imposed by the size and shape of the particle framework, and it is proposed that consideration of these results, together with the identification of any anisotropy, should help in defining more precise geophysical–fluid flow relationships. In turn, laboratory results under controlled conditions may be applied with caution to *in situ* results to interpret the field geophysics measurements in terms of porosities and permeabilities (e.g. Barker & Worthington, 1973); estimation by such methods may involve certain assumptions such as pressure and temperature effects at depth, and pore-fluid conductivity distribution.

Previous work

The remote assessment of permeability by geophysical means has in the past relied largely on empirical formulae relating particular geophysical measurements to the permeability for a specific geological locality. Using such techniques a wide range of rock types have been fairly successfully evaluated within narrow confines (Archie 1942, consolidated sandstones; Barker & Worthington 1973, Bunter Sandstone; Brace 1977, crystalline rock). Each of these particular examples concentrated on electrical resistivity

From Cripps, J.C., Bell, F.G. & Culshaw, M.G., (eds), 1986, *Groundwater in Engineering Geology,* Geological Society Engineering Geology Special Publication No. 3, pp. 335-340

measurement as a predictor of permeability although Barker & Worthington also made sound speed measurements. In each case the technique relied on laboratory measurements for evaluating empirical relationships which could then be attributed to that rock type. Hamdi & Taylor Smith (1982) successfully predicted the permeability of a range of marine sediments in the laboratory using a theoretical model based on original work by Biot (1956). This required a fairly detailed array of input information, particularly the sound speed, sediment rigidity, and porosity. They suggested the last two of these were best determined through measurement of the shear wave velocity and the electrical resistivity formation factor respectively; thus three individual geophysical measurements on the same sample enabled permeability to be predicted to within an order of magnitude. Within the oil industry, the geophysical prediction of permeability has been a much sought goal, but its accurate evaluation for a wide range of geological occurrences has been generally disappointing. At best, local equations can be established for a given field situation (e.g. using electrical logs, Brown & Husseini 1977).

Experimental studies on unconsolidated sediments

Unconsolidated sediments represent perhaps the simplest of porous media since they may be considered as assemblages of discrete particles, the pore spaces between which are filled with a pore fluid which may consist of one or more discrete phases. For some clays the individual particles can form lattice arrangements linked by electro-chemical bonding. The role of the pore space and particularly its spatial relationship to adjoining voids is of primary importance in determining the permeability. For clean sands the nature of this pore space is controlled by the particle shapes and the packing arrangements, whilst for clays the pore space may be additionally influenced by the absorbed water layers at the particles surfaces — this also reduces the amount of 'free' water able to move easily through the structure. Partly for these reasons, but also due to their different requirements in laboratory testing (sands can be easily redeposited whereas clays generally suffer irreversible deformation) unconsolidated sediments have been tested as two distinct groups: sands and clays.

Sand samples representing a range of particle shapes, sizes, and size distributions have been tested in a number of novel laboratory cells enabling porosity and permeability to be measured with electrical formation factor. Clay samples from the North Atlantic have been tested using a fixed-ring oedometer cell, modified to enable the simultaneous measurement of P-wave and S-wave velocities, and electrical resistivity formation factor (the electrical resistivity of the sample normalised with respect to the electrical resistivity of the pore fluid); porosity and permeability values are deduced from uniaxial consolidation theory. Details of these cells are available elsewhere (Lovell 1983, 1985; Lovell & Jackson 1985).

Results

Porosity

Jackson (1975) originally developed a novel laboratory cell for monitoring the electrical resistivity and porosity of an unconsolidated sand as it was compacted from its loosest to its densest state. Fig. 1 shows similar results for four natural sands of differing particle shapes; the increase in shell content corresponds to a decrease in sphericity. Each sample adheres to Archie's (1942) empirical law relating electrical formation factor to porosity

$$FF = n^{-m}$$

with the exponent m increasing with decreasing sphericity of the particles. These are plotted together with results for a series of nine deep-sea clays (Lovell 1983) and the overall trends confirm those produced by Jackson *et al.* (1978).

Permeability

Porosity, however, while defining in simple volumetric terms a most important facet of the physical state of the material, remains a static scalar quantity. Permeability, which refers to the ease with which a fluid flows through a medium under an applied pressure gradient, extends the concept of the pore space in that it depends also on the distribution and interconnection of that space.

Measurements of vertical permeability and horizontal electrical resistivity formation factor for the range of sands produces Fig. 2. For each individual sample a linear relationship results on a log–log basis:

$$K = c\,(FF)^{-x}$$

This is similar to that proposed by Brace (1977) for crystalline rock. In studying a range of sand samples characterised by different particle shapes and sizes it is possible to suggest some causes for the variations in the constants c and x (see Table 1): this is similar to the analysis for a series of artificial sands (Lovell 1985) and confirms those earlier results. Generally the coefficient c appears to be a function of the pore size; c decreases with increasing spread of sizes and increases with decreasing sphericity (corresponding to decreasing and increasing porosities respectively). The exponent x, however, increases with both increasing spread of sizes and decreasing sphericity.

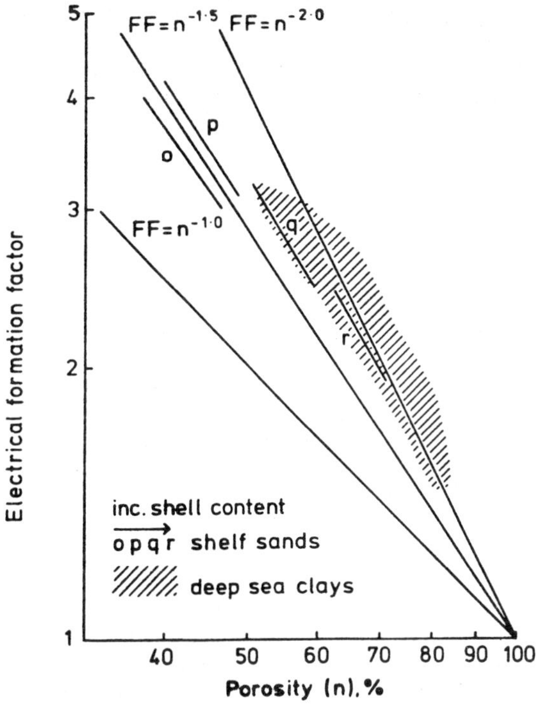

FIG. 1. Formation factor – porosity plot for 4 natural sands and a series of 9 deep-sea clays.

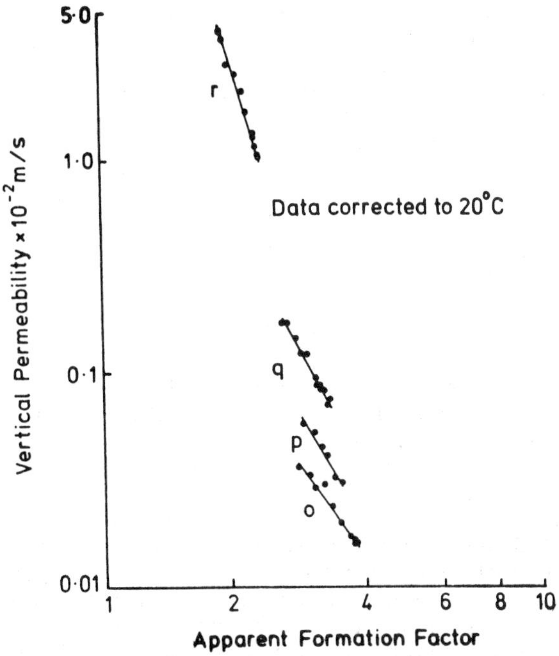

FIG. 2. Permeability – formation factor plot for 4 natural sands.

TABLE 1. *Permeability-formation factor regression lines for natural sands.*

Sample	$K = c\,(FF)^{-x}$		
	c	x	R
o	0.0105	3.1598	0.98
p	0.0221	3.3622	0.98
q	0.0876	3.9623	0.98
r	3.9104	6.7278	0.98

Brace (1977) attached a dependence of x upon the exponent in Archie's law; Fig. 1 shows this to be a dependence on the shape of the particles. Similarly Archie (1942) showed clear relationships to exist between formation factor and porosity and porosity and permeability. Combining these two together produces a relationship between formation factor and permeability of the form:

$$FF = bK^{-y}$$

where both b and y are partially dependent on the exponent in Archie's law (Lovell 1983). This equation is similar to that of Barker & Worthington (1973) for the Bunter Sandstone.

The results suggest some overall relationship between particle shape and the nature of the formation factor permeability relationship for unconsolidated sands, although the absolute magnitude of the permeability is controlled predominantly by the particle size and size distribution (see Fig. 3, after Lovell 1983). Extending this to consolidated rocks or crystalline rocks the relationship may be expected to be determined by the pore structure or crack size and shape.

Similar results pertain for the deep-sea clays with linear fits on a log–log plot (Fig. 4). The variation in each fit presumably corresponding to differences in pore shape and size, though this is probably effected by the mineralogy in controlling the free water, pore size and shape through the pore water absorption mechanism as much as by the solid grain parameters.

Cross-plotting

Problems do exist in the application of geophysics for determining physical properties. With respect to permeability assessment one of the major problems is the lack of uniqueness in the relationship between porosity and permeability. As an example, a high porosity clay and a high porosity shelly sand will exhibit similar electrical formation factors, but will have inherent permeabilities at opposite ends of the spectrum (10^{-8} and 10^{-2} m/s respectively). Thus while electrical formation factor may show a unique relationship with permeability for any one sample, it may be difficult to resolve the precise nature of the sample, solely on the basis of a single geophysical measurement. Even within one sediment type problems exist over smaller ranges of magnitude where the ranges of permeability values corresponding to an electrical formation factor may be less than an order of magnitude (Lovell & Ogden 1983).

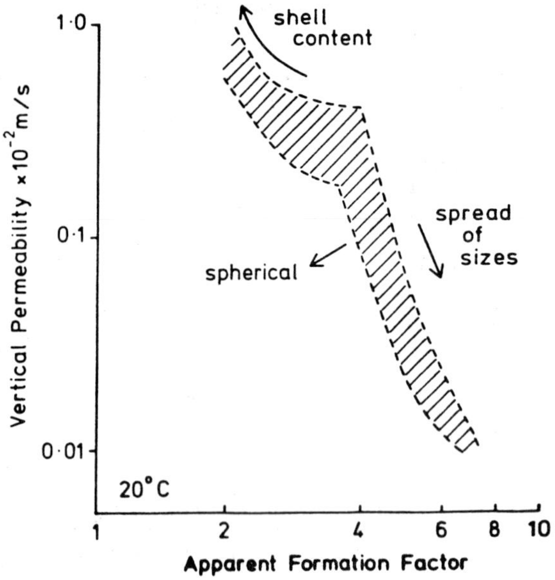

FIG. 3. Composite permeability – formation factor plot for sands of different particle shapes and sizes (after Lovell 1983).

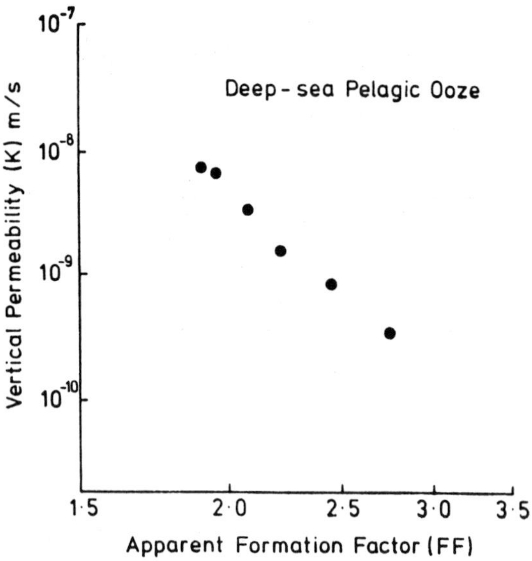

FIG. 4. Permeability – formation factor plot for a deep-sea pelagic ooze.

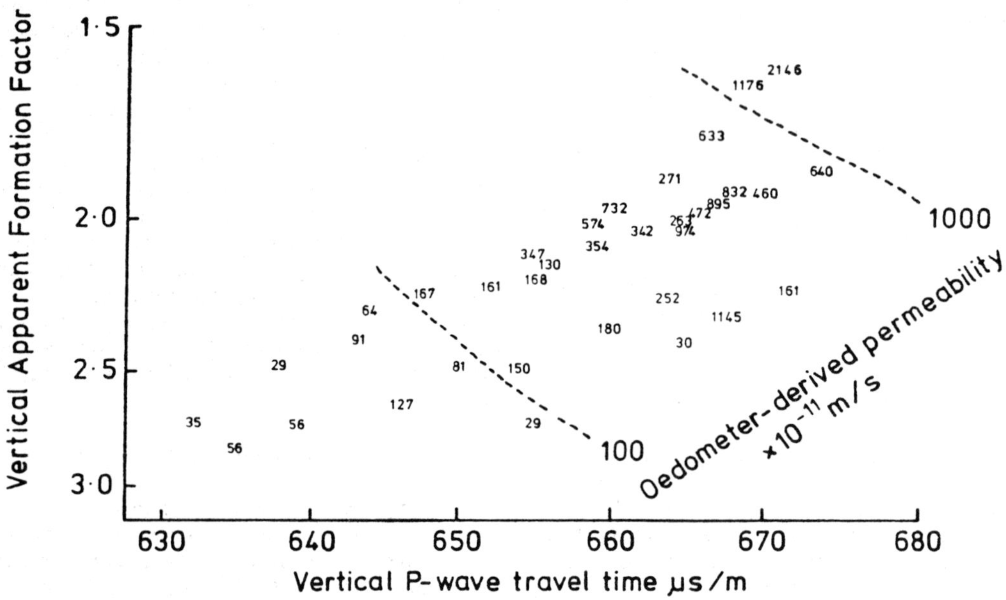

FIG. 5. Cross-plot of formation factor and P-wave travel-time showing permeability variations for 9 deep-sea clays tested in the modified oedometer.

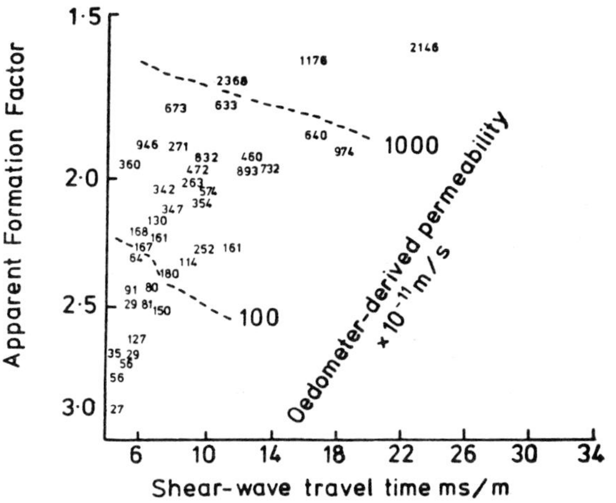

FIG. 6. Cross-plot of formation factor and S-wave travel-time showing permeability variations for 9 deep-sea clays tested in the modified oedometer .

One possible solution here may be to use the technique of cross-plotting. This form of cluster analysis is now used throughout the hydrocarbon exploration industries in the analysis of wireline logs (McFadzean, 1973) and has also been applied to geological engineering (Cratchley 1977; Jackson *et al.* 1981). Here, the technique is applied to the deep-sea clay results. Fig. 5 is a cross-plot of electrical formation factor against S-wave travel time. In both cases the permeability generally plots in fairly distinct zones enabling assessment to within an order of magnitude. It is worth noting that substantial errors may be present in the velocity data due to the very short propagation paths; these are discussed in detail elsewhere (Lovell 1983) and together with the errors in the permeability values (derived from uniaxial consolidation theory) must be considered in viewing the plots. Certainly, within this one sediment type the cross-plot appears to work reasonably well and should perhaps be tested further for different sediments and rocks.

Anisotropy

In defining a dependence of permeability on pore shape, the results effectively identify differences in the sediment structure and hence different distributions of pore geometry. These are in turn related to depositional environments and the fabrics produced are often antisotropic (Pettijohn *et al.* 1972). For the sands studied here, the major fabric is a result of the vertical deposition and compaction, though less well developed horizontal anisotropic fabrics may arise through deposition under horizontal flow regimes (Lovell & Jackson 1975).

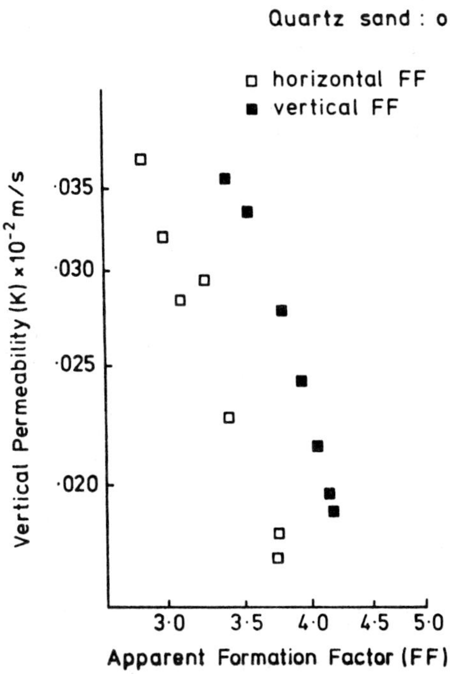

FIG. 7. Permeability - formation factor plot showing anisotropy of the electrical measurements.

Geophysical measurements in consolidated sediments rely on the propagation of energy through the sample. For electrical and compressional wave energy the majority of the received signal travels through the pore fluid, whereas for shear-wave propagation energy is generally transmitted through the particle structure. In clays, P-wave velocity anisotropy is known to exist (Lovell & Ogden 1983) , the horizontal velocity being greater than the vertical, while Fig. 7 shows electrical resistivity anisotropy for sand sample O deposited vertically under water in the laboratory. The links already demonstrated between permeability and geophysics both here and elsewhere in the literature suggest that anisotropy of geophysical measurements should imply permeability anisotropy (Lovell & Jackson 1985). By identifying and measuring this anisotropy for any given situation it may be possible to improve the geophysical-permeability prediction.

Conclusions

1. Empirical relations between electrical resistivity formation factor and permeability have been shown both here and elsewhere in the literature for a variety of different rock types. The precise nature of each individual relationship appears to be governed by the shape and size of the pore structures. These are in turn dependent on the depositional environment and ultimately, for consolidated

rocks, the diagenetic and lithification processes involved.

2. Cross-plotting of different geophysical measurements can provide an integrated approach to the medium and for deep-sea clays allows the permeability to be assessed to within an order of magnitude. The extension of this technique to a range of materials is suggested.
3. Anistropy exists for the geophysical measurements in near surface marine clays and laboratory redeposited sands. The close link between the geophysical parameters and permeability suggests permeability anisotropy may also exist, and at depth in the sediment column this may be considerably greater. Identification and measurement of any anistropy should help in defining geophysical-permeability relationships.

ACKNOWLEDGEMENTS: The experimental work on unconsolidated sediments was carried out at the Marine Science Laboratories (University College of North Wales, Bangor) and thanks are due to Professor D. Taylor Smith and many colleagues for their help and encouragement. Financial support was provided by NERC, BP Petroleum Development Ltd., and DOE/EEC (contract 263-81-7 WAS UK). Thanks also to Dr D. McCann who first suggested the use of cross-plots, to Dr P.D. Jackson for reading the original manuscript and to Mrs J. Angell, Mr D Jones, Mrs J. Pearson and Mrs J. Wilkinson for putting the raw manuscript and figures together.

References

ARCHIE, G.E. 1942. The electrical resistivity log as an aid in determining some reservoir characteristics. *Transactions of the American Institute of Mining and Metallurgical Engineers,* **146,** 54-62.

BARKER, R.D. & WORTHINGTON, P.F. 1973. Some hydrogeophysical properties of the Bunter Sandstone of Northwest England. *Geoexploration,* **11,** 151-170.

BIOT, M.A. 1956. Theory of propagation of elastic waves in a fluid saturated porous solid. I Low-frequency range II High-frequency range. *Journal of the Acoustics Society of America,* **28,** 169-191.

BRACE, W.F., 1977. Permeability from resistivity and pore shape. *Journal of Geophysical Research,* **8,** 3343-3349.

BROWN, A. & HUSSEINI, S. 1977. Permeability from well logs, Shaybah Field, Saudi Arabia. SPWLA 18th Annual Logging Symposium 1977. Paper Q.

CRATCHLEY, C.R. 1977. Geophysical measurements in rock mechanics investigations. *Proceedings of Goguel College Orleans,* May 1977. 447-456.

DAVIS, A.M. & SCHULTHEISS, P.J. 1980. Seismic signal processing in engineering site investigation - a case history. *Ground Engineering* May.

GRATON, L.C. & FRASER, H.J. 1935. Systematic packing of spheres with particular relationship to porosity and permeability. *Journal of Geology,* **43** 785-909.

HAMDI, F.A.I. & TAYLOR SMITH, D. 1981. Soil consolidation behaviour assessed by seismic velocity measurements. *Geophysical Prospecting,* **29,** 715-729.

HAMDI, F.A.I. & TAYLOR SMITH, D. 1982. The influence of permeability on compressional wave velocity in marine sediments. *Geophysical Prospecting,* **30,** 622-640.

JACKSON, P.D. 1975. Electrical properties of Irish Sea Sediments. *PhD thesis,* Univ. of Wales.

JACKSON, P.D. TAYLOR SMITH, D. & STANFORD, P.N. 1978. Resistivity-porosity-particle shape relationships for marine sands. *Geophysics,* **43,** 1250-1268.

JACKSON, P.D., BARIA, R. & MCCANN, D.M. 1981. Geotechnical assessment of superficial marine sediments using in-situ geophysical probes. *Ocean Management,* **7,** 189-209.

LOVELL, M.A. 1983. Resistivity-porosity-thermal conductivity relationships for marine sediments. *PhD thesis,* Univ. of Wales.

LOVELL, M.A. 1985. Thermal conductivity and permeability assessment by electrical resistivity measurements in marine sediments. Marine *Geotechnology,* **6,** No. 2.

LOVELL, M.A. & OGDEN, P. 1983. *Remote assessment of permeability/thermal diffusivity of consolidated clay sediments.* Report to the European Atomic Energy Community, published in the EUR-series, Brussels (1984) EUR 9206 EN.

LOVELL, M.A. & JACKSON, P.D. 1985. Anisotropic electrical and acoustic properties of sediments and their implications for the assessment of permeability. *47th Meeting of the European Association of Exploration Geophysicists,* Budapest, Hungary, June 1985.

MCFADZEAN, T.B. 1973. Cross-plotting - a neglected technique in log analysis. SPWLA ICETL Annual Logging Symposium Paper Y.

PETTIJOHN, F.J., POTTER, P.E. & SIEVER, R. 1972. Sand and Sandstone. Springer Verlag, New York.

M.A. LOVELL, Department of Geology, University of Nottingham, Nottingham NG7 2RD, UK.

The representation of time variant free surface seepage using finite difference methods: a review

J.C. Miles

ABSTRACT: Several numerical techniques are available for the representation of time variant free surface seepage, notably finite elements, boundary elements and finite difference methods. This work concentrates on the latter. Over the last twenty years several papers have been published on the representation of free surface seepage using finite differences but unfortunately each paper advocates the use of different approximations and stability criteria. The choice of which variation to use is therefore a problem. In this paper the various methods are reviewed and tested against a variety of solutions and recommendations are made as to which method is the most stable and accurate.

Introduction

The use of numerical methods to represent the way in which water flows through porous media is nowadays commonplace and much work has been published describing various applications. The majority of these applications utilise the Dupuit-Forcheimer assumptions to avoid the complexity of having to represent vertical flows. For many groundwater flow problems this is a valid approximation especially where the horizontal flow path is much greater than the vertical flow path, but there are some problems where vertical flows are important.

In such cases two types of numerical model can be used, these being models which fully represent saturated/unsaturated flows and free surface seepage models. The former require knowledge of the unsaturated flow parameters and involve the solution of non-linear equations. To avoid this the latter method is frequently used. Skaggs & Tang (1976) show that such models give reasonably accurate estimates of flows and head changes despite the fact that the unsaturated zone is not represented.

Various numerical techniques have been used to solve free surface seepage problems but only digital finite difference approximations are considered in this paper. The major contributions in this area were by Todsen (1971), and Verma & Brutsaert (1971). Todsen advocated the use of fairly simply finite difference approximations and a constant mesh shape whereas Verma & Brutsaert suggested using a moving mesh and complex approximations. Subsequent work by Amar (1975) followed the work of Todsen but Singh (1976) used similar techniques to those of Verma & Brutsaert but without the moving mesh. Guvanasen & Volker (1980) whilst mainly concentrating on finite elements did look in detail at the work of Todsen and briefly at that of Verma & Brutsaert and Singh.

In this paper the work of Guvanasen & Volker is extended to fully consider both methods and the various problems which occur with regard to stability are discussed.

Equations

The equation for two dimensional seepage can be written as:

$$\frac{\partial}{\partial x}\left(k_x \frac{\partial h}{\partial x}\right) + \frac{\partial}{\partial z}\left(k_z \frac{\partial h}{\partial z}\right) = 0 \qquad (1)$$

where k_x and k_z are the x and z direction permeabilities respectively and h is the groundwater potential, The equation assumes that the compressibility of the porous media and the water can be ignored.

Assuming that the slope of the water table is small the time variant movement of the free surface can be described by the equation:

$$\delta H = -\frac{\delta t}{S_y}\left(k_z \frac{\partial h}{\partial z} + k_x \frac{\partial h}{\partial x} \tan \alpha - q\right) \qquad (2)$$

where δH is the change in the height of the free surface δt is the time increment over which the change takes place S_y is the specific yield or effective porosity of the media. q is the recharge to the water table expressed as a depth per unit time, and α is the angle of the free surface below the horizontal plane.

An alternative formulation of the equation is:

$$\delta H = -\frac{\delta t}{S_y}\left[k_z \frac{\partial h}{\partial z} - k_x \left(1 - \frac{\partial h}{\partial z}\right) \tan^2 \alpha - q.\right] \qquad (3)$$

From CRIPPS, J.C., BELL, F.G. & CULSHAW, M.G., (eds), 1986, *Groundwater in Engineering Geology,* Geological Society Engineering Geology Special Publication No. 3, pp. 341-345

By using either Equation (2) or (3) in conjunction with Equation (1) and the boundary condition that

$$h = z \tag{4}$$

on the free surface, the movement of the free surface can be calculated. For details of the way in which these equations are combined to obtain a solution the reader is referred to the literature.

Finite difference approximations

The representative of Equation (1) in finite difference form is straightforward. Details of suitable approximations are to be found in Rushton & Redshaw (1979). The difference between the various methods lies in the choice of free surface equations and the finite difference approximations used therein.

The representation of the tan α term in both Equations (2) and (3) is straightforward:

$$\tan\alpha = \frac{HF\,(I+1) - HF\,(I-1)}{2\Delta x} \quad . \tag{5}$$

The accuracy of the equation is improved if a regular mesh spacing is used in the x direction. The notation is given in Fig. 1.

Todsen (1971), Amar (1975) and Guvansen & Volker (1980), use equation (3) and represent the term $\partial h/\partial z$ by the approximation

$$\frac{\partial h}{\partial z} = \frac{1}{\Delta z}\left\{\frac{1+2\eta}{\eta\,(1+\eta)}\,HF(I) + \frac{\eta}{(1+\eta)}\,h(I,J+1) - \left(\frac{1+\eta}{\eta}\right) h(I,J)\right\} \tag{6}$$

where $h(I,J)$ is the head at node (I,J), $HF(I)$ is the free surface head, etc. as shown in Fig. 1.

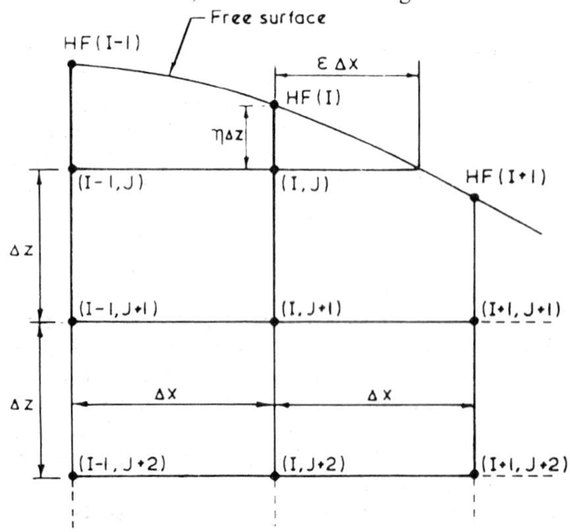

FIG. 1. Mesh Notation.

Verma & Brutsaert and Singh utilise equation (2) and represent δh/δz by the approximation

$$\frac{\partial h}{\partial z} = \frac{1}{\left(\frac{1-\eta}{1+\eta}\right)\Delta z}\left\{\frac{HF(I)}{\eta} - \left[\left(\frac{1+\eta^2}{\eta}\right)h(I,J) + \eta h(I,J+1) + \frac{\eta^2}{2(1+\eta)}\left(h(I,J)-h(I,J+2)\right)\right]\right\}. \tag{7}$$

Also the representation of the $\partial h/\partial x$ term in equation (2) is made using the approximation

$$\frac{\partial h}{\partial x} = \frac{h(I+1,J)-h(I-1,J)}{2\Delta x} + \left[\frac{\partial h(I+1,J)}{\partial z} - \frac{\partial h\,(I-1,J)}{\partial z}\right)\Bigg]\frac{\eta\,\Delta z}{2\Delta x} + \left[\frac{\dfrac{\partial^2 h(I+1,J)}{\partial z^2} - \dfrac{\partial^2 h(I-1,J)}{\partial z^2}}{2\Delta x}\Bigg|\,\frac{(\eta\Delta z)^2}{2}\ .\right] \tag{8}$$

Verma & Brutsaert recommend that the standard forms of finite difference approximations should be used for the $\partial h/\partial z$ and $\partial^2 h/\partial z^2$ terms above unless the distance from row J to the free surface is less than one mesh space, in which case curved boundary approximations of the form (Smith 1965):

$$\frac{\partial h}{\partial z} = \frac{1}{\Delta z}\left\{\frac{1}{\eta(1+\eta)}\,HF(I) - \frac{(1-\eta)}{\eta}\,h(I,J) - \frac{\eta}{(1+\eta)}\,h(I,J+1)\right\}$$

and

$$\frac{\partial^2 h}{\partial z^2} = \frac{2}{\Delta z^2}\left\{\frac{HF(I)}{\eta(1+\eta)} + \frac{H(I,J+1)}{(1+\eta)} - \frac{h(I,J)}{\eta}\right\} \tag{9}$$

should be used.

Time stepping

Equations (2) and (3) contain time variant components and so it is possible to define the heads in the equations at various time levels. Todsen (1971) recommended using a centred formula of the form:

$$\delta H = HG_{t+\delta t} - HF_{t-\delta t} \tag{10}$$

where δt is the time step size and t is the current time. Singh uses the formula:

$$\delta H = HF_{t+dt} - HF_t\,. \tag{11}$$

Both the above result in explicit time stepping schemes and hence limitations have to be placed on the size of the time step δt. Various recommendations have been made as to a suitable upper limit of δt. That of Todsen is too lax and leads to inaccuracies whilst Singh's criterion is too restrictive. The most pragmatic value is that of Guvanasen & Volker:

$$\frac{\delta t\, K}{S_y\, \Delta x} \leqslant \frac{1}{3} \quad . \tag{12}$$

Smoothing

Todsen found that even with upper limits on the time step size, numerical instability can occur. He recommended the use of smoothing to avoid this problem but Guvanasen & Volker shows that this gives erroneous results. If the time stepping scheme of Equation (11) is used the need for smoothing can be avoided provided that the constraints described below are imposed.

Constraints

Where the free surface head is almost coincident with a finite difference mode (i.e. $\eta \rightarrow 0$ or $\varepsilon \rightarrow 0$) various constraints are needed to avoid instability.

When solving the finite difference form of Equation (1) for node (I,J) (using the notation of Fig. 1) if η and/or ε are small, then the terms $\partial^2 h/\partial x^2$ and/or $\partial^2 h/\partial z^2$ will become very large. It has been found that if the minimum values of η and ε are restricted to 0.001, then the accuracy of the solution is not noticeably affected and instability is avoided.

Similarly in equations (6) and (7) if η becomes small then problems occur. Todsen suggests that if the value of $\eta < 0.1$ then the following equation should be used:

$$\frac{\partial h}{\partial z} = (HF(I) - h(I,J+1))/(HF(I) - z(J+1)). \tag{13}$$

This has been found to give good results although if the method of Singh is used $\eta < 0.01$ has been found to be more appropriate.

For Equation (8) Singh recommended that if the free surface intersects row J in the vicinity of node (I,J) then a curved boundary approximation of the same form as Equation (9) should be used instead of Equation (8).

Free surface — seepage dace intersection

On a seepage face the condition $h = z$ applies everywhere as it does on the free surface but unlike the free surface the seepage face is not a flow line because flow passes through it. Therefore at a node where the seepage face meets the free surface a singularity exists. This means the Equations (2) and (3) are inapplicable and so the position of the free surface has to be estimated. The usual method of doing this is by using linear extrapolation from the two proceeding upstream nodes.

This method is reasonably satisfactory except when it is necessary to calculate flows. For example, if it is wished to calculate the discharge to the drain in Fig. 2 the flows across the last two columns of nodes cannot be calculated satisfactorily because the above extrapolation technique causes small errors. In such cases it is best to calculate the flow to the drain by estimating the total change of volume caused by the movement of the free surface.

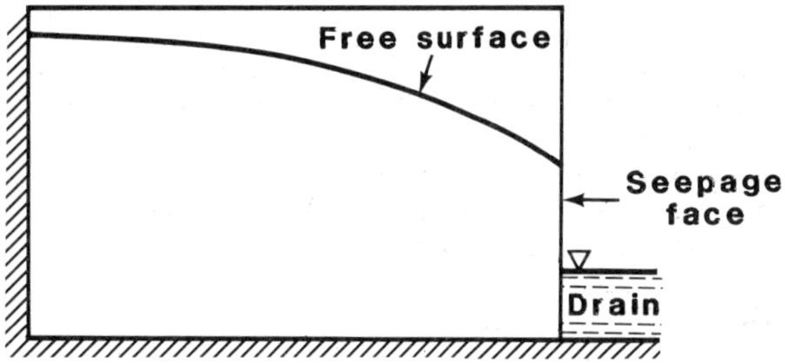

FIG. 2. Flow to a drain.

Comparison of methods

The usual method of checking the accuracy of numerical methods is by comparing the results with analytical solutions. For free surface seepage no suitable analytical solutions are available and so other comparisons have to be resorted to. Amar (1975) and Guvanasen & Volker (1980) used sand tank tests to check the accuracy of their programs but obviously experimental results are subject to errors. Todsen (1971) compared his results with those obtained by Herbert (1968) from electrical resistance analogues.

One of the examples modelled by Herbert is shown in Figs 3 and 4, the latter gives the heads at 2.0 from the left hand boundary. In this problem at times $t<0$ all heads are set to 16.0. At times $t \geqslant 0$ the fixed head on the left hand boundary is set to zero. Herbert used a mesh with $\Delta z = \Delta x = 1.0$ and this spacing has been used to obtain the results given in Fig. 4. The time step used for the numerical results is $k\Delta t/\Delta x S_y = 0.333$.

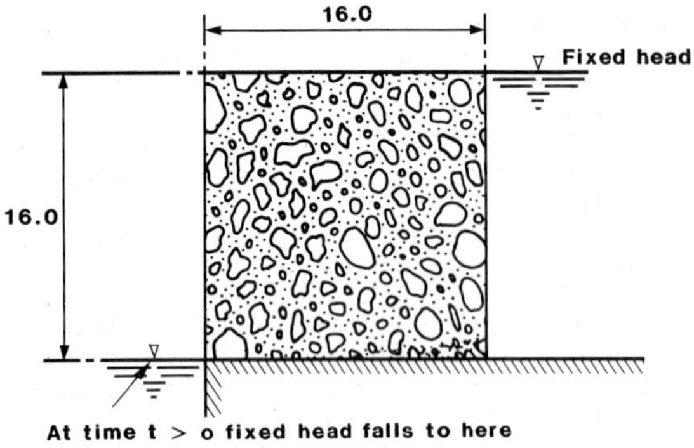

FIG. 3 Herbert's example

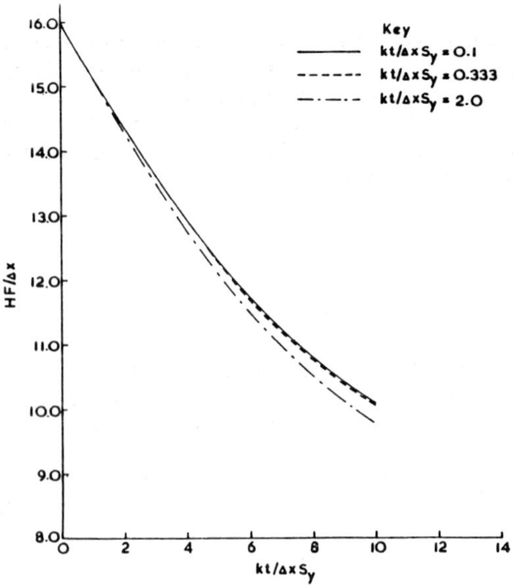

FIG. 5. Effects of time step size.

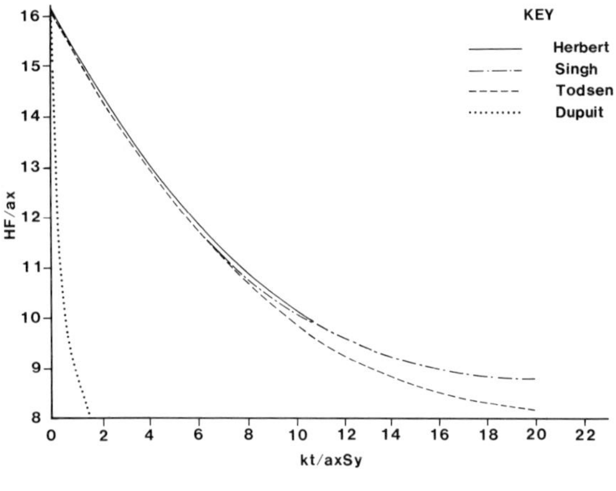

FIG. 4. Comparison of methods.

Fig. 4 shows that Singh's method gives better agreement with Herbert's results than does Todsen's method. This bears out the previously mentioned findings of Guvanasen & Volker about the effects of smoothing. Also included are the results for a solution using a non-linear version of the Dupuit-Forcheimer assumptions (i.e. one dimensional example). The errors caused by not representing vertical flows close to a seepage face are shown to be large.

Fig. 5 shows the variations in the results obtained by the use of different time step sizes. The results for the time step size $k\Delta t/\Delta xSy = 0.333$ agree with the maximum value allowed by Equation (12).

Recharge

It is possible to solve problems involving recharge using the above described methods but two problems arise. Fig. 6 shows the type of instability that can occur if the slope of the water table becomes steep enough to invalidate Equation (5).

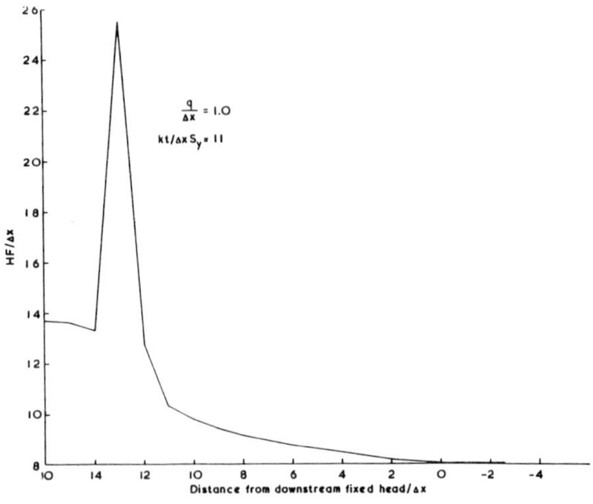

FIG. 6. Recharge example.

The plot in Fig. 6 shows a water table profile for a problem of the same dimensions as in Fig. 3 but with all heads at time $t<0$ set to 8.0 and for times $t \geqslant 0$ the left hand fixed head maintained at 8.0 but the right hand boundary being treated as being impermeable. The instability eventually becomes so great as to stop the execution of the computer program.

The other type of instability that can arise is when the recharge extends across the full width of the problem domain. Then the inaccuracies caused by the use of a linear extrapolation at the free surface-seepage face node can cause unrealistic answers to be generated.

Summary

Two different methods using finite difference approximations to represent free surface seepage have been described. The better of the two methods is that described by Singh (1976). It is less prone to instability and gives more accurate results provided that certain constraints are imposed. The methods described are strictly speaking, only suitable for meshes which have constant values for Δx and Δz. The methods cannot cope with complex situations and it is doubtful for example whether a 3 dimensional problem could be simulated. It recharge is to be included this must be done with great care.

References

AMAR, A.C., 1975. Groundwater recharge simulation, *Proceedings of ASCE, Hydraulics Division,* **HY9,** 1235-1247.

GUVANASEN, V. and VOLKER, R.E., 1980. Numerical solutions for unsteady flow in unconfined aquifers, *International Journal for Numerical Methods in Engineering,* **15,** 1643-1657.

HERBERT, R., 1968. Time variant groundwater flow by resistance network analogues, *Journal of Hydrology,* **6,** 237-264.

RUSHTON, K.R. and REDSHAW, S.C., 1979. *Seepage and Groundwater Flow,* Wiley, Chichester.

SINGH, R., 1976. Predication of mound geometry under recharge basins, *Water Resources Research,* **12**(4), 775-780.

SKAGGS, W.R. & TANG, Y.K., 1976. saturated and unsaturated flow to parallel drains, *Proceedings of ASCE, Irrigation Division,* **IR2,** 221-238.

SMITH, G.D., 1965. *Numerical Solution of Partial Differential Equations,* Oxford University Press, Oxford.

TODSEN, M., 1971. On the solution of transient free-surface flow problems in porous media by finite-difference methods, *Journal of Hydrology,* **12,** 177-210.

VERMA, R.D. & BRUTSAERT, W., 1971. Unsteady free surface groundwater seepage, *Proceedings of ASCE, Hydraulics Division,* **HY8,** 1213-1229.

J. C. MILES Department of Civil and Structural Engineering, University College, Newport Road, Cardiff CF2 1TA, UK.

Analysis of an advanced dewatering scheme at an opencast coal site in Northumberland

S.T. Minett, D.A. Blythe, G.D. Hallam & D.B. Hughes

ABSTRACT: A triangular area of Coal Measures rocks bounded to the south by the major WSW-ENE trending Causey Park Dyke, to the north and west by the limit of outcrop and the east by the North Sea, has been extensively worked by opencast methods over the past twenty years. Due to the close concentration of opencast workings it has been feasible to maintain a regional dewatering scheme over a number of years from Hauxley Shaft. However, during prospecting operations for an opencast coal site it was found that substantial quantities of groundwater existed in old workings horizons, ponded against the Causey Park Dyke in a broad anticlinal structural trap.

It was decided by the Contractor that the area should be dewatered before excavations took place, to avoid the danger of flooding and water induced slope failures, and a submersible borehole pump was, thus, installed.

This provided an opportunity to undertake a pumping test in mining disturbed strata. Due to the high permeability of the old underground workings it was found that the data did not conform to any standard analysis methods. However, it was found that the system could be modelled by treating it as a lagoon of triangular wedge shape. A mass balance calculation was performed and it was predicted that it would take approximately 350 days to dewater the site. The field data conforms closely to the predicted drawdowns and it is concluded that this method may be applicable to many regions with a mining history.

Introduction

The most northerly section of the exposed coalfield in Northumberland forms a triangular area bounded to the south by the major WSW-ENE tending quartz-dolerite Causey Park Dyke, to the north and west by the limit of outcrop and to the east by the North Sea. The area has been worked by deep coal mining methods for over 100 years, the last colliery closing in 1967.

Opencast operations have taken place in north Northumberland for more than 20 years. Initial high regional groundwater levels and the large size and depth of opencast sites meant that some form of dewatering was necessary. Due to the close concentration of opencast workings it has been feasible to maintain a regional dewatering scheme at Hauxley Shaft. The pumping scheme influences a very large area due to the high mass permeability of strata which are affected by old underground coal mining. This has maintained groundwater levels below the pavement seams on most sites at a level approximately 65 m BOD. However, during prospecting operations it was found that substantial quantities of groundwater existed in old workings horizons ponded against the Causey Park Dyke in a broad anticlinal trap.

The geology of the site is simple. The strata form a typical Coal Measures sequence of interbedded sandstones, siltstones and mudstones with intercalated coal seams and seatearths. The sandstones, which comprise approximately 30% of the succession, are generally fine to medium grained, well cemented and laterally impersistant. The strata dip gently to the south each at 2° being truncated to the south by the Causey Park Dyke. There is no major faulting on the site apart from a 60 m north throwing fault, along which the dyke is intruded (see Fig. 1).

The area is covered, as is usual in this region, by drift, which thickens to the south. It is dominated by clayey till with impersistant lenses of sand and gravel. To the south of the Causey Park Dyke there is a major W-E trending drift channel which is over 45 m deep and is mainly sandy in nature.

The hydrogeology of the area is controlled by four main factors:

(a) the pumping at Hauxley Shaft,
(b) the proportion, distribution and type of old workings,
(c) the Causey Park Dyke,
(d) the drift cover.

Before opencast mining operations started on the site in 1983 the groundwater table was approximately 45 m BOD. Initial investigations (NCB (OE) internal reports 1977, 1982), suggested that the Hauxley pump would not lower the groundwater level any further. It was concluded that provided the water level is maintained below 65 m BOD at Hauxley

From Cripps, J.C., Bell, F.G. & Culshaw, M.G., (eds), 1986, *Groundwater in Engineering Geology,* Geological Society Engineering Geology Special Publication No. 3, pp. 347-352

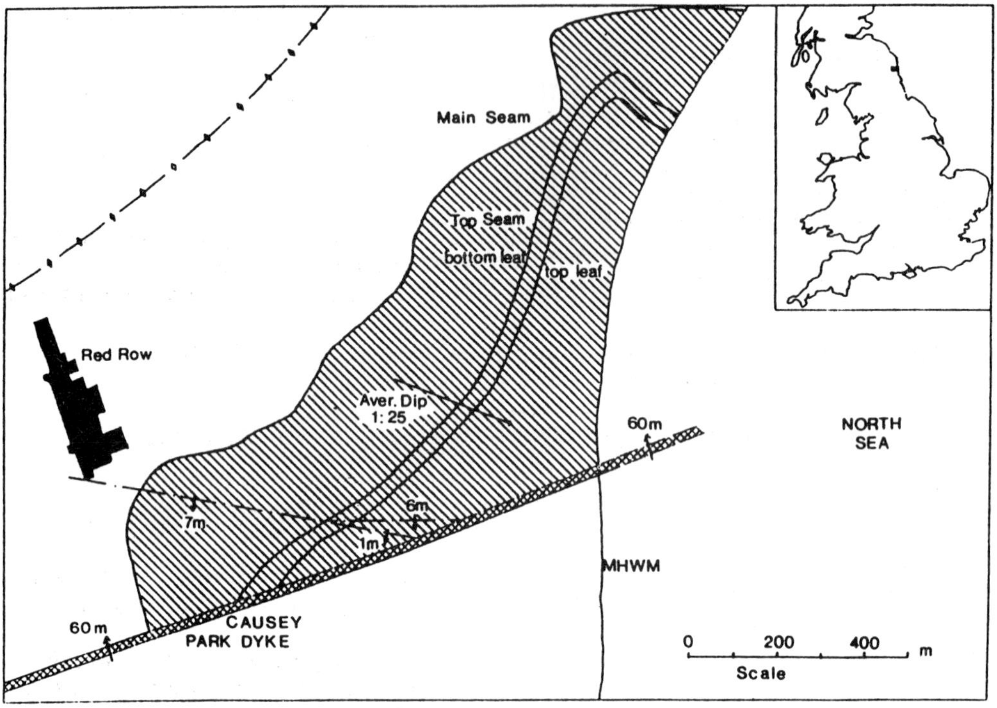

FIG. 1. A location map of the area. The hatched area is the extent of the flooded old workings in the top and bottom leaves of the top of Broomhill Seam and the main of Broomhill Seam.

Shaft, then water levels on the site will remain unaffected. However, if the regional water level was allowed to rise above 65 m BOD then groundwater would flow through old workings horizons over the crest of anticline and water levels would rise on the site.

The major aquifers of this system are the old workings horizons and the broken strata associated with mining activities. Some attempts have been made to estimate the permeability of Coal Measures strata. D.J. Hammond (1977, Unpublished data) and D.A. Blythe and K.G. Walters (NCB (OE) internal report, 1982) have studied permeability in virgin strata on opencast sites, and Hammond (1977, Unpublished data); Sgambat *et al.* (1980); Garritty (1980, 1982, 1983); McWhorter (1981) and Aston & Singh (1983) have attempted to estimate mass permeabilities in old workings and the effects of underground mining on the groundwater regime. In this region permeabilities for typical Coal Measures rocks in areas unaffected by mining are relatively low. In a suite of packer tests Blythe and Walters (NCB(OE) internal report, 1982) found permeabilities of 10^{-6} to 10^{-7} ms^{-1} for mudstones, sandstones and coal seams. Hammond (1977, Unpublished data) estimated that the permeability of the collapsed strata above longwall panels is 5.7×10^{-5} ms^{-1}. However, due to the variable nature of roof measures and the unpredictability of caving above workings a permeability of 5×10^{-5} ms^{-1} is considered a minimum value.

The area of flooded old workings is terminated by the Causey Park Dyke. The strata to the south of the dyke are unaffected by mining and will consequently have lower permeabilities. Piezometer data indicate a standing water level of 25 m BOD, substantially higher than the northern side. It is assumed, therefore, that the dyke is a hydrogeological barrier with the pumping from Hauxley Shaft having little or no affect on the groundwater regime to the south.

The drift largely isolates the system from rapid infiltration from the surface or from the sea. The water from the well has a chloride concentration of 1000 ppm indicating that sea water constitutes approximately 5% of total.

The dewatering well is 82 m deep (71 m BOD) and is placed in an open old workings roadway in the lowest worked seam. A 10 inch (250 mm) Sumo submersible pump was installed which has a measured rate of discharge of $6.26 \times 10^{-2} m^3 s^{-1}$. Water from the well drains to the sea, via a lined ditch and a set of lagoons and, thus, does not return to the system.

Analysis of data

It has been found, while analysing this data, that the system does not conform to any standard aquifer testing methods, Cooper-Jacob analysis (Cooper & Jacob 1946), Theis method (Theis 1935), Theim method (Theim 1906) and Stallman method (Ferris *et al.* 1962) for porous aquifers and the Kozeny-Carman equation and its derivatives (Kozeny 1927; Carman 1956; Sharp 1970; Garritty 1980) for fractured aquifers.

It is possible, however, to conduct a mass balance calculation, if a number of assumptions are made, to predict and rate of drawdown, time taken for total drawdown and recharge rate. These assumptions are:

(i) That the system was in equilibrium before pumping started (i.e. recharge to the system equalled discharge from the system).

(ii) That the system can be approximated to a triangular wedge (see Fig. 2), the top surface (ABC) of which is the regional water level at 45 m BOD and the over dimensions are shown on Fig. 2.

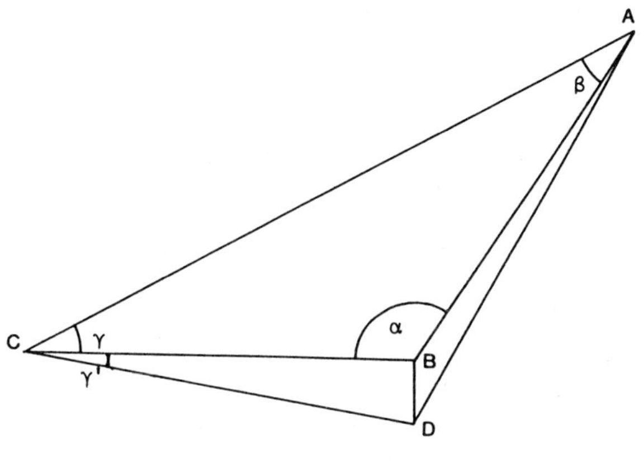

FIG. 2. The approximate shape of the groundwater system ($\beta = \gamma = 30°$; $\alpha = 120°$; $\gamma' = 0.69°$, and before dewatering starts CB = BA = 2150 m; BD = h = 26m).

(iii) That all water emanates from the voids in the old workings. The minimum initial volume of water is 1.02×10^6 m^3 (Blythe, NCB(OE) internal report, 1982). The maximum initial volume of water is equivalent to the total volume of coal removed, 3.81×10^6 m^3.

(iv) That the flow of water into the well is instantaneous and that the upper surface (ABC) drops uniformly across its surface (i.e. that there is no cone of depression and that the groundwater reservoir can be regarded as a lagoon). Thus

$$P = 0.5\,(h \cot \gamma')^2 \sin \alpha \tag{1}$$

and

$$V = 1/3\, Ph, \tag{2}$$

where P is the area (ABC) of the top surface and V is the volume (ABCD) of the wedge. As $\cot \gamma'$ and $\sin \alpha$ are constants, then

$$V = 995\, h^3 \tag{3}$$

and

$$\mathrm{d}V/\mathrm{d}h = 2985\, h^2 \tag{4}$$

so

$$\mathrm{d}V/\mathrm{d}t = \mathrm{d}V/\mathrm{d}h \times \mathrm{d}h/\mathrm{d}t \tag{5}$$

where $\mathrm{d}V/\mathrm{d}t$ is the effective pumping (discharge) rate and $\mathrm{d}h/\mathrm{d}t$ is the drawdown rate.

If the assumption is made that throughout pumping the recharge to the system equals its natural discharge, then

$$\mathrm{d}V/\mathrm{d}t = Qw^{-1} \tag{6}$$

where Q is the measured pumping rate from the well of $6.26 \times 10^{-2}\ m^3 s^{-1}$ and w is the water ratio, which is found from the volume of water (V_w) and the total volume,

$$w = V_w V^{-1}. \tag{7}$$

Thus the water ratio is between 5.8% and 21.8% (from assumption iii) depending on the estimate of initial water volume. Then

$$\mathrm{d}V/\mathrm{d}t_{(max)} = 1.08\ m^3 s^{-1}$$

and $\mathrm{d}V/\mathrm{d}t_{(min)} = 2.87 \times 10^{-1}\ m^3 s^{-1}$.

From Equation (5) and drawdown rate can be expressed:

$$\mathrm{d}h/\mathrm{d}t = \mathrm{d}V/\mathrm{d}t.(\mathrm{d}V/\mathrm{d}h)^{-1}$$

$$= 3.62. \times 10^{-4} h^{-2} \text{ for } \mathrm{d}V/\mathrm{d}t_{(max)} \tag{8}$$

and

$$\mathrm{d}h/\mathrm{d}t = 9.61 \times 10^{-5} h^{-2} \text{ for } \mathrm{d}V/\mathrm{d}t_{(min)}. \tag{9}$$

TABLE 1. *The drawdown rates, d_1; d_2; d_3; for dV/dt $_{(max)}$ dV/dt $_{(min)}$; dV/dt $_{(ob)}$ in ms^{-1} respectively and the corresponding drawdown times, t_1; t_2; t_3;.*

h (m)	d_1	t_1	d_2	t_2	d_3	t_3
25	5.80×10^{-7}	21	1.54×10^{-7}	78	3.95×10^{-7}	31
24	$6.29.\times 10^{-7}$	40	1.67×10^{-7}	150	4.29×10^{-7}	59
23	6.85×10^{-7}	58	1.82×10^{-7}	217	4.67×10^{-7}	85
22	7.48×10^{-7}	74	1.99×10^{-7}	278	5.10×10^{-7}	108
21	8.21×10^{-7}	89	2.18×10^{-7}	333	5.60×10^{-7}	130
20	9.06×10^{-7}	102	2.40×10^{-7}	384	6.17×10^{-7}	150
19	1.00×10^{-6}	115	2.61×10^{-7}	430	6.84×10^{-7}	168
18	1.12×10^{-6}	125	2.97×10^{-7}	471	7.62×10^{-7}	184
17	1.25×10^{-6}	135	3.33×10^{-7}	501	8.54×10^{-7}	198
16	1.42×10^{-6}	144	3.75×10^{-7}	541	9.64×10^{-7}	211
15	1.61×10^{-6}	152	4.27×10^{-7}	569	1.10×10^{-6}	222
14	1.85×10^{-6}	159	4.90×10^{-7}	595	1.26×10^{-6}	232
13	2.14×10^{-6}	164	5.69×10^{-7}	617	1.46×10^{-6}	240
12	2.52×10^{-6}	169	6.67×10^{-7}	636	1.71×10^{-6}	248
11	2.99×10^{-6}	174	7.94×10^{-7}	651	2.04×10^{-6}	254
10	3.62×10^{-6}	177	9.61×10^{-7}	665	2.47×10^{-6}	259
9	4.47×10^{-6}	179	1.19×10^{-6}	676	3.05×10^{-6}	263
8	5.66×10^{-6}	182	1.50×10^{-6}	684	3.86×10^{-6}	267
7	7.39×10^{-6}	184	1.96×10^{-6}	691	5.04×10^{-6}	269
6	1.01×10^{-5}	185	2.67×10^{-6}	696	6.86×10^{-6}	271
5	1.45×10^{-5}	186	3.84×10^{-6}	700	9.88×10^{-6}	273
4	2.26×10^{-5}	186	6.03×10^{-6}	702	1.54×10^{-5}	274
3	4.03×10^{-5}	188	1.07×10^{-5}	704	2.74×10^{-5}	274
2	9.06×10^{-5}	188	2.40×10^{-5}	705	6.17×10^{-5}	275
1	3.62×10^{-4}	188	9.61×10^{-5}	705	2.47×10^{-4}	275
0	-	188	-	705	-	275

The rates of drawdown are presented in Table 1 and it is possible to calculate the time taken for total drawdown from Equation (5):

$$dh/dt = dV/dt\,(dV/dh)^{-1}$$

$$dt = 2985(dV/dt)^{-1}h^2dh.$$

As the effective pumping rate is constant by setting $2985(dV/dt)^{-1} = k$, then:

$$dt = kh^2dh \qquad (10)$$

and

$$t = kh^3/3 + C, \qquad (11)$$

where t is the time taken for drawdown to values of h and C is the integration constant. This equation may be rewritten as:

$$t = (9.95h^3 - 1.75 \times 10^7)/(dV/dt). \qquad (12)$$

It is, then, possible to draw a set of type curves for various values of dV/dt (Fig. 3) and the limits, $dV/dt_{(max)}$ and $dV/dt_{(min)}$, may be plotted, the corresponding times for total drawdown are approximately 200 days and 700 days.

Using data from the field it is possible to predict an improved drawdown rate (see Fig. 3 and Table 1) and the estimated time taken to dewater the system. An estimate has been taken on the first 108 days of the test or the initial 4 m drawdown. The observed effective pumping rate ($dV/dt_{(ob)}$) is 7.37×10^{-1} m^3m^{-1} and the time for total drawdown is 275 days. It is also possible to calculate a water ratio and an initial water volume from the system based on observed data from Equation (6):

$$w = Q(dV/dt_{(ob)})^{-1} = 8.5\%$$

and from equation (7):

$$V_w = w.V = 1.47 \times 10^6 \text{ m}^3.$$

However, it is unlikely that the total recharge will equal the natural discharge once pumping starts. As the system is dewatered drainage from the surrounding strata and leakage from the strata in the system will occur. This effective recharge will increase as the system is further dewatered. It was found that effective recharge started to have a significant effect on the predicted drawdown curve at 14 m (a drawdown of 12 m) and the proportion of recharge increased at a steady rate of 1.6×10^{-2} m^3s^{-1} per metre as drawdown proceeded. Effective recharge to the system can be seen in Fig. 3 as the deviation of

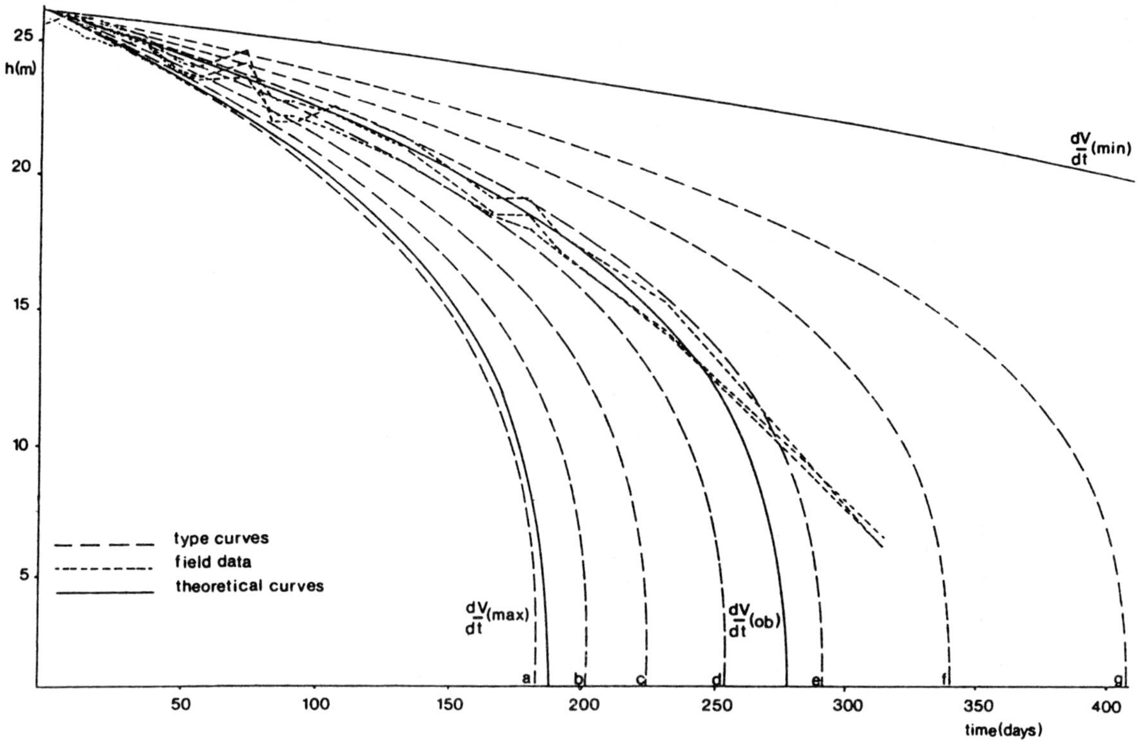

FIG. 3. The type curves for drawdown against time for a range of dV/dt values, a=1.1 ms^{-1}; b=1.0 ms^{-1}; c=9.0 × 10^{-1} ms^{-1}; d=8.0 × 10^{-1} ms^{-1}; e=7.0 × 10^{-1} ms^{-1}; f=6.0 × 10^{-1} ms^{-1}; g=5.0 × 10^{-1} ms^{-1}. Also plotted are the curves for $dV/dt_{(max)}$; $dV/dt_{(min)}$, $dV/dt_{(ob)}$ and the field data.

the field data plot from the calculated drawdown curve. Thus, it was estimated that the flooded old workings would take 375 days to dewater. This situation, however, was not reached as water from a surface pond in the opencast site was pumped into the system after 320 days, thus prolonging the dewatering.

It has been shown that in an area where a number of coal seams have been worked by underground mining methods the resulting highly fractured aquifer behaves a a pseudo-lagoon and limits may be put on the initial volume of water in the system, the drawdown rate and effective recharge to the system.

ACKNOWLEDGEMENTS: We would like to thank the National Coal Board Opencast Executive and Derek Crouch the Contractor for providing facilities with this study. We gratefully acknowledge the assistance of M. Money, D. Panich, N. Ochei, N. Turk, E. Passaris and Miss A. Collings of the Department of Geotechnical Engineering, University of Newcastle upon Tyne. The views expressed in this paper are those of the authors and do not necessarily represent those of the NCB.

References

ASTON, T.C.R. & SINGH, R.N., 1983. A reappraisal of investigations into strata permeability changes associated with longwall mining. *International Journal of Mine Water,* **2**, 1, 1-14.

CARMAN, P.C., 1956. *Flow of Gases through Porous Media* Academic Press, New York.

COOPER, H.H. Jr. & JACOB, C.E., 1946. A generalised method for evaluating formation constants and summarising well-field history. *Transactions of the American Geophysics Union,* **27**, 526-534.

FERRIS, J.G., KNOWLES, D.B., BROWN, R.H. & STALLMAN, R.W., 1962. Theory of Aquifer Tests. *Geological Survey Water Supply Paper 1536-E* US Geological Survey, Washington.

GARRITTY, P., 1980. Effects of mining on surface and subsurface water bodies. *PhD Thesis,* University Newcastle upon Tyne.

GARRITTY, P., 1982. Water percolation into fully caved longwall faces. *Symposium Strata Mechanics,* Newcastle upon Tyne, 25-29.

GARRITTY, P., 1983. Water flow into undersea mine workings. *International Journal of Mining Engineering.,* **1**, 237-251.

KOZENY, J., 1927. Uver kapillae leitung des wassers in Boden. *Sitz der Wein Akad der Wissenschften,* **136,** 271.

McWHORTER, D.B., 1981. Predicting groundwater response to disturbance by mining: selected problems. *Proceedings of the Symposium on Surface Mining Hydrology, Sedimentology and Reclamation,* Kentucky, 89-95.

SGAMBAT, J.P., LA BELLA, E.A. & ROEBUCK, S., 1980. *Effects of underground coal mining on groundwater in the eastern United States.* Geraghty and Miller, Inc., Annapolis. NTIS PB 80-216 757.

SHARP, J.C., 1970. Fluid Flow through Fissured Media. *PhD Thesis,* Imperial College, Univ. London.

THEIM, G., 1906. *Hydrologische Methoden.* Gephardt, Leipzig.

THEIS, C.B., 1935. The relation between the lowering of the piezometric surface and the rate and duration of discharge of a well using groundwater storage. *Transactions of the American Geophysics Union,* **16,** 519-524

MINETT, S.T., Department of Geotechnical Engineering, University of Newcastle upon Tyne, Newcastle upon Tyne NE1 7RU, UK.

BLYTHE, D.A., HALLAM, G.D., & HUGHES, D.B., NCB Opencast Executive (N.E. Region), Ashfield Towers, Kenton Road, Newcastle upon Tyne NE3 4PE, UK.

Prediction of groundwater pressures and uplift below excavations in tidal limits

J.M.A. Pontin

ABSTRACT: A study to explain the mechanisms of cyclic pore pressure generation in ground below a tidal estuary is described: and how to predict these pressures caused by tidal fluctuations is explained. These estimates are used to examine uplift and instability to be avoided during construction inside deep marine excavations. Failure modes to be expected from predicted hydraulic pressures are reviewed and appropriate safety factors discussed. Results are presented for a range of *in situ* ground parameters. Critical levels within the excavations are used and the range of pore pressure relief required are presented for a particular safety factor.

Introduction

Construction in estuaries can involve temporary excavations of considerable extent and depth below the normal bed level of inter-tidal areas. Hydraulic pressure in the soil and rock formations below the base of these excavations, wherein tidal inundation is prevented, can cause vertical motion and dislocation within them. This uplift or 'heave' can cause general failure and disruption inside the excavations. Indeed, if it were not controlled the hydraulic pressure could force certain methods of construction to be abandoned.

It is important therefore to know exactly how these pressures will be generated below the areas of excavation. The values of pressure needed to provide certain factors of safety against uplift have also to be calculated. If the maximum expected value of pressure is defined the degree of pressure relief required to achieve a certain factor of safety is determined. The practical means of exercising this relief, for example, by wells, then can be evaluated. The latter is not discussed herein.

The purpose of this paper is therefore to provide a proper physical basis for the prediction of these pore pressure phenomena in a tidal estuary; and to apply the results to the basic engineering considerations of how they should be controlled.

Analysis of pore pressure changes produced in formations by tidal loads

A mathematical theory was needed to apply to the problem of how pore pressures, generated by tidal loading, would be propagated through saturated formations below an estuary in a horizontal direction. If the mechanism could be described by a general differential equation of groundwater motion this would afford estimations of the pore pressures below the construction areas.

The interaction between an external load and internal pore water pressure is a direct consequence of a degree of elasticity in the porous framework of soil or rock. This interaction, and the implications of it, forms part of the foundation of groundwater hydraulics, and so it is not a new idea. An important simplification has usually persisted, however, viz that the external load and internal *total* stress remain constant during non-steady state flow of groundwater. This assumption is relaxed into the general theory of elastic soils (Biot 1941), wherein the interactions of stress, stain and pore pressure are described precisely. The conventional soil mechanics applications of this and its approximations viz the Terzaghi theory, also assume that the rates of change of total stress and surface loadings do not change with time.

For the problem of tide created pore pressures and their non-steady state propagation this basic assumption is patently invalid. In the Appendix we show how the tide water stress, introduced in the form of a local surface load, is incorporated into the groundwater hydraulics equation developed by Jacob (1950). The latter is derived according to the infinitesimal strain principle applied to elastic soils. The approach is quite general and applies to soils and rocks. In this manner the physical limitations of the differential equation can be clearly appreciated.

To explore the consequences of the mathematical theory we consider a reduced form of the general differential equation derived in the Appendix, viz:

$$\frac{\partial^2 h}{\partial x^2} + \frac{\partial^2 h}{\partial y^2} = \frac{\partial}{\partial t}(C_1 h - C_2 H) \qquad (1)$$

From CRIPPS, J.C., BELL, F.G. & CULSHAW, M.G., (eds), 1986, *Groundwater in Engineering Geology,* Geological Society Engineering Geology Special Publication No. 3, pp. 333-366.

where

$$C_1 = \gamma_w (\alpha + \theta\beta)/K$$
$$C_2 = \gamma_w \alpha/K$$

are hydraulic coefficients. This equation describes the two-dimensional (horizontal) propagation of hydraulic head (h) or pore water pressure (p) produced by a changing surface load of tidewater head (H).

Use of this equation implies the following assumptions:

(i) deformation by the range of stress increments is small and ideally elastic;
(ii) stress increments cause only vertical strains of the soil or rock framework so confined compressibility (α) is the characteristic elastic property;
(iii) pore water motion is described by the Darcy equation within a formation of continuous permeability (k) and porosity (θ) and these are characteristic hydraulic properties;
(iv) deformations of the elastic framework are so small that changes in the hydraulic properties are virtually zero;
(v) the formation is uniform to a degree that allows the constancy of the coefficients C_1 and C_2 to be reasonable approximation;
(vi) motion of pore water is confined entirely to the formation;
(vii) deformation of the confining formations is zero and so only infinitesmal changes in the position of the upper boundary of the confined formation are allowed.

The hydraulic coefficients C_2 and C_1 control the generation and propagation of pore water pressure in the saturated formation. We can note that if its compressibility (α) approaches zero (rigid formation) or if its permeability (k) approaches infinity then tidal loading cannot cause changes in hydraulic head or pore water pressure in that formation.

The reciprocals of coefficient C_1 and C_2 are identical in form to the coefficient of consolidation:

$$c_v = k/\gamma_w m_v \tag{2}$$

which is used in soil mechanics to predict the rate of consolidation of soils. In groundwater hydraulics the emphasis is on the prediction of pore water pressure of hydraulic head. In the latter context c_v is defined to be 'that change in pore volume equivalent to the volume of pore water which can flow in unit time through unit area when the hydraulic gradient normal to that area is unity'.

The quantity c_v is, therefore, analogous to thermal diffusivity, which controls the propagation of temperature changes through a heated solid, because it controls the propagation of pore water pressure through a loaded soil or rock. For this reason the term 'hydraulic diffusivity is appropriate because this refers to either positive or negative changes (increasing or decreasing) of pore water pressure. The symbol c_v is retained to show the correspondence with soil mechanics. We can note also that c_v is essentially the ratio of permeability to compressibility of the porous framework of a soil or rock formation. Equality of permeability does not necessarily mean, therefore, that pore pressure responses of two formations will be the same when loaded in an equivalent manner.

Equation (1) is equivalent to the basic differential equations of groundwater hydraulics and soil mechanics combined and generalised for changes of superimposed total stress. It is expressed above in terms of hydraulic head but we can write an identical equation in terms of increments of pore water pressure. The latter is not so convenient for the purpose of this paper. We have to emphasise that the formation of this differential equation can be made directly from previous derivations. A detailed discussion was merited because the equation had not, apparently, received expression in the form developed for the problem herein (see, for example, Verruijt 1969).

Mathematical solutions

Our purpose is limited to finding solutions of the differential Equation (1) for computation of porewater pressures below areas protected from tidal flooding after the tide is excluded by dykes and sheet pile walls, Fig 1. The idealised mathematical description of pore pressure generation by tidewater variations outside these barriers and the propagation of these pressures through soils or rock extending inside the excavations is developed by recognising that the sheet pile wall forms the boundary ($x = 0$) between external and internal regions of the uni-directional groundwater flow normal to them. The flow is equivalent to that in one or more 'confined' layers with uniform hydraulic properties of each distinct permeable layers of a stratified sequence.

If we distinguish the external region and the internal region by 1 and 2 respectively the differential equations for horizontal groundwater flow within each region are:

$$\frac{\partial^2 h_1}{\partial x^2} = C_1 \left(\frac{h_1}{\partial t} - \frac{C_2}{C_1} \frac{\partial H}{\partial t} \right) \quad \text{in region 1} \tag{3}$$

$$\frac{\partial^2 h_2}{\partial x^2} = C_1 \frac{\partial h_2}{\partial t} \quad \text{in region 2} \tag{4}$$

the external boundary condition of surface loading is assumed to be caused by an equivalent uniform sinusoidal variations:

$$H(t) = \bar{H} + (u_m/2) \sin \omega t, \quad \omega = 2\pi/\tau \tag{5}$$

where $\overline{H}$ is the mean tide above datum, u_m is the maximum tidewater height above the bed at the boundary, and τ is the tidal period, see Fig. 2.

If we define

$$\phi = h - \overline{H} \quad (6)$$

$$g = H - \overline{H} = U\sin \omega t \quad (7)$$

these functions are deviations from the mean tide. This 'mathematical' tide is noted to be higher than the real tide when the latter rises above local bed level, while the peak values are the same. Equation (5) is a conservative approximation when the inter-tidal areas do not drain sensibly, a usual situation.

The formal description of the boundary value problem is completed by

$$\frac{\partial^2\phi_2}{\partial x^2} = C_1 \frac{\partial\phi_2}{\partial t} \qquad 0 < x < X \ (a)$$

$$\partial\phi_2 \partial x = 0 \qquad x = X \ (\text{b})$$

$$\frac{\partial^2\phi_1}{\partial x^2} = \text{C}_1 \left(\frac{\partial\phi_1}{\partial_t} - \frac{\text{C}_2\,\partial\text{g}}{\text{C}_1\,\partial\text{t}}\right) \quad x < 0 \ (\text{c}) \qquad (8)$$

$$\phi_1 = \phi_2 \qquad x = 0 \ (\text{d})$$

$$\partial\phi_1/\partial x = \partial\phi_2/\partial x \qquad x = 0 \ (\text{e})$$

$$\partial\phi_1/\partial x = 0 \qquad x = \pm\infty (\text{f})$$

The hydraulic head function 0 is denoted by ϕ_1 in the formation external to the sheet pile wall and ϕ_2 internal to it. There is no tidal loading inside the wall so $g = 0$, Equation (8a). The centre line of the excavation area, region 2, is assumed to be equivalent X from each boundary so there is zero flow across that line, Equation (8b). The general equation has to be satisfied in region 1, Equation (8c). The pore pressure and therefore the hydraulic head functions ϕ_1 and ϕ_2 are continuous at $x = 0$ and so equal, Equation (8d). This is true also for groundwater velocity, proportional to the spatial gradients of these functions, Equation (8e). With the uniform tide loading assumed, the net motion of porewater in the formation reduces to zero with indefinite distance from the sheet pile wall, Equation (8f).

The solution of the problem was verified to be

$$\phi_2 = (C_2/C_1)\,(U/2)\,(\text{e}^{-\lambda i(\omega t - \lambda)} + \text{e}^{(\lambda - 2L)}\,\text{e}^{\,i(\omega t + \lambda - 2L)}) \quad (9)$$

where

$$\lambda = x\,\sqrt{(\omega\, C_1/2)},\; L = X\,\sqrt{(\omega C_1/2)}. \quad (10)$$

For practical purposes the complex form was reduced to trigonometric terms and from this the maximum head developed in the formation below the excavation, h_m, derived. This occurs at time t_m.

The expressions were:

$$(\phi_2/U = G(\lambda/L\ \omega t/_m) \quad (11)$$

$$G = (\text{e}^{-\lambda}\sin(\omega t - \lambda) + \text{e}^{(\lambda - 2L)}\sin(\omega t + \lambda - 2L))/2$$

and for the maxima

$$(\phi_m/U = G_m\,(\lambda/L,\, L,\, \omega t_m) \quad (12)$$

where

$$\omega t = \tan^{-1}\left[\frac{\cos(\lambda - 2L)\text{e}^{(\lambda-2L)} - \cos\lambda\text{e}^{-\lambda}}{\sin(\lambda-2L)\text{e}^{(\lambda-2L)} - \sin\lambda\text{e}^{-\lambda}}\right] \quad (13)$$

The dimensional forms of (11) and (12) were

$$h(x,t) = \overline{H} + (H_m - \overline{H})\,G\,(x/X,\, L,\, \omega t) \quad (14)$$

$$h_m = \overline{H} + (H_m - \overline{H})\,G_m(x/X,\, L,\, \omega t_m) \quad (15)$$

for the general and peak values of hydraulic head in the formation, respectively. These are noted to be determined by the tide function and the parameter

$$L = X\sqrt{\omega/2c_v} = X\sqrt{\pi\lambda_w m_v/\tau k}$$

if the variation of L with the dimension X is ignored then L is inversely proportional to $\sqrt{c_v}$ for a tide with certain characteristics.

The solution (11) to Equations (8) provide the changes of head and pore pressure during the tidal variation when this cycle is repeated *ad infinitum*. The solutions are in fact the 'dynamic steady state response'. An important practical question is how long is needed to achieve this ultimate response after the areas of excavation are first closed by the dykes and sheet piles.

We assume that inside those areas the head will correspond initially to hydrostatic below the original bed level. The non-steady state response is thus a transition from the initial head h_o to the mean value $\overline{H}$. The cyclic fluctuations generated by the tide are superimposed on this trend. A solution which reproduces this response can be determined but it is very complicated. An approach, sufficient for practical purposes, is to ignore the effect of the tide *per se* and consider mean values. Thus the non-steady state is approximately described by a mean value function, $\overline{h}$, ($\overline{h}_1$ and $\overline{h}_2$) within each region. These mean values will eventually reach that of the tide in region 1, $\overline{H}$. The boundary value problems is designated by

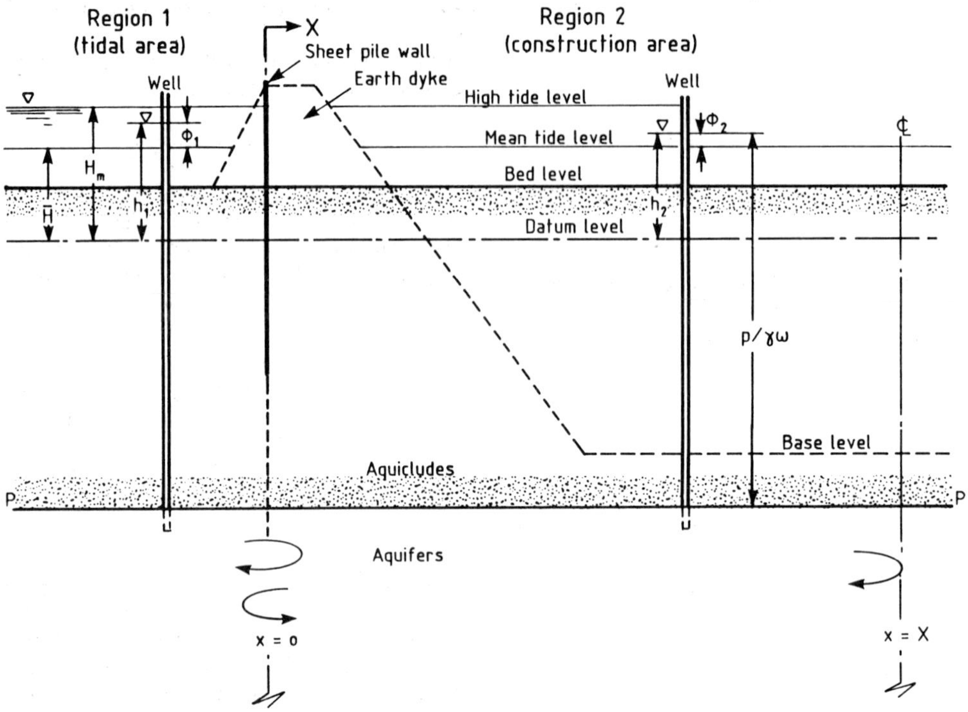

FIG. 1. Section through a marine construction area.

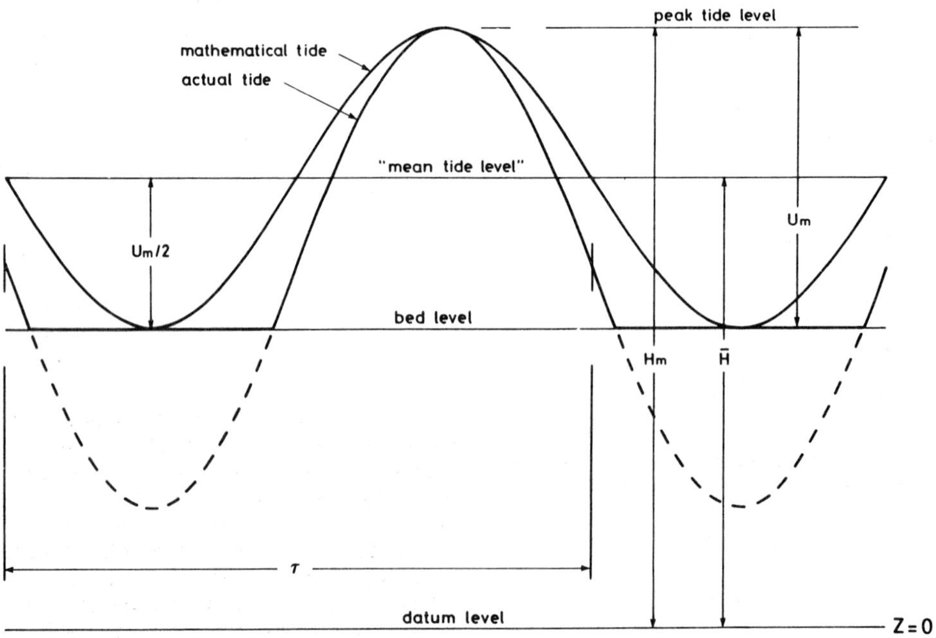

FIG. 2. Tide functions outside excavation area.

$$\frac{\partial^2 \bar{\phi}}{\partial x^2} = C_l \frac{\partial \bar{\phi}}{\partial t} \quad -\infty < x < X,\ t > 0 \qquad (a)$$

$$\bar{\phi} = \bar{\phi}_0 \qquad 0 < x < X, \qquad t = 0 \qquad (b) \qquad (16)$$

$$\bar{\phi} = 0 \qquad x < 0, \qquad t = 0 \qquad (c)$$

$$\partial\bar{\phi}/\partial x = 0 \qquad x = X \qquad t > 0. \qquad (d)$$

The solution to this problem is obtained by a transformation of an analogous problem in the theory of heat conduction. In previous notation

$$(\bar{\phi}/\bar{\phi}_0) = \text{erf}\ (\frac{x/X}{2\sqrt{T}}) + \text{erf}\ (\frac{2-x/X}{2\sqrt{T}})\ /2 \qquad (17)$$

where

$$T = C_l t/X^2 = c_v t/X^2 \qquad (18)$$

is a time parameter. The solution can be expressed in terms of head by

$$\bar{h} = \bar{H} + (h_o - \bar{H})\ E\ (x/X\ T) \qquad (19)$$

where E is the function of Equation (17), which tends to zero with L increasing. At any finite time it has a maximum value when $x = X$, the centre of the excavation. The mean head in the formation here is therefore the minimum value and is expressed by a most convenient form, viz.

$$(\bar{H}_m - \bar{H})/(h_o - \bar{H}) = \text{erf}\ (X/2\sqrt{c_v t}) = E_m \qquad (20)$$

which allows the maximum deviation from the 'steady state' flow (mean) condition to be evaluated in terms of a single time parameter.

Example computations

The uniform external tide was characterised by

H_m = +4.5 m OD (peak)
$\bar{H}$ = +2.5 m OD (mean)
τ = 12.5 hour (period)

so that the tidal amplitude outside the sheet pile wall was assumed to be 2 m. It can be shown that

$$c_v \rightarrow 0,\ Gm \rightarrow 1/2 \quad \text{so} \qquad (21)$$

$$h_m \rightarrow (\bar{H} + H_m)/2$$

With the tide values above this limit is +3.5 m OD. Values of the parameter L were derived for $X = 90$ m, which corresponds to a width of excavation, normal to the centre line between the sheet pile wall, of 180 m.

For illustration we will consider a range of c_v's derived from the expression

$$c_v = \frac{k}{\gamma_w(\alpha+\theta\beta)} \qquad (22)$$

where α, β are the compressibilities of foundation and formation water respectively with the latter unit weight, γ_w, and the former permeability k, and porosity θ. The values:

	'silt' (laminated)	'rock' (weathered)
k(m/sec)	3×10^{-6}	3×10^{-6}
θ -	0.4	0.04
α(m^2/N)	2×10^{-7}	6.5×10^{-9}, 2×10^{-10}
c_v(m^2/sec)	0.0015	0.045, 0.9

were calculated with γ_w equal to 9920 N/m^3 and β equal to 4.5×10^{-10} m^2/N. The range of α for 'fissured rock' was adopted from a recent publication (Van der Kamp & Gale 1983). Laminated silts are universal products of recent lacustrine or estuarine deposition, while rock, exposed to one or more types of weathering, is also quite usual in marine margins. For comparison the c_v's are 'normalised' to a constant hydraulic permeability, which is typical of each lithology in fact. Numerical calculations were made by micro-computer programs developed by W A Pontin for a HP75-IL system.

Profiles of hydraulic head at certain times after high tide are presented in Fig. 3. In each graph there are 6 curves corresponding to:

Profile Curve No	*Time after external high tide hours*
1	0
2	1.5625
3	3.125
4	4.6875
5	6.25
6	9.375

Curve 5 is thus the profile at low tide, while curves 3 and 6 are those at mean tide falling and rising. The c_v values are inversely proportional to formation compressibility so that they increase with incompressibility. A reference point is obviously the c_v selected for the laminated silts. Profiles for this value are shown in Fig. 3a. A 'pore pressure wave' effect is seen clearly by the progress of successive peaks away from the sheet pile wall — with its amplitude so severely attenuated that it is zero beyond c. 30 m from the boundary.

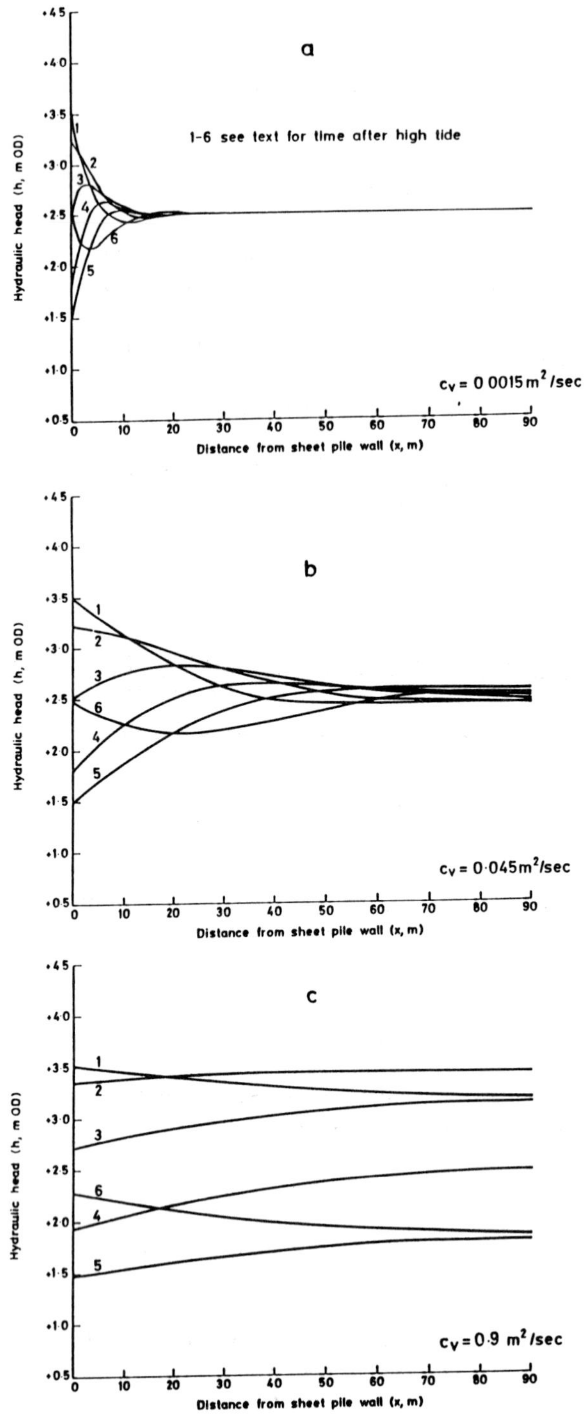

FIG. 3. Hydraulic head profiles during tidal cycle.

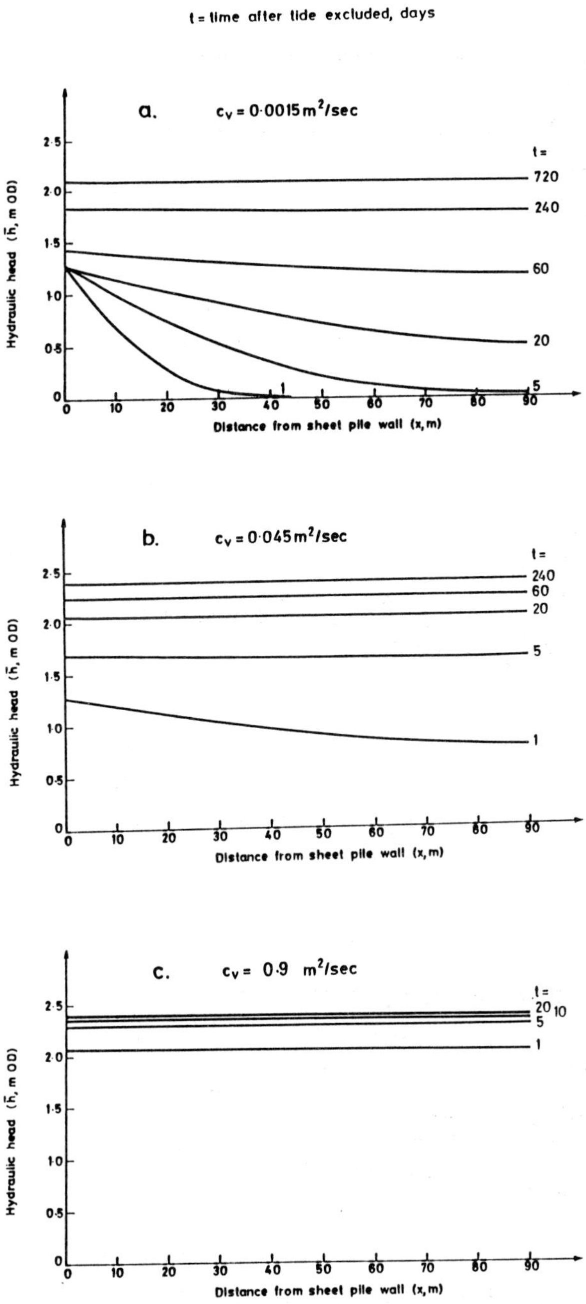

FIG. 4. Mean hydraulic head profiles.

In comparison the profiles computed for the rock c_v's show a 'wave' which penetrates farther into the area of construction and which is less attenuated. For relatively uncompressible rock, $c_v = 0.9$ m²/s, the head profiles fail to develop peak values except at the inner and outer boundaries. The maximum head developed at any x is sensibly the peak value at $x = 0$.

The equilibration of 'mean' hydraulic head for the same numerical values are shown in Fig. 4. The curves are profiles of mean head at increasing times after tidewater is first excluded from the area of construction. The rate of non-steady state change toward the mean tide level, external to the sheet pile wall, is determined entirely by the value of c_v when x is constant.

If c_v is equal to the rock values we can see that the mean head below the protected area is computed to exceed +2 m OD after 3 weeks or so. If c_v is equal to 0.9 m²/s equilibrium is rapid and achieved in a matter of days. The silt formation curves, Fig. 4a, show the time needed to exceed +2 m OD is 8 to 9 months. A point to note is that, for each c_v, after a certain time the mean head is virtually constant across the protected area after which it rises uniformly toward mean tide level.

Engineering considerations

The discussion is limited to a consideration of the stability of the base of construction areas inside the sheet pile walls when these areas are in an unloaded condition during and after excavation. The types of instability we can analyse in a practical manner are related to the magnitude of head and pore water pressure predicted in 'aquifers' below the foundation level. The formations above are assumed to be impermeable 'aquicludes' and so excess pore pressures are not relieved materially by vertical flow of groundwater. (The term 'excess' is defined to be a value of pore water pressure which exceeds the hydrostatic value from the level of pore saturation at that time).

Two major types of formation failure occur in response to 'excessive' pore pressures below the base of marine excavations.

Fluidisation: A 'fluidised' boiling condition is produced in an aquifer when a critical hydraulic gradient of vertical flow is achieved. The condition is well described by experimental situations (Terzaghi & Peck 1967) but remains an unexpected practical effect wherein groundwater flow velocities reach 'piping' values and bearing capacity is reduced to near zero. Fluidisation pre-supposes that the time required to achieve the critical hydraulic gradient is exceeded and that the aquifer has direct hydraulic connection with an external supply of water, viz tidewater in an estuary or groundwater outside it. The process of pore pressure generation developed herein is quite distinct from these provisions and so no further discussion is now merited.

Flotation: The condition wherein a discontinuity of pore pressure exists at a boundary between a soil of low permeability (and diffusivity) relative so that below is normally associated with the phenomenon of 'heave'. The base of an excavation can start to rise and continue to do so until equilibrium is re-established. (We exclude the effect of elastic de-compressibility of an unloaded soil, which is essentially limited in magnitude, when pore pressure effects are absent.)

If the latter are present, vertical translation of large bodies of soil can occur which are terminated by dislocations from the excavated level to the seat of the heaving motion, so relieving the pore pressures at that level. This general failure is not likely in marine excavations because slope failures, sliding or spreading, promoted by a total loss of shear strength at some level below, would occur before it. Irrespective of the actual soils or rock along the sub-horizontal boundary a critical condition is reached when the (vertical) effective stress at that level is zero. There are then no vertical reactions in the soil structure along a horizontal surface at that level and the upper body of soil 'floating' on the lower with motion resisted by cohesion and friction along sub-vertical surfaces. Flotation can be quite general if pore water pressures are nearly in the horizontal direction.

We define a local factor of safety against vertical heave after flotation to be

$$F = \sigma_v/p \tag{23}$$

the ratio of total vertical stress to pore water or pressure. If $F = 1$ at a particular elevation, flotation failure could be prevented by friction due to lateral effective stresses if the condition were localised. However, the equality of the stresses in a vertical direction should not be normally contemplated.

If we designate a certain level P to be a boundary between aquicludes above and aquifers below extending to indefinite depth, then,

$$h_z \leqslant \frac{\mu\,(EL - z)}{F} + z \tag{24}$$

is the hydraulic head in an aquifer at level z which is required for a factor of safety F against local heave when excavation has reached a level EL m OD. The factor

$$\mu = \bar{\gamma}/\gamma_w \tag{25}$$

is the ratio of mean unit weight of soil between EL and z to unit weight of groundwater.

In the series of aquifers at levels Po, P_1 ... Pn below p the peak hydraulic head is assumed to be equal to that developed at P viz Hp. The head or pressure relief of groundwater to be produced at level z is

$$R_z \geqslant H_p - h_z \tag{26}$$

so that estimates of the pressure relief ('drawdown') required at a particular location and elevation below the marine excavations can be calculated if hydraulic heads to design against are determined in the aquifers.

A 'critical depth' below a final level of excavation FEL mOD can be defined below which R_z can be zero i.e. no drawdown is necessary for the selected design values. Then

$$h_z = h_p,\ z = z_c \tag{27}$$

so from Equation (24) we have

$$z_c = \frac{\mu\,(FEL - FH_p)}{(\mu - F)} \tag{28}$$

Hypothetical calculations demonstrate the use of these expressions. Numerical values used are: $\overline{\gamma}$ = 21.0 kN/m^3, γ_w = 10.1 kN/m^3, μ = 2.08, p = -20 m OD, EL = 10 m OD, FEL = -15 m OD. The design peak values derived from the previous section are

	H_p
Silt	+ 2.5 m OD
Rock	+ 3.5 m OD

so that the boundary P corresponds to clay aquicludes above silt or rock aquifers. A factor of safety, F = 1.2, is not unrealistic to avoid failure, and this value is adopted.

The relief or drawdown values required at the level P, which is below FEL, are computed to be

EL (m OD)	*Relief or drawdown value R_p (m)*	
	silt	rock
–10	5.2	6.2
–15	13.8	14.8

and the critical depths

	z_c
silt	38.9 m OD
rock	40.2 m OD

or approximately 25 m below the foundation level of the excavation. The latter calculation is not dependent on the level of the aquiclude – aquifer boundary, which can be supposed to be located at any level between FEL and z_c.

Discussion

The methods and results above have to be applied generally for a particular proposed marine excavation in order to estimate what pressure relief or drawdown has to be produced below them to maintain safety against base heave and uplift. The crux of the exercise is obviously the computation of 'design' pore pressures so it is necessary to mention that the theory used herein is believed to be conservative.

The reason for this conclusion is that there are assumptions intrinsic in this theory which combine to make the computed pore pressures over-estimates. The most important of these are:

(i) absence of vertical components of groundwater flow;

(ii) absence of vertical uplift deformations when static equilibrium has approached or exceeded limited values.

Dissipation of pore pressure excesses by the plastic deformations associated with general uplift is thus ignored. This 'automatic' relief could be anticipated if a limiting condition is allowed to occur but would not necessarily prevent serious failure.

While this paper is addressed to the problem of uplift the prediction of pore pressures has a profound effect on the analysis of slope stability in marine situations. When effective stress methods are used with the conventional limit equilibrium formulations the most critical quantity is most definitely the distribution and magnitude of pore pressure along a potential failure surface. It is suggested that these surfaces will tend to be located along critical levels where excess pore pressures are generated by tidal loading. Simultaneous head profiles during a design tidal cycle can be used in conjunction with non-circular stability analysis to define the locus of safety factors.

In conclusion it has to be stated that a pore pressure relief programme in large scale marine construction engineering is a complex design problem if it is to be optimised. The numbers, types, positions, modes and schedules of operation of wells and wellpoints have to be selected on the basis of the drawdown required at various times and levels during the excavation programme. It is believed that optimisation can be achieved by an appropriate combination of pre-contract design and real time simulation by on-site computer facilities.

References

Biot, M. A., 1941. General theory of three-dimensional consolidation. *Journal of Applied Physics* **12.**

Biot, M. A., and Willis, D.G., 1957. The elastic coefficients of the theory of consolidation. *Journal of Applied Mechanics,* **24.**

Jacob, C.E., 1950. Flow of groundwater. In: Rouse, H. (ed). *Engineering Hydraulics,* 321-386.

SKEMPTON, A.W., 1954. The pore pressure coefficients A and B. *Geotechnique* **4,** 143-147.
TERZAGHI, K, & PECK, R.B., 1967. *Soil Mechanics in Engineering Practice.* John Wiley, New York.
VAN DER KAMP, G. & GALE, J.E., 1983. Theory of earth tide and barometric effect in formations *Water Resources Research* **19.**
VERRUIJT, A., 1969. Elastic storage of aquifers. In: de Weist, R.J.M. (ed). *Flow Through Porous Media* Academic Press, New York.

PONTIN, J.M.A., Groundwater Section, Hydraulics Research Limited, Wallingford, Oxfordshire OX10 8BA, UK.

Appendix: Analysis of pore pressure changes in sub-surface formations due to tidal loads: differential equations of motion

Consider a volume element of the formation, soil or rock, $\Delta x \Delta y \Delta x$, referred to axes Ox and Oy horizontal and Oz vertical.

The loading of the formation by tide water above can be supposed to cause deformation of the volume element. This allows two effects to occur at the same time, the compression of all the pore water within the element and the expulsion of some form it or vice versa.

The element is considered to deform by compression or extension along the directions of the axes. (This is called dilatation or deformation without rotation.). The axial strain increments, with contractions positive, are

$$\delta e_{xx} = \delta \Delta x/\Delta x, \ \ \delta e_{yy} = \delta \Delta y/\Delta y, \ \ \delta e_{zz} = \delta \Delta z/\Delta z \qquad \text{(A1)}$$

and the total volumetric strain increment

$$\delta e = \delta \Delta V/\Delta V = \Delta V - \{(\Delta x - \delta \Delta x)(\Delta y - \delta \Delta y)(\Delta z - \delta \Delta z)\}/\Delta V. \qquad \text{(A2)}$$

If the changes in dimension $\delta\Delta x$, $\delta\Delta y$, $\delta\Delta z$ are assumed to be small the it follows that

$$\delta e = -(\delta e_{xx} + \delta e_{yy} + \delta e_{xx}).$$

Also, with small strains, these are expected to be linearly related to changes in the effective stresses within the framework of the formation. In general terms we can write:

$$\delta e_{xx} = a\,\delta\sigma'_{xx} + b\,\delta\sigma'_{yy} + c\,\delta\sigma'_{zz} \qquad \text{(a)}$$

$$\delta e_{yy} = b\,\delta\sigma'_{xx} + a\,\delta\sigma'_{yy} + c\,\delta\sigma'_{zz} \qquad \text{(b) (A3)}$$

$$\delta e_{zz} = b\,\delta\sigma'_{xx} + c\,\delta\sigma'_{yy} + a\,\delta\sigma'_{zz} \qquad \text{(c)}$$

where a, b and c are arbitrary constants of proportionality. These can be reduced by noting that $b = c$ for the type of deformation we have specified. If elastic constants are now defined according to

$$E = \text{stress/strain, a constant,}$$

then

$$E = \delta\sigma'_{xx}/\delta e_{xx} = \delta\sigma'_{yy}/\delta e_{yy} = \delta\sigma'_{zz}/\delta e_{zz} \qquad \text{(A4)}$$

where E is the Young's modulus for each axial direction, viz the ratio of the effective stress increment to the strain it produces in that direction when the effective stress increments are zero in the other directions, and

$$\nu = -\delta e_{xx}/\delta e_{zz} = -\delta e_{yy}/\delta e_{zz} \qquad \text{(A5)}$$

where ν is Poisson's ratio for each axial direction, viz the ratio of lateral extension to vertical or axial compression of the formation. With these definitions it is found that

$$a = 1/E, \quad b = -\nu/E \qquad \text{(A6)}$$

and the strains are now expressed by

$$\delta e_{xx} = \frac{1}{E}\left[\delta\sigma'_{xx} - \nu\,(\delta\sigma'_{yy} + \delta\sigma'_{zz})\right] \qquad \text{(a)}$$

$$\delta e_{yy} = \frac{1}{E}\left[\delta\sigma'_{yy} - \nu\,(\delta\sigma'_{xx} + \delta\sigma'_{zz})\right] \qquad \text{(b) (A7)}$$

$$\delta e_{zz} = \frac{1}{E}\left[\delta\sigma'_{zz} - \nu\,(\delta\sigma'_{xx} + \delta\sigma'_{yy})\right] \qquad \text{(c)}$$

The total volume strain is therefore

$$\delta e = -\left\{\frac{1\text{-}2v}{E}\right\}(\delta\sigma'_{xx} + \delta\sigma'_{yy} + \delta\sigma'_{zz}) \quad \text{(A8)}$$

or we can write this in the form

$$\delta e = -(1/K\,\delta\bar{\sigma}' = -\alpha_v\,\delta\bar{\sigma}' \quad \text{(A9)}$$

if we define

$$K = E/3\,(1\text{-}2v) = 1/\bar{\alpha} \quad \text{(a)}$$

$$\delta\bar{\sigma}' = (\delta\sigma'_{xx} + \delta\sigma'_{yy} + ds'_{zz})/3 \quad \text{(b)} \quad \text{(A10)}$$

the total volume modulus and mean effective stress increment respectively. the quantity $\bar{\alpha}$ is the unconfined compressibility of the formation framework. If we consider that the lateral strains are zero:

$$\delta e_{xx} = \delta e_{yy} = 0 \quad \text{(A11)}$$

the total volume strain has to be equated with the vertical strain only.

$$\delta e = -\delta e_{zz}. \quad \text{(A12)}$$

In terms of effective stress this is now expressed by

$$\delta e_{zz} = \frac{(1+v)\,(1\text{-}2v)}{E\,(1\text{-}v)}\,\delta\sigma'_{zz} \quad \text{(A13)}$$

or

$$\delta e = -\alpha\,\delta\sigma'_{zz} \quad \text{(A14)}$$

where α is the vertical compressibility of the formation, when an increment of vertical effective stress produces a change in the vertical dimension of the element only. The formation is then considered to be laterally confined. We have the relation

$$\alpha = \frac{(1+v)}{3(1-v)}\,\bar{\alpha}$$

between the compressibilities

The tidal loading superimposed on the formation is equivalent to a change in the total stress. The relation between the latter and the effective stress in its framework is Terzaghi's equation for stress increments,

$$\delta\sigma' = \delta\sigma' - \delta p \quad \text{(A15)}$$

where p is the pore water pressure and σ is the total stress. If $H(t)$ is the tidal head imposed on the seabed vertically above the element the changes in effective stress produced within the element are not determined *a priori*.

When lateral strains are non-zero, from Equation (A9),

$$\delta e = -\bar{\alpha}\,[(\delta\sigma_{xx} - \delta p) + (\delta\sigma_{yy} - \delta p) + (\delta\sigma_{zz} - \delta p)]$$

$$\delta e = -\bar{\alpha}\,(\delta\bar{\sigma} - \delta p) \quad \text{(A16)}$$

if the tidal load can be considered to produce a hydrostatic increment to the total stresses.

When lateral strain is zero, from Equation (A14).

$$\delta e = -\alpha(\delta\sigma_{zz} - \delta p) \quad \text{(A17)}$$

and lateral strain is zero, from Equation (A14),

$$\delta e = \alpha(\delta\sigma_{zz} - \delta p) \tag{A17}$$

and the lateral effective stresses change to prevent horizontal deformation by

$$\delta\sigma'_{xx} = \delta\sigma'_{yy} = \nu\delta\sigma'_{zz}\,(1-\nu) \tag{A18}$$

This means that the total stresses in the horizontal directions change by:

$$\delta\sigma_{xx} = \delta\sigma_{yy} = [\delta\sigma_{zz} + (1-2\nu)\delta p]/(1-\nu). \tag{A19}$$

In fissured rock this is more likely to be so than the hydrostatic increments because we can expect that large horizontal reactions could be produced. If this is accepted, then the tide water loading causes deformation due to changes in total vertical stress:

$$\delta\sigma_{zz} = \gamma_w \delta H \tag{A20}$$

The changes in pore water pressure are expressed in the conservation principle applied to the mass of water in the pores of the volume element during its deformation. The net rate of flow of groundwater expelled from the element has to equal the net rate of change of groundwater stored within the element. The latter has to include the compressibility of pore water if the compressibility of the formation framework, for example, fissured rock, is low. In normal soils this factor is usually ignored, and is a distinction between the 'soil mechanics' description of pore pressure changes and that developed from 'groundwater hydraulics', for aquifers.

A convenient form of the conservation equation is one derived from Verruijt (1969):

$$\left(\frac{\partial}{\partial x} q_x + \frac{\partial}{\partial y} q_y + \frac{\partial}{\partial z} q_z\right) = -\frac{\partial}{\partial t}(\delta e - \delta f) \tag{A21}$$

a formal expression of the principle stated above. The components of the groundwater flow velocity, volume per unit time per unit area normal to the axial directions, q_x, q_y, q_z, are those relative to the framework of the formation. Their gradients in space are equal to the net rate of flow through the faces of the volume element normal to the axes.

The terms in brackets are the volumetric deformations of the rock framework and of the pore water within the volume element. By substituting the appropriate expressions for the total volume strains of each viz:

$$\delta e = \frac{\delta \Delta}{\Delta V} = -\,\alpha(d\sigma_{zz} - \delta p) \qquad \text{(a)}$$

(A22)

$$\delta f = \frac{\delta V_w}{\Delta V} = -\,\theta\beta\delta p \qquad \text{(b)}$$

where

$$\theta = \Delta V_W / \Delta V$$

is the porosity, and α and β are the compressibilities of the rock and water respectively. So

$$\delta e - \delta f = (\alpha - \theta\beta)\,\delta p - \alpha\delta\sigma_{zz} \tag{A23}$$

or, in terms of hydraulic head in the rock and the tidal head,

$$\delta e - \delta f = \gamma_w\,(\alpha + \theta\beta)\,\delta h - \gamma_w\,\alpha\,\delta H \tag{A24}$$

where

$$h = Z + p/\gamma_w. \qquad \delta p = \gamma_w\,\delta h.$$

If we consider that during a time interval Δt the change in tidal loading above the element ΔH occurs without flow from the element — 'undrained loading' —then Equation (A24) can be equated to zero and

$$\Delta h = \frac{\alpha}{(\alpha+\theta\beta)}\,\Delta H = C\,\Delta H \tag{A25}$$

is the relation between changes in pore pressure head and tidal head. The compressibility factor is equivalent to a 'pore pressure coefficient' of the formation (Skempton 1954; Biot & Willis 1957).

The groundwater veolcities are expressed in terms of the hydraulic head by the Darcy equations:

$$q_x = -K_x \frac{\partial h}{\partial x} \qquad q_y = -K_y \frac{\partial h}{\partial y}, \qquad q_z = -K_z \frac{\partial h}{\partial z} \tag{A26}$$

where k_x, k_y, k_z are the permeabilities parallel to the axial directions.

Substituting expressions (A26) and (A24) in Equation (A21) a differential equation for the hydraulic head in a formation subject to variations in tidal loading is obtained:

$$\frac{\partial}{\partial x}\left(k_x \frac{\partial h}{\partial x}\right) + \frac{\partial}{\partial y}\left(k_y \frac{\partial h}{\partial y}\right) + \frac{\partial}{\partial z}\left(k_z \frac{\partial h}{\partial z}\right) = \gamma_w (\alpha + \theta\beta) \frac{\partial h}{\partial t} - \gamma_w \alpha \frac{\partial H}{\partial t} \tag{A27}$$

SESSION 9

MONITORING OF GROUNDWATER CONDITIONS

Automatic monitoring of groundwater conditions

R.C. Weeks & P.Starzewski

ABSTRACT: Recent innovations, particularly in electronics, coupled with the needs of the construction industry have led to the development of a range of equipment which allows reliable, economic and automatic data logging of groundwater conditions in boreholes and piezometers. The paper describes the development, installation, operation and performance of some of the equipment which is currently available for obtaining data from groups of piezometers and also single installations.

Introduction

Piezometers of various types are frequently used to measure groundwater response due to construction activities or some other external factor, e.g. rainfall. As more projects are planned on or near soils whose behaviour is dependent on groundwater conditions, so the need has arisen for more accurate knowledge of the groundwater regime.

To collect sufficient data manually is often difficult and requires considerable manpower — not just to collect the information but also to check, collate and present the data in a manner which can be readily analysed. Often there is a considerable time lag between an event occurring on site and the data being presented for analysis. This has meant that valuable data has been lost due to adverse weather conditions, inaccessibility of remote instruments and misplacement of written data.

Automation would seem an obvious answer especially for large projects such as dams (Penman 1972), road embankments (Insley *et al.* 1982), and bridges, where instrumentation may not only be inaccessible but may also require monitoring for long periods.

One of the first serious attempts at large scale automatic monitoring of groundwater conditions was carried out in Hong Kong as part of a major geotechnical study of the Mid-Levels area of Hong Kong Island (Rodin *et al.* 1982). In 1979 the Geotechnical Control Office of the Government of Hong Kong commissioned Geotechnical Instruments (UK) Limited to design, manufacture, test and install automatic monitoring systems to read 135 standpipe piezometers (Pope *et al.* 1982).

The requirements of these systems provided a basis of the specification of current data logging equipment. Hong Kong with its severe weather conditions also proved a valuable testing ground for new solid state components now incorporated in automatic monitoring equipment.

Principle

There are four basic types of piezometers which lend themselves to automation:

(i) standpipe,
(ii) pneumatic,
(iii) hydraulic,
(iv) electrical.

This paper will not attempt to discuss the merits of the different types of piezometers as they are well documented by many others e.g. (Hanna 1985). In all cases the basic automatic recording system consists of three main components:

Piezometers — Scanner or Multiplexer — Data logger.

Fig. 1 shows a typical automatic system layout, including optional data links.

Automation of standpipe, pneumatic and hydraulic piezometers requires an electromechanical fluid wafer switching mechanism incorporated into a scanner, Fig. 2. Although the details of operation vary with each type of piezometer the scanner allows each one to be sequentially connected to an electronic pressure transducer which senses the pore pressure.

The functions of the scanner are controlled by the data logger. This unit may be programmed *in situ* to vary the number of piezometers to be read in a sequence, the dwell time on each piezometer and the interval time between sequences of readings. The length of time taken to read a sequence is dependent on the number of piezometers (N) and the minimum possible dwell time (T).

$$\text{Sequence time} = N \times T.$$

The dwell time is largely dependent on the maximum lead length — this having a significant effect on the equalisation of pressures throughout the system.

In recent years, the accuracy and reliability of electrical piezometers has improved, thus making

From CRIPPS, J.C., BELL, F.G. & CULSHAW, M.G., (eds), 1986, *Groundwater in Engineering Geology,* Geological Society Engineering Geology Special Publication No. 3, pp. 369-373

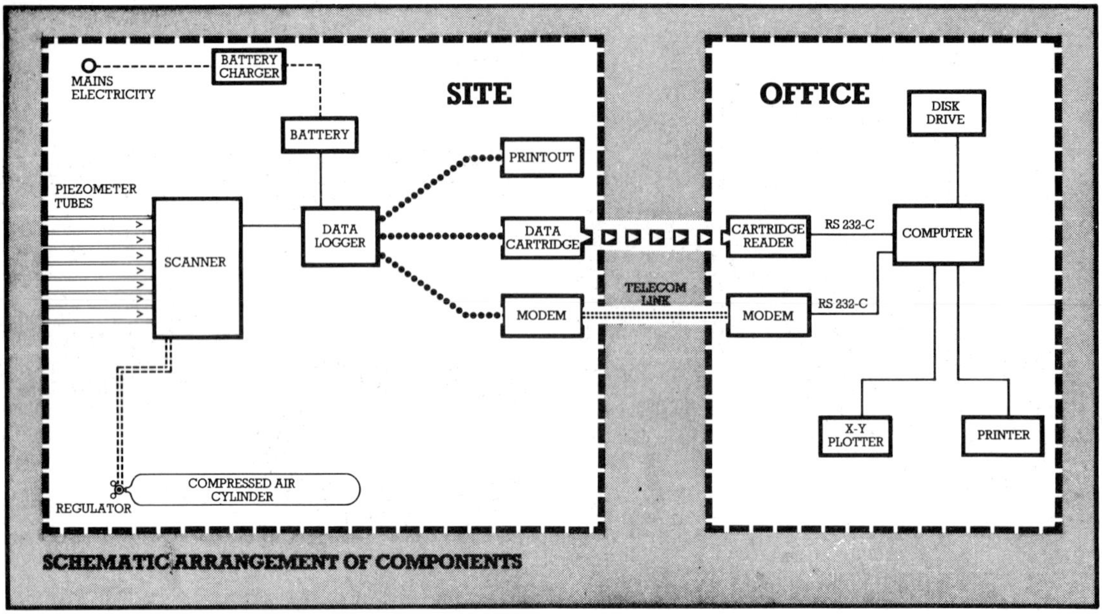

FIG. 1.

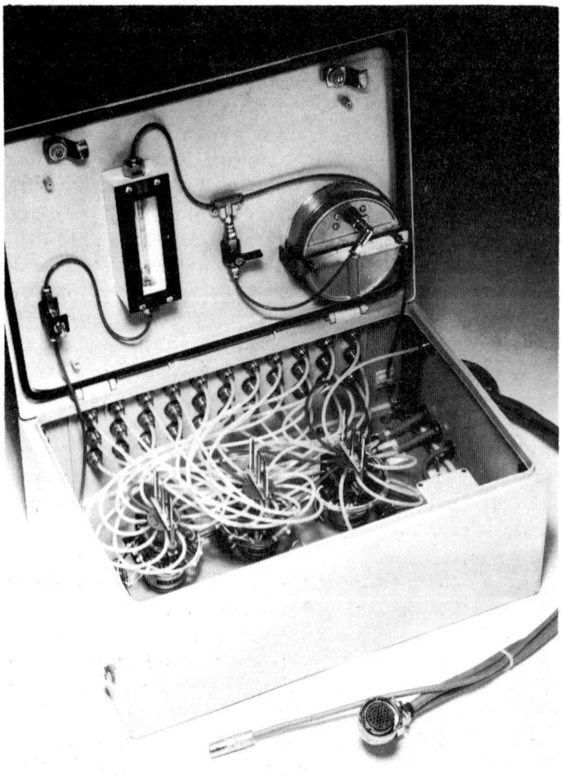

FIG. 2. Scanner

them a viable alternative for long term monitoring. They utilise one of two basic transducers and are available in a variety of shapes and sizes:

(i) vibrating wire transducer,
(ii) DC bridge transducer.

Since only an electrical output has to be measured, then an electrical switching mechanism or multiplexer is required instead of a scanner. The dwell time can therefore be reduced with this system. Electrical piezometers react almost instantaneously to pressure changes. This means rapid scanning is possible which is particularly advantageous in tidal regions or in pumping tests where frequent changes in pore pressure are anticipated.

Description

Data loggers are not in themselves a new concept. However, until the recent revolution in electronics they have been mainly confined to the laboratory. Their size and sensitivity to harsh environments meant they were unsuitable for site use.

The new generation of dedicated data loggers developed for the purpose of piezometer monitoring makes use of the latest solid state technology. They are controlled via sealed pressure-sensitive membrane switch panels and mounted in environmentally sealed cases. Built-in interactive software allows them to be operated by non-specialist personnel.

Information can be recorded by the data logger in a variety of ways. Loggers are available which incorporate lithium battery powered data cartridges, Fig. 3. These are of variable capacity ranging from 4,000 readings up to 30,000 readings. They are sealed and interchangeable on site, requiring only an interface to be interrogated by any microcomputer. Data recorded on these cartridges can be stored for up to 10 years without loss.

FIG. 3. Cartridge data logger.

A second option is to record data directly in tabular form onto an electrosensitive printer, the printer replacing the cartridge module as in Fig. 4. The printers are robust and encased in sealed units well suited for use in harsh environments.

Data loggers can also be connected directly to computers or a modem telecommunications link can be used to transmit recorded readings. This then provides a rapid monitoring system for projects where groundwater response is critical.

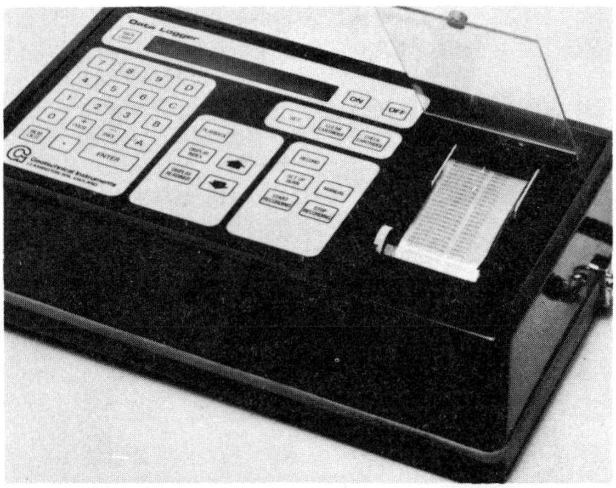

FIG. 4. Printer data logger.

It is not always possible to monitor large groups of piezometers from a central location. To overcome this problem single station data loggers have been developed, Fig. 5. Normally used with electrical piezometers these compact data loggers can be installed directly above the location of the piezometer or borehole.

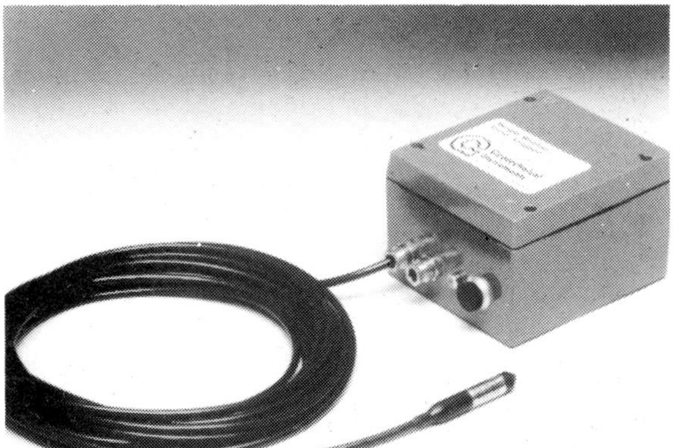

FIG. 5. Single station data logger.

Single station data loggers are programmed by means of an interrogator which is also used to extract data when required from the memory in the logger. They are in effect an extension of the data cartridge technology previously mentioned.

Installation

Even though the correct system may have been selected for a particular project it is all too common for subsequent data to be unreliable because of inexpert techniques used during installation. The best equipment poorly installed will yield useless and highly misleading information.

Although the data logger and scanner or multiplexer are all built in environmentally protected cases they should be installed in lockable gauge houses or cabinets to prevent vandalism. Since such systems can be installed many hundreds of metres from piezometers then a convenient location for the readout equipment can normally be selected. Installation of equipment in manholes should be avoided wherever possible. Access to and maintenance of the equipment is generally very difficult and there is always the risk that the manholes will fill up with water — either by seepage of groundwater or flooding.

Pneumatic tubes, hydraulic tubes and electrical cables should be laid and wherever possible buried in PVC ducts which are sealed against vermin.

Existing manual piezometer systems can easily be converted into automatic monitoring systems since in most cases only reconnection of piezometer leads or wires to a scanner or multiplexer is required. Since the data logging equipment is powered by a 12 volt DC motor, provision of mains power is not a limiting factor.

Operation

Once automatic equipment has been installed its operation is simple. Data loggers are interactive and very little training in their use is required. However, the person who obtains the readings should at least have a basic understanding of the systems, and have some idea of what data is expected.

Electric piezometer systems require the least maintenance. Data collection and battery changes are all that should be needed once the system has been commissioned. Occasionally calibration may also be necessary.

In addition to the above, pneumatic and standpipe piezometer systems also require the changing of an air or nitrogen cylinder. This can affect the actual location of a cabinet or gauge house as good access is required. The frequency of changes is dependent on the number of piezometers and how often they are being read.

Hydraulic piezometers need to be purged or de-aired at regular intervals in order to remove any accumulated gas. This should be carried out regularly to ensure good results and unless portable de-airing equipment is used then a small gauge house is required for permanently mounted devices.

Scanners require occasional routine maintenance such as lubrication of the fluid wafer switches and printers. Latest developments have, however, reduced the number of actual moving components to a minimum and modular design ensures rapid servicing time.

Data processing

Automatic monitoring systems can generate valuable data previously not obtainable by manual methods. The volume of data produced also makes manual analysis a lengthy process and sometimes the value of using automatic systems can be lost. Because the majority of analyses often involve a large amount of routine simple calculation, then the advantages of using computers are immediately apparent. The overall design of automatic data logging systems is such that they can easily transfer data to most types of computer.

Fig. 6 and Fig. 7 give two examples of the advantages of automatic data monitoring. Fig. 6 compares manual results with automatically monitored readings from a standpipe piezometer. These results are taken from a station in the Far East where typhoons are common, bringing with them large amounts of rain which fall in very short periods of time. A significant effect on the groundwater regime results. The dangers due to high winds and flying debris make it virtually impossible to monitor manually in these periods when the critical peak readings occur. In Fig. 6a the manual monitoring gave highly misleading data. In Fig. 6b readings with the automatic monitoring equipment illustrate well the significant and rapid variations that can occur in the water level.

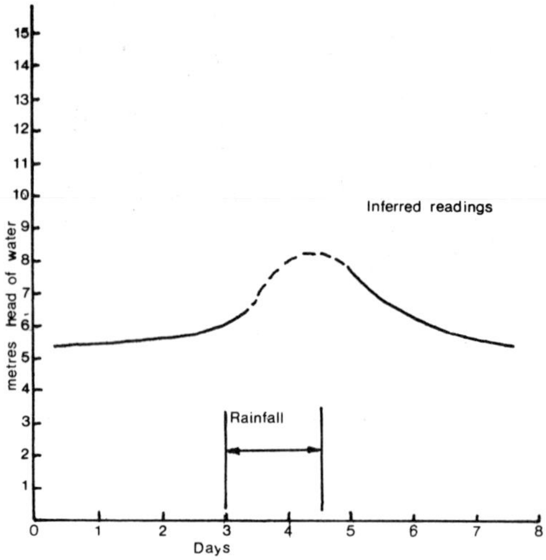

FIG. 6(a). Manual monitoring.

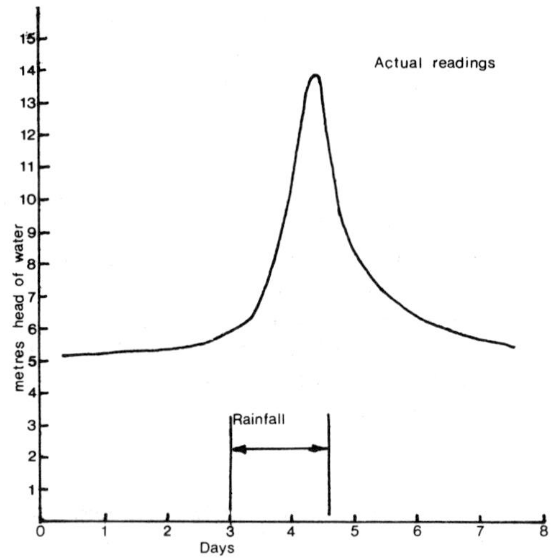

FIG. 6(b). Automatic monitoring.

Fig. 7. demonstrates how an automatic system can be used to monitor groundwater response to tidal conditions. One transducer measures the tide directly and the other the response in a piezometer. Results of this type would have been difficult to obtain manually.

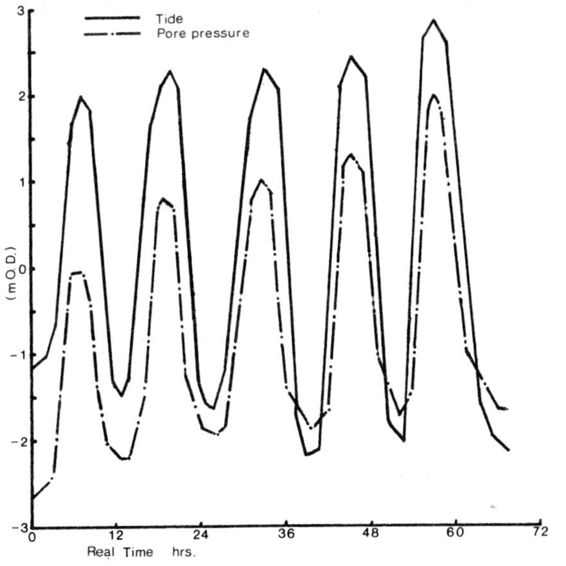

FIG. 7. Tidal response.

Advantages

Automatic monitoring systems have considerable advantages over traditional methods of obtaining data from piezometers. Some of these are summarised as follows:

(1) one system can obtain readings from up to 120 piezometers,
(2) it is possible to obtain readings 24 hours per day in all weather conditions,
(3) data can be regularly obtained from piezometers with limited or difficult access,
(4) printed output is available or direct interfacing to a computer is possible,
(5) systems require little maintenance,
(6) existing standpipe, pneumatic, hydraulic and electrical piezometers can be easily adapted to such systems,
(7) systems can be battery or mains operated allowing greater system flexibility,
(8) systems are fully programmable on site and this can be achieved by non-specialist personnel,
(9) the equipment is rugged, reliable and may be installed in harsh environments.

Conclusion

The function of automatic data loggers is to provide data, and since superfluous data can always be disregarded, most engineers would prefer more, rather than less data.

It has been due mainly to the advances in computer technology that these automatic data logging systems can now be manufactured at relatively low prices. This allows them to be cost effective even on small projects.

They complement the introduction of computers into the civil engineering industry, forming a new and vital link between the need for geotechnical equipment and the power of the computer.

References

HANNA, T.H., 1985. *Field Instrumentation in Geotechnical Engineering* Trans Tech. Publications, Clausthal, Germany.

INSLEY, H., & MCNICHOLL, D., 1982. Groundwater monitoring of a soil slope in Hong Kong. *Proceedings of the 7th South East Asian Geotechnical Conference, Hong Kong.*

PENMAN, A.D.M., 1972. Instrumentation for embankment dams subjected to rapid drawdown. *Current Paper CP1/72,* Building Research Station.

POPE, R.G., WEEKS, R.C., & CHIPP, P.N., 1982. Automatic recording of standpipe piezometers. *Proceedings of the 7th South East Asian Geotechnical Conference, Hong Kong.*

RODIN, S., HENKEL, D.J., & BROWN, R., 1982. Geotechnical study of a large hillside area in Hong Kong, *Journal of the Hong Kong Institute of Engineers*

R.C. WEEKS & P. STARZEWSKI, Geotechnical Instruments (UK) Limited, Station House, Old Warwick Road, Leamington Spa CV31 3NR, UK.

Monitoring of groundwater conditions at West Chevington opencast coal site, Northumberland

D. A. Blythe

ABSTRACT: West Chevington opencast coal site is a productive mine situated near Widdrington, Northumberland and is operated on behalf of the National Coal Board Opencast Executive.

Excavation of a multi-seam sequence of strata mainly within the upper Westphalian A of the Carboniferous is being carried out in a series of benches so that the excavation void proceeds eastwards with increasing depth. A number of piezometers have been installed along sections running across the site, both within the backfill, which follows the slowly migrating excavation void, and in advance of the excavation. The monitoring of groundwater levels has allowed a study of the drawdown and recovery due to the excavation to be carried out. A relationship between water level changes and the position of the excavation benches has been investigated.

Rock samples were obtained from the excavation, as they were removed by faceshovel and dragline, and moisture contents and porosities determined in the laboratory. Estimates of volumes of water removed by earthmoving plant have been calculated to demonstrate the proportions that are mechanically extracted from the excavation.

Introduction

West Chevington opencast coal mine is located approximately 11 km NNW of Morpeth and 1.25 km due east of Widdrington in Northumberland, North-East England. It lies immediately north of the worked out, Sisters opencast site in an area of generally flat lying land (Fig. 1). Operation of these two sites has been undertaken by independent contractors for an on behalf of the NCB Opencast Executive.

Geological succession

Both sites are located within the Northumberland coalfield and coal-bearing strata are covered by a layer of glacial deposits consisting almost entirely of lodgement tills. These vary in thickness from 1.5 to 30 m, the thickest deposits occurring in an area of a former infilled glacial erosion channel which runs approximately NW-SE across the centre of West Chevington site. On Sisters site superficial deposits decrease in thickness in a south-easterly direction.

Coal-bearing strata consist of interbedded coals, mudstone, silty mudstone, siltstones and sandstones displaying typical deltaic sedimentary features. A number of seams are subject to washout and thinning by overlying sandstones some of which display west to east trending channels with lateral accretion surfaces (e.g. above the Widdrington Main, Top Leaf (Q120)). Two seams, the Tilley (O000) and the Top Yard (P200), are overlain by substantial sandstone bodies and are also underlain by commercially valuable underclays used in refractory brick manufacture.

A generalised vertical section for both sites is shown in Fig. 2. The worked succession for the two sites lies within the upper Westphalian A and lower Westphalian B stages. Excavation to underlying seams was prevented by high overburden-to-coal ratios resulting from the presence of washouts and thinnings, former underground workings, and lack of excavation room.

Structure

The strata generally dip in an easterly direction with inclinations ranging from 1.5° to 6°, with minor localised variations adjacent to faults.

The southern part of Sisters site is crossed by the Grange Moor Fault, one of the major structural features in the area. This trends approximately west to east and downthrows to the south from 27 to 34 m. It is associated with a number of other, smaller, subparallel faults.

A SW–NE trending fault also crosses Sisters site, downthrowing to the south-east by 7.4–17 m. West Chevington site is generally free of faulting, though a WNW–ESE fault (downthrowing northwards by 4–15 m) crosses the southern central part of the excavation area. The site is bounded to the north by the Causey Park Fault which downthrows to the north by approximately 80 m. Intruded along this is the Carboniferous doleritic, Causey Park Dyke which has produced heat alteration of the adjacent coal seams.

From CRIPPS, J.C., BELL, F.G. & CULSHAW, M.G., (eds), 1986, *Groundwater in Engineering Geology,* Geological Society Engineering Geology Special Publication No. 3, pp. 375-385

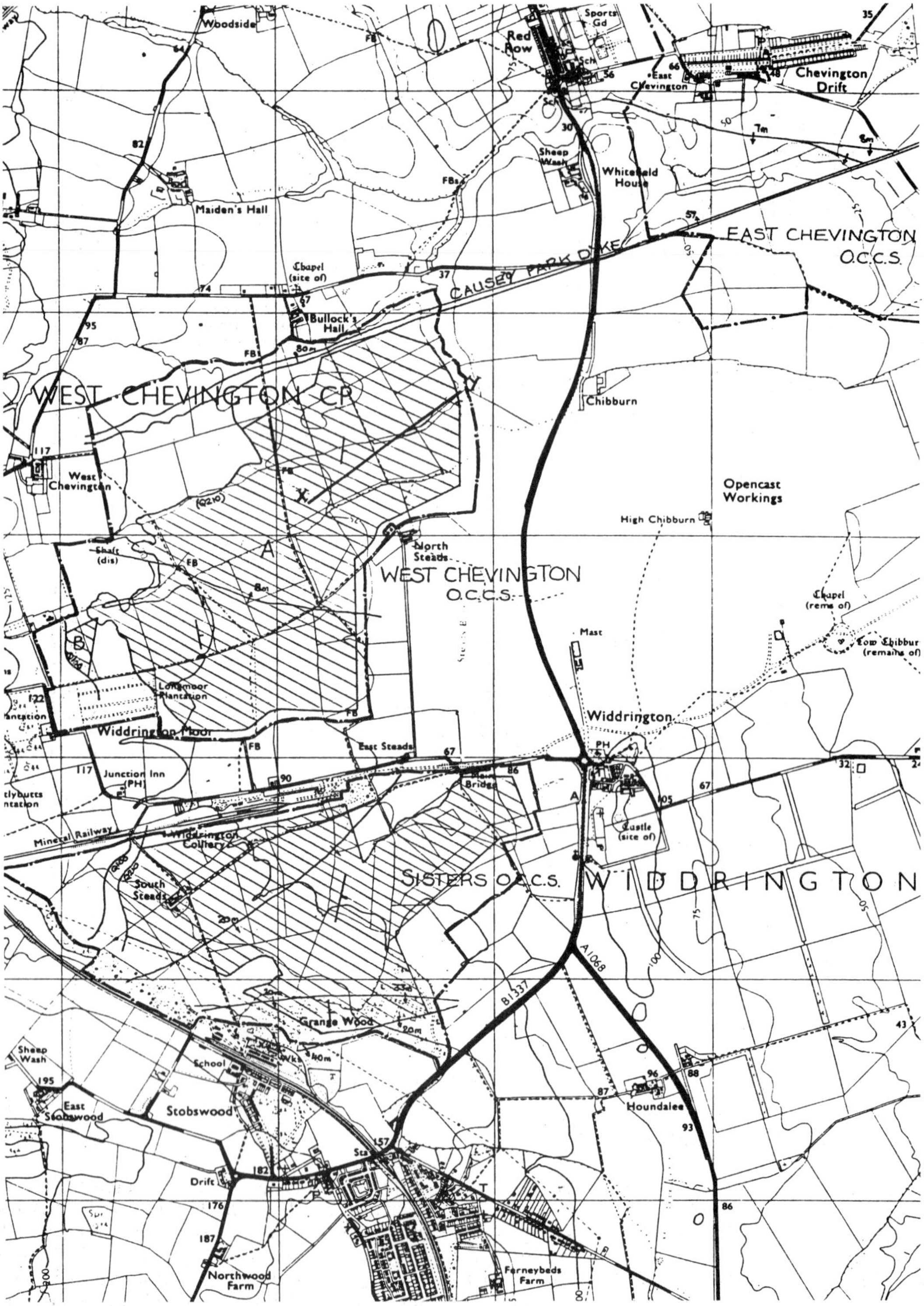

FIG. 1. Location of West Chevington and Sisters site. (Based upon Ordnance Survey 1:25,000 map with the permission of Her Majesty's Stationery Office, Crown copyright reserved.)

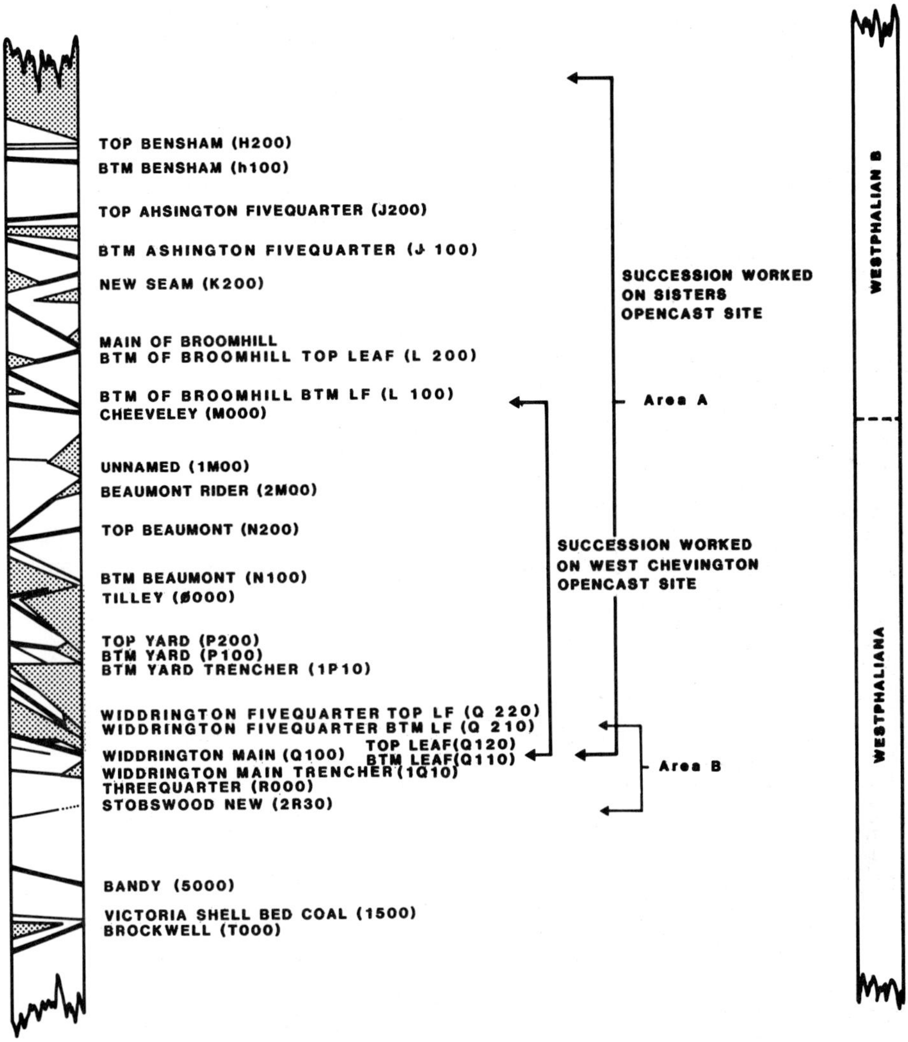

FIG. 2. Generalised vertical succession

Former coalworkings

The nearest working colliery (May 1985) is at Ellington some 5.5 km to the south-east. However, there are considerable, disused, underground workings from the former collieries at Widdrington, closed 1911 and now the site of the modern coal disposal point between Sisters and West Chevington Sites, and Stobswood Colliery south of the Grange Moor Fault, which closed in 1965.

Pillared workings were encountered in the Top Beaumont (N200) and Bottom Yard (P100) seams during the working of Sisters site. Workings in the Widdrington Fivequarter (Q210) and Widdrington Main (Q100) seams are more extensive across the two sites and the degree of extraction has an important bearing on the depth and location of excavation areas on them.

Extraction of coal from the Widdrington Main (Q100) seam appears to have ceased where the seam splits into two separate leaves on West Chevington — this has allowed deeper excavation to the east. Older opencast workings occur both partly within and adjacent to the two sites and are generally restricted to shallow single seam excavations.

Site operation and dewatering

Considerable quantities of water were encountered during prospective drilling on both sites, particularly during penetration of major sandstones (e.g. above the Top Yard (P200) seam) and also when old workings were encountered. The area is unaffected by the Executive's dewatering of old workings from the Hauxley shaft about 7 km to the north-east due to the presence of the Causey Park Dyke which acts as an almost impermeable barrier to strata and workings to the south. Strata to the south of the Grange Moor Fault are dewatered by pumping through old workings to Ellington Colliery to a level of approximately 54 m BOD (Below Ordnance Datum). The Grange Moor Fault acts as another hydrological barrier, though some strata dewatering occurs to the north. Approximately 1.6 m^3 min^{-1} was pumped at Stobswood Colliery to enable mining operations to proceed.

(a) Sisters

Excavation on Sisters site was restricted to one large area with varying pavement seams. Coal extraction commenced in May 1973 with overburden removal undertaken by three faceshovels and a large walking dragline (predominantly the Bucyrus Erie 1550 W, 'Big Geordie').

Dragline operations undertaken on the Widdrington Fivequarter (Q210) and Widdrington Main (Q100) overburden can be divided into three areas. Firstly, a series of north-south strike cuts were excavated from the Fivequarter (Q210) outcrop in the west, proceeding eastwards down-dip. The operations were then switched to the south where excavations took the form of west–east cuts progressing northwards running slightly up-dip. The latter enabled any groundwater to gravitate to pumped sump holes left in previous cuts. Finally excavations were completed by working up-dip along north–south strike cuts in a westerly direction to reach the first area.

To minimise the problems of groundwater ingress the operator decided to dewater the Widdrington Main (Q100) old workings using pumping boreholes equipped with 4 inch (0.10 m) and 6 inch (0.15 m) electric submersible pumps. Two main stations were eventually established one inside and one outside the excavation area. The operation proved successful due to the higher permeabilities of the collapsed

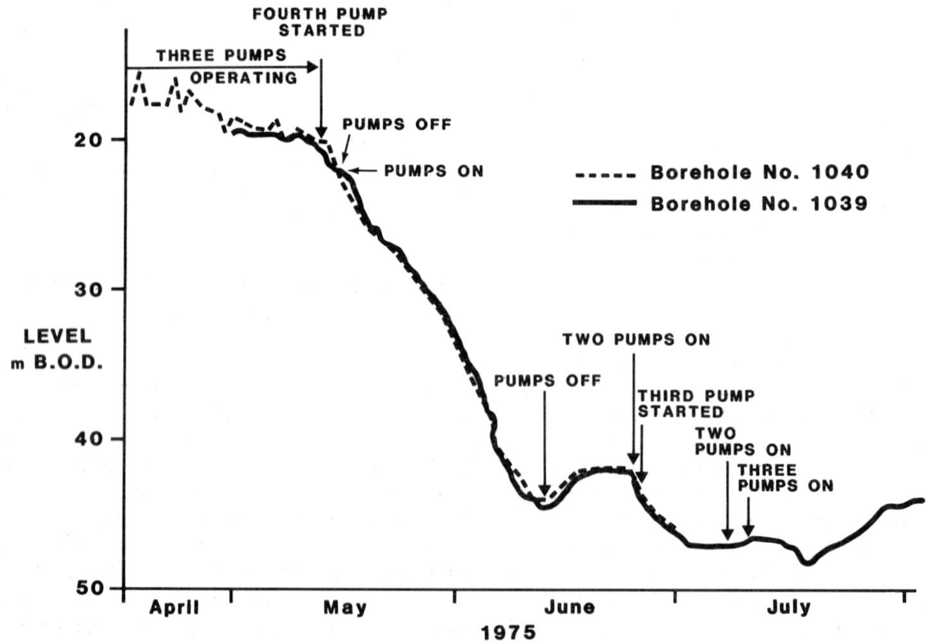

FIG. 3. Sisters site dewatering — water levels from standpipe piezometers.

strata above the old workings. Opening of joints and bed separation also provided interconnections for otherwise independent aquifers. The former was also enhanced by unloading of strata during excavation.

Drawdown curves based on water level data from two observation standpipe piezometers installed in the Main (Q100) seam are shown in Fig. 3. These demonstrate the rapid response to pumping. Records (NCB, O.E.) show that groundwater was only present in the excavations in any appreciable quantities when pumps were inoperative. Problems were encountered during 1978 and until completion of coaling operations in June 1979.

(b) West Chevington

Shortly after June 1979 all groundwater pumping ceased on Sisters Site. The effect of this cessation is clearly shown in Fig. 4 which shows water level data for two standpipe piezometers installed on the north-eastern side of West Chevington site in the Main Trencher (1Q10) horizon. Both are located outside the area of old workings and it is clear that the hydrogeology of the adjacent areas of solid strata are influenced by conditions within the old workings.

On commencement of operations on West Chevington site, therefore, it was clear that the task of dewatering had already reached problematic proportions.

There are two excavation areas on West Chevington (see Figure 1). The larger (Area A) consists of an excavation from Cheeveley (M000) to Main Trencher (1Q10) with the pavement seam governed by the presence or absence of old workings, washouts and coal thinning. A smaller area (Area B) lying to the west, was centred around the Main (Q100) and Threequarter (R000) outcrops with an extension down to the Stobswood New (2R31) seam but is now worked out.

Excavation has been undertaken predominantly by a large faceshovel, working on the upper advance benches and a Bucyrus Erie 1260 W walking dragline working on the lower seams. Individual excavation cuts are approximately 25–30 m wide.

Operations in Area A commenced on the Widdrington Fivequarter (Q210) outcrop with coal extraction in September 1980. The first few excavation cuts ran the full length of the outcrop area. The cuts were then split working the area south of the fault by NNE–SSW cuts and then the northern area by NNW–SSE cuts. Full length cuts were resumed just before completion of the faulted area as operations proceeded in an easterly direction. Maximum excavation depth will be approximately 90 m (= 69 m BOD) towards completion of Area A.

Dewatering of the excavations in Area B did not prove too difficult until the lowest seam was excavated, by which time two, 3 inch (0.08 m) and two 4 inch (0.10 m) pumps had been deployed.

The site diaries (NCB, O.E.) show that few problems were encountered in Area A until the tenth cut was excavated on the Fivequarter (Q210) seam. (The minimum level was approximately 0 m OD). A 6 inch (0.15 m), diesel pump was deployed and water abstracted from the cut. This practice continued for successive cuts, though by the time full length cuts were resumed, two pumps were required and it was decided to excavate sump holes at the southern end into which water would gravitate.

During this time, the fault plane was found to act as a feeder of groundwater and prospecting boreholes which had penetrated the underlying aquifers were also significant sources. When the excavation was extended downwards to the Widdrington Main Trencher (1Q10), the total pumping task was increased but the excavation cuts allowed the creation of more efficient water collecting areas, both for groundwater and surface run-off. Sump areas were usually spared immediate backfilling by the dragline or dumptrucks so that dewatering of succeeding cuts could be aided. Fig. 4 shows that significant drawdown had already been created on excavation of the Main Trencher (1Q10).

One sump area was converted into a permanent pumping station by the insertion of large vertical pipework from the Main (Q100) horizon at 50 m BOD and backfilling the surrounding area. A Hayward Tyer, 8 inch (0.20 m) electric submersible pump was installed which cut off automatically when the head of water became too low. This was later replaced by a 4 inch (0.10 m) type which pumps approximately 0.9 m^3 min^{-1} continuously. Although this undoubtedly aids in the dewatering of the site through the more permeable backfill it cannot satisfy all requirements down-dip to the north–east.

A previous attempt to dewater the site in advance of the excavation, through solid strata by means of a borehole proved unsuccessful as the permeability of the rock was too low to allow a sufficient inflow to feed the pump.

By May 1985 it is estimated that 5.7 m^3 min^{-1} of water are being removed from the site and supplied to the settling lagoons.

Observations from current mining benches (May 1985).

Groundwater is observed to be flowing from strata above the Top Yard (P200) seam, which is mainly sandstone, and from horizons below the Main (Q210). This flow is seen to be influenced by less permeable horizons such as mudstones and by the presence of a well developed joint system.

There are two joint sets present, striking approximately NNW–SSE and WSW–ENE. The former set is predominant and most joints are observed to be subvertical in dip. Joint spacing varies markedly from one lithology to the other and is also dependent on bed thickness. Generally the spacing ranges from about 0.20 m to 1.00 m with more

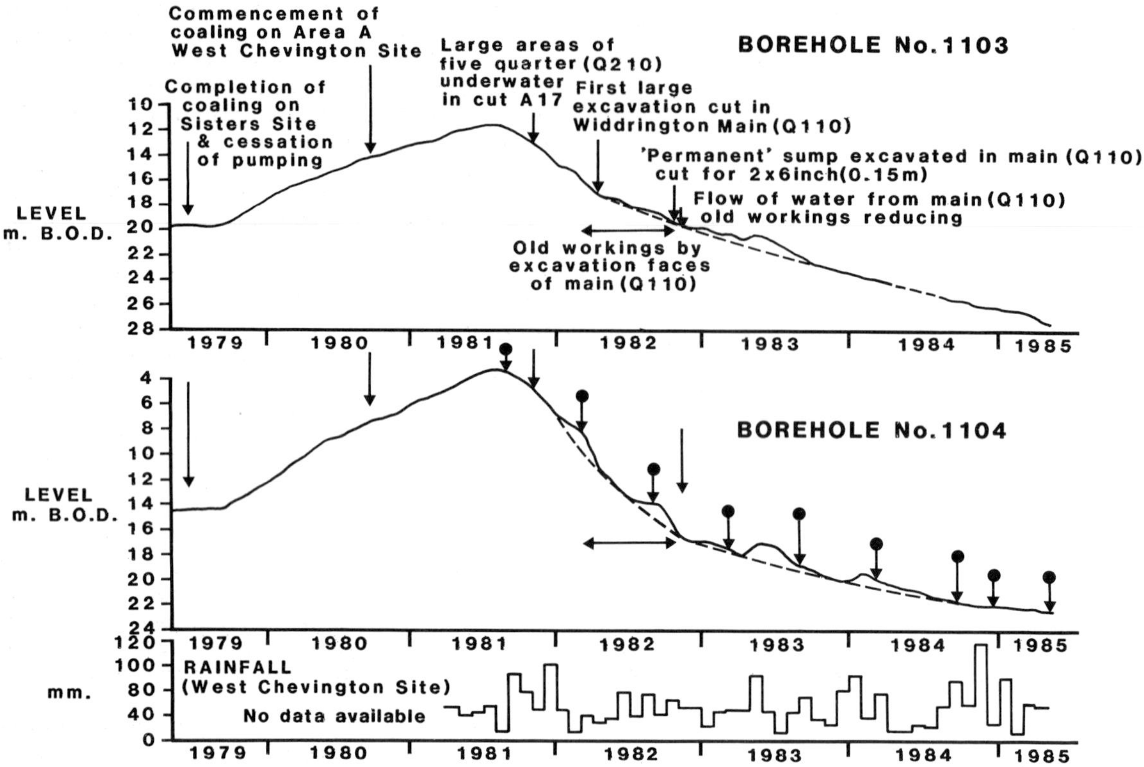

FIG. 4. Standpipe piezometer and rainfall data for West Chevington.

fractures apparent in the mudstone sequences. Estimation of joint openings is difficult because these are influenced by the unloading of strata during excavation. Joints available for measurement in the excavation faces will naturally be more open than those within the rock mass. It is very difficult to assess the distribution of the degree of openness of these joints and hence the permeability due to fractures.

A considerable flow of water is also observed to be emanating from the toe of the loose wall (backfill). However very little water is observed to be issuing from joints beneath the Main Trencher (1Q10) seam except where prospecting boreholes penetrate this and lower horizons.

Piezometric data

Reference has already been made to piezometer nos. 1103 and 1104 (Fig. 4). These together with nos. 1105 and 2675 were all standpipe installations with tips set at the Main Trencher (1Q10) or its immediate vicinity. A broad, long-term drawdown can be observed after making allowances for fluctuations due to rainfall. Data for the latter was obtained from a rain gauge situated on site and is found to be in close agreement with overall data from the Ellington station used by the Meteorological Office.

In March 1984 further piezometric installations were made creating three section lines running orthogonal to the excavation benches and orientated only slightly from the true dip direction. Piezometer nos. 9500, 9501, 9502, 9503 and 9504 were grouted in at the Main Trencher (1Q10) horizon. At the localities for 9501 and 9502, piezometer nos. 9505 and 9506 were installed and grouted into the Top Yard (P200) sandstone. Most piezometers have survived topsoil and subsoil stripping, piezometer nos, 9500 and 9504 have recently (May 1985) been dug through and no. 2675 was destroyed during ancillary work.

In Fig. 5 water level data for piezometer nos. 9500, 9501, 9502, 9505 and 9506 are plotted. A significant change in water level can only be observed in piezometer 9500 during the five months to May 1985 the remainder are fairly static.

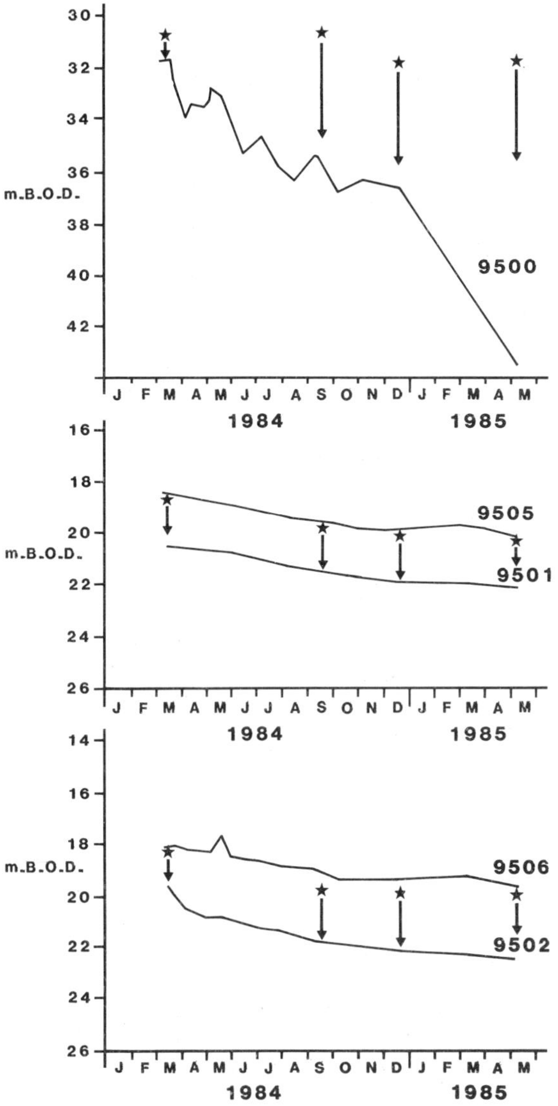

FIG. 5. Piezometric data for West Chevington.

Section X-Y (Fig. 6) shows both the geological details of a cross-section through the actively mined and future excavation area, together with selected piezometric levels observed in the piezometers and the corresponding position of the high wall at that time (these points are marked with asterisks on Figs 4 and 5). Excavation position dates refer to the accounting month end for coaling operations so that they do not represent actual dates of overburden excavation. However, generally there is little delay between excavation and coaling and for the purposes of this exercise the differences are insignificant.

It will be noticed that initially there is little change in the piezometric levels shown for the Main Trencher (1Q10) installations, with only a slight drawdown shown in piezometer no. 9500. The drawdown does not change significantly until the excavation face approaches within about 275 m. Drawdown then becomes fairly rapid (May 1985). Note that for the corresponding period, December 1984–May 1985, there is little change in the other piezometers. Also the positions of the observed seepages below the Widdrington Main (Q210) seam in May 1985 lie above the piezometric level in 9500. (This level was confirmed by subsequent measuring).

The seepages observed above the Top Yard (P200) seam will relate to a piezometric level in the sandstone above. To date (May 1985) the levels recorded in piezometers 9505 and 9506 show no marked change and it is assumed that the excavation in this horizon has not reached the critical distance as for the Main Trencher (1Q10) data.

Piezometric profiles have been extended to the toe of each successive excavation toe although it is by no means certain what may be happening in this zone.

In addition water levels for dates prior to the installation of piezometers 9500 etc. have been plotted on borehole no. 1104 position. It is assumed that the water profiles decreased steadily across the area during this period producing only a very flat drawdown. It is suggested that significant changes do not start until the advance of the excavation into their immediate vicinity.

Piezometers have also been installed in the backfill and in the Main Trencher (1Q10) horizon outside its area of excavation to the west along the same section line. To date (May 1985), none of these piezometers indicate any groundwater rebound. Similarly a piezometer installed off site in the Main (Q100) indicates a total absence of piezometric head. This suggests that the entire strata above the Main (Q100) and Main Trencher (1Q10) have been dewatered.

Water flowing from beneath the backfill on site is considered to be a combination of a very small, vertical flow component through the floor, inflow from the end walls of the excavation and surface water which has percolated through the backfill.

Piezometers installed in horizons below the Main Trencher (1Q10), off site to the north and west, indicate piezometric level above Ordnance Datum, as would normally be expected.

In situ sampling

In order to examine the quantities of water remaining within the excavated rock after dewatering a number of sampling sessions were carried out on all working benches. Sampling was carried out from actively excavated faces during conditions of dry weather. Samples were immediately wrapped in cling film to prevent loss of moisture and generally processed in the laboratory the same night.

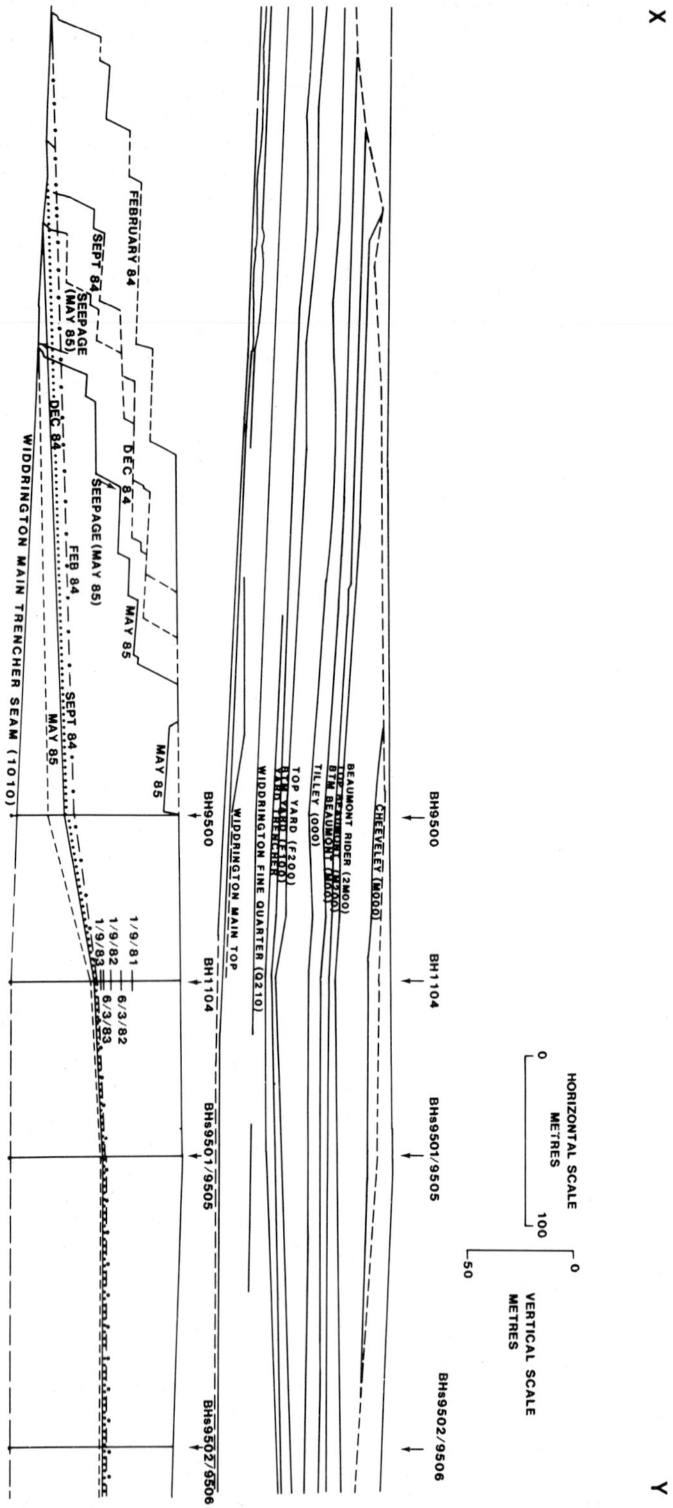

FIG. 6. Cross section of West Chevington showing geology and excavation faces.

Apart from moisture content determination by oven drying a number of specimens were also set aside on each occasion for immediate saturation under vacuum. This allowed the determination of saturated moisture contents and porosities of all rock types (including mudstones which slake if allowed to dry before saturation).

Specimens were classified according to their lithological descriptions and mean values determined for moisture contents on each bench. The degree of saturation was determined by dividing these mean values by the saturated moisture content values for the same rock type. The results are shown in Table 1.

Ideally many more samples are required especially for some of the less predominant rock types within certain benches. However, some conclusions can be drawn from this data:

(i) There is an indication that the fine grained materials release very little of the water which they hold prior to excavation. Sandstones generally release more.

(ii) There is an increase in degree of saturation for the sandstones with increasing depth of excavation.

(iii) The topmost bench material shows a slightly higher degree of saturation. This may result from downward percolation through the superficial deposits and/or water movement at rockhead.

In order to appreciate the quantities of water which remain in the excavated material and are mechanically removed during mining operations, consider the faceshovel operations on the Unnamed (1M00) seam overburden. Estimation of the quantity of water present in the average bucketful of this material must take account of the bucket capacity (12 m^3); the increase in volume of the material upon excavation (the bulking factor); and the degree to which the bucket is filled due to the size and shape of the rock fragments (the fill factor).

Using a dry density of 2.32 g cm^{-3} for mudstones obtained from laboratory data (n = 34, σ_{n-1} = 0.17), which on bulking gives a density of 1.6 g cm^{-3}, it can be shown that the weight of dry material in an average bucketful of material is 20.9 tonnes. This increases by just of 1 tonne when account is made of the average 5.2% moisture content (from Table 1).

The shovel takes about four passes to fill one of the fleet of Caterpillar 777 dumptrucks used to remove overburden to the backfill area. Their load is 87.8 tonnes, of which 4.1 tonnes is water. The capacity of these dumptrucks is quoted as 87 tonnes (Caterpillar 1976).

Similarly for the dragline operation on the Top Yard (P200) overburden, and considering medium grained sandstones only it is estimated that, on average, the bucket (capacity 32 m^3) carries 54.4 tonnes of material of which 3.3 tonnes is water. A dry density of 2.13 g cm^{-3} (n = 15, σ_{n-1} = 0.08) was used, which on bulking gives 1.33 g cm,$^{-3}$.

Dewatering process

Considerable unloading of the joints will occur above the lower seams during excavation of the advance, shovel benches. This will have the effect of increasing the mass permeability of the rock, as only relatively small changes in joint openings can produce significant changes in permeability (Hoek & Bray 1974). Solid, Coal Measures strata have been found to have mass permeabilities of about 10^{-6}m s^{-1} (Hammond, 1977) and more recent work confirms this figure. Joint opening will lead to increases in these values.

It is possible that several zones of rock may be present whose permeability decreases away from the excavation as the influence of unloading diminishes and there is a greater cover of intact rock. The apparently undisturbed zone, where piezometers show little change in level, acts as a reservoir of groundwater from which there is some flow down the hydraulic gradient into the excavation. Appreciable releases of groundwater may not occur until the excavation reaches within about 300 m of any one particular area.

Some of the groundwater flow into the excavation appears to be simply water draining from the joint system, fed by some intergranular flow from the body of the rock material. Observations from a standpipe piezometer installed behind the high wall of an excavated fault block or 'panel' at Butterwell opencast site indicates that this drainage via the joint system continues long after excavation and pumping have been completed (Fig. 7). The recovery of groundwater in this locality is apparently much slower than that witnessed at West Chevington probably as a result of the absence of old workings and the lower permeabilities of the intact rock.

Conclusions

Groundwater ingress into West Chevington opencast site appears to be largely influenced by joints and fractures present within the Coal Measures strata. These fractures open upon excavation and unloading thus increasing permeabilities. It appears that there is a substantial loss of head of the groundwater in a broad zone in advance of the working face and this has implications for slope stability.

Measurement of moisture content and porosity of excavated material suggest that the dewatering processes draw water mainly from the discontinuities, leaving most of the pore water in the intact rock material to be removed by the excavation plant.

ACKNOWLEDGEMENTS: Financial and technical assistance from the National Coal Board Opencast Executive and the cooperation of the Operating Contractor are gratefully acknowledged. The author wishes to thank Mr D.B. Hughes and Dr M.S. Money for their help during the study and members of the site staff for their assistance with sampling, surveying and monitoring. The opinions expressed in this paper are those of the author and not necessarily those of the NCB Executive.

TABLE 1. *Results of in situ sampling from excavation benches and the degree of saturation (σ_{n-1} = sample standard deviation, where n is the no. of tests).*

BENCH	ROCK TYPES																							
	MUDSTONES				SILTSTONES				SANDY SILTSTONE				SILTY SANDSTONE				FINE SANDSTONE				MEDIUM SANDSTONE			
	MC %	n	σ_{n-1}	SAT %	MC %	n	σ_{n-1}	SAT %	MC %	n	σ_{n-1}	SAT %	MC %	n	σ_{n-1}	SAT %	MC %	n	σ_{n-1}	SAT %	MC %	n	σ_{n-1}	SAT %
Unnamed (1M00)	5.2	23	1.3	100	2.5	1	0	72					3.2	5	0.9	98								
Beaumond Rider (2M00)	4.2	5	0.4	89	3.0	8	0.8	84					2.5	5	0.9	79	3.4	2	0.2	56				
Top Beaumond (N100)	4.4	8	1.2	91	3.5	7	0.4	99																
Bottom Beaumont (N100)	4.3	16	0.8	88	3.2	6	0.2	92																
Tilley (O000)	4.5	4	0.3	95																	6.5	22	1.2	69
Top Yard (P200)	5.4	2	0.8	100	2.8	1	0	80									4.1	4	0.4	68	6.5	7	1.5	69
Bottom Yard (P100)	4.5	8	0.3	92	3.2	3	0.5	91									5.4	19	0.9	89	6.9	2	0.9	73
Widdrington Fivequarter (Q210)	4.3	22	1.0	88					3.5	2	0.3	85					5.9	7	0.6	97	7.4	10	1.2	79
Widdrington Main (Q100)	4.5	5	0.2	92	2.3	6	0.2	66	2.3	5	0.2	57	2.7	10	0.3	83	6.1	7	1.0	100	9.1	2	0.8	97

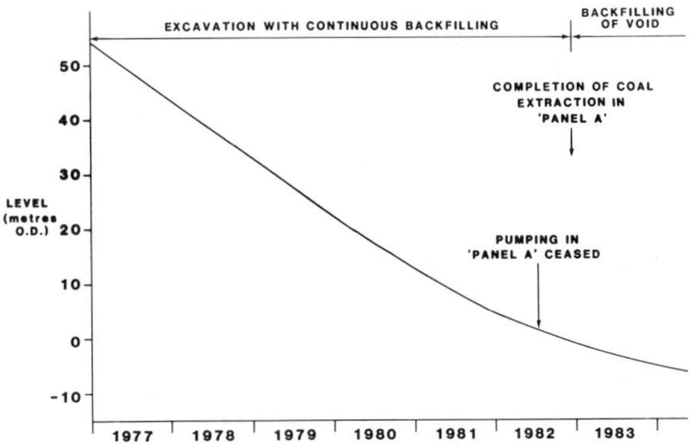

FIG. 7. Standpipe piezometer data from Butterwell opencast coal site.

References

CATERPILLAR TRACTOR CO. 1976. *Caterpillar Performance Handbook, Edition 6*, USA.

HAMMOND, D.J. 1977. An investigation into groundwater conditions likely to be encountered during opencast coal mining of a site in Northumberland. *M.Sc. Dissertation*, University of Durham.

HOEK, E. & BRAY, J.W. 1974. *Rock slope engineering*. IMM London.

NATIONAL COAL BOARD. Northumberland and Durham Division. North Northumberland Area - A Group, July 1965, *Northumberland Underground Water Survey (unpublished)*.

National Coal Board, Opencast Executive Engineers site diaries for Sisters and West Chevington opencast sites (internal reports).

D. A. BLYTHE, NCB Opencast Executive (N.E. Region), Ashfield Towers, Kenton Road, Newcastle upon Tyne NE3 4PE, UK.

Groundwater in Engineering Geology, London, 1986, pp. 387-394 Printed in England

Monitoring groundwater conditions in a highway embankment

G. I. Crabb & G. West

ABSTRACT: The paper describes how a section of 7 m high embankment made of Gault Clay on the Cambridge Northern Bypass (A45) has been instrumented with the aim of elucidating the mechanism of shallow slope failures that have affected adjacent lengths of the embankment. A comprehensive array of automatically recording hydraulic piezometers was installed to monitor pore water pressure changes, an automatic rain gauge was set up to record the rainfall and automatic stereocameras were positioned to photograph the slope daily so that the onset and development of any slip can be followed.

The piezometer installation is described in some detail and the results of the first fifteen months, monitoring of groundwater conditions in the embankment are presented: the results, together with those of laboratory strength tests, enable a simple stability analysis to be made for the embankment slope for different pore water pressure conditions. Comments are made on the operation of the hydraulic piezometers and the automatic logging system used to record their readings. In particular it was found that a diurnal cyclic pressure variation is caused by temperature changes affecting the piezometer tubing. The serviceability of the logger and rain gauge over a full year is also discussed. Finally some improvements to the system that are being incorporated in two new installations on sites on the M26 motorway in Kent, and which were made as a direct consequence of the experience with the Cambridge installation, are briefly described.

Introduction

More than thirty embankment slope failures have occurred on the A45 trunk road and the M11 motorway to the north of Cambridge in the period since the construction of the embankments in 1977–1978. The embankments involved were made from the overconsolidated Gault Clay that occurs in this region. The slope failures were typically shallow translational slides affecting the upper 1 – 1.5 m of the embankment slope and tend to be initiated in the winter or spring, often after high rainfall. More failures occurred on north facing slopes than on south facing slopes during this period. This paper describes a field experiment made to investigate the role that changes in pore water pressure play in shallow embankment slope failures of this kind.

It should be noted that the problem of shallow slope failures in embankments made of Gault Clay are by no means confined to the Cambridge area alone. Parsons & Perry (1985) and Perry (1985) have shown that the problem is more widespread, and in fact is likely to occur wherever highway embankments made of overconsolidated clays are to be found unless their side slopes are very shallow (say less than 1 in 4). These matters have recently been discussed *inter alia* by Garrett & Whale (1985) and by Greenwood *et al.* (1985).

The present paper is a much extended version of a very preliminary report made earlier (Crabb & West 1985).

Experimental site

The site chosen was the north facing slope of a 7 m high embankment made of Gault Clay on the Cambridge Northern Bypass (A45). A section of intact embankment slope having a slope of 1:2 was selected adjacent to a section that failed over a length of 60 m in the winter of 1982–83 and where four experimental repair methods and two experimental preventative techniques are being studied (Johnson 1985). Typical values of soil properties, measured during a site investigation made of the site are given in Table 1.

TABLE 1. *Typical properties of the Gaunt Clay on site at 0 - 1.5 m depth.*

Liquid limit (%)	66
Plastic limit (%)	26
Plasticity index (%)	40
Casagrande classification	CH
Moisture content (%)	28
Vane shear strength (kN/m^2)	80

The instrumentation of the site consisted of (a) a comprehensive array of 26 automatically recording hydraulic piezometers to measure the pore water pressure distribution within the embankment and any changes of it with time, (b) an automatic rain gauge to record the rainfall, and (c) automatic stereo-cameras positioned to photograph the slope daily so

From CRIPPS, J.C., BELL, F.G. & CULSHAW, M.G., (eds), 1986, *Groundwater in Engineering Geology,* Geological Society Engineering Geology Special Publication No. 3, pp. 387-394

that any slope movement can be monitored. The leads from the hydraulic piezometers are brought down the slope to a small instrument house at the foot of the embankment shown in Fig. 1, this contains the hydraulic scanning valves and the associated control and data logging system that utilises replaceable integrated-circuit memory cartridges. The site is visited about once a month to change the cartridge, any piezometers which need de-airing being attended to at the same time. Back at the Laboratory the data stored in the cartridge is transferred via a cartridge reader to a microcomputer system where it is recorded on magnetic disc. The data is processed and output is provided in the form of graphs, shown in Figs 3a, 4a and 5a, of the pore water pressure at each piezometer tip plotted against time. A cartridge from the rain gauge is dealt with in a similar way to give a histogram of rainfall against time shown in Figs 3b, 4b and 5b.

FIG. 1. Piezometer scanner and logger.

The objective of. the work is to monitor the site over several years, in particular to study the changes in pore water pressure near the surface of the embankment slope in the winter and spring so that a better understanding is gained of how the pore water pressure in this upper layer affects slope stability. If a failure should occur, then the piezometric and rainfall data immediately before failure will be invaluable, particularly as the time-lapse photographs should enable the time and rate of slipping to be pinpointed.

In the event, the 1983–84 winter and spring were exceptionally dry so that the likelihood of an embankment slope failure would be low, and in fact none occurred at the site or nearby. A slip, however, did occur on the south facing embankment slope on the opposite side of the road to the experimental site in March 1985.

Installation of piezometer tips

Twenty-six tapered 'Bishop' tips with high air entry pressure ceramic filters were installed, covering a complete cross-section of the embankment slope down to the foundation level (Figs 3c and 3d). The holes were drilled by a lightweight rotary auger rig working on narrow benches formed on the slope. As it was expected that the majority of the piezometers would be measuring pressures below atmospheric, each tip was installed in a separate borehole. It was also initially specified that the tips be pushed into position at the base of the bored hole to ensure good contact with the *in situ* material. However, the clay proved to be too strong to allow this, and it was eventually decided to surround each tip with a weak cement-bentonite grout. This was expected to have a permeability similar to that of the *in situ* material. This technique appears to have been entirely successful, as shown by the system's rapid response to disturbance and otherwise satisfactory performance. The piezometers at the lowest level penetrated a natural layer of gravel underlying the embankment and below the groundwater surface; it was therefore safe to install them in conventional sand-cells.

Results and discussion

The experimental installation briefly described above was commissioned in December 1983. Results for the whole of 1984 and the first quarter of 1985 have therefore been obtained to date. The rainfall and soil moisture deficit (Penman 1950) for the year 1984 are shown in Figs 2a and 2b, and beneath these are shown the pore water pressures recorded by four of the group of piezometers in the second profile from the top of the embankment (Fig. 2c). It can be seen that although the rainfall is relatively uniformly distributed throughout the year, the soil moisture deficit reached a pronounced maximum at the end of August. The pore water pressures at 0.5 m and 1.0 m depth below the embankment slope reflect the soil moisture deficit, becoming increasingly negative as the summer advanced. During October, as the soil moisture deficit fell, the pore water pressures at 0.5 m and 1.0 m depth suddenly returned to their winter value of about zero. The shallow piezometers at the other profile positions in the embankment did not show this effect so markedly, possibly due to variations in the depth of root penetration and surface cracking in the upper layer of the embankment slope. The occasions when the piezometers were de-aired are also indicated on Fig 2c, and the portions of the curves shown by broken lines indicate those periods of the year when the logger was out of action.

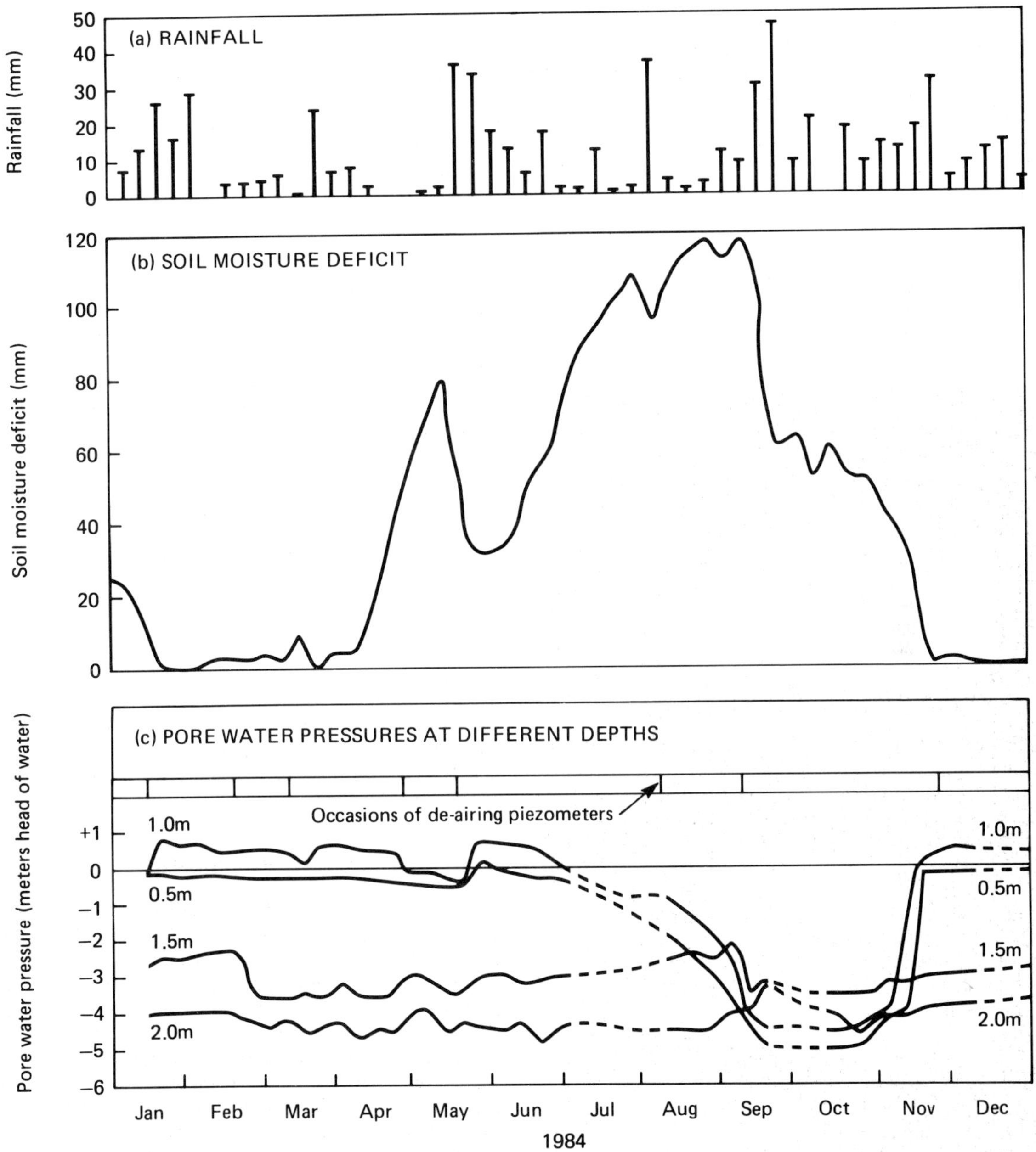

FIG. 2. Rainfall, soil moisture deficit and pore water pressures for experimental site.

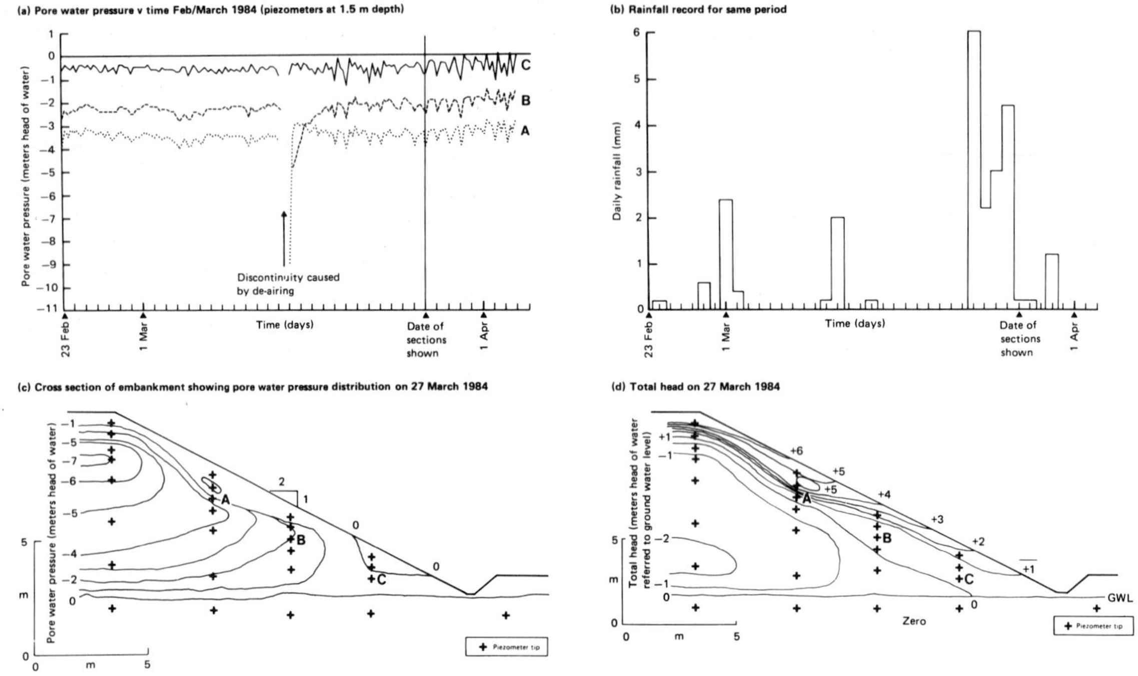

FIG. 3. Results at end of first winter.

Pore water pressure and total head profiles

Fig. 3a and 3b show the pore water pressures recorded by three piezometers and the rainfall for the site for the period 23rd February to 3rd April 1984. The diurnal cyclic pressure variation is caused by temperature changes affecting the system and can be disregarded. Fig. 3c shows the pore water pressure distribution over the instrumented cross-section during this period. It can be seen that the distribution is consistent with the initially dry clay having become wetter by moisture movements, both up from the water table and down from rainfall on the slope, so that high negative pore water pressures remain only in the core of the embankment.

The pore water pressure distribution is required for stability analysis in terms of effective stress, but for the study of seepage the results are required in the form of a total head distribution, and this is shown in Fig. 3d. It can be seen that at this time of year the down-slope component of the hydraulic gradient is small compared with that normal to the slope surface, so that seepage would be mainly into the slope and not down it if the permeability of the embankment was isotropic.

Figs 4a and 4b show the pore water pressure and rainfall for August–September 1984, and Figs 4c and 4d give the pore water pressure and total head distribution over the instrumented cross-section during this period. Comparing these with Figs 3c and 3d, it can be seen that the pore water pressure is more negative in the zone beneath the surface of the slope and that the small regions positive pore water pressures have disappeared. The total head diagram also reflects these changes, the higher negative pore water pressures giving rise to lower values of total head on the embankment slope.

Figs 5a, 5b, 5c and 5d show conditions at the experimental site in March 1985, at the end of the following winter. It can be seen that the pore water pressure and total head distributions have returned to a state very similar to that prevailing in March 1984 (see Figs 3c and 3d).

It can be seen from these results that cyclic annual changes of pore water pressure occur to a depth of about 2 m, with the highest pore water pressures occurring, as expected, at the end of the winter, the time at which slips tend to occur.

Stability analysis

Fig. 6 shows the result of an investigation made by the Northamptonshire County Council Engineering Services Laboratory into the failed section of embankment slope adjacent to the piezometer installation. It indicated that the failure was of the shallow translational kind with a more or less planar slip surface at a depth of about 1.5 m. Fig 6 shows this idealised as a infinite slope together with the forces acting on the potential slip plane imposed by a prism of soil of unit cross section overlying it, the depth to the potential slip plane being z and the unit weight of soil γ. The pore water pressure at point P is u. Consideration of the shear stress that the prism imposes on the potential slip plane and of the shear strength available to resist it leads to the expression for the factor of safety F given in Fig. 7. This form of analysis takes no account of boundary effects which would tend to increase the factor of safety.

Strength parameters for the clay

Fig. 8 shows the results of three preliminary triaxial compression tests on undisturbed 38 mm diameter specimens of clay prepared from a U100 sample taken from 1.5 m depth in the embankment slope, carried out by the City University. The testing was conducted at low effective stresses similar to those acting at this depth and so as to simulate the field situation failure was induced in the specimens by raising the pore water pressure while keeping the confining pressure and axial pressure constant. It can be seen that the envelope to the Mohr circles is curved and intersects the origin. Values of the effective cohesion, c', and the effective angle of friction, ϕ', for any effective normal stress, can be obtained from the tangent to the envelope at the corresponding point.

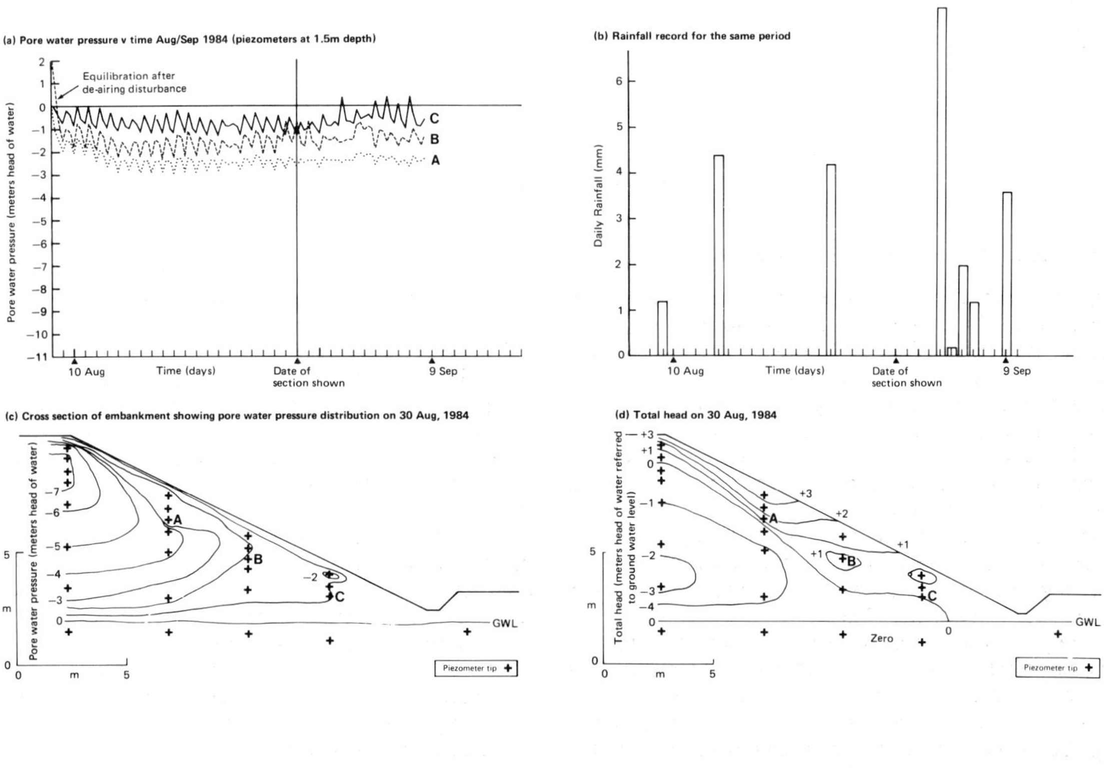

FIG. 4. Results at end of first summer.

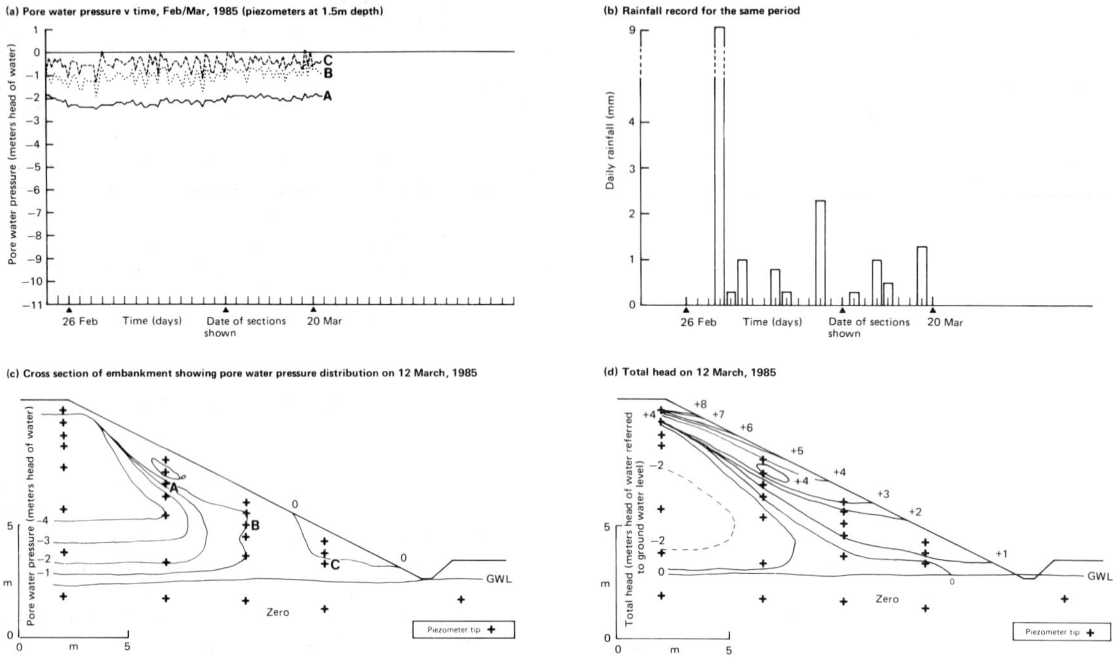

FIG. 5. Results at end of second winter.

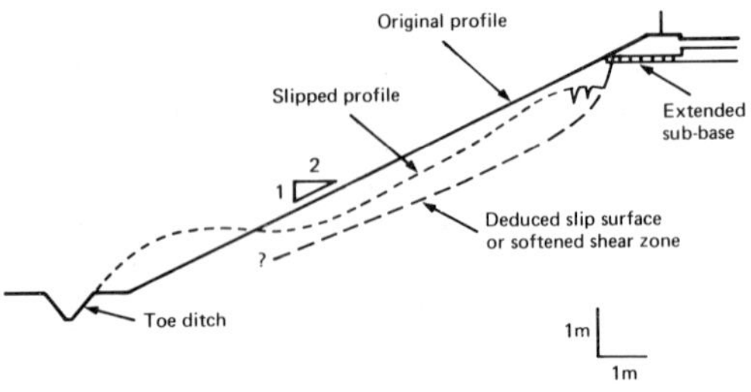

FIG. 6. Cross section of shallow slope failure.

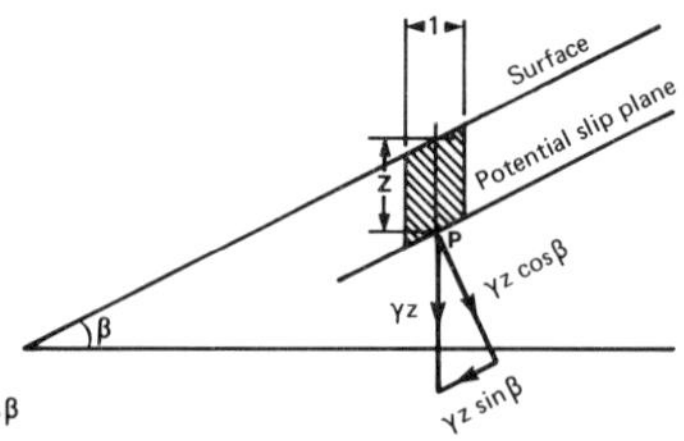

Shear stress = $\gamma z \sin\beta\cos\beta$

Available shear strength = $c' + (\sigma - u)\tan\phi'$

Factor of safety (F) = $\dfrac{\text{Available shear strength}}{\text{Shear stress}}$

$$F = \frac{c' + (\gamma z \cos^2\beta - u)\tan\phi'}{\gamma z \sin\beta\cos\beta}$$

FIG. 7. Stability analysis of embankment slope.

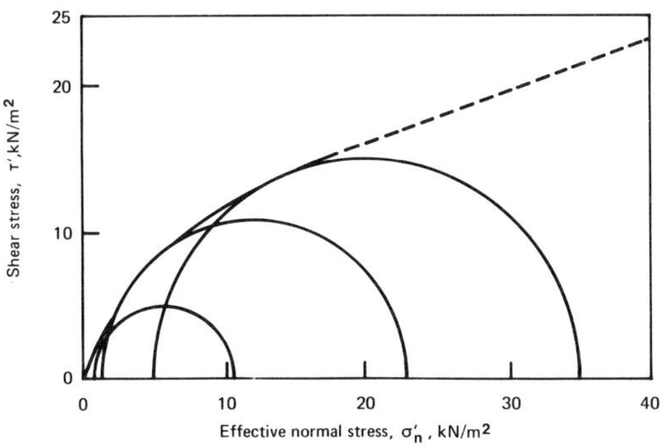

FIG. 8. Mohr's circles at peak failure.

Effect of varying pore water pressure

We now have all the factors we need to make a preliminary simple analysis of the effect of varying the pore water pressure on the embankment slope stability. For a depth of 1.5 m below the slope surface, this is given in Table 2. It can be seen that by raising the pore water pressure from −1.5 m water to +1.5 m water, the factor of safety of the embankment slope is lowered from 2 to just under 1. This is very important because it shows that raising the pore water pressure alone may be sufficient to cause the slope to fail.

There are a number of other considerations that could have the effect the further reducing the factor of safety. It is well known that overconsolidated clays *in situ* have a fissure structure which causes the mass strength to be less than the intact strength (Marsland 1971). Although when these materials are used in embankments, they are remoulded to some extent, it is possible that many of the fissures remain, randomly redistributed. During construction horizontal discontinuities can be formed between compacted layers and this was clearly observed in the trial trenches dug as part of the site investigation of the embankment at Cambridge. Seasonal changes in moisture content causing cycles of shrinkage and swelling, and the roots of vegetation, also are known to produce fissure structure (Anderson *et al.* 1982). To take account of these effects it is necessary to test much larger specimens than those of Fig. 8. Progressive failure, in which it is not necessary for the whole failure surface to reach peak strength simultaneously, is another mechanism by which the factor of safety may be reduced below that arrived at in the simple analyses given above if as is likely the material exhibits a post-peak reduction in strength (Skempton 1964). Seepage forces also have a bearing on stability if seepage velocities are sufficiently high, although this would not appear to be true in the case studied. If seepage has, as suggested by Figs 3d, 4d and 5d, a major component normal to the slope this would, in fact, tend to have a stabilising effect.

Operational comments

It was remarked earlier that some of the piezometer records exhibited a diurnal cyclic pressure variation and this is well shown in Figs 3a, 4a and 5a. It was suspected that this was caused by temperature changes affecting the piezometer tubing and it was found that the coefficient of expansion of the material from which the tubing is made is about three times greater than the coefficient of expansion of water. An increase in temperature, because of the stiffness of the hydraulic system and the low permeability of the clay, causes the tubing to expand more than the water leading to a temporary reduction in pressure. This is corroborated by the fact that the piezometers in the gravel are much less affected. The tubing in the instrument house is the most affected, and has been subjected to a maximum daytime temperature of 43°C, resulting in a day to night variation of at least 20°C. The remainder of the tubing is buried at 0.5 m depth and should be little affected.

In March 1985 a wooden shelter was erected around the instrument house to shade it from the sun and it is hoped this will reduce the effect. In two new

TABLE 2. *Stability at 1.5 m depth in embankment slope. Depth (z) = 1.5 m, unit weight (γ) = 19 kN/m³, γz = 29 kN/m².*

Pore water pressure (u)		Effective normal stress ($\gamma z \cos^2 \beta - u$) kN/m²	Effective cohesion (c') kN/m²	Effective friction angle (ϕ') degrees	Factor of safety (F)
m water	kN/m²				
−1.5	−15	38	9	20	2.0
−1.0	−10	33	9	20	1.8
−0.5	− 5	38	9	20	1.7
0	0	23	9	20	1.5
+0.5	+ 5	18	9	20	1.4
+1.0	+10	13	8	21	1.1
+1.5	+15	8	6	32	0.95

installations on the M26 motorway in Kent, larger thermally insulated instrument houses have been installed equipped with thermostatically controlled heaters and extractor fans in order to restrict the diurnal temperature variation.

The periods when the logger was out of action are shown by the broken lines in Fig. 2c. These were caused by a number of minor faults mainly confined to the power supply circuitry, such as the failure of a voltage regulator. In all cases these problems were swiftly attended to by the manufacturers, thus minimising loss of data.

The occasions on which the piezometers were de-aired also also shown on Fig. 2c. Some piezometers — those measuring high negative pore water pressures — need de-airing about once a month, and at the new installations on the M26 two modifications were made in an attempt to extend the time between de-airing. The first of these was to install the piezometer tip directly in contact with the clay rather than in a sand cell or bentonite-cement cell. To do this a pre-formed hole was prepared at the base of the borehole using a specially-made narrow, tapered auger followed by a mandrel having exactly the same profile as the piezometer tip. The second improvement consisted of fitting the piezometer tip with an internal pipe that brought the reading tube of the piezometer down to the bottom of the piezometer tip. This has the effect of ensuring that any air bubbles which form in the tip find their way into the de-airing tube rather than the reading tube of the instrument and is based on an idea proposed by Anderson & Burt (1977). It is suggested that this modification should become a standard feature of hydraulic piezometer tips.

Conclusions

The monitoring of an extensive array of piezometers at the experimental site for a period of 15 months has enabled the following conclusions to be drawn.

1. After its initial construction from relatively dry Gault Clay, the embankment has been wetting up from the water table at its base and from rainfall on the slope, and the soil in the interior would seem to be a long way from equilibrium at the present time.
2. The embankment slope is wettest at the toe, and failures may well begin here and then extend up the slope, but a small region of positive pore water pressure higher up the slope in winter shows that infiltration, perhaps following cracking of the soil during the previous summer, may occur higher up.
3. During the period of observation, it was only the top metre or so of soil on the embankment slope that was affected by seasonal fluctuations of pore water pressure.
4. A simple stability analysis shows that a rise of pore water pressure in itself may be enough to cause a shallow slope failure.

ACKNOWLEDGEMENTS: The work described in this paper forms part of the programme of the Transport and Road Research laboratory and the paper is published by permission of the Director. The authors would like to thank the Eastern Regional Office of the Department of Transport and Cambridgeshire County Council for permission to make the experimental installation described. Northamptonshre County Council Engineering Services Laboratory is thanked for installing the piezometers and City University for carrying out the triaxial tests. The authors' colleagues are thanked: Mr W.M. Todd for installing and attending to the stereocameras on site and Mr M.P. O'Reilly for contributing to the paper.

References

ANDERSON, M.G. & BURT, T.P., 1977. Automatic monitoring of soil moisture conditions in a hillslope spur and hollow. *Journal of Hydrology,* **33** pp 27-36.

ANDERSON, M.G., HUBBARD, M.G. & KNEALE, P.E., 1982. The influence of shrinkage cracks on pore-water pressures within a clay embankment. *Quarterly Journal of Engineering Geology,* **15**, 9-14.

CRABB, G.I. & WEST, G., 1985. Monitoring pore water pressures in an embankment slope. *Proceedings of the International Symposium on Failures in Earthworks,* Technical Note 4. Institution of Civil Engineers, London.

GARRETT, C & WHALE, J.H., 1985. Performance of embankments and cuttings in Gault Clay in Kent. *Proceedings of the International Symposium on Failures in Earthworks,* Paper 7. Institution of Civil Engineers, London.

GREENWOOD, J.R., HOLT, D.A. & HERRICK, G.W., 1985. Shallow slips in highway embankments constructed of overconsolidated clay. *Proceedings of the International Symposium on Failures in Earthworks,* Paper 6. Institution of Civil Engineers, London.

JOHNSON, P.E., 1985. Maintenance and repair of highway embankments: studies of seven methods of treatment. *TRRL Research Report* RR30. Transport and Road Research Laboratory, Crowthorne.

MARSLAND, A., 1971. The shear strength of stiff fissured clays. *Proceedings of the Roscoe Memorial Symposium,* Cambridge, 59-68.

PARSONS, A.W. & PERRY, J., 1985. Slope stability problems in ageing highway earthworks. *Proceedings of the International Symposium of Failures in Earthworks,* Paper 5. Institution of Civil Engineers, London.

PENHAM, H.L., 1950. Evaporation over the British Isles. *Quarterly Journal of the Royal Metereological Society,* **76**, 372-383.

PERRY, J., 1985. Incidence of highway slope stability problems in Lower Lias and Weald Clays. *Proceedings of the International Symposium on Failures in Earthworks,* Technical Note 13. Institution of Civil Engineers, London.

SKEMPTON, A.W., 1964. Long term stability of clay slopes. Fourth Rankine Lecture. *Geotechnique,* **14**, 77-101.

CRABB, G.I. & WEST, G., Ground Engineering Division, Transport and Road Research Laboratory, Crowthorne, Berkshire RG11 6AU, UK.

Crown copyright. Any views expressed in this paper/article are not necessarily those the Department of Transport. Extracts from the text may be reproduced, except for commercial purposes, provided the source is acknowledged.

The value and interpretation of groundwater level measurements

M.D. Eggboro & S. Walthall

ABSTRACT: The measurement of water levels in the ground is a routine part of groundwater resource assessments and most site investigations. However the techniques used are often inadequate to reveal the true behaviour of aquifers, often resulting in inadequate data or a large number of expensive but useless readings. Continuous recording of groundwater levels using both float-operated chart recording and electric pressure transducers have been made on a wide range of observation wells to determine the most economical frequency of measurement for different aquifers and different purposes. The results of the investigation revealed a number of factors which are relevant in the observation of groundwater levels and these are discussed. Of these the construction of the observation point in the form of an open observation well, tube well, standpipe or one of a variety of piezometers is of fundamental importance. The behaviour of the aquifer as confined, or unconfined, with or without perched water tables, has considerable bearing on the results obtained. Barometric, tidal, fluvial, seasonal and abstraction effects on groundwater levels are discussed with reference to the amplitude of groundwater fluctuation and its rate of change. Finally suggestions are made for the design of observation points and the frequency of readings for different applications.

Introduction

Groundwater levels are used in a variety of ways. Each individual job has its own examples but what exactly is wanted from these measurements? It is important to give this question careful consideration before the value and interpretation of groundwater level measurement can be accurately assessed.

Hydrographs

The record of water level measurements, known as a hydrograph, gives a continuous picture of groundwater level fluctuations at a given site. The shape of the curve is influenced by the frequency of the measurements and in the following examples hydrographs produced by continuous recording equipment are shown. Less frequent readings taken manually would give less detailed and possibly different pictures. The water level in the ground is not static and can indeed vary by tens of metres so the value of a single measurement is limited. Hydrographs can be used to define the extent and rate of groundwater level change.

Fig. 1 shows changes in groundwater due to pumping and seasonal effects.

Fig. 2 shows changes due to tidal effects. The tide can have an effect by not physically pushing water into the aquifer but also by loading the aquifer to such an extent that water is squeezed out of the compressed part into the more dilated part.

Fig. 3a shows the changes in groundwater level in an observation borehole as a result of the change of loading of an aquifer by barometric pressure. Whereas tidal effects usually act only on part of an aquifer, barometric changes affect larger areas. Fig. 3b shows the response of two boreholes some 180 km apart in completely separate aquifers for a period of one month. The similarity indicates that the dominant effect on the groundwater level is barometric and that local effects are both very small and masked. If there is a borehole in the aquifer the full effect of barometric changes will be felt on the water in the open borehole. However, the water in the aquifer itself will only feel a partial effect as the structure of the aquifer will absorb some of the loading change. If high barometric pressure occurs it will tend to push down on the open water in the borehole and because the full effect does not reach the rest of the water in the aquifer, water is effectively pushed back into the aquifer from the borehole thus lowering the water level in the borehole. The ratio of the change in water level in the borehole to the change in the atmospheric pressure is know as the barometric efficiency. This effect also leads to blowing and sucking wells where the rate of ingress or egress of air through the well cap causes the well to whistle. Aquifers which are confined show barometric efficiency to varying degrees. For example, where they are overlain by a thick sequence of clay the barometric efficiency may approach 100%, because the clay absorbs all the loading changes. This then allows the borehole to reflect 100% of any barometric change. A typical value for the Permo-Triassic Sandstone aquifers is around 80%. In unconfined aquifers, where the water is in direct contact vertically in the atmosphere, no change of groundwater level in a borehole should

From CRIPPS, J.C., BELL, F.G. & CULSHAW, M.G., (eds), 1986, *Groundwater in Engineering Geology,* Geological Society Engineering Geology Special Publication No. 3, pp. 395-402

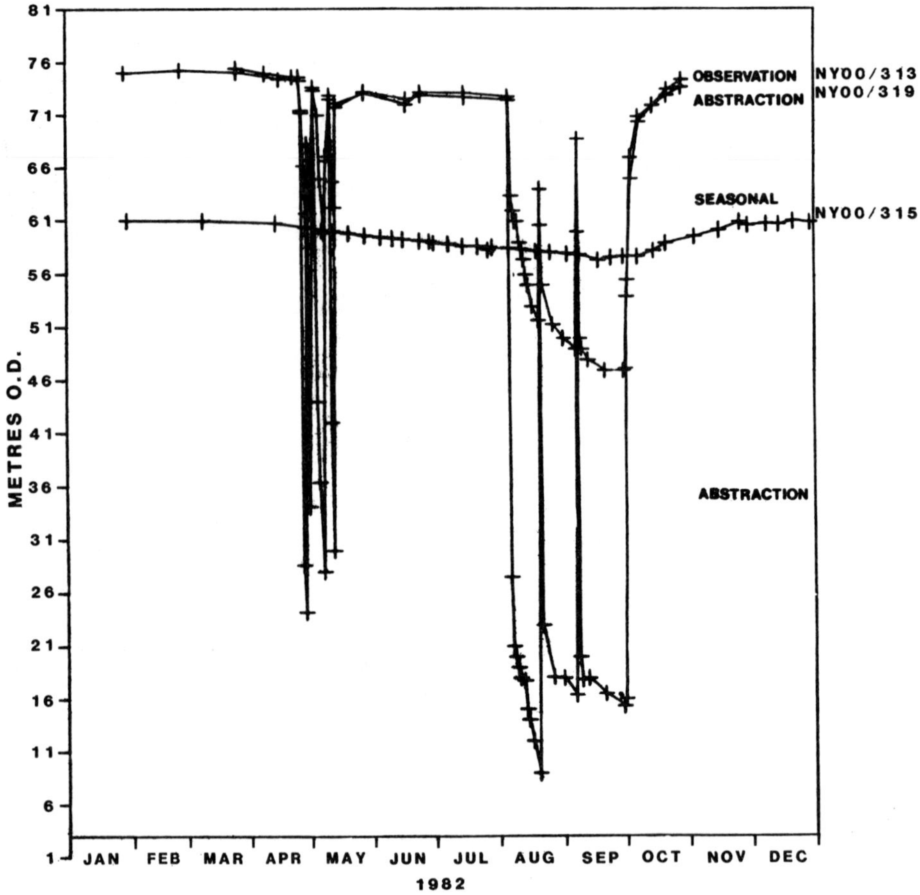

FIG. 1. Well hydrograph from Jan 1982 to Dec 1982

occur but it has been shown by Weeks (1979) that barometric effects can occur in deep, apparently unconfined aquifers. It is of interest to note that when using multiple piezometers, sealed at various levels in an aquifer, the barometric efficiency changes with depth. A further complication is that if sealed water pressure measuring devices are used then differences between these and open standpipes will occur.

Fig. 4 shows the long-term effect of abstraction patterns on an aquifer due to depletion and recovery.

Fig. 5 shows short-term effects of a river on an aquifer in direct contact with it.

A consideration of the hydrographs above shows the average water pressure of all the different layers open to the monitoring point or response length. As the aquifer is not uniform it has a large number of different layers within it such that it resembles a Spam Sandwich! Each of these layers has its own permeability and piezometric head and the observation point and the interpretation of the data therefore must take account of this important fact. These hydrographs also show that a continuous record at a given site is vitally important because spot readings give only part of the truth which can, and does, lead to erroneous interpretation. This can be very costly if an excavation floods or the quantity of water available in an aquifer is markedly less than expected.

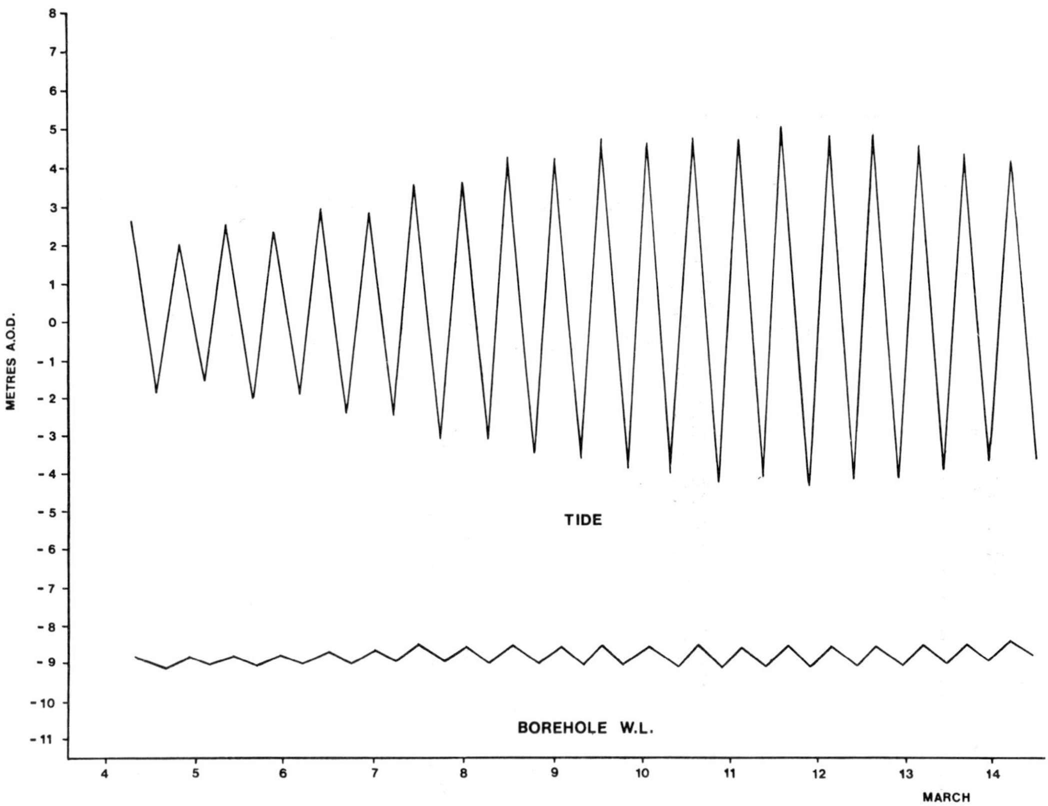

FIG. 2. Tidal efficiency for African Oil Mills borehole.

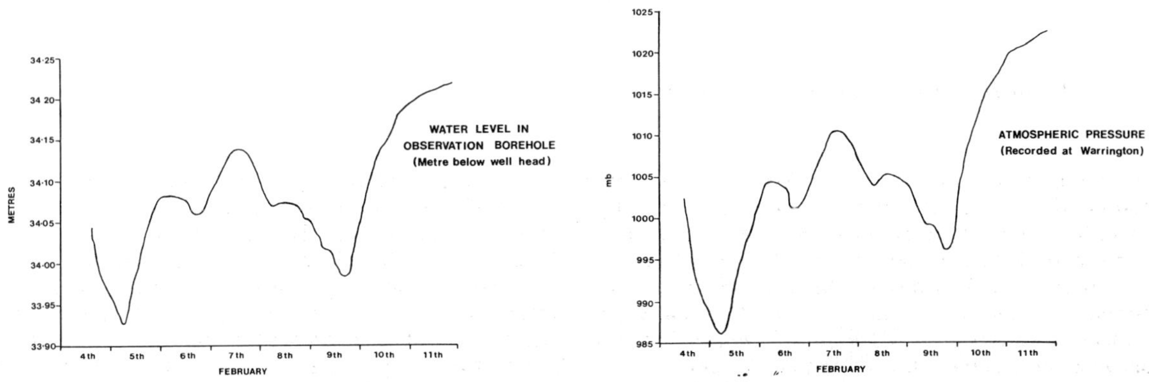

FIG. 3a. Comparison of water level in observation borehole with atmospheric pressure.

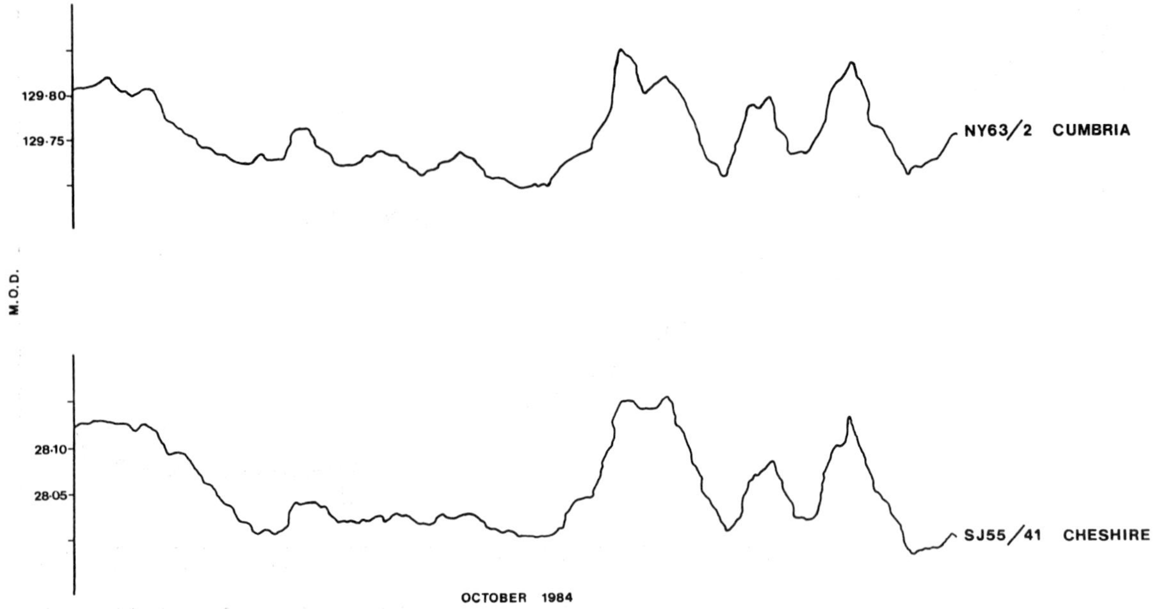

FIG. 3b. Comparison of water levels in mutually remote boreholes.

Applications

Groundwater can either be a resource or a problem. The hydrogeologist is keen to use groundwater as a resource. What could be more elegant than an artesian borehole supplying life-giving clean water completely free of charge; no energy requirements, no pollution, no problems. However, the engineering geologist sees groundwater as a potential problem. So each professional collects groundwater data with a different objective in mind. Consequently, each may need to construct his monitoring point differently and to monitor it at different frequencies. For example, if an observation borehole, is constructed in conjunction with an abstraction borehole, then care must be taken with depths and casings so that the part of the aquifer being exploited by the abstraction borehole is indeed being accurately monitored by the observation well. The frequency of observation needs to be such that subsequent analysis of the pumping test data is possible.

One use of water level measurements is to control local abstractions. Without such controls, over-abstraction could lead to mining of groundwater or to saline intrusion which would spoil a valuable resource. Fig. 6 shows the hydrograph of such a controlling borehole. It is constructed so that it monitors all the aquifer being exploited, being of a similar depth, and therefore accurately reflects any changes so that the aquifer can be managed properly. In this particular case the borehole is monitored continuously so that early action can be taken if necessary.

A second use is to check the long-term regional changes over many years and here a monthly frequency is sufficient (Fig. 4). These changes can be depletion of, or recovery of, groundwater levels. Each can cause a problem. Long-term depletion of an aquifer is not easy to spot because it tends to be slow and therefore changes are not dramatic or noteworthy and can be overlooked. However, the consequences are a problem because the quantity or quality of water from an area can deteriorate with the result that expensive alternative sources need to be established. Again it is necessary to have sufficient well constructed boreholes to define the problem accurately. Long-term recovery of groundwater is being experienced in several cities at the moment. Abstraction of groundwater has diminished in many areas because of the industrial recession and many basements in places such as Liverpool and Birmingham are now awash because groundwater levels have recovered. This problem was predicted using long-term level records but again knowledge of what the borehole is measuring is important to interpret its results.

A knowledge of groundwater levels in civil engineering is of prime importance in the design and construction of any structure. In this case there are

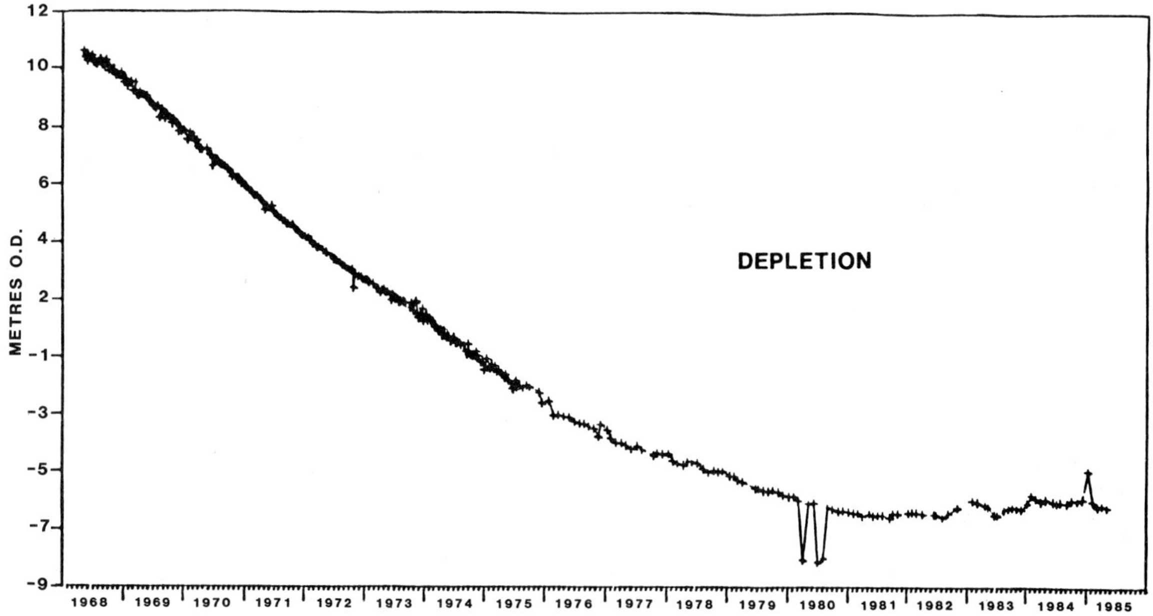

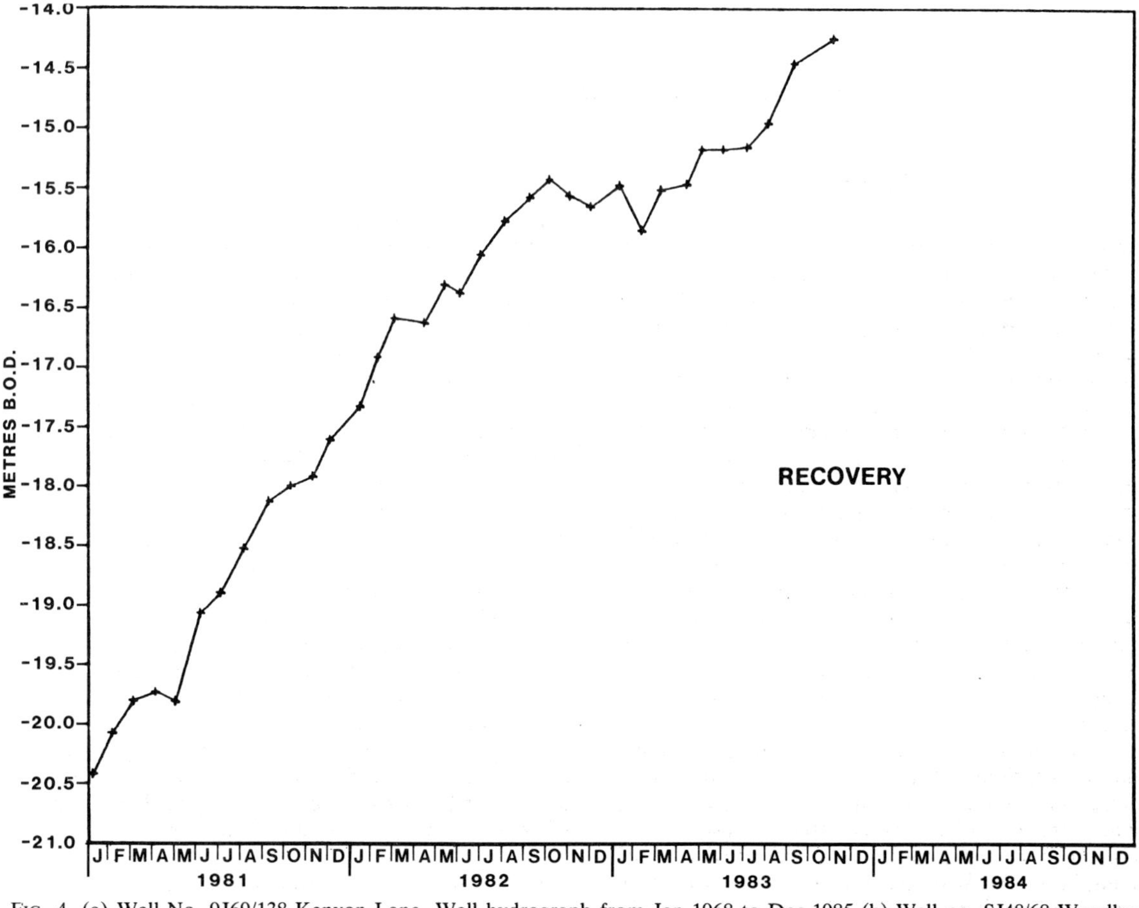

FIG. 4. (a) Well No. 9J69/138 Kenyon Lane. Well hydrograph from Jan 1968 to Dec 1985 (b) Well no. SJ48/68 Woodlands CST24301 133LN30. 48LN: 30.4 Well hydrograph from Jan 1981 to Dec 1984.

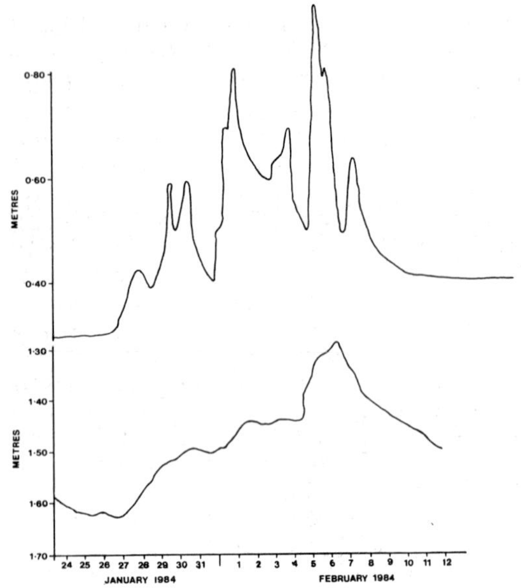

FIG. 5. River level of Broomstair Bridge 1 km upstream of site on River Tame. Groundwater level in pumping test borehole No. 1.

two separate pieces of information required; the first is the groundwater level or pressure within the aquifer and the second is the quantity of water likely to be encountered. The more permeable aquifers, which include sands and gravels and fissued rocks, govern the quantity of water which may enter an excavation, while the lower permeability horizons, particularly the sandy silts, silts and clays, often develop high water pressures. While the quantities of water are small due to the low permeability, any excess piezometric heads may give rise to stability problems. In both cases permeability and/or water pressure are required.

Know your aquifer

It is essential during the investigation of an area to identify the individual elements of the aquifer and to observe the response of each element. Permeable sands and gravels, or rock aquifers are readily identified during the initial boring and instrumentation installed. This usually takes the form of standpipe piezometers or tube wells in the sands

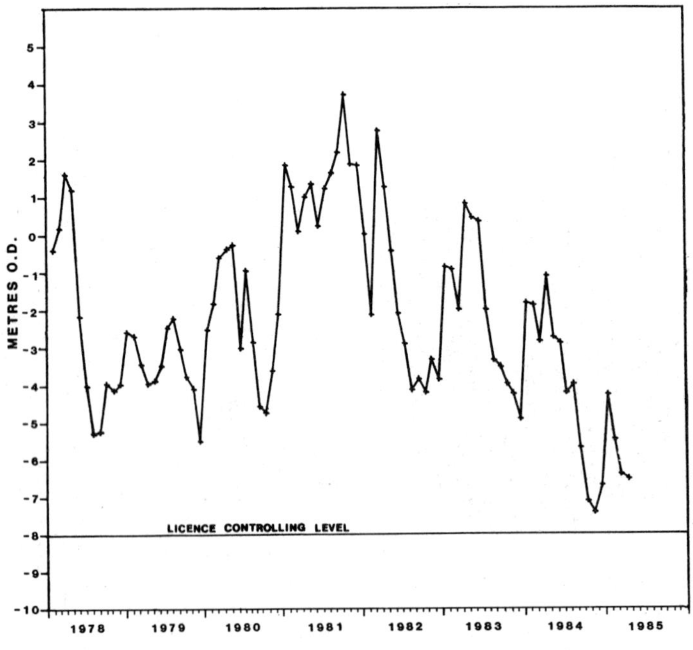

FIG. 6. Well No. SD17/5, Ormsgill. Well hydrograph from Jan 1978 to Dec 1985.

and gravels, or open boreholes in the rock aquifers. The problem with the unreliability of water level readings in open boreholes in fissured aquifers has been discussed by Rushton & Howard (1982) and by Walthall & Ingram (1984) for the Permo-Triassic Sandstone at Kenyon Junction where the different layers within the aquifer were only identified after the installation of multiple piezometers. The presence of different piezometric heads within what at first sight appears to be a uniform aquifer has also been demonstrated by the use of borehole packers (Brassington 1985) and also by a combined multiple packer/piezometer installation. The effects are important in the interpretation of the response of the aquifer to any development, and in particular to interpreting whether it will show an unconfined or confined response. The presence of low permeability horizons across which there is a hydraulic gradient will also give clues to the behaviour of the aquifer. A dewatering test carried out to measure the permeability of peats gave anomalous results in one of six observation piezometers due to the response zone being in direct contact with a high permeability zone into which some of the abstraction wellpoints had been jetted. The use of orthogonal lines of observation points during pumping tests to investigate aquifer anisotropy is more likely to detect local changes than regional variations. In particular, in certain aquifers direct connection through fissues or very high permeability bands may give higher drawdown at more remote locations during pumping tests. It is therefore very important that the hydraulics and the hydrogeology of the groundwater regime are understood before placing great emphasis on the interpretation of any groundwater level measurements.

Instrumentation

The basic elements of different types of water level observation points have been reviewed by Barr (1977) whilst their response times were investigated by Hvorslev (1951). It is worth noting from Hvorslev's work that for highly permeable aquifers i.e. $k > 10^{-3}$ m/s (85 m/day) open observation wells of 150 mm diameter will give a basic time lag as defined by Hvorslev of less than 20 s whilst for permeabilities of less than 10^{-4} m/s (8 m/day) the same observation well could be unsuitable for monitoring the rapid changes in groundwater level during the early stages of a pumping test in a confined aquifer. For lower permeability materials $k = 10^{-3}$ to 10^{-5} m/s (85–0.8 m/day) then a standpipe piezometer with a dip tube of 20 mm diameter will give a basic time lag of the order of 10 s. In materials with a permeability of less than 10^{-6} m/sec (0.08 m/day) standpipe piezometers are no longer sufficiently responsive for the observation of short-term groundwater level changes and null displacement pressure measuring devices should be used. For long-term groundwater level observations much less responsive water level monitoring points may be used and the range over which a particular type of instrument can be used is extended considerably. For monitoring over a period of time in all but the lowest permeability materials conventional standpipe piezometers have been found to be adequate when used in conjunction with 25 mm, 19 mm and 12 mm rising pipes. These have the advantage that they can be cleaned out by air/water jetting should they become blocked and that it is now possible to instal both remote reading and continuous recording instruments at a later date. The installation of water level monitoring equipment has received insufficient attention in the field to date. Often the installation is left to the drilling foreman and very little time allowed for the correct placing of the filter section and the seals. This problem is magnified when more than one piezometer is installed in a single borehole particularly with increasing depth. It is imperative when placing materials below water in a borehole that a tremmie pipe is used to ensure their correct location. This includes filter sands, bentonite seals and any bulk filling between the sections. The time taken for this operation, including any necessary settling time can be long and it may be necessary to take a full shift to place a piezometer below 30 m. The installation at Kenyon Junction, where seven piezometers were placed in a 100 m deep borehole, required a total of six working days spread over three weeks. Under normal circumstances it is likely that it will be more cost effective to instal piezometers in individual boreholes, particularly at shallow depth.

Frequency of measurement

Whilst water levels are normally read with an electric dipper and a calibrated tape capable of reading to an accuracy of 10 mm, the frequency of reading is even more important. Whereas once it would have been almost impossible to collect and process say, hourly readings from ten sites for twelve months, now with the advent of the ubiquitous computer such intensive data collection and processing is possible. It can be seen from the first part of this paper that the reasons for groundwater level changes can be many and varied and that these must be identified as far as possible for each site. The only way to achieve this aim is by the use of continuous recording equipment on one or more observation points at a site, coupled with a series of readings at less frequent intervals on all the other observation points. This technique has been successfully used both in regional aquifers and on small sites. For example, the Permo-Triassic Sandstone aquifer of the Eden Valley in Cumbria has a total of 12 observation boreholes measured on a monthly basis supplemented by two continuous chart recorders. A new sewage treatment works had a total of eight piezometers and one observation borehole with a continuous chart recorder during the site investigation. In both cases it is possible to observe the daily, and seasonal changes, whilst being able to determine the shape of the groundwater surface in the area. In the case of the sewage works this was found to be very closely linked to the adjacent river both in the change in level across the site and in the rate of change.

Conclusions

It is important, as with any physical measurement, that the level to be measured is understood and that the measuring device is appropriate for the project in question. For aquifers this means that the construction of the water level observation point must be carefully designed to monitor the correct horizons. To observe the true effects of climate, both seasonal and daily, and to detect man-made effects it is necessary to have continuous readings at a given site. However, this is not necessary for all the boreholes within a group, provided that frequent readings are made on the remaining boreholes for the local pattern to emerge. In practice once a month is adequate for most applications and may even be stretched to six monthly readings for long-term effects within an aquifer. On a small restricted area, such as a site investigation, weekly readings may be required to establish the site relationships before increasing the interval to once a month. In aquifer resource studies this will lead to a true picture of the resource whilst in engineering the statement that "water levels may be different from those encountered at the time of the investigation" can be replaced with a statement containing numerical values for the maximum and minimum levels, together with the likely effects and response time to climactic and other changes.

ACKNOWLEDGEMENTS: The Authors are grateful to the General Manager and Divisional Engineer of the Rivers Division and the Design and Construction Manager (East) of North West Water for permission to publish this paper. The views expressed are those of the Authors and not necessarily those of the Authority.

References

BARR, M.V., 1977. Downhole instrumentation - a review for tunnelling ground investigation. *CIRIA Report No. 90.*

BRASSINGTON, F.C., WALTHALL, S., 1985. Field Techniques using borehole packers in hydrogeological investigation. *Quarterly Journal of Engineering Geology*, **28**, 181-195.

HVORSLEV, H.J., 1951. Time lag and topsoil permeability in groundwater observations. *Bulletin 36 of the Corps of Engineers. US Army, Vicksburg, Mississippi.*

INGRAM, J.A., 1978. The Permo-Triassic Sandstone aquifers of North Cumbria. *North West Water Authority internal hydrogeological report No. 19.*

RUSHTON, R.R. & HOWARD, K.W.F., 1982. The unreliability of open observation boreholes in unconfined aquifer pumping tests. *Groundwater* **20,** 346-350.

WALTHALL, S. & INGRAM, J.A., 1984. The investigation of aquifer parameters using multiple piezometers. *Groundwater,* **22,** 25-30.

WEEKS, E.P., 1979 Barometric fluctuations in wells tapping deep unconfined aquifers. *Water Resources Research* **15,** 1167-1173.

M.D. EGGBORO, North West Water, Rivers Division, P.O. Box 12, New Town House, Buttermarket Street, Warrington, WA1 2QG, UK.

S. WALTHALL, North West Water, Design & Construction Unit (East), Oakland House, 1 Talbot Road, Old Trafford, Manchester M16 0QF, UK.

Groundwater in Engineering Geology, London, 1986, pp. 403-415 Printed in England

Surveillance of pore water conditions in large urban slopes

D. P. McNicholl & G.W.F. Cho

ABSTRACT: Landslides in Hong Kong are almost always associated with periods of heavy rainfall. Groundwater has a major influence on the strength of soil and the stability of slopes. In the time preceding failure, the pore water pressure can change from being strongly negative to become increasingly positive. Hence the prediction of groundwater levels under critical design conditions is important in the assessment of slope stability.

If it is considered that a slope has an inadequate margin of safety and that the structures above and below the slope should be evacuated or restricted the advice given can often be over-cautious or inappropriate.

A rational approach is suggested to help engineers carry out surveillance of pore water conditions and to give advice on managing urban slopes with low factors of safety. Surveillance implies the collection of information, analysis and initiation of actions. A range of subjects relating to the automatic monitoring and surveillance of hazardous slopes is briefly covered. Emphasis has been placed on situations where it is essential for professionals to react quickly and decisively to changes in the surveillance case. A case study is included to demonstrate the suggested approach.

Introduction

Investigating engineers sometimes advise that a large urban slope has an inadequate margin of safety and that structures above or below the slope should be evacuated or restricted because of the danger of a slip. This advice can often be over-cautious or inappropriate. Many cases encountered are more properly described as lacking the required margin of safety. For instance, if normal practice (defined in Hong Kong by the Geotechnical Control Office 1984) requires a factor of safety of 1.4 and the calculated factor of safety is 1.3 the slope should not be described as 'dangerous' and indeed evacuation and/or clearance would probably be unnecessary. On the other hand there are cases where there is a real risk to life and to property so prompt action must be taken. The engineer must therefore rigourously examine the case to ascertain for himself the basis and correctness of the initial analysis. He should discharge his duty of care by examining what other steps should be taken to improve the situation and he should institute surveillance.

In this paper the word 'surveillance' is defined as the close supervision of a slope with low factor of safety and includes the prompting of any necessary action by the responsible engineer. Surveillance implies the reception of information, analysis and initiation of action.

The word 'monitoring' is defined as observation and reporting back on the status of an individual parameter.

A 'surveillance threshold' is a predetermined bench mark within a range of monitoring results at which the engineer will recommend that action should be taken.

For a large owner of land or property such as the Hong Kong Housing Authority, there are many reasons for initiating surveillance of a slope.

(i) After a failure in a slope, all similar slopes built to the same standard should be checked to determine their condition and probability of failure.

(ii) The engineer may need to exercise proper care for a slope which has been identified as needing reconstruction or repair but is awaiting completion of design and financial approvals.

(iii) It may be necessary to confirm the validity of design assumptions and to monitor the effect of early construction activities.

Monitoring

Choice of parameters for monitoring

A properly designed surveillance system should be based on the monitoring of one or two parameters which must be directly related to the failure process. Ensuing paragraphs give examples of typical approaches to monitoring.

Conventional surveying

Conventional surveys are useful where the expected type of failure is of a slow moving type — perhaps a deep seated rotational circular slip. In that case it is often comparatively easy to set up a base line or laser

From CRIPPS, J.C., BELL, F.G. & CULSHAW, M.G., (eds), 1986, *Groundwater in Engineering Geology,* Geological Society Engineering Geology Special Publication No. 3, pp. 403-415

alignment which can read relative movements in carefully aligned monuments (Fig. 1).

Level surveys are useful, particularly on previously plane surfaces such as flexible roads where early movement shows up. Measurements of crack widths and tell-tales also have their uses.

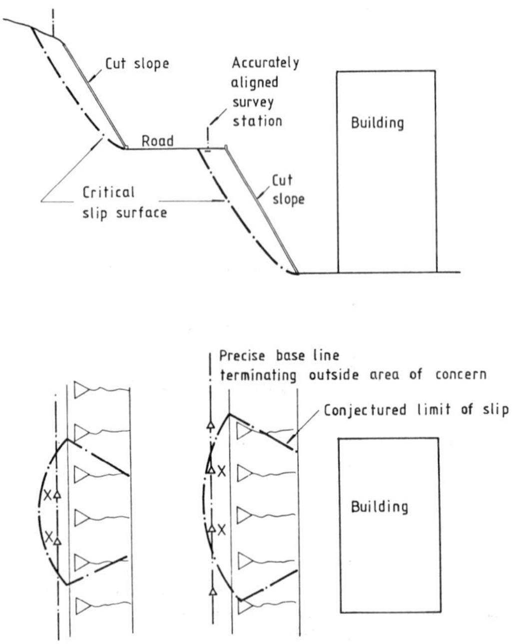

FIG. 1. Monitoring by conventional surveying.

Notes: 1. If slip is initiated or moves further, then stations XX will move out of alignment.
2. Used in conjunction with measurement of crack width and precise leveling.

Vertically installed sensors
Inclinometers are appropriate in some circumstances where a deviation from the vertical may indicate that a movement has taken place. Reduction of readings, however, can be tedious and must be office based. Another vertically installed sensor can be contrived out of dowels or bamboo rods placed in drillholes whereby a slip would disturb the rod. The problem is that once a shear has broken or jammed the rod there is no way of knowing the continuing progress and extent of the slip.

Another simple device known as a slip indicator is also widely used in many forms. This instrument is fairly crude and will not give any indication of the magnitude, rate or direction of movement.

Extensometers have been used in some cases for monitoring slope movements.

Inspection
Slope inspection has a role to play and wherever possible, final decisions should be accompanied by inspection.

The main purposes of inspection are:
(a) to prompt maintenance and urgent repairs;
(b) to look for signs of distress.

Limitations of monitoring system so far described

The above methods are less useful in non-circular slip situations or in soils of low relative density which might collapse on failure and become liquid. In both these types of failures the event occurs rapidly without much warning from physical signs.

Another problem is that failures often occur during times of heavy rainfall when the engineer may not be able to get to site or at night when inspection is restricted.

These arguments point towards the monitoring of a parameter directly related to the cause of failure and not the symptoms.

Monitoring primary parameters of failure

Landslides in Hong Kong are almost always associated with periods of heavy rainfall. Groundwater has a major influence on the strength of soil and on the stability of slopes. The prediction of groundwater levels under critical design conditions is important in the assessment of slope stability. Relevant papers have been published by Endicott (1982), Rodin *et al.* (1982) and Leach & Herbert (1982). Reference can also be made to the Mid-levels Study Report (1982).

The precise links between RAINFALL → INFILTRATION → RISE IN PORE WATER PRESSURE → CHANGE IN SHEAR STRENGTH → FAILURES are not well understood, particularly for unsaturated residual soils (Brand 1982; Brand & Hudson 1982).

For most slope configurations, the calculated factor of safety is most sensitive to minor changes in pore water pressure (u) and effective cohesion (c').

Rainfall monitoring
Rainfall is easy to monitor automatically and is the prime cause of failure. The use of rainfall readings as a predictive tool is, however, in its research phase . It would be necessary to carry out considerable research over several wet seasons in order to link rainfall empirically to the probability of failure at any one site. Some work in this field has been done by Lumb (1962, 1965), Koo & Lumb (1981) and Anderson & Kneale (1980a).

Choice between u and c'

In very crude terms c' would remain more or less constant during the period in which pore pressure reaches a limiting value. It is also difficult, if not impossible, to monitor c' and the most important parameter which is capable of being monitored continuously is pore water pressure (u). This is discussed in more detail later.

Measurement of pore pressure

Pore pressure range

In the time preceding failure, the pore water pressure can change from being stongly negative (pore suction) to become positive (pore pressure). Pore suction can enhance shear strength by increasing effective stress while pore pressure reduces effective stress and hence shear strength.

The range of interest for pore pressure measurement should cover the change in pore pressure caused by infiltration and/or rise in base water table level and should extend from about –60 kPa to +100 kPa depending on circumstances and slope geometry.

Pore suction

There are not many commercially available methods of measuring pore suction. Psychrometers may be used to monitor high values of soil suction but are generally not reliable in the range of most interest to the geotechnical engineer. The most common approach is to use tensiometers. Tensiometers measure directly the negative capillary pressure of soil moisture. They consist of a porous, ceramic cup filled with water and connected by a water filled tube to a pressure gauge (Fig. 2).

When the soil re-wets, water moves out of the soil into the porous pot and pore suction decreases.

Burt (1978) described a tensiometer for shallow depths using a mercury manometer within the water column as a measuring device (Fig. 2). Both positive and negative readings can be registered to ±8 m of water (probably less in practice).

Similar sensitive research based tensiometers have been installed by Anderson & Burt (1977) and Anderson & Kneale (1980b) and in a recent study in Hong Kong, agricultural tensiometers were used in an unfavourable and tropical environment.

Errors have been shown to be ± 0.05 m of water and response time is good.

Agricultural tensiometers are very robust (Jet fill tensiometers Messrs. Soil Moisture Equipment) and respond very rapidly on installation. They are of immediate use in getting a feel for the subsurface suctions and are very convincing, but need frequent adjustment and topping up in dry weather.

The disadvantage of tensiometers is their limited length (max. 2.5 m) which makes it difficult to install them close to a predicted slip surface unless they are installed horizontally in shafts.

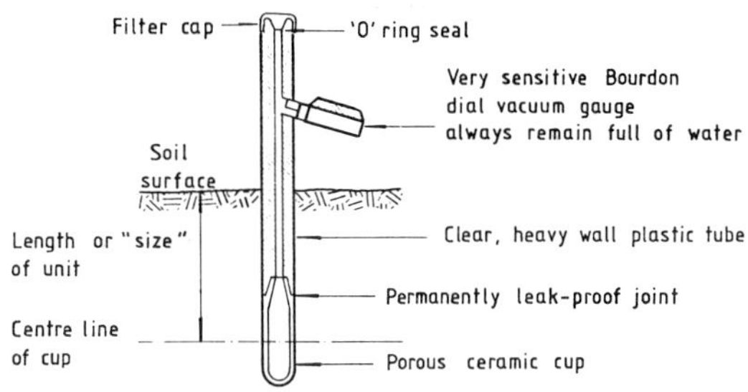

(a)

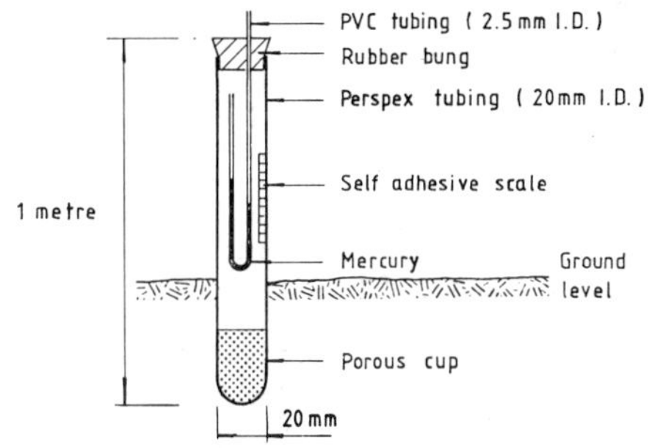

(b)

FIG. 2. Simple tensiometers.
(a) Soil water tensiometer for field use (agricultural tensiometer) (Soil Moisture Equipment Co.);
(b) new tensiometer (Burt 1978)

Pore pressure

Most engineers are familiar with the use of piezometers to measure pore pressure and a good description is to be found in Hoek & Bray (1981). It is usual to install standpipe piezometers (Fig. 3) for manual observations or alternatively, closed hydraulic piezometers, air actuated piezometers or electrically indicating piezometers. A recent technique is the use of an air bubbling system in a standpipe piezometer. This is not as precise or sensitive as a hydraulic piezometer but is very robust and obviates frequent de-airing problems with hydraulic leads.

Only hydraulic piezometers can cater for negative pore pressures but if the piezometer dries out completely or operates for prolonged periods at negative pressures, then considerable problems may be experienced with re-establishing the system and in bleeding of air bubbles.

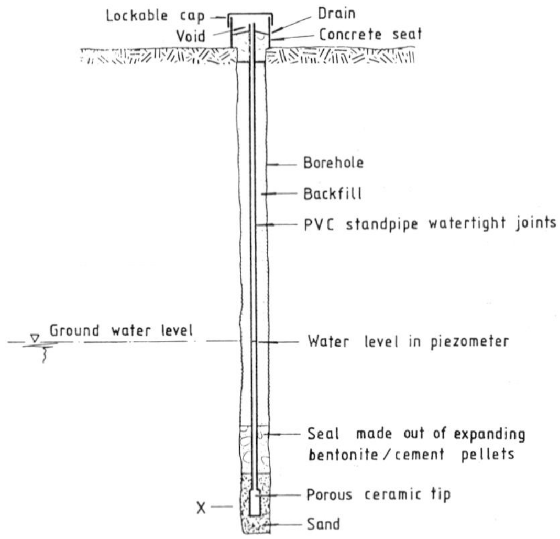

FIG. 3. Simple standpipe piezometer

Notes: 1. Groundwater level and hence p.w.p. at point X is reflected in depth of water in standpipe tube.
2. In some soils long response time is required.

Effect of topography

If there is a hill slope hollow behind the cut slope under consideration, then there is a possibility of saturated conditions or perched water conditions persisting over long periods (Anderson *et al.* 1983).

Automatic systems

Limitations of manual monitoring

It is possible to get reasonable results by regular manual monitoring. Access is of course important and methods of payment should encourage the labour employed to carry out the work conscientiously.

Whilst manual monitoring works well for regular routing purposes, it has several disadvantages for surveillance purposes (Insley & McNicholl 1982).

Even if good access and stairways are provided, problems arise when readings are needed in heavy rain, when berm surfaces are slippery and particularly during typhoons when there are added risks from strong winds and flying debris. It becomes practically impossible to read all piezometers quickly enough under bad weather conditions when most critical peak readings are anticipated. It is possible that the build up to a high perched water table could be missed.

A maximum water level measuring device such as a piezometer standpipe bucket (Halcrow Bucket, British Patent No. 1538487) is useful but does not register intermediate water levels during a storm.

Automatic monitoring

The main purposes of automatic monitoring are:

(a) to highlight peak readings which might be missed by manual monitoring,
(b) to monitor more frequently than would be the case with manual monitoring thus improving slope surveillance,
(c) to build up a well documented record of ground water level fluctuations as a data bank to assist similar cases.

The main requirements are:

(a) the system should be robust and unaffected by weather,
(b) the system can operate through the night,
(c) the system can be left unattended,
(d) readings can be taken at pre-set intervals and stored.

Choice of available systems

Ideally the instruments used for monitoring should require little or no water volume charge at the sensing tip thus assisting instantaneous response to pore pressure changes (Klute & Peters 1969). Early automatic systems used a strain hydraulic fluid system. The link from transducer to each pipeline was through a manifold. Problems were encountered in switching between manifold branches and the whole system had to be kept full of water and de-aired.

Scanivalve system

A better system known as a Scanivalve was developed using a rotary manifold which operated as a fluid switch by rotating a pressure transducer from one connection to another around a ring as shown in Figs 4 and 5 (Burt 1978; Brand 1982). This system has been well adapted to the sensitive differential tensiometer system and has been used successfully for automatically monitoring pore pressures in field conditions (Anderson & Burt 1977; Anderson & Kneale 1980b). Several systems have also been installed in Hong Kong and are the subject of current publications (Brand 1982; Pope *et al.* 1982).

The Scanivalve system uses either water or air as the fluid medium joining the piezometer tip to the valve. Water filled systems give rise to significant de-airing problems and for recent installations an air filled system has been used (Insley & McNicholls 1982; Pope *et al.* 1982). This system, known as the 'bubbler' system, uses compressed air from a dip tube, which is suspended within a piezometer standpipe at a known distance up from the tip. The back pressure developed in this tube is dependent on the depth of water in the standpipe above the base of the dip tube. As the water level varies, the back pressure changes and is measured by means of a single pressure transducer which can 'read' all the drill holes grouped around the Scanivalve station.

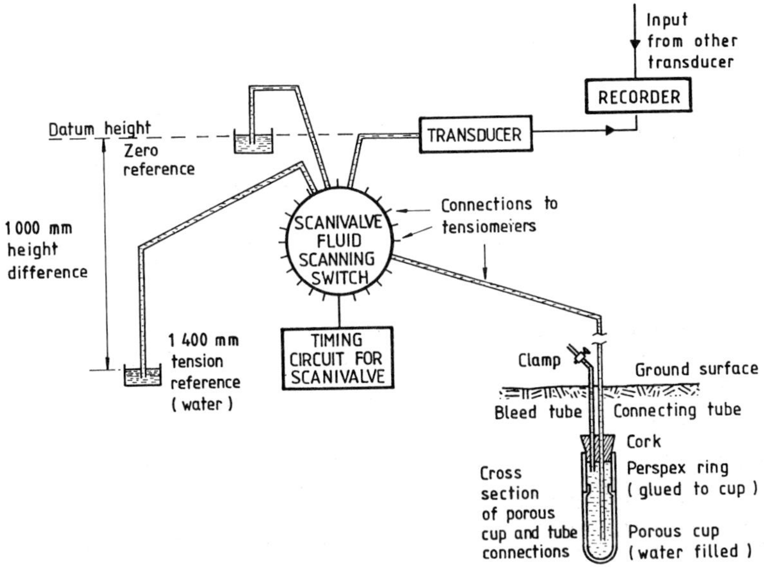

FIG. 4. Scanivalve tensiometer system (Burt 1978).

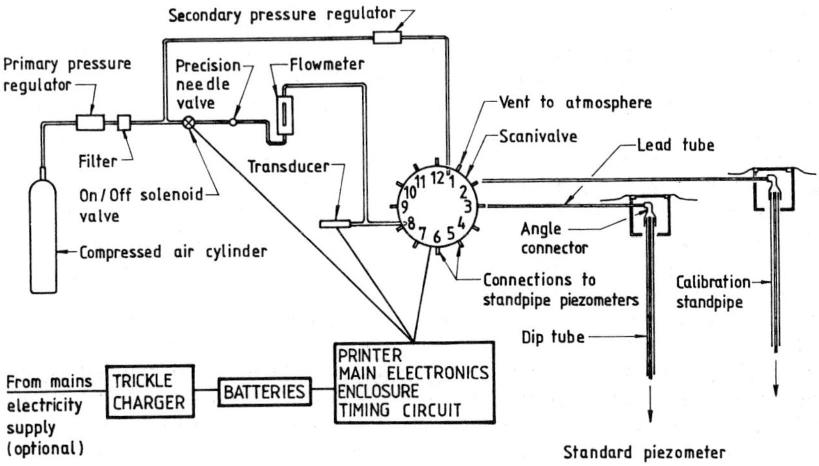

FIG. 5. Schematic diagram of the bubbler system (Pope *et al.* 1982).

This system is simple and robust but relies on a series of vulnerable air tubes with risk of organic growth in the tubes (can be prevented by the use of Nylon 11 for the bubbler tubes) and possibly vandalism. Furthermore, line friction varies with back pressure in line, but can be easily calibrated for a given line type, diameter and length.

Individual transducer system

An alternative system is one which uses electricity as the medium connecting the piezometer tip to recording apparatus. Some types have an electrically monitored diaphragm whilst a recent alternative is to use a conventional piezometer tip which fills with water and then to use mini transducers suspended down each drillhole standpipe. The latter system is said to be intrinsically more robust and has no moving parts. Transducers can be replaced easily and there is no practicable limit to radius of operation providing signals are amplified. Many practicing engineers, however, are somewhat wary about a totally electrical system in exposed, humid and environmentally aggressive situations.

The tried bubble technique has been used more frequently in some recent installations.

Recording of monitoring information

The monitoring information can be recorded conveniently on paper tape (using a converted print set up from a pocket calculator) or, alternatively, transcribed on to magnetic tape using standard tape cartridges/cassettes. Either alternative has advantages. For research purposes the magnetic tape is better. For emergency situations the print read out has instant advantages.

Application of monitoring results

Most critical slip surface

It is necessary to estimate the most critical potential slip surface which will be a function, *inter alia,* of the slope geometry, soil horizons, phreatic level(s) and material parameters.

In cohesive soils, deep seated failure surfaces tend to be the most critical. Frictional soils, particularly saturated residual soils, fail on shallow near planar surfaces which are curved at the transverse limits of the slide (Lumb 1975).

Selection of a most critical slip surface may be done, first, by inspection, taking into account the subsurface geology, and then using a computer suite which generates successive failure surface for a set geometry, material parameters and pore pressure conditions. There are then two alternative approaches to setting a critical pore water pressure threshold.

Stability charts

Having established the most critical slip surface, stability charts can then be compiled by repeatedly calculating the slip surface's factor of safety for a range of water tables and hence pore pressures. This is illustrated in Fig. 6 and the results can be plotted on a graph of the form illustrated in Figs 7 and 8.

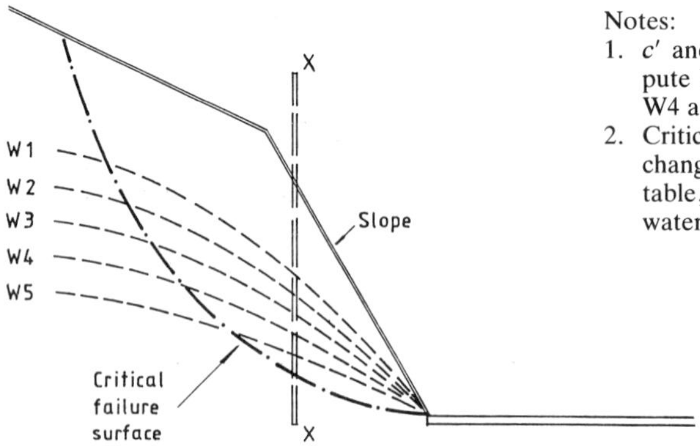

Notes:
1. c' and ϕ' held constant, compute F.O.S. for W1, W2, W3, W4 and W5.
2. Critical failure surface may change depending on water table, particularly perched water table.

FIG. 6. Sketch showing application of a range of water tables to a critical failure surface.

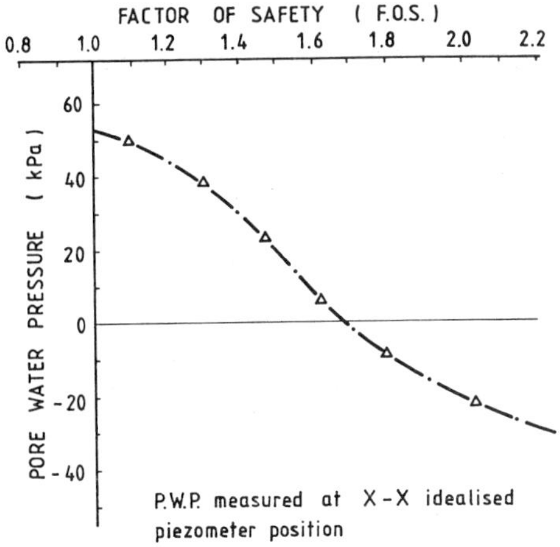

FIG. 7. Sketch showing relationship between F.O.S. and P.W.P. (idealised).

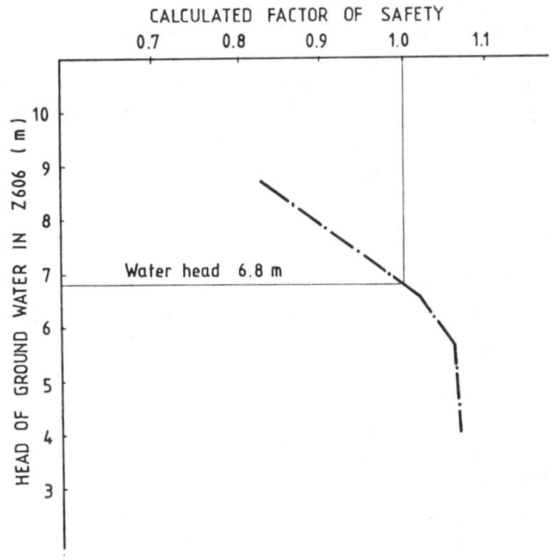

FIG. 8. Monitoring and surveillance of slopes (this sketch is adapted from a case history).

Effective stress and failure envelope approach

In this method, pore pressure is related to effective stress conditions on the most critical slip surface.

A typical slice is considered near the toe of the failure surface or close to an array of piezometers. The failure mechanism is that of rainfall causing infiltration and/or a rise in perched water table which in turn leads to an increase in pore pressure. This results in a decrease in effective stress on the potential failure surface which is reflected in a decrease in shear strength to a point where equilibrium cannot be maintained (Brand 1982). Failure takes place therefore under conditions of almost constant total stress and increasing pore pressure.

Fig. 9 shows the stress path to failure of an element of soil in the critical slip surface. The change in pore pressure from antecedent pressure to pressure at failure can then be related to measured rise in the piezometer readings.

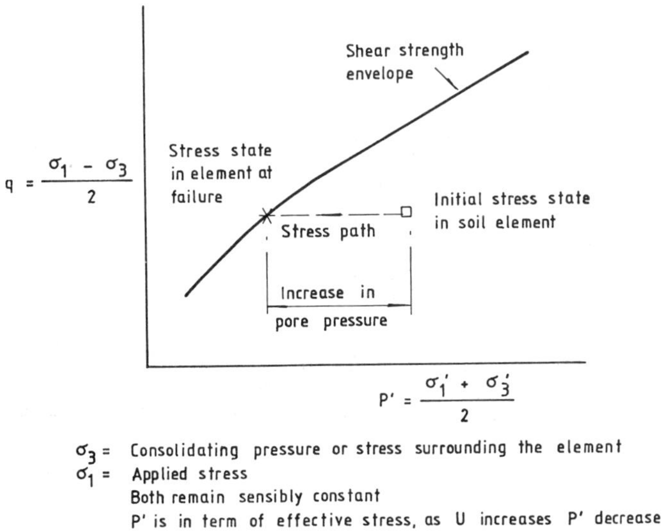

FIG. 9. Stress path to failure of a soil element on a critical slip surface.

Additional records

Whichever approach is taken, it is a good idea to plot pore pressure against time during and after a storm to a suitable scale that will give the engineer a feel for the changes which are taking place.

It is also helpful to plot the derivative of pore water pressure against time.

Hourly rainfall readings should be collected automatically or by telephone from the nearest available rain gauge.

This information helps the engineer to make his judgement. As an example of this in a typical case, it may be found that pore pressure begins to rise about two hours after the onset of a severe tropical storm. In this case even though the critical pore water pressure may be reached it may be possible to defer action because:

(a) rainfall has ceased, and
(b) the rate of change of pore pressure with time has reduced (Fig. 10).

Additional tools

Probabilistic method of analysis
Considerable insight into the uncertainties involved in the design and checking of slope stability can be gained by the application of probability theory, which requires that the mean of a soil parameter be estimated in the form of a probability function.

Normally there will rarely be sufficient data to justify a full probabilistic analysis. But it will always be possible to examine the reliability of stability analysis in the light of engineering judgement. Some work in this field has been done by Lumb (1974).

Sensitivity analysis
As an alternative to accounting for variability of the soil forming the slope by the statistical or probabilistic approach described above, the slope stability analysis may be repeated using differing values for c' and ϕ' and different groundwater conditions. the effect on factor of safety as a result of variations in these

t (hr)	u (m)	Δt	Δu	Δu/Δt
16·00	4·40			
		0·50	−0·10	−0·2
16·30	4·30			
		0·50	+0·50	+1
17·00	4·80			
		0·50	+0·15	+0·3
17·30	4·95			
		0·50	+0·50	+1
18·00	5·45			
		0·50	+0·20	+0·4
18·30	5·65			
		0·25	+0·10	+0·4
18·45	5·75			
		0·25	+0·05	+0·2
19·00	5·80			
		0·25	+0·05	+0·2
19·15	5·85			
		0·25	+0·05	+0·2
19·30	5·90			
		0·25	0·25	−1
19·45	5·65			
		0·25	−0·35	−1·4
20·00	5·35			
		0·50	−0·35	−0·7
20·30	5·00			

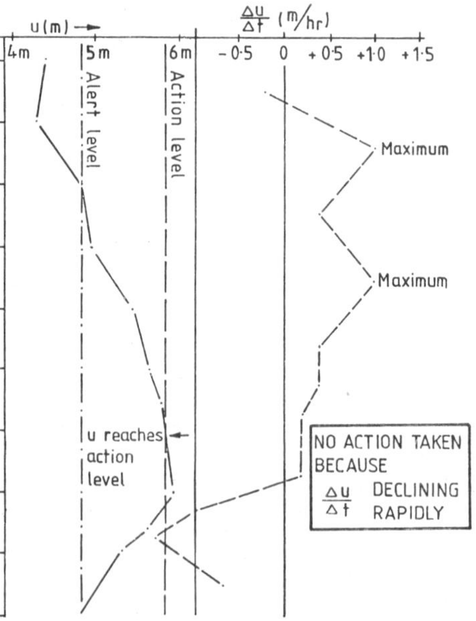

FIG. 10. Example of monitoring plot (reconstructed from case notes).

Notes: 1. Figures have been modified for simplicity;
2. Rainfall figures not shown but rainfall peaked at 16.00 hr and stopped at about 19.00 hr.

these parameters can then be assessed and, if considered necessary, more or less conservative criteria may be adopted for surveillance based upon the frequency distribution of the data obtained during testing and observation.

Various publications by Hoek (1980) and Hoek & Bray (1981) exemplify the approach of grouping different parameters.

Final Review

In situations where critical levels are reached or expected, the engineer should:

(i) arrange to have a second or third opinion available, or

(ii) call for a group discussion and review.

This should result in a corporate engineering judgement that is based on several views of the original assessment.

A site inspection at the start of this final review is mandatory.

Duty of care

Apart for surveillance, it is of the utmost importance that the engineer should continually look for simple steps which will improve the situation. The proposed course of action must be continually reviewed whilst on site to seek ways of improving the situation.

For example, the stability of slopes can, in many cases, be improved cheaply and within 24 hours by the installation of horizontal drains (McNicholl 1984), repairs to channels, slope surface protection and other means which can be carried out quickly. A good temporary measure that is often taken is to prevent water from discharging onto a slope by diverting it, perhaps with the use of sandbags. Measures such as provision of a fence or barrier to direct the public away from the area of risk are equally important.

Surveillance principles

Threshold

The engineer should give serious consideration to the thresholds and type of surveillance to be instituted. It must be stressed that the surveillance system is only a tool and the final decision must rely heavily on the experience and judgement of a responsible engineer.

If thresholds are set too low, then the engineer will be 'on the safe side' in recommending any necessary action. But he must remember that a 'cry wolf' situation will lead to a loss of credibility and future recommendations of evacuation may well be ignored. He should bear in mind the possible anxiety, inconvenience and even dangers which the alarm or evacuation may result in.

On the other hand, if the thresholds are set too high then there is a possibility of an incident occurring before the engineer has had the chance to initiate action.

The preferred approach is to set up a rigourous surveillance programme with thresholds a little on the conservative side.

It is usually advisable to set a Standby Alert level and an Evacuation Alert level, both are dependent on many factors, the foremost of which is the time required to achieve standby readiness and to effect evacuation.

The engineer should gather as much information as possible to assist him in making his decisions and all associated analysis should be done in such a way as to rule out, as much as possible, sources of error or ignorance.

Consequences

Consideration must also be given to the action to be taken should the surveillance results be adverse. An estimate should be made of the possible modes, extent and consequences of failure (Fig 11).

Plans must be well thought out where adverse conditions will result in evacuation of property.

The action for each stage should be predetermined and well understood by all concerned.

As reconstruction of the slopes progress the critical groundwater regime may also alter. Re-analysis of the existing current profile should be carried out during periods of surveillance if necessary.

The probability of a failure is normally reduced as construction work progresses.

Case history

One example has been chosen from a number of case histories. The following sections are abstracted from a paper by Insley & McNicholl published in the Proceedings of the 7th Southeast Asian Geotechnical Conference, Hong Kong. November 1982.

The site

The site was formed in the 1960s by excavation into the Kowloon foothills of a major ridge which attains a maximum level of 500 m above sea level. (Figs 12 and 13).

At the rear of the site, which has been used for housing development (total population 120,000), is a large cut slope 800 m long and 30 to 50 m high. Several failures of significant volume occurred in 1972 and 1973.

Geology and Hydrology

The slope is formed in completely decomposed granite (Cheung Chau Granite corresponding to Jurassic age). At the crest of the cut slope where failures are most likely to occur (Lumb 1975) there is an almost total cover of colluvium about half of which has been classified as showing evidence of inherent instability. During the summer wet season there is a rapid outflow from the upper face of the slope from

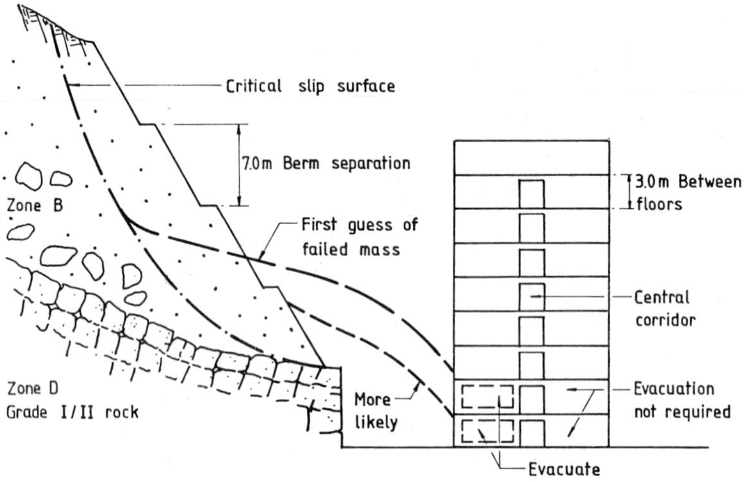

FIG. 11. Estimate consequences of failure.

Notes: 1. Estimation of failure volumes refer to
(a) Lumb (1975) - Table II,
(b) case history,
or (c) analysis.

2. Evacuation criteria:
(a) structural strength of building — whether the whole building is likely to be affected by a failure or only part of building,
(b) possible loss of lateral support and undermining,
(c) type of failure — liquefaction of a fill slope could have a more severe effect than a cut slope.
(d) Any catchment area — e.g. a playground between the building and the slope.
(e) Number of flats or persons involved with respect to time taken to complete evacuation.
(f) Number of existing paths available for evacuation — are some of these potentially in the line of a landslip.
(g) Probable failure volumes.

'perched' water flowing within the cobble and boulder beds in the colluvium. There is a base water table within the decomposed granite.

In the crest area the colluvium has an angle less than 20° but this steepens upslope to 30° or steeper.

Preventive works

The factors of safety against failure of the existing residual soil cut slopes are less than the required standards, under high groundwater conditions in a perched water table in colluvium, or, in the case where a high groundwater table builds up in the

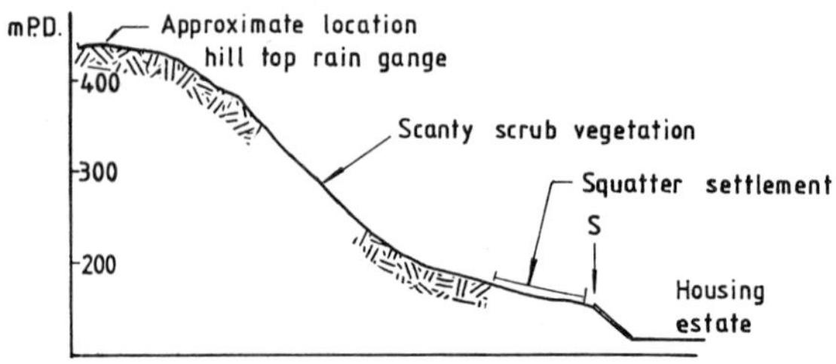

FIG. 12. Cross section from top of ridge to site of case history.

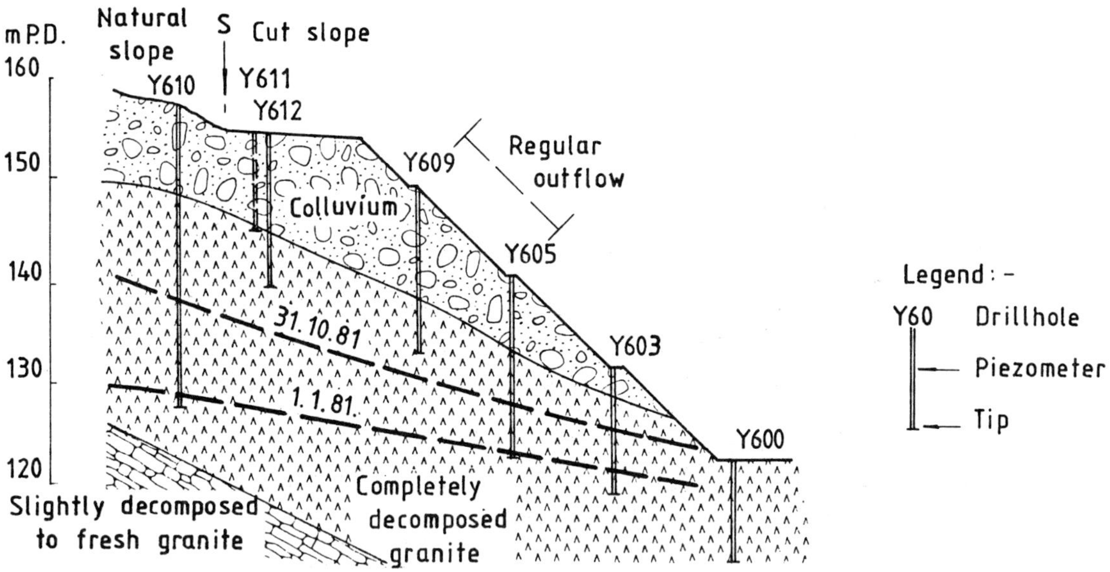

FIG. 13. Case history: section showing geology in outline and position of piezometers.

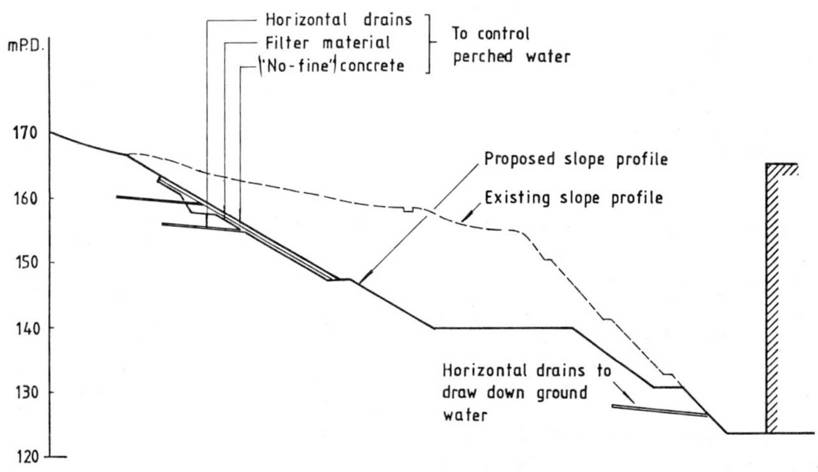

FIG. 14. Case history: cross section with major excavation.

decomposed granite. This can rise by as much as 10 m in a typical summer wet season.

The slopes are being flattened by extensive earthworks (330,000 m^3) in order to improve their stability to current standards (Fig 14). To reduce ground water level to control seepages, horizontal drains and other drainage measures are being installed.

During the final design and early construction period automatic monitoring was for surveillance purposes and for confirming the design parameters employed.

System chosen

Many piezometers were already installed and had been read previously by manual means. For the current exercise three of the steepest cross sections close to buildings were selected for surveillance. Existing piezometers were reinforced by new ones so that there was an array of about 10 piezometers on each of the three sections.

The 'bubbler' system was chosen on a combination of technical, financial and performance grounds.

An air operated Scanivalve was used and a system developed by Messrs. Geotechnical Instruments Ltd., UK and installed in Hong Kong by Omtis Ltd.

A monitoring station was established in a lockable gauge house on each section, each controlled by a microprocesser which was to take readings continuously with the option of a manual over ride. Each transducer was connected to 10 individual piezometers. The Scanivalve fluid switch was used to connect the air supply and transducer to each standpipe in turn. The tubes to individual piezometers were laid in buried PVC ducts which were sealed against vermin and surface boxes locked using an Allen key and stud bolt.

Results

The system worked reliably for about one and a half wet seasons until construction rendered each station obsolete in turn. Fig. 15 shows a typical storm response with a 2.5 m rise in water level in the colluvium over a period of less than one day in July 1981.

Between 28th May and 31st May 1982, 651 mm of rain was recorded, and 537 mm was recorded between 16th and 18th August. On 13th September 1982 with one month of the wet season remaining, the mean annual rainfall of 2250 mm had been exceeded by 15%. Despite the high 1982 rainfall figures there was no serious incident.

The system worked well in providing the engineer with information on which to make judgments.

Observations

There are a few observations which may be made based on the experience of this case history and others. They are:

(a) If a good surveillance system is set up with regular automatic monitoring, the quality of information is such as to give the professional engineer more confidence in making his decisions.

(b) The peaks highlighted by automatic monitoring can be treated after a storm by the urgent installation of extra horizontal drains which may otherwise have been installed in another position or not at all.

(c) It should be re-emphasised that automatic pore pressure monitoring is a tool and an aid to the engineer rather than a decision maker in itself.

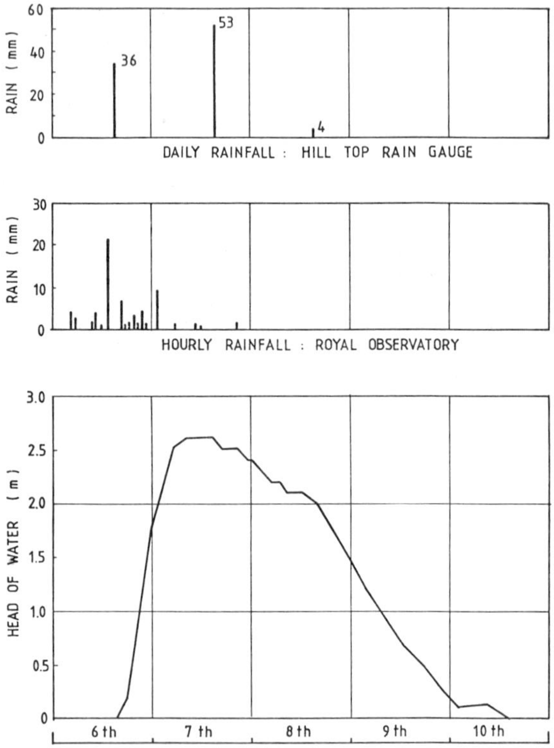

FIG. 15. Case history: variation of groundwater in colluvium (6th-10th July 1981).

References

ANDERSON, M.G. & BURT, T.P., 1977. Automatic monitoring of soil moisture conditions in a hillslope spur and hollow. *Journal of Hydrology*, **33**, 27-36.

ANDERSON, M.G. & KNEALE, P.E., 1980a. An examination of the relationship between storm precipitation and pore water conditions in road cut slopes, St, Lucia, West Indies. *Singapore Journal of Tropical Geography*, **1**, 1-8.

ANDERSON, M.G., & KNEALE, P.E., 1980b. Pore water pressure and stability conditions on a motorway embankment. *Earth Surface Processes*, **5**, 37-46.

ANDERSON, M.G., MCNICHOLL, D.P. & SHENN, J.M., 1983. On the effect of topography in controlling soil water conditions, with specific regard to cut slope piezometric levels. *Hong Kong Engineer*, **11**, No. 11, 35-41.

BURT, T.P., 1978. An automatic fluid scanning switch tensiometer system. British Geomorphological Research Group, *Technical Bulletin No.* **20**, 1-30.

BRAND, E.W., 1982. Analysis and design in residual soils. *Proceedings of ASCE Speciality Conference on Engineering and Construction in Tropical and Residual Soils*, Honolulu, 89-143.

BRAND, E.W. & HUDSON, R.R., 1982. CHASE - an empirical approach to the design of cut slopes in Hong Kong Soils. *Proceedings of 7th Southeast Asian Geotechnical Conference*, Hong Kong, **1**, 1-16.

ENDICOTT, T.J., 1982. Analysis of piezometer data and rainfall records to determine groundwater conditions. *Hong Kong Engineer*, **10**, No. 9, 53-56.

GEOTECHNICAL CONTROL OFFICE, 1982. *Mid-levels Study: Report on Geology, Hydrology and Soil Properties*. Government Printer, Hong Kong.

GEOTECHNICAL CONTROL OFFICE, 1984. *Geotechnical Manual for slopes*. Government Printer, Hong Kong.

HOEK, E., 1980. An empirical strength criteria and its use in designing slopes and tunnels in heavily jointed weathered rock. *Proceedings of the 6th Southeast Asian Conference on Soil Engineering*, Taipei.

HOEK, E. & BRAY, J.W., 1981. *Rock Slope Engineering*. The Institution of Mining & Metallurgy, London

INSLEY, H.T.M. & MCNICHOLL, D.P., 1982. Groundwater monitoring of a soil slope in Hong Kong. *Proceedings of the 7th Southeast Asian Geotechnical Conference*, Hong Kong, **1**, 63-75.

KLUTE, A. & PETERS, D.B., 1969. Hydraulic and pressure head measurements with strain gauge pressure transducers. *Proceedings of the UNESCO Symposium on Water in the Unsaturated Zone, Wageningen*, Netherlands.

LEACH, B. & HERBERT, R., 1982. The genesis of a numerical model for the study of the hydrogeology of a steep hillside in Hong Kong. *Quarterly Journal of Engineering Geology*, **15**, 243-259.

KOO, Y.C. & LUMB, P., 1981. Correlation between rainfall intensities and groundwater levels in the natural slopes in Hong Kong. *Hong Kong Engineer*, **9**, no. **10**, 49-52.

LUMB, P., 1962. Effects of rainstorms on slope stability. *Proceedings of Symposium on Hong Kong Soils*, Hong Kong, 73-87.

LUMB, P., 1965. The residual soils of Hong Kong. *Geotechnique*, **15**, 180-194.

LUMB, P., 1974 Application of statistics in soil mechanics. In: Lee, I.K. (ed.). *Soil Mechanics - New Horizons*, Butterworth, Sydney, 44-111.

LUMB, P., 1975. Slope failures in Hong Kong. *Quarterly Journal of Engineering Geology*, **8**, 31-65.

MCNICHOLL,, D.P., 1984. The assessment of rock mass characteristics prior to excavation of large cut slopes in Hong Kong. *MSc Thesis*, University of Leeds, 174-179.

POPE, R.G., WEEKS,. R.C. & CHIPP, P.N., 1982. Automatic recording of standpipe piezometer. *Proceedings of 7th Southeast Asian Geotechnical Conference*, Hong Kong, **1**, 77-89.

RODIN, S., HENKEL, D.J. & BROWN, R.L., 1982. Geotechnical study of a large hillside area in Hong Kong. *Hong Kong Engineer*, **10**, No. 5, 37-45.

MCNICHOLLS D.P., & CHO, G.W.F., H.K. Housing Authority, 101 Princess Margaret Road, Kowloon, Hong Kong.

Monitoring and investigation of water inflow into a coal mine in New South Wales, Australia

L.M. Whitfield

ABSTRACT: The Dams Safety Committee is responsible for ensuring the safety and integrity of all substantial dams and storages in New South Wales. Coal mining is currently taking place beneath the Sydney Water Board catchment where a system of five reservoirs constitute part of the urban water supply for the cities of Sydney and Wollongong.

Pillar extraction of the Wongawilli seam adjacent to Avon storage resulted in an unexpected inflow of water into the mine workings. Algal analysis indicated that a proportion of this inflow may have originated from a surface source.

The paper outlines the events which occurred, the monitoring initiated and the results of the investigations.

Introduction

The Southern Coalfield of New South Wales is located close to the industrial city of Wollongong, 80 km south of Sydney. The majority of extraction in the area takes place in the vicinity of three major water supply storages which form an integral part of the water supply system for Sydney and Wollongong.

The Permian Illawarra Coal Measures comprise four workable seams varying in thickness from 1.3 to 3 m which dip less than 5° NW, are laterally extensive and are overlain by massive lithic and quartzose sandstones with minor claystone horizons.

The New South Wales Dams Safety Committee was established as a Statutory Authority in 1979 and is responsible for ensuring the safety of all significant dams in the state and making recommendations on proposals for coal mining in the vicinity of such dams, specifying conditions under which such mining should be carried out. The Committee recommended that bord and pillar development be allowed in Blue panels at Wongawilli Colliery under the storage of Avon Dam with total extraction outside a Marginal Zone equivalent to a plan distance of 0.7 times the depth of cover from Full Storage Level to the coal seam (a 35° angle of draw).

The majority of coal extraction in the Southern Coalfield takes place at depths greater than 200 m. However, at the above location, the lower 3 m of the 9 m thick Wongawilli seam was extracted at depths ranging from 90 m under the storage to a maximum of 180 m. The workings are located under the upper reaches of Avon storage, adjacent to the pumping station which supplies water to Wollongong. The stratigraphy consists of the carbonaceous shale which comprises the upper part of the Wongawilli seam overlain by the remainder of the Coal Measures strata approximately 25 m of sandstones and shales containing the Bulli seam. The overlying massively bedded and well jointed sandstones and minor clay/shale horizons of the Triassic Narrabeen Group outcrop in the storage. Massive quartzose Hawkesbury Sandstone forms a capping to the hills outside the immediate area of the storage. A number of major dykes transect the development area but have not been located at the surface. An extensive 2 m thick sill intruded along the seam has severely restricted the extent of development in the area (Fig.1).

The workings underlie the confluence of four arms of the Avon storage. Conditions in the area — viz. deep V-shaped valleys, strata with numerous vertical and horizontal weakness planes and a high horizontal stress field — are conducive to the formation of valley bulging deformation and increased strata permeability.

Because of the complex geological and geomorphological conditions in the immediate area, stringent monitoring conditions were placed on the development. As well as regular underground geological mapping and measurement of surface subsidence and strain, a system of monitoring of water movement within the mine was established. This included measurement of intake and return ventilation moisture, coal moisture, ingoing and outgoing piped or pumped water and water flowing along the roadways, this latter being measured over a number of 90° V-notch weirs.

The Wongawilli Seam is believed to be an aquifer with a vertical permeability of approximately 6 × 10 cm/s (Cantwell & Whitfield 1984). It is inherently wetter than the other seams in the Illawarra Coal Measures but its permeability is laterally discontinuous.

From Cripps, J.C., Bell, F.G. & Culshaw, M.G., (eds), 1986, *Groundwater in Engineering Geology*, Geological Society Engineering Geology Special Publication No. 3, pp. 417-421

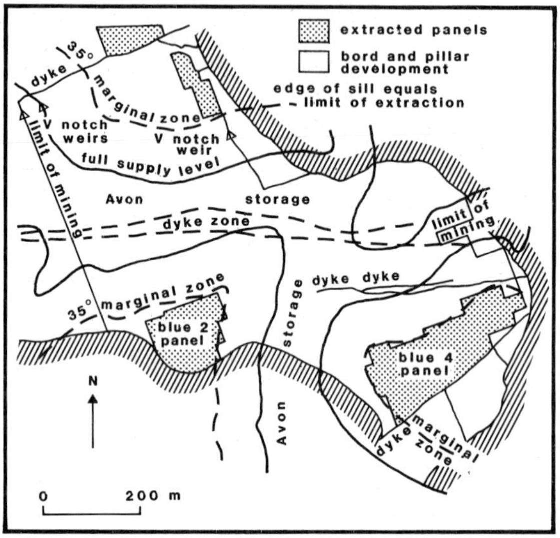

FIG.1. Layout of workings in Blue Panel at Wongawilli Colliery.

Water seepage has been common from the sill-seam contact particularly where the sill has been intruded into the upper part of the seam. Some degree of water inflow was anticipated in the monitoring requirements. A number of substantial water inflows have been noted from the seam in other areas of Wongawilli and the adjacent Nebo collieries but details of the nature and magnitude of the inflows have not been recorded (Wilson *et al.* 1983).

The major water inflow at Wongawilli colliery commenced in Blue-2 panel in mid December 1982 during extraction of pillars adjacent to the sill (Fig.1). The inflow was estimated to be approximately 100,000 l/hr (2.4 Ml/day) subsequently stabilising at 80,000 1 /hr. The inflow caused considerable damage to belt and track roads before being contained behind a breezeblock dam and piped from the area. Measurement of the flow over 4 V-notch weirs, 3 in Blue-2 panel and 1 in Blue-4 panel, recorded a staged increase associated with periods of pillar removal.

The extraction of pillars in Blue-4 panel adjacent to a 2 m thick dyke resulted in an increase in water make from 6,000 l/hr to 12,000 l/hr. On completion of extraction in September 1982, water make increased to 29,000 l/hr. It remained at this level until mid-December 1982, when it rose dramatically to 100,000 l/hr on extraction of pillars adjacent to the sill in Blue-2 panel.

The flow is gradually decreasing and is currently 44,500 l/hr. However, increases in water inflow have been recorded immediately following periods of heavy rainfall, notably in February 1984, when 704 mm of rain fell in 48 hours. This deluge was reflected in an increase in flow over the V-notch weir measuring water make from Blue-4 panel with a time lag of approximately 4 days. No such increase was recorded in Blue-2 panel.

Previous instances of heavy rainfall evoked no such response in the mine suggesting that the water was emanating from a stable source, either a surface storage or a rechargeable aquifer. However, since the above mentioned incident, the inflow from Blue-4 panel whilst continuing to decrease, has also continued its behaviour of retarded fluctuation in response to major rainfall events supporting the hypothesis that the source is a groundwater system which is being sustained by recharge from rainfall. On the other hand, the rate of decline of the inflow via Blue-2 panel has been fairly consistent and suggests that the dominant control is that of a confined groundwater system draining to a sink (Williamson 1984).

Methods of investigation

Investigations to determine the source of the original water inflow commenced in late December 1982. Three sources of water were considered:

- flooded old working (pre-1950) up-dip from the Blue panels on the southern side of the sill;
- groundwater in the seam and surrounding strata;
- surface water from the overlying Avon storage.

Algal, chemical and tritium analyses of inflow water, storage water and water from the old workings were conducted. Tritium analyses reflected the hypothesis that the old workings were the source of the water but suggested that the inflow contained components of surface water and groundwater.

Chemical analysis

Chemical analysis indicated that all three samples had markedly different chemical composition (Table 1.). The high sulphate content of the water from the old workings effectively ruled them out as a source of inflow, there being no known mechanism whereby the sulphate could be almost totally removed during transit of 400–800 m. The chemistry of the storage water did not correlate absolutely with that of the inflow water, notably in the former's lower concentration of cations. It was concluded that the inflow water could be a composite of water from the overlying Avon storage and the groundwater/Wongawilli seam aquifer. Chemical analyses of the inflow water continued to be conducted at 3 monthly intervals but to date, no significant variation in chemical composition has occurred.

TABLE 1. *Chemical composition of waters (analyses by Metropolitan Water Sewerage and Drainage Board). Quantities expressed in mg/litres.*

Analyses	Old workings	Avon storage	Mine inflow
pH	7.6	6.43	7.43
Calcium	40.7	1.7	27.5
Magnesium	32.4	1.8	12.5
Sodium	38.0	8.2	18.9
Potassium	4.05	0.94	2.18
Iron	0.53	0.95	0.09
Chloride	27.4	14.3	19.1
Sulphate	118.9	1.8	0.2
Nitrate	0.04	0.02	0.45

Algal analysis

Algal analyses found a number of species of algae in the mine water in small quantities and various states of preservation which correlated with those found in Avon storage (Table 2). Algae are photosynthetic and require the presence of sunlight or strong light to survive and develop. Such a situation is not found in a coal mine therefore it was concluded that they were derived from a surface storage. Analyses of water from the overlying Avon storage and a nearby swamp, the only other surface source in the area, found a correlation between algae from Avon storage and the inflow water but none from the swamp algae. The presence of a number of live algae in the mine water indicate that the travel time from surface to mine could be as little as 4 days (Cheng 1983).

TABLE 2. *Analysis of algal content (cells/1000 m) (Analyses by Metropolitan Water Sewerage and Drainage Board).*

Alga	Mine Sites (Blue-2 Panel)			
	Avon Storage	Goaf	Impounded water	Goaf
Cryptomonas	50	1016	230	4
Staurastrum	7	4		
Cyclotella	365	8		
Cymbella			10	
Melosira granulata	17	176		304
Navicula	2	4		
Synedra	100	4		4
Peridinium	25	16		4
Dactylococcopsis	36	4		

In later analyses a reduction in both species and numbers of algae was noted suggesting either that the leakage path is being restricted by siltation or that an alternative source of water is diluting the storage water.

Doubt has been cast on the validity of the algal analyses, the possibility of contamination of samples being expressed (Jones 1984). As far as is known, no extensive study has been undertaken of the presence and viability of algae in coal mines although it is known that their longevity is variable. The Committee has therefore commenced a study collecting and analysing water from goafed areas and roof drippers. To date, seven mines other than Wongawilli have been sampled both underlying and remote from water storages. Water from only one mine has been found to contain algae, the source of which is believed to be ponded water in an adjacent opencut area of the same seam.

The possibility of algae being introduced into the mine via airflow or transport of men, materials or mains water has not been investigated. There is some limited evidence to suggest a gradual decrease in algal content of waters from the portal to the interior of the mine (Jones 1984). However, Blue panels are approximately 4 km from the mine entrance whereas samples taken from other mines at distances of only a few hundred metres from the portal have not contained algae.

Other investigations

A number of other methods of determining the source of the inflow were considered including:

- high resolution seismic reflection;
- tracers;
- drilling of boreholes;
- thermometric analysis.

High resolution seismic reflection was rejected as it was not considered sufficiently sensitive to locate a zone of increased fracturing in strata which is believed to be inherently well jointed.

A number of different tracers were considered including fluorescein dyes, sodium chloride, potassium sulphate and lithium chloride in an attempt to indicate a possible leakage route from the storage. All were rejected on political or social grounds. The Avon storage is the major source of water for the city of Wollongong and the pumping station is adjacent to the mined area, over the probable location of the dyke suspected of being the focus of the leakage.

Drilling of boreholes to obtain permeability data and rock samples for testing was rejected as being too costly in view of the inhospitable terrain and the limited information to be gained from such an exercise.

Determination of the temperature variation within Avon storage, thermometric analysis, in an attempt to locate a fracture or zone which could serve as the point of inflow of water was rejected as similarly being too expensive. It has been estimated to cost approximately A$50,000 for a number of traverses and subsequent analysis with no guarantee of success. The presence of the pumping station may result in disruption to the thermal profile of the storage adding to the difficulty of analysing temperature variation.

Mechanism of inflow

Investigation of the mechanism whereby water could be transmitted from surface to seam over a short time period has been very limited. As noted previously, water make increased upon extraction of Blue-4 panel on the northern side of a dyke. Pillars were not extracted on the southern side. A horizontal tensile strain of 1.41 mm/m was measured over the dyke associated with a surface fracture of similar trend (Whitfield 1984). The dyke corresponds with a strong surface lineament as deduced from aerial photographs and is therefore believed to extend under Avon storage. It is thought possible that this dyke could act as the point or zone of leakage of water from the storage to mine. Increase in water inflow restricted to Blue-4 panel following periods of heavy rain similarly supports this theory. However, it is difficult to envisage a mechanism which could cause disruption of strata over approximately 180 m depth so as to allow such an inflow yet only result in a surface tensile strain of 1.41 mm/m.

The quantity of water entering the mine is at variance with that to be expected purely as a result of the strata permeability. If one assumes a bulk permeability of 10^{-7} cm/s for the overlying strata, the source to the mine would be 0.8 Ml/day. The original measured inflow of 2.4 Ml/day suggests that the strata has been more severely affected than surface subsidence and strain measurements and seam level observations would suggest. The limited areas of water inflow also suggest that such disturbances may be highly localised.

The unique topography and the relative shallowness of the mining could combine to produce a localised concentration of stresses which would not necessarily manifest as large surface strains. It is known that stress is concentrated at the base of steep valleys and this, exacerbated by the regional high horizontal stress field, could result in a much increased localised stress regime. The magnitude of operations, although on a small scale, might then have been sufficient to open fractures and generally increase strata permeability at depth without any concomitant major surface deformation.

Remedial measures

Remedial measures were considered briefly, the most likely being the plugging of the leak at the surface. However, this would depend on the accurate location of the leakage point and determination of the seepage path. As noted previously, there are no known methods by which the leakage point can be accurately located in this particular area of the Avon storage.

The geology of the area is complex and though it appears that the dykes and sill are inter-related, their role in the movement of water from surface to seam is not fully understood. Without this knowledge, similar situations cannot be foreseen and preventative or remedial measures established.

Discussion

This situation at Wongawilli Colliery emphasises the inherent problem in underground extraction that the full geological and hydrogeological situation cannot be ascertained prior to mining. Even the development phase may not foresee future problems. In this particular area, headings were driven to and along the crop of the sill with only minor water inflows being experienced. Although an inflow occurred during pillar extraction in Blue-4 panel, it was not until pillars were extracted in Blue-2 panel immediately adjacent to the sill that the major inflow occurred. The exact location of the inflow is not visible, being obscured by the goaf.

The quantity of water involved in the inflow at Wongawilli Colliery is manageable in the mine workings and does not impose a hazard to the overall development of the mine. However, the Dams Safety

Committee is responsible for maintaining the integrity of other storages where similar mining has been permitted. It is concerned that such a situation should not occur again and is therefore continuing and extending its current investigations as well as reviewing its criteria for the approval of coal mining in the vicinity of dams and their storage.

It is stressed that the views expressed are those of the author and not necessarily held by other parties involved.

References

CANTWELL, B.L., & WHITFIELD, L.M., 1984. Underground mining near large Australian dams. *Water Power and Dam Construction,* April Issue, 20-24.

CHENG, D., 1983-4. Biological examination of Wongawilli mine water. *Dams Safety Committee Internal Reports.*

HEATH, C.W., 1983. Wongawilli mine - investigation of increased water flow. *Metropolitan Water Sewerage and Drainage Board, Internal Report.*

JONES, V., 1984. Review of evidence of algae in South Coast mines, particularly Wongawilli. *The BHP Co. Ltd., Internal Report.*

WHITFIELD, L.M., 1984. Ground movements from coal extraction in the vicinity of dams and storages in New South Wales, Australia. *Proceedings of the 3rd International Conference on Ground Movements and Structures,* Cardiff Pentech Press, London. 298-313.

WILLIAMSON, W.H., 1984. Wongawilli-1 Water Inflow. *Dams Safety Committee Internal Report.*

WILSON, R.G., DOYLE, J.F., POOLE, G.R., 1983. Wongawilli Colliery -Water Inflow - Blue Panels. *The BHP Co., Ltd., Internal Report.*

WHITFIELD, L.M., NSW Dams Safety Committee, Chief Secretary's Building, 121, Macquarie Street, Sydney, 2000, Australia.

Discussion Session 9

Written contribution further to verbal discourse:

L. THREADGOLD (Consulting Geotechnical Engineer)

The paper by MCNICHOLL & CHO (this volume) provides a welcome case history showing the day to day practical implementation of geotechnical instrumentation and monitoring which can be seen to directly affect the lives and safety of an urban community.

If instrumentation is to be of value its results must believed unless they can be proved to be in error. It is of course essential that the correct instrument is installed in the correct manner and in the correct location, but to disregard the readings because they do not tie up with pre-conceived concepts of what results are to be anticipated is dangerous arrogance.

It must be appreciated that in relation to certain analyses it is not conservative to assume a groundwater level higher than that which has been monitored. This is particularly so in relation to back-analyses of slope stability since an over-estimate of water level can lead to an over-estimate of the frictional parameter appropriate to the potential failure surface. Furthermore, the improvement resulting from any drainage measures is likely to be over-estimated.

D.W. COX (Polytechnic of Central London) commented on the paper by Mr. G. CRABB and Dr. G. WEST in respect of the effects of undercompaction at the edges of an embankment and permeability anisotropy due to layered construction. Would the compaction of the sides of embankments improve matters?

G.I. CRABB (Transport and Road Research Laboratory) pointed out that although there could be a less well compacted zone at the edges of embankments, which prior to about 1969 may have been removed during construction there is no evidence that this is a major factor precipitating failures. This matter is the subject of continuing investigations.

Horizontal compaction planes and shears were observed in trenches in the embankment studied but there was no evidence of water ingress or softening along these planes. Probably there are zones of relatively high horizontal permeability but little water gains access to the material owing to the sealing afforded by the road construction.

D.N. HOLT (Freeman Fox Group of Consulting Engineers) pointed out that it is feasible to compact embankments right up to the edges and that also the sides can be compacted since in practice the slopes generally are less than 1 on 1 which allows them to be traversed by compacting plant.

J. PONTIN (Hydraulic Research Ltd) pointed out that the presence of negative pore water pressures near to the pavement would appear to confirm that there is little ingress of water.

G. CRABB responded to a question put by Mr PONTIN by stating that none of the slips studied affected the pavement. The back scarps remain stable for a long period (up to a year) but, of course, they need repair eventually.

Written Contribution:

J.A. CHARLES (Building Research Establishment)

CRABB and WEST (this volume) have presented a stability analysis of a shallow slope failure in a 7 m high embankment made of Gault Clay. The failure envelope of the clay fill showed marked curvature at low effective normal stress. In the analysis, stability on a potential slip plane 1.5 m below the surface was examined and the problem was simplified by assuming an infinite slope. Effective normal stresses were calculated for different pore water pressure assumptions. Effective cohesion and friction angle were determined for every value of effective normal stress by taking a tangent to the failure envelope at that value of effective normal stress.

An alternative approach to the stability analysis is to represent the failure envelope by an equation of the form,

$$\tau = A\,(\sigma_n')^b \qquad (1)$$

and analyse the slope stability in terms of the shear strength parameters A and b. In Equation (1) τ is the shear stress, σ_n' is the normal effective stress and A and b are constants. The failure envelope shown in Fig. 7 of Crabb and West can be represented by:

$$\tau = 3.6\,(\sigma_n')^{0.50} \qquad (2)$$

where τ and σ_n' are in kN/m^2.

From CRIPPS, J.C., BELL, F.G. & CULSHAW, M.G., (eds), 1986, *Groundwater in Engineering Geology,* Geological Society Engineering Geology Special Publication No. 3, pp. 423-425

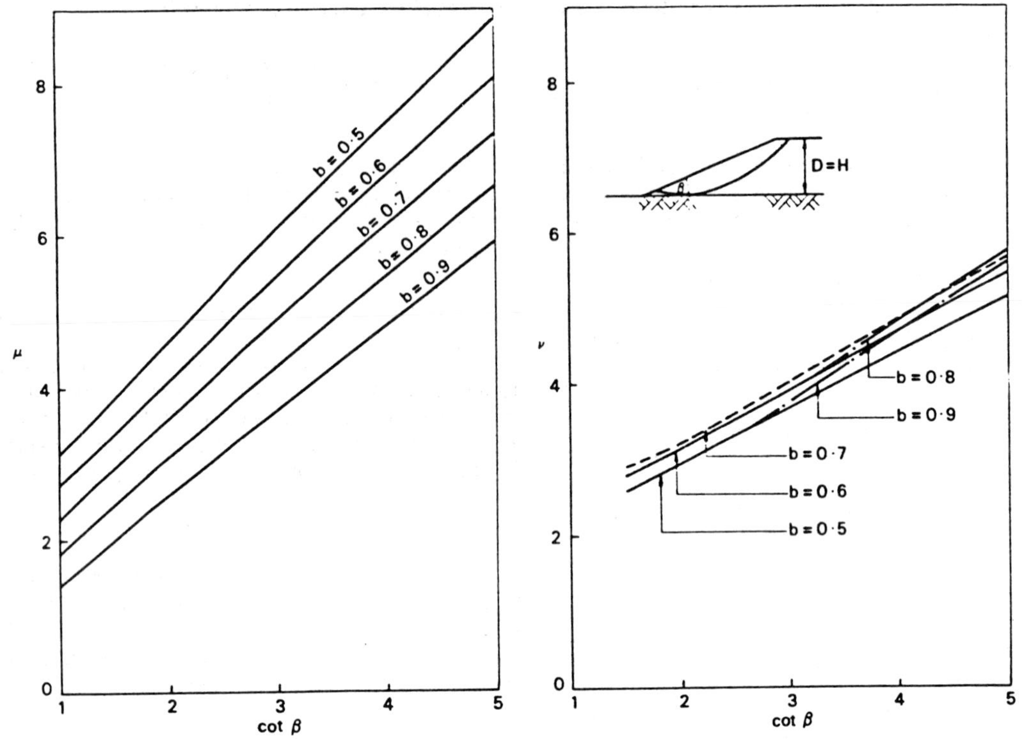

FIG. 1. Stability chart for *D/H* = 1.

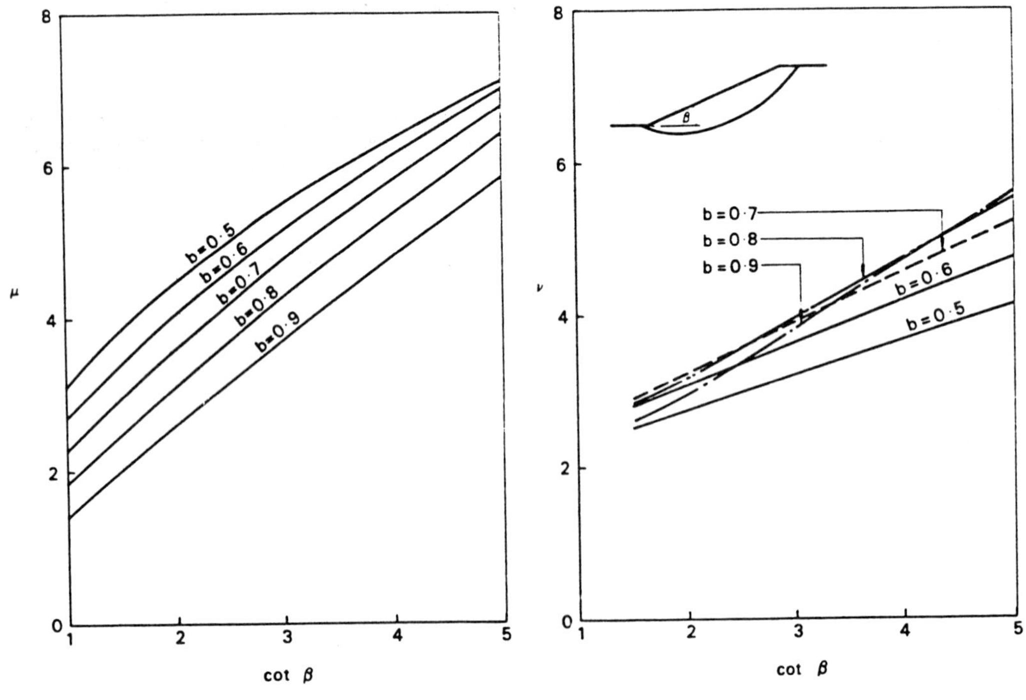

FIG. 2. Stability chart for *D/H* = ∞

For a translational slide in an infinite slope it can be shown that

$$F = \frac{A}{\gamma^{(1-b)}} \quad \frac{(\cos^2\beta - r_u)^b}{\cos\beta\sin\beta} \quad \frac{1}{z^{(1-b)}} \qquad (3)$$

where F is the factor of safety against stability failure on a plane at depth z below the surface of a slope at an angle β το τηε ηοριζονταλ, ϱ_θ ισ τηε πορε πρεσσθρε ρατιο

$\frac{u}{\gamma z}$, γ is the bulk density of the soil.

For the case presented by Crabb and West, A = 3.6, b = 0.50, γ = 19 kN/m^3, cot β = 2 and therefore

$$F = 2.06 \left(\frac{0.80 - r_u}{z}\right)^{0.50} . \qquad (4)$$

If, following Crabb and West, we examine the stability on a potential slip plane 1.5 m below the surface (i.e. z = 1.5 m), then

$$F = 1.68\,(0.80 - r_u)^{\,0.50} \qquad (5)$$

From Equation (5) it is found that F = 1 when r_u = 0.45

The analysis of Crabb and West indicated that F = 1 when u = 13 kN/m^2 i.e. when r_u = 0.45. Thus the two approaches do give the same answer. The advantage of the method shown above is that Equation (4) makes it possible to calculate rapidly the factor of safety for a range of values of r_u and z.

Charles & Soares (1984) have presented stability charts for soils with failure envelopes of the type described by Equation (1). The charts are based on circular slip analyses and are in terms of coefficients μ and ν which are related to a stability number Γ by

$$\Gamma = \mu - \nu r_u. \qquad (6)$$

The stability number Γ is defined by

$$\Gamma = \frac{F}{A}\left(\gamma H\right)^{(1-b)}, \qquad (7)$$

where H is the slope height.

For the case presented by Crabb and West, H = 7 m, γ = 19 kN/m^3, A = 3.6 and b = 0.50 and hence from Equation (7),

$$\Gamma = 3.20\,F, \qquad (8)$$

Fig. 1 which is taken from Charles & Soares (1984) applies to a situation in which there is a hard stratum at the base of the slope. For b = 0.50 and cot β = 2, μ = 4.61 and θ = 2.96. Substituting these values in Equation (6) and combining with Equation (8) we obtain:

$$F = 1.44 - 0.92\,r_u. \qquad (9)$$

Equation (9) indicates that F = 1 when r_u = 0.48. Fig. 2, also from Charles & Soares, is applicable where no hard stratum is present at the base of the slope. The corresponding values of the coefficients are μ = 4.46 and ν = 2.74 and hence,

$$F = 1.29 - 0.86\,r_u. \qquad (10)$$

Equation (10) indicates that F = 1 when r_u = 0.45. It is interesting that the stability charts based on the most critical circular slip surfaces give a similar result to the translation slide at 1.5 m depth.

References

CHARLES, J.A. & SOARES, M.M., 1984. The stability of slopes in soils with nonlinear failure envelopes. *C-anadian Geotechnical Journal* **21**, 397-406.

Crown copyright. Reproduced by permission of the Controller, HMSO.

SESSION 10

CONTROL OF GROUNDWATER BY EXCLUSION

Control of groundwater by exclusion

F.G. Bell & J.K. Mitchell

ABSTRACT: The movement of water into an excavation can be impeded by placing a barrier around the excavation. Such excavation barriers include sheet piling, contiguous bored pile walls, slurry trenches, diaphragm walls, grout curtains and panels, and ice walls. Ideally an exclusion barrier should extend into an impermeable stratum. If this does not happen, then upward seepage of water into the excavation may occur which, in turn, may give rise to instability at excavation level. To prevent such instability, the exclusion barrier should be extended horizontally. Of the exclusion techniques mentioned, only grouting and freezing can be used to form a horizontal exclusion barrier.

Introduction

Groundwater frequently provides one of the most difficult problems during excavation. Not only does water make working conditions difficult, for example, vehicles may become bogged down, muck becomes more difficult to handle etc, but piping, uplift pressures and flow of water into an excavation can lead to erosion and failure of the sides. Collapsed material has to be removed and the damage made good. Some of the worst conditions are met in excavations which have to be taken beneath the water table.

Artesian conditions, in particular, can cause serious trouble in excavations and therefore if such conditions are expected it is essential that both the position of the water table and piezometric pressures should be determined before work commences. Otherwise excavations which extend close to strata under artesian pressure may be severely damaged due to blow-outs taking place in their floors. Such action may also cause slopes to fail and in fact has led to the abandonment of the site in question. Sites at which such problems are likely to be encountered should be dealt with prior to and during excavation, by employing exclusion or dewatering techniques.

In addition, it may be necessary to exclude water from a site because of the effect it has on some ground materials. Obviously the absorption of water can lead to the weakening of soils and the breakdown of some weakly cemented rocks. Water also can lead to some clay soils expanding, whilst the inundation of loess soils can bring about collapse of their metastable structure. Water can be responsible for the solution of some materials, for example, gypsum, and the hydration of others, for example, anhydrite (the latter also can be dissolved). The reaction of water with minerals can cause their alteration and breakdown. Pyrite, for instance, undergoes fairly rapid breakdown, and, like the dissolution of gypsum and anhydrite, this gives rise to sulphate ions in solution which can attack concrete.

One of the techniques of controlling groundwater is by using some type of cut-off. In order to help select the type of cut off required for groundwater control a number of basic questions have to be answered. First, is a cut-off the best method of groundwater control at the site concerned? Secondly, if so, then which cut-off method is best suited to the site conditions? Thirdly, can the cut-off be temporary, that is, remain in existence for more or less the construction period only, or has it to be permanent? Fourthly, can the cut-off serve any further functions?

Methods of forming such a barrier, include sheet piling, contiguous bored pile walls, bentonite cut-off walls, geomembrane barriers, concrete diaphragm walls, grout curtains and panels, and ice-wall barriers. The economy of providing a barrier to exclude groundwater depends on the existence of an impermeable stratum beneath the excavation to form an effective cut-off for the barrier. If such a stratum does not exist or if it lies at too great a depth to be practical to use as a cut-off, then upward seepage may occur which, in turn, may give rise to instability at excavation level. In these circumstances the barrier will not be effective unless it can be extended horizontally beneath the excavation. The only methods of forming a horizontal barrier are by grouting or freezing. Consequently it is necessary to carry out a site investigation before decisions on the type of wall to be adopted are made.

In urban areas the presence of existing buildings and high price of land frequently dictate that construction has to be carried out in the least possible space. Where surrounding property has to be safeguarded it will usually be more appropriate to provide a barrier or cut-off about an excavation in order to prevent inflow of water, whilst maintaining the surrounding water table at its normal level, rather than to adopt a dewatering technique. Maintenance

From Cripps, J.C., Bell, F.G. & Culshaw, M.G., (eds), 1986, *Groundwater in Engineering Geology*, Geological Society Engineering Geology Special Publication No. 3, pp. 429-443

of water at its usual level does not give rise to settlement problems which could occur when a groundwater lowering technique is used.

Perimeter exclusion walls should be capable of construction through soft, wet ground, should be economical in installation cost and should be capable of being installed without any adverse effects on the stability of adjacent buildings. If possible, they should eliminate the need for temporary work during construction of the permanent works. Any movement of the barrier can adversely affect both the structure to be incorporated within the excavation and the stability of existing buildings immediately adjacent to it. Accordingly measurements should be taken on the constructed walls and adjacent properties to check for any movements.

Sheet piling

Sheet piling may be composed of either steel, timber or reinforced concrete piles, each pile being linked to the next to form a continuous wall. Indeed steel sheet piling is perhaps the simplest method of forming an impermeable barrier, particularly if the piles can be extracted for re-use, and so they are the most widely used (see Cornfield 1975). By contrast pre-cast reinforced concrete sheet piles are nowadays seldom used.

Steel sheet piles are used in temporary structures such as strutted excavations and cofferdams, as well as for permanent use, notably in retaining walls and bulkheads. With all sections of pile or types of steel, higher stresses are permissible when used in temporary works rather than for permanent use. Sheet piling, when installed before excavation takes place, provides an effective cut-off, as well as affording ground support in that the piles are used to resist horizontal pressures developed by soil and pore water. The piling derives its stability from the horizontal support provided by the ground into which it is driven and from any anchors, ties or struts located near the top of the piles.

Steel sheet piling is most effective as a cut-off when driven into an underlying impermeable bed of clay. Should a thick deposit of sand exist below the clay, the differential head should be monitored by a piezometer (installed outside the piling) to ensure that the thickness of the bed of clay can offer adequate resistance. If it cannot, then pressure relief wells must be provided.

Steel sheet piling generally is not used in ground containing boulders or in rubble fill, because of the difficulty in driving and the risk of torn piles. One technique that has proved effective in such materials involves excavating the boulders in a slurry trench, backfilling with sand, and then driving the piling. If the sand is thoroughly mixed before placement, the effectiveness of the piling as a cut-off is enhanced.

The pile section to be selected depends on height of ground to be retained, soil and groundwater conditions, and on whether the piling is tied back or strutted. The shape of the pile cross-section also is designed to make the wall capable of resisting bending. The section has to be capable of being driven into the soil to the desired depth and in this context the shape of the cross-section provides stiffness.

The most critical parts of a sheet pile wall for groundwater exclusion are the interlocks which provide the connection between piles, thereby forming a continuous wall. Such a wall generally is sufficiently watertight for most practical purposes. However, if piles come out of interlock during driving this means that the effectiveness of the cut-off is impaired. Moreover when the sheet piling remains interlocked, the cut-off is still of limited effectiveness until the sheet is stressed, wedging the adjacent piles into tight contact. In fact leakage can be quite high when a row of steel sheet piles acting as a cut-off, is unstressed. It is therefore usually necessary to seal the interlocks by using heavy grease before installation or by grouting.

Since a much larger amount of penetration below excavation level is required for a cantilever wall than an anchored wall, it is imperative to ensure that the section of piling finally calculated on bending moment considerations is large enough to withstand the vertical stresses developed during driving. In the former type of wall the depth of penetration should be at least equal to the height of the sheet piling above excavation level.

Steel sheet piling is used most frequently for anchored retaining walls. These normally range in height between 4.5 and 12 m, depending on the soil conditions.

When steel sheet piling extends into undisturbed soil, even though it goes below the water table, corrosion tends to be very slight. Hence it is generally unnecessary to apply a protective coating to the piles. Some corrosion may occur above the water table or in a disturbed soil. In such circumstances it may be necessary to apply a protective coating. Normally two coatings of a bituminous paint are given to these steel sheet piles before installation.

Contiguous bored piles

Contiguous bored piles frequently are associated with both shallow and deep excavations for basements to buildings and with cofferdam work. This is particularly the case in urban areas where noise or the effect of installation on adjacent property are important, as well as in industrial complexes where access, headroom or restrictions on vibration may make other exclusion methods such as steel sheet piling or diaphragm walling less suitable.

Contiguous piles can be designed to cantilever vertically, to be strutted or to be anchored back. It is usual to employ ground anchors, either temporary or as part of the final design, to eliminate the need for struts which limit the working space within an excavation and also to reduce the required pile diameter and amount of reinforcement within the pile.

Bored piles possess a number of advantages. For example, their length can be varied to suit the ground conditions, they can be installed with large diameters, in cohesive soils their ends can be enlarged by up to three diameters and their use involves no risk of ground heave. Unfortunately bored piles are susceptible to 'necking' in squeezing ground. Under artesian conditions water may pipe up the pile shaft and in so doing remove cement. Bored piles may be impossible to install in very soft or loosely packed silts and sands unless the boreholes are cased to prevent caving. A loss of ground may occur when sinking piles in groups in non-cohesive soils, leading to settlement of adjacent structures.

When contiguous bored piles are used for retaining walls, the piles are installed either in a single or double row and are positioned such that they are touching or are in very close proximity to each other. The piles are reinforced to resist the applied bending moments and shear forces. There is much flexibility as far as the choice of pile diameter is concerned. However, the circular section of the pile generally means that more main reinforcement per metre run of wall is required than in the constant section of a diaphragm wall in order to produce an equivalent moment of resistance (see North-Lewis & Lyons 1975).

Presetting a number of casings in advance of the piling ensures the correct spacing of the piles in the line. The depth to which the casing extends depends on the ground conditions. For instance, where the piles are in clay, only a short length of casing generally is necessary to provide a seal in the top of the clay. Boring can then continue in an uncased hole. The use of temporary casing leads to gaps between piles. Such gaps normally should not exceed 75 mm at the head of the pile, and the verticality of the pile shafts should remain within the commonly accepted tolerances for bored piles (for example, 1 in 75).

The concrete of contiguous bored piles is placed directly into the pile shaft through a short trunking and hopper for dry conditions, and by the tremie technique under water. Alternate piles are constructed and the concrete allowed to harden before boring the intermediate piles. The holes are excavated by a grabbing rig or power auger.

It is often possible to chisel through an obstruction during boring operations. However, in soils, where the boring is slow and difficult, due for example, to the presence of boulders, the amount of overbreak resulting from driving and boring out casings at very close centres in a line is likely to be greater than that arising from the installation of piles at wide spacing. Concrete can flow into weak ground or cavities as the casings are withdrawn.

Where water cannot be removed from a borehole either by baling or pumping, then the hole has to be cased from top to bottom and a concrete plug is formed, beneath the water, at the base of the hole. The plug is placed by a tremie pipe. Water is pumped from within the casing when the concrete plug has hardened sufficiently. Tomlinson (1980) indicated that the casing should then be gently turned and lifted slightly in order to release it from the plug. The casing is now filled with concrete. Next the casing is removed. As an alternative to plugging the bottom of a borehole, concreting under water from a tremie pipe can take place over the complete length of the hole. However, according to Tomlinson, such construction may not be able to avoid the formation of weak zones in the pile.

Where contiguous bored piles pass through water bearing strata various grouting techniques can be employed to seal the joints between adjacent piles. Grouting of joints is carried out behind the piles and is completed before excavation begins. The water-tightness of the structure can be further improved by grouting the joints on the exposed face after excavation.

Watertight walls may also be obtained by using interlocking secant piles. Again alternate piles are installed first but at a closer spacing than in the contiguous method. The intermediate piles are installed by boring out the soil between alternate pairs, followed by cutting a groove down the sides of the pile shafts while the concrete is still green (see Anon 1985). Lastly, concrete is placed in these holes, including the grooves, thereby forming a fully interlocking watertight retaining wall. If, however, individual holes drift out of plumb or fail to achieve continuous interception, the integrity of the wall is violated.

Slurry trenches

Slurry trenches are used extensively as a means of groundwater cut-off. For example, they have been used to control seepage beneath dams and to contain groundwater pollution, notably from sanitary landfills and hazardous waste impoundments. Slurry trenches can achieve permeabilities of the order of 10^{-9} m/s.

There are two types of slurry cut-off walls, namely, soil bentonite and cement-bentonite. In the soil-bentonite cut-off process, the bentonite slurry is displaced by a soil-bentonite mixture similar in consistency to high slump concrete. The soil backfill and slurry forms a low permeability highly plastic cut-off wall. In the cement-bentonite cut-off process cement is added to a fully hydrated bentonite slurry. The addition of the cement causes the slurry to harden, giving a strength comparable to that of stiff clay.

Permeability and wall integrity are the most important characteristics of a cut-off wall in that the major reason for its construction is to minimize the passage of groundwater. Consequently the continuity of slurry trench excavation is important. The depth of a slurry trench in controlled in many cases by the ground conditions, notably the depth to an impermeable formation, and the type of barrier to be constructed. The minimum depth of penetration into an impermeable formation at the base of the cut-off wall depends upon the nature of the formation. For instance, if this is competent impervious rock, then only a very small penetration may prove satisfactory. On the other hand an excavation into clay may need to go to a depth of 1 m or so. The width is governed by the required permeability of the cut-off wall, the head of water across the wall, the size of the excavation equipment available, and the materials which form the wall. In the case of soil bentonite backfill slurry walls, Millet & Perez (1981) suggested that the walls should have thicknesses varying from 1.5 to 2.3 m. The increased shear strength of the backfill forming cement-bentonite slurry trench cut-off walls, allows them to be thinner, their thickness ranging from 0.6 to 0.9 m. Such widths are satisfactory for depths of up to 30 m of hydrostatic head. Beyond this the ability of the cement-bentonite cut-off wall to withstand hydrofracturing should be determined. The deviation from verticality is only important insofar as it affects the continuity and integrity of the wall.

Excavation is made under slurry using backhoes, clamshells, draglines and special devices. The head of slurry should be kept to 1 to 1.5 m above the maximum anticipated groundwater level in the trench. The method of excavation which is chosen is based primarily on the width and depth of trench, and the type of soil More than one method of excavation can be used, for instance, a dragline can be employed to dig the upper part of a trench (down to around 21 m), greater depths being removed by clamshell. Speed of construction is increased if several alternate panels can be formed during the initial hardening period, the grab returning to dig out the intermediate panels, which being of lesser width are cut into the primary panels. There are no joints between panels and stop-end tubes and guide trenches are not required.

As excavation proceeds, sands and silts become suspended in the slurry. When a sample of slurry obtained from the trench exceeds a relative density of about 1.6, the slurry should be circulated through settling ponds or separators. This improves the quality of the filter cake on the sides of the trench and enables the backfill to settle into position satisfactorily.

During excavation a filter cake of bentonite forms on each side of the trench, resulting from the filtrate as fluid escapes into the surrounding soil. This may be around 3 mm in thickness. The impermeability of the trench is attributable to the two filter cakes on either wall as well as the backfill material. In fact the permeability of the filter cake may be as low as 10^{-11} m/s.

Backfill is mixed with slurry at the surface, using bulldozers or front-end loaders, before being placed in the trench. The initial backfilling operation for a soil-bentonite cut-off wall involves the backfill being placed at the bottom of the trench by a clamshell bucket. This continues until the backfill material reaches the top of the trench. Then the remainder of the backfill at the surface is moved by a bulldozer onto the backfill exposed in the trench. This action forces the material to slide down into the trench under its own weight, and continues until the entire excavation has been backfilled. There is always some consolidation and settlement of a soil-bentonite cut-off wall. The effects of consolidation on the backfill material at the top of the soil-bentonite cut-off wall may lead to the development of seepage paths.

In the beam method a cut-off wall is constructed by forming a narrow cavity in the soil by repeatedly driving or vibrating an I-beam section within it. As the I-beam is extracted an impervious material, usually a mixture of bentonite and cement, is pumped into the cavity. Successive penetrations of the I-beam are overlapped to develop a continuous membrane (see Boyes 1975). In the driven pile system, a clutch of H-piles is forced into the soil along the line of advance, the end pile being withdrawn and redriven at the head end. Grouting, by a tube fixed to the web, is carried out during withdrawal. Walls are limited to a maximum of 20 m depth with these systems due to the danger of deviating during driving.

When slurry trenches are used to control groundwater pollution from hazardous waste sites, it may be that some of the chemicals present are incompatible with bentonite. In such situations high density polyethelene sheeting is placed in the trench prior to backfilling. Sheeting also can be used to form an impermeable membrane when excavated material is not readily suitable for use as backfill yet cannot be economically disposed of. The permeability of high density polyethelene sheeting can be as low as 10^{-14} m/s. When a slurry trench cut-off is required with a permeability of less than 10^{-9} m/s, then such sheeting is used to enclose the backfill.

Ressi & Cavalli (1984) described the construction of a sheeting-enclosed cut-off wall (the Envirowall). First, a narrow trench is excavated and filled with bentonite slurry to prevent the trench collapsing. The geomembrane sheeting is fabricated to form a U-shaped envelope fitting the dimensions of the trench and then ballast is placed within it so that it can sink into the slurry trench. After initial submergence into the trench, the envelope is filled with wet sand. A system of wells and piezometers can be installed in the sand to monitor water quality and piezometric pressure respectively. If the sheeting is damaged, then the system will detect the leakage. Wellpoints can be used to abstract this water and to maintain a net gradient into the liner.

Specifications for soil-bentonite backfill have involved a wide range of particle sizes from coarse to fine, with the bentonite content in the range of 2 to 4% by weight, and of the order of 10 to 20% fines. Permeabilities of such material can be as low as 10^{-9} m/s. Silty sand provides the ideal material for this purpose. Clays are suitable, except that hard clays tend to remain as blocks and therefore may give rise to voids in the fill. Cobbles and boulders, as well as roots or other organic material, should be removed before placement of fill.

The principle factors affecting the permeability of a cement-bentonite slurry are the cement-water ratio, the bentonite-water ratio and the mechanical procedure in making panel connections between the fresh cement-bentonite slurry and the set cement-bentonite. It is important that the bentonite by fully hydrated before any cement is added. This means that the cement-water ratio is the factor which controls the ultimate permeability of the backfill. The permeability of a cement-bentonite slurry cut-off wall can be as low as 10^{-9} m/s.

A major factor controlling the deformability of cut-off walls is the properties of the materials of which they are composed. Soil-bentonite walls are quite deformable and do not have problems with regard to cracking. The cement-water ratio has a dramatic effect on deformability of cement-bentonite backfill. For instance, the higher the cement-water ratio, the higher the strength, the more rigid and hence less deformable is the eventual cement-bentonite wall. On the other hand the higher the bentonite-water ratio, the more flexible, the more deformable the wall may be.

Diaphragm walls

Diaphragm walls represent a method of constructing walls in the ground in a narrow trench filled with bentonite slurry. They compare favourably in terms of watertightness, stiffness and mechanical strength with cut-offs formed of steel sheet piling, pre-cast piles or cast-*in-situ* piles. In such cases the wall may be rigid (concrete) or plastic (concrete and bentonite mixture) and load carrying capacity is not required. Diaphragm walls also may be used as load bearing and retaining walls for a variety of applications such as underpasses, deep basements, underground stations, tunnels, docks and pump houses (see Sliwinski & Fleming 1975). Diaphragm walls usually are more economical when used as part of the permanent load bearing structure.

A diaphragm wall allows a small area to be isolated for excavation through water-logged ground by using the diaphragm wall as a sort of cofferdam. In such instances the panels are constructed to form a polygon. Diaphragm walls are composed of panels which are formed successively or alternatively. In the first case panels are formed next to each other in line, one after the other. In the second case primary panels are formed at regular intervals and then the wall is completed by constructing secondary panels between those already in place. Obviously watertight joints are required. In ground where artesian pressure exists, even a slight tilt of the panels relative to each other or inadequate concreting could cause disastrous flooding in deep excavations. This is even more important if the situation is aggravated by tidal fluctuations.

Diaphragm walls possess several features suitable for use in built up areas (see Boyes 1975). For example, they can be constructed immediately adjacent to columns subject to live loadings, to multi-storey buildings and to old foundations. Structural waterproof walls may be formed in advance of the main excavation, which permits above and below ground construction at the same time if necessary. In many applications the interior portions of concrete diaphragm walls can be left exposed as in subway excavations or basement car parks. Furthermore the technique is virtually vibrationless and quiet, and so offers minimum disturbance.

The subsurface conditions should be thoroughly investigated so that any potential problems during excavation attributable to the presence of impervious layers are revealed. In particular the groundwater conditions around the trench should be understood. This includes seasonal variations, as well as potential for dramatic changes in groundwater levels due to anomalous weather conditions. Artesian pressures, springs and artificial sources of water (such as potential broken sewers) should be recorded to avoid instability of the trench during excavation.

As far as the construction of a diaphragm wall is concerned, first, a trench is excavated to a depth of a metre or so. The guide walls are then constructed. The guide walls enable excavation of the main trench to begin below levels where made ground, drains or footings and other surface obstructions might be met. They support the trench against heavy construction surcharge pressure at levels over which the slurry rises and falls, and protect the sides against wave action set up during digging or during the introduction of fresh slurry.

A panel of predetermined dimensions, is excavated between the guide walls (Fig. 1). The width of a panel usually varies between 0.5 and 1 m, although greater widths are excavated when required. Lengths of up to 5 or 6 m represent the maximum that can be concreted conveniently with one tremie pipe (see Millet & Perez 1981). Problems of supply and coordination sometimes arise when more than one tremie pipe is used.

Slurry is introduced into the panel and maintained at as high a level as possible without spilling onto adjacent ground, thereby offering maximum support to the sides (see Anon 1985). Individual panels are excavated by first digging to the full depth at each end, then the central section is excavated. Each grab or drill excavates a full panel width. The depths required in most structural work present little difficulty and depths over 122 m have been reached.

It may prove impossible to construct diaphragm walls by normal methods in soils with very low strengths. This is because the internal pressure of the bentonite occupying the trench may be less than the horizontal total pressure of the adjacent soil. For example, instability commonly occurs in soft marine clays and fresh hydraulic fills. However, as Sliwinski & Fleming (1975) pointed out stability also is influenced by the length and shape of the panel, and arching effects. They recommended that any soil in which $\phi = 0$ and the cohesion is less than 10 kPa should be treated cautiously. In such cases panels should be short and uncomplicated.

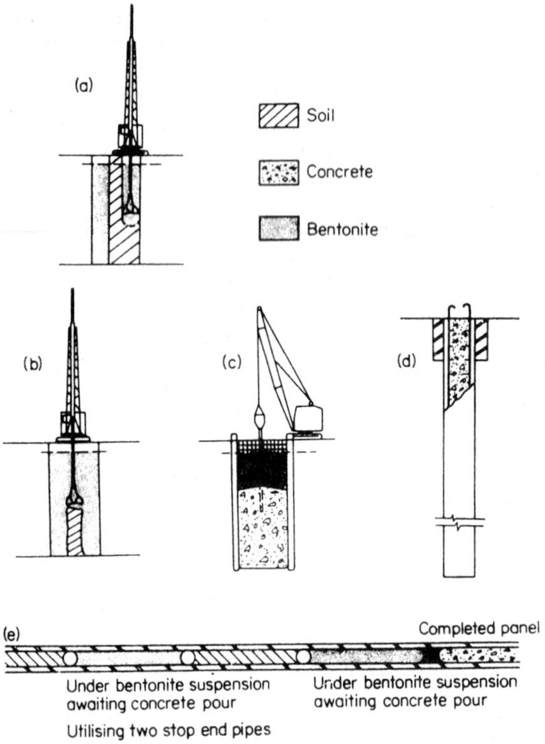

FIG. 1. Diaphragm wall construction. General procedure for the excavation of panels during which time the excavation is kept filled with bentonite suspension. (a) First one end, then opposite end of panel is excavated to full depth. (b) Third and last stage is excavation of centre of panel. General procedure for concreting of panels when the bentonite is displaced by the concrete. (c) Steel stop-end pipes and reinforcement cage positioned and concrete placed through tremie pipe. (d) Section through completed panel showing guide walls and steel. (e) Series of panels in plan. Illustrating variety of uses of steel stop-end pipes as required.

Sinking diaphragm walls into rock frequently poses problems. This is because a standard clamshell grab cannot penetrate hard rock, indeed its efficiency is reduced significantly even in soft rock. Heavy chisels are used to break through occasional hard layers, the rock fragments being reduced to a suitable size to be picked up by the grab. A special percussive reverse circulation drilling machine can be used for penetrating hard rock but its output, according to Sliwinski & Fleming (1975), is painfully slow.

Another disadvantage as far as the construction of a diaphragm wall is concerned is a high water table. The latter hinders the casting of guide walls. It also reduces the differential between the pressure inside the trench and the active pressure of the adjacent soil. Not only does this have a detrimental effect on stability conditions but it also adversely influences the formation of filter cake. Furthermore the unavoidable fluctuation of the level of the bentonite in the trench during digging must be taken into account. It is generally recommended that the level of the slurry in the trench must be at least 1.25 m above that of the surrounding water table. If this cannot be realised, then the trench may be dug within a raised embankment, the water table may be lowered by installing wellpoints or the slurry may be weighted by adding sand or powdered barite.

After the panel has been excavated, any slurry-soil sediment which has accumulated at the base of the trench is removed and replaced with fresh slurry or treated with chemicals. Frequently, cleaning the base of the trench by removing the sludge with a grab proves adequate.

In a reinforced diaphragm wall the reinforcement cage is lowered into place through the slurry and supported from the concrete guide walls of the trench. Reinforcement is made up into cages prior to placement.

The concrete is placed using a tremie pipe. Slurry is pumped from the trench as the concrete is placed, being removed at a similar rate. Pours should be completed in the minimum time (up to 30 m^3/h) and with minimum discontinuity in order to avoid embedment of stop-end tubes, blockages in the tremie pipe and flotations of the cage due to upward drag from stiff concrete. All the concrete for a panel must be poured before setting or significant stiffening occurs. In practice this generally means 3 to 3.5 h.

As the concrete for each panel is tremied in, it moulds against the concave end of the previously cast panel at one end and against a circular steel stop-end placed temporarily at the other end. The knuckle-joint between adjacent panels provides sufficient connection for normal diaphragm wall work. Nevertheless this simple butt joint between panels cannot prevent water penetration, however, significant leakages are rare. This is probably due to the presence of a thin layer of contaminated bentonite at the edges of the joint. Where leaks take place they are associated with differential deflections between wall panels, and these are worst near corners.

Because bentonite cannot penetrate impervious clays filter cake does not form in such ground conditions, hence the suspension is totally displaced by concrete and a perfect soil-concrete contact is obtained. Conversely in pervious sand or gravel, all the free bentonite is displaced but the advancing concrete cannot displace the soil impregnated with bentonite. Furthermore, the harder bentonite filter cake is not removed, nor any soft filter cake which occurs in cavities.

For practical purposes concrete used for diaphragm walls can be regarded as impermeable. Because of the humid conditions in which curing takes place shrinkage cracks due to drying are almost eliminated.

The commonly accepted tolerance on verticallity is 1:80. However, the control which can be achieved is governed by the soil, the plant and the early detection of any deviation. Obstructions such as boulders, or hard strata, especially if inclined, increase the difficulties appreciably and correction normally involves the use of long heavy chisels. Millet & Perez (1981) recorded that walls have been built to depths exceeding 122 m with less than 150 mm deviation from the vertical, a tolerance of 1:800.

Whether successive or alternative panels are constructed, a diaphragm wall consists of a series of independent panels, with a break in continuity at each joint despite tongue and grooving one panel to the next. In order to overcome this disadvantage a special type of form (the CWS form), which consists of a caisson beam made of high strength steel, has been developed (see Dupeuble 1985). This is extracted from the side after the adjacent panel is excavated, not before as are stop-ends in conventional construction. The technique makes use of a rail, which is temporarily fixed in the concrete of the previous panel, to act as a guide for the excavating grab. This ensures continuity of the wall and good quality of joint. Furthermore lateral extraction of the CWS form allows waterstop blades to be installed in the joint, the form being fitted with an additional grooved caisson in which is inserted one or more plastic or rubber waterstops. The free half of these water stops is sealed in concrete when the panel is concreted. Sideways removal uncovers the other half of the plastic or rubber strips which subsequently are sealed in the concrete of the adjacent panel. This method of construction also allows rams to be used to link reinforcement cages in adjacent panels and thereby improve the mechanical integrity of the wall.

The character of bentonite slurry

Bentonite slurries which are used to give support to excavations can vary widely in their properties, however, according to Hutchinson *et al.* (1975) they must:

(1) support the excavation by exerting hydrostatic pressure on its walls;
(2) remain in the excavation and not flow into the soil;
(3) suspend detritus to avoid sludgy layers building up at the base of the excavation;
(4) be easily pumped;
(5) be capable of clean displacement by concrete, with no subsequent interference with the bond between reinforcement and set concrete;
(6) be able to undergo screening or hydrocycloning to remove detritus and enable recycling.

Usually the first three items require thick dense slurries, whilst the latter three are best achieved using very fluid slurries. As a consequence there is a conflict of requirements which has to be resolved in order that an acceptable specification regarding slurry properties can be developed.

One of the basic functions of a bentonite slurry is to form a filter cake layer on the exposed soil as the trench is excavated. The filter cake only develops if the fluid pressure in the trench exceeds any external groundwater pressure and if the permeability of the soil is within certain limits (see Sliwinski & Fleming 1975).

The bentonite slurry is prepared by adding a measured quantity of the powder to clean fresh water. It is well mixed and allowed to hydrate before it is introduced to the trench. Mixing and hydration are necessary if the slurry is to form a good filter cake. The time taken for hydration, after which the bentonite suspension has attained the desired rheological properties, depends on the method of mixing. For instance, a strong shearing action leads to rapid hydration and good performance.

If a bentonite slurry is to exert a stability pressure on a trench with permeable walls, then it must seal the surface with which it comes into contact. This prevents slurry penetrating into the soil.

Surface filtration occurs when a filter cake starts to build up by hydrated particles of bentonite bridging across the entrances to pores in the soil, without penetrating the soil to any great extent. Water continues to percolate from the slurry through the filter cake into the soil during and after the filter cake has formed.

Deep filtration occurs when a slurry slowly penetrates soil, blocking the pores and developing a filter cake within them. In such instances the seal may extend several centimetres into the soil. The concentration of bentonite in the filter cake in surface as well as deep filtration, is greater than in the slurry (typically 15% for a slurry containing 5% bentonite).

In this context, determination of the swelling period, that is, the time which elapses between the end of mixing and the introduction of the slurry into the trench, is important (see Veder 1984). For example, when used in highly permeable soils the swelling period should be short so as to allow the suspension to continue to swell in the trench, thereby preventing deep penetration into the soil with accompanying significant loss of bentonite.

The concentration of bentonite in a slurry has to take into account the often conflicting requirements of trench support, prevention of losses to adjacent strata, reduction of drag and flotation of the grab during digging and the need to prevent too much soil being suspended in the slurry. Normally the concentration of bentonite in slurry varies from about 3% in stiff clays to about 8 to 10 % in dry open ground. For instance, a 4 to 6 % concentration is retained by soil with a permeability up to about 10^{-5} or 10^{-6} m/s. If the permeability of the soil exceeds this, then a denser suspension, ranging up to about 12%, may prove necessary. In exceptional cases, where even this dense fluid is not retained, a number of additives may be tried. However, none of these methods is likely to be very effective if permeability exceeds about 0.5^{-3} m/s. In open ground (k = 0.01 m/s), the sealing mechanism can be changed by including small quantities of fine sand in a slurry, with a resultant significant lowering in the initial loss of fluid. Bentonite entering the trench should contain less than 1 % fine sand to give a good performance. As far as the formation of a filter cake is concerned the presence of fine sand can counteract, to a certain extent, incomplete hydration. The loss of slurry to open dry ground also can be avoided by increasing the viscosity by adding cement (up to about 17 kg per cubic metre of slurry) or by the addition of sawdust, cellulose flakes or shredded clay.

The presence of water in the ground adjacent to a trench can impede the penetration of the surrounding strata. Consequently slurry concentrations may have to be reduced to a level at which they provide sufficient support to the sides only, usually to between 6 to 8% bentonite concentration. Salty and limey water cause slurry to thicken and in extreme cases may lead to bentonite settling out of suspension. In practice, however, these have not proved serious problems. They can be taken care of by adding a phosphate water mix (up to 1% phosphate) to thin the slurry to its original viscosity. Ground consisting of chalk can give trouble by reducing the effective concentration of bentonite during excavation thereby diluting the slurry. Peaty water also has the effect of thinning the slurry. In such cases it may be necessary to increase the bentonite concentration to restore the viscosity. In fact the presence of organic compounds can render the method impracticable.

The pH value is particularly significant where the chemistry of the soil excavated or of the groundwater can give rise to a dramatic change in the pH value of the bentonite slurry. Millet and Perez (1981) proposed that the most desirable range of pH value of slurry is of the order of 6.5 to 10. If the pH value exceeds 10.5, the clay particles in the slurry will have a tendency to flocculate and settle out. If this happens, then a deflocculating agent has to be added to ensure continued effectiveness of the slurry.

From the point of view of wall stabilisation, the higher the density of the slurry the better. The density of freshly prepared slurry varies from about 1.02 t/m^3 for a mix with 4% bentonite to 1.05 t/m^3 for one with an 8% bentonite concentration. The minimum allowable density should be, according to Hutchinson *et al.* (1975), 1.034 t/m^3. In other words it is slightly above that of the groundwater. Millet & Perez (1981) maintained that specified maximum density typically is of the order 1.04 t/m^3 to 1.2 t/m^3. This is to ensure that the tremie process is not impeded. In fact if slurry is to be displaced by tremied concrete its density should not exceed 1.3 t/m^3 and its plastic viscosity should not be above 20 cP (equivalent to a maximum bentonite concentration of about 15%). Fine sand is added to increase the density of a slurry. However, the amount should be limited, again so that the subsequent tremie concreting operations are not impeded. For example, it generally is specified that the sand content should not exceed 5%.

During excavation detritus inevitably becomes mixed with the slurry. Veder (1984) remarked that incorporation of about 5 % fine particles in the slurry does not impair its effectiveness but a higher percentage leads to a number of shortcomings. Not only does incorporation increase the average density of the slurry, which makes excavation by the grab in the slurry more difficult, but it can generate slowly settling layers of sand and silt. These, in turn, produce density gradients in the slurry and lead to build up of sludge in the bottom of the trench. Unfortunately this means that tremied concrete is unable to displace this material from the base of the excavation. Moreover such slurry tends to have a high viscosity, with adverse effects on displacement of bentonite by concrete past the reinforcing cage. Indeed incomplete displacement of bentonite by concrete can give rise to the formation of cavities.

If the amount of detritus in the slurry slows down the rate of excavation significantly, then the slurry must be replaced. It certainly must be replaced before the reinforcement is placed or concrete poured.

If the bentonite slurry is to be re-used, then it must have sandy or silty material removed.

By far the most common form of contamination is by cement, which occurs to a varying degree during tremieing. Contamination by cement leads to the formation of thick permeable filter cakes, very high fluid losses and high viscosities. Hutchinson *et al.* (1975) showed that in a standard bentonite slurry, the effects become noticeable at 0.3% cement contamination, which is the equivalent to a slurry with a pH value of about 11.7. They therefore proposed that this figure should be the maximum pH value of a slurry.

Grouting

Penetration grouting involves the injection of fluids and suspension into pores in soils and discontinuities in rocks. Such penetration grouting is the commonest type of grouting used in construction. However, grout may be used to displace soils as in compaction grouting or to replace soils as in jet grouting. The process is widely used to reduce the permeability or to increase the strength of soil or rock masses. If grouting is to be effective for water cut-off, then it must reliably penetrate the ground the required distance and arrive in a satisfactory state.

The grout used to reduce the permeability of the ground must be able to develop sufficient strength to withstand the hydraulic gradient imposed. Normally a clay-cement grout is used in coarser soils or rock with large fissures and clay-chemical or chemical grouts are used in finer grained soils and fissures (Fig. 2). In the case of chemical grouts, especially on large jobs, cheaper high viscosity grouts may be used to fill the larger voids and more costly low viscosity grouts to fill voids too small to be penetrated by the initial grouting. While it is not possible to state with any certainty what will be the result of any given grouting programme, it is fairly certain that a given degree of watertightness can be achieved if sufficient grouting is performed. In general terms a cut-off with a permeability of 10^{-6} m/s is alluvium and of 10^{-7} m/s in rock is attainable, although in individual cases an improvement of one or more orders of magnitude over these figures has been claimed.

Grouts can be grouped into two basic categories, that is, the suspension or particulate grouts, and solution or non-particulate grouts. Particulate grouts consist of cement-water, clay-water or cement-clay-water mixes. Other materials may be added such as sand, gravel or pulverized fuel ash, to provide bulk when large openings require filling. In the case of cement grouts, water-cement ratios varying from as low as 0.5:1 up to 10:1 or over have been used. Obviously grouts with low ratios are more difficult to inject but they undergo less segregation and give a higher strength than those with higher ratios. It has been suggested that the inclusion of 3% bentonite in a cement grout helps to keep cement in suspension (Little 1975). A deflocculating agent, such as monosodium sulphate, may be used in clay grouts, or a hardener like sodium silicate may be included.

The most common classes of chemical grouts are silicates, lignins, resins, acrylamides and urethanes. The silicate grouts are the most widely used (over 90% of present use). Use of others is limited because of cost or toxicity.

The ability of particulate grouts to penetrate a formation depends upon the particle size of the suspended material in that the particles of the grout must be smaller than the voids which require filling. This has been indicated in terms of a groutability ratio which has been defined as

$$N = \frac{(D_{15})\text{ soil}}{(D_{85})\text{ grout}} .$$

This ratio should not exceed 25 if a grout is to penetrate the formation concerned successfully. Grouting is not possible if the ratio is less than 11. The suggested groutability ratio for fissures in rock has been given as

$$N = \frac{\text{Width of fissure}}{(D_{95})\text{ grout}} .$$

If the ratio is above 5, then grouting is consistently possible while it is not possible if the ratio is less than 2.

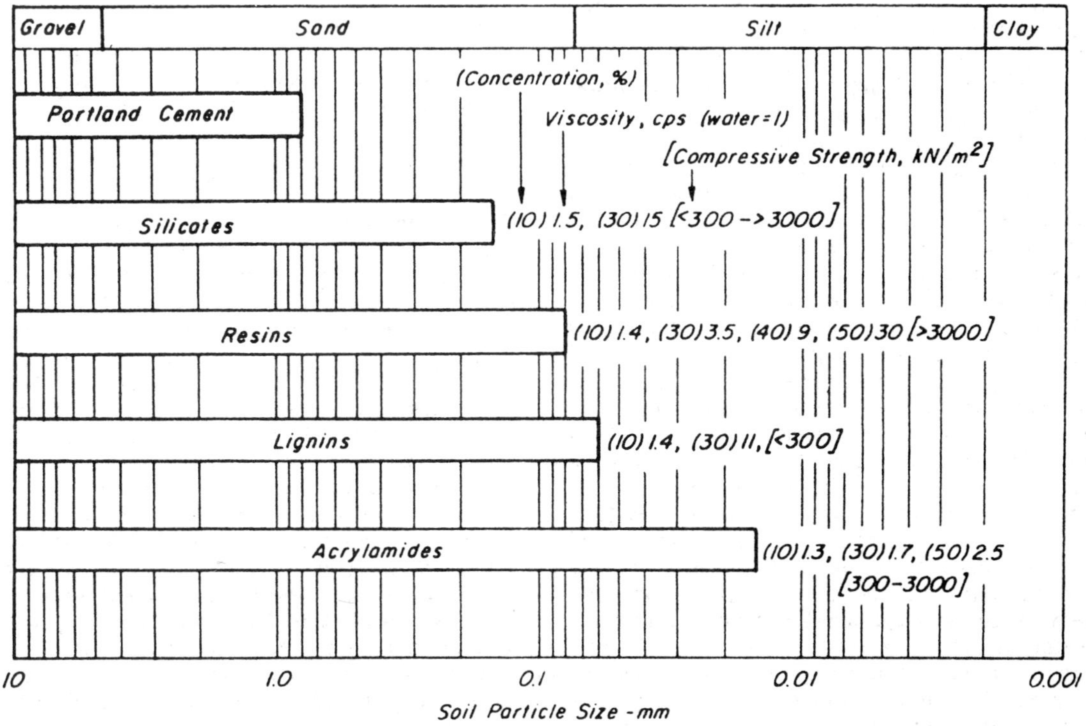

FIG. 2. Soil particle sizes for different grout types and grouted soil properties (Mitchell 1981).

Alternatively permeability may be used to assess groutability. For example, cement grouts cannot be used for treating soils with permeabilities less than 10^{-3} m/s as cement particles would be filtered out during injection. As far as clay grouts are concerned the small sizes of the particles permits them to penetrate voids in soils with a permeability of 10^{-5} m/s. For permeabilities less than this silicate grouts are used, whilst resins may be used in yet finer grained soils. Initial estimates of the groutability of ground frequently have been based on the results of lugeon tests, in which water is pumped into the ground via a drillhole to assess permeability.

Some chemical grouts posssess viscosities only 2 to 5 times higher than that of water. Their gelation times are usually short (10 to 90 mins) and can be controlled. According to Mitchell (1981) soils containing less than 10% fines (< 74 μm particles) usually can be permeation grouted with chemicals. However, effective permeation chemical grouting may prove difficult if the fines content exceeds 15% and when the fines content is greater than 20% such grouting is impossible.

Grout curtains

Grouting has proved an effective method of reducing the percolation of water through the ground. This can be accomplished by forming a grout curtain which involves drilling the grouting lines of holes in the ground to create a barrier or cut-off against excessive seepage. The spacing of groutholes varies from about 1 to 2.5 m apart and grouting takes places either from an open-pipe or a sleeve-pipe (*tube-a-manchette*). The latter method provides better control in terms of the placement of grout.

The rate at which grout can be injected into the ground generally increases with an increase in the grouting pressure, but this is limited since excessive pressures cause the ground to fracture and lift. The safe maximum pressure depends on the weight of overburden, the strength of the ground, the *in situ* stresses and the pore water pressures. However, as there is no simple relationship between these factors and safe maximum grouting pressure, engineers may use hydraulic fracture tests, especially in fissile rocks, to determine the most suitable pressures or simply relate the pressures to the weight of overburden. For example, the pressures used, as measured at the top of the hole, may start at 70 kN/m^2 for the first 3.1 m stage and increase by 70 kN/m^2 in each successive 3.1 m stage, whilst not exceeding 350 kN/m^2 for the fifth and lower stages.

Once the standard of permeability has been decided, for the whole or a section of a grout curtain, it is achieved by split spacing or closure methods in which primary, secondary, tertiary etc. sequences of grouting are carried out until water tests in the groutholes approach the required standard. Houlsby (1976) recommended that in multiple row curtains the outer rows should be completed first thereby allowing the innermost row to effect closure on the outer rows. The upstream row should be the tightest row, tightness decreasing downstream. Single row curtains are usually constructed by drilling alternative holes first and then completing the treatment by intermediate holes.

Ideally a grout curtain is taken to a depth where the requisite degree of tightness is available naturally. This is determined either by investigation holes sunk prior to the design of the grout curtain, or by primary holes sunk during grouting.

Jet grouting

Jet grouting represents a significant advance over conventional grouting and offers an alternative means of forming an impermeable barrier (Coomber 1985). Measured reductions in permeability due to jet grouting have ranged from factors of 100 to 10 000, depending on the permeability of the soil replaced. Replacement jet grouting involves the removal of soil from the zone to be treated, by a high energy erosive jet of water and air. The water and air are jetted under very high pressure from closely spaced nozzles at the base of a triple fluid phase drill pipe, the operation proceeding upwards from the base of a borehole. The grout is emplaced simultaneously. The soil which is removed in brought to the surface by air lift pressure. In granular soils some of the coarsest particles are not removed and are incorporated in the grout. Treatment has been completed successfully to depths in excess of 40 m.

There are three methods of jet grouting, namely, column, panel or wing grouting. In column grouting, the columns are formed by rotating the jet pipe as it is raised. A barrier is produced by overlapping the columns. In panel grouting, jetting takes place from the pipe in a single vertical plane as it is raised. Interconnecting panels form cut-offs of low permeability. Wing jetting uses two nozzles, without rotation, to form a fan-shaped grouted mass.

The use of pre-bored guide holes, usually 150 mm in diameter, facilitates discharge of spoil, helps to maintain verticallity and provides a visual check on the continuity of adjacent grouted areas. Grouting can be terminated at a given level, without disrupting the overlying soil. The pipe is simply withdrawn, restricted grout flow only being maintained to top up the guidehole.

The grouts used must possess a very low viscosity so that pumping can be maintained at the required rate. In addition, abrasive fillers such as sand must be avoided. Apart from these two factors the constituents of a grout mix can be varied to meet the requirement. For example, if low permeability is the principal criterion, then cement-bentonites (5% bentonite) normally are used.

Jet grouting offers the following advantages:

1. Most types of soils can be grouted including silts and clays, and weak rocks.

2. Poor soils in relatively inaccessible layers at depth can be replaced.
3. Low permeability and high compressive strength can be achieved.
4. Grout quantities can be predicted accurately and there is no heave of the ground.

Conversely there are limitations. For instance, the finished product is weaker than concrete, it is difficult or impossible to reinforce rows of columns and panels cannot be regarded as free-standing or cantilever retaining walls. They lack tensile strength and bending resistance.

Horizontal grouted diaphragms

Tausch (1985) described the construction of horizontal grouted diaphragms which are connected to impermeable vertical cut-offs. Such diaphragms are needed when excavations extend below the water table and groundwater lowering techniques, for one reason or another, cannot be used and an impermeable formation does not exist at a suitable depth for the vertical cut-off to be keyed into. Tausch suggested that such diaphragms should be 1 to 2 m in thickness. Because of uplift pressure, they must be constructed at a level significantly below the proposed base of the excavation in order to prevent blow-out into the excavation.

Two methods of construction were outlined by Tausch (1985), one making use of the *tube a manchette* to emplace grout at the required depth, the other employs a valved grout pipe within a drive pipe. The latter is vibrated into the ground and when the chosen depth is reached the drive pipe is withdrawn, again under vibration, exposing the grouting valve. Groutholes may be spaced at 1.1 to 1.5 m, depending upon ground conditions, and are usually set out in a triangular pattern (Fig.3). The grout injected from each hole overlaps with that injected from the adjacent holes. Thick diaphragms are formed by grouting in more than one layer.

Tunnel grouting

Grouting in water bearing ground is frequently carried out ahead of a tunnel face in order to reduce the quantity of water entering the tunnel to readily manageable amounts. Such grouting has been referred to as aureole or umbrella grouting and involved drilling groutholes in advance of the tunnel heading which fan out to form a series of concentric grouted cones. The tunnel is then excavated through the grouting zone. The length of individual holes depends upon the type of ground on the one hand and the quantity and pressure of water on the other. Nonetheless Anon (1962) suggested that in most cases 9 m should prove sufficient. The type of grout used to form the aureole is governed by the ground conditions.

Difficulties frequently arise when tunnels are excavated beneath rivers which contain buried channels. For example, the major problems encountered during the construction of the new Blackwall tunnel beneath the Thames occurred when it went through the buried channel occupied by sands and gravels. A grouted aureole was formed around the periphery of the main tunnel to reduce the permeability of the sands and gravels (which varied from 10^{-3} m/s to 10^{-6} m/s) to 10^{-8} m/s. In addition to the grouted aureole, a series of transverse cut-offs were formed.

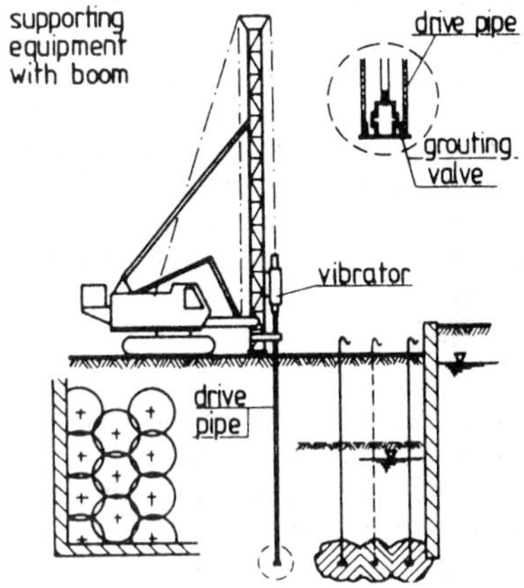

FIG. 3. Method of construction of horizontal membranes.

One of the most serious problems which has to be faced in tunnelling is that of carrying a heading through a fault zone of intensely brecciated rock under a high head of groundwater (Glossop 1960). Again resort can be made to aureole grouting. Grouting often is carried out with cement grout following by single fluid silicate grout. By using suitable mixtures and very high grouting pressures, all but the finest sandy material can be grouted. Even so the finer material is left as isolated bodies which have been sufficiently compressed by the pressure grouting to be more or less impermeable.

Grouting also has been used as a means of reducing or eliminating the flow of water into shafts when the ground conditions are suitable for accepting grouts. Grouting was used in shaft No. 2 at Boulby Mine, North Yorkshire, where saline water in the Bunter Sandstone was at a high pressure. Cement grout was initially used to seal fissures and joints and then, because of the extremely fine nature of the pores in the sandstone, chemical grout of very low viscosity was injected. Grouting took place from one, two or three rings of holes, depending upon the permeability of the rock, which were drilled from the floor of floor of the shaft. The depths of the groutholes varied between 36.5 and 52 m below floor level. Similar grouting operations were associated with shafts sunk at Wistow and Riccall in the new Selby Coalfield, North Yorkshire (Black *et al.* 1982).

Ground freezing

Ground freezing involves the artificial lowering of ground temperatures so that pore water is converted into ice. As such its design is basically a problem in heat transfer and, in the absence of moving groundwater, transfer in the ground is by conduction. Design also requires consideration of stress deformation and creep properties of the frozen ground. The thermal conductivity and the heat capacity of the ground governs the rate at which freezing proceeds. Fortunately the values of thermal conductivity for all types of frozen soil fall within quite narrow limits. This is one of the reasons why artificial ground freezing is a versatile technique, being able to deal effectively with a great variety of soil and rock types (Jones 1981).

Artificial freezing can be employed as an exclusion technique to stop the flow of groundwater into excavations. The frozen ground provides a clear area which is free of bracing and earth supports within which excavation can occur. The barrier to groundwater provided by an ice wall means that there is no need for drawdown of the water table which could given rise to settlements detrimental to adjacent structures.

According to Jumikis (1979) the following factors influence artificial freezing of the ground:

1. The geological and hydrological conditions at the site and their thermal regimen.
2. The type of soil to be frozen, its porosity, water content and amount of ice. Although complete saturation of the soil is desirable, freezing may be carried out safely when the degree of saturation is as low as 10%.
3. The amount and type of clay minerals in the soil. Incomplete freezing of pore water in cohesive soils has been attributed to intersection with the active surfaces of mineral particles (Tsytovich 1959).
4. The position and fluctuations of the water table.
5. The velocity and temperature of groundwater flow. Attempts at ground freezing may be doomed to failure if there is significant groundwater flow.
6. The physical and thermal properties of soil, water and ice. Thermal conductivity varies vertically with each change in water content and lithology. It also changes horizontally from the frozen to the unfrozen ground.
7. The various cryogenic textures (arrangement, shape and size of ice inclusions) of the frozen soil.

The mechanical properties of frozen ground are much more dependent on time and temperature than on the geology of the strata involved. Consequently artificial ground freezing is less sensitive to geological prediction when other methods of ground treatment and may be used in any moist soil or rock formation, irrespective of its structure, grain size or porosity. Weak, running or bouldery ground, and ground in which the pore pressures are high, lend themselves to freezing. Soil and rock profiles, too heterogeneous to grout predictably, also are amenable to freezing. However, in heterogeneous ground the frozen zone tends to be irregular in shape. Khakimov (1966) noted that when clay soils are frozen, vertical cracks may form in the soil between the frozen columns and the soil may bulge near the zone being frozen. The fibrous structure of organic soils tends to be altered permanently by freezing.

Design and construction of an ice wall

Artificial freezing of ground is usually carried out in two stages, which are referred to as the active and passive stages of freezing. Active freezing involves freezing the ground to form the ice wall whilst passive freezing is that required to maintain the established thickness of the ice wall against thawing. The refrigeration plant has to operate at a much higher capacity during the active, than the passive stage of freezing.

The first step in the design and construction of an ice wall is to find out whether the site is suitable for the freezing process. This entails carrying out a site investigation. Then, if the site proves satisfactory, a structurally stable configuration is chosen for the ice wall. When the depth of artificial ground freezing has been finalised, the diameter of the freeze pipe circle is usually determined by surface construction requirements. The frozen wall then has to be designed to ensure safety in the various water bearing formations. Most ground freezing contractors have evolved their own method of calculating the thickness of a frozen wall. When the thickness of the frozen wall has been decided, the number of freeze holes and their spacing; the refrigeration capacity; the diameter of inner tubes, brine mains and ring mains; and capacity of pumps required, is ascertained.

A frozen ground-support system is a massive, relatively rigid structure, which is formed generally before excavation commences. Because of the relatively high compressive and low tensile strengths of frozen soil, curved arched walls, particularly circular walls, provide the best structural solutions and should be chosen whenever possible. An elliptical shape may be used in the case of a rectangular structure if its length-breadth ratio is less than 2.0. Other structural alternatives are a gravity wall, an anchored wall or a braced wall (Braun *et al.* 1979). Although a gravity wall may be constructed in a straight line, the volume of ground which must be frozen varies from two to more than five times as much as that required for a comparable curved wall. Moreover the design of a gravity wall may be governed by overturning and sliding criteria, not stress in the frozen wall. An anchored wall also may be constructed in a straight line but its design is more sophisticated and requires careful field control. The frozen wall may be damaged by inflow of water when the anchors are placed through the frozen wall. This, together with localised thawing, may result in unloading of anchors and unacceptable movement of surrounding frozen ground. As a consequence the use of anchors generally is considered unreliable for most projects. A braced wall is much less sensitive to installation procedures. In addition it can be readily packed or rebraced if required. On the other hand it is more difficult for the contractor to deal with during other subsurface work.

The two principal types of freezing configurations used in tunnelling operations are, firstly, a circular or elliptical frozen zone in which the freeze pipes are arranged horizontally around the perimeter of the tunnel and, secondly, an arched configuration where the freeze pipes are placed vertically or inclined in relation to the axis of the tunnel. The freezing system must avoid bottom heave. The thickness of the frozen ground has to be sufficient to provide an adequate factor of safety against structural failure or excessive deformation. Hence thickness is largely governed by the loading conditions and freezing temperature. The likelihood of frost heave developing with subsequent damage to adjacent structures or underground services also has to be considered in the design of the tunnel system.

The primary plant with pumped loop secondary coolant is the system most frequently used to freeze ground (Shuster 1972). An ammonia or freon refrigeration plant provides the primary source of refrigeration. The advantage of ammonia is that leaks can be detected easily by smell and it is considerably cheaper than freon which is odourless. The system is worked by either electricity or diesel motor and the condensers may be air or water cooled. These plants have a wide range of capacities. Two stages are required to produce temperatures below −25°C. Several different types of coolant have been used with this system. However, the most common type of refrigerant is calcium chloride brine. Other brines have been made from chlorides of sodium, magnesium or lithium. The crystallisation point of the chosen brine should be at least 5°C lower than the ultimate temperatures at which ground freezing will proceed.

Typical brine temperatures for commercial refrigeration plants are around −20°C to −40°C. A temperature of −5°C can normally be attained throughout the zone to be frozen in reasonable freezing times and freeze-pipe spacings.

Heat is removed from the ground by way of probes placed in boreholes. Each probe consists of an external pipe, closed at the lower end, and containing an open-ended inner tube of slightly shorter length. Cooling fluid is normally introduced through this inner tube. Large quantities of coolant have to be circulated to bring about freezing.

The use of thermocouples or other sensors will identify inefficient freeze pipes and areas which are under- or over-cooled. Such situations sometimes can be corrected by varying the flow of brine. The frozen wall should be checked repeatedly to make sure that there is no leakage of brine from the freeze pipes.

All piping above the surface in a freezing system must be insulated to minimise heat loss. Insulation is also necessary in large shallow excavations. The exposed frozen surface should be covered as soon as possible with foam blankets, fibre glass or plastic which is reflective on both sides, because no refrigeration plant can fight solar radiation or open-air convection. A temporary cover is desirable for large excavations which may be open for months. However, in a deep excavation of small diameter such as a shaft, the air is cooled appreciably and as a result protection of the exposed surface of the frozen soil may be unnecessary.

Expendable refrigerants are very attractive for individual projects which only last a few days or for projects where the cost of delay is high (for instance, for emergency situations and initial water cut-off). Although no refrigeration plant is required, the cost of refrigerant is high. Liquid nitrogen (LN_2) provides the fastest and thermally most efficient means of freezing the ground. The use of carbon dioxide is thermally less efficient and normally is more difficult to control.

Indeed, the use of expendable refigerant to freeze ground often proves difficult to control. In other words venting of liquid nitrogen in vertical or horizontal freeze probes produces an irregular frozen zone as well as wasting refrigerant. The irregular nature of the frozen zone is due to the variation in the heat transfer coefficient, it being governed by the quality and velocity of the liquid-vapour mixture. A supply and exhaust manifold with appropriate valves at each end of a series of freeze probes allow a reversal of flow and so help smooth irregularities in freezing.

It is possible to combine brine and liquid nitrogen freezing. In this way the freezing capacity can be raised to meet peak demands, or liquid nitrogen can be used as a standby in the event of failure. Another possibility involves freezing the ground with liquid nitrogen and then keeping it frozen with cooled brine. Liquid nitrogen can be very quickly applied again if unforeseen difficulties arise. Depending upon the circumstances, a decision will have to be made as to whether it is better to inject both the liquid nitrogen and brine into one system of pipes, or into two separate systems.

Ground conditions and artificial freezing

Hydrogeology is one of the most important factors in an evaluation of the feasibility of ground freezing. In particular, the position of the water table plays a significant role in ground freezing because of the importance of water content as far as the mechanical properties of the frozen ground are concerned. Some coarse grained soils which are unsaturated may remain pervious when they are frozen. Hence in such a situation if the water table rises outside of the excavation after freezing has been completed, water may seep through and cause thermal erosion of the frozen wall. A fluctuating water table may occur near rivers or lakes which have significant seasonal variations in level, or along coasts where the rise and fall of the tide are appreciable.

The ground beneath the base of excavation is extremely important. Ideally the base of a frozen wall should be located on an impervious layer which forms a seal preventing the incursion of groundwater into the excavation. Hence there is no need for any significant pumping to control groundwater. Where a frozen wall is not founded in an impervious horizon an excavation has an open base. Consequently extreme caution must be exercised to minimise water movement beneath the frozen wall and to avoid piping or heaving of the floor. Ground freezing is usually much more risky when an open base condition is likely to be met with beneath the water table. This is particularly the case in pervious soils. Braun *et al.* (1979) mentioned that continuous pumping from an open sump should not be used within a frozen wall because of the free water and relative lack of control. Wells provide a more reliable means of disposing of seepage water. The drain should be located at more than three times the thickness of the frozen wall from its outer circumference.

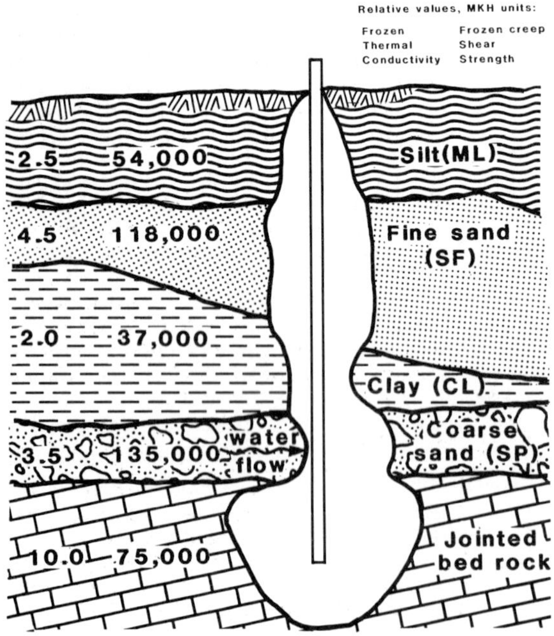

FIG. 4. Effect of heterogeneous ground conditions on shape of frozen zone

Movement of water normally leads to an uneven development of an ice wall in the direction of the flow (Fig. 4). In addition, if the velocity of groundwater flow is too great, the frozen soil columns will not merge thereby leaving windows in the wall. Obviously it becomes impossible to freeze soil if the amount of heat conveyed to the freeze zone by flowing groundwater exceeds that being removed by the freezing system.

It is generally accepted that the normal freezing process is not likely to be successful where the flow of water exceeds 1.5 m/day. In such situations special provisions have to be taken either to lower the flow rate or to cope with the excess heat. If groundwater flow is greater than 1.5 m/day but less than 3 m/day, Braun *et al.* (1979) recommended that the spacing between freeze pipes should be reduced or a second row of pipes should be placed on the upstream side. When the flow exceeds 3 m/day, either the groundwater gradient or the permeability of the formation must be reduced. This is brought about by grouting before or during the installation of the freeze pipes; or by intercepting the flow with wells. Generally the latter method provides a more reliable solution. Liquid nitrogen has been used successfully where groundwater flows of more than 50 m/day have been encountered, but this is expensive.

Where dissolved salts are present in the groundwater, chemical analysis should be made and the freezing temperature should be ascertained. Sea water presents no special problems in this respect since its freezing point is only about 3°C below that of fresh water. However, when a volume of saline water in granular soils has been confined by a perimeter array of probes, further inward advance of the frozen wall causes a progressive increase in the salinity of the unfrozen fluid. In narrow excavations the salinity of the water in the unfrozen core is commonly several times that of the original groundwater.

Frost heave may occur in some soils when they are frozen, depending on their permeability, and *in situ* confining pressures. Such movement is due to the conversion of pore water to ice on the one hand and to secondary frost expansion which results from migration of pore water on the other. However, if the soil drains at the same rate as the progress made the freezing front, expansion does not occur. Usually clean free-draining sands and gravels are not susceptible to either type of frost expansion.

Both forms of expansion occur at the same time, although they differ in amount. Expansion due to freezing of pore water amounts to a maximum of 9% of the volume of the water in the pores if it is completely frozen, and it does not increase with time. The second and more important cause of expansion in frozen soil in attributable to pore water migration under capillary action which leads to the development of segregated ice at the freezing front or in the frozen zone. This does not give rise ground movement if the confining pressure is greater than that developed by the freezing soil-water system.

Further causes of movement associated with artificial ground freezing are a consequence, firstly, of stress relaxation upon excavation and, secondly, of thaw consolidation. If the soil was previously relatively dense, then the amount of consolidation is small. This situation only occurs in the coarse grained frozen soils containing very little segregated ice. On the other hand some degree of segregation of ice is always present in fine-grained soils. For example, lenses and veins of ice may be formed when silts have access to capillary water. Under such conditions the moisture content of the frozen silts exceeds significantly the moisture content present in their unfrozen state. As a result when such ice-rich soils thaw under drained conditions, they undergo large settlements under their own weight. As ice melts and settlement occurs, water is squeezed from the ground by overburden pressures or any applied loads. Excess pore water pressures develop when the rate of ice melt exceeds the water discharge capacity of the soil.

ACKNOWLEDGEMENTS: We wish to extend our thanks to Mrs Anne Geraldine Blades who assisted in the preparation of this document by typing the script.

References

ANON, 1962. Cement grouting. Progress Report of the Task Committee on Cement Grouting. *Proceedings of the American Society of Civil Engineers; Soil Mechanics and Foundations Division,* **88,** SM2, 49-98.

ANON, 1985. Dock wall at Middlesbrough formed by secant piles. *Ground Engineering,* **18,** No. 5, 33-35.

ANON, 1985. Specification for cast in place concrete diaphragm walling. *Ground Engineering,* **18,** No. 6, 11-19.

BLACK, J.C., POLLARD, A.C., & DAW, G.P., 1982. Hydrogeological assessment and grouting at Selby. *Proceedings of the Conference on Grouting in Geotechnical Engineering,* New Orleans. Ed by Baker, W.H., American Society of Civil Engineers, 665-679.

BOYES, R.G.H., Dec 1975. The versatility of diaphragm walling. *Civil Engineering,* 57-59.

BRAUN, B., SHUSTER, J.A., & BURNHAM, E., 1979. Ground freezing for support of open excavations. *Proceedings of the 1st International Conference on Ground Freezing,* Bochum, Ed by Jessberger, H.L., Engineering Geology, Special Issue, **13,** 429-454.

COOMBER, D.B., 1985. Groundwater control by jet grouting. *Preprints Conference on Groundwater in Engineering Geology, Sheffield,* Engineering Group of the Geological Society, 485-498.

CORNFIELD,, G.M., 1975. Sheet pile structures. *Foundation Engineering Handbook,* Ed by Fang, H.Y., and Winterkorn, H.F., Van Nostrand Reinhold, New York, 418-444.

DUPEUBLE, P., 1985. C.W.S. system provides load-bearing and watertight joints between diaphragm wall panels. *Ground Engineering,* **18,** No. 6, 22-26.

GLOSSOP, R., 1961. The invention and development of injection processes. *Geotechnique,* **11,** 255-279.

HOULSBY, A.C., 1976. Routine interpretation of the lugeon water test. *Quarterly Journal of Engineering Geology,* **9,** 303-314.

HUTCHINSON, M.T., DAW, G.P., SHOTTON, P.G., & JAMES, A.N., 1975. The properties of bentonite slurries used in diaphragm walling and their control. *Proceedings of the Conference of Diaphragm Walls and Anchorages,* London Institute of Civil Engineers, 33-40.

JONES, J.S., 1981. State-of-the-art report, Engineering practice in artificial ground freezing. *Proceedings of the 2nd International Symposium on Ground Freezing,* Trondheim, Ed by Frivik, P.E., Janbu, N., Saetersdal, R. & Finborud, L.I., Engineering Geology, Special Issue, **18,** 313-326.

JUMIKIS, A.R., 1979. Cryogenic textures and strength aspects of artificially frozen soils. *Proceedings of the First International Symposium on Ground Freezing,* Bochum, Ed by Jessberger, H.L., Engineering Geology, Special Issue, **13,** 75-85.

KHAKIMOV, K.R., 1966. Artificial Freezing of Soils: Theory and Practice. *Israel Program for Scientific Translations,* Jerusalem.

LITTLE, A.L., 1975. Groundwater control by exclusion. *Methods of Treatment of Unstable Ground.* Ed. by Bell, F.G., Newnes-Butterworths, London, 37-68.

MILLET, R.A. & PEREZ, J.Y., 1981. Current USA practice: Slurry wall specifications. *Proceedings of the American Society of Civil Engineers, Journal of the Geotechnical Engineering Division,* **107,** GT8, 1041-1056.

MITCHELL, J.K., 1981. Soil improvement — state-of-the-art report. *Proceedings of the 11th International Conference on Soil Mechanics and Foundations Engineering,* Stockholm, **3,** 509-565.

NORTH-LEWIS, J.P. & LYONS, G.H.A., 1975. Contiguous bored piles. *Proceedings of the Conference on Diaphragm Walls and Anchorages,* London, Institute of Civil Engineers, 189-194.

RESSI, A & CAVALLI, N., 1985. Bentonite slurry trenches. *Engineering Geology,* **21,** 333-339.

SHUSTER, J.A., 1972. Ground freezing for temporary ground support. *Proceedings of the 1st North American Conference on Rapid Excavation in Tunneling.* Chapter 49, 863-879.

SHUSTER, J.A., 1981. Engineering quality assurance for construction ground freezing. *Proceedings of the 2nd International Symposium on Ground Freezing. Tronheim.* Ed by Frivik, P.E., Janby, N. Saetersdal, R. & Finborud, L.I., Engineering Geology, Special Issue **18,** 333-350.

SLIWINSKI, Z & FLEMING, W.G.K., 1975. Practical considerations affecting the construction of diaphragm walls. *Proceedings of the Conference on Diaphragm Walls and Anchorages,* London, Institute of Civil Engineers, 1-10.

TAUSCH, N., 1985. A special grouting method to construct horizontal membranes. *Proceedings of the International Symposium on Recent Developments in Ground Improvement Techniques,* Bangkok, Ed by Balasubramanian, A.S., Chandra, S., Bergado, D.T., Younger, J.S. and Prinzl, F., Balkema, Rotterdam, 351-362.

TOMLINSON, M.J., 1980. *Foundation Design and Construction.* Pitman, London.

TYSTOVICH, N.A., 1959. The fundamentals of frozen ground mechanics (new investigations). *Proceedings of the 4th International Conference on Soil Mechanics and Foundations Engineering.* London **1,** 116-119.

VEDER, C., 1984. The construction of diaphragm walls -some causes for failure and proposals for their avoidance. *Ground Engineering,* **17,** No. 7, 13-17.

BELL, F.G., Department of Civil and Structural Engineering, Teesside Polytechnic, Middlesbrough, UK.
MITCHELL, J.K., Department of Civil Engineering, University of California, Berkeley, California, USA.

Groundwater control by jet grouting

D.B. Coomber

ABSTRACT: Control of groundwater using familiar injection or permeation grouting techniques has experienced a chequered history, with technical success and cost control remaining difficult to predict. This results mainly from the excessive dependence of grout acceptance and therefore treatment continuity, on the permeability of the ground requiring stabilisation.

Over a decade in Europe generally and relatively recently in Britain, jet grouting has emerged as a tried and tested means of minimising these uncertainties and increasing confidence in the grouting of soils. Soil structure is destroyed using high energy erosive jets while simultaneously either mixing grout with the disturbed soil particles *in situ* or largely *replacing* eroded ground with tremmied grout.

The process has been used to (i) seal cofferdams against or around buried structures or installations, (ii) seal joints between closely spaced but non-contiguous piled walls, (iii) exclude water from basement excavations, (iv) stabilise shaft bases against piping and (v) treat unstable ground for tunnelling purposes.

Following a brief description of the technique this paper illustrates these applications, with appropriate reference to examples.

Introduction

Treatment of soils by grouting, to minimise permeability, to enhance inherent strength, or both, may be broadly classified as follows:

(a) *Permeation grouting* –
where natural pore spaces in the ground are injected with cement slurries or chemicals.

(b) *Displacement or compaction grouting* –
increasing the strength of soils by displacement with a thick mortar under high pressure from the base of a borehole casing.

(c) *Replacement or jet grouting* –
simultaneous controlled injection of simple cement grouts to replace soils eroded by high pressure jets.

Permeation grouting covers the familiar methods long established throughout the world, though less prominent in Britain, where geological soil conditions are frequently unsuitable, than in many other countries. Compaction grouting is seldom concerned with groundwater control and is therefore outside the scope of these remarks. The principles and applications of jet grouting are described below. Practical experience demonstrates that in moving through the above list from (a) to (c) the user of the technique becomes markedly less dependent on soil properties and variations in the ground requiring treatment.

Achievement of continuous treatment between grout positions, so essential in controlling groundwater, together with prediction of grout take and therefore cost, is determined by the natural three-dimensioned permeability of the ground. This property, particularly its potential for variation, is arguably the most difficult to predict even from rigourous site investigation in all soils, but especially glacial deposits.

The success of jet grouting in recent decades has been in minimising technical and economic uncertainty produced by soil conditions and increasing confidence of both specialist and non-specialist engineers in soil grouting.

The jet grout method

Jet grouting evolved around the erosive capabilities of jets of water under high pressure. The principles are allied to those of jet cutting and blast cleaning technologies, regularly reviewed by the British Hydraulics Research Association e.g. Brown (1981).

The technique takes two main forms, the most striking differences being noted below:

(a) *Mixing in situ* uses only a single grout jet both to break down soil structure and to force mixing of grout with the natural soil particles. Column shaped grouted masses are formed, with a diameter usually less than one metre. Work is carried out through a borehole of the same diameter as the grout rods with little slurried spoil exiting to the surface, but because fluid pressures are temporarily trapped in the ground, heave at the surface can sometimes develop. Large specialist pumps are required to pressurise grout to several hundred bars.

From Cripps, J.C., Bell, F.G. & Culshaw, M.G., (eds), 1986, *Groundwater in Engineering Geology,* Geological Society Engineering Geology Special Publication No. 3, pp 445-454

(b) *Partial replacement* is achieved using compressed air and water jets to erode and remove soil for simultaneous replacement by grout through a third jet beneath. Column, panel and 'wing' shaped grouted elements (see below) are formed for differing applications. Column diameters in excess of three metres can be obtained in good conditions. Work is performed through an oversize borehole using the annulus as exhaust channel for substantial volumes of slurried spoil containing the eroded soil. This exhaust facility, if properly maintained, prevents the risk of ground heave.

The following method description deals with the second of these categories, which is more flexible in application, includes and exceeds the technical capability of the first and is currently prevalent in Britain.

This method utilises a 90 mm o.d. grout stem containing three concentric tubes carrying water in the inner, compressed air between and grout in the outer tube. At the base of the stem is mounted an adaptor or 'monitor' fitted with nozzles to convert flows into horizontally directed jets (Fig. 1). With the exception of the monitor, grout stem and triple fluid swivel assembly, together with a high pressure water pump, all plant and equipment used is interchangeable with that required for medium or large permeation grouting projects.

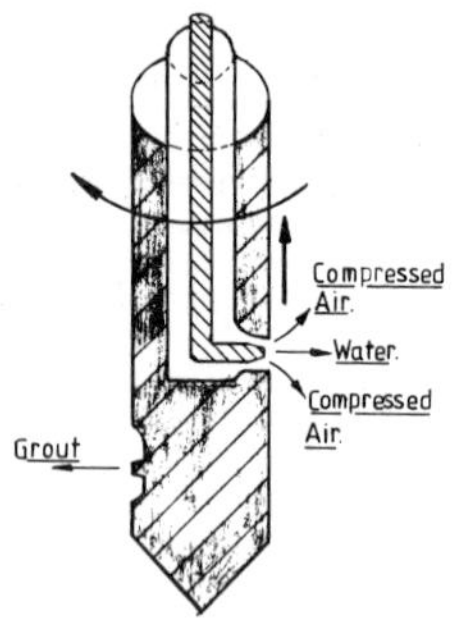

FIG. 1. Diagrammatic view of jet grout monitor.

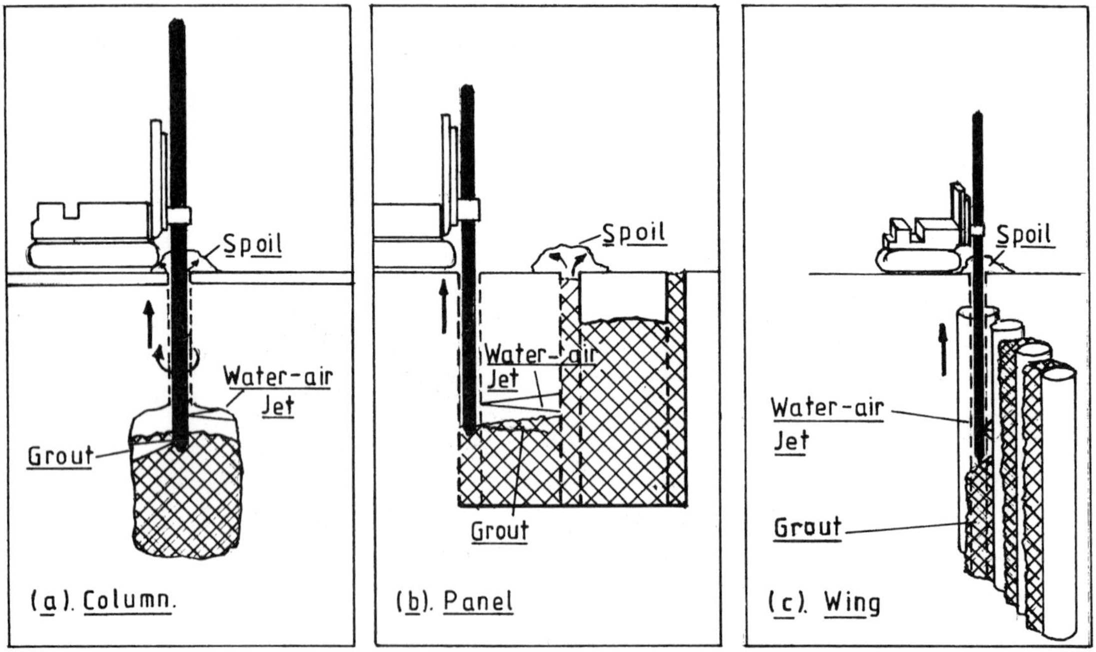

FIG. 2. Jet grout construction.

The high pressure water jet, expelled from a nozzle some 2 mm in diameter and surrounded by a collar of compressed air through a concentric annulus, is used to break down soil structure and flush out soil particles to the surface. Simultaneously with erosion, grout is tremmied in through an 8 mm to 9 mm nozzle below, to replace the soil removed. The process is outlined in Fig. 2 and described in greater detail by Yahiro & Yoshicki (1973), Anon (1982) and Yahiro *et al.* (1982).

The commonest form of jet grouted element constructed takes the shape of a column, built by controlled continuous withdrawal and rotation of the grout stem with the jets active. Columns (Fig. 2a) are formed through a single pre-bored hole which also provides an exhaust channel for disposal of the slurried erosion spoil to the surface. Columns are usually interlocked to form continuous two or three dimensional grouted blocks to a required size. The leaflet also describes formation of a continuous cut-off using inter-connecting panels (Fig. 2b) of grouted ground in slots cut between adjacent carefully spaced pre-bored holes, without rotation of the monitor. A further variation known as 'wing' jet grouting (Fig. 2c) uses twin water jets set 90° apart in plan, with a grout jet centrally below, again without rotation. The result is a wedge shaped grouted mass sealing buried vertical joints to prevent groundwater movement e.g. gaps between closely spaced piles in bored pile retaining walls.

With each form, grouting can be terminated at any level by switching off the erosive jets and topping up the pre-bored hole with grout. Treatment need not therefore extend to ground level unless demanded by the application in question. Pre-bored holes are often formed separately in advance and stabilised with a weak bentonite slurry or, occasionally, a temporary casing, prior to grouting. In some circumstances, however, a drill bit may be fitted to the grouting stem, such that the complete operation becomes continuous.

Importance of soil type

Detailed soil properties inevitably become relatively unimportant when considering jet grouting feasibility, owing to the inherent capacity of an erosive jet to break down soil structure *in situ*. If it is accepted that any soil structure can be destroyed when subjected to sufficient passes of a jet of sufficient axial energy, it follows that virtually all soils are, at least technically, suitable for treatment. (Figs. 3 and 4).

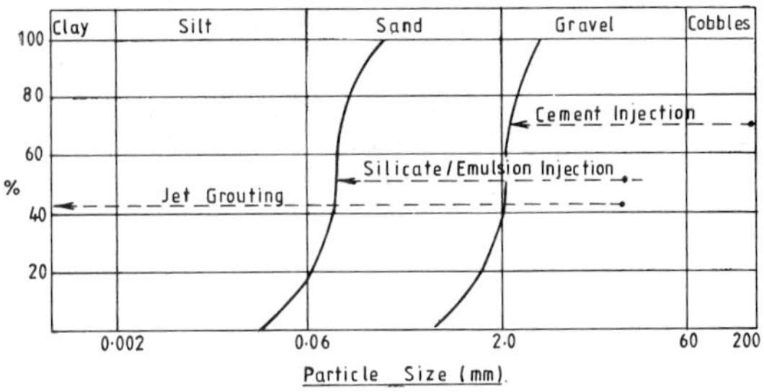

FIG. 4. Limits of grout acceptance by particle size.

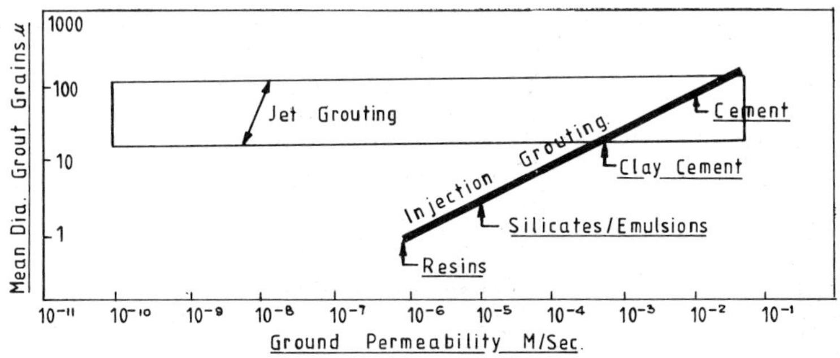

FIG. 3. Limits of grout acceptance by permeability (Modified after Cambefort 1977).

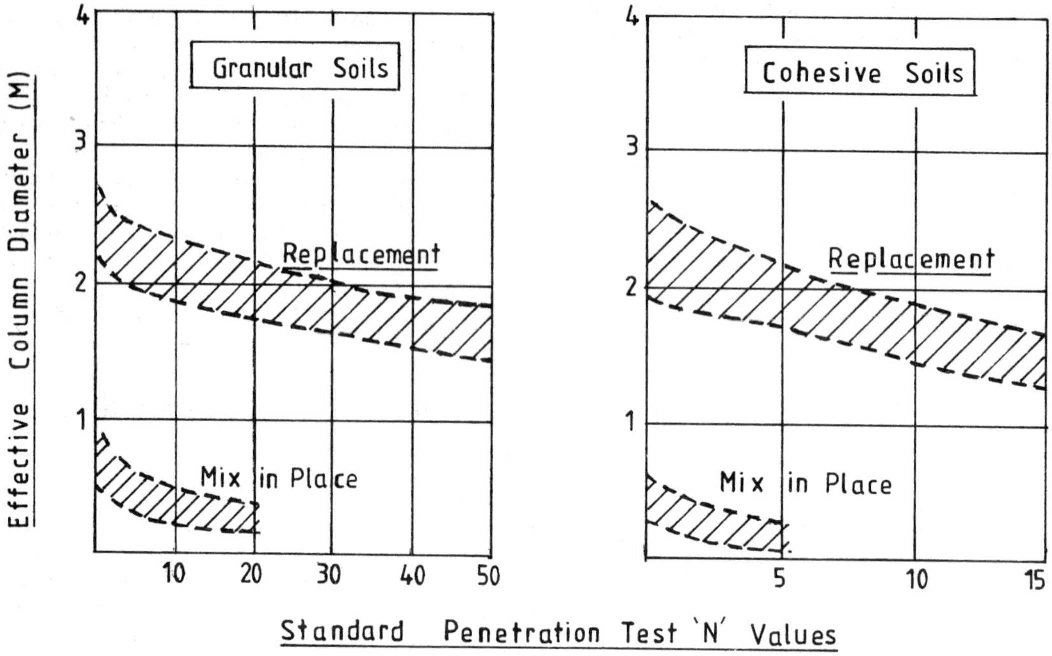

FIG. 5. Column diameter by *in situ* density (after Miki, 1984).

Nevertheless accurate qualitative soil description and, to a lesser extent, quantitative assessment of *in situ* density in a predominantly granular soil, or sheer strength in a cohesive material, remains the key to achieving optimum jet grouting *design*. This is never more true than when treatment is aimed at groundwater control, where attainment of continuity between adjacent grout positions is usually paramount. These properties dictate the penetrative capacity of a jet of given axial energy exposed to an element of soil for a given time period, and hence the size of cavity created for simultaneous replacement by grout. Fig. 3 however, demonstrates that *in situ* permeability, so fundamental to the success of historical grouting methods, yet so difficult to predict in advance, has little bearing.

Miki & Nakanshi (1984) publish empirically derived data on typical jet grout column diameter, relative to standard penetration test 'N' values and demonstrate the increase in diameter attainable by the triple jet 'replacement' method, by comparison with single jet 'mix in place' techniques (Fig. 5).

A multiple of 2 to 4 kN/m^2 for each blow of 'N' value used in normally consolidated cohesive soils, provides a rough approximation to undrained shear strength.

These curves, however, relate to constant, though unspecified, construction working parameters. It is evident that the latter play as great a part as soil properties if not greater, in determining the diameter of an individual jet grout column, or length of each jet grout panel. A full list of working parameters might read as follows:

(a) water jet energy (flow and pressure),
(b) compressed air jet energy,
(c) grout jet energy,
(d) nozzle sizes and shapes,
(e) grout rod (or monitor) withdrawal and rotation speeds.

Experience shows that, of these, water jet pressure and monitor withdrawal rates are by far the most significant. Extensive exposure of completed columns used in practice for temporary works applications has allowed derivation of a series of empirical design curves, relating pressure and distance from the nozzle to the time of exposure to the jet of an element of ground (withdrawal speed). An example is shown in Fig. 6, for medium dense sands or loose clayey silts, at typical water jet pressures. Curves of a similar nature can be used for other broad soil categories (e.g. loose sands, soft clays, dense silts etc.).

Increased cohesive proportions in the soil produce smaller diameters for the same jet pressures and lifting speed, or require increased energy and/or decreased withdrawal rates, if the same diameter is to be achieved. At a typical working water pressure of about 400 bars, most of the energy is absorbed in breaking down the cohesion of soils having shear strengths greater than 40 kN/m^2 and column diameters of more than 1.4 m are seldom achieved. At lower shear strengths, breakdown is rapid and secondary erosion takes place to produce larger diameters.

Particle size distribution has little effect on jet penetration (diameter) in sand or silt dominated mixtures with uniformity coefficients (D_{60}/D_{10}) of 8 or more. Lower coefficients denote the most easily eroded soils of all, with diameters of 3 m or more becoming feasible at normal operation parameters.

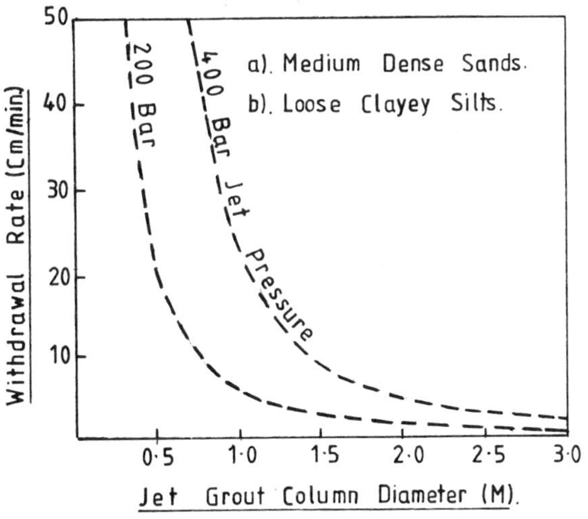

FIG. 6. Typical design curve.

Exceptions to the general guidelines set out above may occur if the proportion of medium to coarse gravel, cobbles, boulders etc., exceeds about 50%. In these circumstances the average particle size may be large in relation to jet diameter. Reflection of the jet may result before attaining its maximum penetration, with reduction, or irregularity, in column diameter. At worst, soils of this type, which may be better treated by historical grouting methods, may preclude use of the thin elements of panel jet grouting for cut-offs.

In any event treatment by jet grouting of gravel dominated soils, fills containing random very coarse ingredients, peats and very prominently structured stiff clays require individual consideration with respect to the objectives envisaged.

Grout mixes

Mixes used are very simple slurries, usually without additives, aimed at high volume usage at minimum unit cost and requiring little comment here. Some features worth noting are listed:

- Almost exclusively cement-based, although use of chemicals is feasible.
- Cement/PFA combinations predominate in ratios from 1:1 to 1:10 by weight.
- Cement/bentonite often preferred for flexible cut-offs etc. where low permeability is principal criterion.
- Fillers coarser or more abrasive than PFA are avoided because of wear on equipment and nozzles.
- Water to solids ratios usually in the range 0.6 to 1.0, depending on application, or up to 4.0 with cement/bentonite.
- Grout cube compressive strengths usually selected from 0.5 to 15.00 N/mm^2 at a given age, with generous tolerance for limitations of cube testing on grouts.
- *In situ* strength is enhanced in granular soils, as residual soil particles are absorbed as aggregate. Conversely remnant clays absorbed in the grouted mass can reduce *in situ* strength by up to 50%.
- Typical set grout permeabilities range from 10^{-6} to 10^{-9} cm/s.

Groundwater control applications

General

The broad field of groundwater control, particularly as a form of temporary works, covers the greatest frequency of use of jet grouting. All three forms, panel, wing and column grouting are used according to requirements, with the latter being most frequent. Occasionally control of lateral groundwater entry into excavations may be combined with other functions of the grouted mass, such as side support. An example of such a case is described by Coomber & Wright (1984).

In most cases, applications may be classified as joint sealing; shaft bases; soft ground tunnelling or cut-off related.

Joint sealing

Either wing or column elements are used to seal vertical joints between closely spaced but non-contiguous tubular or bored piles in retaining walls (Fig. 2c) or gaps in cofferdams, due to interruption by existing services (Fig. 7). The first of these, with the piles providing structural retention and the jet grout seal minimising groundwater entry, overcomes many of the practical difficulties with secant bored piling and usually stands favourable cost comparison, both with secant piling and diaphragm walling. Wing jet grouting is normally employed when gaps, after allowing tolerance, are about 350 mm or less, with columns substituted for wider spacings.

The same approach offers a convenient means of sealing an intersection between a cofferdam and existing underground installations or structures, following construction of the cofferdam by conventional means. This concept is amplified by a British project in Fig. 8, with sheet-piled walls linked to previously installed precast concrete ring shafts. The soils treated were an interesting feature of this work. They included an awkward transition, at depths up to 16 m, from fine granular alluvium to the weathering products of the underlying mudstones and sandstones of the Millstone Grit Series, which would have defeated conventional permeation grouting techniques.

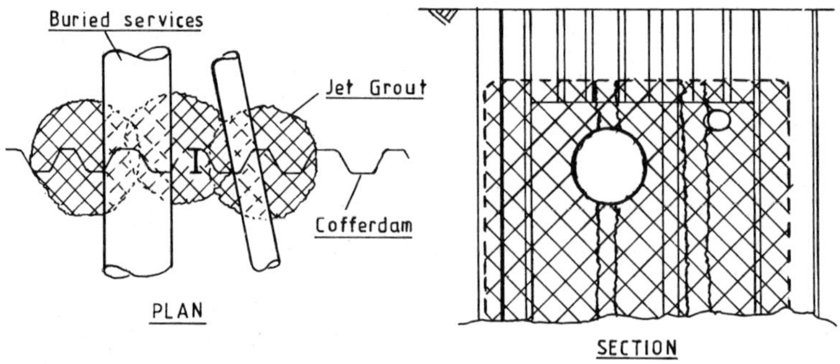

FIG. 7. Joint sealing around existing services.

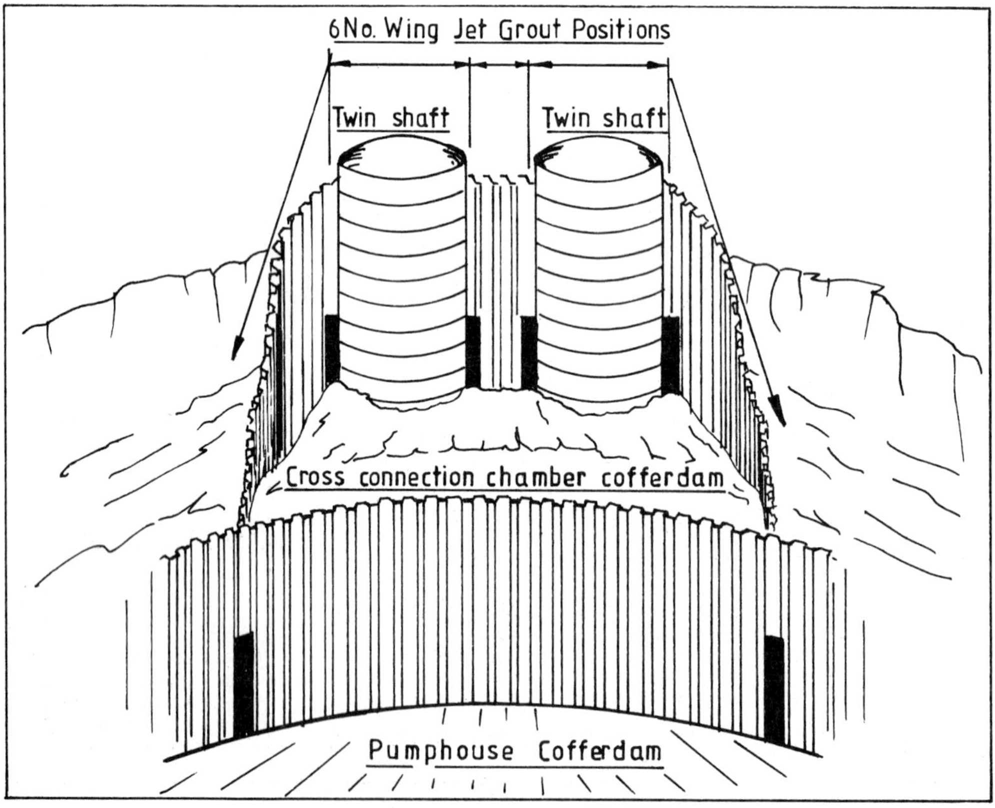

FIG. 8. Wing jet grouting to seal ends of cofferdam.

It is possible to theorise at length, using an assumed or measured value for permeability of the soils and a design value for grouted ground, about the extent to which water entry to an excavation will be reduced. In practice virtually all, if any, lateral seepage after treatment, will be concentrated at the critical interface between grouted mass and pile, or buried structure.

For this type of application grouting is usually carried out from a position only some 400 mm from the interface in question. This places little demand on the penetrating capability of the erosive jet and allows much of its power to play in the interface itself, regardless of soil type. This results in the interface being scoured clean by the jet. Examples in which rust or latence have been removed in this way have been visually observed. Consequently the bond achieved at the interface by the simultaneously placed grout is particularly positive.

Shaft bases

Frequent use has been made of column jet blocks to counteract boiling, piping or heave of the base of shafts associated with soft ground tunnelling projects, or installation of deep machine pits. In the process groundwater entry through the floor of the excavation is reduced to readily manageable proportions and is sometimes exceeded by entry through the sides, at joints in steel sheet or secant piles, if pile installation conditions have proved difficult at shallow depth. In many cases the length of embedment of side supports may be reduced as a result of the propping effect of the grouted slab. Recent experience, however, shows that it is often attractive to leave a limited thickness of unexcavated ground between formation level and the top of the grouted zone. Apart from gaining some increase in stabilising weight, this provides useful protection for the grouted ground against damage during excavation and a level in which sumps can conveniently be installed to deal with seepages.

Fig. 9 summaries the dimensions of a selection of such problems treated on this principle in Britain, Germany and Japan.

Brief study of Fig. 9 quickly establishes that the weight of the grouted plug and overlying soil to formation level alone, would be quite inadequate in resisting hydrostatic uplift forces. The majority of the resistance is generated by vertically downward forces around the periphery of the grouted plug. Opinion is divided on whether these are frictional forces generated by active earth pressures on the side walls, or a function of shear strength bond between the grouted mass and the side walls themselves. The latter strength is potentially high, for reasons explained above. Both analyses are often checked in practice.

Soil Treated	Z (m)	S (m)	H (m)	h (m)	d (m)
Alluvial silts/ sands	10.5	14.5	7.0	2.0	2.0
	13.0	10.5	9.5	2.2	2.0
	7.5	10.5	4.0	Nil	1.5
Alluvial silty sand	10.6	3.5	6.9	Nil	2.0
Thanet sands	13.0	6.0	9.2	Nil	2.0
Gravelly sands	7.0	2.0	6.1	1.6	1.5
Alluvial clayey sands	10.0	8.0	9.0	Nil	2.6
Very soft clays	12.9	10.0	11.9	Nil	2.9
Loose sandy silt	14.7	5.0	12.4	Nil	3.0
Alluvial sands	13.4	7.5	9.6	Nil	2.5

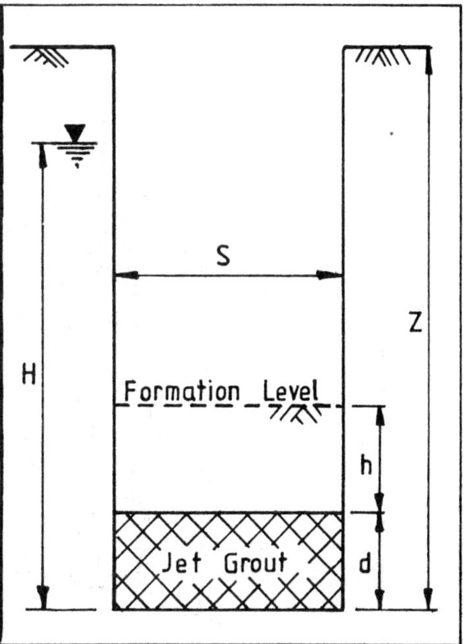

FIG.9 Jet grout treatment of shaft bases.

For this approach to be valid, the grouted mass must be considered as a monolithic slab, achieved by the replacement characteristic of the grouting process. The slab thickness can then be checked in lateral compression and bending due to uplift forces.

Soft ground tunnelling

Even after recent technical advances in soil tunnelling methods and tunnelling machines, a recurring difficulty is presented by unstable ground in the critical starting or reception zones immediately adjacent to shafts. The shield must be some distance from the shaft wall with the eye of the tunnel broken out, before sufficient bentonite slurry or compressed air stabilising pressures can be established, in the case of a starting section, or maintained in a reception section. Consequently stabilised grouted blocks are used for short distances along the axis from the shaft wall, preferably bonded to the back of wall (Fig.10). Blocks are constructed after installation of the shaft walls, but before they are excavated.

In Japan, for over a decade, this application has provided the more frequent single outlet for column jet grouting, while similar examples internationally are now becoming numerous.

The blocks are created using fully interlocking arrangements of jet grout columns, of a diameter suited to soil conditions in each case. The grouted mass is then sufficiently monolithic to allow its dimensions and design strength to be predicted from a geotechnical assessment of the forces acting upon it. Minimum all round cover (t) to the external tunnel diameter and length of the block (l) are related to over burden pressure and lateral earth pressures respectively, bearing in mind a selected grout strength. Analysis of the resistance required on the potential failure plant AA may be included, for starting blocks. The lateral earth pressures and, therefore, the length of the block are less demanding for reception blocks as the shield will have entered the treated ground before the shaft walls are breached.

A survey of completed examples, on tunnels constructed at depths of between 10 m and 30 m, showed that cover to the tunnel diameter (D) in range $0.35D$ to $0.65D$ was used, with lengths of starter blocks from $0.75D$ to D+2 m. Reception blocks were often about 60% of the length of starting blocks. It is possible that dimensions of the treated zone may also be influenced by particular tunnelling machine characteristics, or by the occasional secondary use of the treated ground as a thrust block, while tunnelling away through the opposite shaft wall.

A balance must be struck between dimensions of the grouted block and grout strengths. Increased cover and length may be preferable to the use of grout compressive strengths higher than the range 2 to 5 N/mm^2 at 28 days. At higher strengths, tunnelling progress may be seriously impaired, albeit only over short distances.

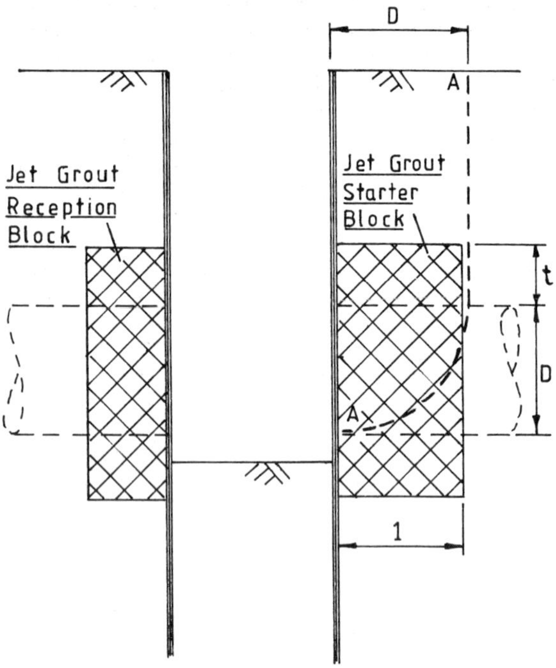

FIG.10. Jet grout blocks for tunnelling.

If sheet piling is used to form the shaft walls, it should be borne in mind that deflection of the piling during excavation may develop. At worst, this may produce a narrow gap between the rear face of the piles and the grouted block previously bonded to it. If suspected this gap is best filled by simple injection grouting through the face after shaft excavation and before breaching of the wall. With more rigid wall construction, of course, the possibility does not arise.

It is arguable whether control of groundwater is the principle objective of this application. Provided a very high proportion of the volume of ground treated is stabilised, quite substantial water flow through it can normally be handled. Nevertheless the presence of groundwater is of course the paramount source of instability at the outset. In practice treatment design is usually aimed at 100% stabilisation. The principle has been less frequently adopted for other tunnelling problems, such as short connecting lengths of tunnel between neighbouring shafts, or tunnel faces partly in rock and partly in overburden, the latter as reported by Tornaghi & Perelli Cippo (1985).

Cut-offs

Grouted cut-offs or barriers to underground movement of fluids or gases are often specified for temporary works purposes, repairs or extension of existing dam cores, effluent containment, flood control, leachate control etc.

Aschieri *et al.* (1983) reported on the use of rows of interlocking jet grout columns to provide a cut-off around a newly constructed cooling water outlet basin on a power station development. Spring lines had developed in the basin on flooding of the outlet channel, despite inclusion of a diaphragm wall in the original construction. Quite extensive laboratory and site testing of the work ended with seepage flows into the basin, after the second successful de-watering, measured and maintained at 0.5 to 0.8 l/s under an external head of 6 m. This represented a decrease by a factor of about 700 to 800 in the bulk permeability of the loose silty sands with organic inclusions in which treatment was performed.

It is also mainly in this category of application that panel jet grouting (Fig.2b) occúrs. Adjacent panel elements are linked, often on a slightly staggered pattern in plan, to form a continuous flexible diaphragm of low strength and very low permeability usually less than 350 mm wide. In European countries where the use of king piles and close boarding is popular for excavation support, panels may be combined with dewatering to allow new basements to be formed without influencing neighbouring shallow foundations. (Fig.11).

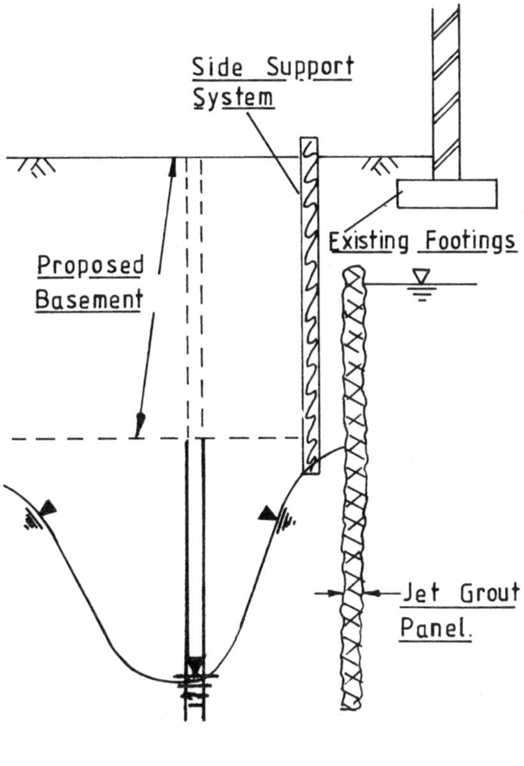

FIG. 11. Panel jet grout for basement excavation.

Yahiro *et al.* (1975), on the other hand, illustrate 150 mm thick panel jet grouting used to intercept underground water movement naturally flowing towards excavations for a nuclear fuel plant. In this case the cut-off was located 2 m behind the crest of a 5 m high side slope formed in sands. The cut-off extended through sands and gravels to key into low permeability mudstones below.

In all cases where sufficient monitoring has been performed after panel jet grouting reductions in permeability by factors of between 100 and 10,000 have been recorded, the greatest occurring, of course, where initial permeabilities are highest. If initial permeabilities are very high, however, resort to column jet grouting for continuous cut-off construction may be necessary because of high gravel contents (see above).

Few completed applications have required the cut-off to withstand hydraulic gradients greater than 100, despite the very limited width of the panels. It has been suggested by Jefferis (1981) that carefully designed cement/bentonites should be capable of resisting gradients well over 500 and possible as high as 2000. Early results from recent work with panel jet grouting in weathered jointed Keuper sandstones beneath a high dam in Germany appear to support this contention.

Conclusion

Jet grouting techniques, now thoroughly established internationally, are capable of decreasing technical and financial uncertainty associated with historical soil grouting. The process has been widely used for many forms of both temporary and permanent groundwater control problems.

References

ANON, 1982 Jet Grouting Comes to Britain. *Ground Engineering,* **15,** No.2., 32-33.

ASCHIERI, F., JAMOILKOWSKI, M. & TORNAGHI, R., 1983. Case History of a Cut-off Wall executed by jet grouting. *Proceedings of the 8th European Conference on Soil Mechanics and Foundations Engineering, Helsinki.*

BROWN, R. (ed), 1981 Jet Cutting Technology - a review bibliography. *BHRA Fluid Engineering Series* Vol.9. BHRA. Cranfield, Beds.

CAMBEFORT, H., 1977. The principles and applications of grouting. *Quarterly Journal of Engineering Geology,* **10,** 57-95.

COOMBER, D.B. & WRIGHT, P.W., 1984. Jet grouting at Felixstowe Docks. *Ground Engineering,* **17,** No. 5.

JEFFERIS, S.A., 1981. Bentonite-Cement Slurries for Hydraulic Cut-offs. *Proceedings of the 10th International Conference on Soil Mechanics and Foundation Engineering,* Stockholm 435-440.

MIKI, G & NAKANASHI, W., 1984. Technical progress of the jet grouting method and its newest type. *Proceedings of the International Conference on In-Situ Soil and Rock Reinforcement,* Paris, 1, 195-200.

TORNAGHI, R. & PERELLI CIPPO, A., 1985. Soil Improvement by Jet Grouting for the Solution of Tunnelling Problems. *Proceedings of the 4th International Symposium on Tunnelling*. Paper 10 Institution Mining and Mettalurgy.

YAHIRO, T. & YOSHICKI, H., 1973. Induction grouting method utilising high speed water jet. *Proceedings of the 9th international Conference on Soil Mechanics and Foundation Engineering*, Moscow.

YAHIRO, T,, YOSHIDA, H. & NISHI, K., 1975. The development and application of a Japanese grouting system. *Water Power and Dam Construction* Feb

YAHIRO, T,, YOSHIDA, H. & NISHI, K., 1982. Soil improvement method utilising high speed water and air jet. *Proceedings of the 6th International Conference on Jet Cutting Technology*. BHRA Guildford.

COOMBER, D.B. GKN Colcrete, Thorp Arch Trading Estate, Wetherby, West Yorks LS23 7BJ, UK.

Some aspects of groundwater control by the ground freezing and grouting methods

J.S. Harris & C.A. Pollard

ABSTRACT: Prior to the construction of a new mineshaft, hydrological and geophysical tests are performed on a shaft centreline borehole and the recovered core which permit water-bearing zones to be located, and potential shaft inflows to be estimated. An assessment of this information enables the most appropriate method of groundwater control to permit safe and controlled shaft sinking to be selected. The three methods discussed in this paper are grouting, freezing and pressure relief.

Grouting is the process of reducing the permeability of water-bearing strata by the injection of appropriate materials under pressure, thereby reducing the shaft inflow to an acceptable level.

The geometry of the injection hole pattern required to form a grout cover, and the selection and subsequent injection of a suitable grout are discussed. An in-shaft test, performed from the advancing shaft sump into the water bearing strata, is described. This test gives a final check on the local aquifer characteristics, and allows the grouting scheme to be modified as required. A similar test performed after grouting enables the efficacy of the treatment to be assessed.

Introduction

In mining works, particularly when shaft sinking, it is frequently necessary to deal with the hazard of groundwater at depth. The threat this poses depends upon the transmissivity, pressure, and nature of the water bearing strata as these dictate the potential inflow, and determine whether the water can simply be pumped away, or has to be dealt with by appropriate ground treatment methods.

This paper briefly discusses methods of assessing the location and possible magnitude of groundwater inflows, and then describes several of the methods of ground treatment used in the mining and civil engineering industries to facilitate excavation in wet and/or weak ground conditions. These methods are:

- artificial ground freezing (AGF), which by refrigeration of the ground, converts the pore-water to ice to bind together the soil particles to form a structural cofferdam. The effect is temporary, and the hydrological regime is restored after the thaw;
- grouting, which is the injection of fluids, with or without fillers, into a rock or soil to fill the voids and thereby reduce its permeability. The effect will normally be permanent;
- pressure relief, which involves the installation of relief wells extending beyond the excavation line to reduce the regional pore water pressure. This technique is often employed in conjunction with grouting.

The various geotechnical processes available are shown in Fig. 1 related to the particle size across which they may be applied. As can be seen, AGF by virtue of its range of application might prove particularly useful in mixed ground conditions.

Hydrogeological assessment for ground treatment

Prior to any major mining activity, it is important that exploratory boreholes are drilled from the surface to enable a detailed hydrogeological investigation of the ground to be carried out. Information derived from:

- the borehole drilling record;
- geological log of recovered core;
- laboratory tests on recovered core samples;
- borehole geophysical logs;
- borehole hydrological tests;

is used to establish

- the location and extent of any water bearing formation, together with their respective transmissivities, water pressure, and the nature of the water flow paths within them (i.e. through pores or cracks/fissures or both);
- the natural strength of each formation;
- the chemical composition of any ground waters;

and an appraisal of this data, together with

- the maximum allowable water inflow (and hence the required reduction in permeability);
- the required strength of each zone;

and factors relevant to treating the ground by freezing or grouting/pressure relief, such as

- availability of a grout suitable for treating each zone;
- required thickness of frozen ground;

From CRIPPS, J.C., BELL, F.G. & CULSHAW, M.G., (eds), 1986, *Groundwater in Engineering Geology,* Geological Society Engineering Geology Special Publication No. 3, pp. 455-466.

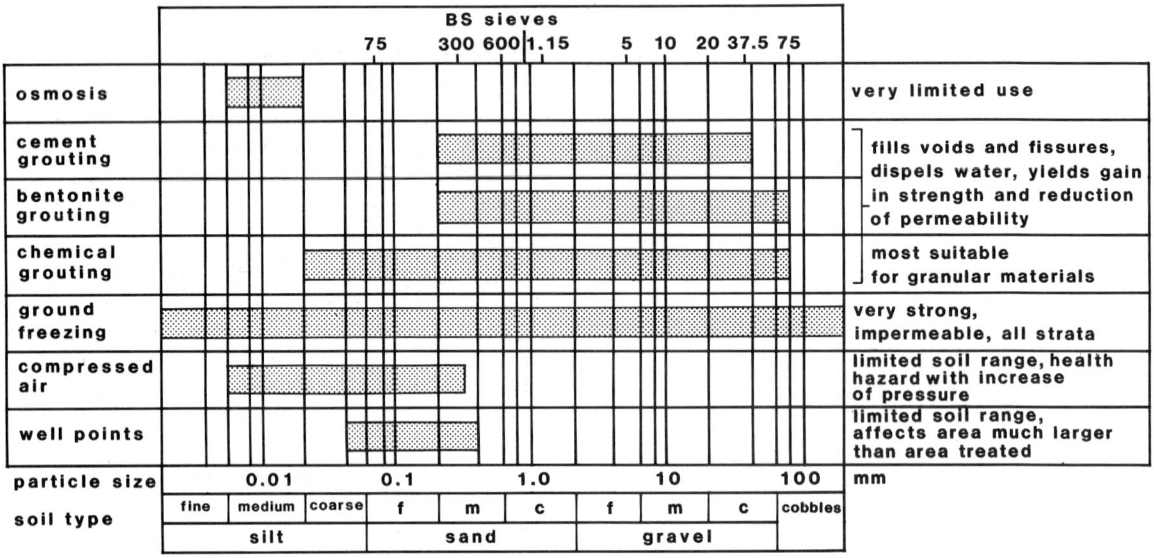

FIG. 1. Applicability of geotechnical process according to soil type.

— estimated total time for ground treatment and subsequent excavation;
— estimated total cost;

allow the appropriate method of ground treatment to be chosen.

These borehole investigation methods, the information which may be derived from them, and the choice of ground treatment method are discussed in more detail below.

Borehole drilling records and core logs

Careful observation of losses or gains in drilling flush returns and penetration rates give preliminary indications of dry open zones and wet surcharged zones. The geological log of recovered core, together with core loss, RQD, and fracture index will enable zones of high permeability arising from, for example, coarse grained sandstones, and fractured vuggy limestones, to be identified.

Laboratory testing of recovered core samples

Core samples will be subjected to a range of tests to provide data for use in the design of both the underground excavation, and the ground treatment method which will be required to enable the rock mass to withstand the hydrostatic stress after excavation, before the permanent lining is installed. These tests will include, for example, the unconfined compressive strength (UCS) of virgin and of grouted rock, vertical and horizontal permeabilities, porosities, also thermal properties such as thermal conductivity, specific heat capacity, and creep tests at various deviator stresses performed at sub-zero temperatures.

Borehole geophysical logging

Geophysical logging provides a continuous downhole record of certain parameters which are related to the physical properties of the rock and the fluid it contains. Radiation and sonic logs provide an indication of density, from which the porosities of the various lithologies may be interpreted. The flowmeter log (which responds to minute flows into and out of the borehole) can be used to locate water-bearing fissures, and the results may be processed to yield permeability coefficients. Competent areas of rock and friable washout zones may be confirmed by the caliper log, and the temperature log yields data of use in freeze design and/or the selection of chemical grout.

Borehole hydrological testing

Of the many hydrological test methods available for a single borehole, those methods which monitor the rate of recovery of aquifer pressure after a period of controlled flow have emerged as the more suitable for testing the deep, multiple, confined aquifers which are likely to be encountered in mining operations. These tests include the drill stem test (DST) as developed by the petroleum industry service companies, and the pump-packer pressure recovery test Daw (1984).

The drill stem test uses a single or double inflatable packer system to isolate a test section of borehole, and by remote operation of a test valve from surface, permits groundwater flow to occur into a previously empty drill stem. After a measured flow period, the test valve is closed and the rate of groundwater pressure recovery in the isolated test section is monitored by a downhole pressure gauge. Analysis of the resulting pressure recovery rate by a variety of methods yields the aquifer transmissivity and undisturbed piezometric head.

The interpretation of DST data becomes intractable when very strong aquifers are encountered at relatively shallow depths, due to rapid filling of the drill stem and large variations in the inflow rate. To overcome this, the pump packer system was developed which enables a sensibly constant flow rate to surface to be maintained during the flow period by using a submersible pump. The aquifer pressure is relayed back to surface via 2-wire downhole electronic pressure transmitters. The pressure recovery data is treated in a similar manner to DST data.

Correlation of, and assessment of, borehole data

The information derived from these borehole investigations should be correlated, and the characteristics of those zones which may require ground treatment for water control and/or stabilisation prior to excavation, should be identified.

The transmissivity and groundwater hydrostatic pressure in the aquifer zones are used in the estimation of the inflow rates to the underground excavation. For preliminary estimates of inflows to shafts, the equations governing radial horizontally confined flow into a well from a laterally extensive aquifer may be used. For more complex excavations, or dipping or anisotropic aquifers, then estimates of inflows can be made using an electrical analogue, where permeability is analogous to electrical conductivity, and piezometric head is analogous to electrical potential. These predicted inflows may then be used to determine whether the aquifer zone is likely to give rise to:

— nuisance water, which can be ignored, or
— water which can be controlled by collection and pumping to surface, or
— water which must be controlled by ground treatment.

The correlation of the geophysical logs with zones of rock with low UCS or high core loss will identify those regions of incompetent strata which require ground treatment to enable the rock mass to withstand the imposed geostatic and hydrostatic stress after excavation has occurred.

Selection of ground treatment process

A flow chart is included (Fig. 2) to illustrate the logic followed when selecting the most technically appropriate ground treatment process.

As noted a costing exercise should be carried out, in parallel with the technical appraisal, to estimate the eventual total expenditure that will be incurred in completing construction of each section of the works requiring ground treatment. Such an exercise should place a value on any time savings/excesses that can be foreseen, as well as the more obvious direct costs. The suitability and merits of each method can then be made on technical and financial criteria, and a choice made.

Artificial ground freezing

Artificial ground freezing (AGF), originally practiced to aid vertical excavations (open pits and shafts) in weak and/or saturated strata, is also a reliable expedient to aid tunnel construction in similar circumstances.

Despite its apparently high unit cost, it has been found that when economies elsewhere on a contract are taken into account a nett saving results.

The advantages of the method are significant and include:

— high strength;
— impermeability;
— non-contamination of groundwater;
— no induced vibration;
— stable groundwater level.

Frozen conditions are created by circulating a cold medium through a series of freeze tubes; thus if they can be so positioned that, when the ice-column associated with each merges with its neighbours, an enclosing ice-wall is formed then the space to be excavated will be protected by a strong cofferdam.

Circular ice-walls are preferred, but the configurations of the site and the desired structure will influence the shape of the ice-wall chosen. Some examples are given in Fig. 3.

This principle has been applied to shafts, drifts and tunnels with equal facility. Frozen shafts have been sunk to over 600 m and pits of up to 40 m diameter. Service tunnels of 1.5 m diameter to road tunnels of up to 14 m diameter have been constructed within frozen ground.

All strata can be frozen. A natural moisture content (m.c.) of at least 11% is normally required to achieve impermeability coupled with a significant improvement in strength. Dry zones can be irrigated

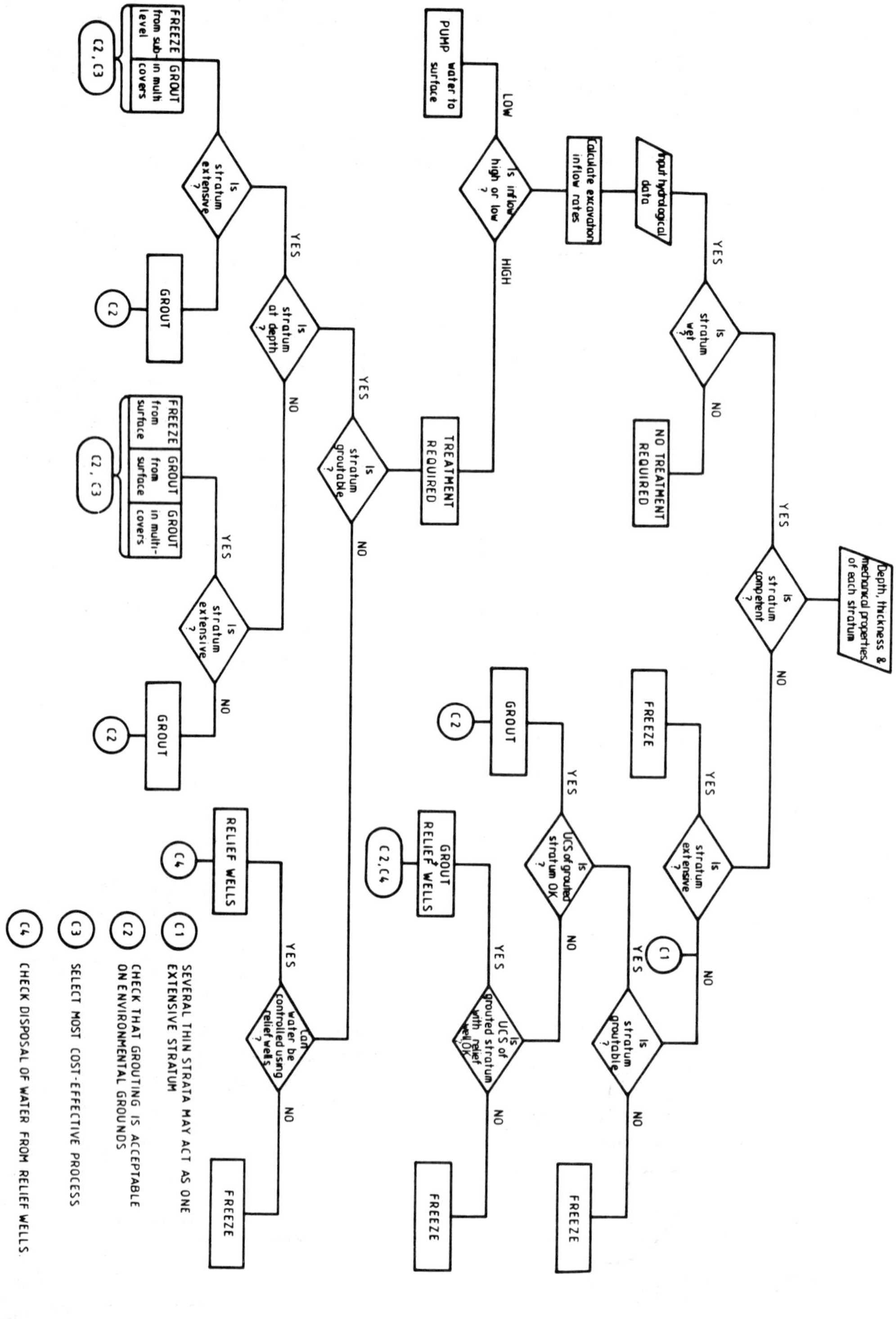

FIG. 2. Selection of process.

to raise the moisture content to an appropriate value. The frozen strength of any soil or rock type varies according to the moisture content and the temperature below freezing point to which it is subjected as illustrated in Fig. 4. Frozen ground behaves in an elasto-plastic manner, although in many excavations the plastic element is negligible within the temperature-timescale being practiced.

A combination of high ground pressure and small negative temperatures may, in some frozen soils, lead to measurable time dependent creep. A high rate of creep dictates special techniques and monitoring.

The strength and creep properties can be measured in the laboratory on specimens recovered from *in situ* frozen locations and on specimens frozen in the laboratory under simulated conditions (Gardner *et al.* 1982). Constant rate of strain refrigerated triaxial compression tests are performed to measure the shear strength parameters c and ϕ at sub-zero temperatures to −20°C. If appropriate, constant stress tests can also be performed to measure the creep properties. A specially designed apparatus to measure these parameters is shown in Fig. 5.

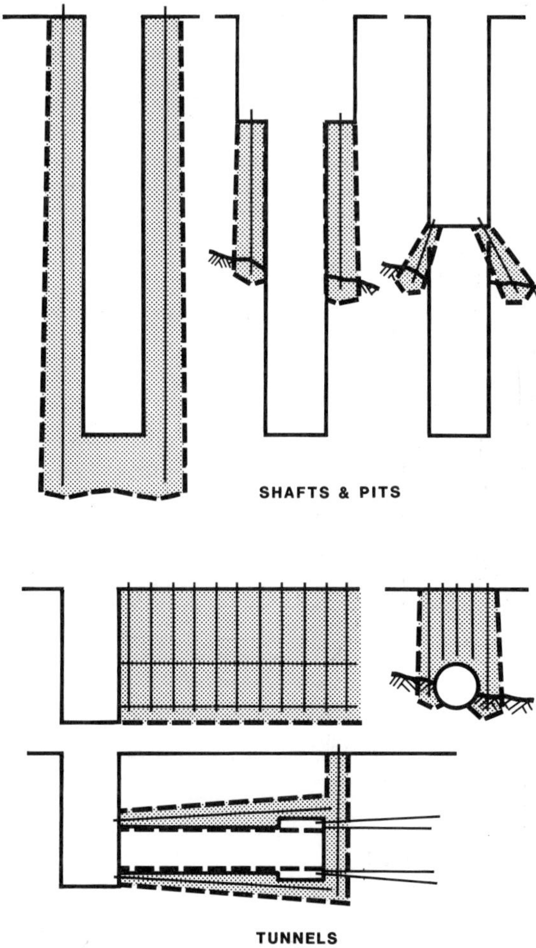

FIG. 3. Alternative ground freezing schemes.

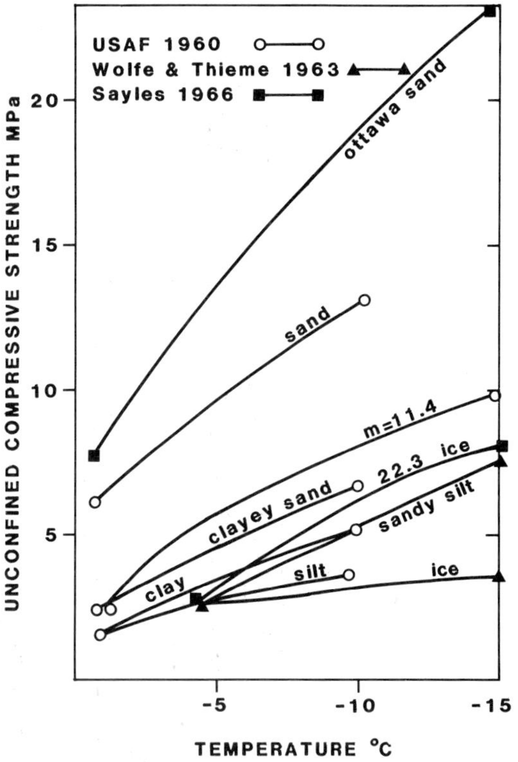

FIG. 4. Dependence of unconfined compressive strength on temperatures for various soils.

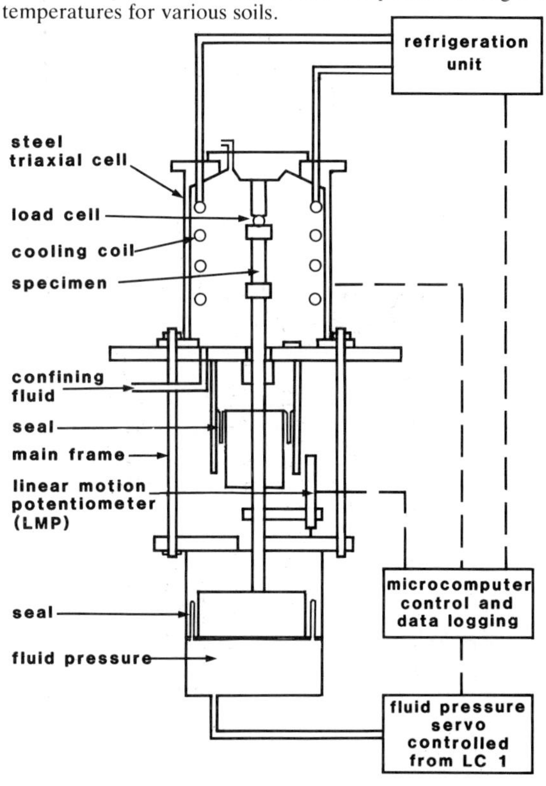

FIG. 5. Refrigerated triaxial compression apparatus.

When the frozen strength and creep properties have been established the ice-wall thickness necessary to withstand the ground pressure loading can be calculated. For shallow vertical cylindrical excavations one of the strength equations based on thick cylinder theory proposed e.g. Domke (Lame) or Vialov (Mohr-Colomb) may be adequate. Structural analyses for deep shafts in weak soils and tunnels are based on the finite element method which can deal with complex temperature distributions, elastic stress behaviour and plastic deformation (time related creep). The design can therefore take account of the length of time that the excavation will remain unlined. A design flow chart is given in Fig. 6.

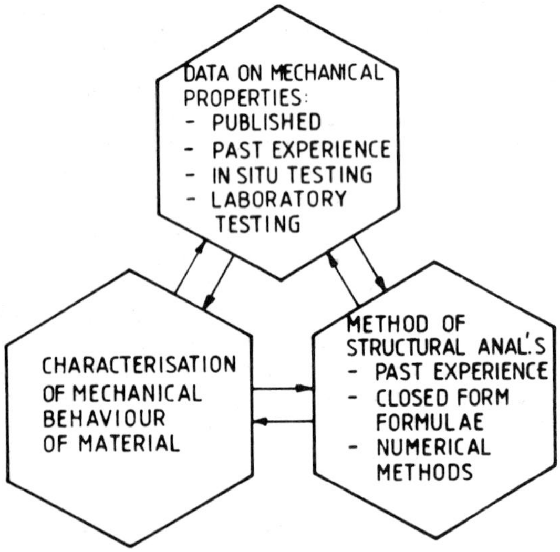

FIG. 6. AGF design considerations.

Design of system

Flowing water constitutes a heat source/cold sink if it passes through or past a refrigerated body. Natural flows associated with rivers or tidal fluctuations must be identified and evaluated; artificial flows associated with wells are best avoided or located remotely from the site. At best sites subject to small flows will require greater refrigerative capacity and/or an increased cooling period; at worst the quantity of heat to be extracted may exceed practical refrigerative effort, and thus preclude use of the method.

Strata containing dissolved salts will freeze at depressed temperatures. In extreme cases either liquid nitrogen or an alternative secondary refrigerant may have to be employed.

The geothermal gradient will affect the refrigerative capacity and time required to achieve the frozen state. Near ground-surface the temperature is commonly around 15°C in the UK, whereas at depth it can reach 25°C in the Middle East and up to 45°C in the Indian sub-continent.

The change in the volume of water as it is chilled is well known. Much of this volume increase is dissipated by expulsion of water ahead of the freezing front, and is a function of the degree of saturation and the permeability of the ground. Such heave as occurs at this stage is of very small, usually negligible, order.

A more significant explanation for groundheave, of an order likely to cause distress, is when a plentiful supply of water has access to the ice-front in a frost susceptible soil.

At the interface between frozen and non-frozen conditions a pressure difference arises (which has been explained in terms of capillary theory) where the pore water pressure is less than the ice pressure which remains sensibly at atmospheric. The resulting suction draws water continuously towards the freezing front. In appropriate circumstances the supply of water results in ice lenses which displace the ground above and heave is observed (Jones 1982).

Distress can be in the form of distortion of the foundations of very close neighbouring structures and services, or of stretching of the freeze tubes themselves. It has been shown that, even in heave susceptible ground careful control of the amount of refrigeration supplied to the ground once the required volume of frozen ground has been created virtually eliminates such effects (Bebi & Mettier 1979). It is, of course, particularly important to recognise a potential heaving condition and make appropriate provision in the design of the associated works and in the manner of applying and controlling the freezing process.

Procedure

A procedural flow chart is given in Fig. 7.

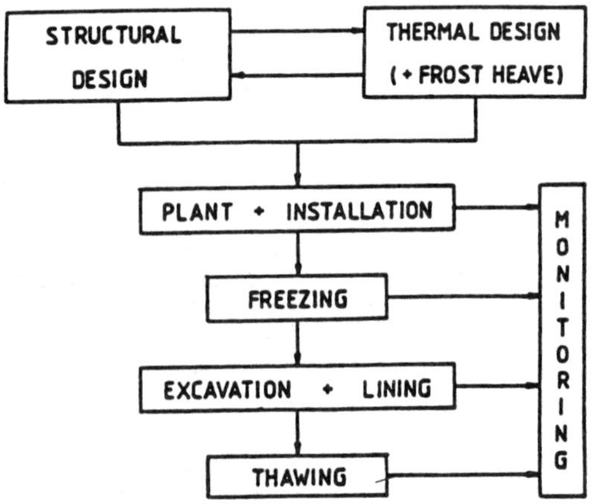

FIG. 7. Artificial ground freezing procedure (after Jones 1982).

Refrigeration is usually supplied by one of two systems, which are compared in Table 1.

- On-site mechanical plant comprising motor driven compressor, condenser and evaporator, and associated floats and pumps. The compressor may be a two-stage reciprocating unit or of the screw-type. Such plant comprises two substantial 'packages' and requires electrical and water service connections. Ammonia or freon are used as the primary refrigerant and calcium chloride brine as the secondary refrigerant or (chilled) heat-transfer medium.
- Off-site production of cryogenic liquid, e.g. liquid nitrogen, liquid carbon dioxide. Such refrigerants are delivered to site storage vessels, and serve as an expendable source after one cycle through the freeze-tube system. No service connections are required.

TABLE 1.

	Mechanical e.g. ammonia –brine)	Cryogenic e.g. liquid nitrogen
Services	Electricity Water	None
Noise	Significant	Negligible
Typical freeze period	3 to 12 weeks	3 to 7 days
Temperature: freeze tube	–25 to –35°C	–100 to –196°C

Installation of freeze tubes

Shell and auger or overburden drilling may be employed for shallow freeze hole construction; for freezing depths greater than 20 m or so a rotary method of drilling is normally chosen. The freeze tube will be 100–150 mm diameter and sealed at its lower end. Thus drillholes of 150 mm or more, each following an accurate course, are required.

Surface casing may first be cemented into place to safeguard the upper part of the hole from erosion/collapse, to isolate the hole from the groundwater, and to act as a guide for the subsequent open-hole mud-flush drilling.

To ensure a stable hole which remains open below the cased level, at sufficient diameter to receive the freeze tube, a good quality mud (bentonite) is circulated. This drilling fluid also removes the cuttings, and its condition has to be monitored regularly. Recycling is possible for several holes by the use of vibratory screens and desanders associated with a multiple mud tank system. The mud type chosen will take account of any expected aggressive strata/water conditions, and be of sufficient density to support the wall and counterbalance any water pressures.

For depths to 250 m it is usually possible to maintain verticality by careful balancing of the drillstring weight to create a pendulum (plumb-bob) effect while maintaining sufficient weight on the bit to ensure that the ground is cut efficiently.

Beyond 250 m, or when angled holes are required, directional drilling techniques are used. This system relies on a down-hole motor (DHM) activated by the flow of the drilling mud. As rotation is restricted to the very lowest part of the drill-string, the azimuth of a bent sub or knuckle of 0.5° to 2° mounted above the DHM can be adjusted to control the subsequent course of the drillhole. The fundamental difference between conventional and directional drilling is shown in Fig. 3.

When drilling commences from below the water table, e.g. horizontal freeze tubes for a tunnel or inset, it is necessary to limit the movement of water and soil grains from the strata into the drilling chamber. An effective and practical method is to incorporate a non-return valve (to pass drilling fluid) in the cutting head and utilise the outer element of the freeze tube as the drill pipe (Harris & Norrie 1982).

The combined assembly then constitutes drill rod/temporary casing while being drilled through a stuffing box; following placing of a sealing-plug to isolate the non-return valve, the tube can be furnished as a conventional freeze tube.

Freeze tubes

The outer casing of the freeze tube assembly must be both leakproof — since loss of the cold medium to the strata would create an anti-freeze condition — and strong enough to withstand tensile stress due to contraction of metal and possible ground heave. Steel quality J55 or better, and API type long thread connections, will usually be specified when freezing for deep excavations such as mineshafts.

The freezetube material represents a significant cost which justifies its recovery for further use whenever possible. However, hydrogeological conditions may preclude this option, particularly when the freeze tubes penetrate more than one aquifer. The permanent linking of aquifers is to be avoided.

A frequent solution is to seal the freeze-tubes into place by cementing the annulus between the drillhole and the freeze tube. The integrity of the impermeable stratum separating the aquifers is thereby restored.

In other circumstances it is adequate to fill the freeze tubes or freeze holes with sand following completion of the freezing process.

Monitoring

Development and growth of the ice-wall is checked by observing temperatures at selected levels in strategically placed observation holes. In addition a central hole lined with perforated casing is normally installed; once closure between all freeze tubes has occurred, the volume change associated with chilling water results in expulsion of water from this 'pressure relief' hole as inward ice-wall growth occurs.

4. Grouting

Introduction

In parts of the world where the major water bearing formations are very heavily fractured or solutioned, treatment of these zones has, on occasions, been successfully carried out by injection of clay-cement grout from boreholes drilled from surface. In general, the nature and depth of the water bearing formations in the UK are such that these pre-cementation techniques are little used, and most grouting operations for shafts or drifts are carried out underground from the advancing excavation face. It is these operations which are discussed in this paper.

The borehole investigations will provide details of the depth and thickness of each aquifer zone, together with its transmissivity and groundwater pressure, and the nature and distribution of the water carrying passages within each aquifer rock, i.e. whether water flow occurs within fractures and fissures, within rock pores, or a combination of these. This will follow from a comparison of the transmissivity based on the intrinsic permeability measured from recovered core samples with the transmissivity determined *in situ*, and enables the relative contribution of fissure and intergranular flow to the aquifer transmissivity to be estimated. This is an important factor in the selection of the main grout to be used.

Fractures, fissures and coarse passages in excess of 200 μm are treated by injection with cement based grouts, whereas chemical grouts are used to permeate and treat fine fissures and pores.

Preliminary grouting design

Following an appraisal of the information available from the surface borehole investigation, a preliminary grouting design is proposed.

The aquifer depth and thickness will determine the number of grout covers that will be required to enable the shaft or drift to be sunk through each water bearing zone. Typical grout cover designs are shown in Fig. 8. The length of a grout cover is restricted to about 35–40 m by both the limited accuracy of rotary-percussive drilling, and by the increased spacing between injection holes as they fan out with increasing depth. The examination of recovered core samples from above and below the aquifer, together with the aquifer rock itself, assist in the selection of the elevations from which the grout covers will be injected.

Grout is injected into the aquifer via a steel standpipe, which is usually sealed in place with cement grout. The length of these injection standpipes will also be considered at this stage. They must be of sufficient length to enable the excavation injection face to withstand the grouting pressures needed to overcome the groundwater pressure, and to inject the rock at an acceptable rate. However, with increasing length they become more difficult to install, and the injection face is held further from the aquifer which is to be treated.

The transmissivity of the aquifer and the proportion of fissure to intergranular flow indicate whether the grout treatment will be primarily cement or chemical. Many factors influence the selection of the optimum chemical grout for a particular situation, and will include properties of both the fluid and the set grout (such as injectability, pH, gel time control, gel permeability, gel permanence, and the UCS of grouted rock; the toxicity of the ungelled components, of grouted rock, and of materials which may be leached from the gelled grout), together with the chemical composition and temperature of the aquifer rock and groundwater.

The width of the grout cover required to achieve the necessary degree of water control is also determined, giving due consideration to the strength of the grouted rock, the groundwater pressure, the method of excavation (e.g. drill and blast or hand jigger), and the length of time for which the excavation will remain unlined. The grout hole spacing required to give this width of cover can then be calculated, taking into account the grout injectability, the net injection pressure and the grout gel time, followed by the general configuration (number of holes, pattern, angles of dip and spin; hole length) of the grout cover.

The preliminary grouting design is based on information obtained from a surface borehole, and may well be modified in the light of additional information supplied by probe holes drilled ahead of the advancing excavation face. These probe holes provide a final opportunity to search for undetected aquifers before they are intersected by the excavation. Rather than simply relying on the discharge rate of the probe hole as evidence of the strength of the aquifer ahead, which can be misleading, an *in situ* test based on the pressure recovery principle has been developed.

The test equipment comprises an accurate means of pressure measurement, such as an electronic pressure transmitter with digital readout or a high precision bourdon tube test gauge, and a test valve additional to the probe hole standpipe valve. The test valve is opened and water allowed to flow from the aquifer, producing a depletion in pressure.

The valve is then closed after a known 'flow period' and a measurable volume of water has been collected, and the recovery of pressure monitored with time for a period typically three times the flow period.

The subsequent data analysis is identical to the conventional pressure recovery tests from surface boreholes.

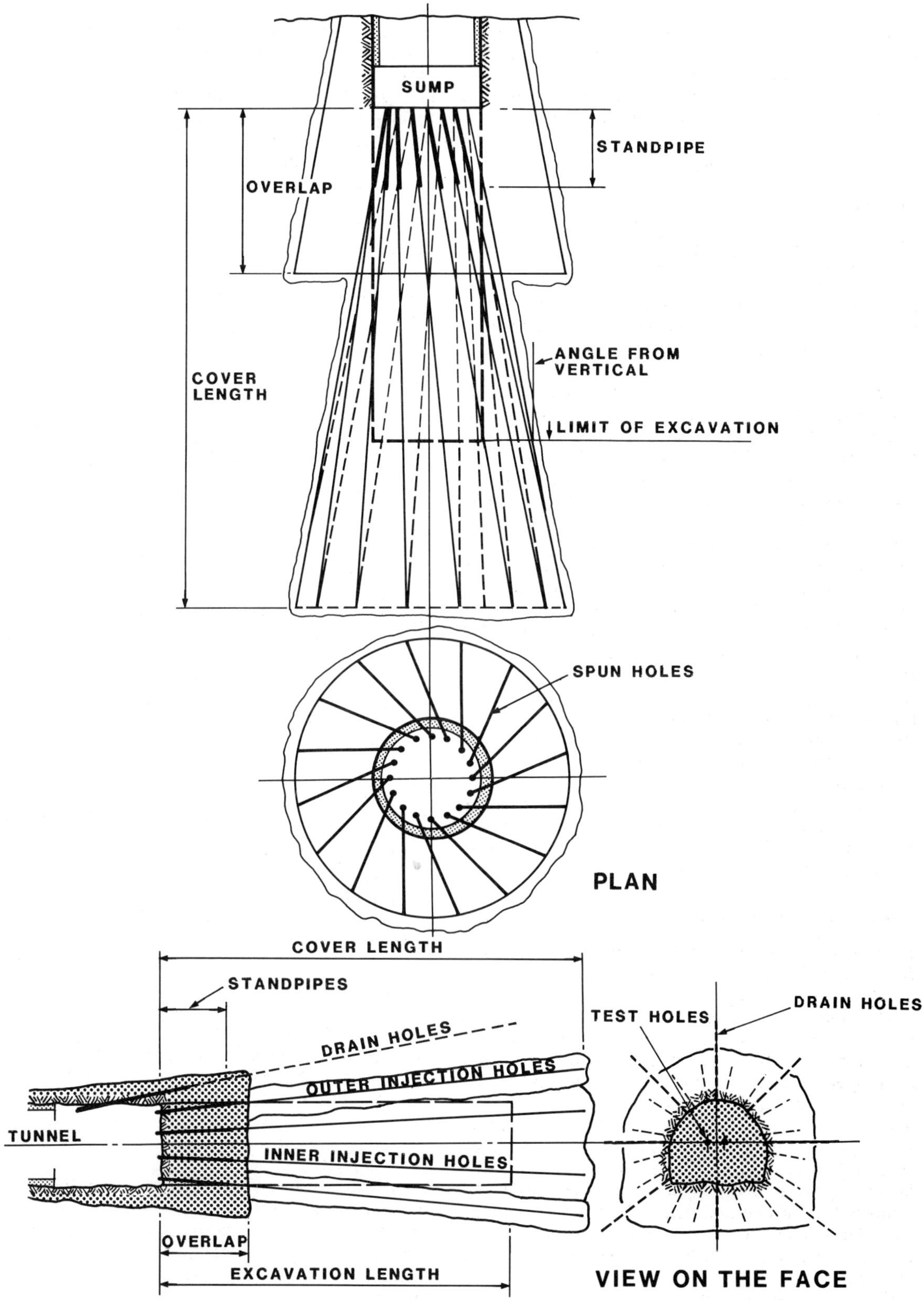

FIG. 8. Typical grout covers.

Grouting procedure

The excavation face is advanced to the elevation from which the grout cover will be formed, and the injection standpipes are installed. A hole, typically 89 mm diameter, is drilled to the required depth, and in dry ground the hole is flushed out and thick cement grout tremied in. The standpipe, typically a length of 2 inch nominal bore high pressure steel pipe, is then lowered into place. If, however, the hole is making water, a mechanical packer-type standpipe is used. It is placed in the hole, then thick cement grout is injected through the standpipe. When a return of thick grout is obtained around the annulus, the packer is expanded to seat and the standpipe grouted in under pressure. The lengths of standpipes may vary considerably; 3 m has been found adequate in competent limestone with injection pressures up to 3.45 $N.mm^{-2}$ (500 psi), whereas in weak Coal Measures mudstones/sandstones where injection pressures of up to 8.45 $N.mm^{-2}$ (1200 psi) were used, required 15 m long standpipes.

Once installed, the standpipes are drilled out to 0.5 m beyond their ends, and then pressure tested. If secure, the grouting works will then commence. The usual procedure is to advance the cover in a number of descending intermediate stages. The length of each stage is defined after close examination of both recovered core samples from the exploratory boreholes from surface, and recorded water inflows obtained from probe holes drilled ahead into the zone to be grouted.

Each injection hole is drilled through a stuffing box, which is used to control any sudden large water inflows. Drilling continues, with the water inflows being checked and recorded every metre, until either an arbitrary maximum water inflow (say 50 gpm) or the stage depth has been reached. The hole is then injected with grout. Cement grouting will commence with very thin mixes which are made progressively thicker until the maximum injection pressure is reached. The rate at which the grout mix is thickened is governed more by the behaviour of individual holes rather than by a set regimen. However, chemical grouting proceeds with the injection of a pre-determined quantity of grout (based on the grout injectability and the porosity of the aquifer rock) unless the maximum injection pressure is attained prior to this quantity being injected. The holes are drilled out and injected in a strict sequence to avoid drilling a hole adjacent to one which is either currently being, or has just been, injected.

Upon completion of the grout cover, its efficiency is checked by test holes drilled ahead into the proposed volume of excavation, and the reduction in transmissivity and hence potential shaft inflow assessed by a further *in situ* pressure recovery test.

Injection pressures

In general, grouting operations which are carried out underground involve drilling and subsequent injection of grout into rock containing groundwater at high pressure. The injection pressure exceeds this back pressure by the net injection pressure, and this governs directly the grout injection rate for a given rock/grout combination. However, it is important when attempting to increase the injection rate by increasing the injection pressure, that the injection pressure is not allowed to rise to the point at which hydrofracture can occur of either the stratum in which the standpipe is installed, or the aquifer being injected. This would possibly result in new leakage channels being open up, and lead to uncontrolled groundwater flow back to the excavation injection face.

A number of factors must therefore be taken into consideration when deciding upon the maximum injection pressure to be used during a particular grouting operation. These will include the mechanical properties of the aquifer and of the stratum in which the standpipes are installed; the geostatic head at the base of the standpipe; the hydrostatic head against which the grout is to be injected; and most important of all, previous experience.

Pressure relief

The ground treatment methods discussed so far aim to control groundwater by excluding it from the underground excavation. A third method, pressure relief, creates a number of preferential paths along which groundwater flow occurs, and so reduces the regional pore water pressure and hence reduces the uncontrolled groundwater flow into the excavation (Scott & Daw 1983). The preferential paths may comprise vertical pressure relief or de-watering wells sunk from surface, with the water pumped by submersible pump (Fig. 9); or a steep conical array of relief wells installed from a temporary sump above the aquifer, with water from the wells collected in a manifold system then pumped to surface from an underground pump lodge (Fig.10). In either case, the total amount of water that has to be disposed is not reduced, but most of it is collected, pumped to surface under control, and so diverted from flowing into the excavation.

Due to the build up of pore water pressure behind a grout curtain, it is difficult to obtain a substantial reduction in the potential groundwater inflow into an excavation unless the grout treatment has been extremely effective. For example, to reduce the water inflow to an excavated shaft by a factor of ten, the grout treatment has to be 99% effective. If, however, a combination of grout cover and relief wells is used, as shown in Fig. 11, for example, the grout cover makes the relief-wells more efficient by retaining a high pore water pressure in their vicinity, and the relief wells reduce the pressure difference across the grout cover, so reducing the flow through the cover. The net result is that the combination becomes more effective than either system could become alone.

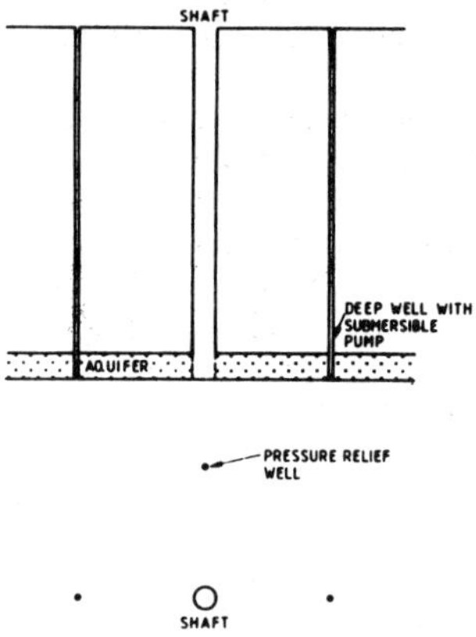

FIG. 9. Vertical pressure relief wells from surface with submersible pumps.

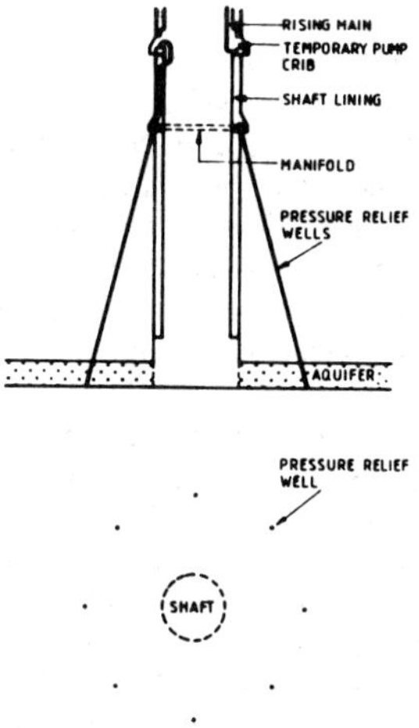

FIG. 10. Pressure relief wells from temporary sump close to aquifer.

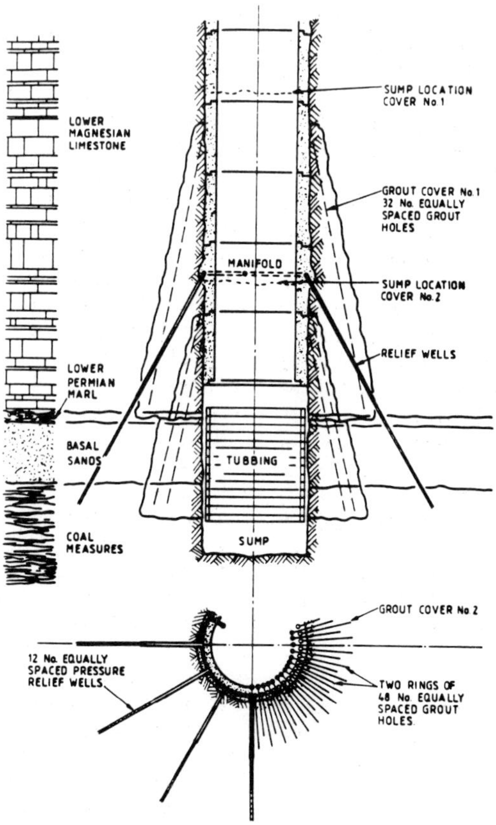

FIG. 11. Combination of grout covers and pressure relief wells.

This technique has been used in situations (Fotheringham & Black 1983)

- where the rock mass face would not withstand the hydrostatic stress after excavation;
- where a shaft lining would be subject to excessive hydrostatic stress during backwall grouting.

6. Summary

It has been shown that many factors are considered before the decision to adopt a particular ground stabilising and/or water excluding process is reached.

Investigatory methods for the ground freezing and grouting processes are common, and yield data at an early stage which dictate the testing programme necessary to enable the design for each method to proceed.

Frozen ground is both strong and impermeable. Since all strata encompassed by each freeze tube array are affected, the design has to cater for the most critical stratum through which the freeze tubes pass. Low thermal conductivity yields a slow ice-wall growth rate which must be allowed for in the programme; low frozen strength/high susceptibility to creep demands a combination of low freezing temperature, thick ice-wall and/or a short excavation and lining cycle, particularly when high loading stresses are to be controlled.

Adverse effects, either on creation of the frozen state, or of the frozen-ground body on nearby property/services, are few. They can be recognised, and generally can be allowed for or avoided.

Freeze tubes are usually placed into purpose-drilled boreholes; most of the recognised drilling methods may be used according to strata conditions, subject to a high degree of accuracy being achieved.

The preliminary grouting design based on data obtained via surface boreholes may be modified by additional information from probe holes drilled ahead from the advancing excavation face. Cement and/or chemical grouts, depending upon the rock formation being treated, are injected via steel standpipes previously sealed into the injection face. The progress of the grout treatment may be checked using an *in situ* pressure recovery test. In certain situations, the combination of grout treatment and pressure relief wells can be particularly effective.

For successful ground treatment by freezing or grouting, it is essential that the process is carefully monitored.

ACKNOWLEDGEMENTS: Figs 1, 4 & 5 are reproduced from Thorburn & Hutchinson (1985) 'Underpinning', Blackie, with permission.

References

BEBI, P.C. & METTIER, K.R., 1979. Ground Freezing for the construction of the 3-Lane Milchbuck Road Tunnel, Zurich, Switzerland. *Proceedings of Tunnelling '79,* Institute of Mining & Metallurgy, 248-255.

BLACK, J.C., POLLARD, C.A. & DAW, G.P., 1982. Hydrogeological assessment and grouting at Selby; *Proceedings of the Conference Grouting in geotechnical engineering.* ASCE New Orleans, American Society of Civil Engineers, 665-679.

DAW, G.P., 1984. Application of aquifer testing to deep shaft investigations. *Quarterly Journal of Engineering Geology,* **17,** 367.

FOTHERINGHAM, J.B. & BLACK, J.C., 1983. Ground water pressure relief in shaft sinking; The Mining Engineer, August, 85-91.

GARDNER, A.R., JONES, R.H. & HARRIS, J.S., 1982. Strength and creep testing of frozen soils. *Proceedings of the International Symposium on Ground Freezing,* 53-60.

HARRIS, J.S. & NORIE, H., 1982. Construction of two short tunnels by AGF. *Proceedings of the International Symposium on Ground Freezing,* 1982, 383-388.

JONES, R.H., 1982. Ground movements associated with AGF. *Proceedings of the International Symposium on Ground Freezing,* 1982, 295-304.

SCOTT, R.A., & DAW, G.P., 1983. Ground water pressure relief wells in shaft sinking. *International Journal of Mining Engineering,* **1,** 229-236.

J.S. HARRIS, Foraky Ltd., Colwick, Nottingham, Nottinghamshire, UK.
C.A. POLLARD, Cementation Mining Ltd., Doncaster, South Yorkshire, UK.

Discussion Session 10

L. McGarry (Groundwater Control Ltd.) asked Mr Coomber whether it would be advisable to continue extraction when jet grouting in groundwater control schemes in which both techniques are being employed.

D.B. Coomber (G.K.N. Keller Foundations) replied that little experience had been gained where both techniques are used in combination but if the well pointing is carried out some distance from the grouting, then it would probably be beneficial to keep dewatering. In answer to a question put by Mr Pontin, Mr Coomber went on to explain that jet grouting was a suitable treatment for most soil conditions except where there is an existing large groundwater flow.

From Cripps, J.C., Bell, F.G. & Culshaw, M.G., (eds), 1986, *Groundwater in Engineering Geology,* Geological Society Engineering Geology Special Publication No. 3, p.467

SESSION 11

CONTROL OF GROUNDWATER BY REMOVAL

Groundwater in Engineering Geology, London, 1986, pp. 471-486 Printed in England

Groundwater control by groundwater lowering

F.G. Bell & P.M. Cashman

ABSTRACT: When an excavation is to extend beneath the water table the ground conditions may be improved by lowering the level of the water table beneath the proposed base of the excavation. Groundwater lowering is accomplished by removing water from the ground at a faster rate than recharge can occur. The most appropriate type of system of groundwater lowering depends mainly upon the soil conditions present at the site but the size of the excavation, the amount by which the water table has to be lowered and the length of time during which the excavation has to be maintained in a stable and workable condition also have to be taken into account. Settlement may occur as a result of groundwater lowering. If a site is surrounded by buildings then measures should be taken to prevent settlement taking place.

Introduction

Problems encountered by engineers working in soils are frequently attributable to the presence of water. Hence when construction operations are to take place in water bearing soils, measures must be taken to control groundwater flow and pore water pressure gradients in order that serious troubles may be avoided.

Dewatering is usually some temporary measure taken to lower the groundwater level so that excavation can take place in stable workable conditions. Groundwater can be controlled by one or more types of dewatering systems. A correctly designed, installed and operating system makes construction work easier as a result of:

1. Lowering the water table and intercepting seepage, which would otherwise emerge from the slopes or base of the excavation.
2. Improving the stability of the excavation slopes and preventing material being removed from them by erosion due to seepage. Hence slopes can be steeper and the area of excavation reduced.
3. Preventing heave, blow-outs or quick conditions from developing in the floor of the excavation.
4. Improving the density, excavation and haulage characteristics of the soils involved.
5. Reducing the lateral loads on any temporary support such as sheet piling.

However, the change in the position of the water table brought about by drawdown, particularly around a deep, wide excavation, may adversely affect the area surrounding the excavation. Of special concern is any settlement likely to occur as a result of withdrawal of groundwater.

The problem of unstable ground

Soil has a skeletal structure of solid material with an interconnecting system of pores which are either wholly or partly filled with water. As a soil is loaded, the (total) stress is carried by both the soil skeleton and the pore water. The pore pressure acts with equal intensity in all directions and the stress carried by the soil skeleton alone is thus the difference between the total applied stress (σ) and the pressure set up in the pores (u). This is termed the effective stress (σ') and is written ($\sigma - u$). Also, as water has no strength, the soil skeleton deforms while the pore water is displaced. This action continues until the resistance of the soil structure is in equilibrium with the external forces, the rate of dissipation of the water being dependent on the permeability of the soil mass and the physical drainage conditions. The shear strength of the soil is dependent on the effective stress and at failure (τ_f) is equal to:

$$\tau_f = c' + \tan \phi'$$

where c' is equal to the apparent cohesion of the soil and ϕ' is equal to its angle of shearing resistance in terms of effective stress.

Thus as the groundwater pressure is reduced the effective pressure is increased with a corresponding increase in the strength of the soil. This in turn increases (i.e. improves) the stability of the ground.

The fundamental requirement of a groundwater lowering installation is that it produces stable and workable conditions throughout excavation during construction. An examination of the conditions which create unstable ground will illustrate what not to do and therefore indicate what should be done.

From CRIPPS, J.C., BELL, F.G. & CULSHAW, M.G., (eds), 1986, *Groundwater in Engineering Geology,* Geological Society Engineering Geology Special Publication No. 3, pp. 471-486.

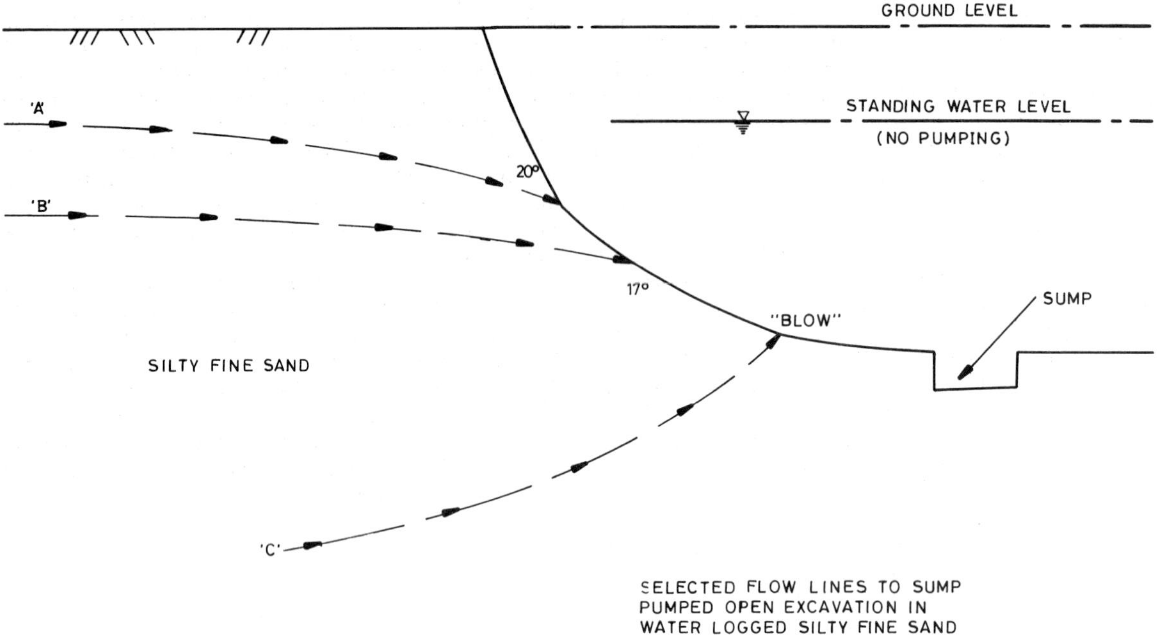

FIG. 1. Water table conditions open sump pumping.

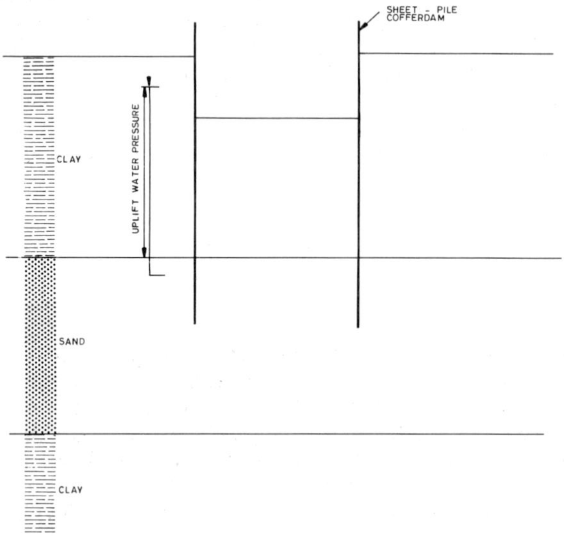

FIG. 2. Confined aquifer condition potential for 'blow'.

There are two cases to consider:

1. Excavation into a sandy soil below the standing water level — unconfined aquifer or water table condition.
2. Excavation within a sheeted excavation into an impermeable bed underlain by a sand stratum charged with water under a pressure head — confined aquifer condition.

Consider the first case (Fig. 1). If a hole is dug in a bed of fine sand to below the level at which water is struck and the inflowing water is pumped out from within the excavation the sides will start to slump in before a depth of about 1 m has been reached. As digging proceeds the situation will get worse and the edges of the excavation will recede. The bottom will soon fill with a sand slurry in an almost liquid condition and this will be constantly renewed by erosion of the sides.

The collapse of the sides is the result of instability of the slopes due to the excess pore water and seepage pressures which are developed in the ground by the flow of water to the excavation. In the case of the top flow line (phreatic surface) which emerges almost parallel to the surface of the excavation, a sand with an angle of internal friction of 35° will only stand at an angle of 20° (about 1 in 3); at the emergence of the intermediate flow line the sand will stand at 17° and in the case of the bottom flow line the sand at the point of emergence will be subject to zero effective stress and may actually be 'boiling'.

Where soil structure is layered or laminated another mechanism will be at work. The sides of the excavation will show gully structures from which sand slurry can be seen to flow into the excavation. This is due to internal erosion. The water flowing along the upper surface of a layer of lesser permeability carries fines with it and thus tunnels backwards beneath the overlying sand, which has some strength due to capillary tension. Before this tunnelling effect has gone far the top collapses and the collapsed material is also carried away. This phenomenon is quite well known in geology under the name of 'spring sapping' or 'seepage erosion'. The curious thing about this process is that it takes place when the flow of water is very small and the gradient very low but it is continuous and so in time very large quantities of material are removed. A case history of spring sapping on a large scale near Newhaven, Sussex, was described by Ward (1948).

The second case is relatively simple to understand (Fig. 2). If the cofferdam is excavated through the clay without reducing the pore water pressure in the sand aquifer, there will be a sudden inrush of water and sand when the weight of the remaining clay is less than the upward water pressure confined in the sand aquifer. This may lead to the collapse of the excavation. This can be prevented either by having a sufficient penetration of the piles or by reducing the artesian head by means of pumping from the confined aquifer.

Generally all of these troublesome conditions can be avoided by draining the ground from water abstraction points outside the excavation such as wellpoints or deep wells, though there will be occasions when water abstraction points have to be installed within the construction area.

Assessing of groundwater lowering requirements

Once this principle has been accepted, it becomes necessary to calculate, with reasonable accuracy for any degree of lowering, first, the number of wells that will be required, their depth, their yield and hence the cost of installation and, second, the capacity and running cost of the pumps. These calculations are, of course, based on the permeability of the ground to be drained, the thickness of the aquifer and the size and depth of excavation required.

The theory of the flow of water through permeable ground originated with the researches of d'Arcy (1856), who investigated the purification of water by filtration through sand, and developed an equation of flow through a granular medium based on the earlier work of Pouseuille on flow in capillary tubes. d'Arcy's contemporary, Dupuit (1863), approached the problem through an analogy with flow in pipes and open channels and examined the case of flow towards a well. Dupuit distinguished between flow to a gravity well and flow to an artesian well.

Dupuit made two simplifying assumptions that:

1. the gradient is the same at all points in a vertical section; and
2. the gradient at the phreatic surface (or at the piezometric surface in artesian flow) is equal to the slope of the surface at that point.

Dupuit, having stated these simplifying assumptions, apparently felt no need to be limited by them; unlike d'Arcy he did not use all possible opportunities to check his theoretical conclusions against experimental data and field observations. No doubt this explains why Muskat (1937) dismissed Dupuit in so peremptory a manner when he wrote "The extension of the Dupuit-Forchheimer theory which has been frequently used to treat such problems involves so many questionable assumptions that reproductions of the analyses based on this theory hardly seem to be warranted".

Bolton (1951) was more tolerant of Dupuit and stated that for all practical purposes the discharge Q can be assessed with sufficient accuracy by using Dupuit's formula and the authors would agree that for most practical purposes the Dupuit equations provide an acceptable means of assessing the amount of water which has to be removed from the ground by pumping. The Dupuit equation for the confined aquifer (artesian condition) is:

$$Q = \frac{2.\pi.k.b.\,(H - h_o)}{\ln (R/r)} ,$$

where Q is the quantity of water; H is the elevation of the original piezometric surface above the impermeable base; h_o is the elevation of the operating level of the pumping well above the base; R is the radius of influence; r_o is the radius of the well borehole; k is the coefficient of permeability; and b is the thickness of the confined pervious stratum (Fig. 3). For the artesian well, the drawdown (H - h_o) at any distance (r) from the well can be derived from.

$$H - h_o = \frac{Q}{2.\pi.k.b} \ln (R/r)$$

If the screen of an artesian well partially penetrates the stratum which is to be dewatered, then the flow pattern around the well will deviate from that developed by a full penetrating well. In such cases the flow required to produce a certain drawdown is governed by the depth to which the well screen penetrates the formation and the equation for flow is as follows:

$$Q = \frac{2.\pi.k.b.\,(H - h_o)\,G}{\ln (R/r)}$$

where G is a correction factor for partial penetration, which is equal to the ratio of flow from the partially penetrating well to that for a fully penetrating well for the same drawdown at the periphery of the wells.

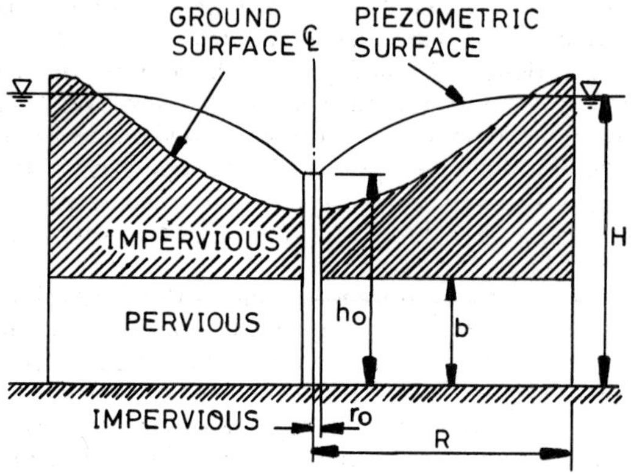

FIG. 3. Dupuit confined aquifer or artesian well condition.

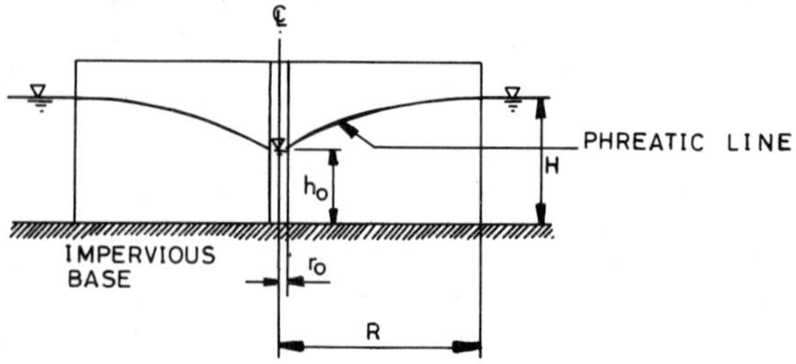

FIG. 4. Dupuit water table or gravity well condition.

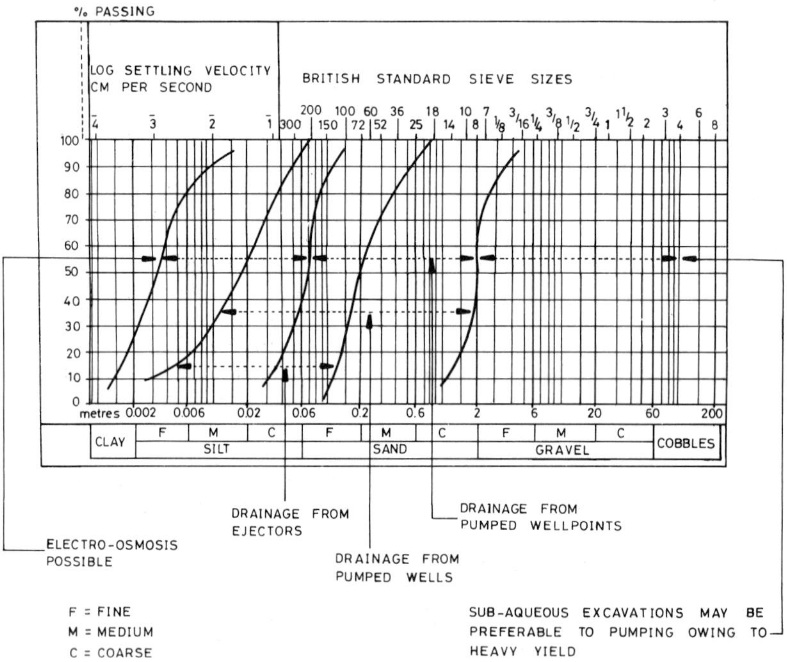

FIG. 5. Tentative limits for groundwater control systems — pumping methods.

The Dupuit equation for the gravity (water table) condition is (Fig. 4):

$$Q = \frac{\pi.k\ (H^2 - h_o^2)}{\ln\ (R/r).}$$

In the case of a gravity well the drawdown (H -h_o^2) at any distance, r, from the well cannot be expressed as readily as it is for the artesian well. However, the term (H^2 - h^2) can be used instead, hence:

$$H^2 - h_o^2 = \frac{Q}{\pi.k.} \ln\ (R/r)$$

The Dupuit equations were derived for a single well but are often used for estimating total flow to a system of wells using the 'equivalent single well' concept.

Cashman & Haws (1970) referred to the limitations of the empirical approach regarding the dewatering of large excavations and multiple excavations of differing depths. They reported that in such situations the use of electric analogue models had proved extremely valuable for design purposes. The Laplacean equations for water flow through porous media are of the same mathematical form as the equations for flow of electricity through a resistor mesh. The groundwater flow problem is converted to an analogous electrical problem; the electrical problem being solved using a resistor mesh or time variant model and the electrical solution is then converted back to an hydraulic solution.

Field tests were carried out by Debindin & Lee (1979) at a trial excavation in semi-confined fine-grained sands in order to assess seepage and drawdown. The results were compared with those obtained from a finite element simulation. The latter were found to be generally compatible.

Investigation of soil and groundwater conditions

The selection of a suitable method of groundwater control and assessment of the equipment needed depends not only on size and depth of excavation below ground level but also on a number of factors of which the most important are:

(1) level of standing water or water pressure;
(2) soil types and permeabilities;
(3) soil structures;

A thorough knowledge of the soil characteristics of the site concerned is required prior to the design and installation of a dewatering system. This means that borehole data, extending to a reasonable depth below the deepest subgrade, should be obtained for the working area. The boreholes should be close enough to each other to indicate any significant variations in the soil conditions. Although the site investigation attempts to record an intimate picture of the succession and character of the soil conditions, careful observation should continue as the dewatering system is being installed so that any variations in soil conditions which could cause problems may be revealed.

The position of the water table(s) and information regarding piezometric pressures are of utmost importance when planning a groundwater lowering installation. Ideally, the position of the water table and the piezometric pressures should be observed over a period of time as they are likely to vary. In practical terms, this frequently is not possible.

An initial assessment of permeability can be obtained from samples taken from boreholes and from rising and falling head tests carried out in boreholes during the site investigation. However, before designing a large dewatering system, Cashman & Haws (1970) advised that a pumping test should be performed to confirm the permeability values to be used for design. In addition, each completed well should be tested for capacity and drawdown, and periodic short tests of drawdown should be made in the aquifer. The design should be adjusted if the situation turns out to be different from that assumed from the site investigation, even if this involves suspending installation temporarily.

In addition to a dimensional plan of the excavation, sections illustrating the geological conditions of the area to be drained, in relation to depth of excavation should be studied.

If the pumping period is expected to be extensive the chemistry of the groundwater should be studied to evaluate the risks of corrosion and encrustation of the wellscreens, filters, pumps and pipework.

Methods of groundwater control by pumping

There are eight systems for achieving groundwater control by pumping:

(1) sump pumping, including garland drains and sumps;
(2) wellpoint systems;
(3) horizontal drainage;
(4) deep wells;
(5) shallow wells;
(6) ejector systems;
(7) electro-osmosis;
(8) electro-chemical stabilisation.

In addition there are two further systems of control which may not entail pumping:

(9) pressure relief wells;
(10) recharge wells.

These will be dealt with in the section on Avoidance of Settlement and Heave.

Sump pumping

This is a simple technique and acceptable provided that 'fines' are not continuously withdrawn thereby encouraging instability of the side slopes below original standing water level and causing springs in the bottom of the excavation (see Fig. 1).

Sumps are widely used, and in many cases are the most economical method of lowering the groundwater level. Commonly pumping from a sump is the first method to be tried when an unexpected problem with groundwater is encountered. Each sump must be deep enough to ensure that the entire excavation is drained as a result of pumping. The open sump method is capable of lowering the water table by not much more than 6 to 8 m.

The sump(s) is located below the general base level of the excavation at one or more corners or sides. It should be surrounded by a graded filter in order to avoid the removal of fines from the soil. A ditch or garland drain which falls towards the sump, is dug around the base of the excavation and conveys water from the slopes of the excavation to the sump. This also avoids the floor of the excavation becoming covered with standing water. The drainage ditch should be lined or filled with gravel in order to prevent erosion taking place. Perforated pipe can be placed within the gravel to increase the rate of removal of water. Lining the ditch with a filter fabric helps to impede the migration of fines.

The pumping capacity needed to establish the initial lowering of the water table is about twice the steady state or equilibrium rate of pumping. Therefore if double the required steady rate capacity is installed then all pumps can be used in the initial drawdown period. Subsequently half the pumps can be shut down and used for emergency standby and maintenance purposes.

Wellpoint systems

A wellpoint system comprises a number of small wells installed at close centres around an excavation. An individual wellpoint usually consists of a gauze screen 0.5 to 1.2 m long and from 40 to 75 mm in diameter, which surrounds a central jetting riser pipe. Various sizes of screen openings are available to correspond with the predominant grain size of the soil. Cashman (1975) recommended that where wellpoints are required to remain in the ground for more than a few weeks, it may be more economical to use disposable plastic wellpoints. These consist of a nylon mesh screen surrounding a flexible plastic riser pipe. The wellpoint and riser are simply cut to various lengths to suit ground conditions and depth. At the end of the dewatering period the riser pipe is cut off at or below the ground surface. In fact plastic pipes now tend to be used instead of metal pipes.

The individual wellpoints are connected by vertical riser pipes to a header main which is under vacuum from a pumping unit. The water flows under gravity to the wellpoint and is drawn by the vacuum to the header main and discharged through the pump. The wellpointing system has the advantage that water is drawn away from the excavation, which means that the sides are stabilised thus permitting steeper slopes.

The system is most suitably employed when the water table in a shallow aquifer does not need to be lowered more than about 5 m. Multi-stage systems are required beyond that depth because of the limitation of suction lift (Fig. 6).

Prugh (1960) considered that the chief advantage of a wellpoint system is its versatility. In particular, a wellpointing system can be readily adapted to deal with local trouble spots. For example, wellpoints can be added at offending locations and additional pump capacity can be speedily connected to the system. Header pipes can be relocated to avert construction difficulties.

Installation of wellpoints

The wellpoints may be either self-jetting (recoverable and reusable) or disposable. The installation of the disposables requires a placing tube (a jetted temporary casing tube) and a crane with a free fall to handle the placing tube. It is prudent to 'sand-in' wellpoints, that is, to provide a vertical sand column around wellpoint and riser. Both types are generally installed using a water jetting pump. It is important therefore to locate an adequate source of jetting water.

Both types of wellpoints are suitable for installation in sands, gravels and silts. In stratified soils a jetting chain is recommended (Powers 1981). This produces a hole with a large diameter, 150 mm being usual in silts and clays, thus providing more room for filter sands. The placing tube needed for the installation of disposables fulfills a similar purpose.

Various drilling methods have been used to install wellpoints when jetting is slow due to the soil conditions. For example, drilling with augers, followed by either plain jetting or installation within a casing is common. Tomlinson (1980) suggested that if a bed of clay overlies the formation to be dewatered, then it is usually more practical to auger than jet through the clay.

On average, around 50 wellpoints can be installed in a day. However, under ideal conditions in clean sand, up to 100 may be put in place. Conversely, with difficult jetting conditions output may drop significantly.

Filter sand is placed around the wellpoint and riser. Where necessary, a larger more effective filter may be formed by the hole being reamed. The filter increases the effective diameter of the wellpoint which, in turn, increases the yield. It also decreases entrance loss and reduces the likelihood of clogging. The filter column above the screen provides vertical drainage through any layers of silt or clay which are present.

Testing and evaluating the performance of a dewatering system as it is installed and operated are advisable. This is because due to normal variations in soil conditions it usually is impossible during the design stage to determine with accuracy the permeability of strata, the radius of influence or distance to the effective source of seepage, and losses of head in the wellpoint system.

The capacity of a single wellpoint with a 50 mm riser pipe is in the range 2.5 to 25 l/min depending on the permeability of the soil. Their spacing about an excavation is governed by the permeability of the soil and the time available to effect drawdown. Where the aquifer extends 3 m or more beneath subgrade, the selection of wellpoint type and spacing is based on the

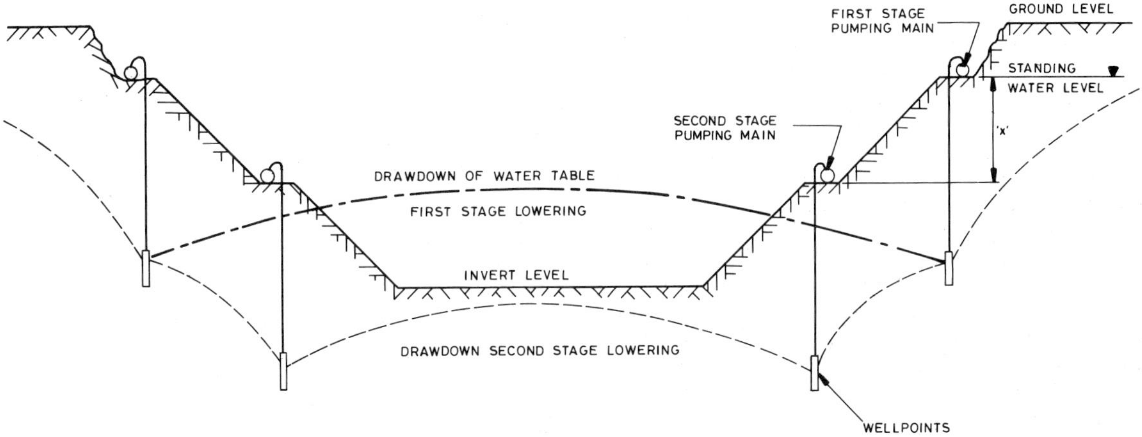

FIG. 6. Two stage lowering with wellpoints; 'x', lowering achieved per stage varies from 3 m to more than 6 m, depending on soils and efficiency of the total pumping system.

quantity of water to be removed. The total flow of the system is estimated and divided by the length of the head main to obtain the flow per unit length. The wellpoint spacing is then chosen.

Normally wellpoints are spaced between 1 and 4 m apart. Wells probably should be considered if the spacing is greater than 5 m. Dewatering an unconfined bed of sand or sandy gravel of moderate permeability above an impermeable bed represents one of the best situations for the employment of wellpoints. Under special conditions, however, closer spacings may be required. For instance, thin layers of clay often occur in water bearing sands, and because they are impermeable they give rise to interruptions in the drawdown curves around wellpoints. In such situations adequate sanding-in is important. Alternatively holes can be positioned outside the line of wellpoints and filled with coarse sand. Such columns of sand provide a vertical path to the zone of sand being drained and enable water to move more readily to the wellpoints than towards the sides of the excavation.

Conversely difficulties can also be brought about by layers of highly permeable material. In this case water tends to by-pass the wellpoints. The problem can be overcome by jetting at close intervals in a line around the excavation.

Wellpoints using vacuum assisted drainage are effective in silty sands and silts, their spacing usually ranging between 1 and 2 m. A plug of bentonite or cement is placed at the top of the filter column to seal against the entrance of air. Vacuum is applied through the filter column to the soil, thereby accelerating drainage.

Amount of groundwater lowering by wellpoint system

The amount of lowering which can be achieved by suction pumping is limited by the amount of suction lift. Also it is related to the permeability of the soil concerned and to the hydraulic efficiency of the whole system.

In practice the suction lift of a single stage wellpoint system is limited to about 5 m. If the total drawdown required is around 6 to 7 m, it may be possible to install a temporary wellpoint stage. This means that the main system can be placed at a deeper level thereby reducing the lift to a manageable 4.5 m. If greater total drawdown is required it is necessary to use a multi-stage wellpoint system (Fig. 6), or a combination of deep wells and a single stage of wellpoints or an ejector system.

There is no theoretical limit to the depth of drawdown if a multi-stage system of wellpoints is used. However, the overall width of the excavation has to be increased in order to accommodate this type of lay-out. Sometimes the upper tiers become dry after the lower ones have been brought into operation. If an excavation is taken down to the water table before the header main and pump are installed, then fewer tiers may be used as the amount of suction lift required is reduced.

Where more than a single stage system is necessary, it is advisable to observe the groundwater level immediately prior to and during pumping from the upper stage and to record the discharge from the wellpoint system.

Unless the wellpoint screens are installed at significant depth, they may begin to draw air as the water table is lowered. The intake of air will not be the same at each wellpoint because of variations in soil conditions. Significant intake of air causes a reduction in vacuum and may result in failure to obtain the drawdown required. The offending wellpoints should be located and tuned by means of the individual trim valves. Tuning refers to the balancing of the flow of water from the wellpoints, so that each one draws its maximum potential yield, without an excessive amount of air.

Because the water level can regain its position very rapidly in highly permeable soil, in fact, within minutes, wellpoints should be shut off one at a time when the work concerned has been completed. In this way the water table rises more slowly to its former position.

Horizontal drainage

Horizontal 'wellpoint' drainage, is particularly suited to groundwater lowering for pipe trenching where fast forward progression is required. According to Cashman (1975) a long perforated pipe is laid horizontally in the soil at the desired depth by a special trenching machine. This excavates a trench 225 mm wide at depths between 1.8 m and 6.0 m and lays thin walled convoluted perforated plastic pipe of 80 or 100 mm diameter in the bottom of the trench as the trencher tracks forward. The pipe is wrapped with a mesh of woven nylon or coco matting. The machine throws the dug spoil back into the excavated trench on top of the perforated pipe. The perforated pipe is cut at given lengths and connected to plain pipe which is brought to the surface at the end of each run. This pipe is then connected to a wellpoint pump. Cashman pointed out that the rapidity with which this system can be installed means that the main contractor can progress speedily. This is particularly relevant to the laying of natural gas pipelines in open country.

Deep bored filter wells

A typical arrangement for a deep well pumping test is shown in Fig. 7. The annulus between the borehole and the installed perforated screen is filled with filter media (Cedergren 1975). The grading of the filter media should be matched to the grading of the soil from which water is to be abstracted, such that, after initial development of the well, clear water and no 'fines' are pumped. The width of wellscreen slot opening should not be such as to allow filter media to be continuously withdrawn when pumping.

The bore of the wellscreen to accommodate small output pumps must be at least 50 mm greater than the maximum outside diameter of the electric submersible pump for the well and the clearance should be greater for larger flow pipes.

Mansur & Kaufman (1962) indicated that deep wells, of large diameter, have been used for lowering the water table where:

1. The soil conditions become more pervious with depth. Deep wells have proved particularly worthwhile where it has been necessary to remove large quantities of water from thick, highly permeable formations.
2. The excavation extends into or is underlain by sands or gravels, indeed deep wells are most effective where the water bearing soil is clean, uniform and high permeable.
3. There is sufficient depth of pervious soil beneath the level to which the water table is to be lowered for adequate submergence of well screen and pump.
4. The bottom of a large excavation is underlain by a relatively impermeable bed beneath which there is a permeable formation under excessive artesian pressure.

For dewatering purposes deep wells commonly are located at 6.5 to 65 m centres, depending upon the permeability of the ground, the quantity of water that must be removed, the source and direction of seepage, the amount of submergence available and the width of the excavation. Such wells usually have diameters of 150 to 450 mm although at times the diameter may exceed 1 m.

The hydraulic efficiency of a deep well system may be increased by some 10 to 15% by applying a vacuum line connected to the tops of the wells. The wellhead and annulus must be airtight.

A deep well system is installed by boring or drilling operations. The selection will depend upon the soil conditions expected and availability of equipment. Holes extending to 30 m in depth with diameters of up to 600 mm have been excavated by bucket augers. Heavy duty augers can excavate holes with diameters up to 1.2 m. Bucket augers are effective in gravels and sands, and in soft to moderately stiff silts and clays. However, cobbles and boulders present problems and very stiff clays are difficult to penetrate. Excavation by bucket auger can yield a hole of good quality if caving is prevented by using a head of water. A minimum of about 3 m of head is recommended (Powers 1981). Subsequently the hole requires little development. In loose permeable sands it may be necessary to mix additive with the drilling fluid so as to form a temporary seal around the hole, thereby preventing caving. Bentonite is not recommended because the filter cake it forms is difficult to remove from the hole but bio-degradable muds are acceptable.

Dewatering wells are occasionally constructed by continuous flight augers. At times heavy slurry is

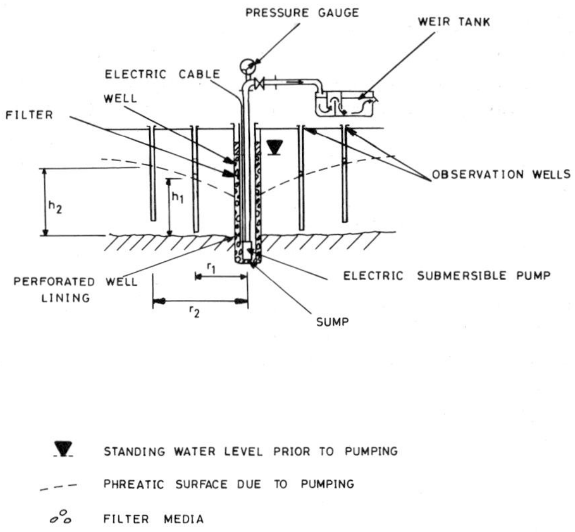

FIG. 7. Typical deep well with observation wells etc. for pumping test in either soil or rock.

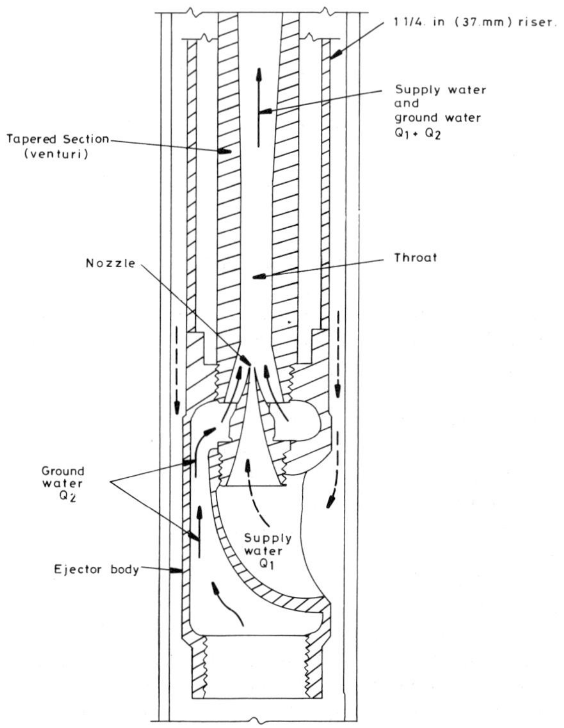

FIG. 8. Section through single pipe ejector.

required to stabilise the hole and to aid the retention of sand on the auger as it is being removed. The slurry, together with the mixing of soil and the smearing of the sides of the hole, results in a hole of poor quality.

Rotary drills, where the cuttings are moved by circulating fluid, are effective for holes of small to moderate diameter. For example, a heavy duty rig can produce a hole 450 mm in diameter.

Reverse circulation rotary drilling is sometimes employed in dewatering programmes. Holes with diameters of 600 mm or more can be sunk. This type of drilling generally results in a cleaner hole than the conventional rotary methods. The method is best suited to loose sands and gravels and soft clays. Stiff clays prove difficult to penetrate and also cobbles may cause difficulty since they tend to become lodged in the drill pipe. A head of water is required to provide support for the sides of a hole when using reverse circulation drilling, a minimum of 3 m from the drilling surface to the water table being recommended. Where the water table is close to ground level the drilling rig can be placed on a platform or alternatively a small wellpoint system can be operated whilst the hole is being bored.

Wells for dewatering fissured rock masses may be drilled with down-the-hole drills. The cuttings are removed from the hole by air flush, together with groundwater. Air flush is used for holes up to 200 mm in diameter. Above that size, additional air supply and special hammers are required. When down-the-hole drills are used through overburden, especially loose sands, the air lift action frequently brings about collapse of the hole. Collapse sometimes can be avoided by using foaming agents. However, where it is necessary to drain water bearing sands above rock, it is more reliable to use mud flush and temporary casing for drilling through the overburden. After the hole has been completed some or all of the casing may be replaced by wellscreen depending upon the competence of the rock. The effectiveness of a well in rock is dependent on the number of water bearing joints or fissues traversed by the drillhole.

Depending upon ground conditions, deep wells either may be used exclusively to dewater an excavation or they may be used in conjunction with another pumping system. For instance, wellpoints may be required at the base of the slopes of an excavation in order to intercept any minor seepage that may by-pass the wells. Alternatively the floor may heave or there may be the possibility of blow-out where the strata beneath are subject to artesian pressure. The artesian pressure can be relieved by the installation of deep wells either around the top of the excavation penetrating into the lower aquifer or at some safe elevation as the excavation is extended downwards. Depending upon the magnitude of the excess pressure and the soil conditions relief wells or 'bleeder' wells may be adequate.

Development of a well is an essential operation in its proper completion. There are three beneficial effects of this development:

1. Development corrects any clogging of the water bearing formation which may have occurred during drilling and it effects removal of boring slurry.
2. Development tends to increase the porosity of the natural formation in the immediate vicinity of the well and so improves its yield.
3. Development stabilises the filter media formation stabiliser in the annulus.

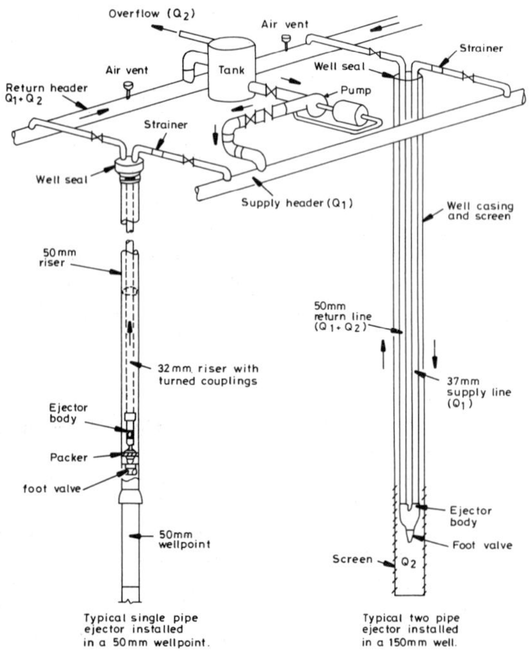

FIG.9. Basic ejector systems.

Also removal of 'fines' during development results in clean water only being pumped when the electro-submersible pump is installed (i.e. no fines in suspension) thereby prolonging the useful life of the pump.

There are two basic techniques of well development:

1. Surge plunger, either solid or valve type.
2. Air lift pumping.

In general the air lift pumping technique is to be preferred although it is not effective for relatively shallow wells, that is less than about 20 m deep.

Shallow bored filter wells

This is a synthesis of wellpointing and deep wells —some refer to this method as 'jumbo' wellpoints. The installation of the wells is the same as for deep bored wells with the same facility to ensure controlled filtering but the wells are pumped by suction pumping either by individual risers to a common suction main or by individual wellpoint pumps at each wellhead.

On a congested site, especially in high permeability soils, this method may be preferred to wellpoints because of the smaller number of risers causing obstruction to the construction operation.

Ejector systems

Ejector systems are most effective when used to dewater deep excavations made in stratified low permeability soils where close spacing is necessary and rate of pumping is relatively small. They are not limited in terms of suction lift, as are wellpoints, and have lower unit installation costs than wells.

The principle of the jet ejector technique is that a very high velocity water jet exiting through a convergent nozzle entrains flow through the water abstraction point installed below the ejector (Fig.8) and in turn causes groundwater flow from the surrounding soils. The total volume of water (high pressure water plus that induced from the ground) is returned to a separate collector main at ground surface (Fig.9).

Nozzle velocities are of the order of 50 m/s to lift from 50 m below ground level and 30 m/s to lift from 20 m depth.

This system is of low mechanical efficiency. Each ejector point resembles a wellpoint assembly having a screen, but with twin riser pipes and twin swing points. The ejector points are installed at close centres — 3 to 5 m. The lift to ground level of ejector systems is generally in the range 10 m to 50 m.

An ejector system may be powered by individual small high pressure centrifugal pumps or by a large pump connected into a common pressure main supplying high pressure water to each ejector point which in turn is connected to a single collector main. The net required input capacity of the pressure pump depends on the drive pressure required to lift from a given depth, input flow, and the diameter and design of the eductor jet nozzle and venturi. With a single pump set up, the water is usually circulated through a stilling tank with an over-flow that is the total induced flow of the system.

The ejector is a self-priming device. It evacuates air from the wellpoint beneath. If one ejector in a system is drawing air, it does not affect the balance of the system, provided air can be vented from the return or collector main. When the flow of groundwater is less than the design capacity of the ejector and the filter column has been sealed properly, then the unit develops a vacuum in the wellscreen and this is transmitted to the soil. The ability to develop a vacuum in the soil is particularly effective in draining fine-grained soils. If the wells are carefully constructed with appropriate filters and bentonite seals are used, the effects on fine sands and silts can be dramatic. Only low volumes of water have to be removed from such soils to achieve stabilisation so that the low efficiency of ejectors is acceptable.

The system design must take due account of the static lift from the ejectors to the water level in the recirculation tank; as well as head loss in the pressure main, the return riser and the return main.

An ejector system can be designed by assuming the line or ring of ejector wells to be equivalent to a continuous drainage slot since they are closely spaced and thereby computing the total induced flow to the system to achieve the required drawdown. A spacing is then selected that requires an acceptable induced flow from each ejector well. Having assessed the required drive pressure at the deepest ejector well head to achieve the required drawdown for the proposed depth of excavation, the rating of the pressure pump can be selected as generally an input flow of about 2 to 2 $^1/_2$ times the induced flow which is required. This ratio depends on nozzle size and design; drive pressure and the depth at which the ejector is positioned.

Careful analysis of the groundwater is advisable since ejectors are more sensitive to clogging by encrustation because of the very considerable pressure drop at the nozzle and venturi. When the groundwater contains more than 0.5 mg/l of iron its precipitation can lead to ejectors becoming blocked. However, nozzles and venturis can be changed.

Avoidance of settlement and heave

When a water table has to be lowered in order to construct an excavation, the effect of such an operation on neighbouring structures must be considered. On occasions it may be necessary to keep the groundwater at its original level in the built-up area immediately adjacent to the excavation so that the existing structures are protected from possible damage.

Lowering the water table by abstracting groundwater leads to an increase in effective overburden pressure, which in turn, can give rise to settlement of the ground surface. Significant

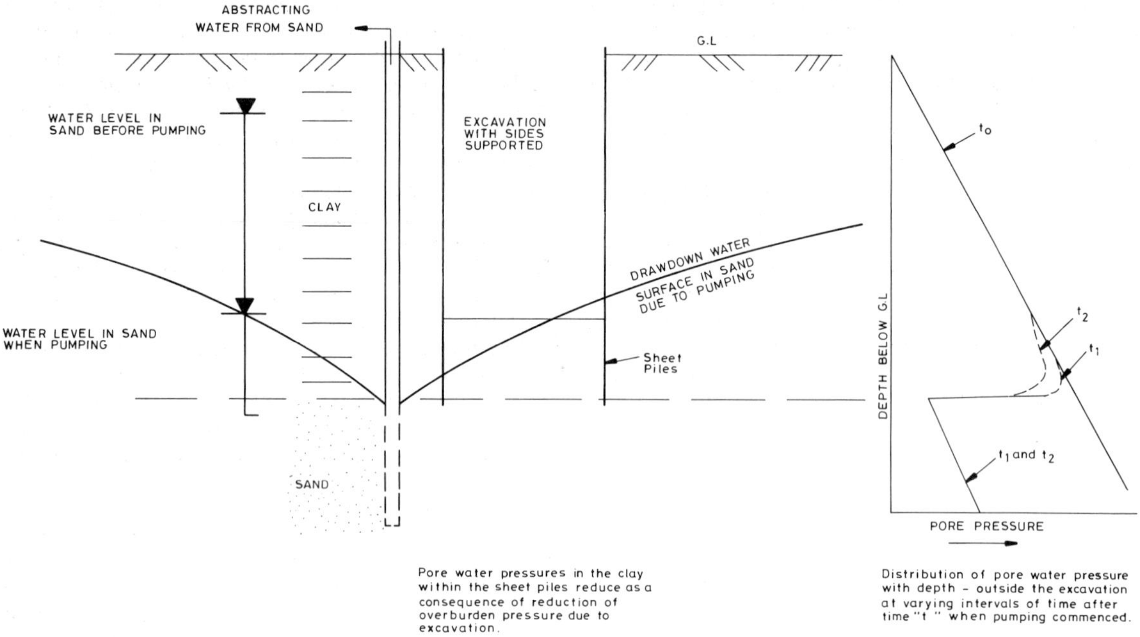

FIG.10. Effects of pumping from a confined aquifer overlain by compressible soils.

settlements also can occur when water is pumped from a confined aquifer overlain by compressible soils. Similarly abstraction of water from an aquifer containing layers or lenses of compressible soils may cause settlement. The amount of settlement which occurs depends on the thickness of the compressible soils and their compressibility, on the amount by which the phreatic surface is lowered and on the length of time over which pumping takes place. The rate of settlement is governed by the permeability of the compressible soil (Fig.10).

Depending on the site conditions, it may be feasible and desirable to use groundwater recharge or other measures to limit the radius of influence, that is, the cone of drawdown and so reduce potential settlement. In such instances the excavation usually is surrounded by sheet piling with the recharge wells located outside the piling. Relief wells within an excavation can be used to reduce piezometric pressures and thereby overcome the problem of heave. This type of relief well is often referred to as a 'bleeder well' system.

Potential problems of settlement outside the foundation area and of heave within had to be resolved prior to the construction of the Tower Latino Americano building in Mexico City. The foundation required a deep excavation into lacustrine sands and clays derived from a volcanic source. Zeevaert (1957) decided that if both settlement and heave were to be made inconsequential, then it was necessary to avoid a large change occurring in the prevailing effective stresses in the soil during excavation. Accordingly the area was first surrounded by sheet piling to a depth of 16 m. Next, four deep wells, going to a depth of 35 m, were installed within the excavation to reduce the piezometric level. The water obtained from these wells was injected, under pressure, into eight recharge wells located outside the excavation. The latter maintained the pore water pressure in the layers of sand.

A more recent case history where measures were pre-planned to minimise potential problems of settlement outside the building area and of heave within the building area, was the sub-basement construction for a new prestigious tower building in the congested business area of the Kasr el Nil thoroughfare in Cairo. The 75 m by 40 m basement area was surrounded by sheet piles 16.5 m in length. The formation level was 7 m below ground level in a clay stratum having confined water pressure beneath (Fig.11). Wellpoints were installed in two levels to reduce pressure in the confined aquifer by pumping. The upper level wellpoints were installed at 1 m above the level of the toes of the sheet piles. These were pumped on as much as was needed to prevent heave and 'blow', whilst the rate of pumping from the lower level wellpoints was restricted as much as possible to minimise drawdown effects outside the excavation area but was just sufficient to assist the upper level wellpoint pressure control. Pumping was continuous from November 1982 until April 1984 at an average rate of about 100 m^3/hr. Since the existing sewers of Cairo are surcharged 24 hours per day, 7 days per week, the pumped water could not be disposed of other than by means of recharged wells constructed so as to discharge at a lower horizon than those from which water was abstracted. This was achievable because of the stratified structure of the Nile alluvials, that is, vertical permeability was significantly less than horizontal radial permeability. The individual recharge wells were flushed out regularly to maintain their acceptance rate. Whereas an abstraction well, if properly designed and constructed tends to be self-cleansing, a recharge well tends to be self-clogging. Also it is important to have the facility at each recharge wellhead to bleed off air.

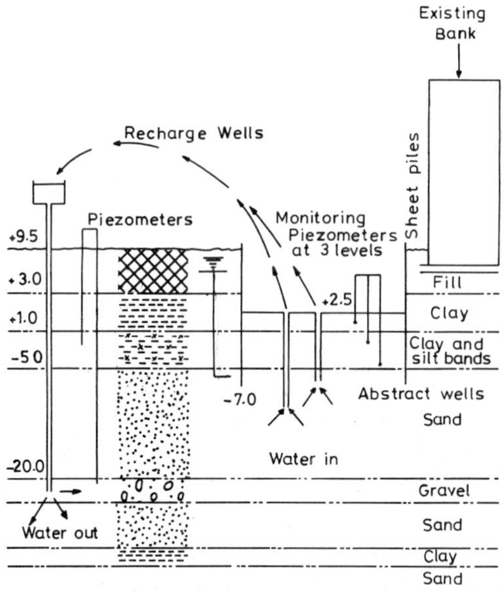

FIG. 11. Section through pressure relief pumping installation.

Electro-osmosis and electrochemical stabilisation

Electro-osmosis was originally developed as a means of dewatering fine grained soils (see Casagrande 1947). It consists of passing a direct current from anodes to cathodes positioned at predetermined locations in the soil which is to be stabilised. As the current passed through the soil it causes water to migrate from the anodes toward the cathodes where it is collected and then removed. The mechanism responsible for bringing about migration is not completely understood but it appears that absorbed water is removed from clay particles, and that ion transfer through the pore water causes it to flow. The anodes can be quite crude such as steel reinforcement rods or old tram rails.

According to Casagrande *et al.*, (1961) not only does electro-osmosis reduce the water content of the soil but it also directs seepage forces away from the surface of an excavation as well as developing pore water tension in the soil. Morris *et al.* (1985) found that the liquid limit of a soil is increased by electro-osmotic treatment and that the reduction in moisture content improves the strength of the soil. Moreover, base exchange occurs in the soil during electro-osmosis, which further enhances its strength. In addition, ion migration, electrolysis and chemical reactions occur leading to the formation of new irreversible compounds.

The fundamental disadvantage of electro-osmosis is that when used solely for dewatering, it is a decelerating process, becoming progressively less efficient as the soil water content is reduced and subsequently rehydration frequently may reverse the process rapidly.

Where electro-osmosis has proved successful, it usually has been as a result of using the process to introduce a chemical (usually silicate of lithium, calcium or ammonium) into the soil, either through anode solution or by direct electrolyte replacement. This improves the stability of the soil either by ionic replacement occurring in the clay mineral content, or by cementitious material being deposited in the pore space. This adaption of electro-osmosis is know as electrochemical stabilisation.

Electro-osmotic flow

The quantity of water (Q_e) moved in a unit time by electro-osmosis in a single capillary, according to Casagrande (1952), can be obtained from the Helmholtz expression:

$$Q_e = \frac{E D r^2 \zeta}{4 \eta L}$$

where
E = the electric potential
D = the dielectric constant of the liquid
r = the radius of the capillary
ζ = the electrokinetic potential, or zeta potential, which is defined as the potential existing between the rigid and the movable parts of the double layer
η = the viscosity of the liquid
L = the length of the capillary between the electrodes.

Unlike the coefficient of soil permeability (k), and therefore the rate of hydraulic (laminar) flow, the coefficient of electro-osmotic permeability (k_e) and the rate of electro-osmotic flow are independent of the size of the capillaries. Hence the electro-osmotic permeability is more or less independent of the soil pore size. In other words, it is about the same magnitude for sands, silts and clays, provided that the electro-osmotic potential is about the same for most of the mineral matter in the soil. The electrokinetic potential depends on the concentration and valence of soluble ions in the electrolyte. It is high in a dilute solution whereas in a concentrated solution it may be reduced to zero. In fact, the potential does not vary much within the ordinary range of concentration of electrolytes present in soils. However, the electrokinetic potential can fall to zero in some situations and the electro-osmotic flow either ceases or may be reversed in direction. For example, in unsaturated soils the thickness of absorbed layers is reduced and so the concentration of ions is increased.

The velocity of flow of water through soils induced by an electric current can be expressed as:

$$v_c = k_c E$$

where v_e is the velocity, in cm/s, due to an electrical gradient and relative to a cross-sectional area of the soil; k_e is the electro-osmotic coefficient of permeability in cm²/volts/s which is equal to:

$$\left(\frac{e}{1+e}\right) \; C_1$$

in which e is the void ratio, and C_1 is a function of the fluid contained in the soil; and E is the electrical gradient in volts/cm (the ratio of voltage across electrodes to the distance between electrodes). Hence, by analogy with d'Arcy's Law, the quantity of water (Q) which discharges through a given area (A) in a given time (t) is:

$$\frac{Q}{t} = k_c EA.$$

For practical field application the above expression can be rewritten as:

$$\frac{Q}{t} = k_c \, p \, I.$$

where ip the resistance of the soil in ohm/cm and I is the current in amps (Bjerrum *et al.* 1967). The discharge of water from a soil decreases with time until a state of equilibrium is reached between inflows and electro-osmotic discharge.

For practical purposes most saturated soils, according to Casagrande (1952) may be assumed to have a coefficient of electro-osmotic permeability of around 5×10^{-5} cm/s for a gradient of 1 volt per cm, for example:

London Clay = 5.8×10^{-5} cm/s

Boston Blue Clay= 5.1×10^{-5} cm/s

Clayey silt = 5×10^{-5} cm/s

Marl = $0.7 - 2.6 \times 10^{-5}$ cm/s

Fine sand = 4.1×10^{-5} cm/s

Zhinkin (1966) suggested a rather wider range of k_e values between 0.5×10^{-5} and 13×10^{-5} cm/s at a potential gradient of one volt/cm, with the majority of values in the 3.0 to 5.0×10^{-5} cm/s per volt range.

The relative efficiency of electro-osmosis as a dewatering technique may be expressed in terms of the ratio between k_e and k, referred to as the coefficient of electro-osmotic effectiveness. It increases with increasing clay fraction in the soil.

Spacing of electrodes depends on the electrical potential available, the potential gradient required for a given soil and groundwater conditions. Casagrande (1952) maintained that spacings between 3.6 and 4.95 m and potentials varying between 30 and 180 volts had proven effective. As far as individual soil types are concerned Zhinkin (1966) proposed a potential gradient of 0.9 volts/cm for sandy loams, 0.7 volt/cm for loams, 0.6 volt/cm for clays, and 0.4 volt/cm for silts. This implies that for a direct current potential of 100 volts, the electrode separation would range from just over 1 m to 2.5 m. For maximum effectiveness, the potential gradient should be in the same direction as the hydraulic gradient. Generally, however, Casagrande suggested that potential gradients of more than 0.5 volt/cm should not be exceeded for long-term applications because higher gradients lead to energy losses in the form of considerable heating of the ground. However, it might prove advantageous to use potential gradients of 1 to 2 volt/cm during the first day hours. This gives a much faster build-up of tension in the pore water. Economy of power consumption could perhaps be obtained by intermittent operation.

Casagrande *et al.* (1981) described the stabilisation of silt where the electrodes penetrated the full thickness of the formation, that is, a depth varying from 12 to 60 m. This is probably the greatest depth so far subjected to electro-osmotic treatment.

Negative pore water pressures develop as water is removed next to the anodes while at the cathodes the pore water pressures can increase to high positive values. This difference in pore water pressure produces a hydraulic backflow. It also means that the soil around the cathodes may be softened.

Consolidation may occur as a result of the development of negative pore water pressure under controlled electro-osmotic drainage. This in turn, leads to changes in the stress-strain and strength characteristics of the soils concerned (Mitchell & Wan 1977). Indeed Esrig (1968) showed that the Terzaghi equation of consolidation for homogeneous material applies to electro-osmotic consolidation where the voltage distribution remains constant with time. Like ordinary consolidation under direct loading, electro-osmotic consolidation is a mass flow process in response to pressure or potential gradients. The rate of consolidation is controlled by the coefficient of consolidation which, in turn, is governed by the hydraulic conductivity and not by the electro-osmotic permeability. For a given compressibility, the amount of consolidation depends on the magnitude of negative pore water pressure that can be developed at the end of consolidation. At the end of the electrical treatment the effective stress developed during the electro-osmotic consolidation relaxes gradually and some rebound occurs.

Under ideal conditions, and starting with no excess pore water pressure, then at the end of the period of electro-osmotic treatment a maximum pore water tension occurs at the anodes and decreases linearly to zero at the cathode. Hence the soil properties vary between anode and cathode. In order to develop a more uniform stress distribution and to increase the average shear strength of the soil, it may prove useful, according to Wan & Mitchell (1976), to reverse the roles of the electrodes after a certain period of treatment.

Farmer (1975) maintained that two factors mitigate against the use of electro-osmosis on a large scale. Firstly, overconsolidated clays tend to be fissured. If the ability of such soil to transmit water along these fissures is similar to the electro-osmotic permeability, then it is unnecessary to use electro-osmotic treatment. Cheaper methods of drainage such as sand drains can be used instead. More seriously, electro-osmosis only operates efficiently in saturated and near saturated soils. It therefore cannot be used to dewater a soil completely. However, Bjerrum *et al.* (1967) noted that an excessive current density means that the soil around the anodes dries out thereby increasing the anode-soil resistance so that the process becomes increasingly less effective. Cessation of the process, especially in the clay soils, readily permits reabsorption of water. This occurs at low hydraulic rates but nevertheless after a certain period the soil reverts to its original state.

One of the most spectacular known uses of electro-osmosis is to accelerate the dissipation of excess pore water pressures generated during the construction of an earthfill dam was masterminded by Casagrande in the mid 1960s. The 25 m high by 2750 m long West Branch Dam in Ohio was nearing completion when a 300 m length near the centre started to move. Immediately 4 m of fill was stripped off. Some piezometers registered pressure heads 10 m to 12 m above the top of the dam. It was estimated that the pore water pressures developed at West Branch might have taken 10 years to dissipate if no remedial measures had been taken. By the use of electro-osmosis they were reduced sufficiently for construction to recommence after 10 months.

Over 1000 electrodes, roughly two thirds anodes, were installed. The cathodes were pumped using eductors. The total installed generating capacity was 2700 kW.

Because electro-osmosis is a process that has high energy costs, its usage is limited.

Electrochemical stabilisation

The mechanical properties of clay minerals vary, particularly those belonging to the smectite family, according to the type of cations associated with them. If a clay mineral contains a significant amount of weakly bonded cations such as sodium or lithium, then it has a tendency to absorb and retain large quantities of water. Accordingly it possesses a low shear strength. By contrast, if the cations have higher bonding strength such as calcium and magnesium, or more especially iron or aluminium, the clay material absorbs less water and has a more stable structure.

As cations associated with the clay minerals are exchangeable, the introduction of solutions containing an excess of cations with higher bonding strength gives rise to ion exchange, thereby enhancing the soil properties. Introduction of stabilising electrolytes into a clay soil is brought about by either solution and dispersion of the anode material through electrolysis or by direct introduction of additives at the anode.

Electrolysis normally involves decomposition of the electrolyte accompanied by solution of anode material. Electrochemical stabilisation due to anode solution results from initial base exchange reactions on the surface of the clay minerals, weakly bonded cations being replaced by strongly bonded cations, and the formation of soil cementing compounds produced by reaction between the electrolyte and dissolved anode solution. Thus a lattice-like structure is built-up around the clay soil particles. This greatly reduces the amount of settlement which would result from the use of electro-osmosis on its own.

Addition of additives to the anode (in addition to the dissolved electrode material) can increase the base exchange reaction, act as a catalyst in the formation of cementitious material and improve the quality and quantity of cementitious material. To increase the base exchange reaction, most of the additives used are organic and inorganic compounds of aluminium or calcium — iron normally being present in the form of the electrode material. Of these the most commonly used compound is calcium chloride. This intensifies dissolution of the iron anode by the formation of ferrous chlorides, at the same time producing calcium hydroxide which acts as a cementing agent.

An important disadvantage, in the use of additives, is the reduction in flow with increase in electrolyte concentration. This may effectively limit stabilisation beyond a certain level.

Early in 1970 trouble was observed with one abutment of the fairly new road bridge over the river Loire at Angers where soft compressive soils overlie the bedrock. Additional fill placed at the approaches to the abutment surcharged the compressible soils and started to squeeze the abutment towards the other side of the river but the bridge was seeking to act as a strut. The abutment could only rotate.

The French Ministry responsible for roads and bridges authorised and paid for an experimental electrochemical stabilisation system to be installed and operated. The anode pipes were extended to above flood level to guard against dilution of the silicate solution if the river level rose.

The electrodes were at 3 m centres. There were 2 × 55 No drillholes and the system was operated for the first 15 months at 40 volts with energy at 195 kW h. For a further 5 months the voltage was raised to 120 volts and the current consumption stepped up to 580 kW h. The cost, in 1970, was about £10 per cubic metre of soil treated with about 25 % of the total cost being for electricity and about 30 % for silicate.

References

BJERRUM, L., MOUM, J & EIDE, O., 1967. Application of electro-osmosis to a foundation problem in a Norwegian quick clay. *Geotechnique*, **17**, 214-235.

BOULTON, N.S., 1951. The flow pattern near a gravity well in a uniform water bearing medium. *Journal of the Institution of Civil Engineers*. London, 534-550.

CASAGRANDE, L., 1947. The application of electro-osmosis to practical problems in foundations and earthworks. *Building Research Station Technical Paper No. 30*, DSIR, Watford.

CASAGRANDE, L., 1952. Electro-osmosis stabilisation of soils. *Journal of the Boston Society of Civil Engineers*, **39**, 51-83.

CASAGRANDE, L., WADE, N., WAKELY, M. & LOUGHNEY, R.W., 1981. Electro-osmotic projects, British Columbia, Canada. *Proceedings of the 10th International Conference on Soil, Mechanics and Foundations Engineering.*, Stockholm, **3**, 607-610.

CASHMAN, P.M., 1975. Control of groundwater by groundwater lowering. *Methods of Treatment of Unstable Ground*, Ed. by Bell, F.G., Newnes-Butterworth, London 12-25.

CASHMAN, P.M., & HAWS, E.T., 1970. Control of groundwater by water lowering. *Ground Engineering*, Ed. by David, J.S., Institute of Civil Engineers, London 23-32.

CEDERGREN, H.R., 1975. Drainage and dewatering. *Foundation Engineering Handbook*, Ed. by Winterkorn, H.F., and Fang, H.Y., Van Nostrand Reinhold, New York, 221-243.

d'ARCY, H., 1856. *Les Fontaines Public de la Ville de Dijon*, Dalmont, Paris.

DEBIDIN, F., & LEE, C.F., 1980. Groundwater and drawdown in a large excavation. Canadian *Geotechnical Journal*, **17**, 185-202.

DUPUIT, J., 1863. *Etudes Theoreteques et Practiques sur le Mouvement des Eause*, Paris.

ESRIG, M.I., 1968. Pore pressure consolidation and electrokinetics. *Proceedings of the American Society of Civil Engineers, Journal on the Soil, Mechanics and Foundations Division* **94**, SM4, 899-921.

FARMER, I.W., 1975. Electro-osmosis and electrochemical stabilisation. *Methods of Treatment of Unstable Ground*, Ed. by Bell, F.G., Newnes-Butterworths, London, 26-36.

MANSUR, C.I. & KAUFMAN, R.I., 1960. Dewatering. *Foundation Engineering*. Ed. by Leonards, G.A., McGraw Hill, New York, 241-350.

MITCHELL, J.K., & WAN, T.Y., 1977. Electro-osmotic consolidation — its effects on soft soils. *Proceedings of the 9th International Conference on Soil Mechanics and Foundations in Engineering. Tokyo*, **1**, 219-224.

MORRIS, D.V., HILLUS, S.F., & CALDWELL, J.A., 1985. Improvement of sensitive silty clay by electro-osmosis. *Canadian Geotechnical journal*, **22**, 17-24.

MUSKAT, M., 1927. *The Flow of Homogeneous Fluids through Porous Media*. McGraw Hill, New York.

POWERS, J.P., 1981. *Construction Dewatering: A Guide to Theory and Practice*. Wiley-Interscience, New York.

PRUGH, B.J., 1960. New tools and techniques for dewatering. *Proceedings of the American Society of Civil Engineers Journal of the Construction Division*, **80**, CO1. 11-25.

WAN, T.Y., & MITCHELL, J.K., 1976. Electro-osmotic consolidation of soils. *Proceedings of American Society of Civil Engineers, Journal of Geotechnical Engineering Division*. **102**, GT5, 473-491.

WARD, W.H., 1948. A coastal landslip. *Proceedings of the 2nd International Conference on Soil, Mechanics, Foundations Engineering*, Roterdam, **2**, 33-38.

WERBLIN, D.A., 1960. Installation and operation of dewatering systems. *Proceedings of the American Society of Civil Engineers, Journal of the Soil Mechanics and Foundnations Division*, **86**, SM1, 47-66.

ZEEVAERT, L., 1957. Foundation design and behaviour of Tower Latino Americano in Mexico City. *Geotechnique*, **7**, 115-133.

ZHINKIN, G.N., 1966. *Electrochemical stabilisation of soils*. Stroiizdet, Moscow. Abridged C.E.R.E., Translation No. 21.

F. G. BELL, Department of Civil and Structural Engineering, Teesside Polytechnic, Middlesborough, UK.

P. M. CASHMAN, formerly at Wimpey Laboratories Ltd., Hayes, Middlesex, UK

Groundwater in Engineering Geology, London, 1986, pp. 487-496 Printed in England

Interceptor drains for cliff-tops and above the crest of slopes and cuttings

M.E. Barton & R.I. Thomson

ABSTRACT: Three types of interceptor drain considered in the paper are (i) drains for surface water, (ii) cut-off drains in pervious strata overlying a relatively impervious stratum and (iii) vertical drainage wells for draining water into either pervious strata at depth or drains bored (or tunnelled) in from the slope face. Their function is to prevent surface or ground water from reaching an actual or potential landslide: drainage of the landslide itself is not considered in this paper.

Various difficulties which may impair the effectiveness of interceptor drains are listed. Cases from the literature are reviewed and new observations made on interceptor drains of which the authors have direct experience. It is shown that the performance of these drains may often be less than satisfactory and it is concluded that by their nature, interceptor drains require rigourous site investigation and monitoring before, during and after construction. Attention is also drawn to the generally negligible influence of interceptor drains on rates of equilibration of pore-water pressures depressed by unloading in clay soils.

Introduction

The word interceptor is used here to define forms of drainage which are designed to intercept surface or ground water and prevent it getting to an actual or potential landslide. Interceptor drains may be installed on cliff-tops to prevent water from getting to an unstable undercliff or on the uphill side of cuttings and potentially unstable natural slopes. Drainage of a landslide mass itself is not considered in this paper.

The following types of interceptor drains are discussed:

(a) Shallow, open drains intercepting surface water and constructed where relatively impervious strata are present beneath the topsoil.

(b) A cut-off drain for ground water contained within a relatively permeable horizon overlying an impervious stratum. This type may take the form of either (i) a trench drain or (ii) a cut-off diaphragm with drainage outlets. Both forms are usually covered over at the surface.

(c) Sand or gravel drains, installed in boreholes or wells, intercepting perched ground water. Drainage takes place either (i) by downward percolation to a pervious stratum from which discharge does not threaten the slope stability or (ii) to a horizontal or inclined, sealed drain bored, or tunnelled, in from the face of the slope.

A composite and idealised diagram showing the main features of each type is given in Fig. 1.

The construction of an interceptor drain may seem highly desirable as part of a stabilisation scheme where there are either surface flows, or perched ground water flows in pervious layers, contributing to seepage erosion and the generation of high pore pressures in potentially unstable soil masses. This initial view may overlook a number of difficulties which could impair the efficiency of the drain or even make the benefits questionable. Some of the possible difficulties for each type of interceptor drain are listed in Tables 1 to 3 respectively. For convenience the difficulties are broadly classified as geological, geo-hydrological and constructional.

Surface water drains

The water to be intercepted may be as a true surface flow or as a flow through the topsoil. It is assumed that below the topsoil, the material is relatively impervious. Drains for intercepting surface water and preventing it from reaching potential slips are referred to by Seaton (1938), Toms & Bartlett (1962) and Sherrel (1976). They may be open, concrete lined channels (Sherrel 1976) or trenches with pipes, back-filled with free draining material (Toms & Bartlett 1962). Open ditches in clay soils are undesirable as the shrinkage cracks developing in dry weather may later prove a source of weakness. Seaton (1938) claimed that the cause of slips in clay cuttings has, at times, been traced to such open ditches. The re-distribution of surface water by open ditches promoting sliding activity has been recorded by Ward (1948) and Henkel (1982).

From CRIPPS, J.C., BELL, F.G. & CULSHAW, M.G., (eds), 1986, *Groundwater in Engineering Geology,* Geological Society Engineering Geology Special Publication No. 3, pp. 487-496

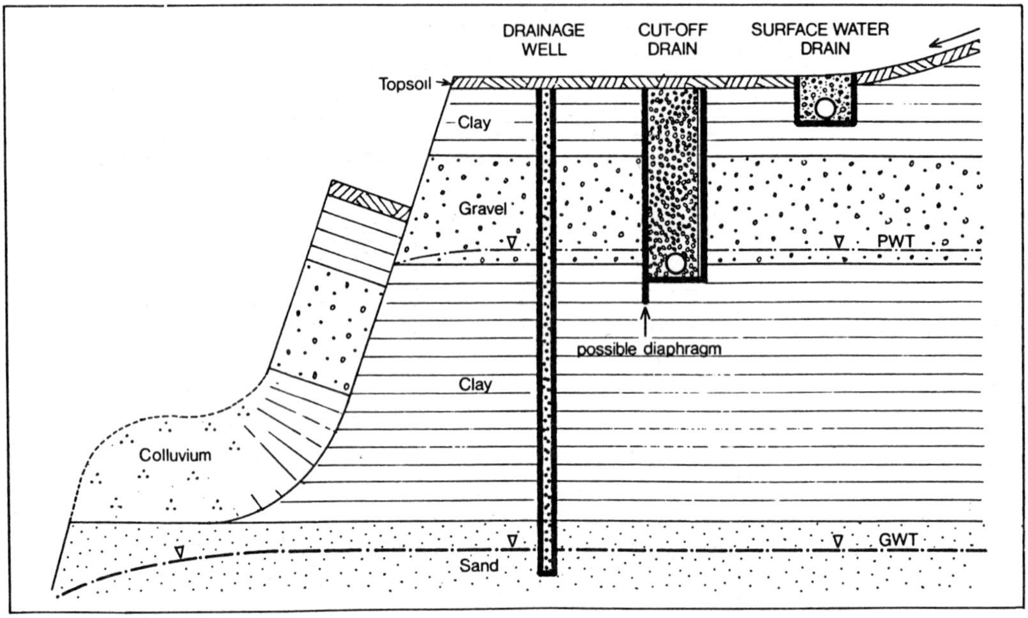

FIG. 1. Composite diagram showing the three types of interceptor drain considered in the paper. A: Surface Water Drain. B: Cut-off in pervious stratum. C: Drainage Well draining to pervious stratum at depth.

TABLE 1. *Some possible difficulties impairing the effectiveness of interceptor drains for surface water.*

Geological	1. Non-uniformity of the clay (or impervious stratum) resulting in more permeable zones occurring at drain invert level.
	2. Presence of fissures or tension cracks at drain invert level.
	3. Presence of fill, or older drains, below drain invert level.
Geo-hydrological	4. Possibility of flash floods overwhelming the drain.
	5. High infiltration capacity in the 'impervious stratum'.
	6. The drain channelling water to zones with a very high infiltration capacity.
Constructional	7. The drain being used as, or extending, a tension crack.
	8. Insufficient gradient to prevent ponding.
	9. Clogging of the drain by silting or the growth of vegetation.

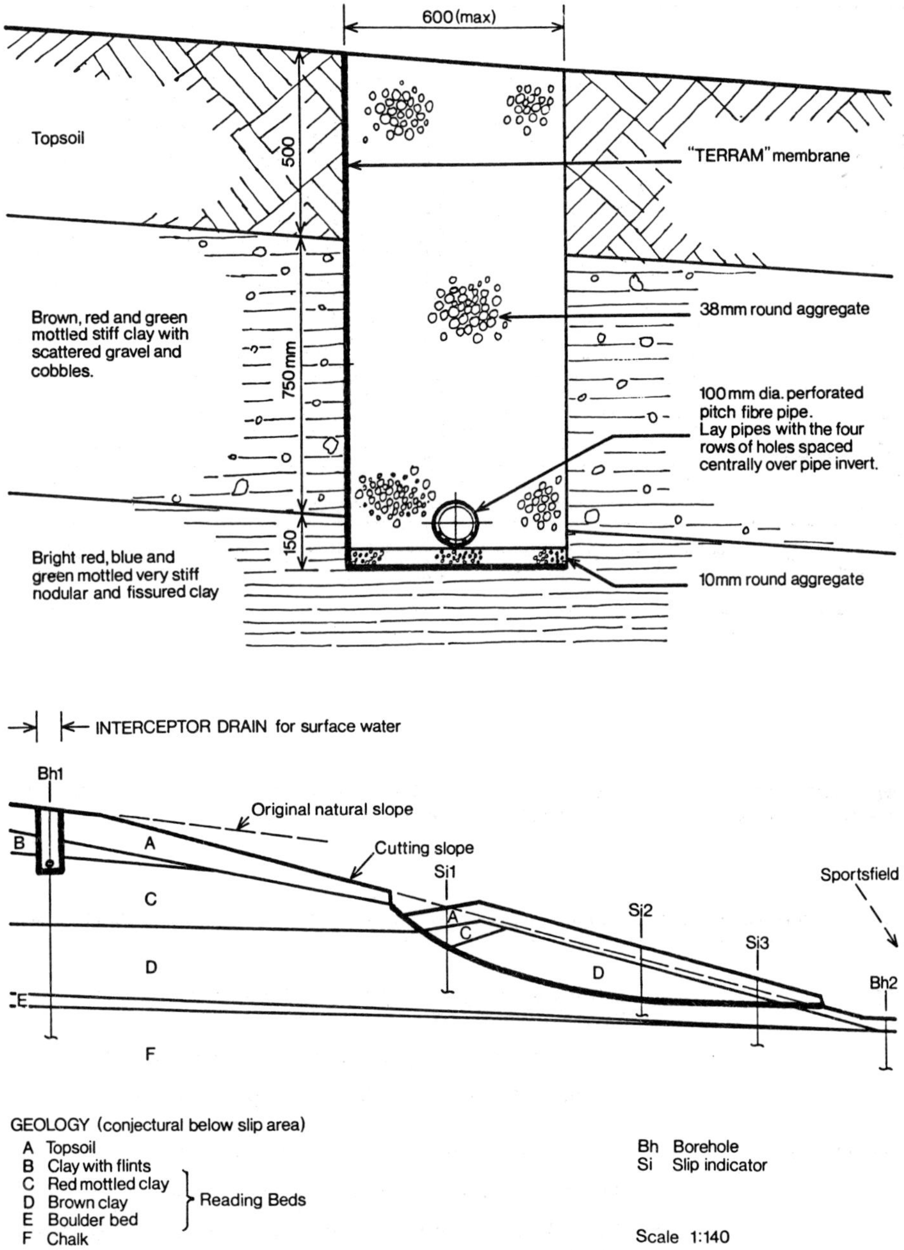

FIG. 2. Above: Interceptor drain for surface water installed above a cutting slope at Hursley, Hampshire. Below: Geological section showing location of drain and slip in cutting.

Open ditches with unpaved inverts were excavated on the up-slope side of all cuttings for the M20 Maidstone By-Pass (Garrett & Whale 1985). Five years after construction, varying degrees of failure occurred in nearly all the cuttings. In the case of the Longham Wood cutting the failure surface led into one of the open, 1 m deep ditches where there was discovered a tension crack extending for at least 1.5 m below the base of the ditch (J.H. Whale pers. comm.).

Fig. 2 shows a backfilled trench, with a perforated pipe at the base, installed to prevent surface water from reaching a 5 m high cutting slope adjacent to a sports field at Hursley, near Winchester, Hampshire. This slope had previously developed two, small compound-type slips. (Fig. 2, section). Apart from regrading the surface, no remedial action was taken on the cutting slope. The base of the interceptor drain was founded in the Reading Bed Clay which could be expected to provide an effective seal to the trench base. Unfortunately at one location, the trench base was above the level of an old drain and its relatively permeable backfill. Much water from these sources continued to reach the cutting slope and further sliding took place during the winter after construction of the interceptor drain.

It is not possible to say whether or not movements would have been renewed either by the pore pressure generation from direct precipitation on the slope or by the continued equilibration of pore pressures depressed by unloading (Vaughan & Walbanke 1973) even if the older drain had been blocked. It can, however, be concluded that with the deficiency, the new interceptor drain could not protect the cutting slope.

Cut-off drains

The terms 'cut-off' drain is used to refer to drains intercepting ground water held up in a pervious stratum overlying a relatively impervious one (Hutchinson 1977). Both strata may be part of the solid sequence (Baker & Marshall 1958: fig. 109) but commonly the pervious stratum is a drift deposit overlying a stiff fissured clay as at Herne Bay (Duvivier 1940), Charmouth (Conway 1979) and Highcliffe (Halcrow 1971; Mockridge 1983).

At Herne Bay the cut-off to intercept ground water in patches of sandy and gravely head overlying London Clay, extended for nearly 1.2 km at a minimum distance of 12 m from the cliff edge. It was constructed as an unlined trench between 1.5 and 3.5 m deep, including a minimum of 0.5 m into the London Clay. The work was completed in 1938. Subsequently considerable landsliding took place in the undercliffs (Bayley 1972; Bromhead 1978; Hutchinson 1965). Regrading of the Herne Bay cliffs destroyed the original cut-off drain (A.G. Roberts, pers. comm.). At the Higher Sea Lane site near Charmouth, a shingle-filled trench drain was installed in 1970 to intercept water flowing from highly permeable Cretaceous Head overlying Liassic Head and Lias Clays. This water had previously contributed to rapid movement of mudslides of the adjacent cliffs. Subsequent to the installation, the mudslide movement ceased although some cliff-top recession continued, albeit at a much reduced rate (Conway 1979).

At Highcliffe, 1.5 to 7.6 m of Plateau Gravel (a sandy, mainly coarse gravel with a saturated permeability (k) of about 1.8×10^{-3} cm/sec) overlies the Barton Clay. As part of cliff stabilisation works, initially undertaken over a length of 0.5 km, a cut-off in the form of a concrete diaphragm was constructed (Halcrow 1971). It was located generally at least 20 m from the cliff edge except at the western end where, owing to restricted access, the distance was reduced in part to about 12 m. Discharge to the beach takes place through sealed drainage outlet pipes provided at intervals along the line of the cut-off. It was intended that the base of the diaphragm would be taken into the clay for a distance of 0.75 m but it was later found that over a length of 20 m, the diaphragm missed the clay by about 0.3 m and a sizeable cliff-top slump which took place after the cut-off construction was ascribed to this deficiency (M.F. Tyhurst pers. comm.) Remedial works to extend the diaphragm were undertaken as shown in Fig. 3 (Mockridge 1983).

From further work at Highcliffe, the following factors relating to the cut-off can be noted:

(i) The actual base of the Plateau Gravel in this area cannot be very precisely predicted owing the irregularities introduced by involutions, frost wedge casts and other periglacial structures (Barton 1984). The frost wedge casts in particular could allow downward percolation into more permeable horizons within the Barton Clay.

(ii) Investigations in the Naish Farm (unstabilised) area at Highcliffe (Barton & Thomson 1984) has indicated that downward percolation from the Plateau Gravel into the Barton Clay measurably increases within about 10 m of the cliff edge. Seepage of water on the cliff face can sometimes be seen emerging from hollows within the weathered zone of the Barton Clay rather than at the gravel/clay contact. Since the diaphragm (in stabilised area) creates additional ponding in the Plateau Gravel (estimated as 0.5 m by Halcrow, 1971) downward percolation into the clay could become a significant factor if the cut-off was located too close to the cliff edge.

(iii) Discharge to the cliff face from the plateau gravel is seasonal at Naish Farm. Provisional estimates (based on Barton & Thomson 1985) suggest that with maximum storage levels during the winter months, about 10 to 20% of the effective rainfall is lost by direct downward percolation into the clay while the rest is discharged from the cliff face. Since the cliff-top catchment is approximately the same size as the undercliff area, then the discharge from the Plateau Gravel cliff-face may add an additional 80–90% to the recharge to the undercliff area. The boundaries of the ground water catchment for the stabilised area at Highcliffe are not known but it may be of the same order.

(iv) The principal benefit of cut-off drains of this type is in reducing outflow to the undercliff. Study of the seepage distribution within the Barton Clay at depth indicates that the pore pressures there will be virtually unaffected by the cut-off and it follows, therefore, that the final level of equilibrated sliding (Vaughan & Walbanke 1973) would also be unchanged by the action of the cut-off. (Bromhead & Dixon 1985 have similarly discussed that 'modest' slope drainage may have no influence on pore pressure equilibration). On the other hand, at Highcliffe (see section, Fig. 3) will certainly have some influence in determining the final equilibrated pore pressures.

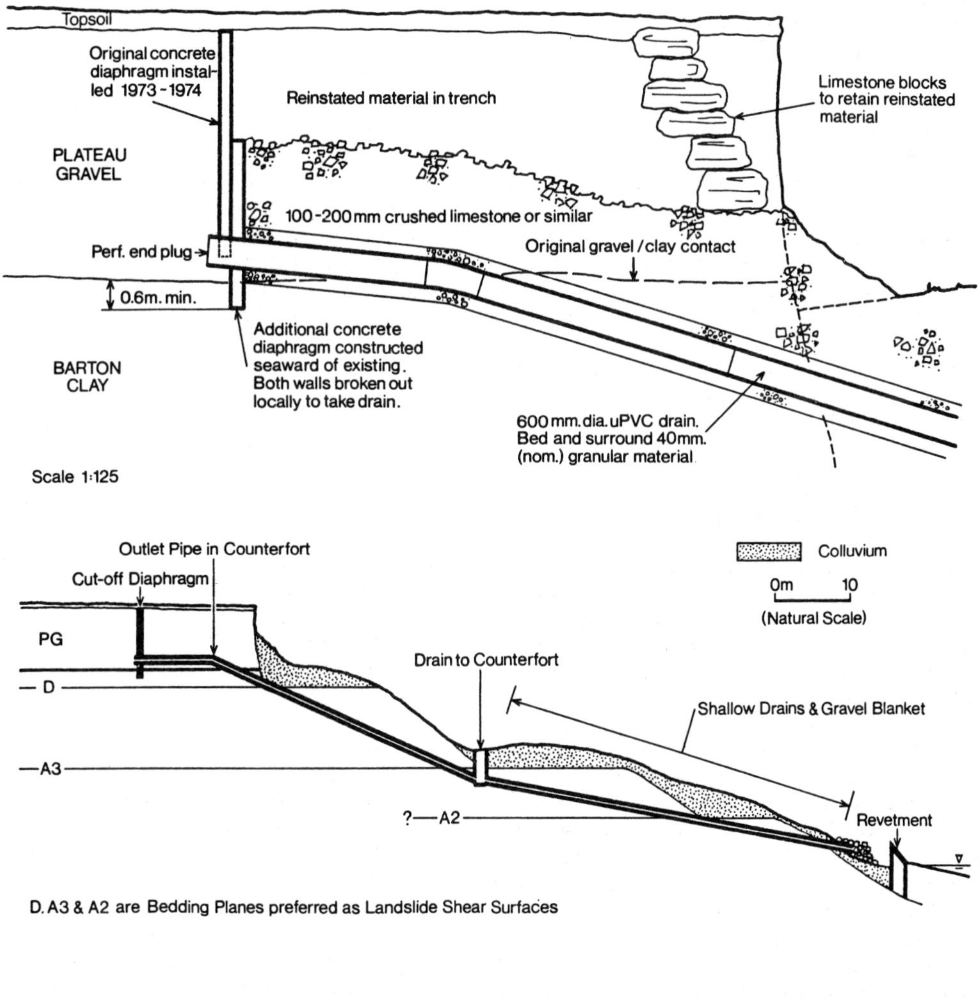

FIG. 3. Above: Cut-off diaphragm installed in the Plateau Gravel to intercept perched water held up above the Barton Clay in the cliff-top area at Highcliffe, Hampshire. The original diaphragm did not fully penetrate the Plateau Gravel and the figure shows the additional work undertaken to make the cut-off effective (from Christchurch Borough Council Engineering Drawing). Below: Geological section trough the stabilised cliffs at Highcliffe. The sealed outlet pipe for the cut-off diaphragm is led downwards through the cliffs in the base of a series of rock-filled counterfort drains running down to the beach.

TABLE 2. *Some possible difficulties impairing the effectiveness of cut-off drains in pervious drift strata overlying solid clay strata in coastal cliffs.*

Geological	1. Irregularity and/or channelling of the drift/solid interface.
	2. Disturbance of the drift/solid contact by periglacial action — especially where deep frost wedge casts may be present.
	3. The lithology and structure of the clay stratum allowing for more permeable horizons to meet the drift landward of the cut-off.
	4. Piping and seepage erosion at the top of the clay due to percolation through fissures or as a result of eluviation in the upper layers.
Geo-hydrological	1. The rate of downward percolation into the clay stratum landward of the cut-off.
	2. The existence of significant seepage flow in relatively pervious horizons within the solid strata.
	3. The recharge to the undercliff by effective rainfall seaward of the cut-off may be sufficient to promote renewed movement.
	4. The cut-off may have negligible influence on the rate of equilibration of pore pressures depressed by unloading in the clay strata due to landsliding.
Constructional	1. Coping with a considerable depth of cut-off and the expense of providing adequate side-support.
	2. Access may be unavailable to allow the cut-off to be located at an adequate distance away from the cliff edge.
	3. The possibility of the cut-off acting as a tension crack or channelling water to a pre-existing one.
	4. The provision of satisfactory engineering-geological supervision of cut-off depth requirements.

Vertical drainage wells

Vertical sand drains designed to intercept water held up over clay bands and seams and to discharge it to a permeable stratum at depth were installed at Bournemouth in 1964 (Anon 1965). Three stretches of cliff with a total frontage of just over 1 km were treated. The drains were generally 2.5 m apart with a double row where seepage on the cliff face was notably large. At East Overcliff Drive the drains were sunk in the footpath at a distance of 6 to 7 m from the cliff edge (Fig. 4).

During January 1965, a slip occurred over a length of 120 m with part of the rear scarp running through the line of sand drains. At the end of April, a second slip occurred taking part of the roadway and exposing the drains at a depth of about 10 m below ground level.

It was claimed that the sand drains notably reduced visible outflow to the cliff face. Apart from the East Overcliff Drive site, it could be that slope movement was imminent (perhaps due to equilibration of pore pressures depressed by unloading) but equally there remains the possibility that the sand drains either (i) induced or extended a tension crack or (ii) redistributed water within the soil profile allowing pore pressures to build up in critical parts of the undercliff.

A notable feature of the exposed sand drain was a transition zone between the backfill material to the well and the surrounding Boscombe Sand (see Fig. 5). The wells were drilled using a rotary bit with compressed air flush and the transition zone appeared to consist of remoulded Boscombe Sand with some impurities giving it a less uniform grain size than the Boscombe Sand (Fig. 5). Permeability tests using falling head tests on the saturated materials gave the following results:

A. Well Core Material $1.2 - 1.8 \times 10^{-3}$ cm/sec.
B. Transition Zone $0.4 - 1.2 \times 10^{-3}$ cm/sec.
C. Boscombe Sand $1.3 - 15.5 \times 10^{-3}$ cm/sec.

It was concluded that the Transition Zone had only a marginally lower permeability than the well core material and would offer no significant impedance to the flow of perched water into the well. Possible segregation of the well core material, as discussed by Powers (1981), could not be explored as only a short length of well was exposed.

Other examples of drainage wells used for vertical drainage to more permeable strata at depth are given by Parrott (1955) and Lefebvre *et al.* (1976). The case described by Lefebvre is not an interceptor in the sense used in this paper since the wells were used to drain an area of superficial sliding, nevertheless, it is relevant in demonstrating a difficulty associated with this form of drainage. The wells were embedded in a stratum not sufficiently pervious to absorb the flow. Consequently there was a measured rise in pore pressure and a reduced safety factor for deep failure surfaces.

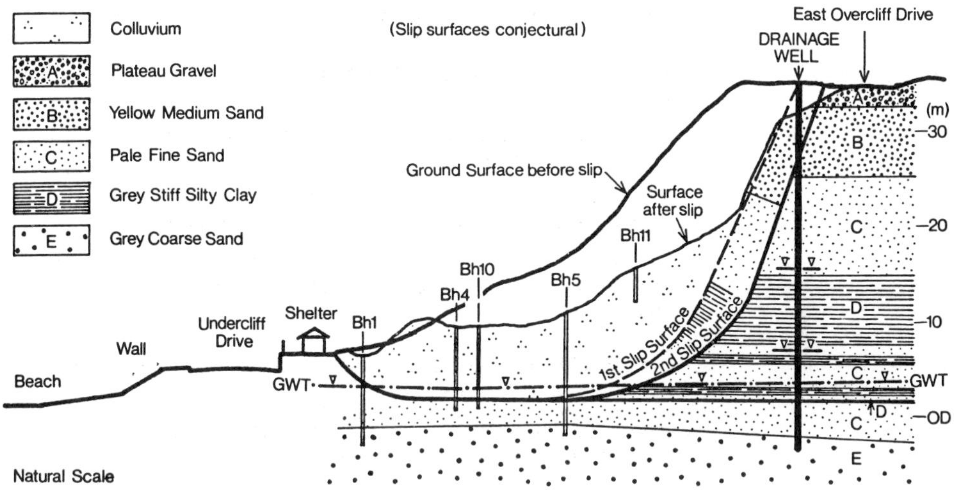

FIG. 4. Geological section of the cliff at East Overcliff Drive, Bournemouth showing the record of strata and ground water recorded in Drainage Well No. 185 (drilled in 1964). Borehole 5 indicated that the surfaces of the two two slips in 1965 extended to the lowest of the clay bands. The location of the slip surface in the toe area is conjectural. There is no record of the permanent (although tidally influenced) groundwater table but it is estimated as the lowest recorded encountered levels in boreholes 1 and 4 and the drainage well (GWT in the figure).

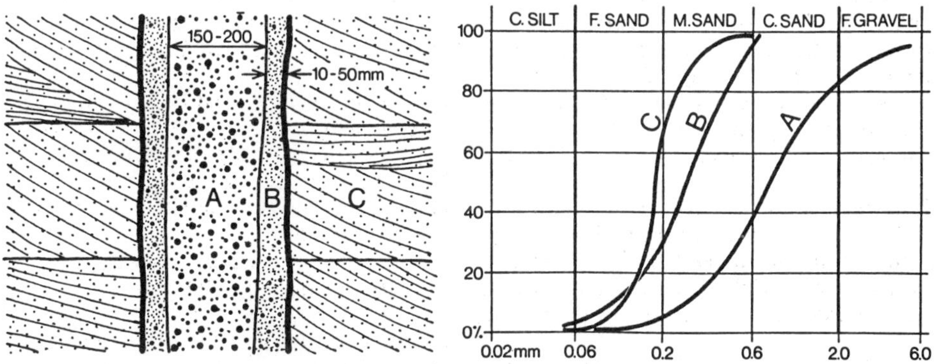

FIG. 5. Left: Section Through drainage well at East Overcliff Drive, Bournemouth as exposed in the rear scarp of the second slip of 1965. Right: Particle size distribution curves. A. Drainage well filter material. B. Transition zone of remoulded Boscombe Sand with impurities. C. Boscombe Sand (Eocene) showing false-bedding.

Drainage wells using bored, inclined drains to intersect vertical sand drains in Sand and Gravel overlying London Clay have been described for a coastal cliff at Clacton-on-Sea, Essex. These were reported as functioning satisfactorily with no large cliff movements occurring in the stabilised cliffs (Harris & Ralph 1980). A system using small diameter raking drains to intersect drainage wells installed as caissons has been described for a site in Hong Kong by Vail & Beattie (1985). Vertical sand drains used in conjunction with horizontal drains have been described by Palmer *et al.* (1950). Systems such as these where part of the drainage system passes through the face of the slope depend on the elimination of sliding movements otherwise they could be seriously damaged as discussed by Baker & Marshall (1958).

Combined forms of interceptor drainage

The drainage system described by D'Appolonia *et al.* (1967) for a site with colluvial and residual soils overlying horizontal rock strata in West Virginia is of interest in combining all three types of drainage previously discussed. It comprises a cut-off trench for both surface water and groundwater contained in the upper highly weathered and fractured rocks overlying a claystone. The cut-off trench is connected to a drainage gallery below by a series of open drain holes inclined at ±5° from the vertical to intersect and collect water contained in the rock joints. The floor of the drainage gallery was sealed with compacted clay to minimise leakage. It was reported that the piezometric levels in the upper part of the slope showed little or no change one year after the installation although it was anticipated that the cut-off and drainage gallery would rapidly decrease pore pressures.

Discussion and conclusions

Interceptor drains to prevent water from reaching a potential, or actual, landslide would normally be used in conjunction with drainage of the landslide area itself. When properly constructed the interceptor drains could prove beneficial in (i) reducing the volume of seepage and contribute to lowering piezometric levels within the landslide area and (ii) eliminate or reduce damage caused by seepage erosion.

Study of the various cases previously described, together with a consideration of difficulties that could be encountered (Tables 1 to 3), demonstrates that interceptor drains may not function as effectively as planned. Hutchinson (1977) has directed attention to the comparative neglect of the assessment of the efficacy of slope stabilisation methods. Interceptor drains are a stabilisation method which need considerable care in their design and construction, together with careful monitoring if they are to repay the cost of installation. The easy solutions suggested by Denness (1972), for instance, for sites with perched water tables can be seen as an over-simplification of the general problem. If an interceptor drain recommends itself as a relatively cheap stabilisation method, it is contended here that

TABLE 3. *Some possible difficulties impairing the efficiency of vertical drainage wells.*

Geological	1.	Lenticular bedding with the clay horizons wedging out and allowing downward percolation.
	2.	False bedding and channelling of beds inhibiting flow towards the wells and allowing significant ponding.
	3.	Presence of a joint or fissure network allowing downward percolation.
Geo-hydrological	4.	The rate of downward percolation of the perched water through the 'impervious' horizons.
	5.	The downward discharge from the wells raising pore pressures to critical levels in the lower horizons.
	6.	Negligible influence of the drainage wells on the rate of equilibration of clay horizons at depth in the soil profile.
Constructional	7.	The drainage wells forming, or extending, tension cracks.
	8.	Zoning of the backfill and smearing the sides of the wells reducing the overall permeability of the latter.
	9.	The backfill not acting as a filter for all the horizons encountered leading to seepage erosion and/or clogging of the well.

the economies proposed could not, with any realistic advantage, be extended to include the site investigation side of the project.

The following conclusions may be listed:

1. Interceptor drains are structures which especially rely on the continuity of some particular geological conditions. They therefore require a very careful and thorough site investigation prior to design.
2. Geotechnical, including engineering-geological, supervision should continue during construction to confirm the results of the site investigation and ensure that all excavations and boring are fully inspected. Any weakness in the system could be accentuated by the drain acting as a channel facilitating entry of water into the landslide mass.
3. It would be unwise to assume that the drains are functioning satisfactorily without monitoring the performance by means of groundwater observations before, during and after construction. Any failure to intercept surface- or ground-water could, in this way, be detected and rectified before slope instability develops.
4. Except perhaps in the case of some very small slopes, interceptor drains are unlikely to influence the rate of equilibration of pore pressures depressed by unloading in clay soils. If the slope is sufficiently steep and the general pore pressures (albeit reduced from pre-construction levels) sufficiently high, then slope failure will occur either during or upon final equilibration.

ACKNOWLEDGEMENTS: The authors wish to thank Mr M.F. Tyhurst of the Borough of Christchurch and Sir William Halcrow and Partners for information on the cut-off diaphragm at Highcliffe and Mr E.C. Whitaker formerly Engineer, Surveyor and Town Planning Officer of the Borough of Bournemouth for information on the Bournemouth sand drains. They also wish to thank Messrs R. Cater, M.R. Cooper, A.G. Roberts and J.H. Whale for helpful discussions. They are also grateful to Mr R.E. Belgrove for soil test results on the Bournemouth sand drain materials and to Mr A. Brookes for drawing the diagrams.

References

ANON, 1965. Cliff stabilisation scheme: East Cliff, Bournemouth, being protected with sand drains. *The Surveyor and Municipal Engineer,* **125,** 25.

BAKER, R.F. & MARSHALL, H.E., 1958. Control and correction. Chap. 8 in Landslides and Engineering Practice. *US Highway Research Board,* Special Report No. 29.

BARTON, M.E. 1984. Periglacial features exposed in the coastal cliffs at Naish Farm, near Highcliffe. *Proceedings of the Hampshire Field Club & Archaeological Society,* **40,** 5-20.

BARTON, M.E. & THOMSON, R.I. 1984. Studies of the water balance in a rapidly degrading soil cliff. *Proceedings of the International Symposium on Landslides, Toronto, Canada,* **1,** 355-361.

BARTON, M.E. & THOMSON, R.I. 1985. A model for predicting ground water level response to meteorological changes. *Engineering Group of the Geological Society,* Annual Meeting at Sheffield. This volume.

BAYLEY, M.J., 1972. Cliff stability at Herne Bay. *Civil Engineering & Public Works Review* **67,** 788-92.

BROMHEAD, E.C., 1978. Large landslides in London Clay at Herne Bay, Kent. *Quarterly Journal of Engineering Geology,* **11,** 291-304.

BROMHEAD, E.N. & DIXON, N., 1984. Pore-water pressure observations in the coastal clay cliffs at the Isle of Sheppey, England. *Proceedings of the 4th International Symposium on Landslides,* Toronto, Canada, **1,** 385-390.

CONWAY, B.W., 1979. The contribution made to cliff instability by Head deposits in the West Dorset coastal area. *Quarterly Journal of Engineering Geology,* **12,** 267-275.

D'APPOLONIA, E., ALPERSTEIN, R. & D'APPOLONIA, D.J., 1967. Behaviour of a colluvial slope. *Journal of the Soil Mechanic & Foundations Division, ASCE,* **93,** 447-473.

DENNESS, B. 1972. The reservoir principle of mass movement. Institute of Geological Sciences Report No. 72, 7. *HMSO,* London.

DUVIVIER, J. 1940. Cliff-stabilisation works in London Clay. *Journal of the Institution of Civil Engineering,* **14,** 412-426 and 439-446.

GARRETT, C. & WHALE, J.H., 1985. Performance of embankments and cuttings in Gault Clay in Kent. Conf. on Failures in Earthworks. Preprint Vol., 83-101. *Institution of Civil engineering, London.*

HALCROW, SIR WILLIAM & PARTNERS, 1971. *Highcliffe: report on the instability of the cliffs and recommended remedial works.* Consulting Engineer's report to the Borough of Christchurch.

HARRIS, W.B. & RALPH, K.J. 1980. Coastal engineering problems at Clacton-on-Sea, Essex. *Quarterly Journal of Engineering Geology,* **13,** 97-104.

HENKEL, D.J., 1982. Geology, geomorphology and geotechnics. *Geotechnique,* **32,** 175-194.

HUTCHINSON, J.N., 1965. A survey of the coastal landslides of Kent. *Building Research Station,* Note No. EN35/65. Building Research Station, Garston, Watford, Herts.

HUTCHINSON, J.N., 1977. Corrective measures for slopes. *Bulletin of the International Association of Engineering Geologists,* **16,** 131-155.

LEFEBVRE, G., LAFLEUR, J. & CHAGNON, J.Y., 1976. Evaluation of vertical drainage as a stabilising agent in a clay slope at Hull, Quebec. *Proceedings of the 29th Canadian Geotechnical Conference,* Vancouver, **V1,** 16-30.

MOCKRIDGE, R.G., 1983. Highcliffe cliffs - the maintenance of coastal slopes. *Proceedings of a Conference on Shoreline Protection,* held at Southampton 1982, 235-242. Telford, London.

MUIR-WOOD, A.M., 1971. Engineering aspects of coastal landslides. *Proceedings of the Institution of Civil Engineers,* **50,** 157-276.

PALMER, L.A., THOMPSON, J.B. & YEOMANS, C.M., 1950. The control of a landslide by subsurface drainage. *Proceedings of the Highway Research Board,* **30,** 503-508.

PARROTT, W.T. 1955. Control of a slide by vertical sand drains. *US Highway Research Board, Bulletin,* **115,** 51-52.

POWERS, J.P., 1981. *Construction Dewatering.* Wiley, New York.

SEATON, T.H., 1938. Engineering problems associated with clay, with special reference to clay slips. *Journal of the Institute of Civil Engineers* **8,** 457-474.

SHERRELL, F.W., 1976. Engineering geology and groundwater. *Ground Engineering,* **9,** 21-26.

TOMS, A.H. &. BARTLETT, D.L. 1962. Applications of soil mechanics in the design of stabilising works for embankments, cuttings and track formations. *Proceedings of the Institution of Civil Engineers,* **21,** 705-732.

VAIL, A.J. & BEATTIE, A.A., 1985. Earthworks in Hong Kong - their failure and stabilisation. Conference on Failures in Earthworks: Pre-print Vol., 15-28. *Institution of Civil Engineers,* London.

VAUGHAN, P.R. & WALBANKE, h.J., 1973. Pore pressure changes and the delayed failure of cutting slopes in over-consolidated clay. *Geotechnique,* **23,** 531-539.

WARD, W.H., 1948. A coastal landslip. *Proceedings of the 2nd International Conference on Soil Mechanics and Foundation Engineering, Rotterdam,* **2,** 33-38.

M.E. BARTON & R.I. THOMSON, Department of Civil Engineering, University of Southampton, Southampton, Hampshire SO9 5NH, UK.

Groundwater in Engineering Geology, London, 1986, pp. 497-508 Printed in England

Estimation of mass permeability for excavations for the Greater Cairo Wastewater Project

R.H. Coe & W.F. Kay

ABSTRACT: Cairo lies at the apex of the Nile delta on fluviatile deposits of Oligocene to Recent age. Groundwater levels are controlled by several factors including the Nile, water supply abstraction, irrigation canals and construction dewatering. The water table on the flood plain is about 2 to 3 m below the ground surface. In many areas the sand aquifer is confined by clays and silts.

The Greater Cairo Wastewater Project involves the design and construction of a network of tunnels, pipelines, culverts and pumping stations to transfer the wastewater to existing and proposed treatment works to the east and west of the city. The structures are up to 32 m below ground surface in the saturated alluvium and therefore groundwater plays a significant part during construction.

Site investigations were carried out to determine the static water levels and quality of groundwater. Permeability of the aquifer was ascertained by means of pumping tests.

Some case histories are given to discuss how the results of the groundwater investigations have been used to design methods of controlling groundwater for the various engineering structures involved in the project. The project is in the early stages of construction and much further experience is likely.

Introduction

The Greater Cairo Wastewater Project is a major public health engineering scheme to expand and rehabilitate the existing Cairo sewage system.

Considerable geotechnical engineering was required at the design stage and is continuing during construction. The purpose of this paper is to describe some groundwater aspects of the scheme pertaining to design and early stages of construction.

The various works, construction sites and other features described are shown on the accompanying figures.

The project

Prior to 1914 Cairo had no sewerage system. In that year the first stage of a plan prepared by C. Carkeet James was implemented.

Over the years the initial system on the east bank has been extended and a scheme to serve Giza and other developing areas on the west bank of the Nile was commissioned.

The present scheme provides for expansion of the system to cope with demand into the 21st century, the main elements of which are shown on Fig. 1.

It comprises 43 km of tunnel in water bearing alluvial sands and silt, 4 km of rock tunnel, 10 major pumping stations, 33 km of twin box culvert, 43 km of collectors and sewers, and 2 treatment works. In addition to this new work there is a significant programme of rehabilitation of the existing system.

Soil conditions

Cairo lies on fluviatile deposits of the apex of the Nile delta and is bound on the east by the limestone cliffs of Gabal Mokattam and on the west by the Pyramids Plateau. Both the meandering of the river and man's occupation of the area for the last 5000 years have had a great influence on the variations to be found in the deposits beneath Cairo.

The approximate extent of the various geological formations where they appear on the surface are shown on Fig. 2.

Stratigraphy and soil properties

The soil within the project area generally consist of alluvium deposited by the Nile. This is overlain by either made ground (fill), topsoil or desert outwash.

The made ground is generally found in the urban areas of the city up to 13 m thick. It comprises typically miscellaneous rubbish in a silty clay matrix but both depth and composition are complicated in the urban areas by the presence of abandoned and infilled canals and birkets [1] which contain rubbish, refuse and construction waste.

In the agricultural areas, the surface soil usually consists of up to 1.0 m of dark brown organic silt and clay topsoil.

Where the valley is in contact with the desert at Abu Rawash and Gabal el Asfar, layers of loose light brown fine and medium sand are found at the ground

1 Lakes or ponds usually formed as a result of changes in the course of the Nile.

From Cripps, J.C., Bell, F.G. & Culshaw, M.G., (eds), 1986, *Groundwater in Engineering Geology,* Geological Society Engineering Geology Special Publication No. 3, pp 497-508

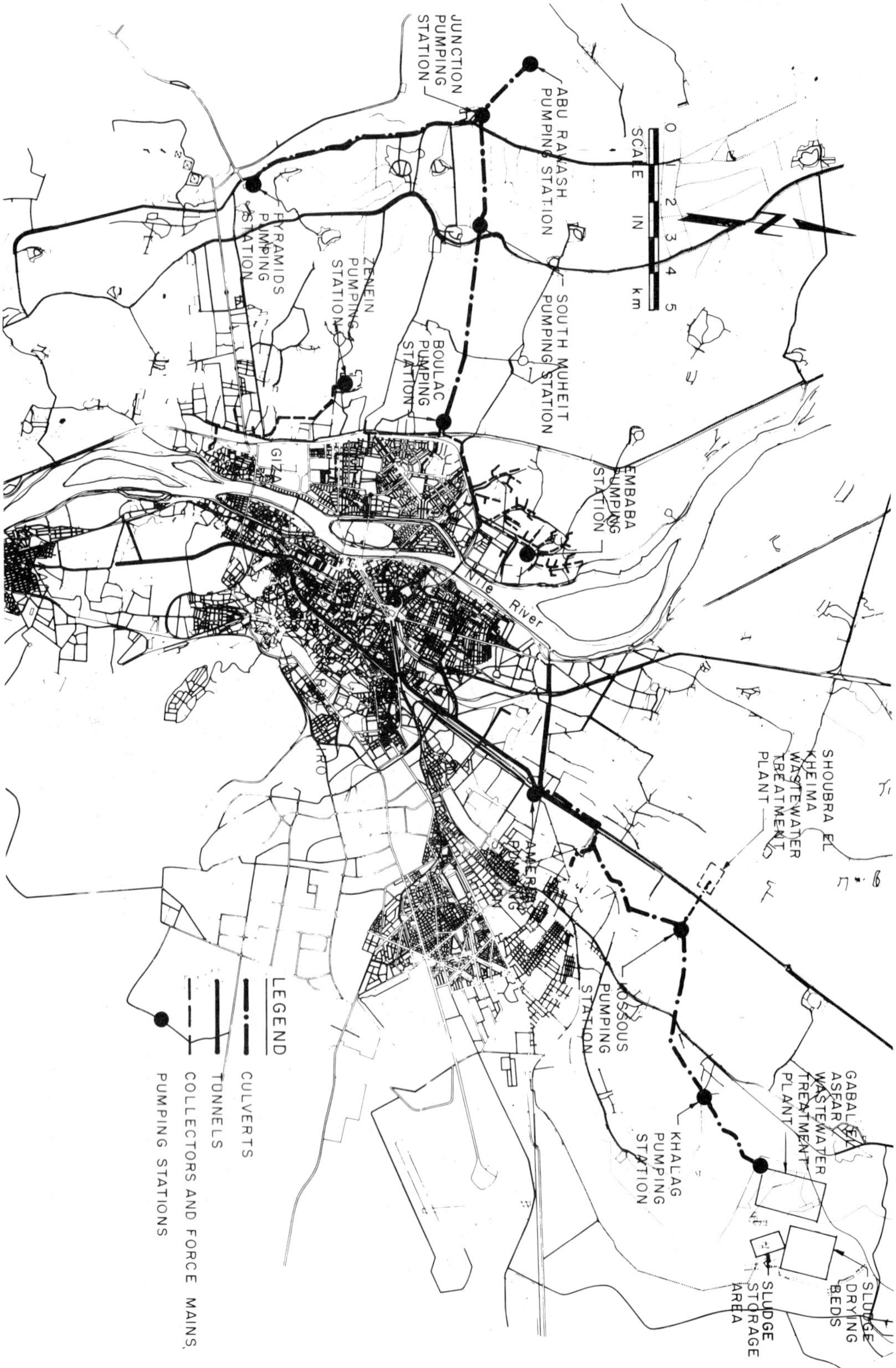

FIG. 1. Cairo Wastewater System first stage works.

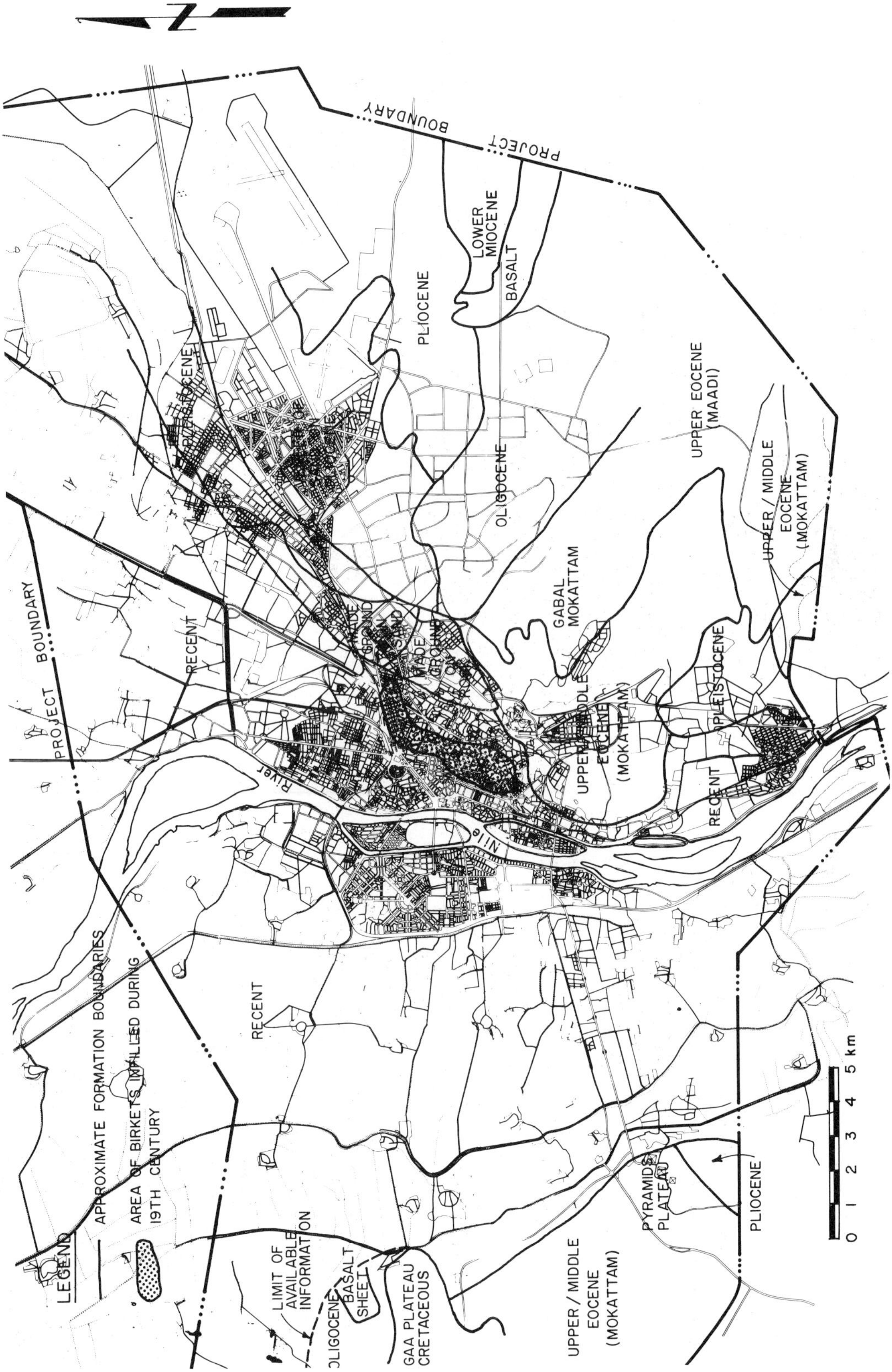

FIG. 2. Geological Map of Greater Cairo.

surface and interlayed with the underlying alluvial clays. It is believed that these are aeolian sands from desert outwash which has periodically overlain the alluvial deposits of the Nile.

From the base of the fill, topsoil, desert outwash sand, or ground surface, the typical stratigraphy of the alluvial soils consists of :

1. Silty clay and clayey silt.
2. Interbedded clay, silt and sand.
3. Sand and gravel.

The silty clay and clayey silt layer generally consists of soft to very stiff grey or brown, often highly fissured, occasionally slickensided, silty clay to clayey silt with small amounts of sand, mica, limestones and calcareous nodules, ferrous stains and occasional sand and silt partings. It was found in thicknesses up to 11 m, with the thickest deposits on the west bank of the Nile and thinning out at the edges of the valley. On the east bank many borings did not encounter this layer due to the depth of the fill material.

The second soil unit is a transitional stratum between the silty clay above and the sands below. Though not always present this interbedded stratum of clay, silt and sand was found to be quite variable and in thicknesses of up to 10 m. A very silty fine sand was usually encountered at the base of these interbedded materials; it usually occurred in thick lenses but in the Abu Rawash area deposits as thick as 13.5 m were encountered.

Beneath the clays, silts and fine sands lies a very thick (greater than 50 m) stratum of sand. This comprises typically a silty fine sand at the top grading to coarse sand with some gravel with depth. Gravel layers also occurred.

Standard penetration tests carried out in this stratum show an increase in density with depth ranging from very loose to medium dense at the top of the layer to dense to very dense at the bottom of the deepest boreholes.

A mean value of permeability of 2.5×10^{-4} m/sec was obtained from laboratory constant head tests and using Hazens Formula based on the D_{10} grain size.

Groundwater

A total of 275 groundwater monitoring observation wells have been installed during the various site investigations. These observation wells have been monitored on a regular basis and the attached groundwater contour map, Fig. 3, was constructed based on these readings. This map shows the groundwater as measured in the latter part of 1984.

Factors affecting the water table in the greater Cairo area include:

— the Nile,
— seepage from Mokattam escarpment,
— irrigation,
— leaking water supply and wastewater pipes,
— water abstraction from wells,
— construction activities including dewatering,
— and low annual rainfall.

Nile water levels are recorded by the Ministry of Irrigation at the southern end of Roda Island. Since completion of the Aswan High Dam in 1965, Nile water levels have remained fairly constant during the months of March to December with a mean of about 16.8 m AD[2] for 1982 and 1983. The January – February mean water levels in 1982 and 1983 were 16.1 m AD and 16.3 m AD respectively. During this time the canals are dredged and cleaned of silt and debris. This is in contrast to the much larger fluctuations prior to completion of the High Dam. In 1964 the monthly mean varied between about 15.9 m in January and 20.0 m above datum in September.

The groundwater levels in the south west portion of the Greater Cairo area are above the Nile water level and the groundwater table appears to be controlled by a source other than the Nile.

Leakage of water supply and wastewater pipes within the city and irrigation outside could have local and widespread affects of the static water table, causing perched watertables within the clay stratum.

The effect of the system of irrigation canals and wastewater drains on the groundwater levels is unknown and probably quite erratic. The authors have observed a number of excavations immediately adjacent to large canals which remained open and dry for long periods of time, even though their depths were below the waterline and in some cases the invert level of the adjacent canal. Some of these excavations were in clay and some in sand.

There are a number of municipal and private wellfields within the Greater Cairo Area (Fig. 3). The abstraction of groundwater from these wells causes a depression in the groundwater levels, sometimes over a significant area, as evidenced by the groundwater contours in the vicinity of the Cairo North and El Ahram wellfields. The El Ahram wellfield in the Pyramids Area has depressed the watertable by more than 7 m. Any change in the pumping rate of these wellfields could cause a change in the groundwater level in that area.

In the Pyramids area the depressed groundwater level is below the invert of the Mansoureya Canal. Borings in the area show sand from the groundsurface to a depth of at least 25 m and still the canal remains full of water. The natural depositing of silt in the canals appears to act as a lining, however, with the annual dredging of the canals the integrity of this lining over a long stretch of canal could be questionable due to possible overdredging.

The groundwater levels are also affected by the dewatering of construction sites throughout the city.

[2] Alexandria datum

Groundwater pumping tests

Three major groundwater pumping tests have been carried out under AMBRIC supervision. Each test has been made at a location where there will be a major excavation during construction of the project. It was considered essential to obtain comprehensive and reliable data about the mass permeability of the soil to facilitate economic and realistic designs, reliable cost estimates and to provide contractors with a good appreciation of the groundwater conditions that they can expect during construction. Furthermore it was anticipated that the data might be reasonably extrapolated to other sites where similar conditions pertain.

The locations of the tests are shown on Fig. 3 and a summary of their details is shown on Table 1.

Site A
This is located on the eastern side of the Nile, while sites B and C are located on the west. Site A is in close proximity to a major municipal wellfield which may have had some affects on the test results.

Site B
This is located in a well populated urban area. However, a number of small agricultural fields are in the vicinity of the pumping test site, one being directly adjacent to it. Irrigation for these fields consists of individual deep pumping wells. Pumping from these may have had some effect on the test results, although the test was run between irrigation cycles of the most adjacent field.

Site C
The third site is located near the western edge of the fertile land of the Nile Valley and the edge of the desert. The hydrogeology of the near surface soils at this site is that of a confined aquifer overlain by a water table aquifer.

The lower aquifer is confined by a layer of silty clay and silt with some sand lenses. Borehole logs in the area of this site show this fine grained alluvial layer thinning out towards the desert and becoming discontinuous at about 2 km from the site.

The pumping test was conducted in the lower sand strata. Piezometers installed in the upper surface sand showed no change in water level during the test as the water levels in the lower piezometers fell. This confirmed that the clay layer was acting as an aquiclude between the two water tables. From the results of the time/drawdown plot for this test shown on Fig. 4 it can be seen that after the test was running for approximately 800 to 1000 minutes the drawdown decreased and began to level off. Extending the distance/drawdown plot for $T = 1000$ min gives a zero drawdown of approximately 2 km from the well.

It is possible that this could have been caused by a recharge into this lower aquifer from the aquifer above, once the zone of influence reached the area where the aquiclude becomes discontinuous.

Values of aquifer transmissibility and storage coefficient have been determined from the plots of the step drawdown and constant rate of pumping aquifer tests for the three sites. These values, together with the values of permeability, based on assumed aquifer thicknesses, are shown on Table 1. Typical time and distance/drawdown data are shown on Fig. 4.

Construction dewatering

Construction dewatering has been carried out at a number of sites in Cairo. This work has been carried out during the early stages of the project concurrent with the groundwater pumping tests referred to in the previous section. The results of those tests were not available for this particular construction work. For this reason and to validate his dewatering designs the contractor carried out his own pumping tests. Five sites are described. Brief details of the ground conditions, dewatering required and techniques employed are given on Figs 5 & 6. Their locations are shown on Fig. 3.

Site 1

An excavation approximately 9.7 m deep was required. Ground conditions comprised clay underlain by silt which was in turn underlain by medium and coarse sand. The groundwater table occurred at about 3.0 m depth.

Sheet piling was driven to a depth of 15 m and initially six deep wells were installed around the external periphery of the cofferdam. Piezometers were placed at the outside corners of the cofferdam to monitor groundwater levels in the silt and the sand.

The Contractor carried out a pumping test in one of the wells and determined that a flow of about 1200 m^3/hr would be required to lower the water table 7 m. The permeability of the sands estimated from the pumping test was 5×10^{-4} m/sec. He also found that maximum discharge from each of the wells was not as high as expected. Accordingly he installed a further seven deep wells giving a total of thirteen.

Disposal of the water was via twin 300 m diameter pipes to a canal. The typical measured discharge was 1135 m^3/hr.

The average drawdown achieved at the piezometers was 6.4 m in the sand and 5.4 m in the silt. Daily and weekly variations in piezometric levels were noted. Whilst some of these were attributable to changes in pumping rate it was also evident that the general groundwater level was fluctuating by at least 0.5 m. It was almost certain that these changes were due to variations in groundwater abstraction rates in public and private wells.

Two large public water supply wellfields each lay about 2.5 km from the site. Piezometer readings on other parts of the Project indicated that their zone of influence extended up to 3 km.

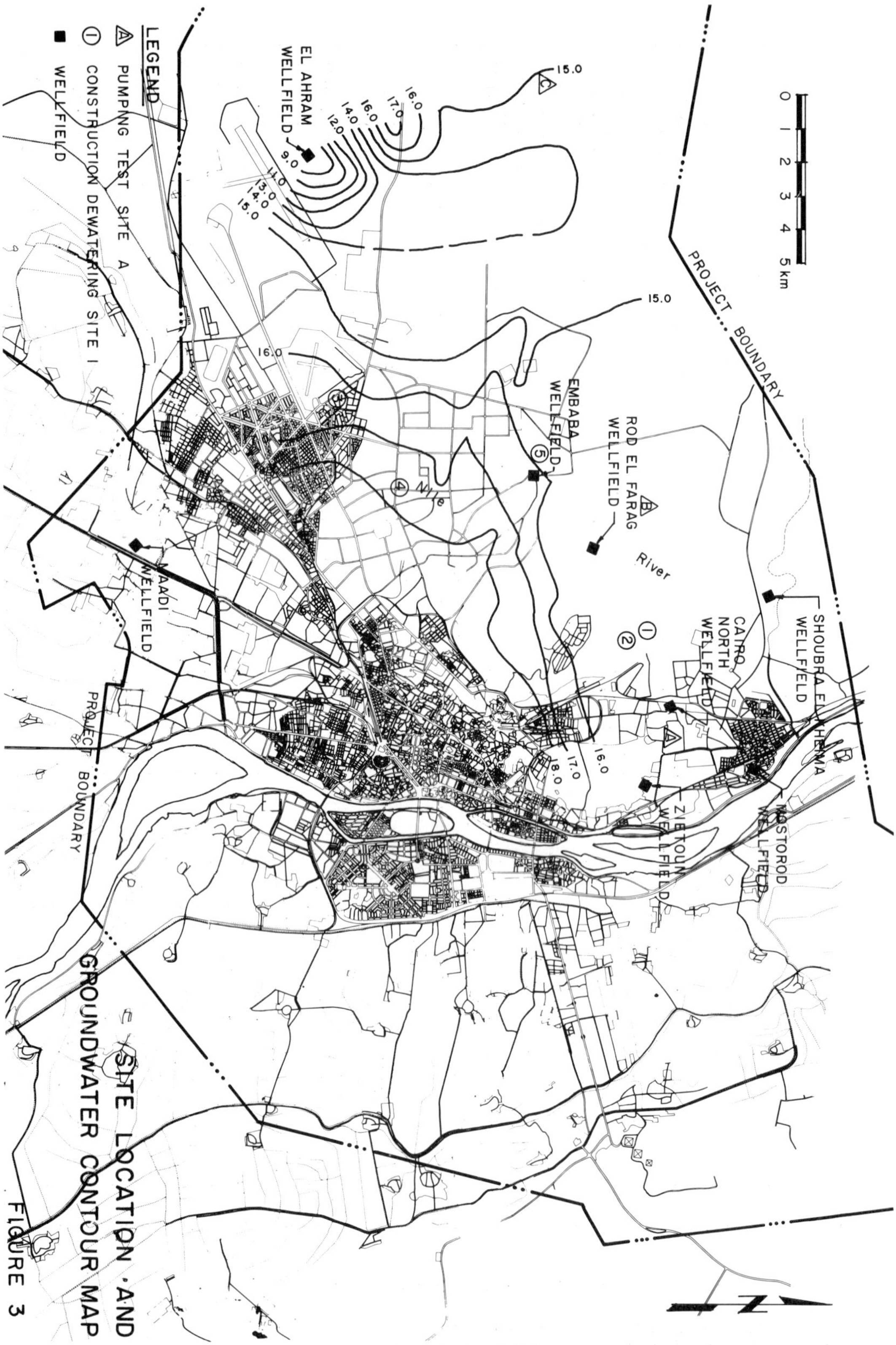

FIG. 3. Site location and groundwater contour map.

SITE	SIMPLIFIED SOIL PROFILE	STATIC GROUNDWATER LEVEL BELOW GROUNDSURFACE	WELL DETAILS: HOLE DEPTH	WELL DETAILS: WELL DIA.	WELL DETAILS: SCREEN TYPE	WELL DETAILS: SCREEN SIZE DIA.	WELL DETAILS: SCREEN SIZE LENGTH	PIEZOMETERS: NO. OF	PIEZOMETERS: ARRAY CONFIG-URATION	PIEZOMETERS: TIP TYPE	PIEZOMETERS: SIZE	PIEZOMETERS: DEPTHS	PUMPING TESTS: STEP DRAWDOWN DISCHARGES (m³/hr)	PUMPING TESTS: AQUIFER TEST DISCHARGES (m³/hr)	PUMPING TESTS: (1) AVERAGE TRANS. T (m²/hr)	PUMPING TESTS: AVERAGE STORAGE COEFF C_s	PUMPING TESTS: ASSUMED AVERAGE PERMEAB. k (m/sec)
A	DEPTH (m) 0–30: CLAY SILT SAND/GRAVEL SILT/CLAY SAND/GRAVEL	4.4m	50m	400mm	WIRE MESH 19mm dia. drilled holes wrapped with No. 28 & 40 wire mesh, approx. 13% open area	250mm	24m (from 26m to 50m depth)	10	90°	2m long Slotted PVC Tip wrapped with No 28 wire mesh	50mm dia.	7@36m 3@25m	DEEP WELL TEST: 109 48 191	191	133 130 142 137	0.0017 0.0007 0.0006 0.0013	$5.3x10^{-4}$ $5.2x10^{-4}$ $5.6x10^{-4}$ $5.4x10^{-4}$ (2)
			30m	400mm			10m (from 19m to 29m depth)						SHALLOW WELL TEST: 38 145 190	188	190 201 211 212	0.0149 0.0071 0.0035 0.0114	$7.5x10^{-4}$ $8.0x10^{-4}$ $8.4x10^{-4}$ $8.4x10^{-4}$ (2)
B	DEPTH (m) 0–25: FILL CLAY/SILT SAND/GRAVEL	2.3m	26.2m	450mm	WIRE MESH 16mm dia. drilled holes wrapped with No. 50 wire mesh, approx. 15% open area	300mm	11.9m	7	90°	2m long slotted PVC tip wrapped with No. 50 wire mesh	38mm dia.	6@14.5m 1@18m	49		139	0.0021	$5.5x10^{-4}$
													113		159	0.0020	$6.3x10^{-4}$
													145		172	0.0020	$6.8x10^{-4}$
														150	181	0.0022	$7.2x10^{-4}$ (2)
C	DEPTH (m) 0–25: SAND SILTY CLAY WITH SOME SAND LENSES SAND CLAYEY SILT SAND	1.0m in the upper surface sand stratum 3.0m in the lower sand stratum	26.2m	450mm	WIRE MESH 16mm dia. drilled holes wrapped with No. 50 wire mesh, approx. 16% open area	300mm	15.1m with 4.1m blank in middle	12	180	2m long slotted PVC tip wrapped with No. 50 wire mesh	38mm dia.	6@ 4.5m 6@14.5m	58		16	0.0029	$8.9x10^{-5}$
													93		17	0.0022	$9.4x10^{-5}$
													120		19	0.0017	$1.1x10^{-4}$
														107	16	0.0025	$8.9x10^{-5}$ (3)

TABLE 1. *Summary of Details of AMBRIC Supervised Groundwater Pumping Test*

(1) Tests performed with two separate well installations.
(2) Calculated from time/drawdown and distance/drawdown plots.
(3) Assumed aquifer thickness t = 70 m for Sites 1 and 2, and t = 50 m for Site 3.

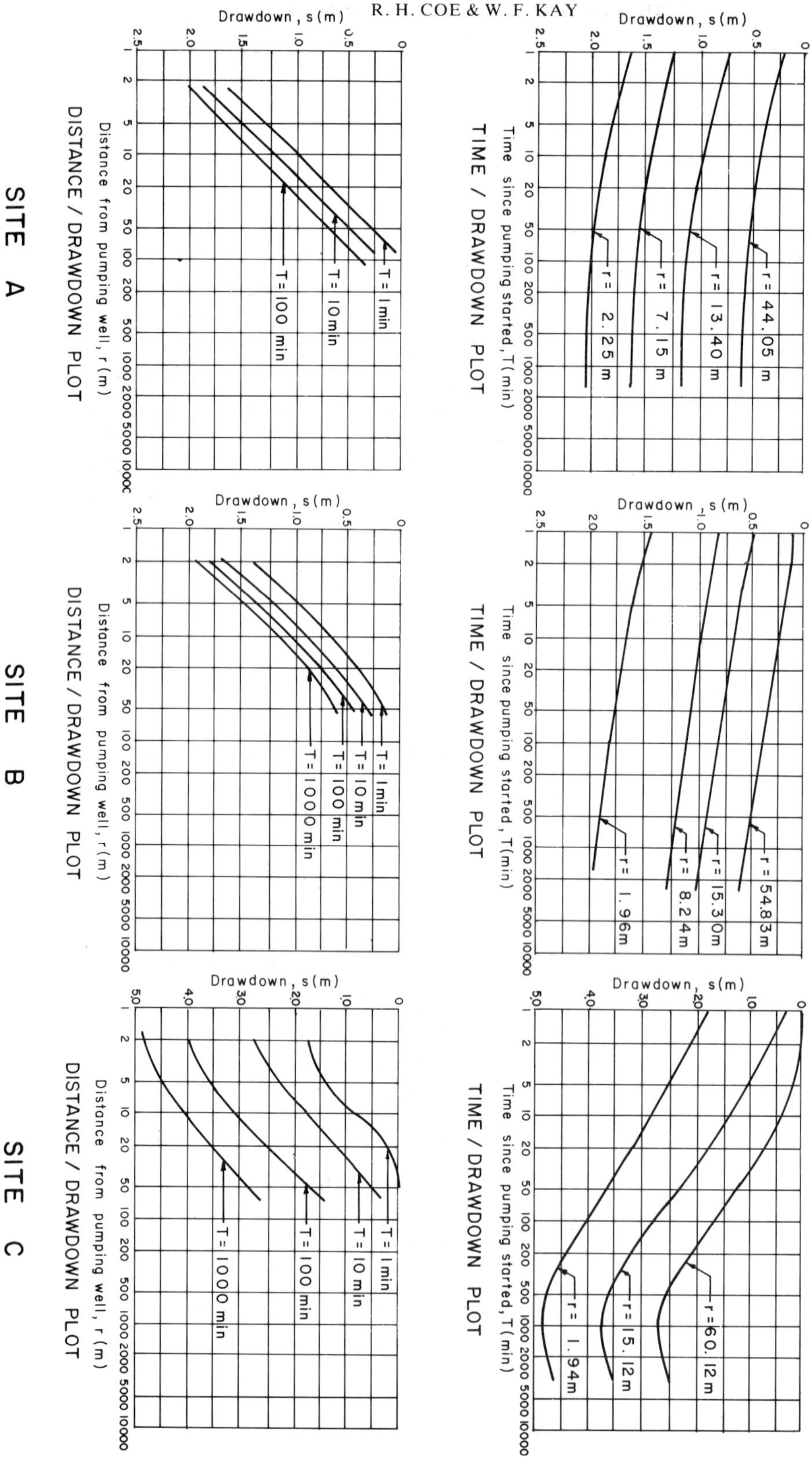

FIG. 4. Pumping test plots. r = Distance from pumping well in meters, T = Time since pumping started in minutes, s = Drawdown in meters.

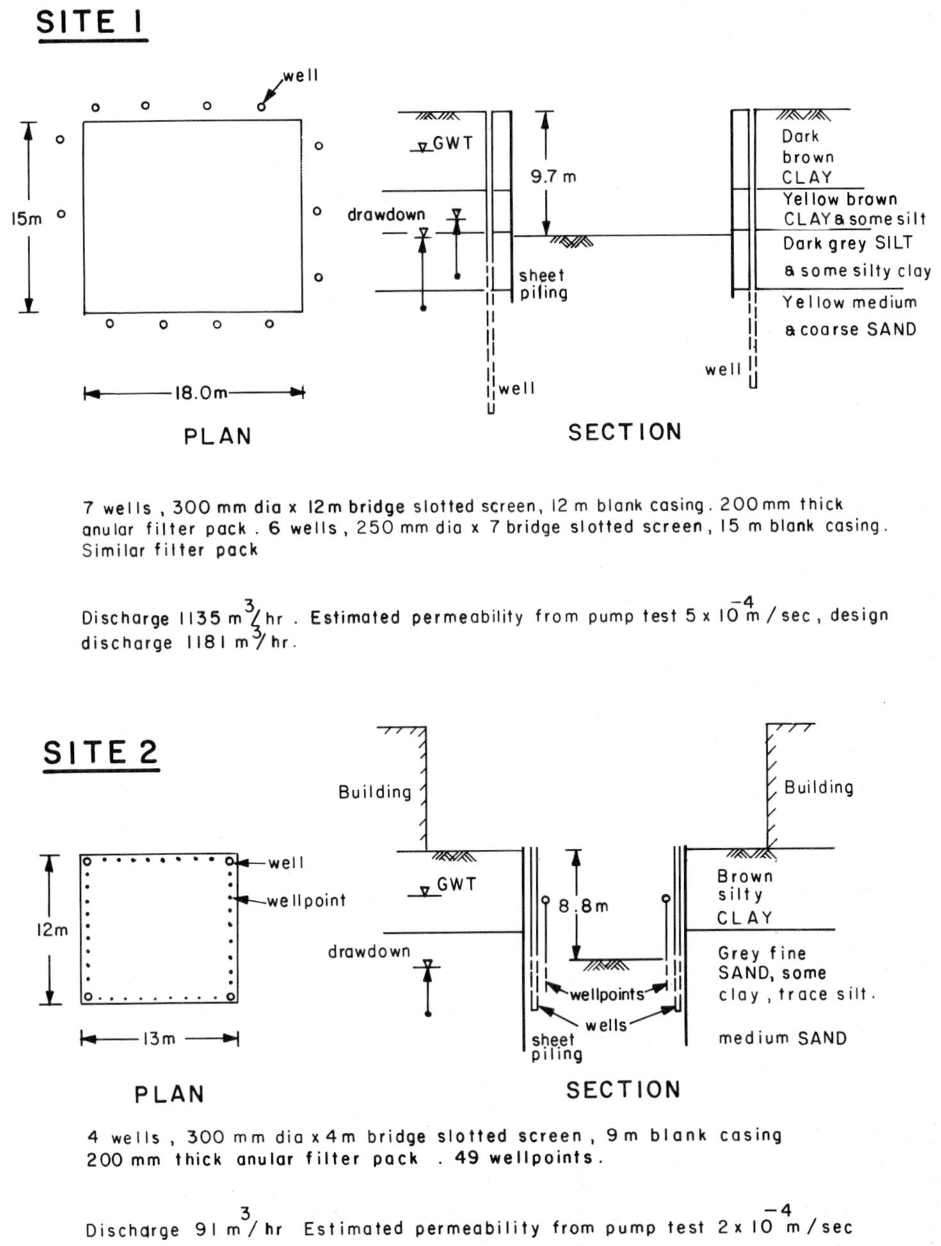

FIG. 5. Details of construction dewatering. Scale = 1:500.

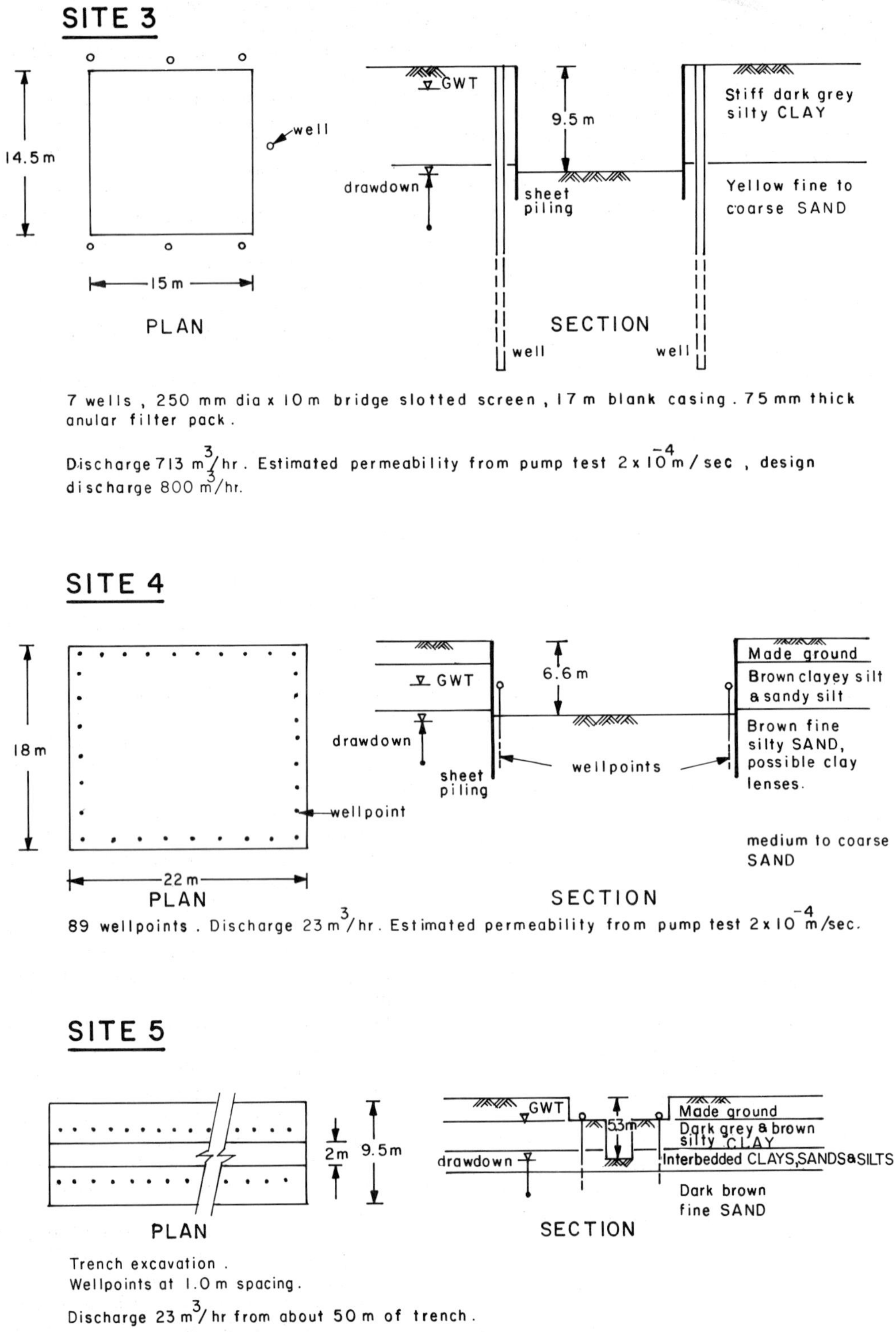

FIG. 6. Details of construction dewatering. Scale 1:500.

Site 2

This site was situated about 0.5 km from Site 1. Adjacent to the site on the north and south sides were four storey blocks of flats. The excavation was smaller and shallower than at Site 1. The proximity of the adjacent buildings was of considerable concern and initially the contractor had contemplated excavation under water. However a further review of construction methods lead to a decision to dewater from inside the cofferdam. By this means it was considered possible to minimise settlement of the buildings.

Ground conditions consisted of clay underlain by fine sand with some clay and traces of silt becoming medium sand at depth.

Sheet piling was driven to 16 m below ground level. Four deep wells were installed inside the cofferdam at the corners. the bottom of the wells did not penetrate below the toe of the sheet piling. Five piezometers were installed around the outside of the cofferdam to monitor groundwater levels adjacent to the buildings. Three piezometers were installed inside the cofferdam. All piezometers monitored groundwater levels in the sands.

A pumping test was carried out by the Contractor in one of the wells from which it was estimated that the permeability of the sands was 2×10^{-4} m/sec.

The external piezometers showed little or no change in water level during the pumping test. Following evaluation of the test results it was decided to install one additional piezometer adjacent to buildings. At a later stage during construction four more were placed inside the cofferdam. This was partly to replace some of the original ones that had become damaged.

Based on the results of the pumping rest the Contractor estimated that to dewater the excavation he would require, in addition to the wells, wellpoints around the internal periphery of the cofferdam. 49 wellpoints were installed but not all were required.

Total discharge from the dewatering system was estimated as 91 m^3/hr for an average drawdown of 5.7 m.

Maximum drawdown beneath the adjacent buildings was 800 mm and inside the cofferdam the water table was lowered to the required elevation.

Settlement monitoring points were installed on the buildings adjacent to the cofferdam but no movement was reported during the 3 months of dewatering.

The substantially lower discharge at this site, compared with Site 1, is believed to be attributable to the finer particle size of the sand at Site 2, the installation of the dewatering system inside the sheet piling, and the increased depth (1.9 m) of sheet piling below the base of the excavation.

Site 3

Site 3 was situated on the West Bank of the Nile and comprised an excavation of similar proportions to Site 1.

Ground conditions consisted of silty clay underlain by fine to coarse sand. The ground water table lay at 2.0 m below ground level.

Sheet piling was driven to a depth of 12 m below ground level.

A pumping test was carried out at the site from which it was estimated that the permeability was of the order of 2×10^{-4} m/sec and a discharge of 800 m^3/hr might be expected.

Seven wells were installed around the exterior of the cofferdam to depths of 27 m. Three piezometers were installed to monitor ground water levels in the sand; two on the exterior of the cofferdam and one on the interior.

Discharge from the wells was collected in a 300 mm dia pipe. The typical measured discharge was 713 m^3/hr.

The drawdown of 7.5 m was sufficient to create a dry formation upon which to commence construction.

Some small fluctuations in water levels were observed but these were very much less than at Site 1.

Site 4

The site was situated on a large island in the river Nile and was immediately adjacent to the river. the location raised concern that the aquifer might be in direct hydraulic connection with the river.

Water levels in the river Nile were approximately the same as static ground water levels on the site.

Ground conditions consisted of 2 m of made ground, underlain by 4 m of clayey and sandy silt. This was in turn followed by fine silty sand becoming medium to coarse with depth.

A pumping test was undertaken at the site prior to any construction work commencing. The results of the pumping test indicated that an aquifer recharge from the river occurred after about 100 mins. A design permeability of 2×10^{-4}m/sec was estimated from the test results and the discharge necessary to achieve the required drawdown was predicted at 380 m^3/hr.

Sheet piling was driven to a depth of 12 m below ground level and 89 wellpoints (38 mm dia.) were installed around the inside of the cofferdam. The tips of the wellpoints were about 0.5 m above the toe of the sheeting. Each wellpoint was surrounded with a sand filter. Discharge from the dewatering system was not measured but was estimated to be no more than 2.3 m^3/hr. At times there was only a trickle.

The reasons for the actual discharge being very much less than predicted from the pumping test are believed to be:

(a) the pumping test was carried out in soils considerably deeper than those in which the wellpoints were installed,
(b) the decision to use wellpoints inside the cofferdam with tips above toe level, whilst sensible in itself, was perhaps not realistically modelled by the pumping test,

(c) inspection of the soils at formation level indicated that there were interbedded sands and clays present which had not been detected during the site investigation. Due to construction exigencies it was not possible to explore further their depth or nature. However such interbedding may well have contributed towards a much lower discharge from the dewatering system than had been predicted.

Site 5

A trench excavation was required for laying a sewer pipe. The groundwater table was about 2.0 m below ground level.

Ground conditions comprised made ground and clay to a depth of about 4.5 m underlain by 2.0 m of interbedded clays, silts and sands. This was followed by fine sand.

Superficial materials were stripped down to the water table level. A wellpoint dewatering system was installed at this elevation with 50 mm i.d. wellpoints at 1.0 m intervals along both sides of the excavation. Excavation then proceeded to the required depth of 5.3 m below ground level. The trench was satisfactorily dewatered throughout its full depth. The discharge was estimated at 23 m^3/hr from about 50 m of trench.

Conclusions

The various site investigations and construction activities have revealed significant variations in the estimation of mass permeability for similar soil conditions.

Permeability has been estimated from

— pumping tests,
— borehole permeability tests,
— Hazen's formula based on grain size,
— and construction dewatering.

Full scale pumping tests have been carried out at three locations together with smaller tests at several construction sites. These are the most reliable method of determining the aquifer characteristics. The cost may initially be high but the authors believe this is more than offset by confidence in the results, a much greater knowledge and understanding of the aquifer, and subsequently reduced construction costs and claims. Inadequate knowledge of an aquifer can result in much wasted time, effort and money during construction dewatering. Permeability of the granular soils determined from pumping tests was typically of the order of 10^{-4} to 10^{-5} m/sec.

Estimations of mass permeability calculated from falling head tests performed in boreholes have typically been in the range 10^{-6} to 10^{-7} m/sec for the granular materials.

Reasons for the variations in calculated permeability between borehole tests and pumping tests are believed to include:

— small volume of soil tested in the borehole,
— disproportionate affects of small local soil variations such as clay or silt pockets and lenses,
— difficulty in cleaning out the borehole satisfactorily before making the tests, and
— variations in vertical and horizontal permeability.

Powers (1981) has reported similar findings.

Hazens formula (k, m/sec = 0.01 D_{10}^2) is applicable over a particle size range of 0.1 to 3.0 mm and a uniformity coefficient less than 5. Permeabilities estimated using this method have been typically in the range 10^{-3} to 10^{-4} m/sec. It appears that estimations of mass permeability based on the D_{10} particle size of a soil sample may be more representative of *in situ* conditions than borehole permeability tests.

Pumping tests carried out at construction sites prior to the installation of dewatering systems have been very worthwhile. In general, the permeabilities estimated from such tests have been consistent with the flows measured from the dewatering systems. Permeabilities measured in borehole permeability tests have not shown such consistency and have generally under-estimated the discharge required for a given drawdown. Where pumping test data is not available and Hazen's formula is applicable this is likely to yield a better estimate of discharge, albeit perhaps a little high, than borehole permeability tests. It must be stressed that pumping tests must adequately model the proposed dewatering system. A deep well pumping test, for example, may not satisfactorily reflect the conditions likely to be encountered by a relatively shallow wellpointing system, because of changes in soil conditions with depth.

Construction dewatering at the locations cited has shown that any dewatering system must be installed with flexibility, adequate monitoring, and willingness to adjust to the response of the system.

ACKNOWLEDGEMENTS: This paper is published with the kind permission of the Chairman of the Cairo Wastewater Organisation and the Board of Control of American British Consultants.

Reference

POWERS, J.P., 1981. *Construction Dewatering, A Guide to Theory and Practice.* Wiley.

R.H. COE & W.F. KAY, American British Consultants, Box 2265, Ataba Square, Cairo, Egypt.

Groundwater in Engineering Geology, London, 1986, pp. 509-511 Printed in England

Groundwater control for a major urban redevelopment project with potential settlement problems and unusual contractual arrangements

U.R. Firth & D.J. Hartwell

ABSTRACT: A dewatering scheme was required to enable the construction of foundations for a large redevelopment complex close the the town centre at Redhill, Surrey. The drawdown beneath nearby buildings had to be limited due to the poor condition of some of the buildings; in many cases, foundation details were also unavailable. A wellpoint scheme composed of various lines was built up and commissioned progressively to allow a flexible approach to the dewatering. Piezometers were installed inside and outside the site so that the drawdown could be monitored and adjustments to the dewatering system made accordingly. In critical areas, investigation boreholes sunk during construction were completed as monitoring points which could also be used as recharge wells if necessary. A thorough dilapidation survey of existing buildings was carried out before dewatering commenced, and the buildings were surveyed at regular intervals during the work to check for settlement or structural distress.

The site

The site, adjacent to Redhill town centre, was to be redeveloped for a large complex of shops and offices, with Mowlem Management acting as Management Contractors. The site is essentially L-shaped, with maximum dimensions of 160 by 90 m, and is bordered on two sides by buildings of various ages. Efforts were made to establish the foundation details of the existing adjacent buildings, but the information was not available. As many of the buildings were relatively old, it was suspected that their foundations would be shallow.

Geology

An initial site investigation consisting of eleven boreholes and a programme of field and laboratory tests had been carried out, showing that the site is underlain by Lower Greensand, described mainly as medium dense and dense fine to medium sand, with alluvial material above. Sieve analyses of the sand showed that it usually consists of uniform medium sand, containing between 10 and 15% silt. The alluvium consists of soft, sandy silty clay, with peat in places, extending generally to two or three metres below ground level, although there appeared to be a buried channel or similar feature running through the site, where up to 6 m of alluvium was present. An additional site investigation consisting of five boreholes was carried out at the beginning of the dewatering contract to define levels of the top of the Greensand close to the existing buildings. These later boreholes confirmed the ground conditions outlined above.

Groundwater

Static water level was found to be between 1.5 and 3 m below ground level in the winter. A pumping test with five observation wells had been undertaken, and analysis of the data produced a permeability value for the Greensand of around 5×10^{-5}m/s.

Dewatering requirements and constraints

Groundwater lowering of about 4 m was required to enable the construction of the foundations for the complex. Although the permeability value obtained from the pumping test indicated that it would be possible to dewater the Greensand, it was obvious that it would be impossible to adequately dewater the site without lowering the water table beneath the nearby existing buildings, some of which may be founded on the alluvial material.

Many of the buildings were only about 15 m from the site boundary, and some were already in a poor state of repair. It was originally suggested that a perimeter cut-off trench outside the dewatering scheme would allow natural water levels to be maintained outside the site; the scheme was, however, rejected as being too costly and impractical, due to the number of service pipes and cables which would have to be moved.

One of the main purposes of the second site investigation was to establish the level of the top of the Greensand beneath the existing buildings in relation to the static water level; obviously the potential settlements caused by dewatering the alluvial material would be greater than those resulting from dewatering within the Greensand. It was

From Cripps, J.C., Bell, F.G. & Culshaw, M.G., (eds), 1986, *Groundwater in Engineering Geology,* Geological Society Engineering Geology Special Publication No. 3, pp. 509-511

established that the static water level was just at or below the top of the Greensand over most of the sensitive area.

It was suggested that during past dry spells, the static water level is likely to have been lower than at the time of this contract (for instance during the drought of 1976). If this were the case, consolidation of the alluvium in confined areas and of the dewatered part of the Greensand in unconfined areas may already have take place. Considerable effort was made to determine the seasonal variations in the water level with little success as the Greensand is not generally used for water supply in the area. A well ten kilometers distant indicated a seasonal range of up to 2.7 m in confined conditions. It was concluded that a drawdown of 1.0 m under confined conditions would have been likely to occur during past droughts, and would have resulted in consolidation of the alluvium under the critical older buildings around the site.

Dewatering both by a system of deep wells or by wellpointing was considered to be feasible, but a wellpointing solution was adopted as the one likely to cause the smallest drawdowns outside the site. Wellpointing has two other advantages: with a large number of wellpoints, greater flexibility is possible around the construction works; deep wells may also prove to be inefficient in medium sand if the grading of the material varies with depth. Instead of merely surrounding the site with perimeter lines of wellpoints, and maintaining it in a dewatered state until all the foundations were completed, several measures were adopted which, it was anticipated, would to some extent safeguard the existing properties by limiting the amount of drawdown produced beneath them and the length of time the drawdown would prevail.

(1) The site was split up into smaller areas, each having its own dewatering system, so that each 'zone' could be operated independently of the rest of the site.

(2) The conventional wellpointing solution would be to intercept inflow to the site by surrounding the site with a ring of wellpoints. Instead of using perimeter lines of wellpoints on the sites bounded by existing property, the lines were moved into the site to reduce drawdown beneath the property. On excavation, it was found that this scheme gave inadequate dewatering at the edges of the site, and perimeter lines of widely-spaced wellpoints were also installed, to be pumped lightly and only when excavation and construction were taking place in the critical areas.

(3) The minimum possible drawdown to allow the work to proceed was maintained. Piezometers were installed inside the site to allow the water levels to be monitored, and the well points were 'trimmed back' either in lines or on an individual basis so as not to exceed minimum requirements. Because each wellpoint was fitted with a valve, a fair degree of control over the water level within small areas was possible.

(4) The water level within each 'zone' was allowed to recover when sufficient weight of concrete had been placed to avoid uplift problems. Again, the 'zones' were operated independently of each other.

The wellpoints were generally installed from ground level to a depth equivalent to the full length of the wellpoint (6.5 m including 0.5 m filter). In a few extremely small areas, for example at the positions of lift shafts or crane bases, a few wellpoints were installed from a lower working level to give additional groundwater lowering for these structures, whilst at the same time confining it to as small an area as possible. The installation and operation of the dewatering system was carried out by Drybase Ltd, under the supervision of Soil Mechanics Limited who were the dewatering consultant.

The Greensand proved to be variable over the site, with medium sand present over the full depth of the excavation in some areas, whilst in other areas, the sand contained layers of silt which were extensive enough to produce local perched water conditions with small areas of running sand forming on excavation. Such problem areas were dealt with locally by pumping from inside specially-constructed sumps with filter mesh and gravel filters.

Extra precautionary measures

Before any work on the contract was started, a thorough dilapidation survey of the existing buildings was carried out in case of later claims; many of the buildings were already in a poor state of repair and it might have been necessary to differentiate previous damage from that attributable to the works.

Several tell-tales were fixed to existing properties and checked at regular intervals during the works. Levels were also take at several locations and were checked weekly concurrent with dewatering.

In addition to the piezometers installed inside the site, several were also installed outside the site and close to existing buildings. Water levels were monitored daily, and continuous plots of the levels were maintained. Using information obtained during the later site investigation together with the measured drawdowns, it was possible to produce a round estimate of the likely settlements of the buildings caused by the dewatering. These settlements were found to be small; differential settlements, which could be caused by differing degrees of drawdown across a building, were estimated to be negligible.

The boreholes constructed outside the site during the later site investigation were completed as recharge wells, using slotted casing and gravel pack within to the Greensand. Should any signs of structural distress have become apparent in the buildings, it was planned to use the recharge wells to bring the groundwater outside the site back to its pre-works level, thus hopefully with the intention of reducing settlements. In the event, the wells were not used.

Contractual arrangements

From the initial site investigation the groundwater problem was identified and a subsequent pumping test undertaken to quantify the problems. It is probably fair to say that the magnitude of the problem in relation to the adjacent structures was not fully recognised until immediately prior to construction, when the full extent of the development was documented. At this stage a number of management options were available for dealing with the problem, but the solutions were closely governed by the programme constraints.

In the traditional contractual arrangement the designer of the structure is only concerned with the water problem once the building has been completed and responsibility for the temporary works dewatering rests firmly with the contractor. In the case where a management contractor is employed, he is intimately involved with integrating a series of sub-contract packages to produce a building to a tight time schedule in the most cost-effective way possible. In the case of the Redhill development, there were over 100 sub-contract packages and the dewatering sub-contract had probably the greatest potential for disrupting the works, particularly given the related geotechnical problems with respect to adjacent buildings; some of these problems were still emerging as the works were beginning. The management contractor engaged a groundwater consultant to assess the sub-contract tenders and monitor and advise on the dewatering and related geotechnical problems. Hence a situation emerged where the dewatering contractor installed and ran the wellpoint system and the management contractor and his specialist consultant were responsible for planning the design, monitoring, and related geotechnical problems. This somewhat unusual arrangement, whereby the responsibility and execution of the dewatering was split between two parties, has worked well in this situation where potentially difficult geotechnical problems existed. Whilst other contractual arrangements might have been equally satisfactory, the benefits in this situation have been considerable.

Conclusions

A wellpointing system provides a flexible approach to dewatering a large site where older buildings of uncertain foundation type and level are situated nearby. For consideration of merely dewatering the site, a perimeter dewatering system would have been the simplest and cheapest solution, but it was agreed that the extra expense and work involved in operating a more complicated system were justified under the circumstances.

U.R. Firth, Soil Mechanics Limited, Foundation House, Eastern Road, Bracknell, Berkshire RG12 2UZ, UK.
D.J. Hartwell, Groundwater Management Consultant (formerly of Soil Mechanics Limited), 48 Hazel Avenue, Cove, Farnborough, Hants GU14 0DW, UK.

Groundwater in Engineering Geology, London, 1986, pp. 513-523 Printed in England

Groundwater control in large scale slope excavations — five case histories from Hong Kong

D.P. McNicholl, W.L. Pump & G.W.F. Cho

ABSTRACT: Slopes in Hong Kong are usually long in relation to their height and, with large quantities of excavation, they commonly require periods of days or sometimes weeks to construct the slope from berm to berm. There is usually sufficient time to implement drainage remedial measures, such as counterfort drains, raking drains, caisson drains, etc. if stability is in doubt due to high groundwater levels or seepage flows. The paper describes five case histories of typical site formation and slope preventive works, which were constructed in different weathering grades and parent rocks. All the schemes entailed extensive excavation and slopeworks. Sketch details of the drainage measure are outlined together with their costs.

Introduction

Hong Kong is situated at the mouth of the Pearl River on the south-east coast of China.

Immediately after World War II, the population of Hong Kong stood at about half a million. By 1951, however, it had risen to 2 million and it is presently 5.4 million. During the early stages of this influx of people, makeshift huts were established on steep, unstable slopes in and around the urban areas. There are currently half a million squatters living in huts of this nature, many of which are on slopes prone to landslides after heavy rainfall.

The government's public housing programme quickly expanded to provide homes for the people of Hong Kong. At the present time, 44% of the population (i.e. 2.4 million people) are accommodated in public permanent housing. Moreover, the annual production of new homes is in excess of 40,000 flats, at an annual capital expenditure of HK$3,460 million (about £346 million).

The majority of sites associated with this extensive capital development programme have been formed in difficult, hilly terrain. During the 1950s and 1960s, because of the pace of development and the scale of the earthworks involved, many of the hillside sites were formed with steep cut slopes, and in some cases with loosely-placed fill slopes. A large number of these slopes would not meet current Hong Kong standards for stability, and a series of preventive works have been undertaken in recent years in parallel with the housing construction programme.

This paper describes a number of case histories where potential groundwater difficulties were a major consideration during construction of new public housing sites in steep hillside urban areas and also in the reconstruction of unstable cut slopes within existing housing estates. All of the the five schemes described entail extensive excavation and slopeworks in complex geology and in different parent rocks.

The work involved installation of a range of groundwater control measures, including raking (or sub-horizontal) drains, counterfort drains and so-called caisson drains. The paper explains the details of the various drain types and discusses the as-built costs at the various sites and observations made of their drainage performance.

Geography and geology of Hong Kong

Background literature

An account of the background conditions is given by Brand (1984a) and McNicholl (1984). The geology of Hong Kong has been described by Ruxton (1960) and Allen & Stephens (1971). A bibliography of geotechnical engineering in Hong Kong has been prepared by Brand (1984b). Other general information on slopes is given by Lumb (1975).

Geography

The Territory has a land area of 1050 square kilometres. It consists of Hong Kong Island, the Kowloon Peninsula, the New Territories which consists of a small piece of the Chinese mainland north of Kowloon and more than 200 islands.

The population is concentrated in a number of distinct geographical locations, dictated to a great extent by terrain. High concentrations of building development and population exist along the north side of Hong Kong Island and the whole Kowloon Peninsula, as well as a number of newly constructed satellite towns in the New Territories.

From Cripps, J.C., Bell, F.G. & Culshaw, M.G., (eds), 1986, *Groundwater in Engineering Geology,* Geological Society Engineering Geology Special Publication No. 3, pp. 513-523

Terrain

The natural terrain of Hong Kong is very hilly. More than 60% of the land area of the Territory is steeper than 15° and about 40% steeper than 30°. On Hong Kong Island and Kowloon, virtually no naturally flat building land exists. The once hilly peninsula of Kowloon has now mostly been levelled to provide filling material for marine reclamation, but isolated hills of 100 m in height still exist. Even in the New Territories, there is little low-lying land, peaks of 400 m being common.

Climate

The average annual rainfall is 2225 mm, with a minimum recorded figure since 1884 of 901 mm (1963), and a maximum of 3248 mm (1982). Nearly 80% of the rain falls during the period May to September. Rainfall intensities can be high with 24 hour rainfalls of more than 250 mm, and one-hour intensities in excess of 50 mm occurring fairly frequently. Extreme rainfall events are often associated with tropical cyclones, or more commonly troughs of low pressures.

Geology

The main rock types are granite and acidic volcanic rocks, which together cover the major portion of the Territory. The small areas of sedimentary and metamorphic rocks are of much less engineering significance.

Granites predominate in those areas of the Territory where building development is most dense. The granite varies somewhat in colour and composition, but the fresh rock is fairly uniform in its engineering properties. Of major significance for slope stability are the joint patterns and the extent of weathering. Joint spacings are typically 0.5 to 2 m. In a few locations, sheeting joints occur almost parallel to the natural ground surface. The granite is extensively weathered in most areas, with depths of up to 60 m of silty sand residual soil being common and containing large corestones in the matrix or exposed on the surface. Below the residual soil mantle, a considerable depth of very mixed material grades exist (fresh rock to moderately weathered material).

The volcanic rocks consist mainly of coarse tuffs, fine tuffs and rhyolite, but some ignimbrites and breccias also exist. These rocks are usually fine-grained and have a blocky structure, with close joint spacings of generally less than about 250 mm. They are more resistant to weathering than granite, the residual soil mantles being only up to 20 m thick. In a volcanic profile, the slightly weathered material usually occurs at a much shallower depth than in a granite profile.

About 20% of the land area of Hong Kong is carpeted with colluvium. This accumulation of semi stable debris from old slips and mass movements commonly covers the lower slopes of most of the hills. It varies in composition from a boulder field, with no soil matrix, to a fine slopewash. It most commonly consists of boulder, cobbles and gravel in a matrix of sand, silt and clay, with thicknesses up to 30 m in places. It is sometimes in a loose state, with a high permeability and frequently gives rise to perched water table conditions. It is also prone to the formation of 'pipes' or 'tunnels' as a result of internal erosion (Pierson 1983; Nash & Dale 1983). These features can be of major significance to the hydrogeology of the area.

Typical site formation conditions

Project outline

The case histories described in this paper consist of three site formation projects and two slope reconstruction projects. These types of projects commonly include large cut slopes, up to 70 m high, and in the case of formation works, involve large volumes of excavation commonly in excess of one million cubic meters per project.

Common hydrological conditions

The principal topic in these case histories is the control of the various forms of groundwater. The paper will not refer to drainage schemes or protective measures to control surface runoff or infiltration.

The maintenance of an acceptable degree of slope stability is dependent *inter alia* on changes in groundwater. Common groundwater features encountered are a fluctuating main water table or short-lived perched water tables. Main water tables typically follow the surface contours of the slope, whilst perched water tables frequently occur above the main water table, usually along the upper surface of a layer of low permeability during periods of recharge from above. Perched water tables are commonly encountered in surface colluvium, which overlies residual soil or completely weathered material. They usually develop and dissipate very quickly depending on seasonal variations in rainfall.

Although the presence of perched water tables would be of immediate concern with regard to the stability of a slope, the numerous shallow slope failures that occur in Hong Kong each wet season are more likely due to increases in pore pressure due to an advancing 'wetting band', following direct infiltration of rainfall (Lumb 1975; Brand 1981c). Deep-seated slope failures are relatively rare in Hong Kong.

Description of case histories

General

Five schemes are described below where groundwater difficulties had been expected. For reasons of time and space these histories are of a narrative nature and do not encompass the mass of site data collected prior to and during the works.

Table 1 sets out the details of the schemes. The cost per square metre of formed platform area ranged between HK$1300 (about £130) and HK$830 (about £83) for the three site formation projects, which serve to illustrate the engineering difficulties and high capital cost of forming land in Hong Kong.

Case history details

1. Chuk Yuen site formation

Background. The Chuk Yuen scheme is set in the Kowloon foothills below Lion Rock peak. The topography is typical of weathered foothill slopes in Hong Kong, being a series of gullies and spurs indented into a general slope of 30° to 45°. The gullies and some flatter areas of slope are filled with colluvium. Underneath the colluvium is weathered granite, which becomes less weathered with increasing depth. The site covers 18.4 hectares and is split into five phases. The first phase of the scheme experienced three slope failures during construction, with serious financial and programme consequences. In each case, the failure occurred in the same place where a road cut intercepted a colluvium-filled gully. The first two failures were during the original excavation, whilst the third failure took place during the construction of the remedial works, which took the form of a caisson wall.

In addition, a nearby high cut slope showed signs of distress: a growth of extra-luxuriant vegetation, bulging of areas of the slope, and seepage from certain areas above the berms. In this last case, trouble was averted by drilling shallow angle raking drains (Tong & Maher 1975) with a normal site investigation rotary rig.

Direct costs in walls and drainage works exceeded HK$6 million (about £600,000) or over 10% of the original contract. That does not include extra consultants fees, client's staff time and reprogramming costs.

These experiences were enough to provoke extreme concern regarding adjoining Phases 4 and 5 which were let in a single contract with slopes up to 70 m high.

Solution. A scheme of vertical drainage wells using hand-dug caisson methods (Evans *et al.* 1982) were proposed, which were excavated by hand into the berms and connected by a long horizontal drain to a berm drain (Fig. 1). Slotted PVC pipes were inserted on the up-slope side of each caisson into pre-drilled holes to intercept channels and flow paths in the colluvium. The total number of caisson drains is 88 and their total length is 1452 m. The total cost of this special drainage (caisson drains and raking drains) was HK$6.8 million. (about £680,000).

Assessment. The drains have functioned for about 18 months and have experienced one and a half wet seasons. They have functioned well and no site problems have been reported; the main contractor is ahead of programme at this stage. The cost per square metre of slope was HK$155 (about £15.5), or about 7.4% of the cost of the scheme. If the cost of the proposed housing tower blocks is included, the percentage cost would reduce to about 1% or thereabouts.

The solution can be said to be satisfactory on the evidence so far.

2. Ma Chai Hang site formation

Background. This site is adjacent to the problem area of Chuk Yuen described in the previous paragraph. It shared several common features, such as a significant gully partly filled with colluvium and having a large up-slope catchment area. The water table was high in places (as determined from the results of pre-construction piezometers). The cut slopes were partly in colluvium and partly in weathered granite and were over 45 m high. There were fixed road levels at top and bottom of the slope which acted as a further constraint, in addition to the need for 250,000 cubic metres of excavation to be completed in about 10 months.

TABLE 1 *Summary of case histories.*

Scheme no	Project description	Platform area (hectares)	Contract sum (millions HK$)	Slope area (hectares)
1	Chuk Yuen site formation, phases 4,5	8.68	92.05	4.4
2	Ma Chai Hang site formation	5.00	41.53	2.9
3	Lam Tin North site formation	5.20	67.02	0.7
4	Tin Wan slope works	N.A.	8.06	1.3
5	Tsz Wan Shan slope works	N.A.	54.00	7.0

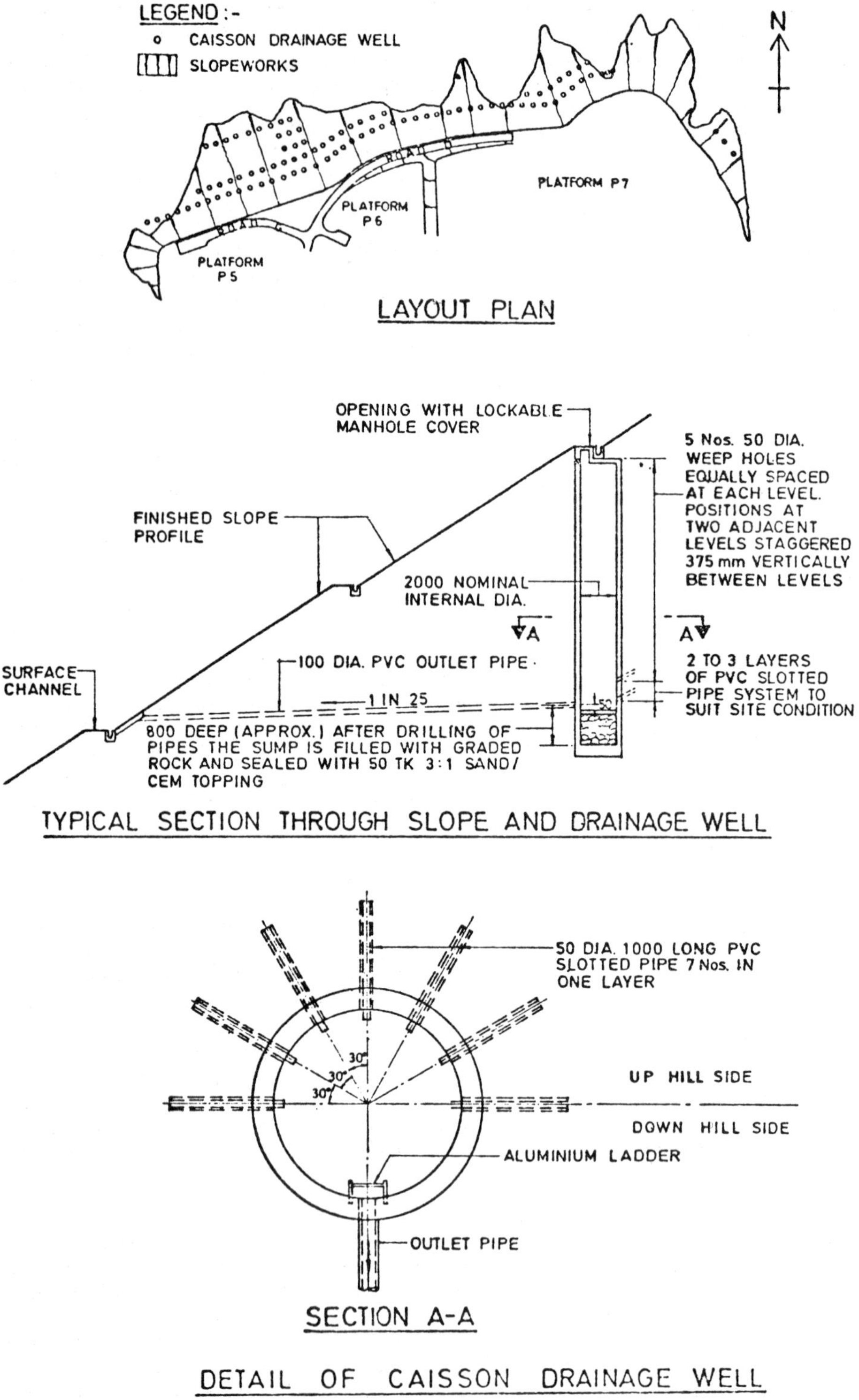

FIG. 1. Chuk Yuen Estate phases 4 & 5 site formation works.

Solutions. The chosen solution (Fig. 2) was to employ caisson drains, which pass through the colluvium into the decomposed granite beneath. The penetration into the granite had to be sufficient to ensure a safe cover of coherent material, as the caisson drains were joined together by 'headings' in the decomposed granite under the colluvium. To assist the construction of the headings and subsequent installation of connecting drains, the 2 m diameter caissons were set at 3 m centres and were provided with two layers, each of 10 in number, of slotted PVC spider drains whose function was to collect water in the colluvium. There were no drainage slots in the semi-horizontal pipes which connected the caisson drains and provided an outfall to a stepped culvert. The position of the drain lines of connected caissons had to be carefully planned to correspond with the low point in the colluvium. The site investigation was thus carried out in four stages, as follows:

(i) feasibility investigation;
(ii) pre-design investigation;
(iii) tertiary investigation to identify localised deepening of the colluvium and to determine the location for the string of caisson drains;
(iv) investigation on site after start of contract to optimise chosen positions.

This procedure was successful in allowing the best planning of the drainage measures. Thirty caisson wells were installed, varying in depth between 3.5 m and 16 m.

In addition to the drainage wells in the colluvium, 35 raking drains were installed with depths of either 10 m or 20 m, depending on circumstances. The total length of drain was 630 m.

Assessment. The site has safely withstood two and a half wet seasons and there have been very few problems.

3. *Lam Tin North site formation*

Background. The site is also located in the Kowloon foothills and covers 5.2 hectares. The works included provision of extensive cut slopes in granite, up to 40 m high, and building platforms for six tower blocks, as well as carparks and schools. About one million cubic metres of material had to be excavated.

Before commencement of work, the site comprised a thin covering of soil with large areas of exposed granite bedrock and core boulders. Local fracture zones up to a few metres wide were encountered in some boreholes and were seen in surface exposures. These zones were closely jointed and chloritised with clay seams, and in places with thin dolerite dykes.

Design at tender stage was based on the first site investigation which was supplemented at the beginning of the contract with a second detailed investigation. The second investigation revealed slightly lower shear strength parameters and the rock head configuration was lower than expected in some areas. Also, the early excavation revealed the presence of unfavourable sheeting joints. Re-analysis showed the calculated factor of safety for failures within the rock mass to be less than the required factor of safety of 1.4. If high water pressure were to occur within the rock joints, then the factor of safety would have fallen to an unacceptably low level.

Solution. In order to avoid reducing the developable platform area, the use of raking drains was proposed to reduce the possibility of transient water levels rising within the rock mass. The drains were set at intervals of 2.5 or 5 m and were inclined upwards at 5° to the horizontal. They were located at a height of 1 m above each intermediate berm. The drains were of 75 mm dia and had internal plastic filters ('Nelton' PDS-65 drainage pipe). A removable 40 mm diameter PDS-40 pipe was installed within the 65 mm pipe. The cost of the installed drains was about HK$850,000 (about £85,000).

Assessment. Very few slope problems have been experienced in the one and a half wet seasons since the drainage works were installed.

4. *Tin Wan Estate slope reconstruction*

Background. This scheme included five slopes which were constructed in the early 1960s. During the construction period, and in the first few years of service there were slope failures (notably in 1966 and 1968) which necessitated remedial works. The failures were attributed to large rises in pore pressure due to infiltration into the extensive catchment above the cut slopes. A comprehensive site investigation revealed a tension crack in one of the slopes, where a thick (3 – 5 m) mantle of colluvium overlies *in situ* weathered volcanics, with moderately weathered material occurring 10 to 20 m below the original ground surface.

Interim measures were initiated for the 1982 wet season. Automatic monitoring of piezometers, of the type mentioned by Insley & McNicholl (1982) and McNicholl & Cho (1985), was instituted and good records of three major rainstorms were obtained, as well as detailed measurements of ground water response and seepage. The records revealed a complex ground water regime with main and perched water tables which responded at different rates and magnitudes to rainfall. Stability analyses, based on shear strength testing and measured water levels, showed that the slopes did not have the required factor of safety.

Solution. A system of raking drains was installed without delay, and were immediately successful in tapping large volumes of water and brought about a lowering of the water table.

A contract was then prepared for permanent reconstruction of the slope. The scheme included extensive drainage which is sketched in Figs 3 & 4. At the head of cut slope, a deep, vertical, caisson cut-off drainage system was constructed similar to that

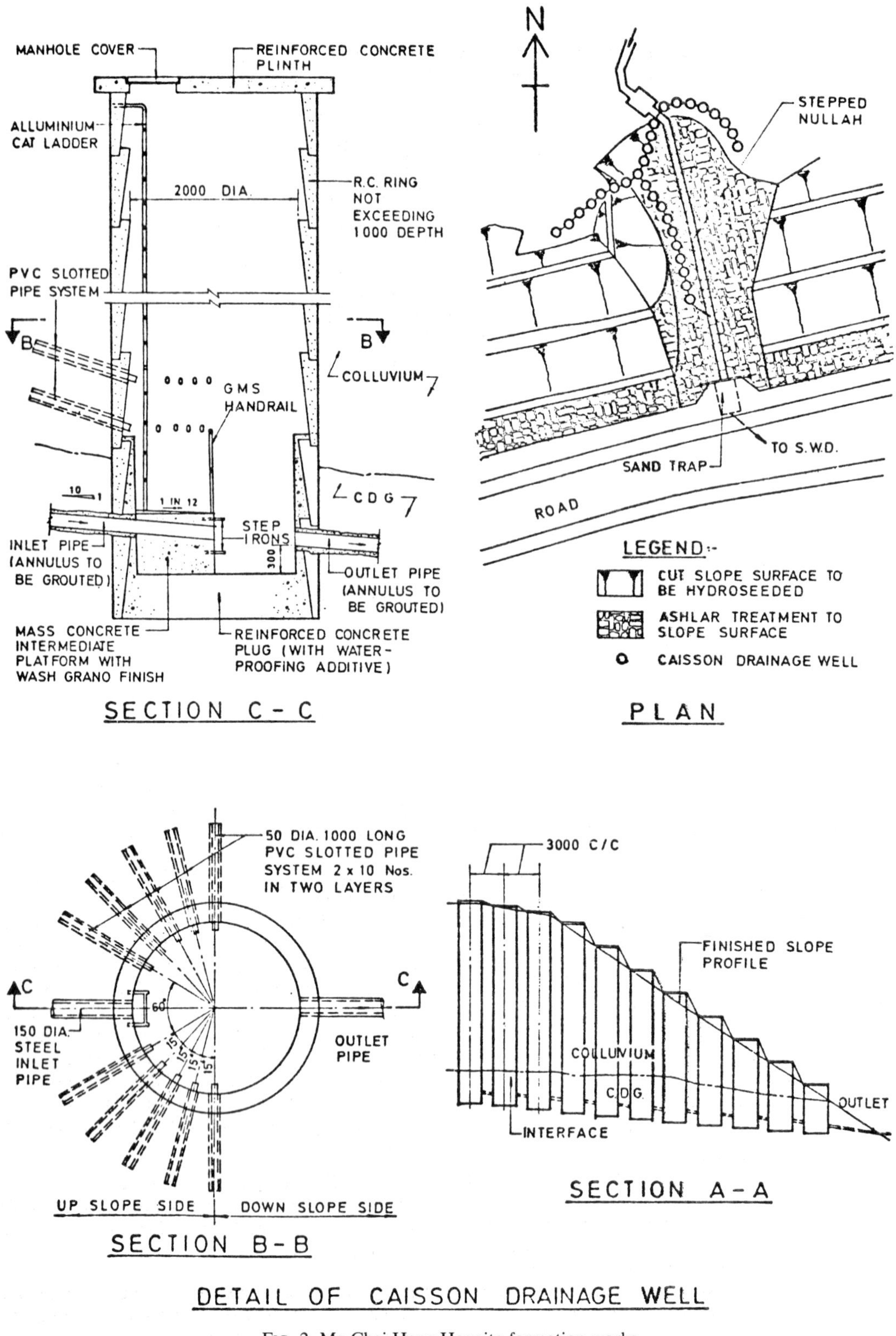

FIG. 2. Ma Chai Hang Hos site formation works.

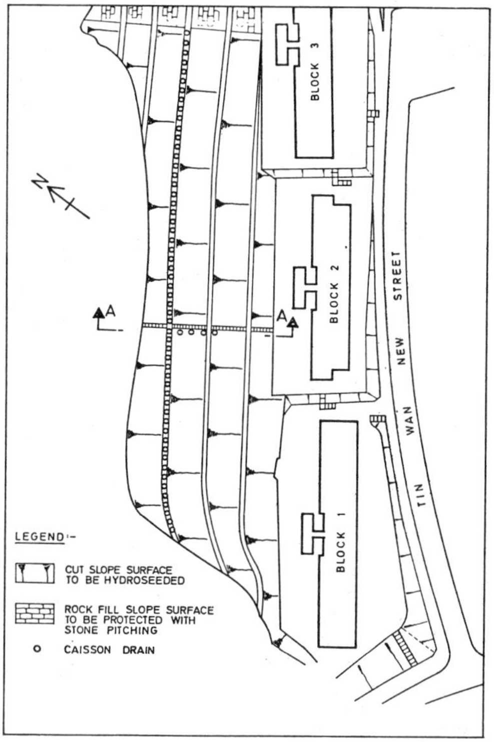

FIG. 3. Tin Wan Estate slope works plan.

installed at Ma Chai Hang and with a discharge drain passing out in a small tunnel to the foot of the slope. The caisson drains were interconnected in turn by small tunnels. All tunnels and vertical drains were backfilled with filter media. 58 caissons were installed with a total length of 585 m. The total cost of sub-surface drainage was HK$2.24 million (about £224,000).

Assessment. The slope works were completed in July 1984 and the drainage measures have successfully coped with one and a half wet seasons.

5. *Tsz Wan Shan slope reconstruction*

Background. The site is located on the northern boundary of the existing Tsz Wan Shan Public Housing Estate. (Case referred to in a paper on the automatic monitoring of pore pressure by Insley & McNicholl (op.cit.)).

Site investigation revealed the soil profile within the slopes to be extremely variable and complex. In very broad terms, two basic soil types were present: colluvium and decomposed granite. Site investigation and automatic pore pressure monitoring recorded that a base groundwater table existed within the decomposed granite and responded relatively slowly to precipitation. Perched water tables, which formed above the interfaces between different soil types, were also detected.

The level of these water tables responded quickly to precipitation, probably because of recharge from direct rainfall on the colluvium and run-off from the steep granite and volcanic slopes to the north. The pressure of a perched water table was confirmed by the extensive seepage visible on the upper benches of the slope in the wet season. Most of this seepage was associated with rapid water flow in subterranean water courses passing through boulder zones in the colluvium. Past failures and stability analysis showed that the factors of safety of the slopes were significantly governed by both the base and perched groundwater tables.

Solution. The adopted scheme used a system of counterfort drains and extensive horizontal drains (over 7 km in total) to reduce the groundwater pressure (Fig. 5).

Assessment. The slope works were completed in October 1983. A considerable amount of seepage has occurred from the horizontal drains and counterfort

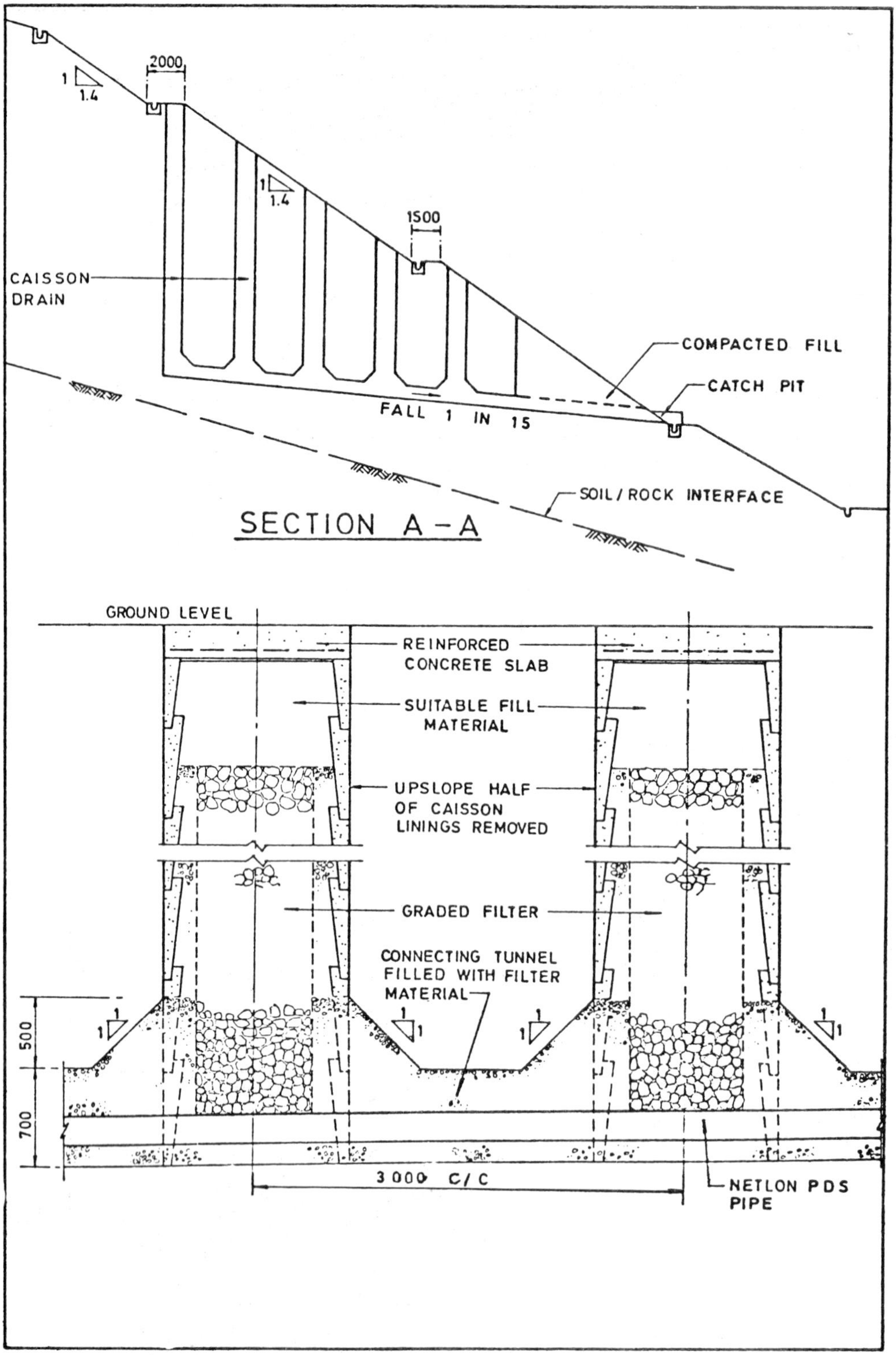

FIG. 4. Tin Wan Estate slope works: detail of caisson drains.

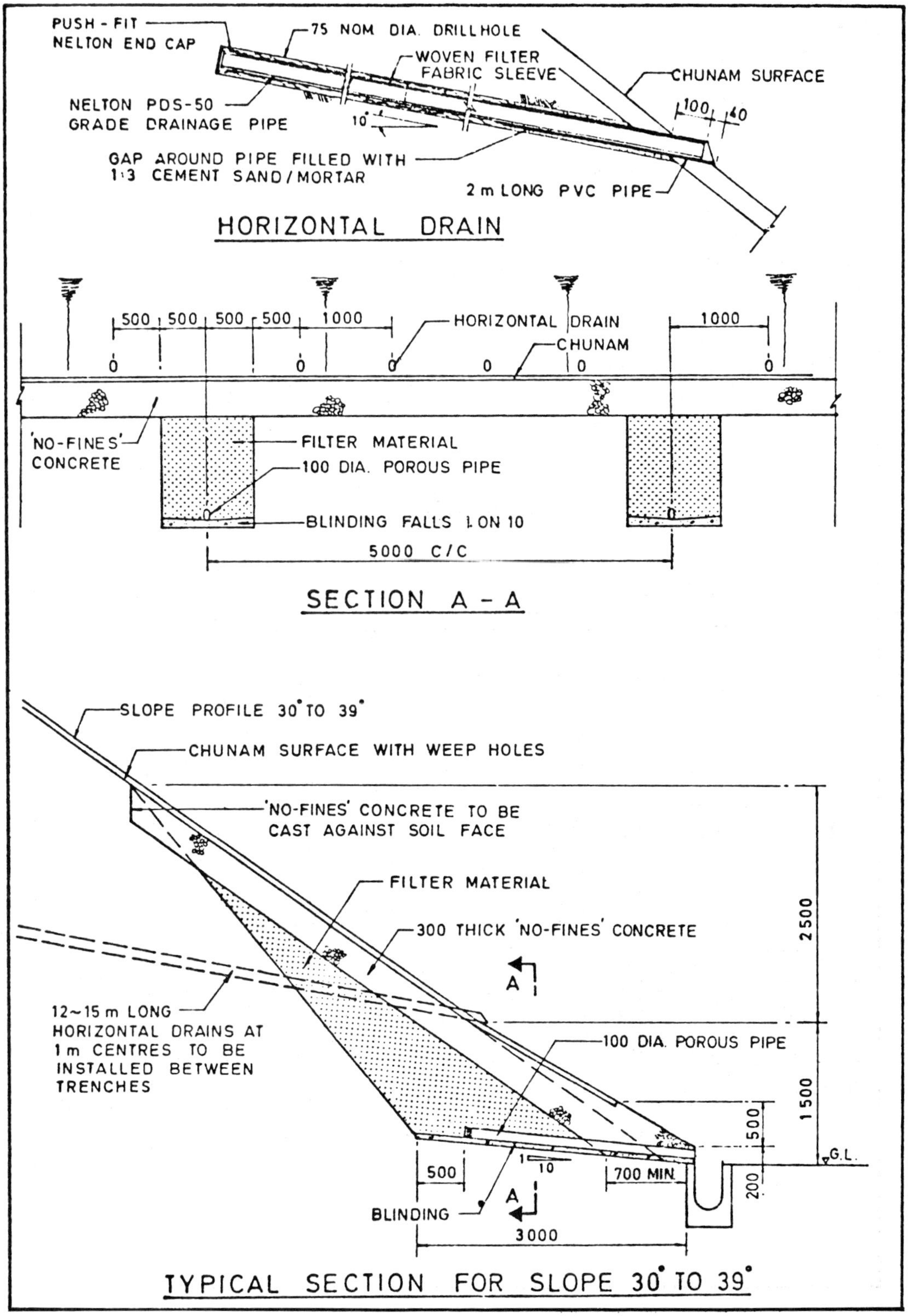

FIG. 5. Tze Wan Shan Estate slope works: detail of counterfort drain.

TABLE 2. *Summary of slope drainage costs.*

Scheme no.	Total length of caisson drains (metres)	Total length of horizontal drains (metres)	Total drainage costs (million HK$)	Cost of drainage per sq.m (HK$)
1	1452	1463	6.80	155
2	440	630	3.86	133
3	-	1560	0.85	122
4	585	1379	2.24	172
5	-	7000	8.00	144
Total/average	-	-	21.75	144

drains. The sub-surface drainage scheme can be said to be satisfactory on the evidence so far.

Costs of completed drainage measures

Table 2 summarises the results of the case history review. The drainage statistics and costs include only sub-soil drainage including raking drains and caisson drainage wells. The usual surface channels installed on slope berms and at the crest and toe of cut slopes have been excluded.

Conclusions

From the review of case histories, the following guidelines and conclusions are evident:

(i) Raking drains (or sub-horizontal drains) can be rapidly and easily installed, even if the need is identified at a late stage in slope excavation works. They are reasonably cheap and effective in the Hong Kong environment;

(ii) Caisson drainage wells are effective in suitable conditions and, based on Hong Kong construction methods, not overly expensive;

(iii) For the Tsz Wan Shan slope works, where extensive excavation took place in weathered granite with a high main water table and transient perched water tables, the extensive use of horizontal drains supplemented by counterfort drains in the colluvium was a reasonably economical solution when compared with other projects. The overall cost was below average;

(iv) A figure of HK$144 (about £14.4) per square metre of slope area is a reasonable figure to allow for subsurface drainage in Hong Kong;

(v) Caisson drainage wells should be excavated down to a level such that the interconnecting drains are located in granite or volcanic rock, i.e. passing through colluvial layers.

ACKNOWLEDGEMENTS: The following firms of consulting engineers were involved in a number of case histories described in this paper and their assistance and cooperation is gratefully acknowledged:

(i) PYS Pun & Associates, together with Leonard Threadgold Geotechnics (Chuk Yuen Site Formation, Phases 4 & 5) — Messrs T. Rogers & L. Threadgold;

(ii) Halcrow Asia Partnership (Lam Tin North Site Formation, Ma Chai Hang geotechnical check) — Messrs N. King, T. Walters, G. Daughton and M. McGuire;

(iii) Binnie & Partners (Tin Wan Slopes Reconstruction) — Mr N.R. Townsend;

(iv) Scott Wilson Kirkpatrick & Partners (Tsz Wan Shan Slope Reconstruction) — Mr H.T.M.I. Insley.

Data from Hong Kong Housing Authority records is included in this paper by kind permission of Mr D Messling, Deputy Director/Construction.

References

ALLEN, P.M. & STEPHENS, E.A., 1971. *Report on the geological survey of Hong Kong.* Hong Kong Government Press.

BRAND, E.W., 1981(a). Landslides in South East Asia: A state of the art report. *Fourth International Symposium on Landslides,* Toronto.

BRAND, E.W., 1981(b). *Bibliography of the geology and geotechnical engineering of Hong Kong to December 1983.* Geotechnical Control Office, Hong Kong.

BRAND, E.W. 1981(c). Some thoughts on rain induced slope failures. *Proceedings of the 10th International Conference on Soil Mechanics & Foundation Engineering, Stockholm,* **3,** 373-376.

EVANS, G.L., MCNICHOLL, D.P., & LEUNG, K.W., 1982. Testing in hand dug caissons. *Proceedings of the 7th Southeast Asian Geotechnical Conference,* Hong Kong, **1,** 317-332.

INSLEY, H.T.M.I. & MCNICHOLL, D.P. 1982. Groundwater monitoring of a soil slope in Hong Kong. *Proceedings of the 7th Southeast Asian Geotechnical Conference,* Hong Kong, **1,** 63-75.

LUMB, P., 1975. Slope failures in Hong Kong. *Quarterly Journal of Engineering Geology,* **8,** 31-65.

McNicholl, D.P. 1984. The assessment of rock mass characteristics prior to excavation of large cut slopes in Hong Kong. *M Sc Thesis,* University of Leeds.

McNicholl, D.P., Cho, G.W.F., 1985. Surveillance of pore water condition in large urban slopes. *21st Annual Conference of the Engineering Group of the Geological Society,* Sheffield, (this volume).

Nash, J.M. & Dale, M.J., 1983. Geology and hydrology of natural tunnel erosion in superficial deposits in Hong Kong. *Proceedings of the Conference on Geology of Surficial Deposits,* Hong Kong.

Pierson, T.C., 1983. Soil pipes and slope stability. *Quarterly Journal of Engineering Geology,* **16,** 1-11.

Ruxton, B.P., 1960. The geology of Hong Kong. *Quarterly Journal of The Geological Society of London,* **115,** 223-260.

Tong, P.Y.L. & Maher, R.O., 1975. Horizontal drains as a slope stability measure. *Journal of the Engineering Society of Hong Kong,* **3,** 1, 15-27.

D.P. McNicholl, W.L. Pump & G.W.F. Cho, Hong Kong Housing Authority, 101 Princess Margaret Road, Kowloon, Hong Kong.

The investigation and control of groundwater at Ysbytty Reservoir, Gwynedd

H. A. Oulton, J. Crowther & P.H. Oldham

ABSTRACT: During early site investigations for an underground water storage reservoir at Ysbytty near Caernarfon, North Wales, groundwater under artesian conditions was encountered.

A subsequent investigation, including pumping trials from two large diameter boreholes, was conducted in order to establish the feasibility of reducing the artesian pressures during and after construction. The parameters thus defined enabled a scheme to be devised for dealing with groundwater during construction, utilising the wells installed during the investigation, and for the detailed design of an under-floor drainage system as part of the permanent works.

A further problem concerned the effects of prolonged draw-down of the water table below the adjacent clay-lined open service reservoir. Recharge wells were installed to maintain the water table and groundwater levels in the immediate vicinity were monitored throughout construction.

The under-floor drainage system for the new reservoir was designed and installed, incorporating porous pipes in a stone filter layer. Its function is to prevent uplift of the reservoir under all working conditions; a gravity outlet drain eliminated the need for permanent pumping.

Introduction

The site of Ysbytty storage reservoir is located at Grid Ref. SH 502 606 some 3 km to the south east of Caernarfon between the A487 and A4085 roads, (Fig. 1). The new reinforced concrete underground covered reservoir holds 8,200 cubic metres (1.8 million gallons of water) and was required to replace an adjacent open service reservoir which supplies the area around Caernarfon.

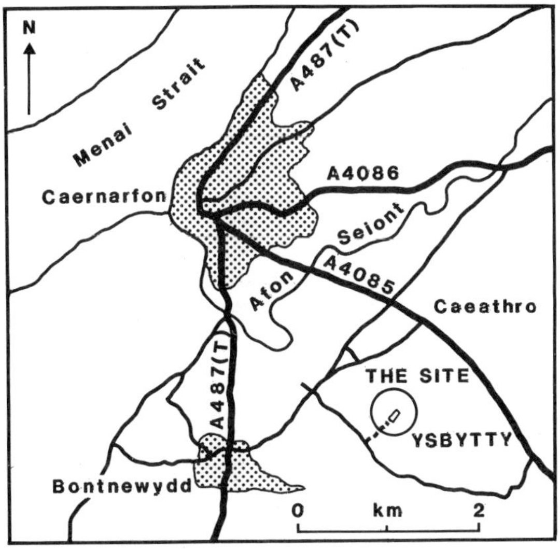

FIG. 1. Location plan.

Following an initial site investigation, carried out in October, 1977, Ward Ashcroft and Parkman were appointed by the Welsh Water Authority to report on the estimated cost of the replacement reservoir. Further investigations were recommended, and a number of alternative sites in the immediate vicinity were considered, some of which were investigated, but ruled out due to land acquisition difficulties.

Detailed investigation of the chosen site followed which enabled detailed design and subsequent construction of the reservoir to proceed.

Geology

The site is located at an elevation of approximately 100 m AOD at the western extremity of a shelf of relatively level land lying above the glaciated valley now occupied by the Menai Straits and shoreline around Caernarfon.

Glacial drift, some 10 m in thickness, overlies acid or intermediate lavas and tuffs; in the vicinity of the existing reservoir the surface is frequently water-logged, whilst immediately to the east, as the land rises slightly, drier conditions allow agricultural use.

Earlier work carried out by Embleton (1964) on the presence of glacial (or sub-glacial) melt-water channels in the Arfon area suggested the possibility of such a channel passing across the site, being formed from discharge water from the Seiont valley, immediately to the north, when the Seiont itself was impeded by ice in its valley floor. This could result in a boulder filled channel carrying appreciable

From CRIPPS, J.C., BELL, F.G. & CULSHAW, M.G., (eds), 1986, *Groundwater in Engineering Geology,* Geological Society Engineering Geology Special Publication No. 3, pp. 525-531

quantities of groundwater. The site investigations were carried out with this in mind.

Investigations

The initial investigation of a site immediately adjacent to the eastern side of the existing reservoir, Area 1 on Fig. 2, was carried out by Norwest Holst Soil Engineering Limited in October, 1977 and consisted of two boreholes and five trial pits. Both boreholes encountered soft becoming firm to stiff (from about 2 m depth) brown silty sandy clay with much gravel cobbles and boulders. This extended to 5 m depth where dense coarse sub-rounded gravel was encountered. In both cases groundwater was encountered at this horizon and in one of the boreholes this rose to ground level in 20 minutes.

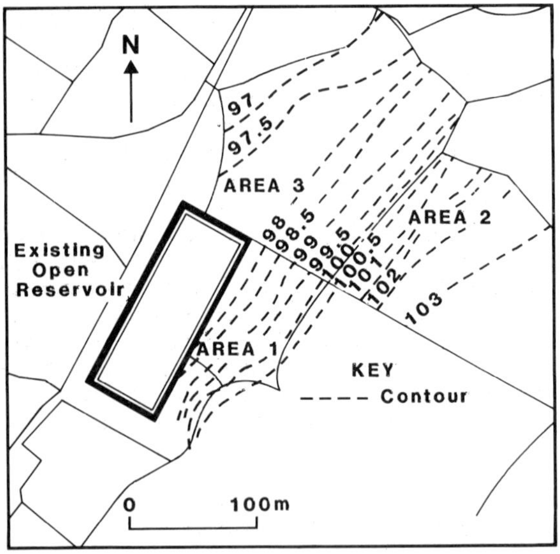

FIG. 2. Investigation of alternative sites.

The results of this investigation were used to carry out a feasibility study including an outline design of the new reservoir with cost estimates. This concluded that a large proportion of the cost of the scheme would be expended in overcoming the anticipated artesian groundwater and the unsuitablity of excavated material for use as backfill to retaining walls. As a result of a recommendation for further investigation of the site for the purposes of detailed design, a further borehole and three trial pits were excavated by Soils and Materials Testing Limited, in September, 1978. The borehole encountered water bearing gravel at only 3.7 m depth. Water did not rise immediately to the surface but a piezometer installed within the gravel recorded a groundwater level above the existing ground level. Water was also seen to rise gradually from the bases of the trial pits which were only 2 m in depth.

Ground conditions were thus shown to be even worse than indicated by the initial investigation. This fact together with the immediate proximity of the existing reservoir, whose stability might be endangered, and the confined dimensions of the site, led to a search for alternative sites which could satisfy the hydraulic constraints of the reservoir.

A suitable site was identified further to the east of the existing reservoir on higher ground, Area 2 on Fig. 2. An investigation of this site was carried out in August 1978 by Soils and Materials Testing Limited which consisted of seven boreholes and three trial pits. Although the conditions were found to be more favourable, as the artesian condition was generally absent, the site is on prime agricultural land. Consideration was therefore given to a third, lower lying but larger field, to the north east of the existing reservoir, Area 3 on Fig. 2.

Investigation of this field was carried out in February 1980 also by Soils and Materials Testing Limited and consisted of six boreholes and five trial pits. Conditions were found to be similar to those in the first site, the depth to the water bearing gravel varying considerably across the field (from 3 to 8 m). The site sloped gently downwards from SE to NW with a gradient of about 1 in 100. The stratum overlying the gravel was mostly silty towards the higher SE end of the field and more clayey in the lower parts. In almost all of the holes water rose to ground level within minutes of the strike within the gravel. Simple perforated standpipes were installed in four of the boreholes, some of which recorded final levels 0.3 m above ground level.

It was concluded that it would be feasible to construct the reservoir towards the higher part of the field at the northern most end i.e. remote from the existing reservoir, provided that groundwater could be controlled during and after construction.

Accordingly, a recommendation was made for a full scale pumping trial to maintain the effects of dewatering from large diameter boreholes.

Pumping trials

The pumping trials were carried out by Fairclough Civil Engineering Limited and fourteen piezometers and two wells were installed at the site during the period 2 April to 27 May 1981. Pumping trials and monitoring of water levels were carried out between 28 May and 26 June 1981.

Initially there was no information relating to the thickness of the aquifer. The first boring was drilled at the centre of the proposed valve chamber to establish the thickness of the aquifer. Rock was encountered at an unexpectedly shallow depth of 7.5 m. The borehole was extended to 9.60 m by open hole drilling and a core extracted from between 9.60 m and 9.90 m. The coring and open hole rock drilling were carried out to establish that bedrock and not a boulder had been encountered.

The weathering profile, shown by flush returns and rock cores, combined with the fractured nature of the rock encountered indicated that bedrock was present rather than boulders. Accordingly it appeared that rockhead was shallowest at the south-east corner of the site. The two wells and their associated piezometers were installed in that area. The location of the wells and piezometers installed in previous investigations are shown on Fig. 3.

In accordance with advice given in BS 5930 (1981) the wells were placed approximately twice the depth to rockhead apart and the piezometers arranged in orthogonal arrays at geometrically increasing distances from the wells. However, the bouldery nature of the ground, especially at shallow depth, did not allow rigid adherence to the geometric spacing, but a close approximation was achieved. Groundwater levels were recorded before the pumping trials started and showed that the artesian groundwater rises to a piezometric level of approximately 99 m OD.

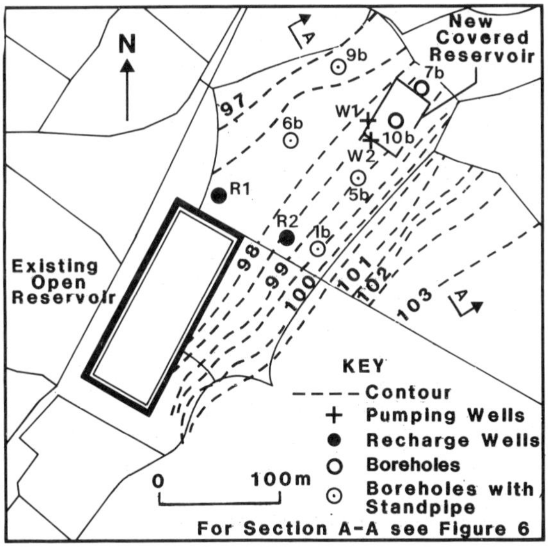

FIG. 3. Pumping wells and outlying piezometers.

The piezometers were installed in cased boreholes drilled with a Craelius B40L tractor mounted rotary rig. The same machine was also used to take NX and 412 rock cores where these were required. The piezometers were built up from 20 mm diameter UPVC pipe. Casagrande type porous elements were fixed at the lower end and the piezometer installed so that it extended one meter above ground level. The piezometer pipe was protected above ground level by a larger diameter plastic pipe. Generally the piezometer elements were surrounded with filter sand and sealed into the aquifer by bentonite pellets. The pipe was grouted close to ground level with bentonite cement grout and the piezometer pipe and protective pipes concreted into position.

The wells were installed in boreholes sunk by means of a Sykes piling winch. The borings were 430 mm diameter and were sunk into bedrock. The well casing was 320 mm exterior diameter UPVC Demco Terra Screen with 3 mm slots giving a free open area of 14% over a perforated length. The wellscreen was surrounded by a gravel pack comprising 10 mm single size gravel. The well screen and gravel pack was sealed into the aquifer with bentonite pellets, bentonite-cement grout, and completed with concrete to ground level. The wells were pumped by two Sykes Univac UV100 pumps. The suction was placed at the bottom of the well casing and the discharge water passed through a 90° 'V' notch 200 mm deep and thence by a 150 mm diameter discharge pipe to a stream 215 m to the north-west.

Testing sequence and summaries of tests

The pumping trials were carried out to establish the mass permeability of the confined aquifer and to assess the drawdown and recovery characteristics of the piezometric head. Tests were carried out in both wells to establish if any significant non-uniformity existed.

A total of five pumping trials were carried out and are summarised in Table 1. As a steady flow could not be achieved it was decided to allow tests numbers 3, 4 and 5 to run to equilibrium, or until very small changes in drawdown were recored over a 24 hour period. What may be considered as an equilibrium condition was achieved after five days pumping when a 24 hour change in drawdown of less than 100 mm occurred.

Problems encountered during the trials

Retrospectively it appeared that the variation of well yield with pumping time and increasing drawdown was a characteristic of the aquifer as it occurred in all the tests, with the exception of the short term low discharge, Test No. 1. This trend may have been masked and in some cases emphasised by the tendency for the UV100 pumps used during the trials to become less efficient over a period of time due to slackening and wear of the packing seal resulting in a gradual loss of vacuum. Consequently it was concluded that maintenance on a daily basis would be necessary if similar pumps were to be employed during a full scale dewatering operation.

Test results and analysis

The test data have been used to consider three aspects of dewatering the aquifer, viz permeability, drawdown and recovery.

TABLE 1.

Test no	Well pumped	Duration (hours)	Pumping rate (m^3/s)	Remarks
1	1	24	3.3×10^{-3} constant	Achieved drawdown below excavation level
2	1	10	1.0×10^{-3} reducing 1.6×10^{-3} to	Possibility of filter disturbance
3	1	120	6.3×10^{-3} reducing 4.5×10^{-3} to	Equilibrium
4	2	120	6.3×10^{-3} reducing 3.8×10^{-3} to	Equilibrium
5	1 & 2	142	6.3×10^{-3} reducing 1.6×10^{-3} to	Equilibrium

Permeability

Steady state analyses were carried out at what was defined to be equilibrium conditions for the data obtained from Test Nos 1, 3 and 5.

Where discharge rates appeared constant at the beginning of the tests the Theis non-equilibrium method has been applied to the data from the piezometers with the most detailed drawdown records. It was established that the data was unsuited to the Cooper-Jacob approximation of the Theis analysis because of the high values of 'u' measured by the latter and the desired relationship between distance-drawdown and time-drawdown plots were not present; i.e. the gradients of the two curves were not related by a factor of 2, as required by BS 5930 (1981).

The results of the Theis analysis are generally consistent with values of permeability in the range 1.2 to 1.5×10^{-4} m/s. They are generally based on test data from two to three hours of testing and probably relate to the ground in the immediate vicinity of the site.

The results of the steady state analysis show a large spread of values from 9.3×10^{-3} to 5.5×10^{-4} m/s. This spread may be attributable to the fact that a true steady state had not been achieved together with the effects of boundary conditions such as the presence of an aquifer of a different permeability.

Drawdown

The trials resulted in extremely wide cones of depression over the site resulting in a more or less consistent

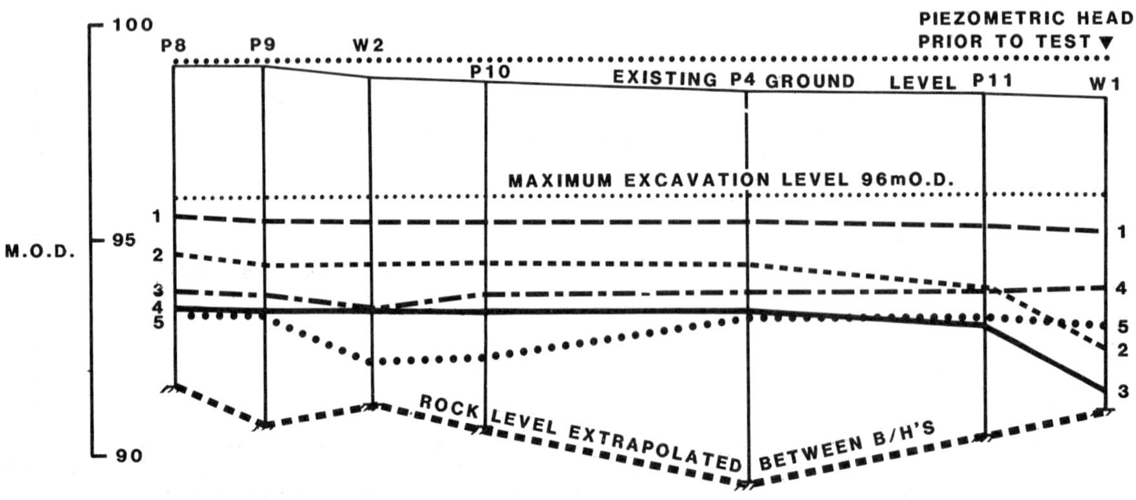

FIG. 4. Section through wells nos. 1 and 2 showing maximum drawdown.

drawdown level except locally in the vicinity of the pumped wells and in piezometer nos. 9B, 6B and 1B. The maximum drawdowns achieved during the trials are illustrated by sections in Fig. 4.

The longer term pumping trials nos. 3, 4 and 5 show drawdowns of generally 2 m below the maximum excavation. However, it should be noted that detailed information is only available in the vicinity of the wells as shown on Fig. 4, but piezometers installed in the peripheral boreholes show similar levels of drawdown due to the extensive cone of depression noted above. Thus it may be inferred that similar levels extend beneath the area of the proposed excavation.

In the absence of information relating to ground conditions beneath the existing reservoir it is not possible to assess the effect of drawdown over a longer period of pumping. In order to minimise the possible detrimental effects of the dewatering operation on the existing reservoir it was proposed that recharge wells located between the existing and proposed reservoir locations be incorporated in the final scheme.

Recovery

The recovery of the piezometric head was monitored after Test nos. 1, 3 and 5. To enable this data to be compared, the ratio of recovery water level and maximum drawdown has been plotted against the ratio T/t where t is the time since pumping ceased and T is the time from pumping starting, Fig. 5. This shows the recovery curves close together and it is possible to make some general observations:

1. The degree of initial recovery varies from test to test. Of particular significance is the large variation between Test nos. 3 and 5 which possibly relates to the greater maximum drawdown achieved in Well no. 1 during Test no. 3.
2. In all three recovery curves the 'semi steady-state' period persists until T/t approximates to 100 after which the rate of recovery increased in an exponential manner. All three curves tend toward the common point.
3. Recovery exceeding 75% of the maximum drawdown did not occur prior to $T/t = 20$ i.e. after five days pumping 25% recovery would not occur prior to six hours from the cessation of pumping.

On completion of the pumping trials, the wells were protected for re-use in the main construction contract.

The conclusions to be drawn from the pumping trials are:

1. The permeability of the confined aquifer lies in the range 9.3×10^{-3} to 5.5×10^{-4} m/s. Materials of these permeabilities are not normally capable of being dewatered other than by wells.
2. Pumping from either well will reduce the head to a level to permit construction with the other well maintained as a standby in case of break down of the pump or deterioration of the filter in the pumped well. An initial pumping capacity of 6.3×10^{-3} m/s will be required to initiate the dewatering but the yield will decrease with time to a lower value.

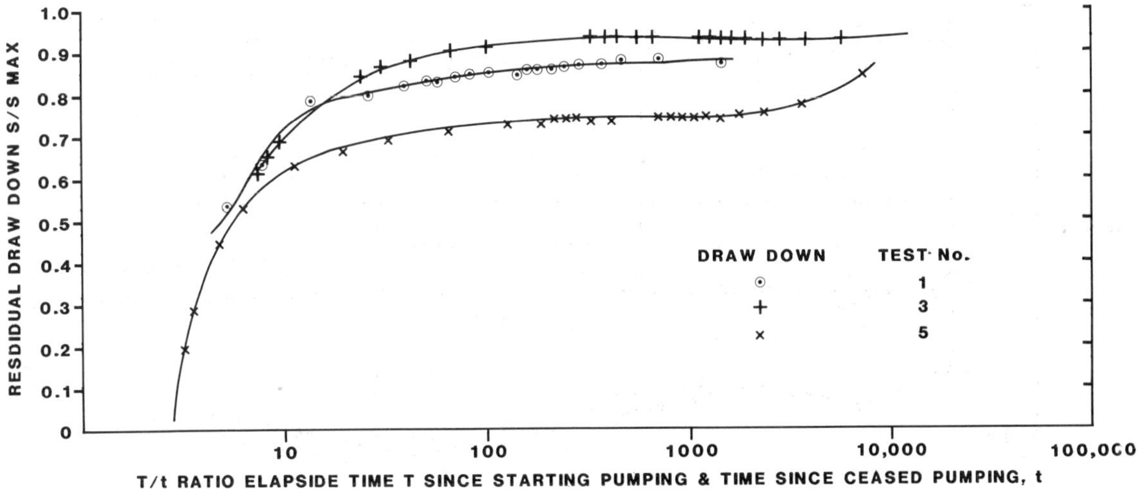

FIG. 5. Recovery curves.

3. To minimise the effects of prolonged drawdown of the piezometric head in the confined aquifer on the existing reservoir, re-charge wells are required between the excavation and the existing reservoir.

Detailed geology of the reservoir location

The site investigations revealed that the reservoir location was underlain by clay with boulders over a thick deposit of very permeable gravels and boulders containing appreciable groundwäter under artesian head. A detailed investigation of all the ground records obtained allowed the preparation of a section across the reservoir, shown in Fig. 6.

The conjectured separation of the two gravel/boulder deposits by a thickness of clay is supported by the evidence that they are not in hydraulic continuity; only the deposit beneath the reservoir site itself revealed artesian groundwater conditions. Further, it is suggested that this deposit represents the position of a glacial melt-water channel, rising in a north-easterly direction, thus fed by groundwater from a higher level, the artesian conditions being provided by a cover of clay.

Design aspects

The reservoir structure was planned to be built partly below ground level to allow for a number of factors which included the need for balancing excavation and fill material, and to reduce the impact of the finished structure on the landscape.

Clearly, with this situation, the return of the high water level (to 99 m AOD) after construction in relation to the flood level of the reservoir structure (at 96.95 m AOD) would cause uplift at the 'reservoir empty' condition unless a permanent underdrainage system were incorporated in the design. The system used includes a 350 mm bed of stone filter material laid on a 'Lotrak' 16/15 membrane under the whole structure with a herringbone network of 225 and 100 mm diameter porous concrete pipes leading to a catchpit, overflowing at 97.30 m AOD and thence to a nearby stream; the filter material complies with the Type B material requirements of the DTp Specification for Road and Bridgeworks -Supplement No. 1 of 1978.

From the results of the pumping trials and the need to reduce the groundwater levels below the reservoir to the order of 93.50 m AOD, it was apparent that there could be a rapid initial draw-down of artesian

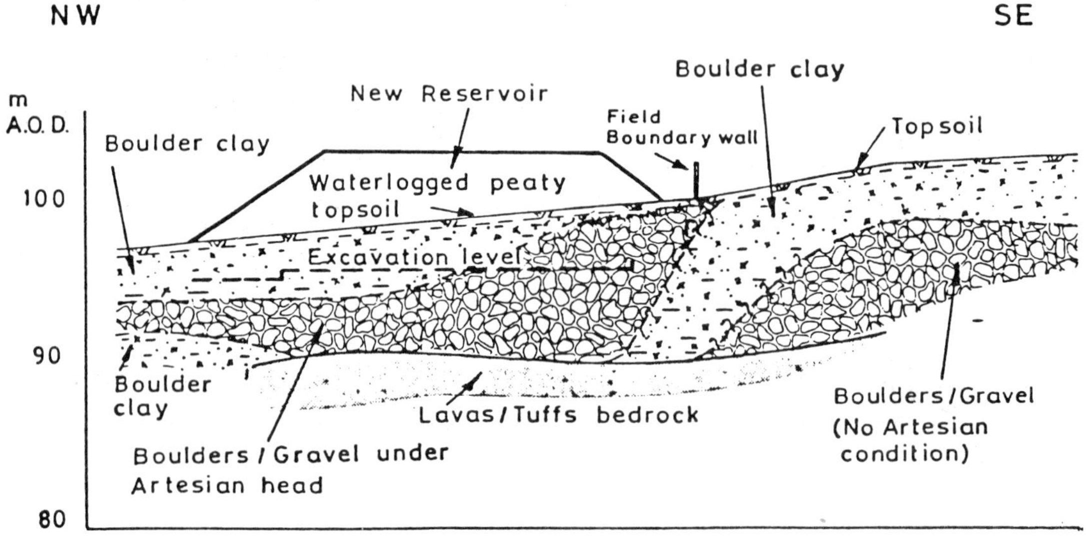

FIG. 6. Conjectured Geological section A-A.

head by some 1.5m below the existing clay-lined open reservoir, which might increase with prolonged pumping during construction over a period of up to 12 weeks. This would cause settlement, possibly leading to leakage and damage to the embankments.

To obviate such an occurrence, it was considered prudent to specify the installation of two recharge wells in the vicinity of the existing reservoir at locations shown on Fig. 3.

Construction aspects

The contract for construction of the reservoir was awarded to McTay Construction Limited and the works were completed over a period of 50 weeks, starting in January 1983.

Use of existing wells

Prior to excavation operations, a further pumping trial was carried out to assess the condition of the wells. The drawdown and yield were found to agree closely with the earlier investigations. Continuous pumping commenced at week 8 of the Contract and was discontinued at week 20 when the underdrainage system was allowed to become effective. Weekly monitoring of water levels was carried out throughout this period.

Excavation and construction activities were carried out without any groundwater problems, the water level below the new reservoir site being maintained at about 93.50 m AOD — some 1.5 m below the deepest excavation — the well discharge being about 3.3×10^{-3} m^3/s. After cessation of pumping, the discharge from the underdrainage system was measured at 0.75×10^{-3} m^3/s.

Water from the pumping wells was fed into lined ponds which acted as settling tanks and maintained a recharge head of 98 m AOD. Observations indicated that about 50% passed into the ground as recharge water, the remainder overflowing to the drainage outfall. It was difficult to assess the effect of recharge but monitoring of stand-pipe water levels did not indicate additional long-term lowering of the water table which would almost certainly have occurred with prolonged pumping; also regular visual inspections of the existing reservoir did not reveal any damage. The recharge wells were installed as a precaution. Whether or not they were really effective is open to question.

The standpipe water levels were generally in close agreement with anticipated conditions throughout groundwater lowering operations. In only one case did an anomaly arise; this occurred with a stand-pipe installed in the initial investigation and was thought to be due to faulty installation in the difficult ground conditions.

Conclusions

The benefits of a phased site investigation procedure have been demonstrated by, *inter alia,* an initial desk study, followed by investigations to assess the anticipated problems and further investigations to define the extent of such problems. Finally, a full scale trial demonstrated the effectiveness of the proposed solution, provided the data for design, and the means of implementation during construction.

The extent and shape of the drawdown curve during pumping from the wells has not been easily determined. It is suspected that it may not be purely radial because of a possible linear channel effect in the aquifer.

ACKNOWLEDGEMENTS: The authors wish to thank the Divisional Manager, Northern Division, Welsh Water Authority for his kind permission to publish this paper.

References

EMBLETON, C., 1964. The Deglaciation of Arfon and Southern Anglesey, and the Origin of the Menai Straits, *Proceedings of the Geologists Association,* **75,** 407-427.

B.S. 5930 (1981) *Code of Practice for Site Investigations.* British Standards Institution. London.

Department of Transport Specification for Road and Bridge Works. H.M.S.O., London.

H.A. OULTON & J. CROWTHER, Ward Ashcroft and Parkman, Consulting Engineers, Cunard Building, Liverpool, UK.
P.H. OLDHAM, Fairclough Civil Engineering Limited, Chapel Street, Adlington, Lancashire, UK.

Groundwater control by drainage gallery at Aberfan, South Wales

H.J. Siddle

ABSTRACT: The Aberfan (Merthyr Vale) tip, which failed with disastrous consequences in 1966 (HMSO, 1969), was constructed above rocks subjected to artesian pressures. Two-thirds of the tip were removed after the failure and sub-surface drainage measures have been installed to secure the remainder. These include a drainage gallery, the construction and performance of which are the subject of this paper. The 'factor of safety' of a critical slope of the tip has been improved to an acceptable value by a reduction in the artesian pressures within the rock. Investigations before and during construction indicate that water moves towards the seat of the 1966 failure in a 20 m wide flexure in the rock adjacent to a hitherto unrecorded minor fault. It appears that tensional strains created by past mining have been concentrated within this structure the existence of which must be added to the other previously established factors which contributed to the failure. The profound anisotropy within the aquifer caused by this geological structure and mining history is demonstrated by the fact that an estimated 25% of the water entering the potential groundwater catchment is released from a length of aquifer equivalent to less than 1% of the available spring line.

Introduction

The failure of Tip No. 7 at Aberfan in 1966 was caused by a rise in pore pressure at the base of the tip due to an increase in water pressure in an underlying fissured sandstone aquifer (Bishop *et al.* 1969). Movements on a shear surface, formed during an earlier failure, were reactivated and, during the failure, the superficial deposits were ruptured releasing groundwater from a new spring at an estimated rate of 505 to 630 l/s, before equilibrium was reached (Moore 1969). Between 1969 and 1971, about two-thirds of the tip were removed to a new site and the remainder was reprofiled. The new spring was protected by an 'arched culvert' consisting of steel arches and sandbags, lined with sprayed cement. Several relief wells, including four within the culvert, were drilled in 1970 in an attempt to increase free drainage of the rock. This paper describes more recent measures taken to control the water table.

Geology and hydrogeology

Boreholes drilled after the failure established that the uphill part of the tip is underlain by Pennant Sandstones of the Brithdir Beds, overlying the Brithdir Seam in the Upper Coal Measures (Fig. 1). The succession dips at six degrees towards the south-east, obliquely into the Taff Valley. The sandstone is an aquifer which carries water towards the tip complex from a potential groundwater catchment of about 2 square kilometers on Mynydd Merthyr, up dip to the north-west. The base of the aquifer, defined by the argillaceous strata underlying the Brithdir Seam, is confined by a tongue of less permeable lodgement till, which fills an embayment in the rockhead under the south-eastern edge of the tip.

As shown in Fig. 2, flowing artesian conditions are present on and just uphill of the crop of the Brithdir Seam and it was from this part of the hillside that the spring appeared as a result of the 1966 failure, some 15 to 20 m higher than the seam.

Even after the relief wells had been drilled, it was known that a reservoir of water remained within the sandstone, confined by the lodgement till and that the spring acted as an overflow for that body of water (Moore 1969). The rapid transmission of appreciable quantities of groundwater towards the tip foundations is favoured by the local dip, together with, it was thought, an interconnected open fissure system within a narrow zone of tension caused by past mining (Gray 1969).

A series of south-easterly trending faults, parallel to the major Kilkenny fault, crosses the area. A number of these faults are shown on the Aberfan Geological plan (Woodland 1969). However, a detailed examination of the records of the boreholes drilled after the failure, some of which had not been completed at the time of the Inquiry, suggested that a minor fault, Fault 'C' downthrowing about 5 m to the south-west, trends towards the area where the spring appeared (Halcrow 1975). It was considered in that study that the fault may be a control on the movement of water.

From CRIPPS, J.C., BELL, F.G. & CULSHAW, M.G., (eds), 1986, *Groundwater in Engineering Geology,* Geological Society Engineering Geology Special Publication No. 3, pp 533-540

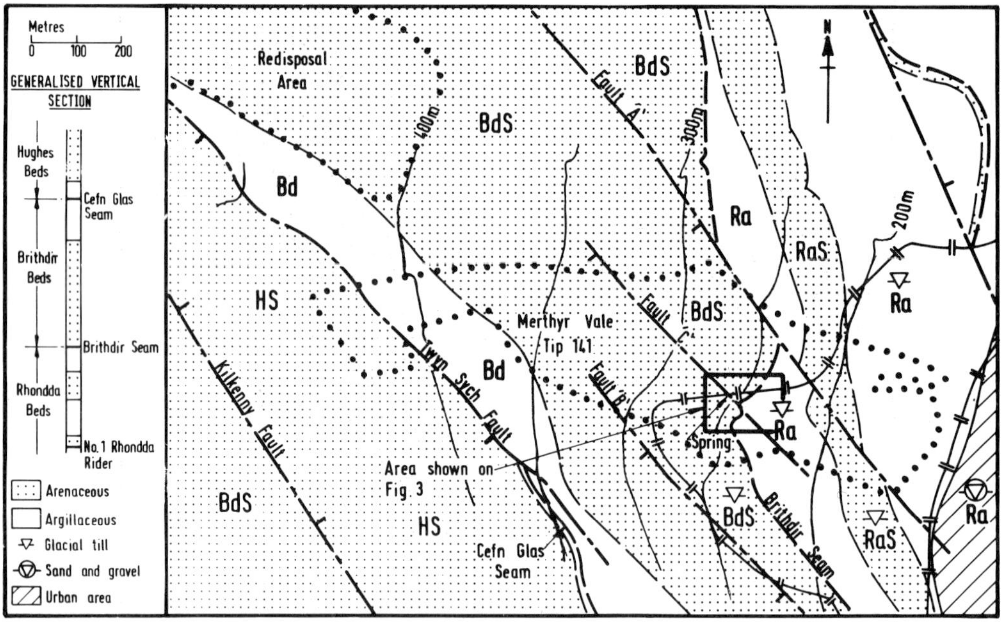

FIG. 1. Geological plan.

Investigation and assessment of remedial measures

In a statutory report on the tip complex (Halcrow 1975) prepared to comply with Regulation 18 of the Mines and Quarries (Tips) Regulations (1971), stability analyses of the regraded slopes of the tip were performed. Given a minimum desired 'factor of safety' of 1.20 for a 'closed classified tip', the main slopes of the reprofiled tip were shown to be secure. However, the slope above the arched culvert had an inadequate margin against failure on the basis of piezometric data dating from about 1970. The assumed piezometric surface was subsequently confirmed by four boreholes MG 1 to MG 4, flowing artesian condition being found in borehole MG 3.

In order to achieve the required minimum factor of safety, two alternative schemes were considered:

(i) Earthworks, involving about 30,000 m^3 of cut and fill to reduce the slope gradient, together with an extension of the arched culvert in the area of fill.

(ii) Drainage to reduce the piezometric surface by at least 2 m either by a system of pressure relief wells or by a drainage gallery.

The earthworks alternative was rejected because it would involve placing spoil on ground subject to artesian pressures and the more positive approach of improving the inherent adverse hydrogeological conditions was considered advisable. Additionally, it was felt that relief wells might not be sufficiently effective if they failed to intercept critical fissures within the zone of extension. Consequently, a drainage gallery became the preferred option, constructed from a new portal downhill of the existing arched culvert. The gallery was designed to enter the base of the aquifer within the zone of tension and to intercept the conjectured fault.

Construction of Gallery

A plan and section of the gallery which was constructed in the spring of 1980 are shown on Figs 3 and 4.

The thin layer of spoil and glacial till were machine excavated in open cut for a distance of 25 m, beyond which the heading continued in tunnel, excavated by pneumatic hand tools. On reaching rockhead, excavation continued by drill and blast using between 3.6 and 14.5 kg of explosive in 11 to 30 charge holes. The tunnel was supported by pre-cast concrete segmental lining in the superficial deposits and by steel manhole arches in the rock. The gallery was completed in eight weeks with an average rate of

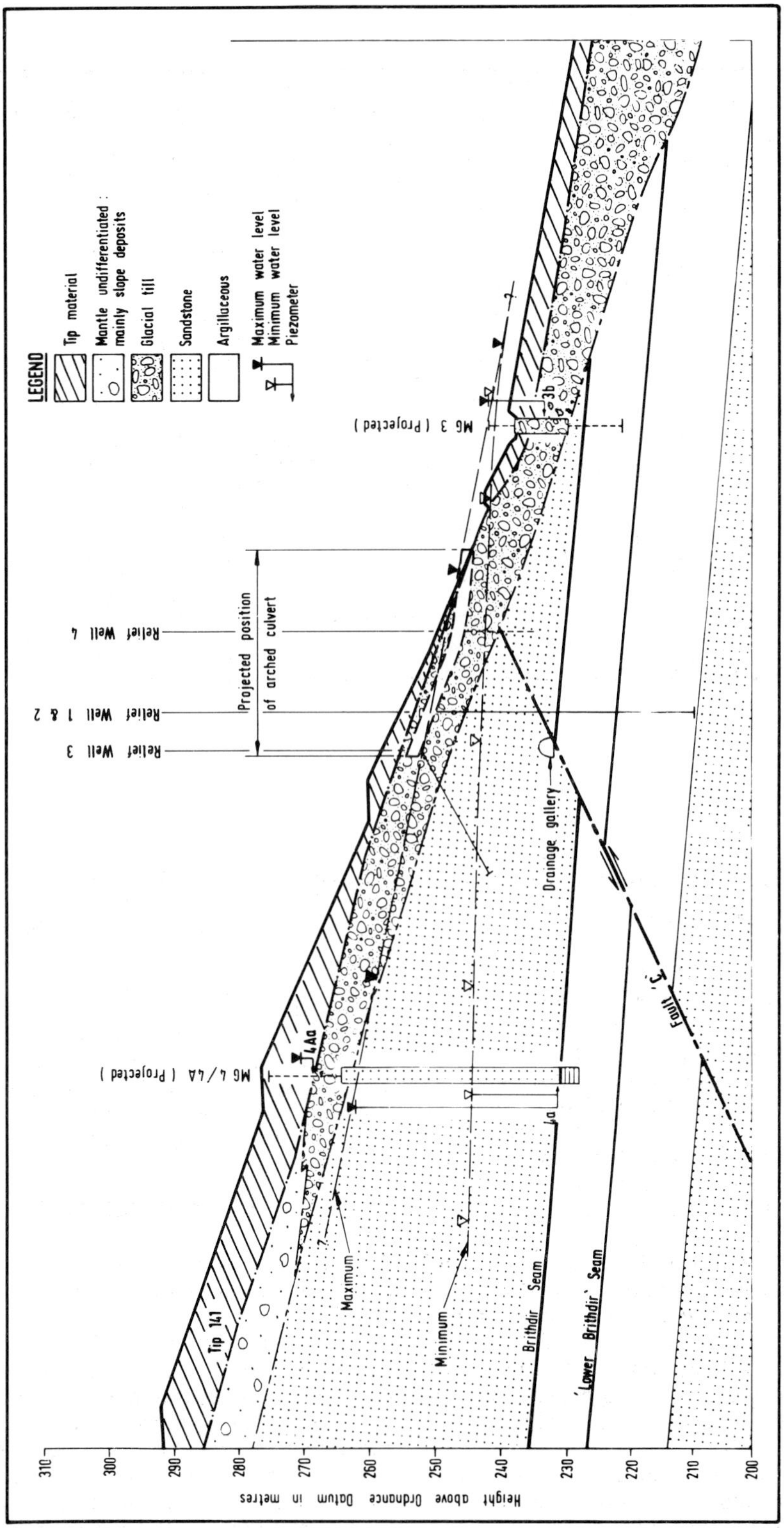

FIG. 2. Geological section.

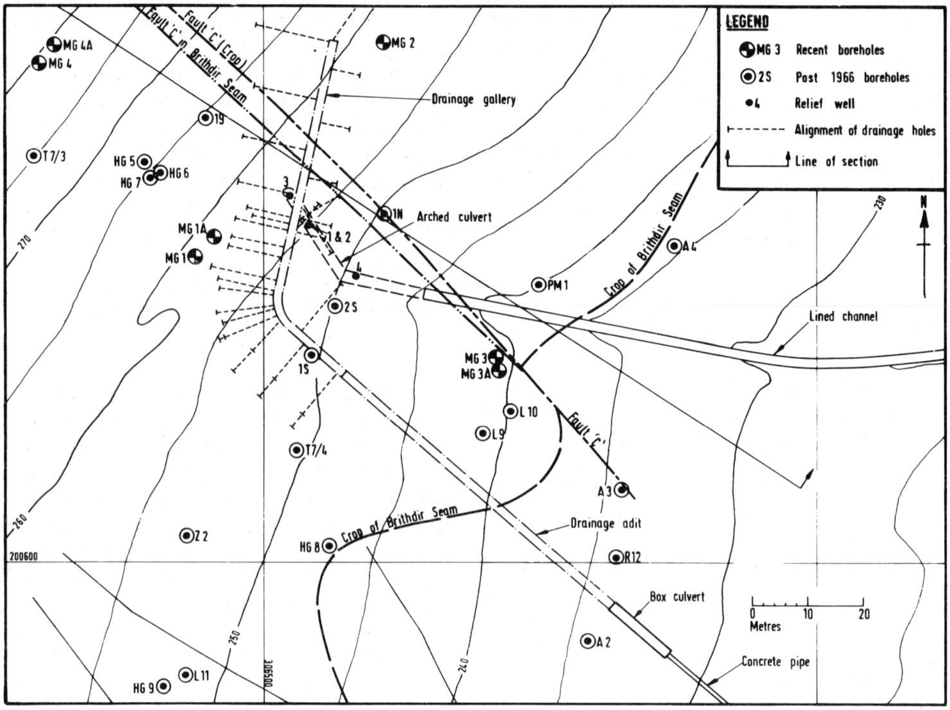

FIG. 3. Plan of drainage gallery.

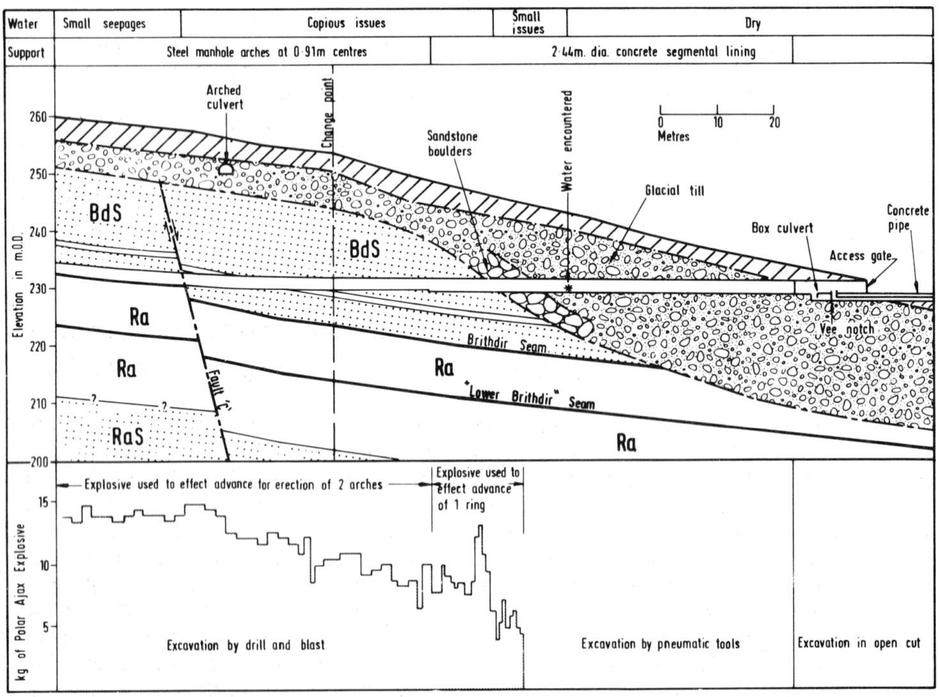

FIG. 4. Section along drainage gallery.

advance of about 1.5 m per 12 hour shift. On completion of the drive, 28 drainage holes between 6 and 19 m long were drilled upwards through the sides of the gallery to enhance drainage of the rock.

At the mouth of the gallery, a reinforced concrete box culvert was constructed, in which a steel vee notch weir and autographic recorder were installed to provide a continuous record of flows. More detailed descriptions of the construction methods of this and other galleries in South Wales are given by Siddle (1984).

Observations during tunnelling

The drive through the glacial till was dry as expected, but a small inflow of water was encountered in a layer of boulders just above rockhead. This layer, which is in hydraulic continuity with the underlying sandstone, is thought to be a deformation till of locally derived rock. Water levels in the aquifer, as measured in piezometers 1a, 2a, 3a and 4a were falling during this period (Fig. 5) but soon after water appeared at the face, a sudden fall of pressure occurred in the confined part of the aquifer as measured by piezometer 3a. The quantity of water released by the gallery increased steadily as it was extended into the sandstone and flows responded sharply to rainfall in mid-June 1980.

The fault which had been conjectured from borehole information was proved and consisted of 150 m of clay gouge between planar, locally slickensided surfaces. Layers of silty mudstone on the downthrown, south-western side of the fault displayed a conspicuous south-westerly dip in contrast to the regional south-easterly dip. A plan showing the local geological structure determined from borehole data and observations in the gallery (Fig. 6) indicates that there is a 20 m wide flexure adjacent to the fault on its downthrown side.

Most water entered the north-western side of the gallery within the flexure associated with the fault and the majority of the drainage holes were drilled in this section. Beyond the fault, the gallery was noticeably drier.

Effect of gallery on water levels and flows

A representative period of water level and flow measurements is shown on Fig. 5, but weekly records are available between 1976 and 1981. Large rainfalls have occurred both before and after construction of the gallery and it is considered that a meaningful comparison of maximum and minimum water levels can be made. The minimum water levels were generally reduced to a greater extent than the maximum, resulting in a greater range of fluctuation than existed prior to the construction. This reflects the deeper drainage now induced within the aquifer and the extension of unconfined conditions to parts of the aquifer where confined conditions used to prevail.

The greatest reduction occurred in borehole MG 3 within the confined part of the aquifer soon after the gallery punctured the base of the lodgement till. The maximum water level in piezometer 3a was reduced by 5.9 m, appreciably more than the requirement described above. Relief well 4 ceased to overflow and wells 2 and 3 now overflow for much shorter periods than before the gallery was constructed. Some 15 m uphill of the gallery, the maximum water level in piezometer 1a was also reduced by more than 5 m but, further uphill, piezometer 4a was little affected. Similarly, the water level in piezometer 2a, only 11 m away from the adit (7.6 m higher and 8 m in plan) was not much reduced.

The gallery did not appear to reduce the piezometric surface to the elevation of the gallery itself, despite the drilling of drainage holes. This is due to the greater permeability of the rock parallel to the bedding than orthogonal to it and suggests that, in general, water moves along bedding surfaces rather than in sub-vertical joints. It is considered that a greater draw-down of the piezometric surface could have been achieved by drilling more drainage holes but this possibility was not pursued because sufficient draw-down was attained to achieve the required factor of safety for the slope concerned.

The total discharge from the 'arched culvert' and gallery between September 1980 and April 1981 was 4.8×10^5 m^3. Assuming that run-off could account for about 30% of the rainfall on the catchment concerned (which is made especially receptive by mining subsidence fissures) and that evaporation losses are negligiblc in this Winter period, the volume of water is equivalent to the effective infiltration over an area of about 0.5 km^2 of catchment. The exceptional nature of the hydrogeology of the Aberfan tip site is therefore illustrated by the fact that above 25% of the water entering the Brithdir Sandstones on Mynydd Merthyr is drained by the drainage gallery (and to a lesser extent by the arched culvert), which were constructed on less than one per cent of the spring line associated with the Brithdir Seam.

Influence of geology and mining subsidence on groundwater flows

An analysis of past mining by Wardell & Piggott (1969) demonstrated that a pillar of coal left unworked adjacent to Fault 'A' is likely to have produced a south-easterly trending narrow zone of tension in the sandstone beneath the tip (Fig. 7). It was suggested that this zone would be more transmissive and would help to explain the concentration of water towards the spring.

This theory was supported by tracer tests (Mather *et al.* 1969) in which brine tracer injected into the sandstone at the north-western end of the 'tensional corridor' moved rapidly towards the relief wells and spring in the 'arched culvert'. It was concluded that

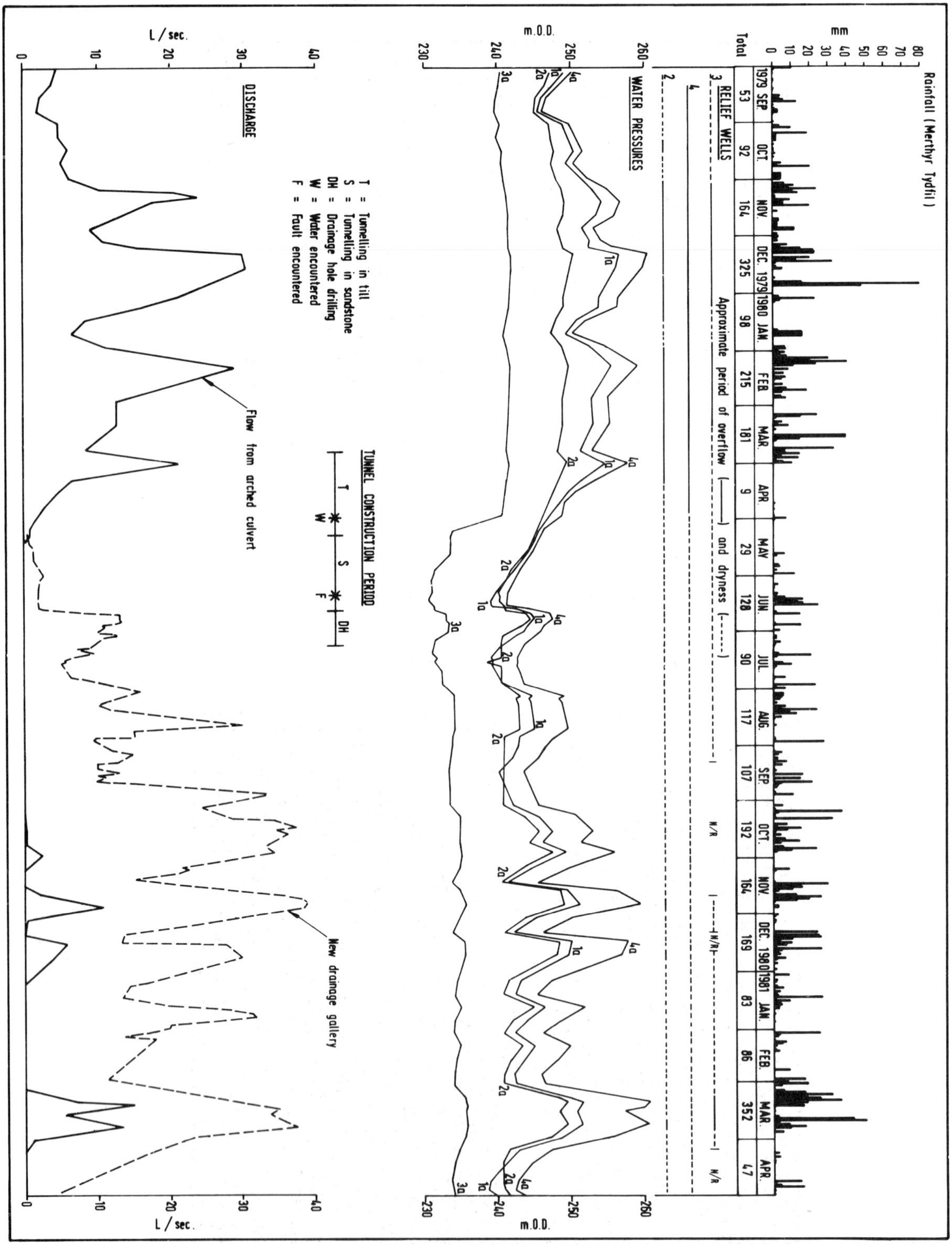

FIG. 5. Effect of gallery on water levels and flows.

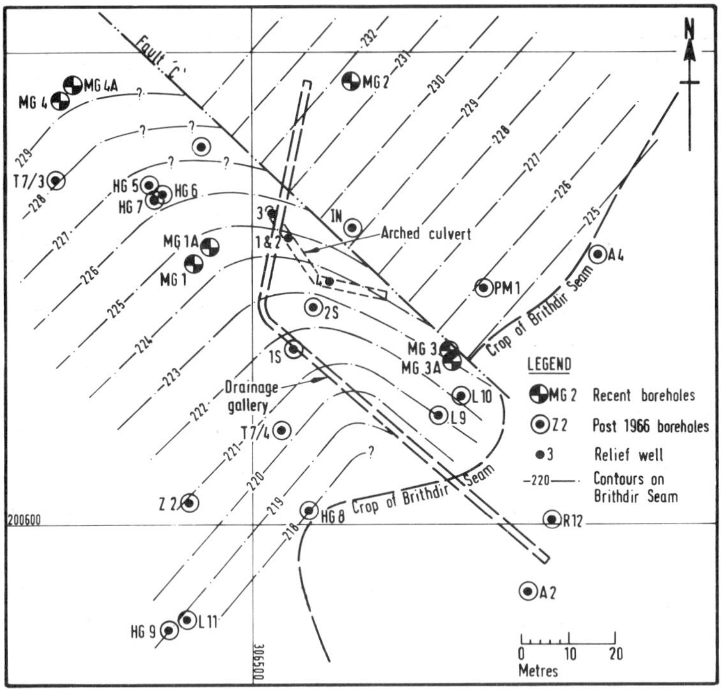

FIG. 6. Geological structure.

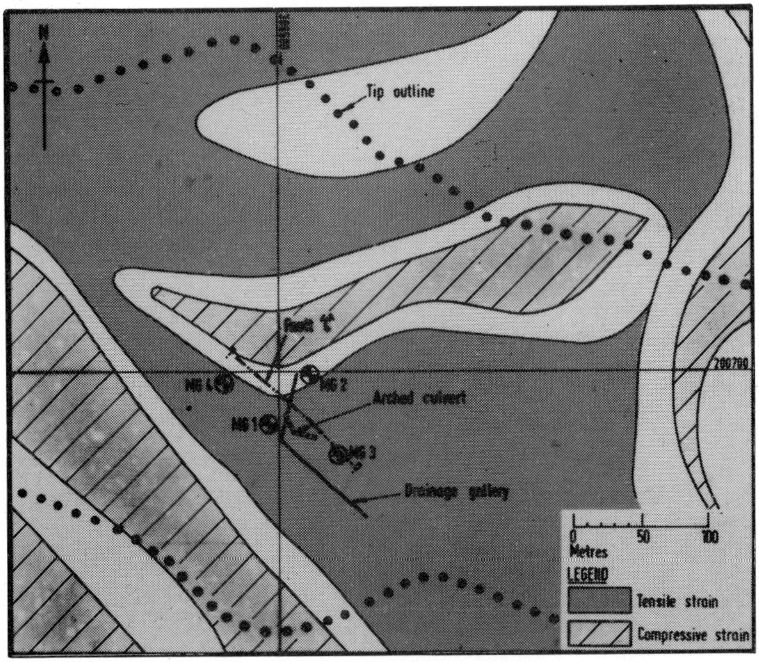

FIG. 7. Mining subsidence tension and compression zones (after Wardell & Piggott 1969).

groundwater moved in an open fissure system. It was also suggested that zones between tensile and compressional strains might by hydraulic boundaries.

The present investigations showed that Fault 'C' lies close to the tensional corridor. Most of the water enters the gallery within the flexure on the downthrow side of the fault and it seems that this structure, controls the movement of water. A dilated vertical joint system was not encountered in the gallery and observations made during probe-hole drilling indicated that most of the water in the vicinity of the gallery is moving along bedding surfaces, albeit interconnected by sub-vertical joints. It was not possible to systematically compare the dilation of discontinuities on either side of the fault proved in the gallery because of the loosening effects of blasting. However, it would appear that the rocks are more transmissive within the flexure where it is interesting to note significantly less explosive was required to advance the gallery, suggesting looser rock conditions. Tighter rock conditions away from the tensional corridor are indicated by the drier nature of the rock north-east of Fault 'C' and the lack of response of water levels in borehole MG 2 to driving the gallery. It is well known that subsidence and strain produced by mining can be concentrated at faults and discontinuities. At Aberfan, it would appear that subsidence and tensional strain were concentrated not on the fault plane itself, which was tight and non water-bearing, but within the flexure on its down-thrown side. It may even be that some of the flexuring may result from subsidence rather than frictional drag.

Conclusions

Construction of a drainage gallery was found to be an effective means of controlling groundwater pressures in a Pennant Sandstone aquifer subjected to residual tensile strains induced by past mining activities. The piezometric surface was reduced sufficiently to achieve the desired improvement in the factor of safety of a critical slope of the tip, but was not drawn down to the level of the gallery itself. This is attributed to the greater horizontal permeability than vertical. Observations during tunnelling and later monitoring also suggest that there is less opportunity for groundwater control in less transmissive zones of compression in the sandstone.

Groundwater movement at Aberfan was shown to be concentrated within a 20 m wide flexure adjacent to a hitherto unrecorded minor fault within the 'corridor of extension' produced by mining. The anisotropy within the aquifer produced by this structure and the effect of mining is demonstrated by the fact that above 25% of the water entering the potential geological catchment is released from the gallery, over less than one per cent of the available spring line.

ACKNOWLEDGEMENTS: The investigations at Merthyr Vale Tip 141 from 1975 onwards were carried out under the overall direction of Mr N.J. Cochrane, then Principal Partner of Sir William Halcrow and Partners.

Thanks are expressed to Mr K. Lisk, Area Civil Engineer of the South Wales Area of the National Coal Board for permission to publish this paper and to my colleagues in the Cardiff regional office of Sir William Halcrow and Partners.

References

BISHOP, A.W., HUTCHINSON, J.N., PENMAN, A.D.M. & EVANS, H.E. 1969. Geotechnical investigation into the causes and circumstances of the disaster of 21st October 1966. *A selection of technical reports submitted to the Aberfan Tribunal,* Welsh Office, London. H.M.S.O., 1-80.

GRAY, D.A., 1969. Groundwater conditions in the Pennant sandstone. Appendix 1 to Woodland, 1969.

HALCROW, SIR WILLIAM & PARTNERS, 1975. Merthyr Vale Colliery. *Regulation 18 Report on the closed classified Tip 141 known as Aberfan Tip,* unpublished report for National Coal Board.

HALCROW, SIR WILLIAM & PARTNERS, 1979. Merthyr Vale Colliery. *Supplementary Regulation 18 Report on the closed classified Tip 141 known as Aberfan Tip,* unpublished report for National Coal Board.

H.M.S.O., 1969. *Report on the Tribunal appointed to inquire into the disaster at Aberfan on 21st October 1966,* H.M.S.O., London.

MATHER, J.D., GRAY, D.A. & JENKINS, D.G., 1969. The use of tracers to investigate the relationship between mining subsidence and groundwater occurrence at Aberfan, South Wales, *Journal of Hydrology,* **9,** 136-154.

MOORE, L.R., 1969. Geological report on the tipping site and its environs at Merthyr Vale and Aberfan. *A selection of technical reports submitted to the Aberfan Tribunal,* Welsh Office, London, H.M.S.O., 147-185.

SIDDLE, H.J., 1984. South Wales spoil stabilisation tunnels in sandstone. *Tunnels and tunnelling* (December), Morgan-Grampian Press, 17-20.

WARDELL, K. & PIGGOTT, R.J., 1969. Report on mining subsidence. *A selection of technical reports submitted to the Aberfan Tribunal,* Welsh Office, London, H.M.S.O., 187-204.

WOODLAND, A.W., 1969. Geological report on the Aberfan tip disaster of 21st October 1966. *A selection of technical reports submitted to the Aberfan Tribunal,* Welsh Office, London, H.M.S.O., 119-145.

H.J. SIDDLE, Sir William Halcrow & Partners, 25 Windsor Place, Cardiff CF1 3BZ, UK.

Groundwater in Engineering Geology, London, 1986, pp. 541-545 Printed in England

Advance land drainage as an aid to construction in areas with high groundwater levels in Warrington New Town

P. M. Wilde & J. M. Crook

ABSTRACT: Warrington New Town, now part of the combined Warrington and Runcorn Development Corporation, was designated in 1968 with an area of approximately 7 500 hectares to be comprehensively developed by 1990. The task has included the development for employment uses of several large low lying areas of land underlain by saturated alluvial and fluvioglacial deposits. When building in these areas, an effective method of controlling the groundwater must be chosen to facilitate the efficient and economic construction of excavations for all groundworks. If the upper soils are permeable sands and/or gravels, land drainage installed in advance of construction to lower groundwater levels can provide such a method. A case study is presented demonstrating the effectiveness of this system illustrated by the results of monitoring of groundwater levels over a 24 month period punctuated by the installation of land drainage.

Introduction

Warrington New Town was designated in 1968 for the purpose of expanding the population from approximately 120 000 to 160 000 by 1990 and within an area of approximately 7 500 hectares surrounding the existing town (Warrington Development Corporation 1972). A vital key to this growth has been the provision of new jobs by attracting companies to take advantage of the communications crossroads at Warrington. Four strategically located employment areas are now available adjacent to the motorway network and to date, in addition to privately built factory accommodation, the Development Corporation has provided 250 000 m^2 of floorspace in Advance Industrial Units for rent. Three of these employment areas are located on low lying and flat land with a history of problems due to the very wet nature of the ground.

Geological background

Warrington is situation on the northern rim of the Cheshire Basin and is underlain at depth by sandstone rocks of Triassic age. These are generally obscured by superficial drift materials which vary from a few metres to in excess of 60 m thick within the Mersey and Sankey Brook sub drift channels.

On the higher ground, Boulder Clay type deposits or sandstone bedrock are predominant but within and adjacent to the Mersey and Sankey Brook channels, alluvial and fluvioglacial sands and gravels predominate in the upper layers. Typically these lower lying areas show between 2 and 4 m of loose sand deposits on top of soft/firm silty clays with groundwater levels within 1 m of the surface throughout a large proportion of the year. This sequence is typical of the ground conditions present in the three employment areas shown on Fig. 1 which also indicates the simplified geological setting of Warrington.

The problem

The programme for the provision of Advance Industrial Units has entailed a succession of construction contracts within each of the employment areas. At an early stage, it was realised that unless an effective method of controlling groundwater could be found, building works were not going to be possible within the tight construction schedules and budgets. Ideally, the system of groundwater control should reduce standing water levels across a whole site prior to any building works to avoid the situation of the initial site strip producing 'paddy field' conditions. Whilst proven techniques existed for reducing groundwater levels for specific excavations, e.g. well point dewatering for drainage runs, none were considered suitable for use to prepare a site for development by way of an advance works contract other than the installation of a network of deep land drains. This was a particularly attractive proposition with the development of combined trenching and pipe laying equipment, commonly used for agricultural field drainage, capable of laying at depths of up to 3 m below ground level. With this system, the machine cuts and supports a narrow trench to the required depth and then lays the drainage pipes with associated backfill in a continuous operation.

From CRIPPS, J.C., BELL, F.G. & CULSHAW, M.G., (eds), 1986, *Groundwater in Engineering Geology,* Geological Society Engineering Geology Special Publication No. 3, pp 541-545

Design philosophy

As soon as a site is earmarked for development, a thorough site investigation is carried out with particular emphasis on assessing groundwater conditions. Standpipe piezometers are installed to establish standing water levels and carry out *in situ* permeability tests if required. Laboratory grading analyses are conducted on typical granular materials. If the results of this work indicate an upper layer of sand and/or gravel in excess of 2 m thickness of minimum permeability 10^{-5} m/s then the site is considered suitable for pre-drainage.

The system of land drains must satisfy the following criteria:

(a) be capable of reducing groundwater levels to allow 'dry' excavation conditions;
(b) be installed far enough in advance of main construction to allow drainage to occur;
(c) be designed so as not to conflict with the proposed construction and remain operative during groundworks.

In practical terms, the proposed layout of building foundations and foul and surface water sewers often restricts the routes available for the land drains to avoid underlying buildings or crossing permanent drainage runs.

Obviously to achieve 'dry' excavations, the land drainage system needs to be below the level of the proposed groundworks with, if possible, a safety zone of 0.5 m to accommodate variations in the reduced water-table. By careful design of the permanent system of foul and surface water sewers and building foundation levels it is usually practical to achieve this aim although on certain occasions, where permanent drainage is below 3 m, recourse has had to be made to a limited site strip along land drain runs prior to installation to achieve additional depth. Occasionally, particularly deep excavations have to be dealt with separately with provision made for well point dewatering for example. A land drainage system typically comprises a grid iron pattern of drains covering a site. The drainage pipes are usually 100 mm diameter slotted/perforated UPVC to allow ingress of the groundwater. To facilitate the efficient transfer of groundwater from the natural ground into the pipe, the land drainage system includes the following features:

(a) the pipe surround/trench backfill is 'designed' to discourage the infiltration of fines which can cause clogging and reduce efficiency. (Spalding 1970);
(b) the pipe is wrapped in a geotextile sock chosen to prevent material entering the pipe and causing a blockage.

A system is laid at very shallow gradients to:

(a) achieve maximum depth;
(b) be capable of draining in either direction. To assist in this aspect, at least two outfalls into the permanent surface water main drainage system are provided and at intersections on the system, catchpit manholes are installed. The result is that if one length becomes inoperative for any reason, the majority of the land drains will still function.

Although a particular site layout may restrict the potential pattern of land drains, it is interesting to compare achievable drain spacing with theoretical spacing. Spalding (1970) refers to work in connection with filter drains adjacent to roadways. In a typical case with the objective of reducing the water table from 1 m to 2 m below surface with parallel drains at 2.5 m depth, then

$$\frac{tkD}{yW^2} \approx 0.1,$$

where

t = time since beginning of drainage
k = coefficient of permeability
D = depth of drains below initial water-table
y = volume of drainable water per unit volume of soil
W = distance between parallel drains.

Assuming a sand deposit with values of

k = 10^{-5} m/s
y = 0.3

and a pre-drainage period of 1 month, W can be estimated to be in the order of 50 m. This in practice is an average installed drain spacing on the majority of sites.

Case study

The Development Corporation proposed to construct factory units at a site known as Phipps 102. Beneath topsoil, the ground conditions comprised between 2.5 m and 3 m of loose saturated sand with some gravel overlying firm silty clay. In order to assess the effectiveness of land drainage in lowering groundwater levels, it was decided to install standpipe piezometers at key locations and monitor standing water levels before, during and after installation of the drains.

The proposed site layout of the structures and sewer system allowed a pattern of land drains to be formulated with between 25 and 70 m between runs at between 2m and 3 m deep. The closely spaced land drains were laid in the sewer corridors where excavations up to 2 m deep were proposed and the more widely spaced land drains between structures where only foundations 1 m deep were proposed. Five monitoring points were chosen and Fig. 2 illustrates the results of recording groundwater levels over a 24 month period punctuated by the installation of the land drainage. Fig. 3 shows in more detail the relationship between ground conditions, drain spacing and depth together with the development of the reduced standing water level profile due to the effect of the drains.

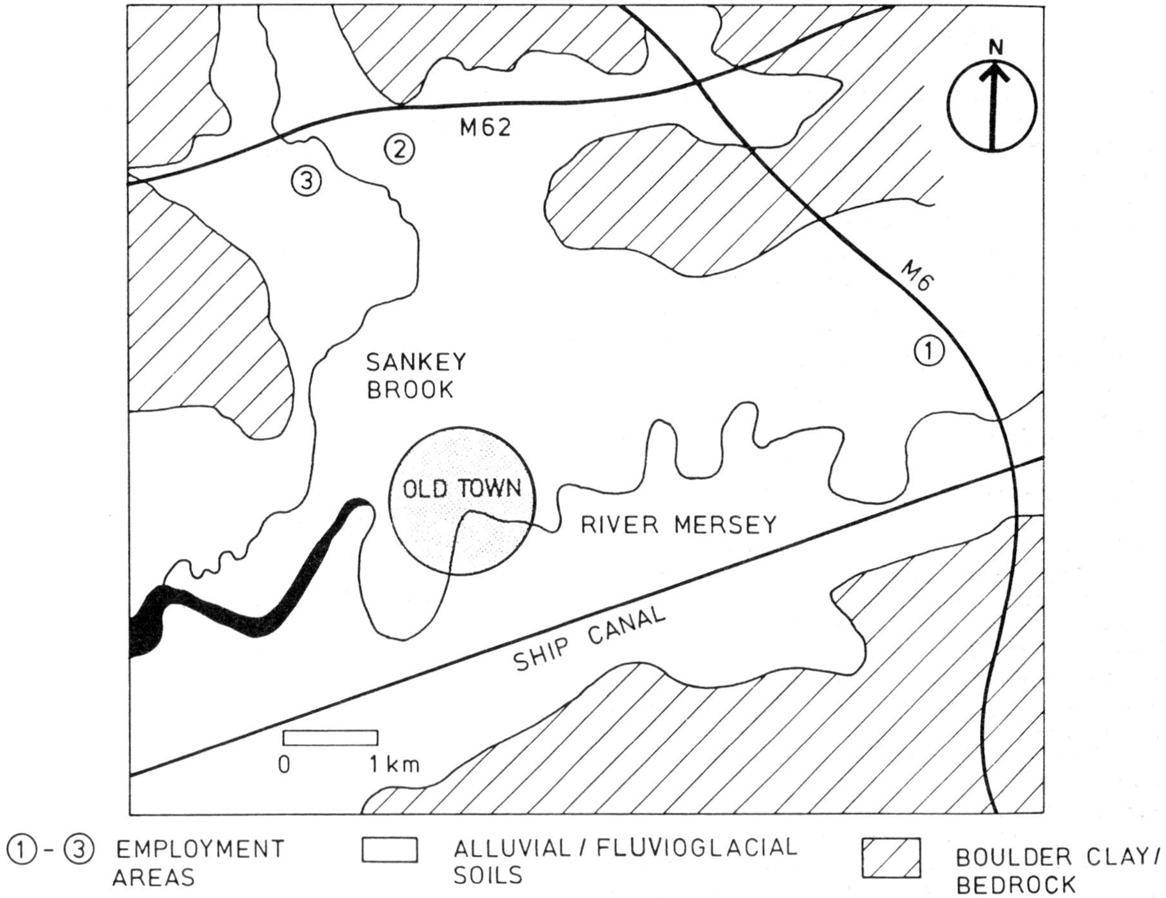

FIG. 1. Geological setting of Warrington.

It can be noted that:

(a) the land drainage system was very effective in reducing the water levels to the required levels to suit the construction of the proposed permanent drainage and building foundations;
(b) as would be anticipated, the maximum drop of 1.5 m was adjacent to the closely spaced drains with a reduced effect of a 1 m drop at the midpoint between more distantly spaced drains;
(c) due to the permeable nature of the ground, it took only 12 days after installation to reach a 'steady state' condition.

Conclusions

Experience with the numerous projects which have included the provision of advance land drainage indicates the success of the system in reducing groundwater levels prior to main construction with the following advantages:

(a) Need only be installed weeks in advance of main construction in permeable soils i.e. when building positions have been fixed and drains can be located to avoid conflict with permanent works.
(b) Cost effective. For typical Corporation factory developments, the cost of the land drainage is around 1.5% of total cost. It should also be noted that this total cost reflects the fact that the site has been pre-drained and without this benefit, an even higher construction cost would have been likely as excavation rates below the water table are commonly double the comparable rates above the water table in 'dry' conditions.
(c) Eases site construction problems. By allowing the majority of groundworks to be carried out in the 'dry', the works are not only completed more speedily but without provision of sophisticated trench shoring systems and pumping to control groundwater.

To date 14 schemes have been successfully completed including advance land drainage and the system has proved its value for dealing with the particular soil and groundwater conditions commonly found in Warrington.

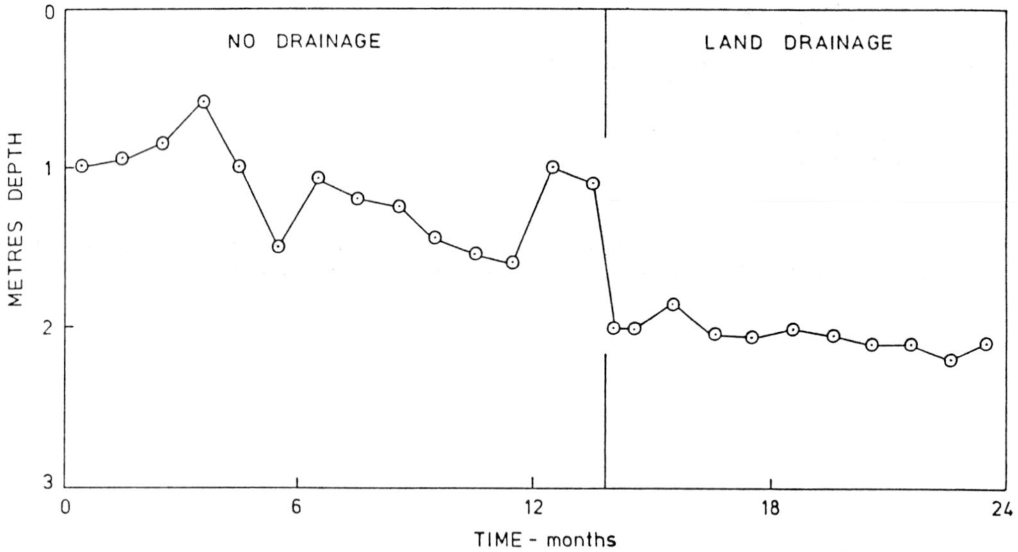

FIG. 2. Typical record of standing groundwater level with time before and after installation of land drainage.

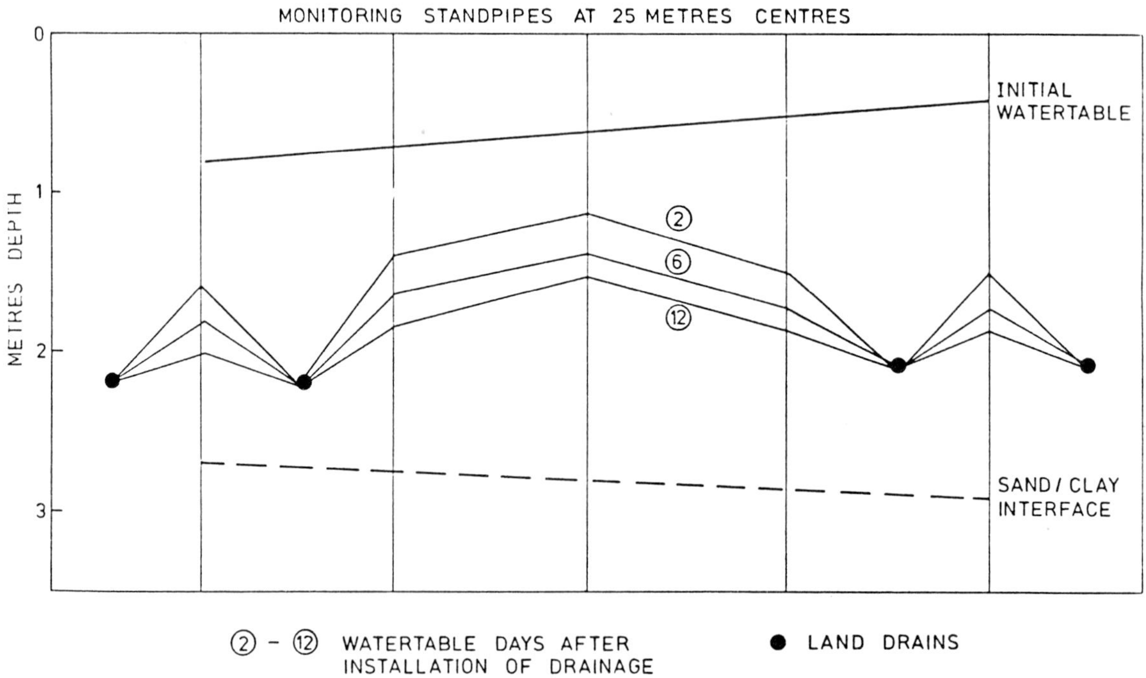

FIG. 3. Change in groundwater profile due to installation of land drainage.

ACKNOWLEDGEMENT: This paper has been written with the encouragement of T. Walsh, BEng, CEng, FICE, FIHT, Chief Engineer, Warrington and Runcorn Development Corporation.

References

SPALDING, R., 1970. *Selection of materials for subsurface drains,* Road Research Laboratory Report LR 346, Ministry of Transport.

WARRINGTON DEVELOPMENT CORPORATION, 1972. *Warrington New Town Outline Plan.*

WILDE, P. M. & CROOK, J. M. Department of Engineering, Warrington and Runcorn Development Corporation, New Town House, Buttermarket Street, Warrington, Cheshire WA1 2LF, UK.

Discussion Session 11

P. CASHMAN (Wimpey Laboratories Ltd.) referred to experience in Cairo where pore water control was achieved by dewatering within a partial cut-off. Thus effects of drawdown were limited. Some may question the practicalities of recharge. However, since the Cairo sewers are surcharged 24 hours per day, the pumped discharge could only be disposed of by recharge at a lower level. About 100 m^3 per hour was pumped down 30 recharge wells.

Mr CASHMAN went on to ask Dr CROOK to give details of the ground conditions at the site in Warrington since, if the sands were homogeneous, there would be no need to import granular backfill.

Written contribution further to verbal discourse:

P.M. WILDE and J.M. CROOK (Warrington and Runcorn Development Corporation).

In response to the questions put by Mr CASHMAN and Dr CROOK, the following comments are appropriate:

(a) Although the drains have been installed in 'sand' material, these deposits are commonly silty in nature and also discretely layered hence exhibiting anisotropic permeability. In most cases, therefore, it has been the practice to backfill the narrow trenches with a selected granular medium of equivalent permeability to the natural sand as opposed to the disturbed excavated material which would possess a reduced permeability so affecting groundwater transfer into the drains.

(b) However, in certain instances where the drains have been installed in coarse, uniform, highly permeable sands, the pipes had only been wrapped in a continuous geotextile sock to protect against material entering them without imported granular backfill. In terms of end result, this appears to have been equally effective in lowering groundwater levels without any reduction in drain efficiency.

H.B. PHILLIPSON (Engineering Development Department, Hong Kong) made reference to experience with the extensive use of horizontal drains in controlling groundwater in cut slopes in Hong Kong. He went on to describe the installation and operation of a drainage system. (See written contribution.)

E.N. BROMHEAD (Kingston Polytechnic) pointed out that a situation similar to that described by Barton & Thomson in which an apparently small scale extraction of groundwater was sufficient to have a beneficial effect on cliff stability on the Isle of Wight. Cessation of brewing activity of the Old Star Inn Brewery (4 tied houses) was associated with a large increase in the recession of nearby cliffs. He suggested that continued support of Burts Brewery, Ventnor, might have a beneficial effect on land stability in that area.

Written contribution further to verbal discourse:

H.B. PHILLIPSON (Engineering Development Department, Hong Kong)

Groundwater lowering using long horizontal drains

Background

McNicholl *et al.* (1986), in their paper for this Conference (this volume) describe the extensive use of horizontal drains in controlling groundwater in cut slopes in Hong Kong. An additional case history from Hong Kong is described as follows, which shows further development and use of horizontal drains in lowering groundwater in a large natural slope. The scheme was designed, and constructed supervised by the Geotechnical Control Office (GCO).

As a result of an extensive study of the Mid-levels area of Hong Kong Island (Geotechnical Control Office 1982), it was decided to improve the stability of a large natural slope by lowering both the main groundwater table and local perched water tables. Lerner (1986) describes in his paper for this Conference the hydrogeology of the Mid-levels, and in his Figure 2, a photograph of the natural slope area is shown. The portion of the slope stabilised is approximately 40,000 m^2 with an overall elevation change of 110 m.

Initially a drainage tunnel scheme, not unlike the one described at this Conference by Siddle (1986), was proposed. But further assessment of the site showed that if horizontal drains up to 90 m long could be constructed, a scheme approximately one-fifth of the tunnel cost could be used. The final scheme comprised a total of 77 drains, installed at four levels. Of these approximately half were 20–50 m long, and half 50–90 m. Spacing of drains was generally 10 m.

From CRIPPS, J.C., BELL, F.G. & CULSHAW, M.G., (eds), 1986, *Groundwater in Engineering Geology,* Geological Society Engineering Geology Special Publication No. 3, pp 547-553

Design and construction details

McNicholl *et al.* (1986) describe the use of 65 mm diameter 'Netlon' drains with removable 40 mm diameter internal drains. Such a drain type works well for shorter drains up to 20 m in length, but recent GCO experience with longer drains showed that construction of the outer drain of this material, and removal of the inner drain for maintenance flushing, presented practical difficulties. It was therefore decided to use non-removable single 40 mm diameter stainless steel tubes for the drains. The screen sections of the drains comprise 1 mm slots perpendicular to the tube axis. A diagrammatic sketch showing details of the longer horizontal drains used to drain the main water table is shown in Fig. 1.

The downhole 'Odex' drilling system was used to facilitate installation of the long drains. This allows temporary casing, ranging in size from 115 to 190 mm diameter, to be installed continuously as the hole is advanced. Without such a technique installation of the drains below the water table would have been extremely difficult if not impossible.

Alignment specification required that the drain endpoint be located within a circle of 1 m radius of the design location for drains less than 30 m long, 2 m radius for drains between 30 and 60 m long, and 3 m radius for drains over 60 m long. A borehole survey device was used to check this requirement. There were, however, no problems in achieving this specification.

In order to guard against the possibility of drains recharging the perched water tables in the colluvium with water drained from the main water table in the decomposed volcanics, it was decided to grout the drain in front of the screen section. Grouting operations proved difficult in certain locations, and proved to be the major construction problem. Techniques required modification and development as the works progressed.

Taps have been installed on the end of all but four of the drains. These are to facilitate flushing of the drains, and also to investigate the effect of drain spacing on the water table drawdown by shutting off individual drains.

The drilling of drains and construction of associated drainage channels took nine months, and was completed in August 1985. The approximate cost of the scheme was HK$4 million (£400,000). The resulting cost of approximately £10 per square metre of slope treated compared well with the £14 quoted by McNicholl *et al.* (1986).

Monitoring and performance

Having constructed the drains it is now extremely important that their performance is monitored; initially to confirm design assumptions, and thereafter for maintenance purposes. Since the purpose of the drains is to lower groundwater tables, contrary to popular belief, drain flow is of secondary importance to piezometric monitoring. Due to the background of

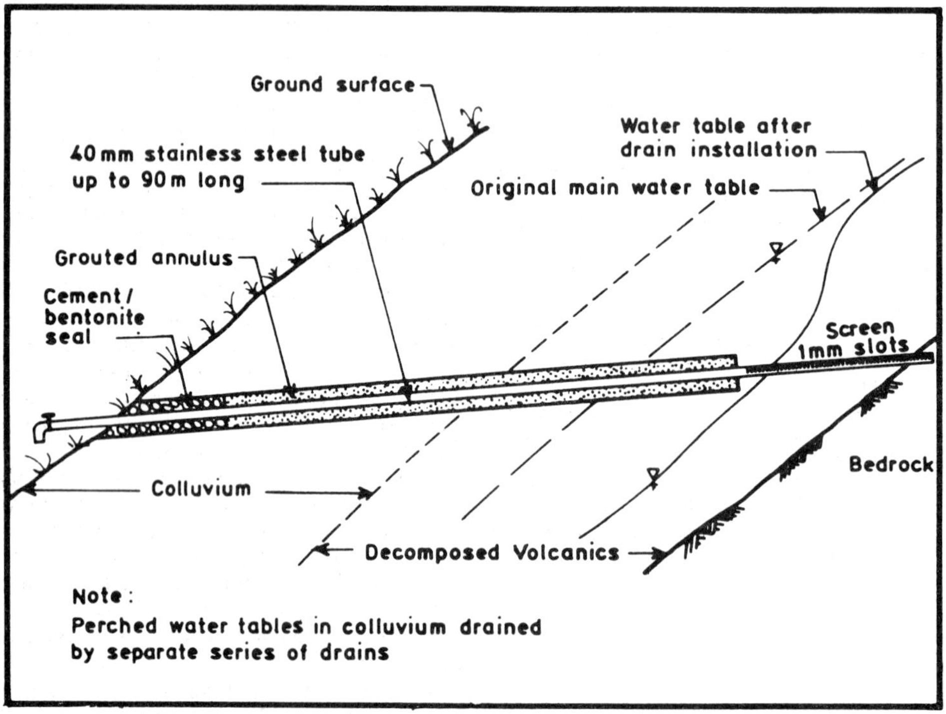

FIG. 1. Horizontal drain details.

intensive study of the area, there are approximately 70 piezometers in the area stabilised. Initially these are being automatically monitored using equipment similar to that described at this Conference by Weeks and Starzewski (1986). Frequency of piezometric measurement is three times per day in the wet season (April to September) and once per day in the dry season (October to March). Drain flows are recorded twice a week in the wet season and twice a month in the dry season. Once design verification has been obtained monitoring of piezometers will be significantly reduced.

Within several weeks of installing the drains, the main water table was lowered, an average of 3 m, and up to a maximum of 10 m, thereby achieving the design objective. Perched water table in the colluvium at this stage do not, however, appear to have been significantly lowered.

Additional information

Craig & Gray (1985), the designers of the drainage scheme, describe in detail the installation of a trial series of long horizontal drains at this site, together with an assessment of their short term performance. The results of this work formed the basis for the design of the scheme described. This Geotechnical Control Office publication also gives a review of international practice with regard to horizontal drain installations, which is thought to be the most comprehensive and up-to-date of its kind.

References

CRAIG, D.J. & GRAY, I., 1985. *Groundwater Lowering by Horizontal Drains*. Geotechnical Control Office, Hong Kong, (GCO Publication No. 2/85).

GEOTECHNICAL CONTROL OFFICE, 1982. *Mid-levels Study: Report on Geology, Hydrology and Soil Properties*. Geotechnical Control Office, Hong Kong,.

LERNER, D.N., 1986. Predicting piezometric levels in steep slopes. *Proceedings of the 21st Regional Conference of the Engineering Group of the Geological Society*, Sheffield, this volume.

McNICHOLL, D.P., PUMP, W.L. & CHO, G.W.F., 1986. Groundwater control in large scale slope excavations -five case histories from Hong Kong. *Proceedings of the 21st Regional Conference of the Engineering Group of The Geological Society*, Sheffield, this volume.

SIDDLE, H.J., 1986. Groundwater control by drainage gallery at Aberfan, South Wales. *Proceedings of the 21st Regional Conference of the Engineering Group of the Geological Society*, Sheffield, this volume.

WEEKS, R.C. & STARZEWSKI, P., 1986. Automatic monitoring of groundwater conditions. *Proceedings of the 21st Regional Conference of the Engineering Group of the Geological Society*, Sheffield, this volume.

Written contribution:

D.A. GREENWOOD (Cementation Piling & Foundations)

In their paper, Coe & Kay conclude that deep well pumping tests gave the most reliable determination of aquifer characteristics providing they accurately modelled the dewatering system proposed for excavations. In contrast small scale falling and constant head borehole tests generally gave results of two orders lesser permeability, and the authors suggest some reasons for the differences.

Flow to wells (or dewatered excavations) is through substantial volumes of soil, many times larger than that of the excavation; thus general aquifer characteristics are most significant for the resulting yield. For grouting applications, however, which are usually localised around structures to fulfil their engineering and economic objectives, it is important to know local permeability and its variation within comparatively small volumes to be treated. The scale of detail required is altogether different, and it is necessary to know how permeability varies in relation to a single injection which may be influenced by a volume of soil of only about 300 to 700 litres. Conceivably grouting may be required to stabilise and make impermeable ground at the 'tight end' of the range of aquifer properties. Such tight zones even if fairly substantial have comparatively little effect, unless they coalesce, on well test results.

Thus for grouting, well tests are not generally appropriate and it is important to use carefully conducted 'short cell' tests which sample the local permeability. The test cell may be only about half a metre long in a 150 mm borehole, and the major part of head lost is within a metre or two of the bore. If practicable, it is desirable to contour the variation of permeability within the zone to be grouted. A variation of 10 times (one order) on permeability means ten times on the grout pumping rate. This is very significant in terms of adherence to programme and costs of grouting works.

Results of borehole tests from the Cairo Project SI Report provided to Grouting Contractors, reproduced in Fig. 1 show marked differences between *in situ* tests and various forms of laboratory tests. This highlights the extreme difficulty of technique when measuring *in situ* permeability. It is suggested that in fine grained soils the more accurate tests are those in the laboratory because leakage around *in situ* borehole casings, necessarily established with some disturbance, becomes a significant part of total measured flow during the test. In coarser soils this is not so significant, and *in situ* tests are probably the most accurate for them as it is impracticable to reconstitute cohensionless soil properly in the laboratory.

To reduce potential for leakages under large test head differences, the apparatus shown in Fig. 2 was developed and used. It retains accuracy of flow

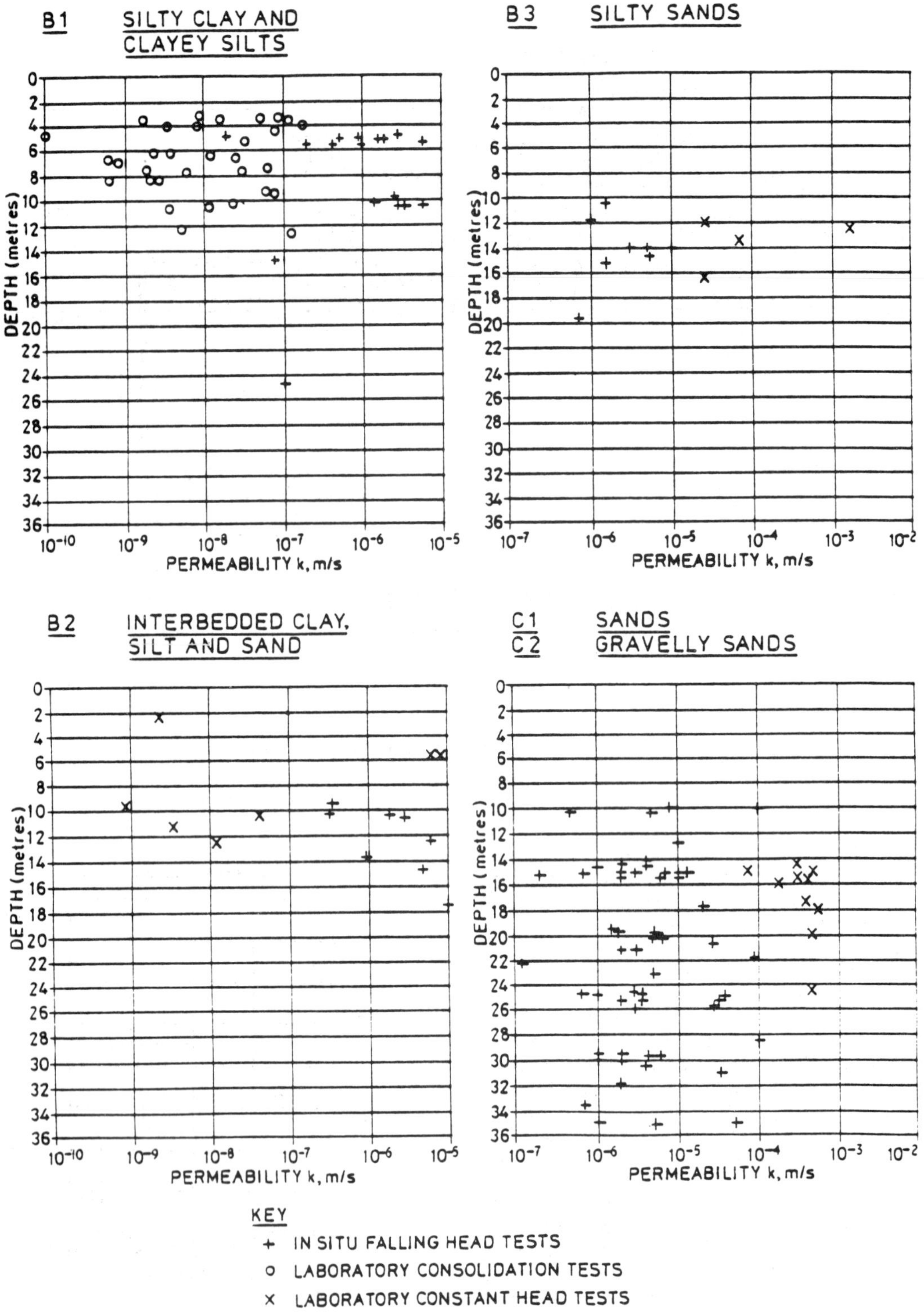

FIG. 1. Results of *in situ* borehole falling head and laboratory permeability tests for differing soil categories; Cairo Waste Water Project.

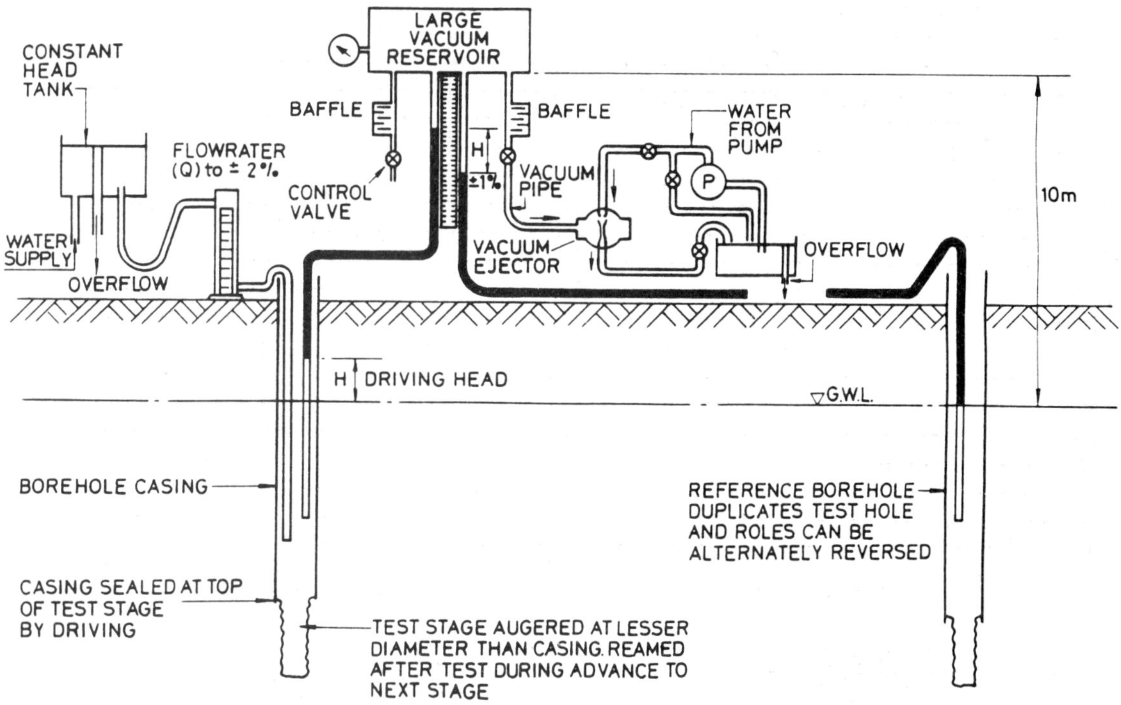

FIG. 2. Low head borehole permeability test apparatus.

measurement even at head differences of only a few centimetres. Results produced as a precursor to the first chemical injections on the current Cairo Contract 3 are reproduced in Fig. 3. They compare well with the authors' well tested results. However, difficulty was experienced initially with rapidly accumulating sediment from borehole water on top of the filter and erroneously low permeabilities were recorded. Not until test water was introduced within the body of the filter through a central perforated pipe to give a positive head of only a few centimetres, was a reliable result obtained. The results were confirmed by water injection through *tubes à manchettes* (sleeved injection tubes) very accurately and continuously recorded by chemical injection apparatus which also displays the pressure flow rate ratio versus time on a site VDU after computing processes (Fig. 4). Subsequent chemical injections also confirm these results. To date the permeability apparatus (Fig. 2) has not been used in ground of tighter than 10^{-5} m/s. It seems however, that tests at this early stage of the contract works confirm the authors' conjecture that the differences between borehole and well tests recorded in the original SI could be primarily due to problems of cleaning out the borehole sediment satisfactorily before making tests.

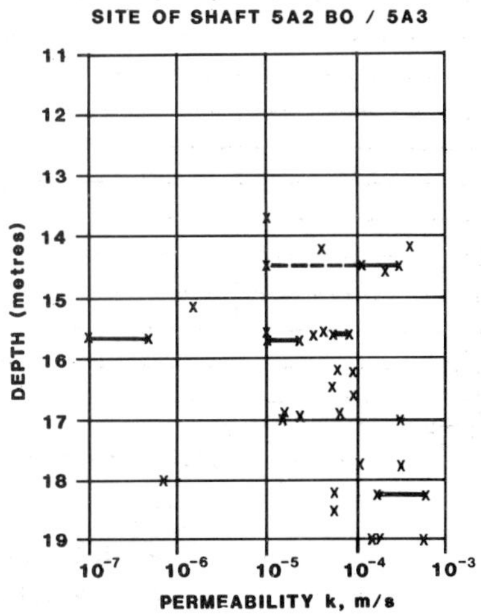

FIG. 3. *In situ* constant head tests by apparatus shown in Figs 2 and 4.

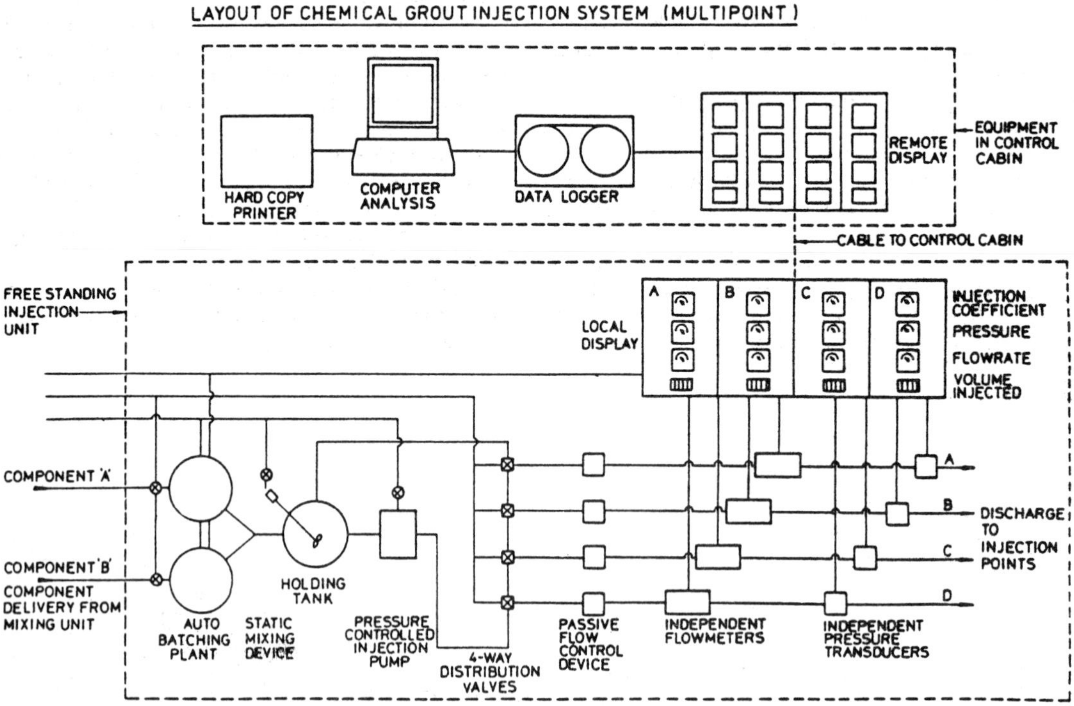

FIG. 4. Multi-point chemical injection, data display and recording apparatus.

It is concluded that the need for short cell tests for assessment of grouting prospects is accompanied by a need for great care in testing technique if misleading results are to be avoided. Implications for technique, cost and programme of subsequent grouting are very significant, and the importance of such borehole test results cannot be over-emphasised.

Written contribution:

T.O. POWER (Laing Design and Development Centre)

The use of gravel filled drainage wells described in Barton & Thomson's paper should be noted. Some years ago similar deep soakaway drains were specified on a contract. These drains would carry water from an upper aquifer through a layer of boulder clay to lower permeable strata. The initial drain, however, produced no noticeable drawdown in the aquifer which was fortunately monitored, and it was concluded that a pressurised air pocket had probably formed in the gravel pack in the clay thus blocking flow. The design of the drain was modified to a tube well system which would vent air and these worked very satisfactorily.

Although it is possible that the slip, may have still occurred even if the drainage had worked properly, there is little doubt in the mind of the writer that the type of drainage wells installed along the cliff top had little affect on the seepage in the cliff.

Written contribution:

L. THREADGOLD (Consulting Geotechnical Engineer)

I wish to refer to the paper by MCNICHOLL, *et al.* (this volume pages) concerning groundwater control in cut slopes in Hong Kong. Since I was directly involved in the design of the slopes and associated drainage at the Chuk Yuen Phases 4 and 5 site to which they refer. I would like to give some background to the selection of the measures adopted.

1. The cut slopes on this site are amongst the highest in Hong Kong.
2. The evidence of high groundwater levels in

these slopes indicated that deep drainage would be required to achieve an adequate Factor of Safety.

3. The evidence of perched water tables required puncturing of the frequently sub-horizontal aquicludes.
4. In order to maintain slope stability it was necessary to lower the groundwater table before excavation.
5. The system needed flexibility in order to deal with local conditions exposed during excavation.

The system adopted achieved the desired effects (Fig. 1). After running field trials to support the design, the caisson wells were installed where required at 12 m centres from each berm level as it was reached and extended to some 16 m depth. On completion, pumps were installed in the wells to remove the groundwater on a temporary basis prior to installation of near-horizontal outlet pipes. These were installed by drilling at a gradient of 1 in 25 from the level of the second berm below that from which the well was installed. This drilling required careful alignment but in the event, no difficult in hitting the wells within 300 mm of their centre was experienced.

This system is flexible in that additional wells could be introduced between the original locations to deal with local conditions without influencing the functioning of the remainder.

Local flows from 'pipes' within Colluvium at the slope toe at one location have been dealt with by local counterfort drainage.

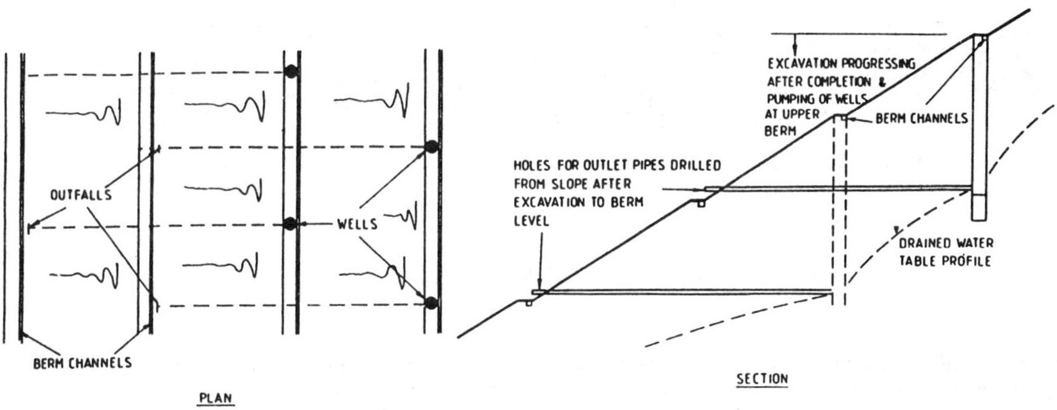

FIG. 1. Drainage system used for cut slopes in Hong Kong.

SESSION 12

CONTRACTUAL AND LEGAL MATTERS

Contractual arrangements for geotechnical work

J. Uff Q.C.

ABSTRACT: The question of appropriate forms of contract for geotechnical processes has received little attention. Such contracts may involve quite unacceptable levels of risk to employers. It is in the interest of all parties that proper contractual terms are set down so that risks and responsibilities may be appreciated.

Introduction

Ground investigation and its procurement has recently been the subject of investigation and recommendation (ref. 1). Many investigation contracts use the ICE Conditions (ref. 2) or some variant. Under this system the investigatory work is billed and specified as though it is construction work. The result is often that, while the required quantities are carried out the quality of the work, the main purpose of which is to obtain reliable data, is poor. But ground investigation can be carried out successfully with traditional forms of contract.

With geotechnical processes, such as grouting, the interests of the parties cannot be stated adequately in terms of the quantities and quality of materials and work. The intention is usually to achieve an objective such as the reduction of flow or permeability. But the specialist contractor is never able (or willing) to give any guarantee of the result, or that the stated processes can be carried out in the particular ground conditions or that, if carried out, they will achieve the desired objectives or any benefit at all.

Nature of the problem

Where forms of contract such as the ICE conditions are employed, work is carried out under the direction of the Engineer. He may have no specialist geotechnical knowledge and even where he does, he will rarely match the expertise of the specialist contractor. Yet the engineer must decide upon the need to order variations, and must account to his client for additional cost. The result is that some contracts turn into cost-plus arrangements with the end result falling far short of expectation. This is not the fault of the specialist contractor, who is usually careful to qualify the results which are achievable.

Where geotechnical work is carried out in conjunction with construction work, both the employer and the main contractor may be exposed to substantial financial risk. Where work such as grouting does not achieve its intended purpose, the main works may be seriously delayed and rendered more costly; and the employer may have to meet claims for unexpected ground conditions. The result is equally serious whether the geotechnical work is carried out independently of the main works, or as a sub-contract.

The Engineer attempting to control work such as grouting under the ICE conditions may be faced with impossible decisions. Quality control, other than by checking physical measurements, is often impossible, so that in practical terms it is rarely possible to reject sub-standard work. The concept of substantial completion has little relevance to geotechnical processes, and the engineer is often faced with the choice of accepting work which has not achieved its objective, or of continuing to pay for extra work with no knowledge of the quantities needed or the end result.

Tentative solution

There is no ready solution to these difficulties, but it is suggested that a starting-point is to avoid the use of standard forms of contract designed for use with construction work. In drawing up a suitable contract for geotechnical work, the following points should be considered:

(1) The employer and designer should be made aware of the degree of uncertainty involved in any geotechnical process to be carried out.

(2) Full use should be made of the expertise and experience of specialist contractors who are asked to tender for such work.

(3) Where the consulting engineer or design professional for the main work does not possess geotechnical expertise sufficient to appreciate fully the likely problems, an independent geotechnical specialist should be employed to advise and if necessary to control the work.

From Cripps, J.C., Bell, F.G. & Culshaw, M.G., (eds), 1986, *Groundwater in Engineering Geology,* Geological Society Engineering Geology Special Publication No. 3, pp 557-558

(4) Geotechnical work should be treated as exploratory, at least in the initial stages of any contract.
(5) There should be enforceable means of detecting and rejecting any sub-standard work.
(6) Work which is to be measured and paid for on a time and materials basis should be closely supervised and monitored by an appropriately experienced geotechnical specialist.
(7) Alternatively, specialist contractors may be invited to quote for the achievement of specified objectives, either on the basis of lump sums or on some form of target payment.
(8) A contract for construction work which is dependent on geotechnical processes should take into account the possibility that the process may not be successful and that it may require substantial variation of method, price and timing.

References

[1] Draft report into methods of procurement in the ground investigation industry: BRE/CIRIA
[2] Conditions of contract for use in connection with works of civil engineering constructions 5th edition revised January 1979

J. Uff Q.C., 10 Essex Street, Outer Temple, London, WC2R 3AA, UK.

Engineering contracts: a legal point of view

M.R. Ludlow

As I am not even a trainee geologist, it seems appropriate to begin this paper defensively with a Biblical quotation:

> 'So likewise ye, except ye utter by the tongue words easy to be understood, how shall it be known what is spoken? For ye shall speak into the air.
>
> There are, it may be, so many kinds of voices in the world, and none of them is without signification.
>
> But if I know not the meaning of the voice, I shall be unto him that speaketh a barbarian, and he that speaketh shall be a barbarian unto me.'

Despite the illustrations of engineering failures provided by Professor Hanna this quotation, which comes from 1 Corinthians, Ch. 14, v. 9-11, is not intended as the beginning of a sermon.

Unlike the distinguished speaker (Dr J. Uff) who preceded me, I have no engineering qualification of any kind, and in some ways it is important to my job as a solicitor advising on engineering contracts generally that this should be the case. The solicitor enters into the field of engineering contracts at one of two possible stages, in both of which he has to do his best to translate the technical into the simple for others who may have even less expertise. The first stage is at the point where different disciplines within the engineering industry enter into contractual arrangements with each other relating to the design and construction of the different parts of an engineering project which go together to make up the completed works. The industry, here, has a choice. The first, obvious, question is:

If the engineering contractor or consultant goes for advice from a solicitor on the terms of a contract which he is negotiating, will the solicitor take so long to understand the technical problems which are relevant to the proposed contract, in view of the language difficulties of an engineering project, that the experience will be too harrowing for the solicitor, and grossly cost ineffective for his client? If there is a doubt, is it on balance preferable to avoid lawyers, and have standard conditions of contract which apply throughout the industry to particular types of contract or sub-contract, or terms of engagement?

Conversely, do standard terms give rise to so many potential loopholes in the definition of your particular rights and obligations, that accepting them without legal advice creates too big a risk?

On the whole, experience suggests that the industry opts for standard conditions, worked out between the various engineering institutions and employer groups, and only infrequently looks for separate legal advice on contractual terms as they affect specific projects.

The second stage when the solicitor becomes involved is if the project turns sour, either because the contract involves more work, time and cost than was expected, or because the works are defective in either design or construction, or both. It is at this stage that legal advice most frequently becomes necessary, although even here the industry has done its utmost by arbitration clauses to create a framework within which disputes can be resolved without recourse either to the courts or to lawyers.

On a purely money claim relating to the cost of the works, it was, I suspect, more often ten years ago than today that lawyers were avoided, since on the whole, claims tended then, more than now, to be a one against one dispute, for example in claims under Clause 12 or 52 of the ICE Conditions of Contract. But in the last decade or so, the industry has undergone a substantial change in attitude to claims. If Clause 12 conditions are encountered, which allow extra claims by the contractor, this begs the question whether the design of the works was careless, in failing to take account of ground conditions which a more detailed site investigation might have revealed. Similarly, if the design of the works is varied under Clause 52, why was the variation necessary? Was the client warned of the potential risk of variation of the original design?

It is not uncommon today when an engineering dispute arises between two parties for the disputants to want to hedge their bets. Take, for example, the following possible series of events, relating to the design of foundations for a building.

Consulting engineers appointed by a client engage site investigation contractors to carry out a borehole survey. From their report, there is reference in the borehole records to soft ground conditions overlying a hard layer, but of insufficient bearing capacity. There is reference to groundwater encountered above the hard layer overlying bedrock. The report advises that vibration would cause damage to adjoining buildings, but that it is feasible to provide piled foundations by continuous flight auger of a given diameter, down to bedrock, and that neither dewatering nor casing of piles is necessary.

From CRIPPS, J.C., BELL, F.G. & CULSHAW, M.G., (eds), 1986, *Groundwater in Engineering Geology,* Geological Society Engineering Geology Special Publication No. 3, pp 559-562

Nominated sub-contractors tender for the piled foundations, offering without qualification piles of the same diameter and the same piling method as the site survey contractors have indicated are feasible. The method of forming piles is by extruding mortar as the auger is withdrawn, into the hole created by the auger, so that at no time is there a void between the base of the auger as it is withdrawn and the pile which is being formed. The consulting engineer recommends the client to accept this tender, and the main contractor has no objection. When the foundations' sub-contract commences, however, serious cracks appear in an adjoining building. The works are stopped and delayed. The piling sub-contractor refuses to continue with his piling scheme, alleging that the soils conditions are not as he was entitled to expect and, hence, the diameter of pile is too wide for further damage to be avoided. He says that the screw of the auger is not penetrating the hard layer at a sufficient speed. Material surrounding the auger, softened by groundwater, is being sucked into the screw, and up the auger removing support from the adjoining structure. Although this is not admitted, ultimately, a variation order is issued on the recommendation of the consulting engineers, to use a pile of narrower diameter. The works recommence and are completed without further damage.

This model problem might result in all or any of the following consequences:

1. a claim for damages, including loss of use, by the adjoining owner;
2. claims for extensions of time and extra payment by the main contractor and the sub-contractor on the grounds either that the works have been varied or that the ground conditions were not reasonably foreseeable;
3. claims by the client against the contractor and sub-contractor for damages, including loss of use, for delay in completion of the works, on the grounds of bad workmanship;
4. claims by the client against the consulting engineers for similar damages for negligent design or negligent supervision of the construction;
5. claims by the client for indemnity in respect of any successful claim against him by the adjoining owner;
6. a claim by the consulting engineers for damages against the site investigation contractor for failure to warn of risks as to the feasibility of this type of foundation in the ground conditions encountered, or for failure to identify the true ground conditions;
7. direct claims by any of these parties against any other, in tort, for indemnity or contribution.

We can ignore the confusion and further permutations that might arise if the works were controlled by a architect under an RIBA contract.

Arbitration provisions under the present law are not adequate to deal with complex combinations of possible claims by many different parties, such as these. For the most part, arbitration agreements cater only for one to one disputes. So you would have to have lots of separate arbitrations to deal with these various claims, with a risk of conflicting findings. The industry is therefore driven to lawyers and the courts so that the totality of the differing disputes can be resolved by a single tribunal, by findings which cut the cake finally as between the claims and liabilities of the various parties. As yet, it does not often happen in practice that the soils or structural investigations specialist is sued. Recently, for example, I advised insurer clients on the possibility of claiming contribution against a University (and its Professor) whose structural stress analysis for consulting engineers of a screen for separating liquid from solid sewage contained fundamental arithmetical errors. The University and Professor might well have been found liable but, on the whole, insurers who have settled claims for defendants seem reluctant to spend money when then standing in the plaintiff's shoes on the risks of continuing the litigation. Also, for the most part, it may be that the consulting engineer has the primary responsibility to direct the investigation and assess the adequacy and interpretation of the findings. However, the climate is such that it is common for more and more disciplines to be dragged into disputes today than yesterday, and it only needs a few successes for a newly widened ambit of risk and liability to become commonplace.

In disputes in the High Court, judges, barristers and solicitors are not often qualified engineers or even mathematicians, and the issues and evidence need to be translated into their own language. There are not many John Uffs around.

This brings me back to the Biblical quotation with which this talk began.

The job of the solicitor when assisting his client in the civil engineering industry is threefold.

First, before he can advise he must learn about the problem in words which he can himself understand. It is my personal starting principle with any client in the engineering industry to confess total ignorance, and not to shrink from showing it, even if this may tend to create a sense of lack of confidence on the client's part. Better to seem stupidly slow in understanding than to hide one's stupidity.

Secondly, the solicitor needs to discuss with his client every possible choice of solution that they can together imagine, and its risks and possible cost. It is only by this two-way process between the solicitor and his client that the client can himself be put in a position of making well-informed decisions about courses of action, and that both can become aware of the risks that have to be either overcome or accepted.

Lastly, when the relevant problems have been fully considered and choices have been made, the solicitor can then set about putting into effect whatever is required of him from the client, in either contractual terms or in the conduct of litigation or other disputes, based on the solicitor's advice.

All three of these principles could, I suggest, apply between different disciplines in the construction industry, as well as between client and solicitor. Many construction defects, whether of design of workmanship, occur because of a lacuna in understanding of what is intended, or of who is to be responsible for what in co-ordination, detail, and so on.

In all three elements of the solicitor's job, it is an advantage to be familiar with the language of engineers, but it is not necessarily a disadvantage to have no qualification as an engineer. The more simply and clearly it becomes necessary for the engineer client to define his problems and objectives to the solicitor, the less chance there is of misunderstanding in communication through the solicitor to his ultimate audience. This principle applies just as much in the creation of contractual terms for a civil engineering project as, for example, in persuading a lay judge when you are accused of breach of duty that your defence is sound.

When I was invited to give this paper, my only brief was to make it specifically relevant to groundwater in engineering geology. I am going to have to admit to you that I have interpreted my brief rather widely! Legal problems on the whole are not special to a particular section of a particular industry. They apply across the board. For the construction industry, the main problem seems to me, today, to be the market. Heavy competition, both national and international, has imposed enormous risks on the industry. You get the job by economising on price. To economise on price, you have to design new methods and materials, calculate factors of safety to narrow margins, and employ less labour. Ten years ago, I was involved in an arbitration between a County Council and a consulting engineer about a primary school which was abandoned after piled foundations had failed. The claim was for £60,000. In excavating for a retaining wall without temporary drainage, groundwater filled the excavations, causing local instability in the ground and movement of some of the piles. It was alleged that the engineer had not carried out a sufficient site investigation. Had he done so, it was said, he would have found that the site was in an area of landslip where the stability of the soils were not much above unity. There was generally a high water table; in one part of the site, a perched water table and, in others, sub-artesian groundwater conditions, and he ought to have arranged as a matter of design for piezometers to measure water pressures and for well points and other specialist drainage measures both as temporary works and as permanent design features. The water table should have been lowered prior to construction. So went the argument for the Council, but after an appalling 90 days of trial and much high-powered expert evidence, the arbitrator found that the loss had been caused wholly by the building contractor's incompetence in excavating without temporary drainage. In the course of the trial, various types of laboratory tests carried out for one of the experts for the Council were produced, which were said to prove either slip planes or slickensides, and for the purposes of the trial, it was necessary for the solicitor to become familiar with the various testing procedures in order to criticise their reliability. The 1981 CP2001 provides plenty of ammunition of its own. For the solicitor practising in this field, however, this type of knowledge is only an *ad hoc* requirement for the particular dispute. The use of the somewhat narrow knowledge acquired, as being relevant to the particular dispute, tends often to be misleading knowledge relating to some other dispute even if the dispute bears some similarities. Thus, rather like working for an exam, a solicitor like me tends to forget the limited knowledge which he acquires for the purpose of a particular trial, or contract, hoping to re-learn it rather more easily if occasion again arises, but in its new context. Until then, only somewhat meaningless phrases remain in his memory which may sound knowledgeable to the unknowing.

In the dispute concerning the primary school, the building contractor was not a party to the arbitration. If the dispute were heard in the High Court today, the engineer would, I believe, have been found liable. In a report of a symposium on liability in the latest arbitration journal, Judge Newey, who is one of the Official Referees who specialise in engineering disputes in the High Court in London, is quoted as saying that, whilst the law of tort is growing day by day, he did not see the importing of the principle of strict liability into the law of negligence in the construction industry. The recent increase in litigation was, he believed, in part due to declining standards of workmanship and in part due to the use of new techniques and new technology. But although Judge Newey may well have been right that strict liability will not be introduced in form; in practice, the principle is very nearly applied as the test of negligence. When civil engineering works fail, the question is usually not whether someone is to blame, but who is to blame, and, if more than one, in what proportions. You must prove your innocence.

In order to prove your innocence, it is important today to think ahead before you even enter into contractual terms and to define clearly and precisely what you agree to do for your money, and to warn the client or employer of any risks relating to work which you do not propose to undertake, or relating to any unproven elements of your designs, or methods of work, or materials to be used. When risks are to be undertaken, you must both draw attention to them, and make clear that the risk is to be taken by your employer, or client, not by you. Unless you do so, you will be liable if the risk of damage turns into a fact of damage.

The difficulty is that in a buyer's market, there will always be someone around the corner who will be prepared to take the risk in order to beat you for the job. The same principle applies in other spheres. For example, institutional landlords require business tenants to undertake full repairing and renewal obligations in leases, even in the extreme eventuality of the building becoming uninhabitable. The lessee must take the risk if he wants the lease, and in times of scarcity, he may feel compelled to do so. So far as civil engineering projects are concerned, the cost of insuring against risks undertaken is often high, and sometimes unobtainable, and even contracting under limited liability does not protect you against liability for your personal negligence.

The mutual interests of the disciplines within the construction industry are to achieve the successful completion of projects. Nothing succeeds like success. However, the present facts of life are that success can only be ensured at a cost which the client may not be prepared to pay. He prefers to rely on his right to sue if things go wrong, presumably on the basis that the risk is sufficiently small that, on average, this approach works out cheaper or him.

With regard to the terms of my brief, I am sure that no one here would dream of embarking on the preparation of a site investigation report, or give instructions for such a report, based only on a statement that it must be relevant to groundwater in engineering geology. To do so would be asking for trouble. This talk should therefore be treated as part of stage one of a solicitor's job. The second stage is detailed discussion for which you will later be allowed 15 minutes, which is only half the time allowed for a preliminary discussion under the Legal Aid Green Form scheme. You may even be hoping in that time to go to the third stage of actually being advised.

Recently, when wondering, perplexed, how to deal with this talk, I mentioned my uncertainty about the nature of my audience to a barrister friend who has more experience than me of this sort of occasion. He told me that you would be a highly intelligent and learned group, who would expect to be not merely educated but entertained. I doubt whether old chestnuts can be described as any more entertaining than I can claim you have been educated. I suppose it depends on how you tell'em, but as I started on a Biblical note, I may as well end on one, to show by repetition that the slowness of the law and difficulties in legal negotiations go back even to the time of Moses. It relates to the creation of the Ten Commandments. Laws in those days of course had to be carved in stone. After days of climbing up and down Mount Sinai and hard bargaining with God, Moses finally came down to the people of Israel bearing the tablets of stone on which the commandments were carved, and told them: 'I have done a deal with God, but there's good news and bad news. The good news is that He has agreed to reduce the commandments to ten. The bad news is that adultery is still forbidden'.

So far as you are concerned, in your future negotiation of contractual terms within the construction industry and resolution of disputes, I hope that whatever terms you achieve, the news will be good.

M.R. LUDLOW, Beale & Co., Solicitors, 22 Great Smith Street, Westminster, London SW1P 3DF, UK.

Discussion Session 12

P.M. CASHMAN (Wimpey Laboratories Ltd.) suggested that there were benefits in the French contractual arrangements in which the specialist contractor is directly invited by the client to carry out work within an agreed budget. The actual method and extent of site investigation etc. is left to the discretion of the specialist. This approach contrasts with the hide-bound British philosophy.

J. UFF (Consultant Engineer) pointed out that there are a number of alternative contractual systems now in use in the United Kingdom. The Construction Industrial Research and Information Association (CIRIA) is currently investigating alternative contractual arrangements for ground investigations. It is logical in such work to remove the distinction between the direction and carrying-out functions in construction work and use a lump sum basis. Frequently a lack of adequate supervision by the engineer and financial pressures under which contractors often operate give rise to an inadequate standard of work. These problems are reduced by using a contract which places all responsibility on one party.

P.M. CASHMAN explained that in France consulting engineers are employed directly by the specialist contractor for the work.

D.A. GREENWOOD (Cementation Piling & Foundations Ltd.) said that disasters arise most often through misunderstandings at contractual interfaces. There is generally no need for interposing a non-specialist main contractor between client and specialist. He pointed out the benefit of a common engineering philosophy between the initiator of work and those carrying it out which would accrue from 'design and construct' contracts. The client's interests can be protected by adopting the German method in which a consultant checks the proposals. It would also be advantageous to both contractor and client in respect of stimulating innovation of new methods of work. Unfortunately, a lack of time often precludes proper attention to the negotiation and drawing up of the contract in the best interests of all parties.

J. UFF in answer to a question about the implications in 'all-in' contracts for long term performance pointed out that, from a lawyer's point of view, determining who was responsible for faults would be simplified compared with the usual type of contract in which both the contractor and the supervisor share responsibility. A saving in litigation would result.

M.R. LUDLOW (Beale & Co) pointed out that until not many years ago, it had been assumed in civil engineering works that the contractor was wholly responsible for bad workmanship, the designer and supervision duties not carrying any real risk. Now, due to the law of tort, liability for bad workmanship is shared between the contractor and the consulting engineer if the consulting engineer fails to spot the defect and to call for it to be rectified. This increases legal costs because it encourages the client to sue both and causes disputes over apportionment. This could be avoided if the contact placed the whole of any liability for bad workmanship on the contractor, the consulting engineering being excused from liability save in cases where (for example, if the contractor went out of business) recovery against the contractor proves impossible.

J.M.A. PONTIN (Hydraulics Research Ltd.) questioned the appropriateness of making somebody 100% responsible for an unintentional event. Mr LUDLOW appears to be criticising the profession for not giving precise performance specifications which can be used for measurement purposes.

J. UFF asserted that the problem did lie with the specification. It is invariably the 'unwritten' terms that cause difficulty. All the conditions may be complied with but the performance may still be unsatisfactory, so that the parties are left to argue whether further terms are to be implied into the contract. The overall obligation and purpose needs to be included in the specification. Dr UFF went on to point out that it is important to appreciate that responsibility is not necessarily apportioned according to the amount of original input. This may be grossly unfair and the Danes have introduced a statutory system which limits responsibility. In the United Kingdom, unless limited within the contract, any degree of responsibility can attract liability for the whole of the resulting loss.

Written contribution:

D.N. HOLT (Freeman Fox & Pts)

To avoid restricting contributions from the floor when time was running out the Chairman (Mr HOLT) refrained from commenting himself but wishes to take the opportunity to stress in writing the role of the Consulting Engineer both in geotechnical work and in major construction.

From CRIPPS, J.C., BELL, F.G. & CULSHAW, M.G., (eds), 1986, *Groundwater in Engineering Geology,* Geological Society Engineering Geology Special Publication No. 3, pp 563-564

Traditionally the Consultant advises the Client (who is assumed to have little or no professional expertise), designs and commissions work on behalf of his Client, sees that this work is correctly carried out and fairly paid for, and takes the responsibility for all these things. This implies a threefold division of activity between Client, (purchaser of the goods and services), Consultant (specifier and guarantor of the appropriateness and quality of the goods and services) and Contractor (provider of the goods and services). This is a division which is clear-cut and implies a clear-cut practical moral and ethical responsibility on the Consultant, even if legally the situation is somewhat less well-defined. This is the basis of Consulting Engineering as we have known it in this country since Victorian times.

However, the position has become confused recently because Clients (and particularly major ones such as Government Departments) have been taking on even greater numbers of specialised professional staff. As a result they inevitably feel more and more that they can act as their own advisers and guarantors.

In this situation it is also inevitable that there should be an increase in the design-and-build turn-key type of agreement for major works. However commissioning this type of work is not a simple matter, and doing it places a very heavy burden of responsibility directly on the Client who has to see that he gets all his requirements right first time and without outside advice.

It will be very interesting to see if those in the public sector who will have to face this responsibility more and more in this new order of things will be prepared to do so and answer for their actions.

Until we can be assured of this, there will be a continuing need for the Consulting Engineer in his traditional role, for commissioning and supervising both geotechnical and construction work for all major projects, even though it has to be admitted that for minor geotechnical contracts this role can be omitted or side stepped.

Index of authors and contributors to discussion

From CRIPPS, J.C., BELL, F.G. & CULSHAW, M.G., (eds), 1986, *Groundwater in Engineering Geology,* Geological Society Engineering Geology Special Publication No. 3, pp. 565-571

CARDIFF
UWCC LIBRARY